Lecture Notes in Networks and Systems

Volume 299

AF880517

The series "Lecture Notes in Networks and Systems" publishes the latest developments in Networks and Systems—quickly, informally and with high quality. Original research reported in proceedings and post-proceedings represents the core of LNNS.

Volumes published in LNNS embrace all aspects and subfields of, as well as new challenges in, Networks and Systems.

The series contains proceedings and edited volumes in systems and networks, spanning the areas of Cyber-Physical Systems, Autonomous Systems, Sensor Networks, Control Systems, Energy Systems, Automotive Systems, Biological Systems, Vehicular Networking and Connected Vehicles, Aerospace Systems, Automation, Manufacturing, Smart Grids, Nonlinear Systems, Power Systems, Robotics, Social Systems, Economic Systems and other. Of particular value to both the contributors and the readership are the short publication timeframe and the world-wide distribution and exposure which enable both a wide and rapid dissemination of research output.

The series covers the theory, applications, and perspectives on the state of the art and future developments relevant to systems and networks, decision making, control, complex processes and related areas, as embedded in the fields of interdisciplinary and applied sciences, engineering, computer science, physics, economics, social, and life sciences, as well as the paradigms and methodologies behind them.

Indexed by SCOPUS, INSPEC, WTI Frankfurt eG, zbMATH, SCImago.

All books published in the series are submitted for consideration in Web of Science.

More information about this series at http://www.springer.com/series/15179

Mostafa Al-Emran · Mohammed A. Al-Sharafi ·
Mohammed N. Al-Kabi · Khaled Shaalan
Editors

Proceedings of International Conference on Emerging Technologies and Intelligent Systems

ICETIS 2021 (Volume 1)

Springer

Editors
Mostafa Al-Emran
The British University in Dubai
Dubai, United Arab Emirates

Mohammed A. Al-Sharafi
Universiti Teknologi Malaysia
Skudai, Johor, Malaysia

Mohammed N. Al-Kabi
Al Buraimi University College
Al Buraimi, Oman

Khaled Shaalan
The British University in Dubai
Dubai, United Arab Emirates

ISSN 2367-3370 ISSN 2367-3389 (electronic)
Lecture Notes in Networks and Systems
ISBN 978-3-030-82615-4 ISBN 978-3-030-82616-1 (eBook)
https://doi.org/10.1007/978-3-030-82616-1

This Springer imprint is published by the registered company Springer Nature Switzerland AG
The registered company address is: Gewerbestrasse 11, 6330 Cham, Switzerland

Preface

During the last decade, nobody denies the role of emerging technologies and intelligent systems in building digital societies and reshaping the knowledge-based economy. The number of experimental and theoretical findings is increasing rapidly due to the large number of successful emerging technologies. The role of those technologies has been witnessed across various domains, including health care, education, tourism, and industry, among others. The aim of the International Conference on Emerging Technologies and Intelligent Systems (ICETIS 2021) is to provide a forum for academics, scholars, and practitioners from academia and industry to share and exchange the recent developments in the domain of intelligent systems and emerging technologies. ICETIS 2021 also aims to shed light on the emerging research trends in intelligent systems and their applications. Further, it attempts to concentrate on the recent design, developments, and modifications of intelligent systems and their applications. It also aims to provide a holistic view of the main theoretical models used to understand the adoption, acceptance, and continued use of intelligent systems and their applications.

ICETIS 2021 was able to attract 249 submissions from 30 different countries across the globe. From the 249 submissions, we accepted 139 submissions, which represents an acceptance rate of 55.8%. Out of the 139 accepted submissions, 57 were selected to be published in this volume. The accepted papers under this volume were categorized under three different themes, including *Software Engineering*, *ICT in Education*, and *Management Information Systems*. Each submission is reviewed by at least two reviewers, who are considered experts in the related submitted paper. The evaluation criteria include several issues, such as correctness, originality, technical strength, significance, quality of presentation, interest, and relevance to the conference scope. This volume is published in *Lecture Notes in Networks and Systems Series* by Springer, which has a high SJR impact.

We acknowledge all those who contributed to the staging of ICETIS 2021. We would also like to express our gratitude to the reviewers for their valuable feedback and suggestions. Without them, it was not possible to maintain the high quality and the success of ICETIS 2021. As gratitude for their efforts, ICETIS 2021 is partnered with Publons to give an official recognition for the reviewers' contribution to peer review. This partnership means that reviewers can opt in to have their reviews added to their Publons profile.

Mostafa Al-Emran
Mohammed A. Al-Sharafi
Mohammed N. Al-Kabi
Khaled Shaalan

Organization

Scientific Committee

Conference General Chair

Khaled Shaalan	The British University in Dubai, UAE

Conference Organizing Chair

Mostafa Al-Emran	The British University in Dubai, UAE

Program Committee Chairs

Mohammed A. Al-Sharafi	Universiti Teknologi Malaysia, Malaysia
Mohammed N. Al-Kabi	Al Buraimi University College, Oman

Publication Committee Chairs

Mohammed A. Al-Sharafi	Universiti Teknologi Malaysia, Malaysia
Mostafa Al-Emran	The British University in Dubai, UAE

Conference Tracks Chairs

Vitaliy Mezhuyev	FH JOANNEUM University of Applied Sciences, Austria
Garry Tan Wei Han	UCSI University, Malaysia
Ibrahim Arpaci	Gaziosmanpaşa University, Turkey
AbdulRahman Al-Sewari	Universiti Malaysia Pahang, Malaysia
Baraq Ghaleb	Edinburgh Napier University, UK
Abdullah Nasser	Universiti Malaysia Pahang, Malaysia

Members of Scientific Committee

Ismail Ahmed Al-Qasem Al-Hadi	Universiti Malaysia Terengganu, Malaysia
Joseph Ng	UCSI University, Malaysia
Joshua A. Abolarinwa	Namibia University of Science and Technology, Namibia
Kamal Ali Alezabi	UCSI University, Malaysia
Kamal Mohammed Alhendawi	Al-Quds Open University, Faculty of Management, Palestine
Kauthar Mohd Daud	Universiti Malaysia Kelantan, Malaysia
Majid Abdulateef Abdullah	University of Hafr Al Batin, Saudi Arabia
Majid Salal Naghmash	Dijlah University College, Iraq
Mohamed Elwakil	University of Cincinnati, USA
Mohammed Ahmed Talab	Almaarif University College, Iraq
Muaadh Mukred	University Kebangsaan Malaysia, Malaysia
Muhammad Akmal Remli	Universiti Malaysia Kelantan, Malaysia
Mustafa Musa Jaber	Dijlah University College, Iraq
Nejood Hashim Al-walidi	Sana'a University, Yemen
Noor Akma Abu Bakar	Universiti Malaysia Pahang, Malaysia
Noor Al-Qaysi	Universiti Pendidikan Sultan Idris, Malaysia
Noor Suhana Sulaiman	University College TATI, Kemaman, Terengganu, Malaysia
Nooraida Samsudin	University College TATI, Kemaman, Terengganu, Malaysia
Nur Ilyana Binti Ismarau Tajuddin	UCSI University, Malaysia
Nur Sukinah Aziz	University College TATI, Kemaman, Terengganu, Malaysia
Osama Al Sayaydeh	Universiti Malaysia Pahang, Malaysia
Osama Mohammad Aljarrah	University of Massachusetts Dartmouth, USA
Osamah A. M. Ghaleb	Almustaqbal University, Saudi Arabia
Qasem Al-Tashi	Universiti Teknologi PETRONAS, Malaysia
Qasim AlAjmi	A' Sharqiyah University, Oman
Rayan Alanazi	Jouf University, Saudi Arabia
Redhwan Al-amri	Taylor's University, Malaysia
Safwan Mahmood Yahya Ghaleb	Universiti Malaysia Terengganu, Malaysia
Salim M. Zaki	Dijlah University College, Iraq
Samer Ali Alshami	Universiti Teknikal Malaysia Melaka, Malaysia
Shahad Al-Yousif	Dijlah University College, Iraq
Sinan Salih	Dijlah University College, Iraq
Suhaila Zainudin	Universiti Kebangsaan Malaysia, Malaysia
Sura Khalil Abd	Dijlah University College, Iraq
Syahrul Fahmy Abdul Wahab	University College TATI, Kemaman, Terengganu, Malaysia
Taha Hussein Rassem	Universiti Malaysia Pahang, Malaysia

Tuty Asmawaty Abdul Kadir	Universiti Malaysia Pahang, Malaysia
Vitaliy Mezhuyev	FH JOANNEUM University of Applied Sciences, Austria
Waheed Ali Hussein Mohammed Ghanem	Universiti Malaysia Terengganu, Malaysia
Yahya M. Al-dheleai	Universiti Sains Malaysia, Malaysia

Publicity Committee

Redhwan Al-amri	Taylor's University, Kuala Lumpur, Malaysia
Noor Akma Abu Bakar	Universiti Malaysia Pahang, Kuantan, Malaysia

Contents

Software Engineering

An Overview of Sentiment Analysis with Dialectical Processing

Jenophia Nelci[1], Said A. Salloum[2,3](✉), and Khaled Shaalan[1]

[1] Faculty of Engineering and IT, The British University in Dubai, Dubai, UAE
[2] School of Science, Engineering, and Environment, University of Salford, Salford, UK
ssalloum@sharjah.ac.ae
[3] Machine Learning and NLP Research Group, Department of Computer Science, University of Sharjah, Sharjah, UAE

Abstract. Sentiment analysis (SA) is one of the most emerging field of study in natural language processing (NLP) domain that primarily focuses on the opinion extortion of people's lingual and penned articulation. Considerable number of studies about sentiment analysis over Arabic language are probed to label its polarity into positive or negative since this becomes very common dialect to Arabian internet users to depict their views. The prime understanding of this systematic review over Arabic sentiment analysis gives an insight about the classic, modern and colloquial flavors of its high morphological richness. Potential documents were filtered from different data sources published from 2015 to 2020 as well relevant 10 research articles were analyzed carefully and findings were captured based on the questionnaires framed with methods used for data selection and approached algorithms. The key finding is that 89% of the analyzed sample resulted in positive outcomes. Positively research papers that originated from the Kingdom of Saudi Arabia has contributed to the above said achievement. In this aspect, this study is an initiative to demonstrate and detail the progressive advancement made in Arabic NLP based SA processing, eventually this work can suffice as an essential reference especially a clear insight about the polarity of the Arabic words used in the social media conversation to researchers of natural language processing domain.

Keywords: Sentiment analysis · Natural language processing · Arabic · Polarity

1 Introduction

In the natural language processing field sentiment analysis is an inevitable area of study, the popular methods used for sentiment identification are data detection and data classification. The data samples for this research are mainly collected from various sources including digital documents, social media data streams and so on [1]. The recent exploration of social media elevated this digital medium as the primary source of information.

In reality, the primary advantage of the social media platform is that it allows its users to express their viewpoint, speculations, assumptions and emotions with their

M. Al-Emran et al. (Eds.): ICETIS 2021, LNNS 299, pp. 3–12, 2022.
https://doi.org/10.1007/978-3-030-82616-1_1

native language itself as humans feel more comfortable to converse their thoughts with one another using their mother tongue. For example, until the last decade, the mechanism of collecting feedbacks across the supply chain or service delivery was minimum for companies, mainly institutions relied on their own surveys or sales book data. Within a decade, robust growth of social media networks such as Instagram, Twitter, YouTube and Facebook made it very easy for the customers and companies to share reviews, comments, likes and dislikes about their products. The choice of making opinions in public brought the information available free and open, on the other hand cases like articulated public opinions circulated to boost companies sales by indirectly influencing customer buying behavior [2]. The key technological renaissance happened in fields like Artificial Intelligence, Machine Learning (ML), Deep Learning, computational linguistics, and NLP made the hard reality of analyzing highest data volumes to be true and practical. Advanced tools with feature engineering known as supervised learning method that in turn use ML algorithms, unsupervised learning techniques that use rule-based methods and exploit sentiment lexicons are the key improvements made in the regional language based analysis especially with Arabic [3].

This review work systematically reports and incorporates the methodologies and algorithms that related to Arabic sentiment analysis of the possessed studies. This study precisely concentrates about the polarity of the Arabic words used in the social medium based the collected papers, thus it extends the below research questions to achieve the goal:

RQ1. What are the main research techniques or algorithms used to achieve Arabic sentiment analysis in the selected studies?

RQ2. How the datasets are collected to process the various data mining operations over the collected papers?

RQ3. What are the main research methodologies used to perform sentiment analysis in the chosen studies?

RQ4. Which countries contributed more researches in Arabic sentiment analysis? Mention the year of publication?

RQ5. What are the active databases in the framework of the Arabic sentiment analysis?

2 Literature Review

Sentiment analysis methodologies mainly deal with extraction of polarized sentiments in the supplied text data. Initially the raw data stream derived from the data source is converted as text, subsequently the sentiments articulated within the texts is detected and classified under three sentiment heads known as positive, negative and neutral [4]. The key process involved in text detection is called as text mining. In fact, it is relatively a new field used across data mining, ML, Information retrieval processes as well computational linguistics those are linked to distinct research areas [5]. Sentiment polarization detection is the potential result expected from this proposed study, in such a way the effective sentiment mapping between sentiment identification (i.e., positive, negative, or neutral) and actual hidden sentiment (i.e., anger, happiness, disgust, and joy) mapping eventually derive the purpose [6].

The process of document analysis is entirely different from sentence level analysis, the main difference is that documents across the flow reflects multiple contradicting sentiments, however sentences usually exhibit a single sentiment [4]. Sentiment weight calculation is inevitable to understand the polarization, hence typical use case classifiers are used to label multiple sentiment classes, and its associated predefined score to each sentiment class, and finally to calculate the overall weight [6]. Arabic language is one of the ancient language that hails its root from the Sami language family, as well people from North Africa and the Middle East region are the highest spoken around the world [7]. The writing nature of the language is cursive, on top of that the cursive letters are inter connected with one another which complicate the data extraction process. Arabic language is principally classified under three broader spectrum known as Modern Standard Arabic (MSA), Classical Arabic (CA), and Dialect Arabic (DA) [7]. Text preprocessing, identification, feature extraction and classification are high level steps performed in Arabic data opinion mining system [2]. The final step is classifiers application on the tweets to group under positive, negative, moderate positive, moderate negative and neutral sentiments categories [8].

3 Research Methodology

The potential literature review consists of four important phases knows as the data sources and search strategies, exclusion criteria and identification of inclusion, data coding and analysis and finally quality assessment [9].

3.1 Inclusion/Exclusion Criteria

See Table 1.

Table 1. Inclusion and exclusion criteria.

Inclusion criteria	Exclusion criteria
Should involve sentiment analysis with dialectical processing	Papers that are not focused on sentiment analysis with dialectical processing
Should involve processing techniques or algorithms of sentiment analysis	Processing techniques or algorithms that are used in contexts other than sentiment analysis
Should be written in English	Articles that use languages other than English

3.2 Data Sources and Search Strategies

The popular research article databases such as Google scholar, Science direct, Springer, IEEE, Wiley and Emerald are involved in this systematic review. The data search was undertaken in JAN 2020, and the search terms include the keywords ((“sentiment analysis” OR “sentiment processing”) AND (“dialectical” OR “processing” OR “analysis”)).

The retrieval of articles are purely based on the keyword identification hence keyword selection is very significant for any systematic review (Costa and Monteiro 2016) [9]. The search results with the above mentioned filter condition retrieved 726 articles out of which 21 articles were found as duplicates and removed from the study, thus the result is further reduced to 705. All the resulted samples were carefully read and further filtered by applying the inclusion/exclusion criterion. Accordingly, 10 research articles were identified as qualified samples, and included for this research. In short, the applied search and refinement stages is known as Preferred Reporting Items for Systematic Reviews and Meta-Analysis (PRISMA) (Moher et al. 2009) as shown in Fig. 1. PRISMA flowchart for the selected studies.

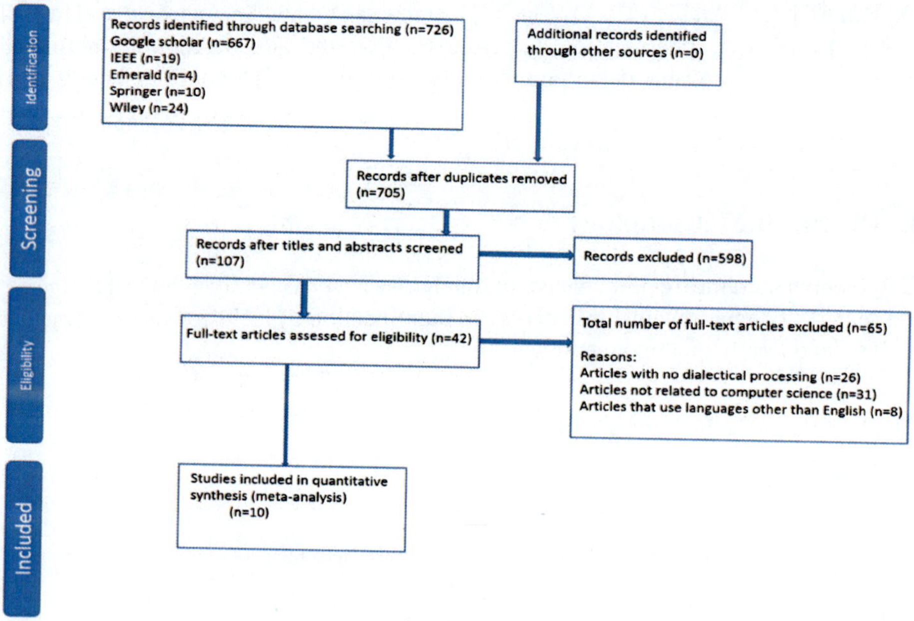

Fig. 1. PRISMA flowchart for the selected studies.

3.3 Quality Assessment

In this review, a quality assessment checklist with 7 questions were prepared and used to filter out the irrelevant research articles in order to assure the quality of study (N = 10). Table 2 illustrate all the quality assessment checklist.

Table 2. Quality assessment checklist.

#	Question
1	Is the article clearly specify its research aims?
2	Is the sentiment analysis study process certainly described?
3	Is the study precisely cover the techniques or algorithms used?
4	Is the study included scenarios such as collection of data and the methods used?
5	Is the study clarify the reliability or validity of the measures?
6	Do the outcomes count to the literature?
7	Does the study deepen your domain understanding?

The checklist was adapted by the popular review guide submitted by Kitchenham [9]. Each question is mapped in a three point scale such that, every "yes" is mapped to 1, "No" is attributed to 0 and neutral outcomes are referred as 0.5. Hence, for each study the value score varies between 0 and 7, with this background the authors presumed that the score of study is directly proportional to the score a study attains. Table 3 lists all the 10 assessments' results that assures that the selected articles passed through the quality assessment.

Table 3. Quality assessment results.

Study	Q1	Q2	Q3	Q4	Q5	Q6	Q7	Total	Per.
Study 1	1	1	1	1	1	1	1	7	100%
Study 2	1	1	1	1	1	1	0.5	6.5	92%
Study 3	1	1	1	1	0.5	1	1	6.5	92%
Study 4	1	1	1	1	1	1	0	6	85%
Study 5	1	1	1	1	0.5	1	1	6.5	92%
Study 6	1	1	1	1	0.5	1	1	6.5	92%
Study 7	1	1	1	1	1	1	0	6	85%
Study 8	1	1	1	1	0.5	1	1	6.5	92%
Study 9	1	1	1	1	0	1	0	5	71%
Study 10	1	1	1	1	0.5	1	1	6.5	92%

3.4 Data Coding and Analysis

Research methods such as reviews, data analysis and survey along with research techniques/algorithms with the research outcomes are correlated in addition to that the country of study also included. As a matter of fact, this detailed analysis on the collected articles of study is completely done manually by the first author.

4 Result and Conclusion

This section covers the studies that related to SA and its results in full view. The sample used in this studies differ in various aspects such as the methods of pre-processing, polarity analysis, and the design techniques [5]. The carefully selected 10 research articles about SA published between 2015 and 2020 are analyzed based on the below five research questions.

4.1 What are the Main Research Techniques or Algorithms Used to Achieve Arabic Sentiment Analysis in the Selected Studies?

There are three different types of classification techniques used in sentiment analysis known as Naive Bayes (NB), K-Nearest Neighbor (K-NN) and Support Vector Machine (SVM) [10]. The popular open source java software used for data mining is Rapid Miner which includes tokenization, stemming and filtering stop words to deal the Arabic language. First, the inclusion of deep learning techniques in SA achieved significant results with the expected accuracy. Next, the bidirectional long and short term memory approach as well the deep learning are used to obtain context sensitive comment representation [11]. The ML algorithms either supervised like Bayesian network, naive Bayes, neural network support vector machine (SVM) and decision tree or unsupervised algorithms such as K-means are used for grouping. Third, Lexicon techniques are efficient alternates, here the dictionary based model and the corpus based approach use either semantic or statistical methods to produce positive results. The prime advantage of domain independency, the single lexicon is built whereas hybrid techniques combine both ML algorithms and lexicon-based methods [4]. Fourth, task classification and regression use decision trees (DT) known as non-parametrically supervised learning algorithm. The random forest (RF) is based on the ensemble method where bagging and boosting techniques are for the classification of trees [8]. Last, the combined approach is for its robust performance, highest accuracy, stability and readability via texts and domains, specifically their hybrid methods in data analysis reached 82.30% accuracy [2].

4.2 How the Datasets are Collected to Process the Various Data Mining Operations Over the Collected Papers?

The data mining operations studied here based on two data sets that contains texts collected from twitter that is preprocessed and manually annotated with seven different sentiment classes grouped into three pairs of opposite sentiments [Fun, Anger], [Love, Hate], [Happiness, Sadness] and a [Neutral] sentiment class. Different studies used separate data samples, depending on their requirement, thus the data set used in each case is carefully documented [6]. Another case analyzed is based on lexicon model which starts from the first step of building own lexicon consists of 2500 words from Sudanese dialect. The data set is prepared manually from different social media and Arabic websites thus the composed data has different sizes and categories for training and testing. Table 4 illustrate a complete overview of the operations over the collected studies.

Table 4. Operations over the collected studies.

Source	Data mining	Sentiment identification	Sentiment classification	Sentiment mapping	Sentiment polarization	Machine learning approach	Lexicon approach
[10]	x	X	x	x	X	x	x
[3]	x	X	x			x	
[7]	x		x	x	X	x	x
[6]	x	X	x		X	x	
[2]	x			x	X	x	x
[1]	x	X	x		X	x	x
[8])	x	X	x		X	x	x
[4]	x	X	x	x	x	x	x
[11]	x		x	x	x	x	
[12]	x		x	x	x	x	x

The training dataset consists of 608 positive and 1246 negative words [10]. Primary data sources used in the surveyed articles are from amazon reviews, We Chat, Corpus OCA, product reviews, video reviews, mobile reviews blogs, hotel reviews, opinions review, tweets, Yelp and IMDB [1]. The datasets are publicly available in the Natural Language Toolkit (NLTK) corpora resource. Indeed, 10000 twitter posts are the total corpus size, out of which 5000 tweets represent positive and the remaining 5000 contribute the negative sentiments. Another dataset studied is the emotion based data model known as AR_EMC used to draw four primary emotions (Anger, Joy, Digest, and Sad) [2]. According to the reviewers, the accuracy of the sentiment detection is achieved very less as the result is directly proportional to the existence of large annotated corpora which is limited for the Arabic [7]. The research revealed the usage of Large Scale Arabic Book Reviews Dataset (LABR), one of the largest Arabic dataset having more than 63,000 reviews suitable for DL techniques to carryout SA.

4.3 What are the Main Research Methodologies Used to Perform Sentiment Analysis in the Chosen Studies?

The primary research methods identified during the study are survey, data analysis, document reviews and literature review, Fig. 2 indicates the distribution of articles by research methods used in the study. According to the result, 50% of the studies were mainly focused on data analysis method (N = 5) for data collection, this is followed by both reviews and surveys (N = 2) respectively. Literature review method is the least of all with 10% (N = 1).

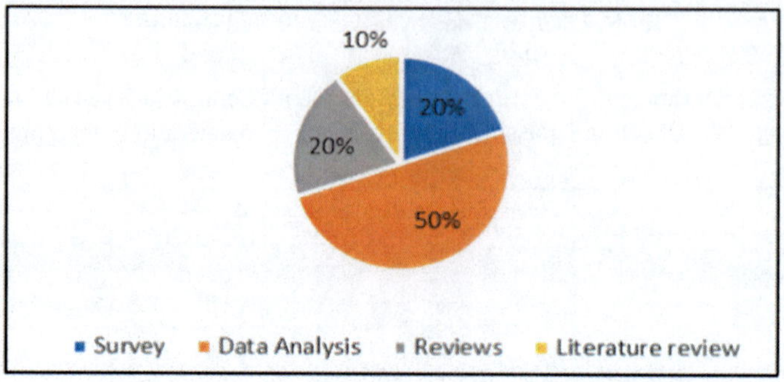

Fig. 2. Distribution of articles by research methods.

4.4 Which Countries Contributed More Researches in Arabic Sentiment Analysis? Mention the Year of Publication?

Another point of research is the period of publication, the number of research work done with the respective year and the publishing country. As given in Fig. 3 the data used for this study ranges from 2015 to 2020. The number articles published was significantly increased from one study in 2016 to 9 studies till early 2020. In addition to that, there is a visible improvement in the number of published work in 2018 and 2019, In fact the growth in terms for Arabic dialect study is very significant during this period.

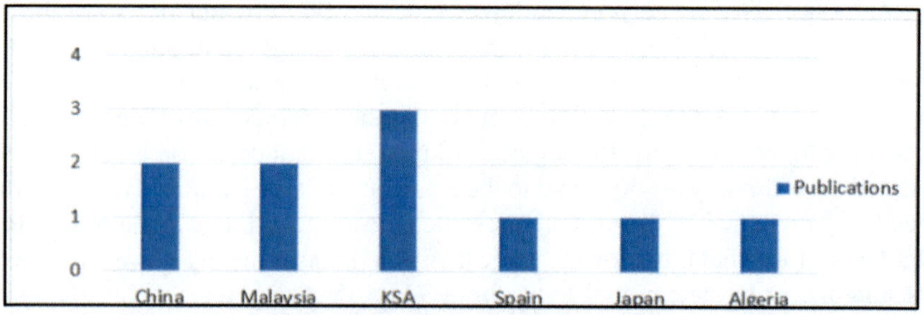

Fig. 3. Countries contribution chart.

4.5 What are the Active Databases in the Framework of the Arabic Sentiment Analysis?

Here, it is evaluated and identified the most active databases that publish studies related to Arabic SA with dialectical processing. Figure 4 shows the databases based on the collected studies distribution. From the graph, it is proved that Google Scholar is the most productive database contributed five article among all the published articles. This is followed by IEEE (N = 3) and Springer with 2 published articles.

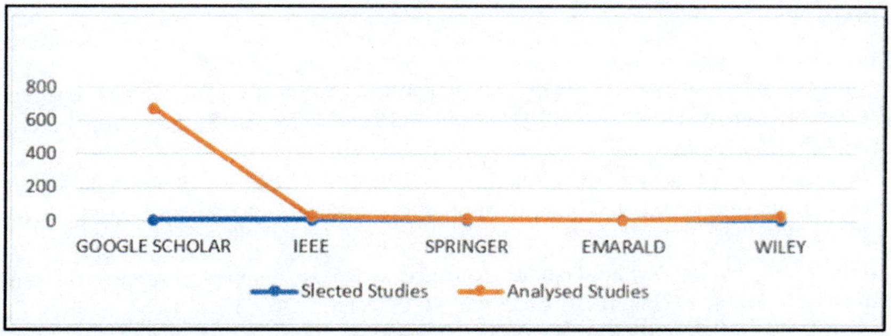

Fig. 4. Distribution of the collected studies in terms of databases.

5 Challenges and Limitations

The critical challenge of sentiment study are the lexicons, since building manual regional language based lexicons are very hard using humans' language pattern. Because people follow different writing styles to express their opinions towards circumstances. The content based sentiment, in reality opinions are context driven thus sentiment polarization should be context dependent. Sentiment analysis methods dependent on sentiment dictionary, hence the availability of standard master dictionary is very vital. As of now, most of the sentiment dictionaries have problems such as insufficient coverage of sentiment words and lack of domain words [11]. Hurdles specific to Arabic language become bottleneck in sentiment extraction such that the word Saleem means Right in English. Saleem as an adjective indicates positive sentiment but as a person name it is neutral as it has no sentiment [10]. The vast reach of social platform introduced another challenge known as linguistic based sentiment analysis. Though there available enough tools and techniques for English language, it requires more research towards other inflated languages for potential exploration.

6 Conclusion

The systematic review for the Arabic sentiments and opinions to identify the polarity of the texts are analyzed over the chosen papers after filtration thus the primary goal achieved. Though it requires more exploring in this regard, especially on the Arabic language dialectical domain. Additional tools and techniques that are capable of processing Arabic dialect should be developed to cater the adherent need of this field. Hence contribution from the research community as well as the development community is very vital at this point of time.

Acknowledgment. This work is a part of a project submitted in fulfilment of PhD Computer Science - Faculty of Engineering and Information Technology at The British University in Dubai.

References

1. Behdenna, S., Barigou, F., Belalem, G.: Document level sentiment analysis: a survey. EAI Endorsed Trans. Context Aware Syst. Appl. **4**(13), 154339 (2018). https://doi.org/10.4108/eai.14-3-2018.154339
2. Abdullah, M., AlMasawa, M., Makki, I., Alsolmi, M., Mahrous, S.: Emotions extraction from Arabic tweets. Int. J. Comput. Appl. **42**(7), 661–675 (2018). https://doi.org/10.1080/1206212X.2018.1482395
3. Al-Twairesh, N., Al-Negheimish, H.: Surface and deep features ensemble for sentiment analysis of Arabic tweets. IEEE Access **7**, 84122–84131 (2019)
4. Alharbi, A., Taileb, M., Kalkatawi, M.: Deep learning in Arabic sentiment analysis: an overview. J. Inform. Sci. **47**(1), 129–140 (2019). https://doi.org/10.1177/0165551519865488
5. Al-Emran, M., Shaalan, K., Salloum, S.A., AlHamad, A.Q.: A survey of Arabic text mining, vol. 740, no. 2, pp. 380–392 (2018). https://doi.org/10.1007/978-3-319-67056-0
6. Bouazizi, M., Ohtsuki, T.: Multi-class sentiment analysis on Twitter: classification performance and challenges. Big Data Min. Anal. **2**(3), 181–194 (2019). https://doi.org/10.26599/bdma.2019.9020002
7. Abo, M.E.M., Raj, R.G., Qazi, A.: A review on Arabic sentiment analysis: State-of-the-Art, taxonomy and open research challenges. IEEE Access **7**, 162008–162024 (2019). https://doi.org/10.1109/ACCESS.2019.2951530
8. Saad, S.E., Yang, J.: Twitter sentiment analysis based on ordinal regression. IEEE Access **7**, 163677–163685 (2019). https://doi.org/10.1109/ACCESS.2019.2952127
9. Al-Emran, M., Mezhuyev, V., Kamaludin, A., Shaalan, K.: The impact of knowledge management processes on information systems: a systematic review. Int. J. Inf. Manag. **43**, 173–187 (2018). https://doi.org/10.1016/j.ijinfomgt.2018.08.001
10. Abdelhameed, H.J., Muñoz-Hernández, S.: Sentiment analysis of Arabic tweets in Sudanese dialect. J. Comput. Sci. **12**(11), 553–563 (2016). https://doi.org/10.3844/jcssp.2016.553.563
11. Xu, G., Meng, Y., Qiu, X., Yu, Z., Wu, X.: Sentiment analysis of comment texts based on BiLSTM. IEEE Access **7**, 51522–51532 (2019). https://doi.org/10.1109/ACCESS.2019.2909919
12. Salloum, S.A., AlHamad, A.Q., Al-Emran, M., Shaalan, K.: A survey of Arabic text mining. In: Shaalan, K., Hassanien, A.E., Tolba, F. (eds.) Intelligent Natural Language Processing: Trends and Applications. SCI, vol. 740, pp. 417–431. Springer, Cham (2018). https://doi.org/10.1007/978-3-319-67056-0_20

Feasibility of Using Attention Mechanism in Abstractive Summarization

Rashed Z. AlMazrouei[1], Jenophia Nelci[1], Said A. Salloum[2,3](✉), and Khaled Shaalan[1]

[1] Faculty of Engineering and IT, The British University in Dubai, Dubai, UAE
[2] School of Science, Engineering, and Environment, University of Salford, Salford, UK
[3] Machine Learning and NLP Research Group, Department of Computer Science, University of Sharjah, Sharjah, UAE
ssalloum@sharjah.ac.ae

Abstract. The Prevalence of information and its magnitude mandates a short description of the core of a document, an article, or legal documents. Abstractive summarization helps to concur with this problem utilizing the evolutions in machine learning and deep neural network. Attention-mechanism has extensively applied in the challenging issue of abstraction a text, in shorter length yet informative. We noticed in [13] after removing the attention layer from their proposed model, the performance only experience soft drawback, even can be ignored. Thus, motivates us to survey the latest models using attention-mechanism and its achievements, and the second objective is to run an experiment to compare standard stacked 3- Long Short-Term Memory (LSTM) layers incorporated with attention layer only (without any other hand-crafted algorithm) to explore how efficient this technique can generate better summarization, then a stand-alone model. The standard proposed model incorporated with attention-mechanism suffered from drawback performance and scored less than a stand-alone model by at least 6 point scores on ROUGE-1&2.

Keywords: Abstractive text summarization · Attention-mechanism · ROUGE

1 Introduction

Nowadays accessing the internet at a person's fingertips. A huge amount of data is generating on a second-basis, which poses a challenge wherein people's minds have become more disintegrated and overwhelmed with the immense aggregated data they find everywhere. Generated data, can take various types, such as audio, video, and textual content. Writing a summary for an article, book or news requires reading the entire content, focusing on key information and items from the text. However, a manual rewriting meaningful succinct text representing what has been read. Adopting manual summarization can be a costly and time-consuming process in a world demanding agility with quick decision-making. Automatic Summarization comes in handy to alleviate this dilemma. Furthermore, summarization is one of the high-level tasks using Natural Language Processing (NLP) mechanisms to output a condensed yet meaningful text, and

M. Al-Emran et al. (Eds.): ICETIS 2021, LNNS 299, pp. 13–20, 2022.
https://doi.org/10.1007/978-3-030-82616-1_2

clearly expressed the core idea of the original text. In the field of NLP, summarization is still a hot topic amongst researchers, striving to enhance the semantic of the output summary not just only structural and syntactical build. Inevitably, summarizers can benefit from prevailing training data which exploded as the online users eagerly become a source of content generation. Unfortunately, specific domains like opinions briefing, discourse summarization, and minutes of meetings are examples of intractable to be labeled.

Automatic text summarization can be classified into two categories: extractive and abstractive. Extractive text summarization generates few exact sentences from the original document or text based on the "TextRank" algorithm or similar statistical technique, where the sentences are ranked based on importance and the amount of information contained [1]. This approach does not change or modify the sentences and keeps them almost intact. Whereas, abstractive summarization reads the entire document or text, and then uses advanced natural language processing (ANLP) to rephrases meaningful text and builds informative sentences [2]. Remarkably, abstractive summarization becomes a trend in the NLP field, with a substantial focus from academia and industrial practitioners alike. Our focus in this paper will be on the latter type of summarization technique. The trend shifting toward abstractive summarization manifested as the former mechanism suffers from redundant information and non-relevant generated text. Thus, reflects on overall performance in terms of accuracy, and consistency. Additionally, the abstractive technique exhibits a better quality of generated summary attributed to several factors. In general, abstractive summarization produces shorter and consistent text compared to the extractive technique using fusion mechanism and compression in the early days [3]. As the abstraction methods evolve by the advancement in technology; new techniques have been introduced, namely extraction, selection, and surface realization.

Furthermore, an elaborated survey [2] divided the types of abstractive summarization - based on used methods - into 1) structured models; 2) semantic models, and recently; 3) neural network models. The structured approach searches for intrinsic information parts in the text and reflects them on pre-defined rules or templates. This approach having already been used in text pre-processing tasks and complements with extractive summarization, or with other two types of abstractive approaches to enhance the quality of output summary. Structured methods to mention a few - with related recent researches and studies- are, tree-based [4]; template-based [5]; ontology-based [6]; and graph-based method [7]. Secondly, the semantic approach analyzes different aspects of text characteristics and relationships between words, for example, object, verb, and subject items in the sentence. Besides, specific items, like time, event, and location are considered to provide richness in the final summary. A semantic graph is considered as a reliable model of semantic approach, which has been harnessed for both abstractive and extractive summarization studies [8]. The third approach –neural network- has drawn an outstanding attention by the researchers in NLP field. Neural network based methods has achieved tremendous results in non-linear relation tasks like in language processing, allowing prediction of complex relations within text structure. A steep increase in conducting studies and experiments for text summarization using neural models, wherein they excel in learning several properties and features from the words in sentences, for instance; co-location, context, similarity and semantic characterstics between them.

Neural capabilities drive the change in handling complex tasks in language processing, then the traditional techniques. Machine translation (MT) complexity inspired researchers in adopting neural mechanisms and they outperform several existing conventional systems. Some of the well-known networks for Neural Machine Translation (NMT) are Convolutional Neural Network (CNN); Recurrent Neural Network (RNN); and Long Short Term Memory (LSTM). Those models apply encoder-decoder methods allowing for better word representation (i.e. word embeddings).

Arguably, NMT systems show better performance and high quality, however, since the source and target texts are not of the same length, a technique namely attention-mechanism incorporated in the network implementation [9]. The purpose of the attention method is to construct a joint alignment between the source and target texts on both sides i.e. encoder and decoder of the network model. Moreover, pioneering work in NMT, motivates other researchers to tackle the problem of abstractive summarization [10]. Preliminary investigation in the efficiency of attention-mechanism in a complex task like abstractive summary is the objective of this work. The second purpose is to run an experiment to investigate how well the attention mechanism incorporated into standard 3-LSTM layers model performances in abstractive summarization task, then similar model stand-alone without attention-mechanism. The study, therefore, is organized as follows. Section 2 -related work- presents an overview of related recent-state-of-the-art work implementing attention-mechanism in the field of abstractive summarization. Following that, Sect. 3 explains conducted methodology. Results and discussion reflect in Sect. 4, proceeded by a conclusion and future research ideas in Sect. 5.

2 Related Word

Attention-based network models have been implemented for single and multiple documents summary tasks to produce fluent target text output in different fields like headlines generating and news summarization. Accessing several news and articles datasets publicly makes it possible for researchers to focus on resolving other issues, then labeling the data. A recent study in news summarization [11], proposed a Word Attention with Convolutional Neural Networks (WACNNs), which integrates the local attention layer using CNN network and expanding the capability of regular seq2seq architecture. The purpose of the suggested system was to focus on subset elements of the input text keywords which incorporated the essence of the document. Three news summarization datasets were used, namely English Gigaword, DUC-2004, and large-scale Chinese short text summarization (LCSTS). The model outperformed most of the base-line models using Recall Oriented Understudy for Gisting Evaluation (ROUGE) scale on English Gigaword.

In another field, customer product reviews and opinions, the attention layer was used in [12] to condense the significant terms into the produced summary. The proposed model implemented Recall-Vocabulary Again (RVA). Besides, to overcome the problem of Out of Vocabulary (OOV) words, a Copy mechanism from the extractive summarization domain is used with LSTM decoder and attended context vector. They judged the model performance against the Amazon product reviews dataset. The model was able to reduce the vocabulary size of the decoder from ~70 to only 4 K and still maintained a good ROUGE score. Moreover jointly combining advantages of attention-mechanism from abstractive summarization side, and auxiliary extractor of information utilizing extractive summarization techniques, [13] were able to strengthen summary generation to a comparable level to state-of-the-art models. Their model adopted hierarchical attention-mechanism as sentence-level, and word-level attention hidden states took into consideration to represent deep semantics from source and target text. Researchers trained their model on CNN/DailyMail dataset. The model scored almost the same in three ROUGE variants once the attention-mechanism was removed in case of testing against the DailyMail dataset.

The work of [14] attempts to solve the issue of abstractive summary of long documents. Their system **g**lobal generative topic fusion with **l**ocal sentence generation **e**ncoding for **a**bstractive su**m**marization (GLEAM), structured a single LSTM decoder layer based on attention-mechanism. This construct gives the ability to focus on specific topics, which is generated by neural generative topics, and produce a hidden state incorporated the global document level important terms, previous output, and context vector. This ability supports fusing a local sequence with document (i.e. global) level topic representation. The model was learned on the English Gigaword dataset and the researchers tested their model on the DUC-2004 dataset. Researchers claimed promising results on the ROUGE metric comparing with other models in the field of abstractive summarization.

From the unpretentious survey on the recent state-of-the-art summarization models that took advantage of attention-mechanism, we were inspired to investigate its relative performance in terms of the ROUGE scale on two deep RNN setups. The main difference between them is being with or without attention-mechanism and no other hand-crafted algorithm or auxiliary techniques were integrated, to detect the impact on the model behavior toward generating a summary quality as we noticed in [13], whereby removing attention-mechanism in one test scenario of that study did not affect the final output of the model.

3 Method

3.1 Dataset and Preprocessing Brief

As the aim of the study to compare the performance and quality of generated summary, based on the RNN model for text summarization with the attention-mechanism layer and without it. Both network models were developed on a seq2seq architecture for both encoder and decoder components. The News summary dataset was downloaded from kaggle.com. The size of the data is (~100 K) news articles with reference headlines. Preprocessing takes care of cleaning the text by; removing duplicates, deleting null values, remove HTML tags, word contraction expanding using a dictionary, and removing noisy characters. The stop words were not removed to find how fluent the summary output would be using both models. Less occurring words were removed (<4 occurrences for reference summary and <6 for article' text).

3.2 Attention-Mechanism

In this experiment, an attention-mechanism was implemented in the second model as suggested in [15]. The main objective of the attention-mechanism is to allow the decoder to focus on certain source word sequences, that is, enabling the decoder to improve the accuracy and preciseness of generated summary, at which it to align both source and target text with attended context vector. The attention layer sets after the embedding layer and it concatenates the context vector of input and LSTM decoder output. In this implementation, the attention-mechanism pays attention to all input words which is considered as 'global' attention. The other type, 'soft' attention not applied in this experiment.

3.3 Model Creation

The created models are normal standard RNN with three stacked LSTM layers for the encoder. This configuration reduces the possibility of long sentences and words diminishing. Then execution of the model took place, proceeded by exploring the pre-processed data to optimize the maximum number of words in sentences. The data has been split as follows (90% for training and building the models and 10% for testing). We considered the following hyper-parameters settings into the experiment, dropout at 0.4 level, 300 hidden states for LSTM, 'softmax' was the activation layer for the model with optimization algorithm namely Root Mean Square Propagation (RMSProp), and 200 dimensions for word-embeddings. The use of RMSPorp to automatically adjust the learning rate according to hyper-parameters setup. The model objective function was the sparse categorical cross-entropy. A single LSTM decoder layer was used for output to the dense layer. The model development uses python Keras and Tensorflow work-frame with Tesla P40 GPU and CUDA version 10.2.

3.4 Performance Evaluation Metric

ROUGE metric, has been accepted widely by researchers community in performing evaluations for machine translation models and automatically generated summaries. In this experiment, the ROUGE metric was applied to 1000 randomly selected samples. ROUGE-1, ROUGE-2, and ROUGE-L were the variants selected from this metric, at which they are used the most in this type of task.

4 Results and Discussions

Before proceeding into the findings and discussions, the created model was based on a standard architecture of the RNN-LSTM network model. There was no manually hand-crafted algorithm or enhanced auxiliary technique, as such, created particularly for this experiment. Moreover, an exception only occurred as of using attention-mechanism in the second model. Demonstrating the difference of use vs. no-use of attention-mechanism and its impact on the summarization model mandated a particular setup.

As depicted in Table 1, the standard seq2seq model without attention-mechanism substantiates interesting results, with comparison to the similar model with the attention

Table 1. ROUGE-1, 2&L metric scores.

Model	3-LSTM + WO/ATT			3-LSTM + W/ATT		
Sample data	1000					
Metric	Recall	Precision	F-score	Recall	Precision	F-score
Rouge-1	0.29	0.23	**0.25**	0.21	0.16	**0.18**
Rouge-2	0.10	0.08	**0.08**	0.06	0.02	**0.03**
Rouge-L	0.28	0.25	**0.26**	0.17	0.18	**0.17**

layer incorporated in it. The former model outperformed the latter with a distinguished higher score. WO/ATT model achieved 6 Rouge-1 (i.e. unigram words) Recall points over W/ATT, subsequently for Precision and F. Moreover, it can be observed the same score received from Rouge-L for both models respectively. This finding raises a question, as to when it is advisable to integrate attention-mechanism into your model for automatic summarization. Further investigation is required, whether the hyperparameters should be manipulated to allow the attention-mechanism playing its role which developed for. Also, it is worth mentioning, that our findings might be in tandem with [13] result when the attention-mechanism has been removed from their model and recorded no differences. In this context, global attention was chosen for this experiment and not the local approach. A part of the preceded outcomes, both models failed to achieve a good score under ROUGE-2 (i.e. bigram words) metric. This outcome necessitates the demand for further research in various areas, i.e. the created model, text preprocessing, text quality, and attention-mechanism. However, we have paid a closer look at the detailed generated summary from two models comparing with the reference headlines. In Table 2, we have selected three different examples, to perform a simple comparison with two folds: information intensity and repeating words. Article #1, both models succeed to generate an informative summary, with repetition of ‘app’ twice produced by W/ATT. However, fluency was achieved at a quite good level by both models. Repetition can be noticed further in Article #3 at word ‘football’ inherited from seq2seq model as mentioned earlier by [16]. Moreover, repetition found also in seq2seq, generated by the same model ‘thai cave rescue’ i.e. 3-g repeated twice. Besides that, in Article #2, the W/ATT produces a bad informative summary, compares to the WO/ATT model.

Table 2. Snippets of resulted abstractive summarization W/O-ATT & W-ATT

Article #1	**Social media giant facebook on wednesday officially launched its video platform watch for users in the us available on facebook mobile apps and desktop site watch allows users to subscribe to their favourite series and offers sections like today spotlight popular now and what friends are watching it also offers special 10 min or more spot for long form videos**
Ref	Facebook launches video platform watch for users in the us
Pred-WO-ATT	Facebook tests feature to let users photos videos
Pred-W-ATT	Facebook launches app for users on live streaming app
Article #2	**A volcano in southwestern japan spread ashes in at least four cities and towns on thursday after an ash plume rose 000 m from its crater the volcano which lies on the border of and started erupting this week for the first time in six years residents have been warned to not approach the volcano**
Ref	Volcanic eruption in japan spreads ash in cities
Pred-WO-ATT	World largest volcano erupts in japan
Pred-W-ATT	World largest nuclear plant kills 8 billion in days
Article #3	**The thai junior football team who were rescued from flooded cave after being trapped for more than two weeks trained with the mls side los angeles galaxy and their striker zlatan ibrahimovic on friday thought was brave but these kids this team is than me this is probably the best team in the world ibrahimovic**
Ref	Thai cave survivors train with zlatan
Pred-WO-ATT	Thai cave rescue boys trapped in thai cave rescue baby rescued
Pred-W-ATT	Football football team catches fire after thai cave survivors

5 Conclusion and Future Researches

This study aims at two folds: explore recent state-of-the-art summarization models, which are integrated attention-mechanism within their functionality. The second fold was comparing the performance of the stand-alone RNN-LSTM (i.e. seq2seq) model, with a similar model incorporate attention-mechanism with it. The results were not as anticipated, whereby the attention-mechanism brings drawback performance to the main model. Also, the recorded score for the model with attention-mechanism less than a stand-alone model by at least 6 point scores on ROUGE-1&2. Fluency of the language for wordings generated by both models, even though W/ATT suffers from bad summarization in some cases. Various results generated by this experiment worth further investigation and research. One inquiry could be about how, when, and where to implement attention-mechanism, and in which circumstances (i.e. hyperparameters configuration) can perform better.

Acknowledgment. This work is a part of a project submitted to The British University in Dubai.

References

1. Mohd, M., Jan, R., Shah, M.: Text document summarization using word embedding. Expert Syst. Appl. **143**, 112958 (2020)
2. Gupta, S., Gupta, S.K.: Abstractive summarization: an overview of the state of the art. Expert Syst. Appl. **121**, 49–65 (2019)
3. Lin, H., Ng, V.: Abstractive summarization: a survey of the state of the art. Proc. AAAI Conf. Artif. Intell. **33**(01), 9815–9822 (2019)
4. Liu, Y., Titov, I., Lapata, M.: Single document summarization as tree induction. In: Proceedings of the 2019 Conference of the North American Chapter of the Association for Computational Linguistics: Human Language Technologies (Long and Short Papers), vol. 1, pp. 1745–1755 (2019)
5. Liu, K., Koyuncu, A., Kim, D., Bissyandé, T.F.: Tbar: revisiting template-based automated program repair. In: Proceedings of the 28th ACM SIGSOFT International Symposium on Software Testing and Analysis, pp. 31–42 (2019)
6. MacAvaney, S., Sotudeh, S., Cohan, A., Goharian, N., Talati, I., Filice, R.W.: Ontology-aware clinical abstractive summarization. In: Proceedings of the 42nd International ACM SIGIR Conference on Research and Development in Information Retrieval, pp. 1013–1016 (2019)
7. Mallick, C., Das, A.K., Dutta, M., Das, A.K., Sarkar, A.: Graph-based text summarization using modified TextRank. In: Nayak, J., Abraham, A., Krishna, B.M., Chandra Sekhar, G.T., Das, A.K. (eds.) Soft Computing in Data Analytics. AISC, vol. 758, pp. 137–146. Springer, Singapore (2019). https://doi.org/10.1007/978-981-13-0514-6_14
8. Ullah, S., Al Islam, A.B.M.A.: A framework for extractive text summarization using semantic graph based approach. In: Proceedings of the 6th International Conference on Networking, Systems and Security, pp. 48–56 (2019)
9. Bahdanau, D., Cho, K., Bengio, Y.: Neural machine translation by jointly learning to align and translate. arXiv1409.0473 (2014)
10. Rush, A.M., Chopra, S., Weston, J.: A neural attention model for abstractive sentence summarization. arXiv1509.00685 (2015)
11. Yuan, C., Bao, Z., Sanderson, M., Tang, Y.: Incorporating word attention with convolutional neural networks for abstractive summarization. World Wide Web **23**(1), 267–287 (2019)
12. Sheela, J., Janet, B.: An abstractive summary generation system for customer reviews and news article using deep learning. J. Ambient Intell. Humaniz. Comput. **12**(7), 7363–7373 (2020)
13. Chen, Y., Ma, Y., Mao, X., Li, Q.: Multi-task learning for abstractive and extractive summarization. Data Sci. Eng. **4**(1), 14–23 (2019)
14. Gao, Y., Wang, Y., Liu, L., Guo, Y., Huang, H.: Neural abstractive summarization fusing by global generative topics. Neural Comput. Appl. **32**(9), 5049–5058 (2019)
15. Luong, M.-T., Pham, H., Manning, C.D.: Effective approaches to attention-based neural machine translation. arXiv1508.04025 (2015)
16. Lin, J., Sun, X., Ma, S., Su, Q.: Global encoding for abstractive summarization. arXiv1805.03989 (2018)

Efficacy of Intelligent Knowledge Management Algorithms on Financial Reporting Quality in Nigerian Listed Companies

A. S. Bappah[1](✉), I. A. Mohammed[2], A. Garba[2], A. A. Palladan[2], S. S. Salihu[2], and H. S. Musa[3]

[1] Faculty of Technology Education, Abubakar Tafawa Balewa University, Bauchi, Nigeria
asbappah@atbu.edu.ng
[2] School of Business, Federal College of Education (Technical), Gombe, Nigeria
[3] School of Vocational Education, Abubakar Tatari Ali Polytechnic, Bauchi, Nigeria

Abstract. This study aims to examine intelligent knowledge management on the audit committee's effectiveness in financial reporting quality. The study utilised data from 101 non-financial firms listed on the Nigerian Stock Exchange (NSE) for the 2011 to 2015 period. The study employed the Generalised Method of Moments (GMM) technique using the system GMM to control potential endogeneity. The results indicate that the process of financial reporting quality has significantly improved. The findings suggest that audit committee shareholders' financial expertise, audit committee shareholder chairman, and external audit provide a significantly negative financial reporting relationship both in the short and long-run. However, the results indicated that the audit committee block shareholders' relationship with financial reporting reduced audit committee performance by increasing earnings management practice. Similarly, institutional ownership's interacting role improves the audit committee's overall effectiveness, particularly on audit committee block shareholder both in the short and long-run. The presence of shareholders with a knowledge management algorithm has provided an enabling environment for an external audit to perform its duty effectively. The finding is also robust in scope concerning unobserved heterogeneity, which prior studies have failed to address. The study offers a contemporary analysis of the audit committee's influence on financial reporting and has contributed to corporate governance and auditing debates in an emerging economy such as Nigeria.

Keywords: Knowledge management · Corporate governance · Financial reporting quality

1 Introduction

In recent time, the financial report received increased attention from regulators, stakeholders, and policymakers following various accounting scandals recorded worldwide [1]. Thus, to address financial reporting, the debate has focused on Knowledge Management mechanisms that might enhance financial reporting quality [2]. Effective corporate

M. Al-Emran et al. (Eds.): ICETIS 2021, LNNS 299, pp. 21–33, 2022.
https://doi.org/10.1007/978-3-030-82616-1_3

governance involves a series of mechanisms that guarantee an effective financial reporting process, accountability, and responsibility [3, 4]. These emphases are indicative of an association between corporate governance (CG) and financial reporting quality (FRQ) that an effective audit committee (AC) will improve on. Accounting scandals have led to the collapse of high-profile businesses across the globe, which raised serious concerns on CG mechanisms' effectiveness and monitoring strategies hitherto employed by firms to protect investors' interest [5]. In Nigeria, these scandals have prompted the release of the 2011 revised Code for Corporate Governance (CCG) and the establishment of the Financial Reporting Council of Nigeria (FRCN) in 2010 and 2011, respectively, to regulate the activities of Nigerian listed firms [6]. The International Financial Reporting Standards (IFRS) were also adopted to strengthen the financial reporting process's reliability and credibility [7]. Hence, investigating FRQ is a good way of assessing AC's effectiveness. Thus, the study examines some attributes regarding AC, Earning Management (EM), External Audit (EXAUD) and Institutional Ownership (IO) based on the agency theory and prior literature related to all variables of this present study. Thus, all variables are discussed to establish a relationship between the independent variables and the dependent variable.

1.1 Statement of the Problem

The financial reporting process and the auditing framework for Nigerian listed companies were generally deficient, and auditor's independence is often compromised, as reported by [8]. As a result of the impairment of auditor's independence primarily due to weak AC, several issues, like EM, financial fraud, and losses, have been reported. It involved some audit firms, which further led to professional misconduct among accountants and auditors in corporate Nigeria [9]. There is also inadequacy in the Statement of Accounting Standards (SAS) issued by the defunct Nigerian Accounting Standards Board (NASB). Such a lack of adequate standards has further created leeway for managers to indulge in more creative accounting, thus causing substantial financial losses. For example, a considerable debt totaling over $1.5 billion was reported by Arik Airline, which led to its takeover by the AMCON [10]. These different gravities of professional misconduct among accountants and auditors in corporate Nigeria have been glaring. Some accountants and auditors are determined to take advantage of the ambiguities that characterise the accounting systems to continue to create escape routes to justify their misdeeds. Hence, it is glaring that the market could not attract sustainable investment following investor confidence loss [11]. These irregularities have led to the collapse of several companies in both financial and non-financial sectors of the economy, regardless of having unqualified reports. These ugly situations have focused attention on regulatory reforms in financial reporting seeking ways to enhance firm levels' financial reporting integrity. Hence, the urgent need to circumvent this problem through the adoption and digital deployment of intelligent knowledge management algorithms based on leading learning models.

2 Literature Review

2.1 Intelligent Knowledge Management Algorithm

Knowledge plays a vital role in sustaining the firm's business. It can be described as one of the essential factors that companies require to maintain their business in the competitive market environment [12]. It is also regarded as a valuable intangible asset. It comprises all the concepts, data, idea technologies kept in the human brain [13]. Therefore, knowledge management (KM) is an important aspect that can influence the growing need for competition, performance standards, monitoring, measurement, flexibility, and control".

Consequently, using Extensible Business Reporting Language (XBRL) can assist in financial reports and business data, allowing these to be accurate, timely, efficient, and of good quality. KM enhances corporate knowledge that will significantly assist shareholders in the AC to create a vista for auditors to perform their job within the regulation time. It can also help them scrutinise effective financial statements and attest to the quality of the financial report [14]. Further, the XBRL method is compatible with the ODM of OMG. In an intermediate technology condition. However, [15] proposed intelligent information integration and pointed out that this method was also applicable to the intelligent incorporation of financial information, query and matching. [16] upgraded the technique proposed by [15] and encouraged using pattern matching and context intermediaries' technologies to solve the XBRL financial reporting taxonomy's semantic ambiguity.

2.2 Financial Reporting Quality

Issues influencing financial reporting quality continue to be of great interest to practitioners, regulators, and scholars alike [17]. Consequently, liberal accounting rules allow managers to use their discretion on what to report; thus, managers, in some cases, use accounting standards opportunistically for personal gain [18]. Thus, to address financial reporting, the debate has focused on corporate governance mechanisms innovation that might enhance FRQ [2]. Effective corporate governance involves a series of mechanisms that guarantee an effective financial reporting process, accountability, and responsibility [19]. These emphases indicate an association between corporate governance and FRQ that an effective AC will improve on. Practically, auditing is one of the methods available to lessen agency problems. The ownership separation from control in today's business arrangement has brought the need for the external examination of the financial report [20]. Such an independent external check of FRQ by a third party is thought to improve accountability and restore confidence and trust in the financial reporting process. However, the auditee and auditors' agreement bring into sharp focus other matters revolving around audit fees. Therefore, the need to evaluate the competitiveness of the audit market and the auditor's independence stimulates empirical investigation into audit pricing [21, 22].

External auditors have the duty of providing a certain level of assurance that a company's resources are being carefully managed reliably and credibly and by shareholders' interests. Thus, audits stewardship utilises agreed-upon standards and fiduciary duties to settle agency problems and, consequently, improve a firm's economic certainty [23].

Evidence from previous research indicates that external auditing offers an essential mechanism for shareholders to monitor managers' activities [24]. In auditing Literature, The Big4 audit firms are referred to as the brand name auditors. This audit firm has specific client quasi-rents, which function as a bond that assures independence for auditors' and its capacity to provide a high-quality audit service [25]. The reputation established by Big 4 audit firms deters them from acting opportunistically. The widespread belief is that such firms offer reasonable assurance outside statutory requirements. Empirical evidence has suggested that the Big 4 audit firm can charge a premium fee due to product differentiation [21]. An enormous investment in a modern audit by Big4 has resulted in acquiring skills and technical know-how, which serve as significant reasons for their services' unquestionable quality. Early research worldwide has adopted Simunic (1980) audit fees model for ascertaining the quality of external audit [26]. [22] work examined the determinants of audit fees in the US audit market. Using survey data of 397 publicly listed firms, he used Ordinary Least Square (OLS) regression. He found that the US audit market competitive irrespective of market segment, with big audit firms showing economies of scale.

2.3 Financial Reporting Quality and Earnings Management

Both developed and emerging nations reacted to poor corporate governance challenges, even before the 2001 Enron collapse, by releasing various corporate governance codes and best practices. The Enron embarrassment and other financial-reporting fraud in the US-led to the passage of the SOX of 2002 [27] a federal law with full corporate governance significance. Further, the Blue-Ribbon Committee's Report of 1999 [28] stresses that the AC must act as a robust monitor of the financial reporting process. Hence, the AC can question the external auditor, internal audit, and management to ascertain whether they serve the firm's best interest. Therefore, one of AC's main functions is to safeguard the external auditor's independence, owing to the strong economic ties inside directors have with the firm or its management.

The growing disappointment with AC performance in Nigeria has necessitated the need for an initiative for identifying the determinants of an effective AC to suit the Nigerian business climate. Consequently, the Nigerian regulators introduced a unique AC composition by including shareholders in the committee. The AC in Nigeria comprises an equal number of two non-executive directors and three shareholders [29, 30]. This is more interesting as the shareholders will know about their investment, which will stimulate their voting power during the AGM should anything go awry in the committee, especially when there is inadequate disclosure [31]. In Nigeria, regulators have enforced that AC members have accounting or financial knowledge [6]. Therefore, the AC should be composed of at least two independent directors (IND) and at least one with financial expertise and three shareholders as representatives of other shareholders. These regulations indicate clearly that the effectiveness of governance mechanisms depends to a large extent. The involvement of shareholders in the AC in Nigeria represents one of the most unique and recent regulatory enforcements that is expected to improve the effectiveness of AC compared to the guidelines for most developed countries, such as the US, the UK and Australia, as well as other emerging economies of Africa and Asia.

Including shareholders in the AC could improve FRQ. Thus, it is believed that as AC members, shareholders can bring their intelligent knowledge and skill to bear when executing AC's roles [31]. Additionally, features that can influence the AC's performance comprises up; members' status, knowledge of accounting and industry expertise [32]. Hence, shareholders are expected to efficiently play dual roles of being owners and display the ability to monitor the financial reporting process leading to FRQ [33]. It is also maintained that shareholders can monitor executive directors (ED) powers regarding financial reporting functions and protect auditor's independence [34].

Shareholders should assume the AC's leadership obligation considering their investment. The AC chair is more responsible for managing the financial reporting process [35] and liable to the FRQ [36]. Further, the AC, management, and external/internal auditors' contact point is the chairman [37]. Previous literature indicates that the AC chairman is one of the firm's top positions and a reliable source of power [38].

Financial expertise in the AC is one of the single most important AC features [1]. Hence, shareholders are expected to have at least first degrees; it's equivalent or higher degrees from a recognised university in Nigeria or abroad in accounting or management-related fields of study [6]. Prior studies argued that Enron's fall was due to the lack of relevant expertise of the AC [39]. Hence, a combination of knowledge, skills, and structural power from shareholders may bring forth AC's effectiveness. Further, individual block shareholders in the AC are likely to provide effective AC performance due to the number of shares held [40]. Consequently, individual block shareholder is better positioned to limit managerial myopia by inspiring managers to invest in long-term and profitable portfolios [41].

In recent time, shareholding across the globe tilted towards identifiable interest groups such as institutions like a pension fund, banks, and insurance [42]. [43] argue that institutional owners reduce information asymmetry and effectively tab management, which lowers EM and enhances quality financial reports. Further, institutional ownership (IO) strengthens CG mechanisms such as firms' ACs for proper supervision of the financial reporting process, reducing agency problem. [44] report that Indonesian firms with IOs suppress real earnings management. Based on the above discussions, the following hypotheses are developed.

Hypothesis 1: AC shareholders with financial expertise can lower EM in Nigerian listed firms.

Hypothesis 2: AC shareholders chairman can constrain EM in Nigerian listed firms.

Hypothesis 3: AC block holders can suppress EM in Nigerian listed firms.

Hypothesis 4: IOs could moderate the relationship between AC shareholders financial expertise, AC chair, AC block shareholder, and External audit.

Agency theory underpinned the present study as the focus of the research is on EM and AC composition. Agency theory emphasised on CG mechanism and human-factor reliant need to realise group corporate goals. However, agency theory believes that all those involved in managing firms' act opportunistically [45]. Further, agency theory advocates suggest that human-factor dependent and CG mechanisms be integrated to lower managers' opportunistic behaviour against shareholders' interest.

3 Methodology

The paper utilised dynamic panel data using GMM. It is more suitable for solving endogeneity issues, omitted variables, error terms, and unobserved heterogeneity. In the model below, ϑ_{it} represent DA, of the i listed firms for t years; α_i is the constant term, and $\gamma\vartheta_{i,t-1}$ indicates the lag value of DA, while Z_{it} is the predictor variable and ε_{it} is the error-term. The Sys-GMM estimations include the firm fixed effects (CFE_{dum}), which is meant to control for the heterogonous effects of unobserved factors across the firms. At the same time, the time (year) fixed effects (Y_j) capture the impact of the time trend for the periods under consideration.

$$\vartheta_{it} = a + \gamma\vartheta_{i,t-1} = \sum_{\rho=1}^{\rho} \beta_\rho Z^\rho it + \sum_{q=1}^{q} \beta_q Z^q it + \sum_{r=1}^{r} \beta_r Z^r it + \sum_{j=1}^{m} \delta j FFE_{dum_j} + \sum_{j=1}^{n} \theta_j Y_j + \varepsilon_{it} \tag{1}$$

Therefore, the modified Jones model (MJM) allows credit sales variations during the event because of EM [46]. The model is as follows:

$$TAC_{it} = EBIT_{it} - CFO_{it} \tag{2}$$

$$\frac{TAC_{it}}{TA_{it-1}} = \alpha_0 + \alpha_1\left(\frac{1}{TA_{it-1}}\right) + \alpha_2\left[\frac{(\Delta REV_{it} - \Delta REC_{it})}{TA_{it-1}}\right] + \alpha_3\left(\frac{PPE_{it}}{TA_{it-1}}\right) + \varepsilon_{it} \tag{3}$$

$$NDA_{it} = \alpha_0 + \alpha_1\left(\frac{1}{TA_{it-1}}\right) + \alpha_2\left[\frac{(\Delta REV_{it} - \Delta REC_{it})}{TA_{it-1}}\right] + \alpha_3\left(\frac{PPE_{it}}{TA_{it-1}}\right) + \varepsilon_{it} \tag{4}$$

$$DA_{it} = TAC_{it} - NDA_{it} \tag{5}$$

Where:

TAC_{it} = Aggregate accruals, TA_{it-1} = the book value of total assets of firm i at the end of year t − 1, $\Delta REV_{it}/TA_{it-1}$ = sales revenues of firm i in year t less revenues in year t − 1 scaled by TA_{it-1}, ΔREC it = the change in accounts receivables, PPE_{it}/TA_{it-1} = gross property, plant and equipment of firm i at the end of year t scaled by TA_{it-1}, $\alpha = \alpha_1\ \alpha_2\ \alpha_3$ = estimated parameters. ε_{it} = the residual. Consequently, the following regression model is used (Table 1).

$$\begin{aligned} DA_{it} &= \beta_0 + \beta_1 ACFE_{it} + \beta_2 ACSCr_{it} + \beta_3 ACBLS_{it} + \beta_4 EXAUD_{it} \\ &+ \beta_5 FRL_{it} + \beta_6 IO_{it} + \beta_7 ACSF{*}IO_{it} + \beta_8 ACSCr{*}IO_{it} \\ &+ \beta_9 ACBL{*}IO_{it} + \beta_{10} EXAUD_{it}{*}IO_{it} + \beta_{11} FSIZE_{it} \\ &+ \beta_{12} LEV_{it} + \beta_{13} PRO_{it} + \beta_{14} IFRS_{it} + \beta_{15} GRWTH_{it} \\ &+ \beta_{16} COMPLEX_{it} + \varepsilon_{it} \end{aligned}$$

Table 1. Variables, measurement, and sources

Variables	Measurement descriptions	Source
LFRL	Log of number of days from year-end to when the financial report is published	Aubert (2009)
ACSFE	Measured by 1 if shareholder is a financial expert 0 otherwise	Abernathy et al. (2014)
ACSCr	Measured by 1 if chairman is a shareholder 0 otherwise	Kibiya et al. (2016)
ACBLS	Measured by 1 if shareholder is a block shareholder, 0 otherwise	Abernathy et al. (2014)
EXAUD	Measured by natural log of audit fees	Mohamed and Habib (2013)
DA	Total Accruals minus Non-Discretionary Accruals	Dechow et al. (1995)
IO	5% or more shares held by investors	Dou et al. (2013)
FSIZE	Measured by the natural logarithm of total assets	Carpenter (2002)
PROF	Net profit divided by year-end owner's equity	Mollik and Bepari (2012)
LEV	Total liability divided by total assets	Hodgdon et al. (2009);
IFRS	Measured by dummy 1 if IFRS is used, 0 otherwise	Yaacob and Che Ahmad (2012)
COMPLEX	Dummy 1if a firm has a subsidiary, 0 otherwise	Abdulmalik and Che-Ahmad (2016)
GRWTH	Measured by market equity value to book value	Mayers (1977); Gavers (1995)
ROA	Measured by net income divided by total assets	Ashbrough (2003)
BSIZE	Total number of board members	Gaver (1995)

Table 1indicates the variable of the study and their measurements as well as their sources.

The study diagnosed issues that are concerned with problems econometric in panel data estimation as shown in below (Table 2);

Table 2. Diagnostic tests

Statistics	F	χ^2	Prob > χ^2	Prob > F	Mean VIF
Heteroskedasticity		(100) = 78496.06	0.0000		
Auto/serial correlation	4.612			0.0342	
Multicollinearity					2.12
Endogeneity DA	118.5386***			0.0000	

The tests results indicate that heteroskedasticity characterises the model, auto/serial correlation and endogeneity. Therefore, the use of GMM is justified. Although the model is free from multicollinearity as the mean VIF is 2.12.

Additionally, the long-run coefficients are generated using only the variables that are significant in the short-run and computed using the formula depicted below;

$$\beta_k \div [1 - \phi]$$

Thus, the following syntax is used: nlcom (_b[β_k])/(1 − b[L1 · DA]) for the DA models based on the number of variables that were significant in the short-run estimations. Consequently, both the short-run and long-run are estimated, and the results are presented in Table 3.

4 Results and Discussions

The result of the descriptive statistics of the continuous and dummy variables as depicted in Table 3 below;

Table 3. Descriptive statistics of continuous variables

Variable	Mean	Std. dev.	Min	Max
LFRL	95	59	10	365
EXAUD	14953	18352	400	145000
DA	0.27	0.18	0.01	0.55
IO	0.15	0.14	0	0.49
BSIZE	9	3	5	20
ROA	0.08	0.05	−0.02	0.33
LEV	0.11	0.03	0.05	0.2
GRWTH	0.08	0.03	0.06	0.21
FSIZE	4.79	0.84	2.54	7.45
PROF	0.14	0.13	−0.07	0.72

Descriptive statistics for dummy variables

	Frequency		Percentage	
Variable	1	0	1	0
ACSFE	435	70	86	14
ACSCr	480	25	95	5
ACBLS	151	354	30	70
COMPLEX	360	145	71	29
IFRS	404	101	80	20

DA is estimated using the MJM and reported based on absolute values. It means that the extent to which managers are involved in creative accounting among firms listed on the NSE is alarming, and therefore, a cause for concern. As indicated by the table, other variables suggested that they are within normal based on their minimum, maximum and standard deviations. Additionally, dummy variables results show that 86% (435 firms) are shareholders with financial expertise, while 14% (70 firms) are without financial expertise. Thus, signifying compliance with SEC regulation. Shareholder's involvement in the AC can bring about fruitful development by lowering EM. The results also revealed that 95% (480 firms) ACs are chaired by a shareholder, while an IND chairs 5% (25 firms) in the AC. Further, the number of block shareholders suggests that 30% (151 firms) of the AC's of Nigerian firms include block shareholders, while 70% (354 firms) do not have block holders. Hence, block holders are expected to enhance the reporting process and provide effective monitoring.

The study estimates a two-step system GMM for EM. The results of the model are shown in Table 4. The two steps are more robust, as argue by (Roodman 2009). The Wald Chi^2 (21) for DA is 454574.78 and statistically significant 1%. This result confirms that all variables are vital and, therefore, must be retained in the model. Thus, the model is found to be fit.

Table 4. Twostep system GMM regression results

	Short-run			Long-run		
Variables	Coef.	z-Stat	p-Value	Coef.	z-Stat	p-Value
DA L1	−0.1266	−7.19	0.000			
ACSFE	−0.2184***	−3.29	0.001	−0.1938***	−3.28	0.001
ACSCr	−0.5660***	−4.35	0.000	−0.5024***	−4.45	0.000
ACBLS	0.0567*	1.79	0.074	0.0503*	1.79	0.073
EXAUD	−0.5476***	−22.09	0.000	−0.4861***	−23.23	0.000
LFRL	0.0555**	2.10	0.036	0.0493**	2.16	0.031
IO	1.9507*	1.93	0.053	1.7315*	1.93	0.053
IO*ACSFE	−2.7687***	−5.48	0.000	−2.4575***	−5.42	0.000
IO*ACSCr	−2.0757***	−5.17	0.000	−1.8424***	−5.31	0.000
IO*ACBLS	−0.5301***	−3.13	0.002	−0.4705***	−3.15	0.002
IO*EXAUD	−0.7828***	−3.77	0.000	−0.6948***	−3.8	0.000
BSIZE	−0.0095***	−2.63	0.009			
LEV	0.0157**	2.10	0.035			
ROA	0.0023***	6.15	0.000			
IFRS	−0.0955	−1.52	0.128			

(*continued*)

Table 4. (*continued*)

	Short-run			Long-run		
Variables	Coef.	z-Stat	p-Value	Coef.	z-Stat	p-Value
FSIZE	0.0418***	3.31	0.001			
_cons	1.2197***	8.32	0.000			
Statistics	**Coefficient**		**p-Value**			
Wald Chi2 (24)	454574.78					
Prob > Chi^2			0.000			
AR2			0.339			
Hansen J.			0.279			
No. of Group	100					
No. of Inst.	96					
Industry	Yes					
Year Effect	Yes					

The results show that a percentage increase in the number of AC's shareholders as ACFE in the AC is related with 22% in the short-run and 19% decrease in EM respectively in the long-run at the 1% significant level, on average *ceteris paribus*. It further suggests that the effect of shareholders in the short-run is better in decreasing EM's extent. Similarly, a 1% increase in ACSCr could lower EM by 57% and 50% in both the short-run and the long-run, respectively. However, the results of ACBLS indicate a similarity of poor performance by increasing EM by 6% and 5% instead of reducing EM *ceteris paribus*. While the result of EXAUD shows a 55% and 49% decrease on EM in the short and long-run at 1% significant level, respectively *ceteris paribus*. Overall, the result indicates that adequate intelligent knowledge management could enhance AC shareholders' performance in both the short-run and long-run. Consequently, interactions of IOs also suggest that an intelligent knowledge management algorithm is an important mechanism that will enhance the effect of IOs, making sure that AC with shareholders provided improvement in reducing the magnitude of EM in Nigeria.

5 Conclusion

This present study explores the moderating effects of IOs in determining the relationship between some of AC's attributes and EM on the Financial Reporting Quality in Nigerian listed firms. Following the accounting scandals leading to several companies' failures due to poor AC performance, factors that can effectively enhance AC performance to suit the Nigerian business horizon have been identified. Consequently, the revised CCG and the CAMA provide a unique AC composition by including shareholders. The AC composition in Nigeria now is made up of two non-IND and three shareholders under section 404 (3) and Part E Article 30. Contrary to what is obtainable in most developed countries, such as the US and the UK, where ownership of shares is dispersed. However, share ownership in Nigeria, and most emerging economies worldwide is tilted towards concentration, where shares are owned by various and identifiable interest groups, such as individuals, foreigners and managers.

Further, Nigeria's political history offers a clear view of how developing shareholder activism as a CG mechanism can succeed in a country with a political background, like Nigeria. The study has confirmed that the Modified-Jones accrual model by [46] is one of the most reliable models to detect manipulative financial reports, using the manager's discretionary right of accounting methods choice estimates. Findings from this study suggested that as sitting members in the AC, shareholders could restore investors' confidence such that foreign and domestic investors can invest in firms listed on the Nigerian capital market without any fear. Additionally, we also hope that the research will benefit the regulators in taking the necessary actions to overcome matters related to scandals, failures and improve financial reporting quality.

References

1. Ghafran, C., Yasmin, S.: Audit committee chair and financial reporting timeliness: a focus on financial, experiential and monitoring expertise. Int. J. Audit. **22**(1), 13–24 (2018)
2. Al-Shaer, H., Salama, A., Toms, S.: Audit committees and financial reporting quality: evidence from UK environmental accounting disclosures. J. Appl. Acc. Res. **18**(1), 1–38 (2017)
3. Haq, I.: Role of the audit committee chair in the financial reporting process. FIU Eletronic Theses and Dissertation, 2212 (2015)
4. Li, J., Mangena, M., Pike, R.: The effect of audit committee characteristics on intellectual capital disclosure. Br. Account. Rev. **44**(2), 98–110 (2012)
5. Khalil, M., Ozkan, A.: Board independence, audit quality and earnings management: evidence from Egypt. J. Emerg. Market Finan. **15**(1), 1–35 (2016)
6. SEC: Code of corporate governance for public companies in Nigeria (2011)
7. Herbert, W.E., Anyahara, I.O., Okoroafor, E.N., Onyilo, F.: Financial reporting council of Nigeria and the future of accounting profession in Nigeria. Int. J. Finan. Account. **5**(3), 146–157 (2016)
8. Adeyemi, S.B., Fagbemi, T.O.: Audit quality, corporate governance and firm characteristics in Nigeria. Int. J. Bus. Manag. **5**(5), 169–179 (2010)
9. Punch, N.: Corporate governance and fraud cases in Nigeria banks (2015). www.punchng.com/?s=fraud+in+bank. Assessed 10 Apr 2015
10. Confidential, T.E.: N300bn Debt: AMCON Takes over Arik Air (2017). http://economicconfidential.com/business/n300bn-debt-amcon-arik/
11. Oteh, A.: A Roadmap for Transforming the Nigerian Capital Markets. Press Briefing, Securities and Exchange Commission Media Center, Lagos, February 5 (2010)
12. Alvarenga, A., Matos, F., Godina, R., Matias, J.C.O.: Digital transformation and knowledge management in the public sector. Sustainability (Switzerland) **12**(14), 2–24 (2020)
13. Rezgui, Y.: Information Retrieval Algorithms for Knowledge Management - The Challenge Continues, December (2015)
14. Omar, N., Amirah, Z.J., Smith, M.: Predicting fraudulent financial reporting using artificial neural network. Knowl. Based Syst. **5**(4), 1–9 (2017)
15. Troshani, I., Lymer, A.: Translation in XBRL standardization. Inf. Technol. People **23**(2), 136–164 (2010)

16. Zhu, H., Wu, H.: Quality of data standards: framework and illustration using XBRL taxonomy and instances. Electron. Mark. **2**(21), 129–139 (2011)
17. Abdulmalik, S., Che-Ahmad, A.: Auditor's Risk Assessment of Independent Directors in Nigeria. SHS Web Conf. **34** 10001 (2017)
18. Li, X.J., Ahmed, R., Than, E.T., Ishaque, M., Huynh, T.L.D.: Gender diversity of boards and executives on real earnings management in the bull or bear period: empirical evidence from China. Int. J. Finan. Econ. 1–19 (2021)
19. Haq, I. (2015). Role of the audit committee chair in the financial reporting process. FIU Electronic Theses and Dissertation, 2212. https://doi.org/10.25148/etd.FIDC000101
20. Rossi, F., Barth, J.R., Cebula, R.J.: Do shareholder coalitions affect agency costs? Evidence from Italian-listed companies. Res. Int. Bus. Financ. **46**, 181–200 (2018)
21. Hay, D.C.: Audit fee research on issues related to ethics. Curr. Issues Audit. **11**(2), A1–A22 (2017)
22. Simunic, D.A.: The pricing of audit services: theory and evidence. J. Account. Res. **18**(1), 161 (1980)
23. Samaha, K., Khlif, H.: Audit-related attributes, regulatory reforms and timely disclosure. J. Finan. Report. Account. **15**(2), 158–179 (2017)
24. Abdulmalik, S.O., Che-Ahmad, A.: Audit fees, corporate governance mechanisms and financial reporting quality in Nigeria. DLSU Bus. Econ. Rev. **26**(1), 122–135 (2016)
25. Cudia, C.P., Dela Cruz, A.L.C.: Determinants of earnings management choice among publicly listed industrial firms in the Philippines. DLSU Bus. Econ. Rev. **27**(2), 119–129 (2018)
26. Al-Rassas, A.H., Kamardin, H.: Internal and external audit attributes, audit committee characteristics, ownership concentration and earnings quality: evidence from Malaysia. Mediterr. J. Soc. Sci. **6**(3), 458–470 (2015)
27. Sarbanes, P.: Sarbanes-oxley act of 2002. In: The Public Company Accounting Reform and Investor Protection Act (2002)
28. BRC: Report and Recommendations of the Blue-Ribbon Committee on Improving the Effectiveness of Corporate Audit Committees. New York Stock Exchange and National Association of Securities Dealers, New York (1999)
29. CAMA: Laws of the Federation of Nigeria, CAP 59, vol. III (2020)
30. Leech, D.: Shareholder voting power and ownership control of companies. In: Power, Voting, and Voting Power: 30 Years After, pp. 475–498. Springer, Heidelberg (2013). https://doi.org/10.1007/978-3-642-35929-3
31. Kibiya, M.U., Che-Ahmad, A., Amran, N.A.: Audit committee independence, financial expertise, share ownership and financial reporting quality: further evidence from Nigeria. Int. J. Econ. Financ. Issues **6**(7S), 125–131 (2016)
32. Carol Liu, M.H., Tiras, S.L., Zhuang, Z.: Audit committee accounting expertise, expectations management, and nonnegative earnings surprises. J. Account. Publ. Policy **33**(2), 145–166 (2014)
33. Sunusi, M.S.L.: The impact of the global financial crisis on the Nigerian capital market and the reforms. In: The Impact of the Global Financial Crisis on the Nigerian Capital Market and the Reforms. 7th Annual Pearl Awards and Public Lectures, Muson Centre, Onikan, Lagos (2011)
34. Enofe, A., Aronmwan, E., Abadua, H.: Audit Committee Report in Corporate Financial Statements: Users' Perception in Nigeria, 1 March (2013)
35. Schmidt, J., Wilkins, M.S.: Bringing darkness to light: the influence of auditor quality and audit committee expertise on the timeliness of financial statement restatement disclosures. Audit. J. Pract. Theory **32**(1), 221–244 (2013)
36. Engel, E., Hayes, R.M., Wang, X.: Audit committee compensation and the demand for monitoring of the financial reporting process. J. Account. Econ. **49**(1–2), 136–154 (2010)

37. PricewaterhouseCoopers, L.L.P.: Audit committees: good practices for meeting market expectations (2003)
38. Sharma, D.S., Tanyi, P.N., Litt, B.A.: Costs of mandatory periodic audit partner rotation: evidence from audit fees and audit timeliness. Audit. J. Pract. Theory **36**(1), 129–149 (2017)
39. Breeden, R.C.: Report of Richard C. Breeden, New York (2003)
40. Arowolo, R.O., Che-Ahmad, A.: Block-holders and monitoring mechanisms in Nigeria. Asian J. Account. Govern. **78**(8), 71–78 (2017)
41. Habbash, M.: Earnings management, audit committee effectiveness and the role of blockholders ownership: evidence from UK large firms. Int. J. Bus. Gov. Regul. **8**(2), 155 (2013)
42. Amin, A.S., Dutta, S., Saadi, S., Vora, P.P.: Institutional shareholding and information content of dividend surprises: re-examining the dynamics in dividend-reappearance era. J. Corp. Finan. **31**, 152–170 (2015)
43. Elyasiani, E., Wen, Y., Zhang, R.: Institutional ownership and earning management by bank holding companies. J. Financ. Res. **40**(2), 147–178 (2017)
44. Susanto, Y.K., Pradipta, A.: Corporate governance and real earnings management. Int. J. Bus. Econ. Law **9**(1), 17–23 (2016)
45. Jensen, M.C., Meckling, W.H.: Theory of the firm: managerial behavior, agency costs and ownership structure. J. Financ. Econ. **3**(4), 305–360 (1976)
46. Dechow, P.M., Sloan, R.G., Sweeney, A.P.: Detecting earnings management. Account. Rev. **70**(2), 193–225 (1995)

Towards Integrated Access Control for Software Product Line Engineering Web Development

Ichlasul Affan(✉), Adrika Novrialdi, Oman Komarudin, Hafiyyan Sayyid Fadhlillah, and Ade Azurat

Universitas Indonesia, 16424 Depok, Indonesia
{ichlasul.affan,adrika.novrialdi,oman.komarudin}@ui.ac.id, {hafiyyan,ade}@cs.ui.ac.id

Abstract. The needs to reduce costs on cloning and adapting software leads to the emerging of Software Product Line Engineering (SPLE), which enables developers to create different software with common requirements and defined variabilities. To support web application generation based on the SPLE approach, a dedicated engine is necessary. We built the SPLELive Engine as our alternative approach to that problem. However, we also need to address another issue of SPLE web generation engine which is multi-product access control. This paper explains how the multi-product access control issue was tackled by doing experimental research using the single sign-on approach with external OAuth 2 authentication services and a decentralized authorization mechanism. This paper also explains and demonstrates how the SPLELive Engine handles authorization configuration variability using delta-oriented programming.

Keywords: Software product line · Access control · Single sign-on · Multi-product deployment

1 Introduction

Companies can employ individual software that is specifically designed to meet their needs. Although several companies have similar types of businesses, each has different characteristics and varying requirements [2]. Clone-and-own approach which clones a software and adapts the clone for other uses could be considered, but the approach requires sophisticated technical skills to modify or adapt the software to meet organizational needs [6].

The costs of adopting cloned software are as high as the costs of procuring new proprietary software, as well as costs for adjustments and improvements necessary to meet the needs of the company. Developing diverse software applications via the clone-and-own process might work for a small amount of products, but when this number increases, this will become a maintenance problem. Developers need to make sure that changes in core components are applied to every single cloned software [11].

Rather than developing a single unique software, Pohl et al. introduced a promising and emerging concept called software product line engineering (SPLE) [12]. SPLE

M. Al-Emran et al. (Eds.): ICETIS 2021, LNNS 299, pp. 34–43, 2022.
https://doi.org/10.1007/978-3-030-82616-1_4

enables developers to create different software with common basic requirements, each with defined variability. The SPLE is a promising method in reducing the complexity of maintaining multiple similar software products. However, this concept still requires support and enabling technology [12] in each phase of the development process.

Developing software based on SPLE has a very different approach in comparison to conventional software. The software developed with SPLE is very likely to change. Slightly different with flexibility during traditional software development, in SPLE product variants are created by adjusting the base artifact to match the requirement of said product variant [12]. This approach needs a mechanism to adjust the base artifact, and that mechanism should be effective, efficient, and easy to reuse. The need of a mechanism that support the SPLE approach led to the need of a specific framework to support the mechanism itself [8].

Software has developed and nowadays, different kinds of software are available. One commonly used nowadays is web application. Many business use web application as one of the "cores" to their business. This fact encouraged us to try applying SPLE in web application.

Like every other application known, security aspect is something that developers must be aware of when developing the web application. The loss because of security vulnerability has proved to be costly. The impact of security failure on customer behaviour shown in [4], when a membership-based Internet shopping mall failed to protect its customer information from a hacker. The compromised shopping site lost a considerable number of customers. Because of how important security issues are, security issues also need to be considered when developing web applications using SPLE. Applying security through access control in a conventional web application can be done through defining the authorization rule between the endpoints and roles available in the system. This rule ensures that a certain endpoint can only be accessed by certain roles.

However, different approaches in developing also means a different approach needed to ensure security in web applications developed using SPLE. The different requirements of product variants led to different needs of each products' variants in securing the application. For example, one endpoint may need a certain role to be able to access that endpoint, but it may not be the same for another product variant. The role or even the endpoint itself could be different for each variant. It is also possible that a role or endpoint from a product variant does not exist in another variant.

This research wants to solve the issue of authentication and authorization in web applications based on SPLE. We created SPLELive Engine[1] as a way to creating web applications using SPLE. To solve this issue of authentication and authorization in SPLELive Engine, external authentication service (OAuth 2) is used and delta-oriented programming is applied to the endpoint routing to solve the requirement change in endpoint routing.

The structure of this paper is outlined as follows. Section 2 explains some related works and the SPLELive Engine that was mentioned earlier. Section 3 shows the reason why authorization-related research needs to be conducted. Section 4 will explain the issue related to multi-product deployment. Section 5 will show how the implementation

[1] https://www.splelive.id.

works. Section 6 provides an example of the SPLELive Engine in action with a case study. Finally, in Sects. 7 and 8 will show possible future works and conclude the discussion.

2 Related Works

The SPLELive Engine is an engine that is able to generate a whole web application, from the application display, business logic, and the mechanism to communicate to the database. SPLELive Engine is able to create web application based on the feature selection from user input and adjusts the generated application based on the input.

The backend layer uses Abstract Behavioural Specification (ABS) and a microservices architecture developed by Naily et al. [10], which complies with the delta-oriented programming (DOP) paradigm. The usage of the DOP enable us to express that a product line is represented by a core module and a set of delta modules [16]. The core module provides an implementation of a valid product that can be developed with well-established single application engineering techniques. Delta modules specify changes to be applied to the core module to implement further products by adding, modifying, and removing code. Later, delta can be applied to the core modules to create a running software based on the selected features.

The result of the customization in the ABS Microservices is then processed by a database schema generator [14] to create the database schema of the generated application. By creating the database based on the result of delta in the ABS Microservices [10], this database schema generator [14] also enables the adoption of the selected feature. The engine then creates the needed display view based on the selected feature. In the end, this engine is able to create a whole working web application from the feature selected by the user.

Our works is not the first effort to provide tools that can automatically generate web applications in SPLE. FeatureJS is an example of a plugin that supports the development of HTML and Javascript in SPLE [5]. Using FeatureJS, variations can be mapped within the source code to a variability model. Later, FeatureJS will compose the target Javascript and HTML source code that corresponds to the selected features.

Horcas et al. integrates a Common Variability Language (CVL) with annotations to manage variability in SPLE for web application [3]. To automatically generate web application, there are various artifacts that need to be generated (e.g. HTML, Javascript, CSS, JSON Config file, and server-side source codes). By integrating CVL, a base model can be used to represent various artifacts in the web application. The variability model in CVL will be used to declare the variation points within a base model that need to be reconfigured. Each variation points represented in the variability model will be mapped to one or more annotations in the artifacts. Later, the engine will compose the target application by processing the annotated artifacts according to the selected variations.

Both of the aforementioned works use annotations in their artifacts (e.g. HTML, Javascript, CSS, JSON Config file, and server-side source codes) as a mechanism that will be later used to generate web applications. In contrast, our works more focused on generating web applications using DOP approach. This research uses ABS as a main language, which is later used for generating various artifacts for producing the web application. Thus, the implementation process can be simplified since there is no need to maintain various artifacts at the same time.

3 Authentication in SPLE

In building software that will be used for multiple users, certain users might not need or even be denied access to sensitive things. As mentioned before, there is still much work that needs to be conducted on the ABS Microservices Framework, one of which is security mechanism [10]. The access control mechanism also acts as one of the crucial gates to the protected resources, so it needs to be prioritized.

Software access control in a single instance is relatively easy, since interactions only happened within that instance. The problem with access control within multiple resource instances is the need of user access restriction to different resource instances. Every service needs to know who is currently accessing their endpoints, and whether they are allowed to access that endpoint. Meanwhile, not every service should know every users' identity, which would otherwise void their privacy violations.

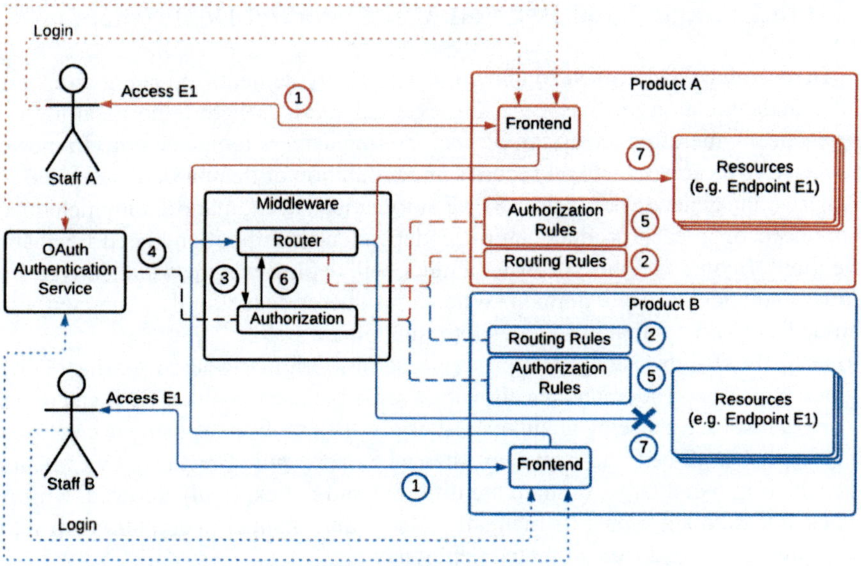

Fig. 1. Modified OAuth workflow for multi-product deployment.

Single Sign-On (SSO) is a prospective approach for solving authentication on multiple software instances. In SSO, user authentication is handled by a different service so that every user can login to any related services by using a single account. We chose the approach where the SSO service only handles authentication, meanwhile access restriction is handled by each respective services [13].

OAuth was used to realize the implementation of SSO in SPLE. We took advantage of the usage of access tokens as a result of authentication process, which represents currently authenticated user. OAuth eliminates the need for resource services to impersonate the user by remembering user credentials [15]. OAuth also enables usage of JavaScript frontend clients.

As illustrated in Fig. 1 by dotted lines, frontend clients will only need to forward user credentials to the authentication service to login. Later, frontend clients can forward user requests on resources in the product using the returned access token. During step number 4, the protected service needs to validate the access token before deciding to allow or deny a request in step number 5 and 6.

OAuth as the authentication protocol also fits the authentication of multiple products. In a SPLE-based web development context, the original role of the authorization server needs to be modified to only handle authentication process. The authorization mechanism will be handled by each product, which in itself contains one or more protected resources. By decentralizing the authorization mechanism, each product can maintain their own authorization rules without altering configurations on shared resources. Thus, the variabilities chosen by a product will not interfere with the authorization of other products, as illustrated by Fig. 1.

4 OAuth Configuration Issue on Multi-product Deployment

ABS Microservices is designed to allow users to easily authenticate using an external OAuth 2 authentication service. By using external service, there is no need to "reinvent the wheel"; therefore, costs can be focused on business logic and core framework maintenance. We can also rely on security and reputation of that particular service.

To derive and generate products in SPLE automatically, automatic deployment mechanism is needed, so each product can run independently without any need for manual interventions. In our case, the SPLELive Engine will deploy every product derived from a product line under a single domain. When a product is successfully generated, users can immediately access that product using a subdomain.

To set up the OAuth 2 authentication service, Cross Origin Resource Sharing (CORS) restriction must be considered when allowing users to request logins. CORS can prevent cross-origin resource abuse by unauthorized origins [9], such as exposing access tokens to other apps. CORS can allow or deny sites when performing requests by evaluating whether the origin and target domain are different and not explicitly allowed. Unfortunately, each trusted sites must be manually added, thus manual interventions to set up OAuth 2 are still needed for automated deployment.

To avoid manual interventions, each generated product will be deployed in a subdomain of certain sites. Later, we can configure the OAuth authentication service to use wildcard subdomain configurations. Since the OAuth authentication service now trusts the whole domain instead of specific subdomains, each generated product will now be allowed to use the service. This approach will be beneficial to programmers because requiring a manual set up could be error-prone and more difficult to adapt due to changes within the product itself.

5 Authentication and Authorization Implementation

Authorization mechanism for the SPLELive Engine is implemented by adding a module inside the ABS Microservices Framework. The `Authorization` module is contained in the Middleware, which acts as set of intermediary modules between incoming requests and other framework layers. Alongside the `Authorization` module, Middleware also contains modules such as `ABSCaller`, which serves as a routing mechanism.

```
# OAuth service definition
auth=auth0
client_id=example.auth0.com$123456abcdef
# Role definition
administrator=alex@example.com
manager=safira@example.com
registered=nadia@example.com, utsman@example.com
```

Listing 1. `auth.properties` example.

Currently, the Authorization middleware reads as a file called auth.properties, as shown in Listing 1. It contains basic OAuth client configurations, such as Client ID. It also contains the relationship between user emails and their roles. The `administrator` role is a mandatory superuser role that can access every endpoint. To simulate the OAuth authentication mechanism, Auth0[2] is used, as it is allows for wildcard subdomains. Currently, support for other OAuth services has not yet been implemented or tested, but the Authorization module itself can be extended to also support those services. JSON Web Token (JWT) standard [1] is used for the access tokens.

As illustrated in Fig. 2, the `Authorization` module will be executed after `ABSCaller` routes user requests to certain resource endpoints. The `Authorization` module will read the access token given by a user. If there is no access token given, the user will be regarded as anonymous. To validate the access token, the `Authorization` module will either contact the chosen OAuth service or validate the signature by itself. If it is valid, the module will read the JWT access token payload to get the user's email. Then, the module will determine eligible roles based on the user-role relationship, defined in `auth.properties`, to compare it with the roles that are allowed to access the target resource. Then, the `auth.properties` will read every user request received by `ABSCaller`, so that the user-role relationships can be changed in real time. The decision will then be returned to `ABSCaller`, which will return `Unauthorized` error message to the user if denied, otherwise the request will be forwarded to the target resource.

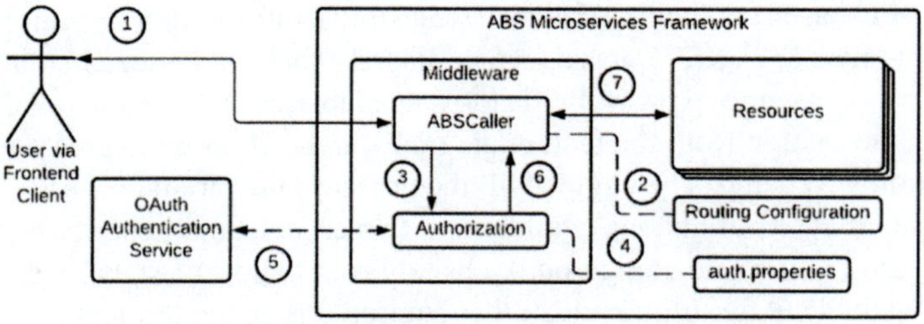

Fig. 2. ABS Microservices Framework structure.

[2] https://www.auth0.com.

The ABS Microservices Framework can manage product variabilities, which includes role to resource access rules. Roles that are allowed to access certain endpoints can be defined via `Route.abs`, which is the routing configuration. Like any other ABS source code, `Route.abs` can also have deltas to modify or add routing functions. For example, as illustrated in Fig. 3, feature `F1` wants endpoint `E1` allowed for role `R1`. Meanwhile, feature `F2` is relying on the same resource as `F1`, with endpoint `E1` now also allowed for role `R2`. By using delta, the routing rule of endpoint `E1` can be redefined to reflect that change. Thus, authorization for multiple products can be easily set up without introducing more code duplication. The next section will present a case study to demonstrate this design.

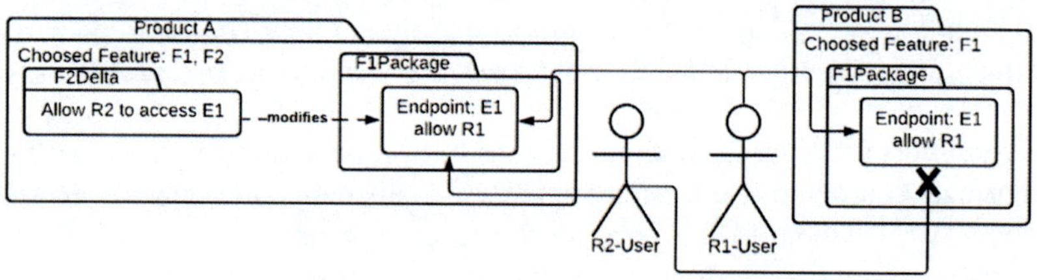

Fig. 3. Example of managing authorization variabilities using delta.

6 Case Study

In this case study we use a Command Line Interface (CLI) tool as a medium to configure and create the application and the ready-to-run jar. The CLI deployment tool takes selected features as input. The CLI deployment tool creates the display (the view layer) and generates the needed backend layer and database. The file named `auth.properties` is where the auth configuration lies. The role and its corresponding email need to be defined in this file.

We use the SPLELive Engine to generate web apps with AISCO as a case study.[3] AISCO is an SPL-based application that realizes adaptive information systems for charity organization domain [7]. AISCO has a feature diagram that can be seen in our website.[4] Feature diagram is used to express the features available, as well as the relationship of the feature. Combination from the feature diagram can be choosed to create variants of the organization's web app. The engine will then create said variant.

To use and create a variant, stakeholders (in this case, a small charity organization owner) can go to feature selection at our website. From there, the stakeholder can pick the features available. After filling out all the required forms, the product is built. When the product creation is finished, a confirmation email will be sent to the stakeholder. Then, the stakeholder will be able to access the application.

3 https://aisco.splelive.id/.

4 https://www.splelive.id/research.

```
module ABS.Framework.Route;
class RouteConfigImpl {
    String routeCreateIncome(String url) {
        String result = case url {
            "/api/income/create.abs" =>
"manager:MIncomeResource.IncomeResourceImpl@create";
            _ => "";
        };
        return result;
    }
}
```

Listing 2. Base module `Route.abs`.

```
delta DConfirmationRoute;
uses ABS.Framework.Route;

modifies class RouteConfigImpl {
    modifies String routeCreateIncome(String url) {
        String result = case url {
            "/api/income/create.abs" =>
"manager,registered:MIncomeResource.IncomeResourceImpl@create";
            _ => "";
        };
        return result;
    }
}
```

Listing 3. Delta for `Route.abs`.

```
productline AISCO;
features Confirmation, ...; //omitted for brevity
delta DConfirmationRoute when Confirmation;
```

Listing 4. Delta assignment configuration.

Listings 2, 3, and 4 demonstrates how the SPLELive Engine manages variabilities of authorization rules using DOP related to Fig. 3. In the base routing configuration, base endpoints and the required roles (if they exist) to access said endpoint were defined. As shown in Listing 2, the endpoint `/api/income/create.abs` requires the `manager` role to access it. Later, the delta will be applied to the base routing configuration to meet specific needs. Listing 3 is a delta that modifies the roles required to access the `/api/income/create.abs` endpoint. If this delta is applied, the `manager` and `registered` roles can access that endpoint.

Configure `framework/ProductConfig.abs` as shown in Listing 4 to assign the `DConfirmationRoute` delta with the `Confirmation` feature. Choosing the `Confirmation` feature will enable access to resources on `/api/income/create.abs` for the `manager` and `registered` role. Otherwise, only the `manager` role can access it.

7 Discussion

Further challenges were introduced when building multiple Representational State Transfer (REST) services for a product, as demonstrated by [17]. We need to synchronize user and role management across those services, while keeping each services loosely coupled. Thus, we need to determine how we can introduce synchronized authorization for user and role management for all instances in the particular product. SPLE web frameworks that was discussed in Sect. 2 are concentrating on building monolithic applications, meanwhile distributed approaches such as Microservices increases the need of improving our current approach.

Based on our current work, we propose to solve the issue by introducing different authentication and authorization layers. Products can keep using the existing OAuth 2 authentication service to authenticate their users. The `middleware` module should then be separated as a different service, which serves the routing mechanism and authorization. The module then acts as a gateway for incoming requests. The authorization process will be done in front, before forwarding the allowed requests to the corresponding service. By using this approach, services no longer need to ask the module to check for permissions.

8 Conclusion

Based on this work, we conclude that using a centralized authentication service is a prospective starting point towards access control on SPLE-based web development, which we implemented using the Single Sign On approach and utilizing an OAuth 2 authentication service that trusts wildcard subdomain as JavaScript request origin. The authorization mechanism is embedded in the middleware of the ABS Microservices Framework, so the authorization process is decentralized to each product's REST API. Our current results show that the fully automated deployment of web products generated using the SPLELive Engine is now possible because we can avoid manual setups, especially regarding authentication services.

Acknowledgement. This work was partially funded by Universitas Indonesia under PUTI Grant Number: NKB-883/UN2.RST/HKP.05.00/2020.

References

1. Bertocci, V., OAuth Working Group: JSON web token (JWT) profile for OAuth 2.0 access tokens (2019). https://tools.ietf.org/html/draft-ietf-oauth-access-token-jwt-02
2. Dubinsky, Y., Rubin, J., Berger, T., Duszynski, S., Becker, M., Czarnecki, K.: An exploratory study of cloning in industrial software product lines, pp. 25–34 (2013)
3. Horcas, J.M., Cortiñas, A., Fuentes, L., Luaces, M.R.: Integrating the Common Variability Language with Multilanguage Annotations for Web Engineering. ACM Press, New York (2018)
4. Lee, M., Lee, J.: The impact of information security failure on customer behaviors: a study on a large-scale hacking incident on the internet (2012)
5. Machado, I.D.C., Santos, A.R., Cavalcanti, Y.A.C., Trzan, E.G., de Souza, M.M.A., de Almeida, E.S.: Low-level variability support for web-based software product lines. In: Proceedings of the Eighth International Workshop on Variability Modelling of Software-Intensive Systems. Association for Computing Machinery, New York (2014)
6. Mateen, A., Kausar, S., Sattar, A.R.: A software reuse approach and its effect on software quality, an empirical study for the software industry (2017)

7. Mauliadi, R., Setyautami, M.R., Afriyanti, I., Azurat, A.: A platform for charities system generation with SPL approach. In: Proceedings of the 2017 International Conference on Information Technology Systems and Innovation, pp. 108–113 (2018)
8. Metzger, A., Pohl, K.: Software product line engineering and variability management: achievements and challenges. Proc. Future Softw. Eng. - FOSE **2014**, 70–84 (2014)
9. Mozilla Foundation: Cross-origin resource sharing (CORS) (2020). https://developer.mozilla.org/en-US/docs/Web/HTTP/CORS
10. Naily, M.A., Setyautami, M.R.A., Muschevici, R., Azurat, A.: A framework for modelling variable microservices as software product lines. In: Cerone, A., Roveri, M. (eds.) SEFM 2017. LNCS, vol. 10729, pp. 246–261. Springer, Cham (2018). https://doi.org/10.1007/978-3-319-74781-1_18
11. Nishiura, Y., Asano, M., Nakanishi, T.: Migration to software product line development of automotive body parts by architectural refinement with feature analysis. In: 2018 25th Asia-Pacific Software Engineering Conference, pp. 522–531 (2018)
12. Pohl, K., Bockle, G., van der Linden, F.: Software Product Line Engineering: Foundations, Principles, and Techniques. Springer, Heidelberg (2005). https://doi.org/10.1007/3-540-28901-1
13. Radha, V., Reddy, D.H.: A survey on single sign-on techniques (2012)
14. Rhizadi, F., Fadhlillah, H.S., Azurat, A., Afriyanti, I., Apriani, N.F.: Database generator to support product derivation in SPL. In: 2019 International Conference on Advanced Computer Science and information Systems, pp. 1–8. IEEE (2019)
15. Richer, J., Sanso, A.: OAuth 2 in Action. Manning Publications, Shelter Island (2017)
16. Schaefer, I., Bettini, L., Bono, V., Damiani, F., Tanzarella, N.: Delta-oriented programming of software product lines. In: Bosch, J., Lee, J. (eds.) SPLC 2010. LNCS, vol. 6287. Springer, Heidelberg (2010). https://doi.org/10.1007/978-3-642-15579-6
17. Setyautami, M.R.A., Fadhlillah, H.S., Adianto, D., Affan, I., Azurat, A.: Variability management: re-engineering microservices with delta-oriented software product lines. In: Proceedings of the 24th ACM Conference on Systems and Software Product Line: Volume A. Association for Computing Machinery, New York (2020)

Finite-Time Adaptive Sliding Control for a Class of Perturbed MIMO Nonlinear Systems

Ha Le Nhu Ngoc Thanh[1](✉), Choong Hyun Lee[2,3], and Sung Kyung Hong[2,3](✉)

[1] HUTECH Institute of Engineering, Ho Chi Minh City University of Technology (HUTECH), Ho Chi Minh City 700000, Vietnam
hlnn.thanh@hutech.edu.vn

[2] Department of Aerospace Engineering, Sejong University, Seoul 05006, Korea
chungh6577@naver.com, skhong@sejong.ac.kr

[3] Department of Convergence Engineering for Intelligent Drone, Sejong University, Seoul 05006, Korea

Abstract. In this paper, we introduce an adaptive sliding mode control (SMC) to deal with a finite-time convergence problem of a multiple inputs and multiple outputs (MIMO) system in the presence of uncertainties and/or external disturbances. The proposed controller is constructed by using the super-twisting sliding mode technique to eliminate a negative effect of the chattering phenomenon on the system. Additionally, an adaptive law is designed to guarantee a rapid adaptation and robust stability of the closed-loop system even without knowledge of the upper bound conditions of uncertainties/perturbations. The stability of a control system is investigated by the Lyapunov theorem. A great efficiency of the proposed algorithm is proven through a numerical simulation of a chaotic system.

Keywords: MIMO system · Sliding mode control · Finite-time convergence

1 Introduction

The conventional SMC and extended SMCs are very popular and efficient robust control methods because of its outstanding capacity to eliminate the undesirable influences of matched/mismatched uncertainties and/or exogenous perturbations on engineering systems [1, 2]. In order to establish a conventional SMC, a nonlinear sliding manifold is formulated by a smooth function of tracking errors and/or system states, and then an efficient controller is designed so that the system states and/or tracking errors drive to the presented sliding function and stay on it [3]. Due to this simple concept and excellent effectiveness in robust control, the sliding mode technique has widely used to solve the tracking control and stabilization control of an engineering system.

For many decades, there are plenty of finite-time stability algorithms proposed by using a terminal SMC [4], second order SMC integrated with a nonlinear disturbance observer (NDO) [5, 6]. However, the accidental singularity may take place when the initial conditions of the system are located in the particular regions causing an enlargement of the control signals [7]. In [8, 9], an advanced sliding mode control technique

M. Al-Emran et al. (Eds.): ICETIS 2021, LNNS 299, pp. 44–54, 2022.
https://doi.org/10.1007/978-3-030-82616-1_5

integrated with a NDO was presented for nth order system in the appearance of uncertainties and/or exogeneous perturbations to enhance the control performance of a nonlinear system. However, these methods cannot insure a stability of the system in a finite time. Moreover, these researches just consider a disturbed single input and single output (SISO) model [10]. This problem is solved by a terminal SMC for MIMO systems described in [11, 12]. However, the upper bound information of uncertainties and/or exogeneous perturbations is strictly required to design an efficient controller. In [13, 14], a robust adaptive sliding mode control algorithm for a nonlinear system in the appearance of matched and mismatched uncertainties is developed to enhance the control performance of a system. However, these algorithms highly concentrated on a second-order nonlinear model. Thus, it may not be appropriate to apply for higher order perturbed nonlinear systems.

The research's motivation of this article is to solve the aforementioned drawbacks of the existent controllers. An essential concept of the proposed algorithm is founded on a super-twisting sliding mode method to construct an effective control law with an excellent capacity to stabilize the MIMO systems in the presence of uncertainties and/or exogenous perturbations. The contribution of this study can be described as follows:

i. An adaptive controller is designed to ensure a finite-time convergence of the system trajectories to zero.
ii. The proposed method can be applied to enhance a stability of the high-order disturbed MIMO systems without prior awareness of the upper bound of uncertainties and/or exogenous perturbations.

The remainder of this paper is arranged as follows. The preliminaries are given in Sect. 2. A robust adaptive sliding controller and its stability analysis are provided in Sect. 3. The simulation results of an example are given in Sect. 4. In Sect. 5, the conclusion is presented.

2 Preliminaries

Consider a disturbed MIMO nonlinear system given by the following expression

$$\begin{aligned} \dot{z}_1 &= A_{11}z_1 + A_{12}z_2 \\ \dot{z}_2 &= A_{21}z_1 + A_{22}z_2 + B_2 f(z,t) + B_2(u + d(z,t)) \\ y &= Cz \end{aligned} \tag{1}$$

where $z = \begin{bmatrix} z_1 \ z_2 \end{bmatrix}^T \in \begin{bmatrix} \mathbb{R}^{n-m} \ \mathbb{R}^m \end{bmatrix}$ is a vector of system states; $u \in \mathbb{R}^m$ represents the controller input; z and u are defined instead of $z(t)$ and $u(t)$ for convenience; $f(z,t)$ is a continuous function of z; $d(z,t)$ is an unknown uncertainty and/or exogeneous perturbation effecting on the nonlinear system; $y \in \mathbb{R}^p$ is the output response; $B_2 \in \mathbb{R}^{m\times m}$, $C \in \mathbb{R}^{p\times n}$; $A_{ij}(i,j = 1, 2)$ are constant matrices with appropriate dimensions.

Let A^* and B^* be the new matrices given by

$$A^* = \begin{bmatrix} A_{11} & A_{12} \\ A_{21} & A_{22} \end{bmatrix}, B^* = \begin{bmatrix} 0 \\ B_2 \end{bmatrix}$$

The uncertain MIMO system (1) are re-written by

$$\dot{z} = A^* z + B^* f(z,t) + B^*(u + d(z,t)) \tag{2}$$

The control objective is to construct an adaptive controller u to ensure that the system states of a perturbed MIMO system converge to zero in a finite time, and also eliminate the negative influences of chattering on the system.

3 Controller Design

3.1 Sliding Surface Design

A novel sliding surface is proposed as the following equation:

$$\sigma = Pz + \Psi \|z_1(0)\| \|z_1\|^{\eta-1} z_1 \tag{3}$$

where $0 < \eta < 1$, Ψ is a gain matrix with an appropriate dimension, $z_1(0)$ is an initial state of the system. Let $P = \begin{bmatrix} \Gamma & I \end{bmatrix} \in \begin{bmatrix} \mathbb{R}^{m\times(n-m)} & \mathbb{R}^{m\times m} \end{bmatrix}$ is a matrix properly chosen, $I \in \mathbb{R}^{m\times m}$ is an identity matrix. Thus, the function (3) is re-written by

$$\sigma = \Gamma z_1 + z_2 + \Psi \|z_1(0)\| \|z_1\|^{\eta-1} z_1 \tag{4}$$

If the sliding surface, σ, drives to zero (i.e. $\sigma = 0$), then the Eq. (4) becomes:

$$z_2 = -\Gamma z_1 - \Psi \|z_1(0)\| \|z_1\|^{\eta-1} z_1 \tag{5}$$

Theorem 1. *The state trajectories z_1 and z_2 of the disturbed MIMO system (1) will drive to zero in finite time if the sliding function (3) is used and the following conditions are also satisfied.*

$$i)\ \textit{Matrices } Q = A_{11} - A_{12}\Gamma \textit{ and } A_{12}\Psi \textit{ are symmetric} \tag{6}$$

$$ii)\ \lambda\left\{Q + Q^T\right\}_{max} \tag{7}$$

$$iii)\ \lambda\{A_{12}\Psi\}_{min} \tag{8}$$

where $\alpha, \varphi > 0$, and $\lambda\{\}_{max}$ and $\lambda\{\}_{min}$ denote the largest and smallest eigenvalues of a matrix.

Proof.
From the Eq. (1), Eq. (5), and Eq. (6), it can be seen that

$$\begin{aligned} \dot{z}_1 &= A_{11}z_1 + A_{12}\left(-\Gamma z_1 - \Psi \|z_1(0)\| \|z_1\|^{\eta-1} z_1\right) \\ &= Qz_1 - A_{12}\Psi \|z_1(0)\| \|z_1\|^{\eta-1} z_1 \end{aligned} \tag{9}$$

A candidate of Lyapunov function is chosen as follows

$$V = z_1^T z_1 \tag{10}$$

Using the Eqs. (9) and (10), the function of $\dot{V}$ can be obtained as

$$\dot{V} = z_1^T \left(Q^T + Q\right) z_1 - 2 z_1^T A_{12} \Psi \| z_1(0) \| \| z_1 \|^{\eta - 1} z_1 \tag{11}$$

It is always able to select the values of Ψ and Γ such that matrices $Q = A_{11} - A_{12}\Gamma$ and $A_{12}\Psi$ are symmetric. Thus, the semi-negative value of $\dot{V}$ can be demonstrated as follows:

$$\begin{aligned} \dot{V} &\le \lambda_{\max}\left\{Q^T + Q\right\} \| z_1 \|^2 - 2\lambda_{\min}\{A_{12}\Psi\} \| z_1(0) \| \| z_1 \|^{\eta + 1} \\ &= -\alpha \| z_1 \|^2 - 2\varphi \| z_1(0) \| \| z_1 \|^{\eta + 1} \\ &\le 0, \quad \forall \alpha, \varphi > 0 \end{aligned} \tag{12}$$

From the inequality (12), it yields

$$\dot{V} \le -\alpha V - 2\varphi \| z_1(0) \| V^{\rho} \tag{13}$$

where $\rho = \frac{(\eta + 1)}{2}$.

From Eq. (13), we can obtain

$$dt \le -\frac{dV}{\alpha V + 2\varphi \| z_1(0) \| V^{\rho}} = -\left(\frac{1}{1 - \rho}\right)\left(\frac{(1 - \rho) V^{-\rho} dV}{\alpha V^{1 - \rho} + 2\varphi \| z_1(0) \|}\right) \tag{14}$$

$$\xrightarrow{d\left(V^{1-\rho}\right) = (1 - \rho) V^{-\rho} dV} \quad dt \le -\frac{1}{1 - \rho}\left(\frac{d\left(V^{1 - \rho}\right)}{\alpha V^{1 - \rho} + 2\varphi \| z_1(0) \|}\right) \tag{15}$$

Let $t_0 \ge 0$, $t_f > 0$ are an initial time and finite time of the control system, respectively, it means that the function $V(t) = 0$ as $t \ge t_f$. Thus, the finite time t_f is obtained by integrating of the inequality (15) with the time interval $t_0 \le t \le t_f$ as follows

$$t_f \le t_0 - \frac{1}{1 - \rho} \int_{t_0}^{t_f} \frac{d\left(V^{1 - \rho}\right)}{\alpha V^{1 - \rho} + 2\varphi \| z_1(0) \|} = t_0 + \frac{1}{\alpha(1 - \rho)} ln \frac{\alpha V^{1 - \rho}(t_0) + 2\varphi \| z_1(0) \|}{2\varphi \| z_1(0) \|} \tag{16}$$

It is very clear that the system states, z_1 and z_2, of the uncertain MIMO system (1) will drives to the origin zero in finite time t_f.

3.2 Adaptive Sliding Mode Controller Design

In this section, an adaptive sliding mode controller based on the super-twisting sliding algorithm is designed by applying the presented sliding surface in Sect. 3.1.

Assumption 1. [15] *the uncertainty/perturbation, $d(z, t)$, influencing on the nonlinear system (1), is a continuous function and bounded by*

$$\|d(z,t)\| \leq \beta \|\sigma\|^{\frac{1}{2}} \tag{17}$$

where $\beta > 0$ is a constant.

Remark 1. In [15], to construct an effective controller, the upper bound of $d(z, t)$ must be prior knowledge via the parameter of β as shown in Eq. (17). However, it is not easy to understand the value of β because of the complexity and variety of the working environment [16]. A classical adaptive SMC is presented in [17] to overcome the mentioned problem. However, it is also difficult to find out the proper controller parameters since there are a lot of coefficients needing to tune. Thus, we propose an adaptive controller with a simple concept and effectiveness to deal with these issues. The proposed method is designed as Theorem 2.

Theorem 2. *Consider an uncertain nonlinear system (1), the trajectories of system state will converge to sliding surface $\sigma = 0$, and achieve the origin zero in finite-time with a controller designed by Eqs. (18), (19), and (20)*

$$u = -(PB^*)^{-1}\begin{bmatrix} PA^*z + PB^*f(z,t) + \eta\Psi\|z_1(0)\|\|z_1\|^{\eta-1}Ez \\ +\left(k_1 + \|PB^*\|\hat{\beta}\right)\|\sigma\|^{\frac{-1}{2}}\sigma + k_3\sigma - \nu \end{bmatrix}, \text{where } k_1, k_3 > 0 \tag{18}$$

$$\dot{\nu} = -k_2\|\sigma\|^{-1}\sigma, k_2 > 0 \tag{19}$$

$$\dot{\hat{\beta}} = \frac{\|PB^*\|\|\sigma\|^{\frac{1}{2}}}{2\delta}, \delta > 0 \tag{20}$$

where $E = \begin{bmatrix} A_{11} \; A_{12} \end{bmatrix}$, $\hat{\beta}$ is an estimate of β.

Proof.
Let $\zeta_1, \zeta_2 \in \mathbb{R}$ denote the new state variables given by

$$\begin{cases} \zeta_1 = \|\sigma\|^{\frac{-1}{2}}\sigma \rightarrow \|\zeta_1\| = \|\sigma\|^{\frac{1}{2}} \\ \zeta_2 = \nu \end{cases} \tag{21}$$

If the states ζ_1 and ζ_2 drive to zero, it means $\sigma \rightarrow 0$. From Eqs. (19) and (21), the derivative $\dot{\zeta}_1, \dot{\zeta}_2$ can be achieved as

$$\begin{cases} \dot{\zeta}_1 = \dfrac{\dot{\sigma}}{2\|\sigma\|^{\frac{1}{2}}} = \dfrac{\dot{\sigma}}{2\|\zeta_1\|} \\ \dot{\zeta}_2 = -k_2\|\sigma\|^{-1}\sigma = -k_2\dfrac{\|\sigma\|^{\frac{-1}{2}}\sigma}{\|\sigma\|^{\frac{1}{2}}} = -k_2\dfrac{\zeta_1}{\|\zeta_1\|} \end{cases} \tag{22}$$

Otherwise, the derivative of function σ can be calculated from Eqs. (1)–(3) as follows

$$\dot{\sigma} = P\dot{z} + \eta\Psi\|z_1(0)\|\|z_1\|^{\eta-1}\dot{z}_1$$

$$= PA^*z + PB^*f(z,t) + PB^*u + PB^*d(z,t) + \eta\Psi\|z_1(0)\|\|z_1\|^{\eta-1}Ez \tag{23}$$

From Eqs. (18), (21) and (23), the first derivative of sliding function σ can be obtained by the following equation

$$\dot{\sigma} = -\left(k_1 + \|PB^*\|\hat{\beta}\right)\zeta_1 - k_3\sigma + \zeta_2 + PB^*d(z,t) \tag{24}$$

Consider a Lyapunov function given by

$$V_2 = \frac{1}{2}\zeta_1^T\zeta_1 + \frac{1}{4k_2}\zeta_2^T\zeta_2 + \frac{1}{2}\delta\tilde{\beta}^2 \tag{25}$$

where $\tilde{\beta}$ is an error $\tilde{\beta} = \beta - \hat{\beta}$.

From the result of Eqs. (22) and (25), it is easy to compute the derivative $\dot{V}_2$ as follows

$$\begin{aligned}\dot{V}_2 =& \frac{1}{2}\left(\dot{\zeta}_1^T\zeta_1 + \zeta_1^T\dot{\zeta}_1\right) + \frac{1}{4k_2}\left(\dot{\zeta}_2^T\zeta_2 + \zeta_2^T\dot{\zeta}_2\right) + \delta\tilde{\beta}\dot{\tilde{\beta}}\\ =& \frac{1}{2}\left(\frac{\dot{\sigma}^T}{2\|\zeta_1\|}\zeta_1 + \zeta_1^T\frac{\dot{\sigma}}{2\|\zeta_1\|}\right) + \frac{1}{4k_2}\left(-k_2\frac{\zeta_1^T}{\|\zeta_1\|}\zeta_2 - \zeta_2^Tk_2\frac{\zeta_1}{\|\zeta_1\|}\right) - \delta\tilde{\beta}\dot{\hat{\beta}}\\ =& \frac{\zeta_1^T}{2\|\zeta_1\|}\dot{\sigma} - \frac{\zeta_1^T}{2\|\zeta_1\|}\zeta_2 - \delta\tilde{\beta}\dot{\hat{\beta}}\end{aligned} \tag{26}$$

From Eqs. (24) and (26), we can obtain a new equation of $\dot{V}_2$ as follows

$$\dot{V}_2 = \frac{1}{2\|\zeta_1\|}\left(-\left(k_1 + \|PB^*\|\hat{\beta}\right)\|\zeta_1\|^2 - k_3\zeta_1^T\sigma + \zeta_1^TPB^*d(z,t)\right) - \delta\tilde{\beta}\dot{\hat{\beta}} \tag{27}$$

From Eq. (21), it can be seen that

$$\begin{aligned}&\zeta_1\|\sigma\|^{\frac{1}{2}} = \sigma\\ \Leftrightarrow\ &\zeta_1^T\zeta_1\|\sigma\|^{\frac{1}{2}} = \zeta_1^T\sigma\\ \Leftrightarrow\ &\|\zeta_1\|^3 = \zeta_1^T\sigma\end{aligned} \tag{28}$$

Substituting Eqs. (17), (20), (28) to Eq. (27), the value of $\dot{V}_2$ can be obtained by

$$\begin{aligned}\dot{V}_2 \le& \frac{1}{2\|\zeta_1\|}\left(-k_1\|\zeta_1\|^2 - \hat{\beta}\|PB^*\|\|\zeta_1\|^2 - k_3\|\zeta_1\|^3 + \beta\|PB^*\|\|\zeta_1\|^2\right) - \frac{\tilde{\beta}}{2}\|PB^*\|\|\zeta_1\|\\ \le& -\frac{1}{2}\left(k_1\|\zeta_1\| + k_3\|\zeta_1\|^2\right)\end{aligned} \tag{29}$$

$$\Rightarrow \dot{V}_2 \le 0, \forall k_1, k_3 \in \mathbb{R}^+ \tag{30}$$

From above discussion, we can see that the state trajectories (z_1, z_2) will converge to the sliding surface $\sigma = 0$ as the Lyapunov theory.

4 Simulation Results and Discussions

In this section, the proposed controller is used to stabilize an example of a disturbed MIMO nonlinear system through a classical chaotic motion of the Lorenz system influenced by uncertainties/perturbations to verify the effectiveness of the proposed algorithm. The Lorenz model is considered by the following equation [18].

$$\begin{bmatrix} \dot{z}_1 \\ \dot{z}_2 \\ \dot{z}_3 \end{bmatrix} = \begin{bmatrix} -q & q & 0 \\ r & -1 & 0 \\ 0 & 0 & -p \end{bmatrix} \begin{bmatrix} z_1 \\ z_2 \\ z_3 \end{bmatrix} + \begin{bmatrix} 0 & 0 \\ 1 & 0 \\ 0 & 1 \end{bmatrix} \begin{bmatrix} -z_1 z_3 \\ z_1 z_2 \end{bmatrix} \tag{31}$$

where q and r are the Prandtl and Rayleigh coefficients, and p is a constant, and its values are selected by $q = 10$, $r = 28$, and $p = \frac{8}{3}$. The initial conditions are given by $z_1(0) = 1$, $z_2(0) = 4$ and $z_3(0) = 1.5$. The chaotic behavior of the uncontrolled system is exhibited in the phase portrait and trajectories performance of z_1, z_2 and z_3 as shown in Figs. 1 and 2. To verify the efficiency of the presented controller, an additional control input is applied to stabilize the Lorenz model with the system matrices and coefficient matrices of the sliding surface given by

$$A^* = \begin{bmatrix} A_{11} & A_{12} \\ A_{21} & A_{22} \end{bmatrix} = \begin{bmatrix} -q & q & 0 \\ r & -1 & 0 \\ 0 & 0 & -p \end{bmatrix} = \begin{bmatrix} -10 & 10 & 0 \\ 28 & -1 & 0 \\ 0 & 0 & -\frac{8}{3} \end{bmatrix}, B^* = \begin{bmatrix} 0 & 0 \\ 1 & 0 \\ 0 & 1 \end{bmatrix}$$

$A_{11} = -10$, $A_{12} = \begin{bmatrix} 10 & 0 \end{bmatrix}$, and; $f(z,t) = \begin{bmatrix} -z_1 z_3 \\ z_1 z_2 \end{bmatrix}$, $P = \begin{bmatrix} 0.45 & 1 & 0 \\ -0.45 & 0 & 1 \end{bmatrix}$, $\Psi = \begin{bmatrix} 0.1 \\ 1 \end{bmatrix}$, and the negative influences of uncertainty/perturbation on the Lorenz system are given by $d(z,t) = \begin{bmatrix} d_1(z,t) & d_2(z,t) \end{bmatrix}^T$, where

$$d_1(z,t) = \left(0.3z_1^2 + z_1 sin(3t) + 0.1z_2 + 0.5t + sin^2 2t\right) \|\sigma\|^{\frac{1}{2}}, \text{ and}$$
$$d_2(z,t) = \left(0.3z_1^2 + 0.7\cos z_1 + 2z_2 z_3\right) \|\sigma\|^{\frac{1}{2}}.$$

The other parameters are chosen as $\eta = 0.9$, $k_1 = 0.51$, $k_2 = 0.2$, $k_3 = 10$, $\delta = 0.85$, $\hat{\beta}(0) = 1.0$. The proposed controller is applied to control the Lorenz model right after $t \geq 0$. In order to easy understanding, we only exhibit the results of simulation in the time interval $t \geq 0$ and $t \leq 2s$. The results of control performance of the Lorenz system are shown in Figs. 3, 4, 5, and 6.

According to the results of simulation in Fig. 3. It can be seen that after controlling, the responses of trajectories, z_1, z_2, z_3, strongly drive to zero even without knowing the upper bound information of uncertainties/external disturbances. The negative effect of chattering on the Lorenz system is also eliminated as shown in Fig. 4. The sliding surfaces of the controller fast drive to zero without any fluctuations as exhibited in Fig. 5. The adaptive gain is automatically adjusted by the proposed control law to ensure an excellent convergence of the system states to zero as shown in Fig. 6.

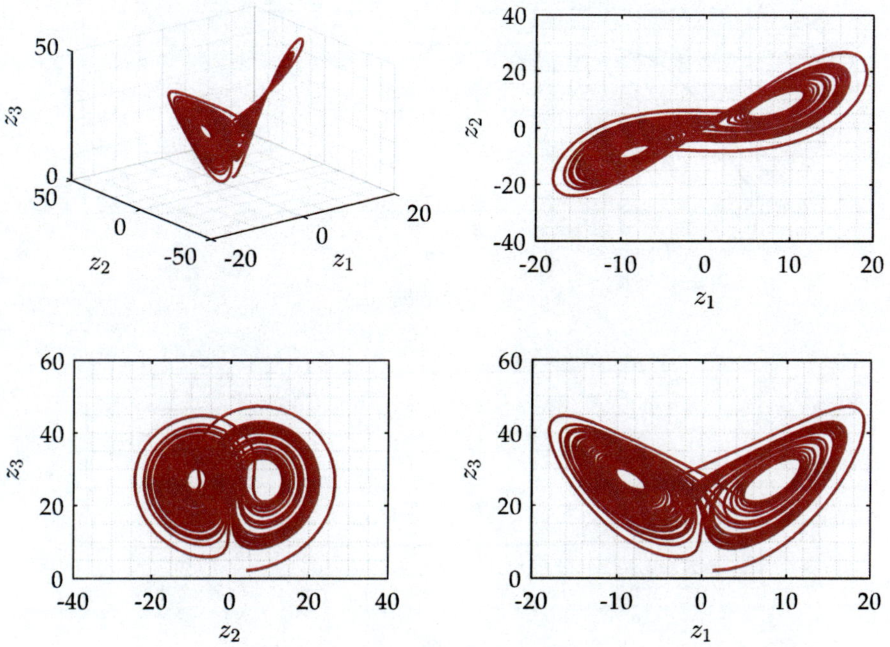

Fig. 1. Chaotic behavior of the uncontrolled Lorenz model.

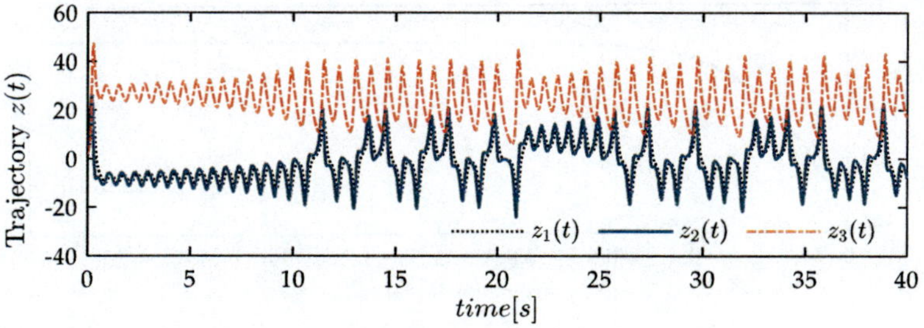

Fig. 2. Trajectories $z(t)$ of the uncontrolled Lorenz model.

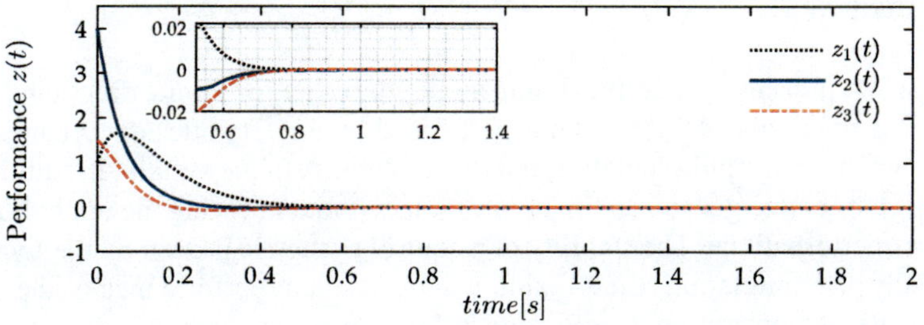

Fig. 3. Performance $z(t)$ of the Lorenz model after controlling.

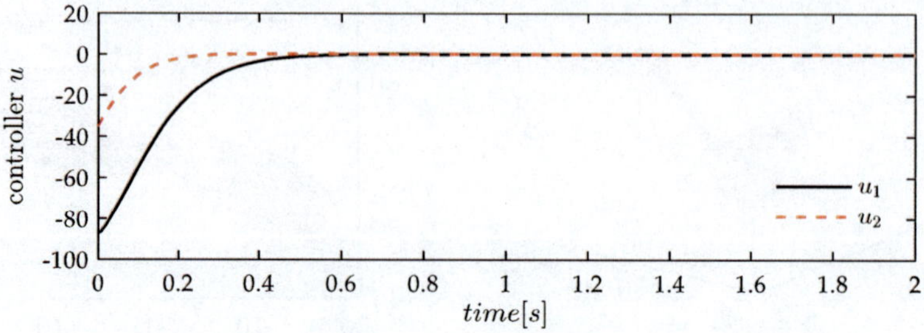

Fig. 4. Controller $u(t)$

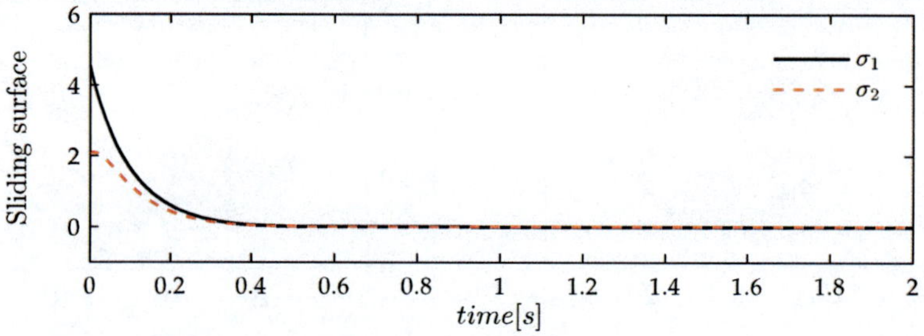

Fig. 5. Sliding surface $\sigma(t)$

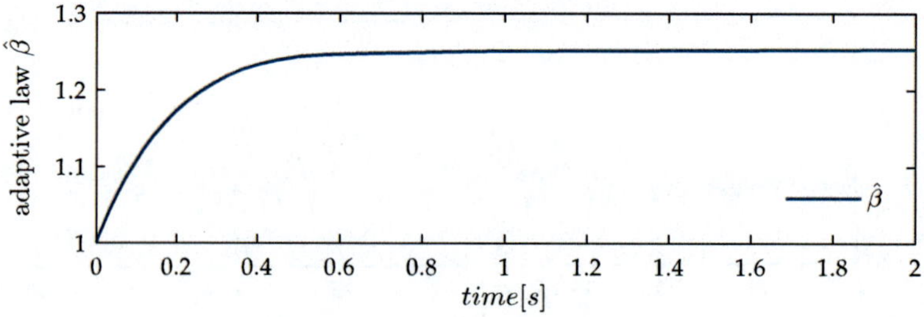

Fig. 6. Adaptive law $\hat{\beta}$

5 Conclusions

In this article, we present an adaptive sliding mode controller for a finite time stabilization problem of the uncertain MIMO systems influenced by uncertainties/disturbances. The controller guarantees a rapid adaptation and strong stability of the system. In addition, the system trajectory converges to zero in a finite-time without knowing the upper bound of uncertainties/perturbations. The stability of a control system is proven by the Lyapunov theory. Finally, the simulation results exhibited an excellent performance of the control system strongly demonstrating the efficiency of the proposed method.

Acknowledgement. This research was supported by the MSIT (Ministry of Science and ICT), Korea, under the ITRC (Information Technology Research Center) support program (IITP-2021-2018-0-01424) supervised by the IITP (Institute for Information & communications Technology Promotion).

References

1. Thanh, H.L.N.N., Hong, S.K.: An extended multi-surface sliding control for matched/mismatched uncertain nonlinear systems through a lumped disturbance estimator. IEEE Access **8**, 91468–91475 (2020)
2. Thanh, H.L.N.N., Vu, M.T., Mung, N.X., Phi, N.N., Phuong, N.T.: Perturbation observer-based robust control using a multiple sliding surface for nonlinear systems with influences of matched and unmatched uncertainties. Mathematics **8**(8), 1371 (2020)
3. Thanh, H.L.N.N., Lee, C.H., Nguyen, X.M., Hong, S.K.: An effective dynamic sliding mode control based nonlinear disturbance observer for a quadcopter UAV. In: 2020 IEEE 16th International Conference on Control & Automation (ICCA) (Virtual), Hokkaido, Japan, 9–11 October 2020, pp. 1553–1558 (2020)
4. Thanh, H.L.N.N., Vu, M.T., Nguyen, N.P., Mung, N.X., Hong, S.K.: Finite-time stability of MIMO nonlinear systems based on robust adaptive sliding control: methodology and application to stabilize chaotic motions. IEEE Access **9**, 21759–21768 (2021)
5. Colorado, R.M.: Finite time sliding mode controller for perturbed second-order systems. ISA Trans. **95**, 82–92 (2019)
6. Mao, J., Li, S., Yang, J.: Design and implementation of continuous finite-time sliding mode control for 2-DOF inertially stabilized platform subject to multiple disturbances. ISA Trans. **84**, 214–224 (2019)
7. Vu, M.T., Thanh, H.L.N.N., Huynh, T.T., Do, Q.T., et al.: Station-keeping control of a hovering over-actuated autonomous underwater vehicle under ocean current effects and model uncertainties in horizontal plane. IEEE Access **9**, 6855–6867 (2021)
8. Yang, J., Li, S., Yu, X.: Sliding-mode control for systems with mismatched uncertainties via a disturbance observer. IEEE Trans. Ind. Electron. **60**(1), 160–169 (2013)
9. Chen, W.-H.: Disturbance observer-based control for nonlinear systems. IEEE/ASME Trans. Mechatron. **9**(4), 706–710 (2004)
10. Chen, M., Wu, Q.X., Cui, R.X.: Terminal sliding mode control for a class of nonaffine nonlinear systems with uncertainty. ISA Trans. **52**, 198–206 (2013)
11. Man, Z., Yu, X.: Termina sliding mode control of MIMO linear systems. IEEE Trans. Circuit. Syst. I: Fundam. Theory Appl. **44**, 1065–1070 (1997)
12. Mobeyen, S., Majd, V.J., Sojoodi, M.: An LMI-based composite nonlinear feedback terminal sliding-mode controller design for disturbed MIMO systems. Math. Comput. Simulat. **85**, 1–10 (2012)
13. Mobayen, S., Tchier, F.: A novel robust adaptive second-order sliding mode tracking control technique for uncertain dynamical systems with matched and unmatched disturbances. Int. J. Control Autom. Syst. **15**(3), 1097–1106 (2017)
14. Thanh, H.L.N.N., Hong, S.K.: Quadcopter robust adaptive second order sliding mode control based on PID sliding surface. IEEE Access **6**, 66850–66860 (2018)
15. Moreno, J.A., Osorio, M.: A Lyapunov approach to second-order sliding mode controllers and observers. In: Proceedings of the 47th IEEE Conference on Decision and Control, Cancun, Mexico, 9–11 December 2008

16. Thanh, H.L.N.N., Hong, S.K.: Robust dynamic sliding mode control – based PID – super twisting algorithm and disturbance observer for second-order nonlinear systems: application to UAVs. Electronics **8**(7), 760 (2019)
17. Shtessel, Y., Taleb, M., Plestan, F.: A novel adaptive-gain super- twisting sliding mode controller: methodology and application. Automatica **48**(5), 759–769 (2012)
18. Yang, S.K., Chen, C.L., Yau, H.T.: Control of chaos in Lorenz system. Chaos Solitons Fractals **13**, 767–780 (2012)

Drone Altitude Control Using Proportional Integral Derivative Technique and Recycled Carbon Fiber Structure

Amenah E. Mohammed Redha[1], Riyadh Bassil Abduljabbar[2], and Majid S. Naghmash[2](✉)

[1] Optics Techniques Department, Dijlah University College, Baghdad, Iraq
amna.emad@duc.edu.iq

[2] Computers Technologies Engineering Department, Dijlah University College, Baghdad, Iraq
{riyadh.bassil,majid.salal}@duc.edu.iq

Abstract. This paper introduces a drone altitude control system using proportional integral derivative techniques and recycled carbon fiber structure. The carbon fiber usage in the fuselage and wings become more important due to numerous mechanical properties that include solidity, lightweight, and effective performance. Hence, the stiffness and mass of the module structure could be increased by using the carbon fiber. To obtain a good sink rate, the weight must be minimized and optimized. In drone designs, the optimization of the drone will break in case of its pressed to the restrictions. The drone design is a vital problem is deflection limitation when gradually insert power to every launch of well trim out the replica. Additionally, the lightweight of carbon fiber allows for smooth and easy takeoff and landing with high and sharp maneuver during operation. The proposed design contains many components for sensors, interface, radios and arduino controllers. The calculation of the control input of the iteration signals and the estimation of the states of the system are done on the chip to allow for command actions. The results show an effective control of the altitude of drone flight, noise effects, and wind performance during the drone flight.

Keywords: Drone altitude control · Integral derivative technique · Recycled carbon fiber

1 Introduction

Nowadays, the Quad Copter is one of the most popular research areas; in the 1800s, researchers began paying attention to drones and small-sized unmanned aerial vehicles (UAV). Early 1900, this type of drone becomes more popular in the world, with the commonest type of UAV being the so-called drones which were favored because of their small size, stable flight, vertical landing and take-off. Additionally, this type of drone could change direction instantly to attain any unreachable place. In military applications, Quad Copter is mostly used to embark on difficult tasks, such as unknown objects identification, enemies and border patrol monitoring [1, 2].

M. Al-Emran et al. (Eds.): ICETIS 2021, LNNS 299, pp. 55–67, 2022.
https://doi.org/10.1007/978-3-030-82616-1_6

Lately, drones are used in transportation of small goods and other trading process. For a subspace system state with the objective of controlling the altitude, the control laws were explored and investigated to reduce the quadratic cost; height controllers were assumed and joint with altitude controllers. One of the most developed technologies in many places around the world is the UAV drone [3, 4]. It is an aircraft that could be used to pilot, control, and manifest wind and aerodynamic. The drone company introduced many types, shapes, and sizes with fixed or rotary wings that could be used for many applications depending on the efficiency and quality [5, 6]. Currently, there is an increasing interest in the rotary type due to the ease of controlling and manufacturing it. The multi-rotor drone is more suitable for high maneuver flight compared to the conventional helicopter. The multi-copter with more than two rotors uses fixed pitch blades not vary as blade rotate situation [7].

To stabilize this type of copter, an electronic control system and sensors are used for the most common configurations, such as trip-copter, quadcopter, and hexa-copter. The application of quadcopter is numerous, ranging from inspection to security, military, and agriculture. Many researchers have focused on the design and manufacturing of drones. For instance, quadcopter pitch and roll moments have been simulated with PID controller by [8]. The results showed the ability of the roll moments and pitch to select the correct controller. By using the PID controller, the quad copter and Kalman filter have been controlled by [9] to reject noise from the IMU sensor. Three algorithms (PID, PD, and back stepping algorithm) have been used by [10] to provide high stabilization response of quadcopter dynamic motions.

To achieve stable and robust performance of the quad copter, the PD auto tuner has been designed and produced by [11]. An adaptive controller was used by [12] to develop the position and trajectory tracking system for drones in the presence of disturbances. The adaptive PID controller has been produced by [13] for stabilizing the motions of quadcopters. With these algorithms, there is no need of using special mathematical studies to derive the appropriate coefficients. A PID controller for steady drone operation has been introduced by [14].

This paper introduced a drone altitude controller and determined the system orientation based on PID controller in MATLAB. This paper consists of the following sections: Sect. 2 presents the materials and methods used in the experimental phase while Sect. 3 presents the discussion of the results and the conclusion.

2 Materials and Methods

The mathematical modeling, design and simulation of the drone in this work was done using SIMULINK block set. The pitch and roll movement of the drone response were obtained via MATLAB. A barometer is a unique altitude sensor of drone used to point the altitude parameters; it is sensitive to the external parameters that cause landing difficulties. The altitude data could be provided using GPS with centimeter level accuracy. The GPS scheme is affected by communication problems between the rover station and the base station. Any breakdown in the transmitter and GPS receiver affects the ability of the rover of the GPS station to receive the modification data that cause drone crash. To overcome these problems, a data synthesis scheme has been introduced for the high data of GPS and barometer. Therefore, the suggested design of the height controller has been given in Fig. 1.

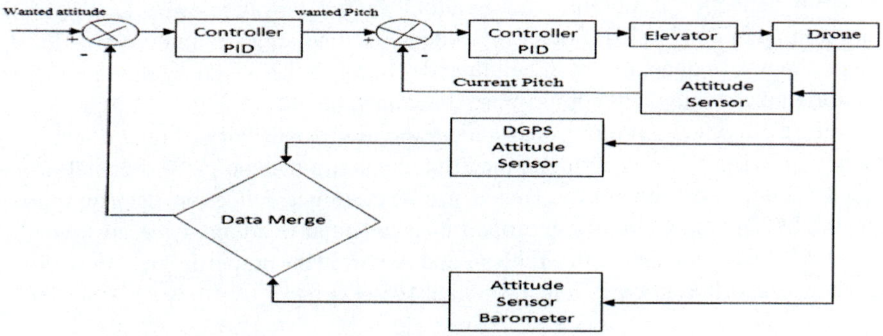

Fig. 1. Altitude controller design of drone

For relative altitude, the rover station could produce 10 Hz at frequency of 10 Hz at normal situation. In case of any failure in the contact between the rover station and base station, there will be no altitude information produced if the GPS is working properly. In this case, one could contract with communication responsibility even though the barometer sensor will produce float for a long time and the instant rate of the barometer altitude will vary. This data will be used to compensate the GPS altitude when there is any breakdown. The fault tolerance could be designed by assuming the autopilot stored the barometer altitude and GPS altitude at the same time. So, when there is GPS breakdown, the estimated real altitude could be calculated based on the standard GPS altitude pointed earlier rather than the communication fault and barometer altitude rate using the following equation:

$$H_{EST} = H_{GPS} + H_{RATE} \times t \tag{1}$$

Where
(H_{EST}) ….is the estimation of real attitude
(H_{RATE})….represent the barometer attitude rate
(H_{GPS})…..represent the GPS attitude before failure is happened
(t) ………is failure time duration

In case of normally working GPS, the normal data of the GPS altitude controller will be used again; here, the first order filter could be used to minimize the influence of switch between the estimated altitude and the GPS altitude. In the modern drone design, the carbon fiber usage in the fuselage and wings become more important due to its numerous mechanical properties, such as solidity, lightweight, and effective performance [15–20]. Hence, the stiffness and mass of the module structure of drones could be increased by using the carbon fiber.

To obtain a good sink rate, the weight must be minimized and optimized. In the drone design, the optimization of the drone will break in case of its pressed to the restrictions or redirect too much that will not talented to carry out as well. The drone design is a vital problem is deflection limitation when gradually insert power to every launch of well trim out the replica [21–25]. Sometimes one could find so as to a number of location additional power reduce the lunch height and will reach the boundary that the model is deflect too much which will compromise the launch height.

In this case, when more power is added, the model will fail or break down. In the principles of essential engineering, if the difficulty is strain incomplete, the final strength of fabric become more imperative. Hence, in case of limited deflection, the new approach is important. The interest in these properties is essential in adding excessive weight to assist in the stiffness/weight ratio. Tables 1 and 2 showed the properties of some materials compared with different types of carbon fiber [25–32].

Table 1. Carbon fiber properties compared with other materials

Material	Density	Strength	Modulus
Carbon fiber	**0.07**	–	–
Boron	**0.21**	**571**	56.700.000
Steel	**0.31**	**211**	30.100.000
Epoxy	**0.039**	**5.19**	590.000

Table 2. Carbon fiber properties compared with other materials

Material	Density	Strength	Modulus
Aluminum	**0.10**	**70**	12.600.000
Titanium	**0.11**	**–**	15.500.000
Balsa wood	**0.003**	**2.2**	300.000
Spruce wood	**0.890**	**12.1**	1.850.000

The pitch fiber is luxurious and hard to implement with PAN fiber due to dissimilarities in the properties; hence, it is typically used in the clubs and golf. The pitch type is usually used in satellite, and most drone model suppliers are selling the PAN fibers. So, many data about carbon fiber is required to study and understand its behavior as a rope or thread. Instead of carbon thread, pieces of carbon tow can be glued together as part of a drone. The carbon fiber and glue then becomes a carbon-fiber composite that can consist of two components. A more complex composition could as well be produced by mixing three or more materials mutually to produce the drone structure or wings of drone. Several factors will affect these compositions, such as the modulus and strength of the carbon fiber, and the glue used; the orientation of the carbon fiber can also affect the composition.

Normally, carbon tow is laid downward in one direction to achieve unidirectional sections. More strength and rigidity can also be achieved in one way arrangement than in other directions. This will give lesser overall strength or stiffness of fraction per unit mass. In the last factor, the influence on the strength or stiffness depends on the amount of every material (carbon fiber and glue) used in the composition. Better performance will be achieved with carbon fiber because it is extremely rigid and burly compared to glue; less glue in the section is also desired. Carbon tow is not a simple task to manage if exposed to water; management of the quantity of glue in the object or drone parts is also tedious. There are two common glue systems used to link carbon tow; one is referred to as damp and room temperature curative devise. In the drone design case, the room temperature glue is used to link the carbon tow. When performing this manually, the quantity of resin in the lay-up is firm to manage and the mechanism can be used to manage the resin quantity. Typically, this process is referred to as pultrusion process; it wets the carbon tow and lay down the carbon fiber tow to manufacture the drone parts using $60 - 75\%$ resin content. Another type of glue system is referred to as prepared which is made without compellation of cured at manufacturing facilities again by use the machine methods.

The carbon tow is laid up into multi directional sheets concerning 12 wide in this work. In the proposed mode, numerous lengths of carbon tow are laid down side by side and held jointly with B-stage epoxy to produce high strength. The B-stage epoxy is semi-cured to be sticky, flexible and soft. These sheets can be purchased from a vendor and laid up using a machine or hand into the desired drone structure. Temperature is applied to cure and harden the composition at 350 F for about 2–3 h; the drone parts are made with less than 20% glue content using an auto-cave. The auto-cave is the stove that provides a stress of 100 psi to the cured fraction, pushing out the glue compared with vacuum begging and work as an auto-cave procedure though the stress is incomplete at 15 psi or complete space. Several model suppliers provide carbon tow, carbon fiber plant, and cured unidirectional carbon fiber.

A number of these suppliers advertise further complex carbon fiber compounds, such as carbon-balsa – glue sheets and carbon-foam – glue sheet. The properties of the supplied carbon fiber dose not includes the epoxy required to grasp every fiber to shape the drone parts. A basic requirement is to ensure the proper glue content that will confer automatic properties to the mixing material of drone parts. Table 3 illustrates the carbon fiber composition with 40% epoxy and 60% carbon fiber, as well as one sheet of carbon tow in multi directional layup case. The most interest in drone structure design is to use stiffness/weight material to calculate the ratios of these properties per density. To reduce the numbers, all values are multiplied by 10^-6 to provide goodness ratio for understanding the material per mass. Table 4 shows that the larger the number, the better the material in terms of stress and rigidity weight.

Table 3. Carbon fiber composition

Material	Density	Strength	Modules
IM7	0.06	450	23.600.000
M46J	0.059	300	30.500.000
K139	0.062	370	65.300.000
K13D	0.063	349	84.850.000
Carbon fiber	–	–	–

2.1 Drone Structure Design Using Recycled Carbon Fiber

Recycled carbon fiber production could offer from 30–40 percent cost saving compared with virgin carbon fiber. The major factor is making sure that the whole thing is ragged to short fibers and still with higher properties. Figure 2 shows the comparison between virgin fiber and recycled fibers in terms of modulus E and strength X. Though, due to its lightweight, the drone and the engine are also low in weight, making it to struggle to take-off when it got windy. That's why all the captured photos are in relatively uninteresting places, like in the outer edge or at the park place. The drone will perform better in stable weather, especially when wind effect is blocked.

Table 4. Several material strength and modulus/stiffness ratio

Material	Strength	Modulus/Stiffness ratio
Carbon fiber	–	–
Boron	7.1	700
Steel	0.7	102
Epoxy	0.1	12

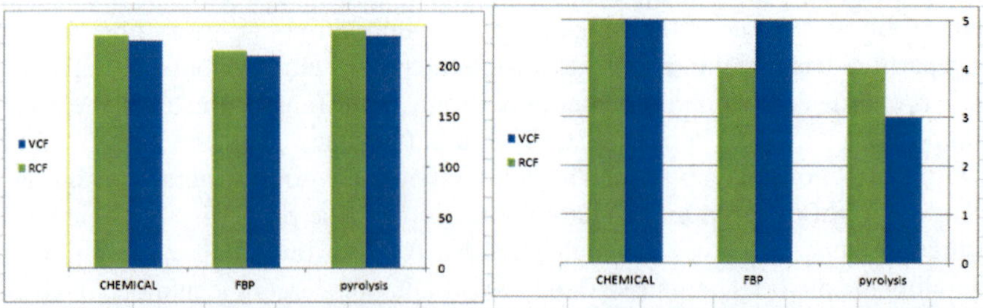

Fig. 2. Comparison of recycle carbon fiber and virgin carbon fiber

2.2 Drone Control Model

The drone control model consists of dynamics, weather, motor, actuators, and sensors. Four actuators channels were used to receive the commands of plant; they are throttle, ailerons, rudder, and elevator; they are used to produce a set of readings of drone sensors. A flat earth approximation has been considered in this model which will only work for a flight range of 20 km. In addition, a preliminary model for the kinematics and dynamics of drone system was developed. To describe the motion of drone, six partial differential equations were found to depend on the following assumptions:

a. In the rotor blades, the flapping does not occur
b. Four thrusts are always in parallel
c. Mass center is located in the meet of arms
d. Drag has been ignored
e. Rigid drone has been considered
f. Same length of arms has been considered

The autopilot controller is a lateral-longitudinal channel flight type that involves control, navigation and guidance based on ground control station commands to control the surface and telemetry report back to the station to keep the drone in stable condition through flight. The autopilot allows the operator to command the drone for fly in the return to base station, waypoint guide, circular guide, and mid-level control. Figure 3 shows the proposed model of drone altitude control.

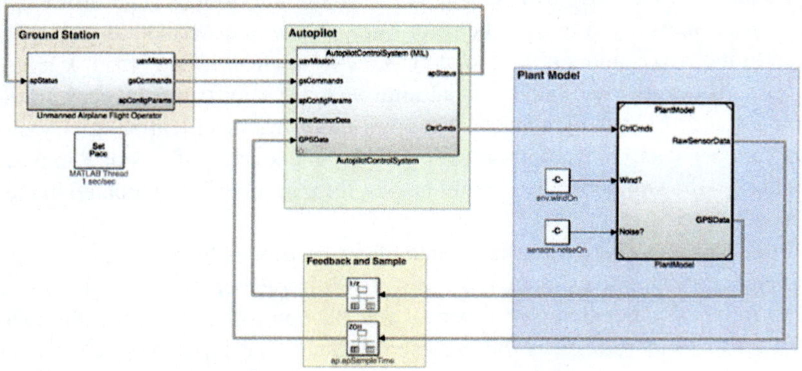

Fig. 3. Drone flight control model

The drone operative boundary represents the soothe operated in trip which sends the command as of grounded station to the drone. This block receives the report of altitude, sensors reading, and all other drone status to allow the description of the desired performance of drone through flight conditions. The communication protocol is used to communicate between the drone and ground station with many common workflows during flight. The principle of this approach is to demonstrate a recommendation to form sanded design procedure and verify the concept of drone flight control based on autopilot in the subsequent steps and workflow:

a. The design of flight controller according to permanent division drone autopilot with behavioral testing in reasonable sensible simulation.
b. The flight control simulation under different conditions and performance assessment of autopilot using local parallel simulation.
c. The characterization of flight envelope of drone routine through using a large variety of ecological situation brush in MATLAB distribution computer attendant.
d. The capabilities of drone assessment demonstration and collect reaction on usable and functional through co-simulation among ground control station and SIMULINK.
e. Code test generation of flight controller in the computer loop development of the designed autopilot.
f. Test the correctness and deployment of flight controller in developed board as represented by Raspier Pi for identical result using SIMULINK computer.
g. The integration of hardware board with SIMULINK for real-time completion of tenuously piloted broadcast control inside drone for flight stabilization and operative boundary to control the drone route.

The body of the drone axis is centered to the gravity point as follow:

a. At gravity center, the x-axis started and pointed in the pathway beside drone nudge.
b. At the gravity center, the y-axis started and pointed to the drone right side.
c. At gravity center, the z-axis started and pointed from drone following the rule of right side.

The complete drone body works as an element and the folder of the drone contain the value for the inertia and mass. The rotor number one is described as positive with admiration to the z-axis and similar situated to the xy-plane with 45° from x-axis. The rotor number 2 is negatively rotated with admiration to the z-axis and placed in similar to the x-y-planes with 135° as of x-axis's. The rotor number 3 has identical rotary motion as in rotor number 1 and located in parallel with x-y-plane and 135° as of x-axis's. The rotor number 4 has a similar rotary motion path of rotor number 2 and located in parallel with xy-axis and 45° from x-axis.

The drone control system uses the Kaman filter to estimate the position and velocity, and complementary filter to estimate altitude which is implemented in the PID organizer for PD controller of yaw, pitch – roll control and PD controller to control the location in the east-north synchronize. In the workspace, the drone input of pitch, roll, yaw, east, down and north coordination is used to change the VSS-COMMAND variable. A set of sensors is used in this model to determine the drone states, such as inertial measurement unit, camera, sonar for angular rate measurement and acceleration, optical flow estimation and altitude measurement.

The generation of drone trajectory tool uses to create a set of pathway navigation. Hence, to produce a route with the list of points, this technique uses a group of pose distinct by location, turn direction, heading and turn curvature. The waypoints describe the poses of trajectory tools required to draw the drone trajectory. The text box used in this window could be defined as: North/East is a position in meters, Curvature is a tuning in meters^-1, Heading is in degree from a North, and Turn is a route clockwise or else anticlockwise. Therefore, to add a pathway point, the poses values could be entered in the correct box and tick (add) option which will appear in the path point list in the panel.

The pathway point's characterization can be added by selecting the point and clicking (Edit) option while clicking the (Cancel) option will cancel the selected point. In the same way, to delete the pathway points in the list, choose it and click delete. The location of drone and its characterization is defined by the panel of no-fly zones as radius for distance in meters, north and east for position in meter, and edge for security edge in meter. The mapping trajectory windows could be used to plot the overall path points of drone during the flight of no-fly region. Hence, to produce the route, the path point with no-fly region individuality could be added to the particular panels and click (generate trajectory) option. The trajectory line shown in red and the black markers shows either a specific pose or any change in the trajectory path. The heading of a specific waypoint is represented by blue line which is accompanied to specific poses. The green circles represent the no-fly zones.

2.3 Drone Controller Model

The pathway controller of drone attitude has been designed and modeled in SIMULINK-MATLAB environments as show in Fig. 4. The drone model is illustrated in Fig. 5.

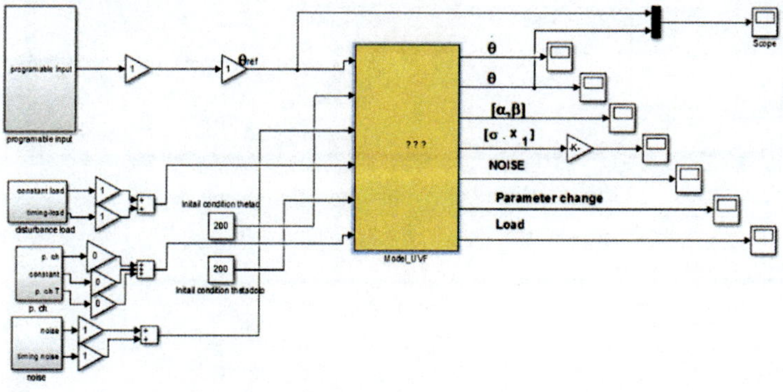

Fig. 4. Drone attitude controller model

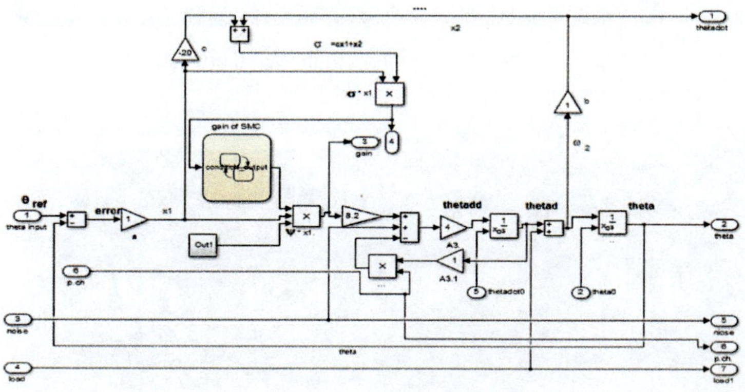

Fig. 5. Drone model

3 Results and Discussion

The finding of this design with low weight structure of drone by using carbon fiber is to increase the maneuver. The carbon fiber materials used in drone structure could affect on the drone behavioral during the flight phase. Low drone weight could allow increasing the sharp maneuver during takeoff or landing stages as well as during photo capturing and fighting. The speed curve shown in Fig. 6 is recorded and the pitch angle curve is in Fig. 7. The random vector Wind in x, y, z is a control loop and has been added with [0, 0, 0] m/s. Figure 8 illustrates the control loop with wind parameters of changeable values. By adding wind vector as distribution in simulation, a reasonably actual environment for ground operation will be achieved. The simulation result presented by curves with typical parameters could show the increase in the instability situation in drone shake frequency in windy environment as close to actual conditions. The simulation results provided by software could satisfy the system purpose and the stability of the system is achieved; wind distribution is considered in this work.

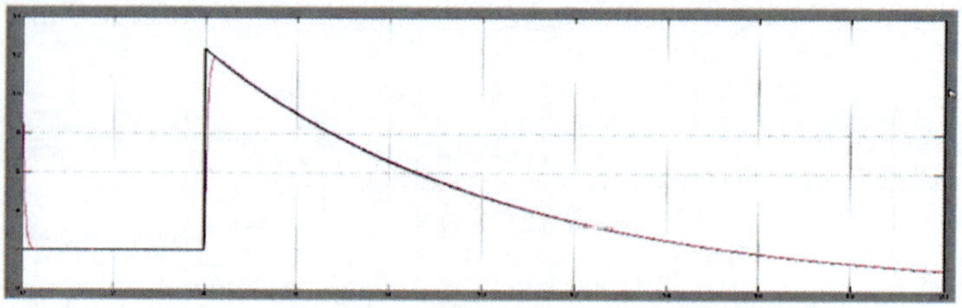

Fig. 6. Speed curve of controlled drone

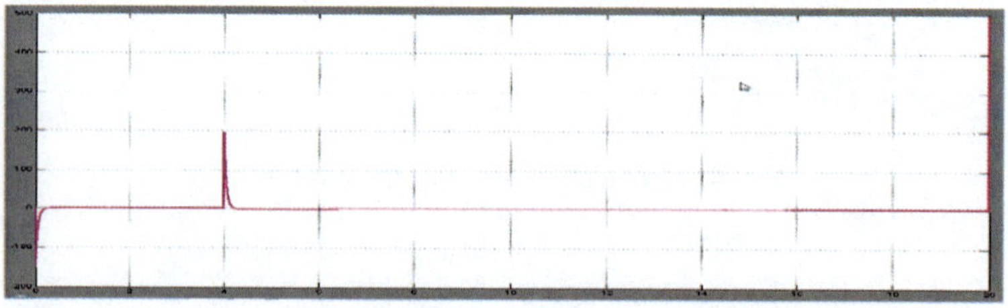

Fig. 7. Pitch angle curve with the wind effects

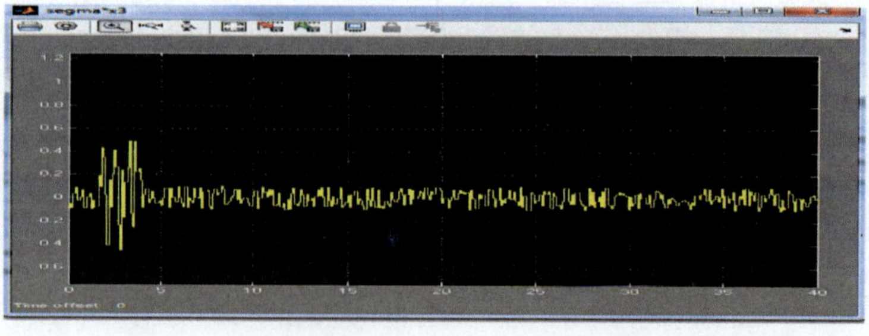

Fig. 8. Pitch angle simulation graph with the Noise

The drone sensors platform from scratch has been developed to include system modeling, control design, state estimation, communications handling, and user interface implementation. By using additive manufacturing methods, the drone frame has been designed and printed. The sensors are coupled via an arduino microcontroller and examined by different altering techniques. The drone motion is modeled with the system model and kalman filter to prove the achievement of the objectives. The organization broadcast matrices and the consequential combinations of latest estimation scheme could be implemented in simple microcontroller designs. Accurate and fast altitude estimation was provided with the aid of extended Kalman filtering principle. To give better stability pitch and roll angles compare to the alternative, a proportional integral derivative controller was used. In the theory of nonlinear controller, enhanced inputs to the system could be provided and thus, guide to presentation improvement with long time consuming through implementation phase. The altitude controller was devised and implemented with stable behavior to allow the user save the drone flight control.

4 Conclusion

This paper introduces the object-oriented method and the application techniques in the flight simulation of drone systems. The core of drone systems in flight simulation is the software of flight simulation. High level design of drone altitude controller with ground station and sensors, coordination, functionality, and communications has been implemented. Different scenarios have been demonstrated to prove the capability of the drone control design following the principle of suggested structure. This design observes that the effective structure of drone controllers, especially from dynamic applications, gives better explanation of the interaction between coordinator, sensing, and communication. In addition, the constraints imposed by these applications at hand are considered as well. The current design focused on advanced modeling and testing a multi drone system, while there are still several open issues to achieve an ideal drone system. The carbon fiber materials used in drone structure could affect drone behavior during the flight phase. Low drone weight could allow increasing the sharp maneuver during takeoff or landing stages, as well as during photo capturing and fighting. More reliable models need to be investigated and developed in future researches to accommodate different noise effects on drone per remark.

References

1. Dukare, H.P., Bondre, M.B., Vidhate, A.R., Tiple, R.I., Kumar, G.: Design and implementation of flying robot. Int. J. Adv. Res. Comput. Sci. Softw. Eng. **5**(2), 458–461 (2015)
2. Ostogic, G., Stankovski, S., Tejic, B., Dukic, N., Tegeltija, S.: Design, control and application of quadcopter. Int. J. Indus. Eng. Manage. **6**(1), 43–48 (2015)
3. Pawar, S.G., Dongare, K., Dalvi, M., Doshi, S., Das, K.K.: Automated quadcopter using android controlling system. Int. J. Eng. Sci. Manage. **5**(4), 115–119 (2015)
4. Nakazawa, A., Jin, B.X. Quadcopter video surveillance UAV. Dept. Electrical and Computer Engineering University of Victoria. Accessed May 2016
5. Prem, K.P., Amirtharaja, S., Harie, S., Kishore, R.S., Kiruba, V.: Quadcopter video surveillance and control using computer. Int. J. Electric. Electron. Eng. **7**(1), 340–345 (2015)

6. Al-sofy, K., Al-Talib, S.A.: Investigation of Cartesian routing for unmanned aerial vehicle networks. In: 2nd International Scientific Conference of Al-Ayen University (ISCAU) (2020)
7. Thanh, H.L.N.N., Hong, S.K.: Quadcopter robust adaptive second order sliding mode control based on PID sliding surface. IEEE Access **6**, 66850–66860 (2018)
8. Thanh, H.L.N.N., Nguyen, N.P., Hong, S.K.: Simple nonlinear control of quadcopter for collision avoidance based on geometric approach in static environment. Int. J. Adv. Robot. Syst. **15**(2) (2018)
9. Nguyen, N.P., Hong, S.K.: Fault-tolerant control of quadcopter UAVs using robust adaptive sliding mode approach. Energies **12**, 95 (2019)
10. Nguyen, N.P., Hong, S.K.: Position control of a humming bird quadcopter augmented by gain scheduling. Int. J. Eng. Res. Technol. **11**, 1485–1498 (2018)
11. Feng, Y., Zhang, C., Baek, S., Rawashdeh, S., Mohammadi, A.: Autonomous landing of a UAV on a moving platform using model predictive control. Drones **2**, 34 (2018)
12. Zhou, Z., Wang, H., Hu, Z.: Event-based time varuing formation control for multiple quadrotor UAVs with markovian switching topologies. Complexity (2018)
13. Xuan-Mung, N., Hong, S.K.: Improved altitude control algorithm for quadcopter unmanned aerial vehicles. Appl. Sci. **9**, 2122 (2019). https://doi.org/10.3390/app9102122
14. Husni, A.F., et al.: Integrated pid controller for altitude control of quadrotor system. In: Proceedings of Symposium on Electrical, Mechatronics and Applied Science 2018 (SEMA 2018), pp. 1–2, 8 November (2018)
15. Rimme, O., et al.: Development and validation of recycled carbon fiber-based binder tapes for automated tape laying processes. J. Compos. Mater. (2019). https://doi.org/10.1177/0021998318820422
16. Uhlmann, E., Meier, P.: Carbon fibre recycling from milling dust for the application in short fibre reinforced thermoplastics. Proc. CIRP **66**, 277–282 (2017)
17. Knobelsdorf, C.: Secondary carbon fibers in large-scale production: sheet molding compounds made with secondary carbon fibers for luxury class cars. Kunststoffe Int. **01**, 105–108 (2016)
18. Wo¨lling, J., Schmieg, M., Manis, F., et al.: Nonwovens from recycled carbon fibres – comparison of processing technologies. Proc. CIRP. 66, 271–276 (2017)
19. Goergen, C., Baz, S., Mitschang, P., et al.: Recycled carbon fibers in complex structural parts – organic sheets made of rCF staple fiber yarns. KEM **742**, 602–609 (2017)
20. Longana, M.L., Ong, N., Yu, H., et al.: Multiple closed loop recycling of carbon fibre composites with the HiPerDiF (high performance discontinuous fibre) method. Compos. Struct. **153**, 271–277 (2016)
21. Ishikawa, T., et al.: Overview of automotive structural composites technology developments in Japan. Compos. Sci. Technol. **155**, 221–246 (2018)
22. Rimmel, O., May, D., Mitschang, P.: Impact of stitching on permeability and mechanical properties of preforms manufactured by dry fiber placement. Polym. Compos. **1**, 56 (2018)
23. Royster, E.: Carbiso-M-TDS: Technical Data Sheet. 49. Sonderheft 2016: Jahresthema recycling. Augsburg: Carbon Composites e.V. (2016)
24. Meng, F., Olivetti, E., Zhao, Y., Chang, J.C., Pickering, S.J., McKechnie, J.: Comparing life cycle energy and global warming potential of carbon fiber composite recycling technologies and waste management options. ACS Sustain. Chem. Eng. **6**, 9854–9865 (2018)
25. Wellekötter, J., Christian, B.: Comparison of fiber-reinforced recycled polyamide in single stage and two stage injection molding. Polym. Compos. **40**, 1731–1739 (2018)
26. Reichert, O., Ausheyks, L., Baz, S., Hehl, J., Gresser, G.T.: Innovative rC staple fiber tapes - new potentials for CF recyclates in CFRP through highly oriented carbon staple fiber structures. Key Eng. Mater. **809**, 509–514 (2019)
27. Wellekötter, J., et al.: Insights into the processing of recycled carbon fibers via injection molding compounding. J. Compos. Sci. 4, 161 (2020). https://doi.org/10.3390/jcs4040161

28. Naqvi, S., et al.: A critical review on recycling of end-of-life carbon fibre/glass fibre reinforced composites waste using pyrolysis towards a circular economy. Resour. Conserv. Recycl. **136**, 118–129 (2018)
29. Wong, K., Rudd, C., Pickering, S., Liu, X.: Composites recycling solutions for the aviation industry. Sci. China Technol. Sci. **60**(9), 1291–1300 (2017). https://doi.org/10.1007/s11431-016-9028-7
30. Pillain, B., et al.: Positioning supercritical solvolysis among innovative recycling and current waste management scenarios for carbon fiber reinforced plastics thanks to comparative life cycle assessment. J. Supercrit. Fluids **154**, 104607 (2019)
31. Bonten, C.: Plastics Technology—Introduction and Fundamentals. Hanser, Munich, Germany (2019)
32. Meng, F., et al.: Comparing life cycle energy and global warming potential of carbon fiber composite recycling technologies and waste management options. ACS Sustain. Chem. Eng. **6**, 9854–9865 (2018)

Design and Simulation of Wind Farm Model Using Doubly-Fed Induction Generator Techniques

Ola Hussein Abd Ali Alzubaidi(✉) and Aya Qusay Dawood

Computers Technologies Engineering Department, Dijlah University College, Baghdad, Iraq
{Ola.hussien,aya.qusay}@duc.edu.iq

Abstract. This paper presents the design and simulation of wind farm model using doubly-fed induction generation (DFIG) techniques and MATLAB platform. To control the output active power of DFIG, the rotor flux oriented vector has been used to improve the vector control strategy instead of stator flux oriented vector control. The rotor flux control technique provide better power tracking accuracy compared with traditional technique represented by stator flux vector control methods. The performance of DFIG under various wind farm grid disturbance has been evaluated and examined in this work. To overcome the stability of voltage in transmission system device, the implementation of dynamic reactive power compensators is used at the tip of ordinary combination to improve the DFIG during disturbances. This includes static VAR and synchronous compensators to improve the transient performance and keeping the turbine in services during the faults of grid. In term of time response, amount of reactive power injection and the cost of applications, the performance of both techniques has been compared and investigated. The suggested solution of integration between energy storage and GFIG grid connection shows better smoothing in the output power during wind rate variation. In addition, the principle of wind farm under twice fed asynchronous generation is connect through 3-phase converter type AC-DC-AC to the network and rotor.

Keywords: Wind farm · DFIG · MATLAB

1 Introduction

To convert the wind energy into useful energy in wind turbine, the wind energy conversion system (WECS) could be used as prime mover of electrical DFIG [1–5]. Basically, the WECS include turbine and electrical DFIG associated with control and converter systems [6, 7]. Figure 1 shows the design of WECS and grid connection system [8].

To enable the power control of wind farm, the reactive power and DC voltage controllers system required to generate a voltage signals and pitch angle to the network and rotor sections [9, 10]. In addition, the measurement of output electrical power at the turbine terminals should be added the losses power which compare with orientation power obtain from characteristics of turbine path [11, 12]. The ability of reactive power

M. Al-Emran et al. (Eds.): ICETIS 2021, LNNS 299, pp. 68–78, 2022.
https://doi.org/10.1007/978-3-030-82616-1_7

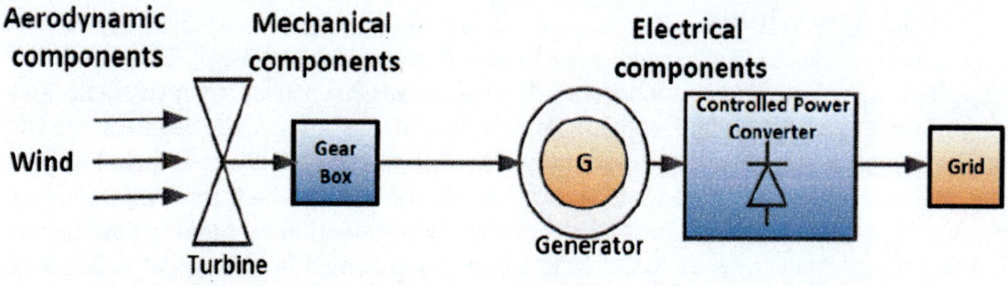

Fig. 1. Virtual diagram of WECS and grid connection system

and voltage modes in the system are synchronized by current flow in converter of rotor [13–17]. In case of hasty power form, the power is synchronized by controller voltage and when the voltage mode is established, the voltage id set up with orientation values pending the current is reached to up values. Figure 2 shows the DFIG principle working which is connected to the turbine [18–20].

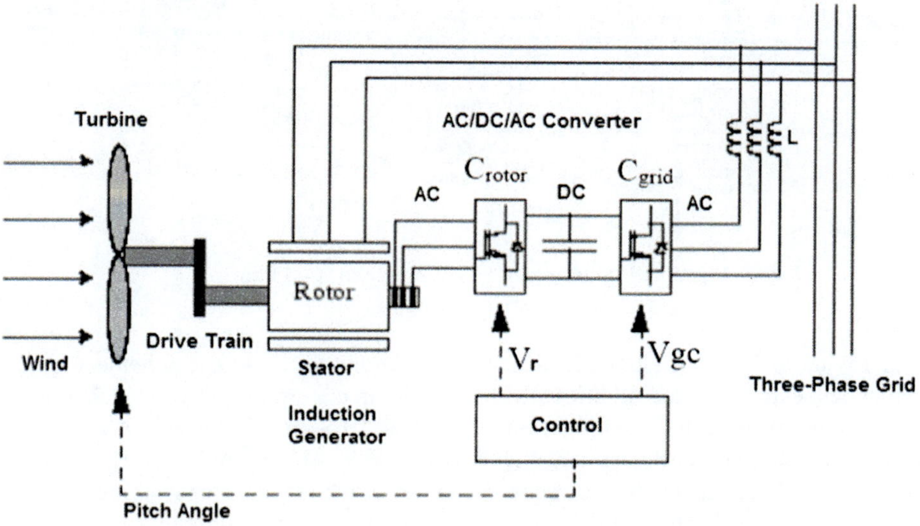

Fig. 2. DFIG work principle connected to the wind turbine

The wind farm turbine use DFIG which consists of AC-DC-AC type IGBT based PWM converter and wound rotor induction generator. The winding of stator is linked straight to the grid system frequency and the rotor is feed at changeable frequency via AC-DC-AC converter [21–25].

2 Wind Farm Model

The software of Simpower structure in MATLAB has been used to investigate and examine the suggested model with dynamic performance and steady state of 9 MW wind farm connected to the distribution networks. The model consists of 6-wind turbines every with 1.5 MW connected to allocation scheme of 25 kV which produce a total of 120 kV grid via 30 km feed. A motor load of 1.68 MW induction motor at 0.93 PF is the specifications of plant with 2300V/2 MVA. The resistive load in this model is 200 kW which is linked on the identical feed at B25 bus. The load of 500 kW is as well linked to the wind farm bus of 575 V. Figure 3 shows the single line model of wind farm linked to the scheme distributors.

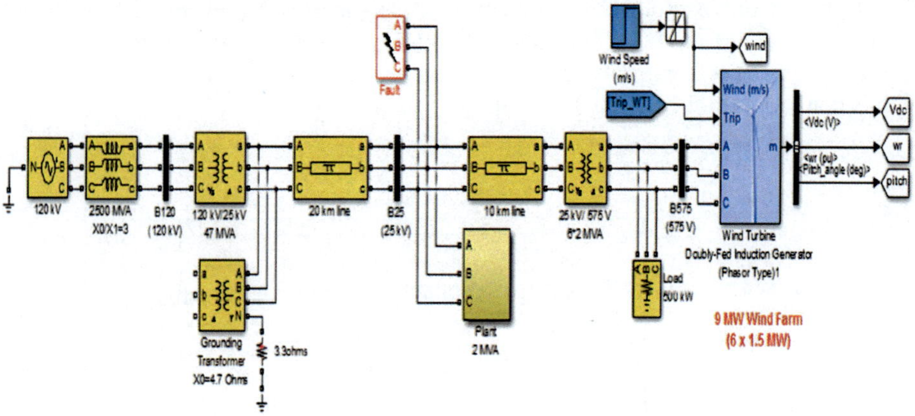

Fig. 3. DFIG phasor model of proposed wind farm

The wind turbine and motor load have guard monitoring voltage scheme with a machine and current speed. Additionally, the DFIG direct current link voltage is monitored also. In this model, the wind turbines are based DFIG method which contains an injury rotor initiation generation and IGBT based PWM AC-DC-AC converter. Then, the rotor is feed at changeable frequency via converter while the stator winding is connecting to 60 Hz grid. The technology of DFIG permit to extract optimal power as of wind to low wind rates by maximizing the turbine rate as minimal mechanical stress on the turbine through wind squall. Maximum mechanical energy could be generated with optimal turbine rate in a given wind velocity which is proportion to the velocity of wind. The DFIG techniques provide another advantage represented by the capability of power electronically converter to produce turbine control characteristic to take up the reactive energy and then eliminate the required capacitor store as in the motor of squirrel cage induction. The rotor is run at sub-synchronous rate for wind velocity less than 10 m/s while it run at super-synchronous velocity for high wind speed in this case.

The automatic energy of turbine is an occupation of turbine speeds which is shown in wind rate range from 5–16.2 m/s. this characteristic is provided with specific parameter in turbine information. Figure 4 illustrate the two MVA place with the shield scheme and the turbine power characteristics is illustrated in Fig. 5.

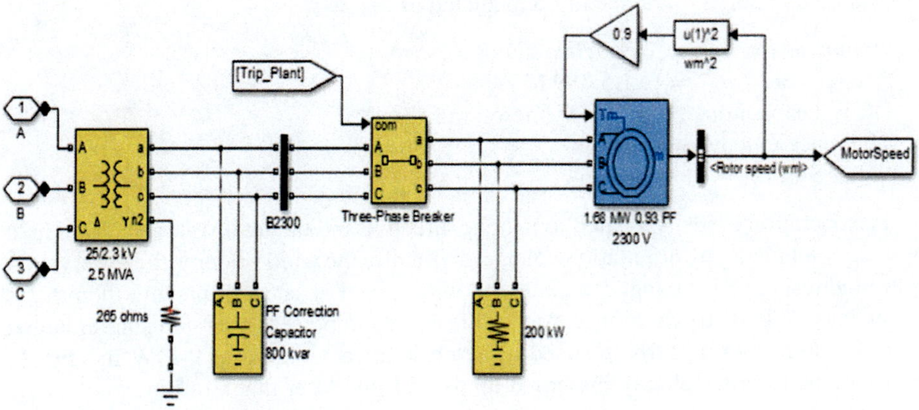

Fig. 4. 2 MVA plant with the protection system

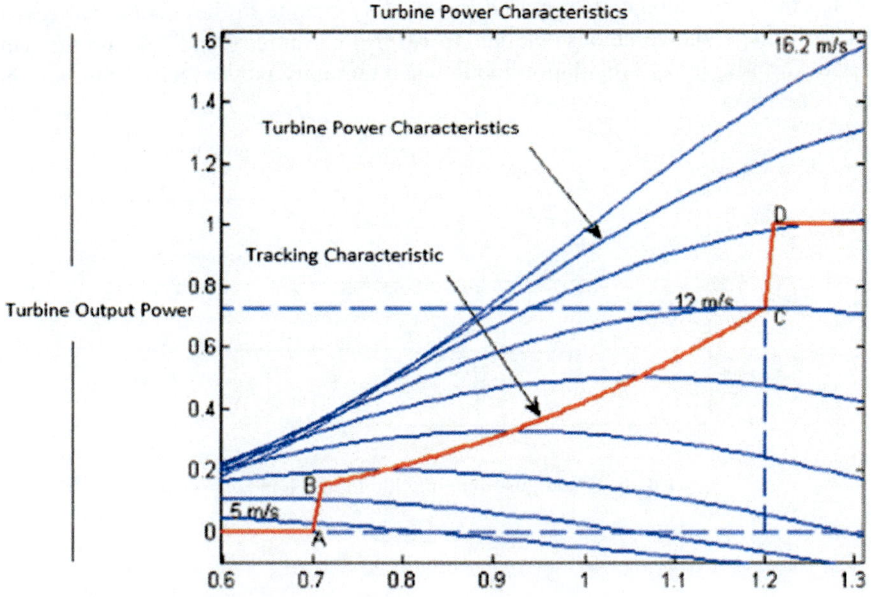

Fig. 5. Turbine power characteristics

To go behind the ABCD arc in turbine power characteristics and menu of data, the DFIG should be controlled. Hence, the optimization of turbine velocity is provided flanked by point B and C in the arc. To stabilize the transient type with long time simulation, the wind turbine model has chosen as a phasor type. Therefore, the system is experiential throughout 50 s and the 6 turbine is replicated by solitary wind turbine chunk and the main 3- parameters is calculated as below:

Output power = $6x1.5e6$ watts
Rate power = $6x1.5/0.9$ MVA
DC bus capacitor = $6x10000$ microfarad
Reference voltage = 1 pu
Voltage droop = 0.02 pu

The operation mode is located to voltage directive and the fatal voltage is forbidden by value compulsory by orientation voltage. To modify the wind velocity, the observation of turbine response is changed. Hence, the wind speed is set to 8 m/s and then to 5 s and increased suddenly to 14 m/s. At 5 s, the energetic power generated start enlarged smoothly proportion to turbine speed to be close in rated value of 9 MW in 15 s. In this time, the turbine velocity increased up to 1.21 pu. After that a turbine pitch angle is initially blade is zero and the operation point goes behind the red color in power characteristic awake to D. After that, the turbine pitch angle is augmented to 0.7 degree to bind the mechanical energy of turbine. Reactive power and voltage have forbidden upholding the 1 pu voltage at supposed power. Any change in the form of process to regulation voltage, the voltage is enlarged to 1.021 pu. Figures 6 and 7 shows the wind farm waveform in voltage regulation mode when the unity power factor with nominal turbine generation.

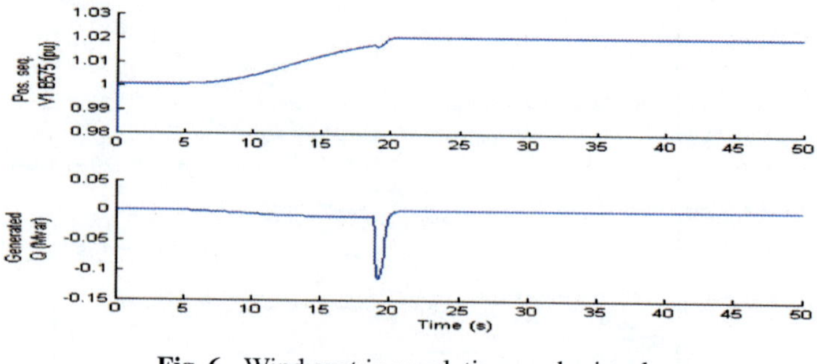

Fig. 6. Wind gust in regulation mode signals

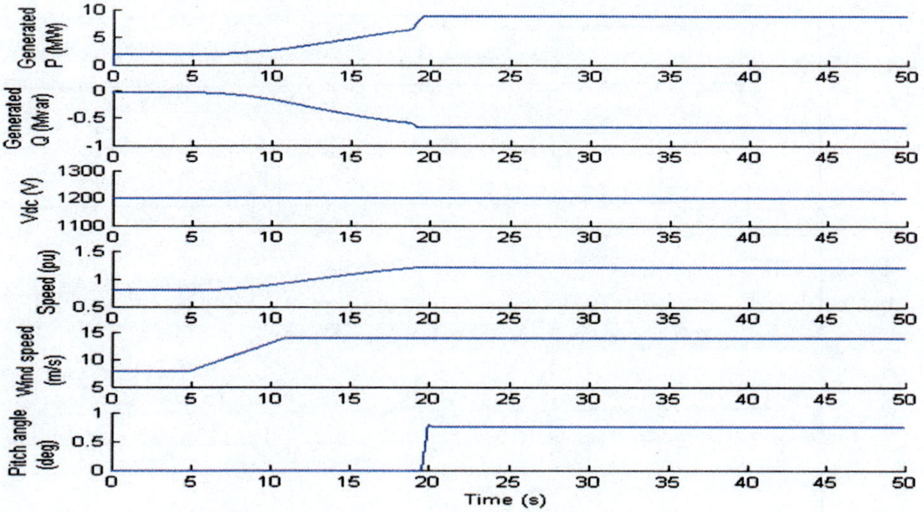

Fig. 7. Wind farm in voltage directive mode signals

The observation of voltage impact resultant from distant liability on 120 kV is initially simulated with zero reference voltage and wind velocity at 8 m/s. In 120 kV source menus, the voltage drop lasting is 0.5 s is automatically in order to happen at t = 5 s. Figures 8 and 9 shows the simulation results of voltage droop on 120 kV. The current, voltage plant and motor speed has been observed to produce 1.8 MW. Hence, at t = 5 s, the electrical energy waterfall under 0.9 pu. At t = 5.22 s, the shield trip the place because of below voltage permanent higher than 0.2 s is detecting. Therefore, the motor speed is gradually decreased and plant current falls to zero. In same time, the wind farm goes on generate at energy altitude of 1.8 MW. Following the tripped of plant, the measured power exported to the grid is found at 1.25 MW. At what time the manage mode of wind turbine is tainted to voltage directive and repeated, the plant dose not tripped owing to the voltage hold providing by five reactive power which produced by wind turbine throughout the voltage droop maintaining the place voltage further 0.9 pu as doorsill guard level.

At 25 kV, the solitary phase to earth liability impact has observed at t = 5 s. Hence, while the wind turbine is in fault applied at phase A, the optimistic succession volts at the terminal of turbine is drooped to 0.8 pu. Consequently, the wind farm stay in repair and the voltage drop under 0.7 pu with the beneath voltage guard trip the wind farm as illustrated in Fig. 10. At t = 40 s, the pitch angle is enlarged to bound the velocity and the turbine speed is increased as illustrated in Fig. 11. Table 1 contains the results of wind farm signals.

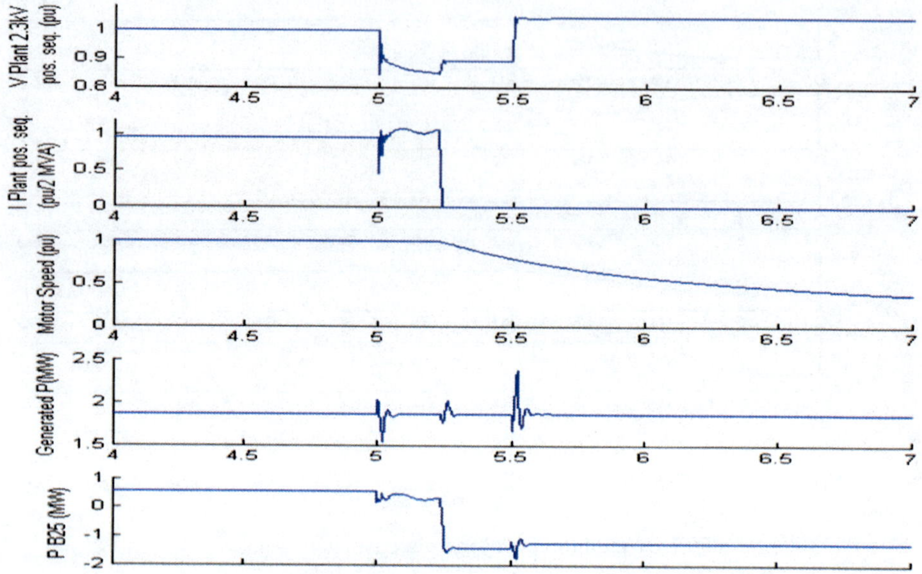

Fig. 8. Voltage droop on the 120 kV scheme

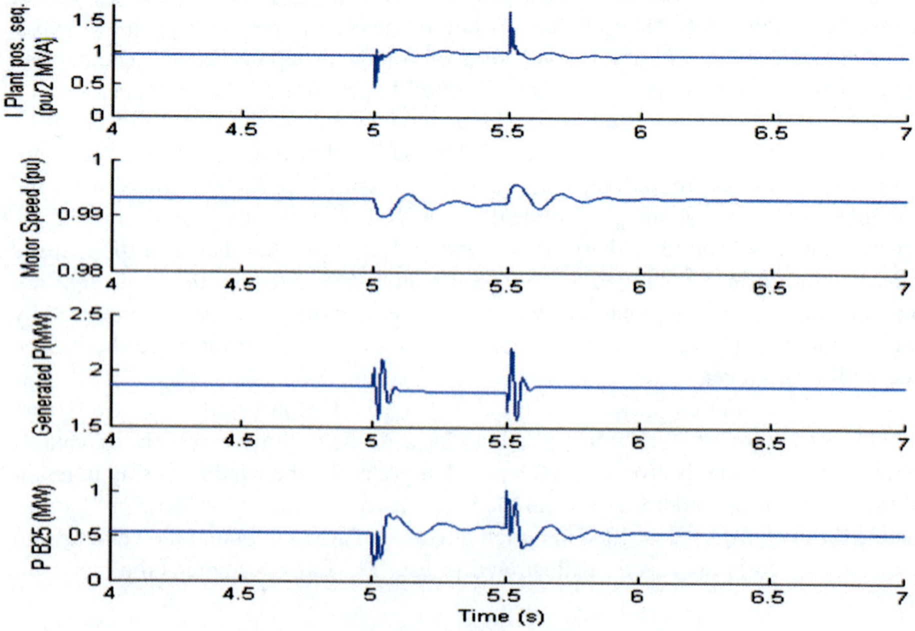

Fig. 9. Voltage droop on the 120 kV System

Table 1. Wind farm signals

Time	Wind farm signals			
	Optimistic succession V1 (pu) at 5s	Optimistic succession I1 (pu)	Speed (pu)	Pitch angle (deg)
Thought fault at 25 kV, t = 5s	0.9	0.18	0.8	0°
During fault at t = 40 s	0.8	0.3	1.2	15°

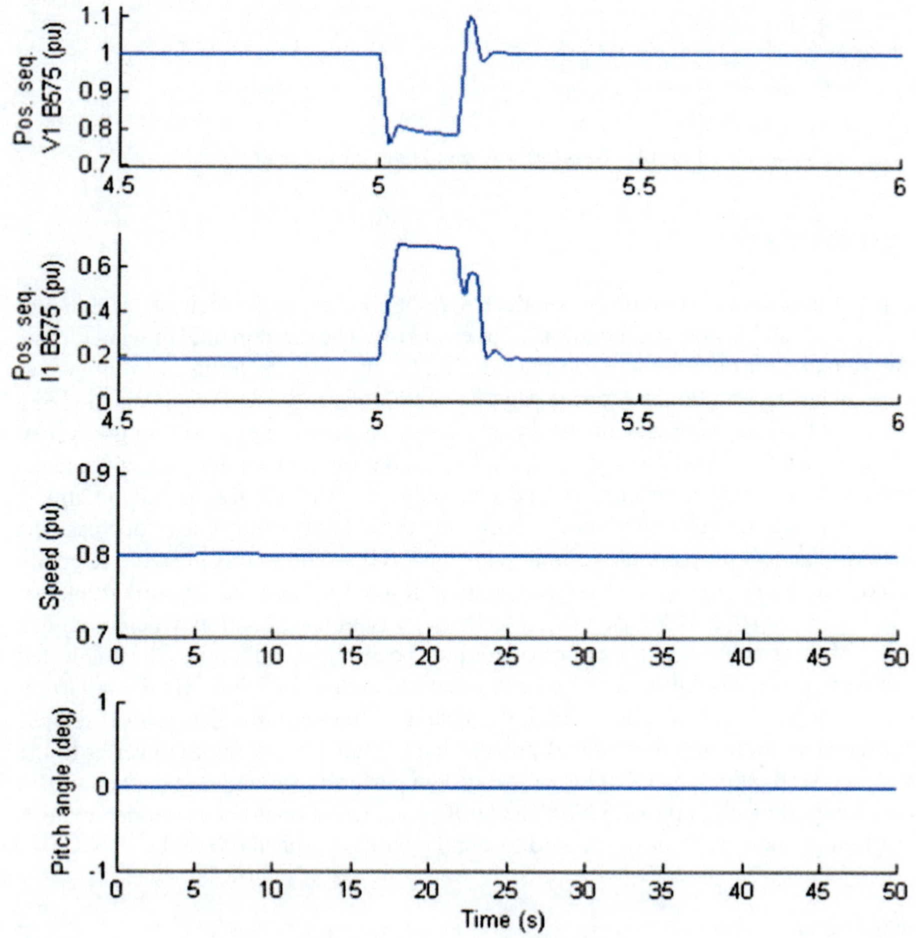

Fig. 10. Wind farm signals throughout fault at 25 kV

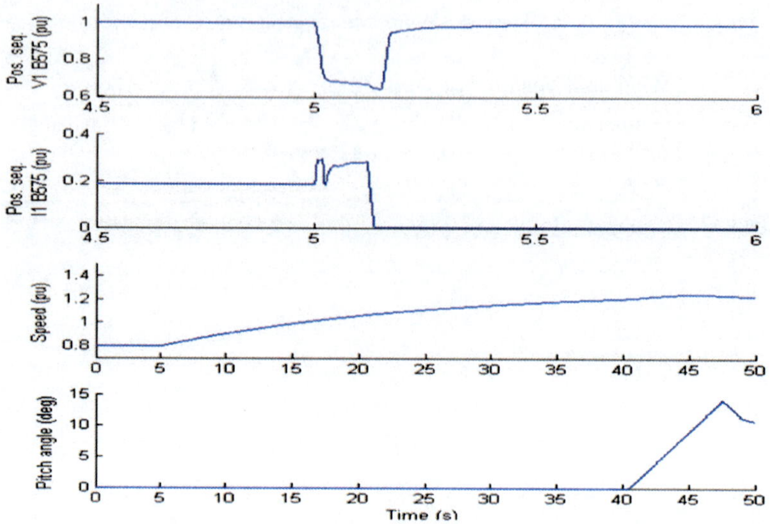

Fig. 11. Waveforms of wind farm during fault

3 Conclusion

This paper introduced wind turbine model under DFIG and power electronic techniques which provide stable power generation as green energy. The fundamental of wind turbine topology and control approaches in phasor simulation methods are used in this work. In case of using traditional squirrel cage asynchronous machines, high magnetizing current was haggard as of grid of energy which recover from nearby fault in the system of power generation. Variable speed capability could provide from this concept to make an optimization in power generation. The technology of DFIG permits to pull out utmost power from the little velocity wind by optimizing the velocity of turbine as decrease the turbine mechanical pressure throughout wind gust. Regarding to this idea, the selection of wind velocity control system is become as well more imperative. Important interest in wind turbine up to 5 MW based DFIG technology is preferred option which minimize the cost of rotor converter in front of operating speed range reduction. The suggested operating map is to use DFIG for usual wind status but when the wind velocity fall to cut off the stator from grid and short circuit the system. Afterward, the generator is relapse to conservative induction methods of process in all channels via rotor converter to the grid. The investigation and examine of the system performance under many conditions has been experimented up to 7.5 kW. The DFIG connection could be forbidden to keep it near to unity power factor downward to small conditions speed of wind.

References

1. Zamzoum, O., et al.: Power control of variable speed wind turbine based on doubly fed induction generator using indirect field-oriented control with fuzzy logic controllers for performance optimization. Energ. Sci. Eng. **6**(5), 408–423 (2018)

2. Bedoud, K., Rhif, A., Bahi, T., Merabet, H.: Study of a double fed induction generator using matrix converter: case of wind energy conversion system. Int. J. Hydrogen Energy **43**(25), 11432–11441 (2017)
3. Mahmoud, M.S., Alyazidi, N.M., Abouheaf, M.I.: Adaptive intelligent techniques for microgrid control systems: a survey. Int. J. Electr. Power Energy Syst. **90**, 292–305 (2017)
4. Moreira, A.B., Barros, T.A.S., Teixeira, V.S., Ruppert, E.: Power control for wind power generation and current harmonic filtering with doubly fed induction generator. Renew. Energy **107**, 181–193 (2017)
5. Wickramasinghe, A., Perera, S., Agalgaonkar, A., Meegahapola, L.G.: Synchronous mode operation of DFIG based wind turbines for improvement of power system inertia. Renew. Energy **95**, 152–161 (2016)
6. Raju, S.K., Pillai, G.N.: Design and real time implementation of type-2 fuzzy vector control for DFIG based wind generators. Renew. Energy **88**, 40–50 (2016)
7. Rahimi, M.: Improvement of energy conversion efficiency and damping of wind turbine response in grid connected DFIG based wind turbines. Int. J. Electr. Power Energy Syst. **95**, 11–25 (2018)
8. Beddar, A., Bouzekri, H., Babes, B., Afghoul, H.: Experimental enhancement of fuzzy fractional order PI+ I controller of grid connected variable speed wind energy conversion system. Energy Convers. Manage. **123**, 569–580 (2016)
9. Aly, H.H.: Dynamic modeling and control of the tidal current turbine using DFIG and DDPMSG for power system stability analysis. Int. J. Electr. Power Energy Syst. **83**, 525–540 (2016)
10. Ganjefar, S., Mohammadi, A.: Variable speed wind turbines with maximum power extraction using singular perturbation theory. Energy **106**, 510–519 (2016)
11. Allouhi, A., Zamzoum, O., Islam, M.R., et al.: Evaluation of wind energy potential in Morocco's coastal regions. Renew. Sustain. Energy Rev. **72**, 311–324 (2017)
12. Ghoudelbourk, S., Dib, D., Omeiri, A.: New techniques to improve the power control in the double-fed induction generator wind turbine, wind engineering, **40**(1) 2016
13. Hossain, M.M., Ali, M.H.: Future research directions for the wind turbine generator system. Renew. Sustain. Energy Rev. **49**, 481–489 (2015)
14. Ghennam, T., Aliouane, K., Akel, F., Francois, B., Berkouk, E.M.: Advanced control system of DFIG based wind generators for reactive power production and integration in a wind farm dispatching. Energy Convers. Manage. **105**, 240–250 (2015)
15. Ananth, D.V.N., Kumar, G.N.: Fault ride-through enhancement using an enhanced field oriented control technique for converters of grid connected DFIG and STATCOM for different types of faults. ISA Trans. **62**, 2–18 (2016)
16. Mehdipour, C., Hajizadeh, A., Mehdipour, I.: Dynamic modeling and control of DFIG-based wind turbines under balanced network conditions. Int. J. Electr. Power Energy Syst. **83**, 560–569 (2016)
17. Mahto, T., Mukherjee, V.: A novel scaling factor based fuzzy logic controller for frequency control of an isolated hybrid power system. Energy **130**, 339–350 (2017)
18. Civelek, Z., Lüy, M., Çam, E., Mamur, H.: A new fuzzy logic proportional controller approach applied to individual pitch angle for wind turbine load mitigation. Renew. Energy **111**, 708–717 (2017)
19. Sudheer, H., Kodad, S.F., Sarvesh, B.: Improvements in direct torque control of induction motor for wide range of speed operation using fuzzy logic. J. Electr. Syst. Inform. Technol. 2017, In press
20. Gupta, N., Garg, R.: Tuning of asymmetrical fuzzy logic control algorithm for SPV system connected to grid. Int. J. Hydrogen Energy **42**(26), 16375–16385 (2017)
21. Yaakoubi, A., Asselman, A., Djebli, A., Aroudam, E.H.: A MPPT strategy based on fuzzy control for a wind energy conversion system. Proc. Technol. **22**, 697–704 (2016)

22. Tir, Z., Malik, O.P., Eltamaly, A.M.: Fuzzy logic based speed control of indirect field oriented controlled double star induction motors connected in parallel to a single six-phase inverter supply. Electr. Power Syst. Res. **134**, 126–133 (2016)
23. Salleh, Z., Sulaiman, M., Omar, R., Patakor, F.A.: Optimization of fuzzy logic based for vector control induction motor drives. In: 2016 8th Computer Science and Electronic Engineering (CEEC), pp. 83–88. IEEE (2006)
24. Ismail, M.M., Bendary, A.F.: Protection of DFIG wind turbine using fuzzy logic control. Alexandria Eng. J. **55**(2), 941–949 (2016)
25. Kumar, D., Chatterjee, K.: A review of conventional and advanced MPPT algorithms for wind energy systems. Renew. Sustain. Energy Rev. **55**, 957–970 (2016)

Text and Web Content Mining: A Systematic Review

Fatima Almatrooshi[1], Sumayya Alhammadi[1], Said A. Salloum[2,3](✉), and Khaled Shaalan[1]

[1] Faculty of Engineering and IT, The British University in Dubai, Dubai, UAE
[2] School of Science, Engineering, and Environment, University of Salford, Salford, UK
ssalloum@sharjah.ac.ae
[3] Machine Learning and NLP Research Group, Department of Computer Science, University of Sharjah, Sharjah, UAE

Abstract. Text and content mining are the subcategory of data mining. This category of data mining is used to extract the information from web or web pages of a website. This mining identifies useful information from web content like web pages, search logs, and other website-related content. The extracted information can be used in many applications, for example, we can extract opinions from online sources and web hierarchy which provides better insights and knowledge. In this paper, we conducted a systematic review that included 18 research papers that are relevant to the topic and matches the inclusion criteria of this study. From these research papers we were able to answer the research questions that we identified. The questions are related to the applications, techniques and issues of the text and web mining. The findings suggest that many research papers made a good foundation for this topic, and gave an informative explanation of each type of techniques used in text and web mining, as well as some issues that can be a future work for researchers who are interested in the topic.

Keywords: Text & content mining · Web mining · Systematic review

1 Introduction

In a world where every action is translated to bits and bytes of data, it is important to extract useful knowledge from the huge number of possibly useless data that occupies millions Terra bytes of space. Text and content mining of websites and applications are the process of extracting and discovering insights and useful knowledge from such data. Data is usually classified into three main categories: structured, semi structured and unstructured. Although the studies stated the importance of textual mining, it was at certain period ignored, and researchers mostly depended on computational and numeric data [1]. Moreover, as the world moved toward digitalizing and converting daily life tasks into electronic website or application, the importance of text and content mining of websites and apps increased drastically. In unstructured data mining, most of the content is unstructured text. To extract the information, computer scientist use a method known

M. Al-Emran et al. (Eds.): ICETIS 2021, LNNS 299, pp. 79–87, 2022.
https://doi.org/10.1007/978-3-030-82616-1_8

as the KDT or the knowledge discovery in text. As the name suggests this technique is usually used to extract the data from the unstructured data like the text data. Sometimes the text mining is also considered as an instance of web content mining. Before the actual applying of the knowledge discovery in text or any technique of data mining, the data need to be first preprocessed, which is basically converting unstructured data to structure. After the preprocessing of the data, the analysis is done on the structured data, then the data is categorization and lastly, the KDT or any natural language technique will be implemented. ("What is Text Mining: Techniques and Applications", 2020) [2]. In this research work, we will analyze the existing works of text and web content mining. The research questions we are interested in investigating are:

1. What are the applications of web content mining and how web content mining works on actual web data?
2. What are the approaches and techniques that are used in the text and web mining for the extraction of the useful information?
3. How is text and content mining used in social networks and is there a difference when dealing with social networks in comparison with usual websites data?
4. What are the misconceptions and current issues faced in text and content mining?

2 Literature Review

This section discusses the different concepts of mining that are researched by different data scientists. More specifically this section will discuss the work that has been done in text and web content mining, as well as giving a brief explanation of its history and its relation with other types of natural language processing. In A. Kumar et al. they proposed their research on the web content mining, and mentioned that nowadays the websites are the main source of collecting information because all the information is available on websites. It is quite difficult and complex to download information from a website as it can take lots of effort and time to complete the whole process. This is how a machine or computer is used to get information about a specific thing that the user wants. In their research, they also presented and discussed several tools that are usually used to extract information from different websites. They also discussed the basis of web content mining and evaluated the performance of the best tools. They also discussed what the merit and demerits of the tools is, relatively. In their research, it was also mentioned how different companies explore the meaningful data that is related to the users of the website. Because of the advertisements, it is the width of any company to get those specific types of users that are having some sort of interest in their products. In this way, web content mining is very beneficial for these companies for different marketing purposes [3].

M.J. Hamid et al. [4], also discussed the data mining related to the web and defined different techniques of web data mining, and implemented several different algorithms related to web data mining. As the internet is nowadays essential for the cycle of daily life, using the internet, the internet users upload and download the bulk of data to the servers of a specific website. The most preferred one where the users mostly upload or download data is known as the worldwide web. On the daily basis, there is tremendous size of data uploading on the internet, which makes it difficult to find the data that is usable and have meaningful information. Because digging into the data that is inappropriate and unstructured it is quite difficult to explore such type of data. The web pages have the number of different web pages, hyperlinks, contents, and the number of different weblogs after retrieving them it extracts some useful results. In their research work, they also described the different types of web mining. For example, web usage mining, web content, and web structure [4]. P. Milic, et al. [5] Explained how the web content mining work. To support their work they used different types of techniques that are usually used for the data mining, using these techniques they explored the information from the web pages. They extracted their respective information from number of web pages that usually a website has. They also discussed the worldwide web, as that is the major source of the information related to the web content mining. The author also thoroughly mentioned that how the web pages are so vast and having a huge amount of the information. That information is very important for companies and different business analysts. Because the data that is provided on these websites belong to several different user, if the company know the behavior of the users it is quite easy for them to attract such types of users toward them. In such cases the companies introduce the specific type of products in the specific area that are according to those specific types if the users. but as explained earlier in this era there are tons of data is downloading and uploading by the huge amount of the users, so in this way it is much difficult for these companies and business analysts to extract the data that is useful for them, means data according to which they can promote their services or products in the market [5].

3 Methods

In this section, we will describe the process, criteria and databases used to choose research papers that will be included in this study. We will also show a PRISMA chart that summarizes the process of screening and excluding irrelevant papers, as well as a critical appraisal chart will show the quality of each paper. The number of studies found in the field of our research was 706. After removing duplicates, the number of studies ready to be screened were 348. After applying the criteria we ended up with 27 relevant papers. Lastly we assessed quality and accessibility to full text, and ended up with 18 papers that are going to be included in our systematic review.

3.1 Inclusion and Exclusion Criteria

To choose papers for this type of studies, a set of rules must be defined to enable a proper selection of relevant papers. These rules are described as following:

3.1.1 Inclusion Criteria

The included studies should match the following criteria:

- The study should be relevant to the chosen topic
- The study should be written in English
- The study should be of high quality
- The study's full text should be accessible.

3.1.2 Exclusion Criteria

The Excluded studies will most likely have one of the following limitation:

- The study is published before the year 2014.
- The study is not related or relevant to the topic.
- The study is not written in English.
- Duplicate studies are excluded.

3.2 Data Sources and Search Strategies

In this systematic review we searched a number of databases, mainly Google scholar (392), Science Direct (271), UCC library (1), and IEEE digital library (215). As shown in Table 1, the main keywords used in the searching process were: Text mining, social networks, websites, text mining techniques and content mining. The results were almost over 800 studies that we later on chose from according to our research inclusion and exclusion criteria.

Table 1. Databases and keywords used

<table>
<tr><th>Database</th><th>Searching keywords</th><th>Results</th></tr>
<tr><td>Google scholar</td><td rowspan="4">Text mining
Content mining
Social networking
Mining techniques</td><td>392</td></tr>
<tr><td>Science Direct</td><td>271</td></tr>
<tr><td>IEEE digital library</td><td>215</td></tr>
<tr><td>UCC library</td><td>1</td></tr>
</table>

3.3 PRISMA Chart

(See Fig. 1).

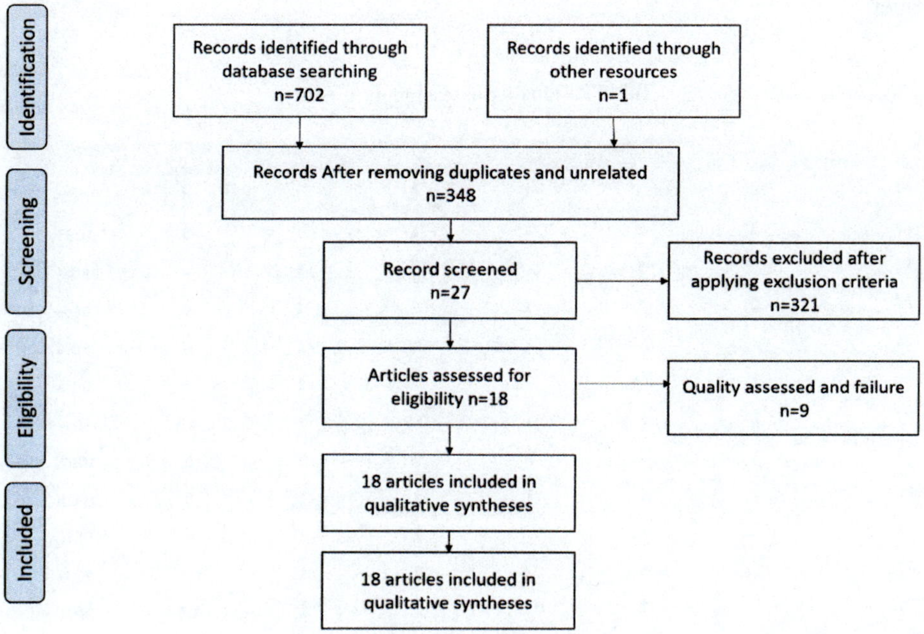

Fig. 1. PRISMA flow diagram (n = 18).

3.4 Critical Appraisal

To evaluate the quality of the chosen papers, we did a quality assessment that will assess and check the quality of the papers. The quality assessment consist of a checklist with standard questions to evaluate the quality. The quality question are:

1. Are the research aims clearly specified?
2. Was the study designed to achieve these aims?
3. Is the study related to text and web content mining?
4. Do the results add to the literature?
5. Does the study add to your knowledge and understanding?

Based on these questions we were able to evaluate each paper and give it a final score that indicates its quality. If the answer is "yes" then the score is 1, if the answer is "no" then the answer is 0. A score of 0.5 indicates that the question is answered partially or does not fully meet the requirements. Table 2 shows the evaluation and score of each paper:

Table 2. Quality assessment (n = 18).

Source	Q1	Q2	Q3	Q4	Q5	Score	%
P1	1	1	1	0	1	4	80%
P2	1	1	1	1	1	5	100%
P3	0.5	1	0.5	1	1	4	80%
P4	0	1	1	1	1	4	80%
P5	1	0.5	1	0.5	1	4	80%
P6	1	0.5	1	0.5	1	4	80%
P7	1	1	1	0.5	1	4.5	90%
P8	1	1	1	1	1	5	100%
P9	1	1	1	1	1	5	100%
P10	0.5	1	0.5	1	1	4	80%
P11	1	1	0.5	1	1	4.5	90%
P12	1	1	0.5	1	1	4.5	90%
P13	1	1	1	1	1	5	100%
P14	1	0.5	1	0.5	1	4	80%
P15	1	0.5	0.5	1	1	4	80%
P16	0.5	1	0.5	1	1	4	80%
P17	1	1	1	1	1	5	100%
P18	1	1	0.5	0.5	1	4	80%

4 Results

The study was designed to investigate three main concepts of text and web content mining which are: applications, techniques and issues. The findings after reading all the included research papers are shown in the following graphs:

Paper	General application	Business	Social media	Academic	Other
P1	x				
P2	x				
P3	x				
P4	x				
P5	x				
P6	x				
P7			x	x	
P8			x		
P9			x		
P10			x		Environmental management
P11		x	x		
P12					TM techniques Selection Criteria
P13	x				
P14	x				
P15		x	x		
P16	x		x		
P17	x		x		Opinion mining
P18	x		x		

Fig. 2. A graph that shows the distribution of application discussed across included research papers

Figure 2 shows the nature of applications mentioned and discussed in the included research papers. As it is shown, we classified the papers accordingly under 5 main categories of applications which are: General, Business, Social networks, Academic and others. The graph shows that the papers we included mostly talk about the social media as a platform to apply text and web content mining. Moreover, Multiple papers explains the concepts of text and web mining without specifying an application which means it can be used in different applications based on the needs of user.

Paper	Retrieval of information	Categorization	Clustering	Summarization	Topic tracking	Information visualization	Other
P1	x	x	x	x	x	x	Mention of other techniques
P2	x	x	x	x	x	x	
P3	x			x			
P4	x						
P5	x						
P6	x	x	x	x	x	x	
P7		x	x				Text mining specific
P8			x				Text mining specific
P9	x		x				
P10							
P11							
P12	x	x	x	x	x	x	
P13							Stemming
P14	x	x	x	x			
P15			x				Stemming and tokenization
P16	x		x				
P17	x			x			Classifiers: SVM and Naïve Bayes
P18							Not focused at

Fig. 3. This graph shows the techniques mentioned in each paper.

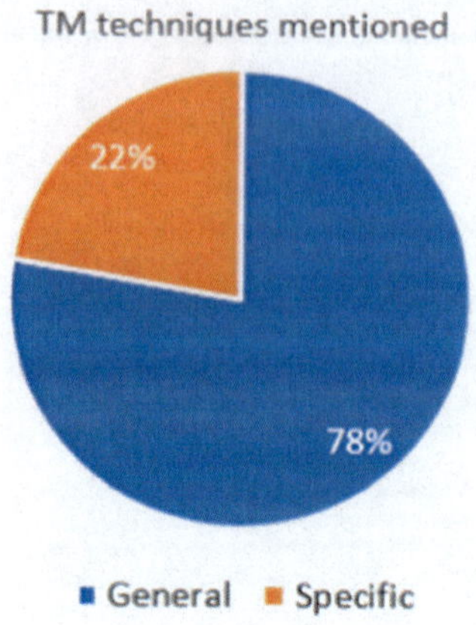

Fig. 4. This graphs shows that the majority of the studies discuss multiple techniques and not a specific one.

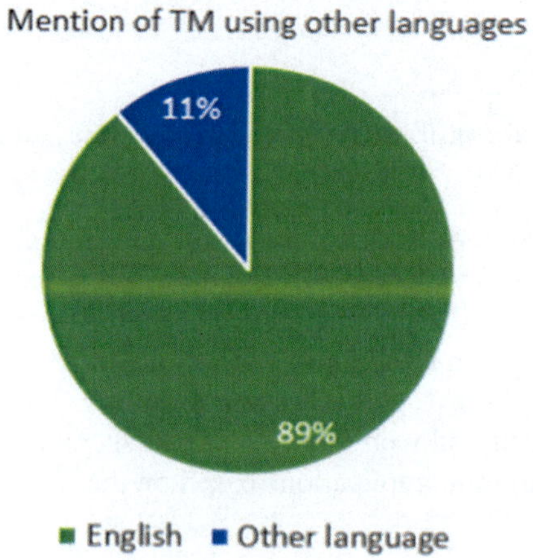

Fig. 5. This graphs shows that the majority of the studies work with English as the language of mining and used tools.

Figure 3 shows the main 6 techniques of web content mining, 4 out of 6 of these techniques are also considered text mining techniques. The figure basically can guides researchers where each technique is mentioned and discussed for each paper. Figure 4 shows that most of the included research papers do not focus on one technique, but they give an explanation of each technique in general. It also shows that 22% of papers actually focus on one technique such as stemmers or clustering. Lastly, Fig. 5 shows the usage of other languages in text mining and how well this issue is recognized by researchers in the field. Only 11% of the papers mention the Arabic language and how limited the advances in this topic is.

5 Conclusion

In this systematic review we had 4 research questions that we wanted to investigate. We applied some searching criteria to choose studies that will be included 18 relevant and related to the topic studies were found and included. The main findings of this paper are as following:

Most papers included were generally explaining the techniques and applications of text and web mining.

- 22% were more specific and focused on one technique.
- We discussed techniques from technical and general perspectives.
- The main techniques used in this research field are: Information Retrieval, Summarization, Categorization, Clustering, Topic Tracking, Information visualization.
- Text and web content mining could be used and applied in many fields such as: Management of risks, social media analysis, Costumer care services, Business Intelligence and fraud detection.
- Applying Text mining in social media is challenging because of the fact that people do not necessary use right sentence structure when writing in media pages.
- The main issues in the field could be summarized in two main points: other languages usage and text errors in social media. Other issues may exist, but these two points are the common ones among the paper we included.

For future perspective: The efficiency of both techniques needs to be improved. For example, web content mining can be improved by adding some artificial intelligence into its algorithm, through this algorithm it can detect the type of the website. If the website's type is an online store then it learns the algorithm based on the template of the store, where only products changed not all of the design of the web page changes. In this way, some load of the processing is reduced which can cause an increment in the efficiency of the processing and generation of the results. On the other side, text mining can be also improved by using any fast learning method. For future work, researchers can also work to solve the limitations and gaps of language usage and dealing with sparse and noisy text.

Acknowledgment. This work is a part of a project submitted in fulfilment of PhD Computer Science - Faculty of Engineering & Information Technology at The British University in Dubai.

References

1. Özyirmidokuz, E.K., Özyirmidokuz, M.H.: Analyzing customer complaints: a web text mining application. In: Proceedings of INTCESS14-International Conference on Education and Social Sciences, Istanbul, pp. 734–743 (2014)
2. What is Text Mining: Techniques and Applications (2020)
3. Sharma, A.K., Gupta, P.: Study and analysis of web content mining tools to improve techniques of web data mining. Int. J. Adv. Res. **1**(8), 287–293 (2012)
4. Hamid Mughal, M.J.: Data mining: web data mining techniques, tools and algorithms: an overview. Int. J. Adv. Comput. Sci. Appl. **9**(6), 208–215 (2018)
5. Milic, P.: Framework for Web Content Mining (2017)

A Survey on the Use of Personalized Model-Based Search Engine

Mohammad Najah Mahdi[1](✉), Abdul Rahim Ahmad[2], Qais Saif Qassim[3], Mohammed Ahmed Subhi[4], and Taofiq Adeola Bakare[5]

[1] Informatics and Computing in Energy, Universiti Tenaga Nasional, Selangor, Malaysia
mehdi.najah@uniten.edu.my
[2] Systems and Network Department, College of Computing and Informatics (CCI), Universiti Tenaga Nasional, Selangor, Malaysia
[3] Ibri College of Technology, Ibri, Sultanate of Oman
[4] Engineering and Built Environment, Universiti Kebangsaan Selangor, Bangi, Malaysia
[5] Faculty of Engineering and Technology, Multimedia University, Cyberjaya, Malaysia

Abstract. Mostly with growth of the information sector, in particular the World wide web and the mobile Internet, the volume of information that we'll have to contend with is rapidly increasing. To a certain degree, users could get the information they need from the Internet. And in context of these vast availability and proliferation of information, users need an efficient way of finding only important and interesting information. How to optimize usability during information search process is still not addressed. Searching was the key feature of a conventional search engine. There has been insufficient customization in search engines to be used by searchers during the information extraction process. This article thus analysed and discussed the personalized recommendations of searches using a tailored framework that supports search engines. We describe a comprehensive review of the study on search engine customization for web-based queries. The study is partially systematic and, also at end of the study, we discuss the difficulties of personalizing search engines.

Keywords: Information retrieval · Model-based search engine · Search engine · Personalized search engine

1 Introduction

The search for information on the Internet has been so popular in recent decades. Online search includes a search engine (SE); however, current search engines are still not personalized as they do not satisfy the individual search needs. They also may not satisfy the personalization criteria. While there is no limit to the growth of information [1], personal interest may be limited to a certain goal. It is essential for the search engine to provide the most reliable and interesting information to users or searchers. Users have different perspectives. These could lead them to having different viewpoints in the same subject or field of knowledge. Although the user can search by keywords or keyword,

M. Al-Emran et al. (Eds.): ICETIS 2021, LNNS 299, pp. 88–98, 2022.
https://doi.org/10.1007/978-3-030-82616-1_9

the SE may generate search results that reflect some kind of basic filtering, that might be partially represent what was searched for. So how do you get search results to be better personalized? Also, how can a query be better to return more important results? These abilities are now very much needed [2–4]. Personalized web search (PWS) is one of the types of general search strategies aimed at achieving improved search results.

While monitoring the search history of registered users, PWS may change search results user's preferences. This takes advantage of the information collected to evaluate the SE actions of the user in relation to other data collected from the internet [5, 6]. The study focused on PWS. We reviewed the literature on personalization by considering the following areas of interest: I problems, challenges and limitations in PWS systems and (ii) performance assessment of personalized SE search application. This study also examined the personalized search by reviewing previous research that focuses on SE using shared resources. In addition to this, the paper also suggests several changes to the implementation of PWS.

The remainder of this paper is structured as follow: Sect. 2 offers a quick overview of interactive visual surveys, and general guidelines explicitly on conventional personalized SE surveys. Section 3 outlines the overall architecture and relevant PWS strategies or techniques. This section also includes a brief analysis of the benefits and limitations of these approaches, the methodologies used to classify the information and application considerations. And the last section of this paper contains the identified problems and conclusion.

2 Research Background

As stated in the introductory section, with the further growth of the information industry, in particular the Internet industry, the quantity of information is growing exponentially. Consequently, the lack of information can no longer be complained about. Whatever bit of information that we like is already available on the Internet. That being said, when presented with so much content, it becomes quite difficult to determine which information, resulting from the query, is important, can be easily identified, and has the information of interest. There is immediate need to overcome this challenge In the new age of Web 4.0, the origins of information have become more varied. Users are not only inactive receivers of information anymore, but can also be the architect to initiate that information. Traditional SEs are also not efficient in managing the bulk of information emerging in this Web 4.0 period, that already has normalization and integration [7–9].

Social media outlets, like Facebook and Twitter, provide a vast volume of user information. These would be a reference point for most users to begin their investigation because browsing with existing user data is more efficient than the conventional SE. For instance, if we had an account on Facebook, we would have already saved a huge volume of information on our own. Which include demographic information, gender, friends, and interests. Such data, along with existing search and browsing history, establishes a personalized profile from which the social networking platform changes the SE result page to match our interests. Conventional SEs are not designed nor positioned to exploit this form of search; they are somewhat limited to enhance user experience. This really is a disadvantage [10–12].

To understand the effect of personalization on SEs, our research answers two questions:

RQ1: Does a PWS enhance users' comprehension of the above goals?

RQ2: Can PWS users build filters to narrow the results of the search phase?

In the context of the SE, we formulated these research questions on the basis of filtering. With consideration to increasing its target exploration capability, PWS is a constructive strategy to enhance SE efficiency by using filters. The field of PWS has evolved in the last two decades and now has a wide range of methods and techniques that can be used for searching [13–15].

Owing to the exponential growth of PWS and its related fields such as ontology based web personalization, SE, Meta Search, human-computer interaction, cross-system personalization, several conventional survey papers aimed at offering a mostly static description of basic approaches cannot be easily modified.

This is attributable to the consistent growth of PWS technologies [16]. Taking into consideration the relevant work for the proposed interactive research method mentioned in this article, we distinguished between general interactive PWS surveys that concentrate on any study field and conventional PWS surveys in SE techniques [17–19]. To the best of our understanding, hardly any interactive PWS relevant to SE techniques has been established specifically for filters to narrow down the searched results. Over 25 papers on PWS have been collected and analyzed for development and adaptation in the management of data filtering [20], along with component analysis, device development, direction and adaptation of PWS. The articles are included in Table 1 below, which contains the title of the articles, the approaches used, the problems and meanings, the contributions reported and the proposed strategy.

Table 1. Components analysis, directions and adaptation of personalized web search

Ref	Methods	Issues and definition	The achievement	Directions and adaptation
[21]	Digital Libraries (DLs)	Investigates the design, compatibility, and model used	Maintains profiles interoperable user models and complain that can be propagated and reconciliation of user data	Six user model were interoperability presented. Focused on reconciliation of various user attributes through which model interoperability
[22]	Ontology based PWS	Presents a general review of ontology based PWS	PWS recommended systems using ontology methods. Also, presented a comprehensive description	Works on context and user profiling. Sophisticated PWS experiences have been provided with web mining combined with ontology

(*continued*)

Table 1. *(continued)*

Ref	Methods	Issues and definition	The achievement	Directions and adaptation
[23]	Search Engine Meta Search	Review of ontology based PWS, challenges that selects the Meta SE are studied	Improve the efficiency of SE, increase the number of result as per the user requirement	PWS focused on various outcomes, user interface, and importance of the SE member sorted
[24]	PWS, User profile	Focused on the survey of many efficient PWS methods	To reduce its applicability that can only operate on repeated queries	It is required to increase the SE accuracy and minimize the time the userspent
[25]	Ontology based PWS	Problems connected that provide a review on present research and user modelling in multi-application environments	To facilitate the methods of understanding the users, their interests and their preferences in terms of providing them with PWS	Directions are: (i) privacy and security in cross-system personalization and (ii) evolutionary, are not sufficiently considered up to now
[26]	Privacy Personalization Human–computer interaction	Technologies that can assist to minimize privacy risks also helped designers and users to contextualize privacy and personalization	Analyses the privacy risks associated with various present and prominent personalization direction, social-based personalization, behavioral	Discussed that frames risks and methods solutions in the intersection between privacy and personalization
[27]	ODP, Semantic web	Presented several methods to personalize Web SE through searcher modelling by analyzing semantic data	To increase in size the query based on searcher intense such model can identify the semantic similarity between the previous SE query	The user interests incorporating to customize the SE results by re-ranking the retrieved results
[28]	Ontology, Personalization, Information	Presented issues and searching methods also covers overall view of ontology model which is mainly utilized for gathering web information	Initialization of information gathering according to user profile and ontology to gather web information based on keywords that may be local database	Discussed the learning concept extract the data in structured format for unstructured input. Also covers basic framework of ontology which focuses on overall data gathering
[29]	Information Retrieval, Semantic Web, Ontology,	Reviewed various techniques of operations to enhance the efficiency of the data retrieval system	Established a variety of techniques used in Web Mining by scientists to accomplish Web Personalization	To minimise brand loyalty and data overload, PWS is a platform that supports consumers with significant competitive advantages

2.1 Search Engine (SE)

A detailed overview of the basic information searched for is given to the search engines. Generally, only two or three keywords need to be provided in the current SEs. More terms or phrases are necessary to support the query method to extend the reach of the query. Similarly, more detailed information, like location, search history, can all be presented side-by-side with the original query, such as the Google search engine. The SE does the following:

a: Accepts the user's submission of requests, including words or sentences, by specifying the precise details in the action to be viewed on the site by the user. b: Tries to search the database to match the request. c: Organizes and delivers a clickable URL array that matches the request. d: Receive the resubmission of the updated query from the customer if any.

The search engine is a programmer designed to search for information found on the internet. It searches for information using keywords or phrases, and presents a set of documents (hits). The targets can be linked or unrelated to web sites, images and files [30–32], and some SEs will mine data that is stored in databases or open directories too. Thus, without SEs, it could be difficult to find details on the Internet without having their unique URL [20, 33, 34]. SEs can be categorized into three types [35]; (a) crawler-based, (b) human-powered type, and (c) hybrid types. Details of the categories are given in Table 2:

Table 2. Types of search engines

SE types	Example and description
Crawler-Based SE	Example: Google. Crawlers access the web, read the details in it, and generate a meta tag. This enable the records to be retrieved by following some hyperlinks to return all the information to the central repository and index the data [35]
Human-Powered SE	Example: Google directory, Yahoo directory, Open directory. Appropriately known as a site directory. More popular due to the dynamic value of human links that are indexed and classified. The content provided is only recorded
Hybrid SE	Example: MSN Search. This intersects between human-powered and crawler-based directories. When seeking for information, both crawler and human-powered results can be accessed while preferring one method of listing to another [36]

2.2 Research Status of Search Technology Based on User Personalization Model

From its inception, the PWS analysis concentrated on the recommendation of search results and user style polymerization. The personalized user model analysis was not enough to attract the interest of the researchers. Nevertheless, with an increment in the amount of users, which raises the network and user criteria for the precision of the searched data, better user classification polymerization techniques have been proposed to satisfy their needs. Recently, scientists have reported some findings by paying closer attention to personalized models. For example, the authors in Leiberman [37], Gourevitch, et al. [38] established a user information agent approach to the design user models [39, 40].

The idea is to track user activity accurately and to automatically construct a user model structure. This method does not require the user to directly include the data; nevertheless, it can be created based on the user's habits. Fab Comprehensive Literature [41–43] was focused on the contents page ranked division to create user models that were developed using the subjective sort of the user.

Balabanovic [44], Nath and Bhushan [45] are already out of the user's perspective; they are concerned with the content from which the user's specialized subject is to be categorized and filtered out. The material available on the page to be categorized uses a user model to predict whether the user has this perception of page relevance. In the SE user model, Wen [46] suggested a rather more systematic understanding of the user model, such as a representation of the vector and a representation of the conceptual level by using Huffman tree to represent the user model.

Hong [47] has focused on tailored customer-side, fine-grained user modelling services. His contribution is no more whatever the consumer is interested in or not. This is clearly classified into two parts; but, from the user's perspective, it is separated into different classes. This is dependent on the user's interpretation of the content. In addition, Zheng, et al. [48] borrowed a model from someone else to refine the system for the consumer model. A further contribution of these studies and methods is also outlined in Table 3.

Table 3. Critical review of personalized search engine

Ref	Methods/Techniques	Description	S: Strengths, B: Weaknesses
[49]	Researched models proposed for web semantics, IR methods, information gathering techniques emotional outputs of SE actions online	The improved search experience, making the users to the to retrieve relevant results easier, was obtained by smaller queries	S: performance of cognitive agent based Web was improved. B: More research is required to be done
[50]	Suggested the evaluation model that describes the aspect of a SE	Presented methods of evaluating SEs to easily suit the user's need by formulating user oriented metrics that compare fresh and duplicate results in SE	S: Performance of SEs were identified by the statistical patterns that further refined the evaluation model. B: A larger scale study involving many users should confirm the validity of the model

(*continued*)

Table 3. (*continued*)

Ref	Methods/Techniques	Description	S: Strengths, B: Weaknesses
[51]	The implementation of an approach that personalized delivery was described by considering the relevance of content that suits the user's need	Using automated personalized delivery, the achievement supports the user to search for new contents	S: Method has the ability to collect and process the stream of new information. B: Implementation of automatic portals in other sectors including culture
[52]	Built a custom context-dependent SE web model utilizing semantic relationships	PWS Context-Dependent named "Sama SE" that provided an optimized model for indexing and matching the terms collected from the retrieved documents and submitted a searches	S: Suggest that the precision, recall, and F-measure of Sama SE are higher, and it is fall-out is lower compared to the other ST method. B: Depends on results collected by Google
[53]	Presented a framework of an academic SE with personalized ranking SEs result mechanism	To improve and enhance the quality of research paper searching, an academic SE capability	S: Result of the experiment proposed that the PWS result ranking mechanism performed better than original result ranking. B: Can be destined to have a limited field of research filed
[54]	Mining user behaviour of characteristics with agent for information retrieval	Puts a meta PWS mechanism with the use of the agent technology on ontology	Mining user behaviour of characteristics with agent for information retrieval
[55]	Using WordNet, a personalized user model, collecting the related sentences from the actual text document, was implemented to form a description	Semantic personalized IR, which encourages the efficient retrieval of semantic information from the Internet, has been proposed for the utilisation of Semantic Web applications	S: Improved the accuracy of the IR since the system depends on the Semantic Web technology. B: Proposed system was not implemented for searching

3 Challenges

Numerous studies on user profile modelling has been designed to enhance personalized familiarization and eliminate unnecessary experiences that have some-times arisen between the user and the system. However, a lot of work also needs to be done to enhance the simulation of the user profile, which will prevent mal-functioning of the device. This could be accomplished by delivering correct services to the customer at the required place and time, although some difficulties persist in achieving these objectives. How users specifically search, from which SEs select information according to their experience and interest, for whatever they need efficiently. How can users, in particular mobile users, be helped to access accurate and relevant information? Most of the suggested approaches have failed, in one way or another, to deal effectively with such challenges. Based on conventional SE technology, the process of obtaining experience with mobile users is not good sufficient. Traditional IR technology typically takes into account only reasonably common requirements; does not effectively meet the more complex query request of the user associated with its context as well as the intent of the search. Personalized IR technology is the answer to this issue. It takes full account of the customer and the disparity between different users of different services by balancing their specific needs [56]. Construction of the user's personalised model, to fit their desires and preferences, has therefore become essential. We further streamlined this by sorting information tools for further classification, acting on the various suggested users, to fit their preferences and information needs.

4 Conclusion

In this study, we addressed personalised user-based SE as well as how to allow users to search efficiently, find references to their own preferences and experience. How to help users navigate their personalised search easily and precisely. Most of the approaches previously used, including traditional SE methods, have not been able to solve the problem. We studied the PWS and various user interface designs. In order to facilitate experimentation, the interface should provide prompt input on the possible actions of the user. The primary contribution of this article is the general view of existing techniques on personalized user-based search engines and the method for filtering the results of the search process. We then highlighted the detailed methods and key characteristics used to define the methodology used to collect and execute the data. The recommendations made in this article may act as a guide to the development of an interface that encourages progress in all aspects and phases of the user search process. Effective filter interfaces that are designed to help users express effective requests, aid them to understand the searched results and, if necessary, encourage query modifications.

Acknowledgment. This work is a part of a project submitted to Universiti Tenaga Nasional (UNITEN) internal grant no J510050783 (2018). Many thanks to the Innovation & Research Management Center (iRMC), UNITEN who provided their assistance and expertise during the research.

References

1. Dao, T.T., Hoang, T.N., Ta, X.H., Tho, M.C.H.B.: Knowledge-based personalized search engine for the Web-based Human Musculoskeletal System Resources (HMSR) in biomechanics. J. Biomed. Inform. **46**, 160–173 (2013)
2. Leung, K.W.-T., Ng, W., Lee, D.L.: Personalized concept-based clustering of search engine queries. IEEE Trans. Knowl. Data Eng. **20**, 1505–1518 (2008)

3. Dongfei, L., Jianguo, D.: The design and implementation of personalized search engine based on bilingual translation. In: 2008 ISECS International Colloquium on Computing, Communication, Control, and Management, pp. 472–476 (2008)
4. Umagandhi, R., Senthilkumar, A.: Time dependent approach for query and URL recommendations using search engine query logs. IAENG Int. J. Comput. Sci. **40** (2013)
5. Xu, J.-F., Zhao, Z.-B., Qiu, T.-R.: Personalized search engine based on granular computing. In: 2008 IEEE International Conference on Granular Computing, pp. 690–694 (2008)
6. Costa, C., Santos, M.Y.: Big data: state-of-the-art concepts, techniques, technologies, modeling approaches and research challenges. IAENG Int. J. Comput. Sci. **43**, 285–301 (2017)
7. Dong-Fei, L., Jian-Guo, D.: The design and research of user interest model in personalized search engine. In: 2009 Asia-Pacific Conference on Information Processing, pp. 639–642 (2009)
8. Chen, X.-D., Huang, L.: The research of personalized search engine based on users' access interest. In: 2009 Asia-Pacific Conference on Computational Intelligence and Industrial Applications (PACIIA), pp. 337–340 (2009)
9. Mahdi, M.N., Ahmad, A.R., Ismail, R., Subhi, M.A., Abdulrazzaq, M.M., Qassim, Q.S.: Information overload: the effects of large amounts of information. In: 2020 1st. Information Technology to Enhance e-learning and Other Application (IT-ELA), pp. 154–159 (2020)
10. Akhlaghian, F., Arzanian, B., Moradi, P.: A personalized search engine using ontology-based fuzzy concept networks. In: 2010 International Conference on Data Storage and Data Engineering, pp. 137–141 (2010)
11. Cao, J., Tang, Y., Lou, B.: Personalized meta-search engine design and implementation. In: 2010 3rd International Conference on Computer Science and Information Technology, pp. 305–307 (2010)
12. Mahdi, M.N., Ahmad, A.R., Ismail, R., Subhi, M.A.: Improving big data technologies with visual faceted search. In: 2020 8th International Conference on Information Technology and Multimedia (ICIMU), pp. 44–49 (2020)
13. Yang, S.H., Wang, F.L.: Study on personalized search engine based on files. In: 2010 International Conference on Internet Technology and Applications (2010)
14. Gong, J.: Analysis the idea of personalized search engine based on user behaviour. In: 2010 International Conference on Computer Application and System Modeling (ICCASM 2010), pp. V5-450–V5-452 (2010)
15. Zhongbao, L.: Research of a personalized search engine based on user interest mining. In: 2010 International Conference on Intelligent Computing and Integrated Systems, pp. 512–515 (2010)
16. Lili, C.: The study of personalized search engine based on Chinese-English bilingual translation. In: 2010 International Conference on E-Product E-Service and E-Entertainment, pp. 1–4 (2010)
17. Annadurai, A., Annadurai, A.: Architecture of personalized web search engine using suffix tree clustering. In: 2011 International Conference on Signal Processing, Communication, Computing and Networking Technologies, pp. 604–608 (2011)
18. Mahdi, M., Ahmad, A.R., Ismail, R.: Similarity search techniques in exploratory search: a review. In: TENCON 2018–2018 IEEE Region 10 Conference, pp. 2193–2198 (2018)
19. Mahdi, M.N., Ahmad, A.R., Ismail, R.: Evaluating search results in exploratory search (2018)
20. Mahdi, M.N., Ahmad, A.R., Ismail, R., Natiq, H., Mohammed, M.A.: Solution for information overload using faceted search-a review. IEEE Access **8**, 119554–119585 (2020)
21. Nika, A., et al.: A survey of context-aware cross-digital library personalization. In: Detyniecki, M., Knees, P., Nürnberger, A., Schedl, M., Stober, S. (eds.) AMR 2010. LNCS, vol. 6817, pp. 16–30. Springer, Heidelberg (2012). https://doi.org/10.1007/978-3-642-27169-4_2

22. Jammalamadaka, K., Srinivas, I.: A survey on ontology based web personalization. Int. J. Res. Eng. Technol **2**, 163–167 (2013)
23. Naval, P., Priyanka, S.: A survey on personalized meta search engine. Int. J. **2** (2012)
24. Akhila, G., Prasanth, R.: A survey on personalized web search. Int. J. Adv. Res. Trends Eng. Technol. (IJARTET) **II** (2015)
25. Viviani, M., Bennani, N., Egyed-Zsigmond, E.: A survey on user modeling in multi-application environments. In: 2010 Third International Conference on Advances in Human-Oriented and Personalized Mechanisms, Technologies and Services, pp. 111–116 (2010)
26. Toch, E., Wang, Y., Cranor, L.F.: Personalization and privacy: a survey of privacy risks and remedies in personalization-based systems. User Model. User-Adap. Int. **22**, 203–220 (2012)
27. Jay, P., Shah, P., Makvana, K., Shah, P.: Review on web search personalization through semantic data. In: 2015 IEEE International Conference on Electrical, Computer and Communication Technologies (ICECCT), pp. 1–6 (2015)
28. Borse, N., Patil, S., Agrawal, N., Pachlor, R.: Survey on A Personalized Ontology Model for Web Information Gathering (2013)
29. Sridevi, K., Umarani, R.: Web personalization approaches: a survey. Int. J. Adv. Res. Comput. Commun. Eng. **2**, 1533–1539 (2013)
30. Schuth, A.: Search engines that learn from their users (2016)
31. Sullivan, D.: How search engines work. SEARCH ENGINE WATCH. http://www.searchenginewatch.com/webmasters/work.html. Accessed 26 June 2001. (on file with the New York University Journal of Legislation and Public Policy) (2002)
32. Chen, G., Lu, Z., Zhang, Z., Sun, Z.: Research on hybrid modified cuckoo search algorithm for optimal reactive power dispatch problem. IAENG Int. J. Comput. Sci. **45**, 328–339 (2018)
33. Goel, S., Yadav, S.: Search engine evaluation based on page level keywords. In: 2013 IEEE 3rd International on Advance Computing Conference (IACC), pp. 870–876 (2013)
34. Leung, N.K., Lau, S.K.: No more "keyword search" or FAQ: innovative ontology and agent based dynamic user interface. IAENG Int. J. Comput. Sci. **33** (2007)
35. Agrawal, H., Yadav, S.: Search engine results improvement–a review. In: 2015 IEEE International Conference on Computational Intelligence & Communication Technology (CICT), pp. 180–185 (2015)
36. Levene, M.: An Introduction to Search Engines and Web Navigation. Wiley (2011)
37. Leiberman, H.: An agent that assists web browsing. In: Proceedings of IJCAI-95, Montreal, Canada. Morgan Kaufmann, San Francisco (1995)
38. Gourevitch, G., Luttinen, T., Boctor, V., De Graaf, W., Zhao, L.: Using e-mail message characteristics for prioritization. Google Patents (2017)
39. Nodelman, U.D., Horvitz, E.J.: Continuous time Bayesian network models for predicting users' presence, activities, and component usage. Google Patents (2010)
40. Mahdi, M.N., Ahmad, A.R., Ismail, R.: Improving faceted search results for web-based information exploration. Int. J. Adv. Sci. Eng. Inf. Technol. **10**, 1143–1152 (2020)
41. Wang, Y., Bu, H., Qiu, Y.: The flexible integration of inference algorithm based on users' preference. In: Proceedings of 2011 International Conference on Electronic & Mechanical Engineering and Information Technology, pp. 2671–2674 (2011)
42. Pazzani, M., Billsus, D.: Learning and revising user profiles: the identification of interesting web sites. Mach. Learn. **27**, 313–331 (1997)
43. Rahimi, S., Gupta, B., Adya, K.: A Novel Page Ranking Algorithm for Search Engines Using Implicit Feedback. Engineering Letters, Advance online publication, Southern Illinois University, USA (2006)
44. Balabanovic, M.: An adaptive web page recommendation service. In: First International Conference on Autonomous Agents, pp. 378–85. ACM (2000)
45. Nath, R., Bhushan, R.: A study of web personalization system: supporting user profiling in web recommendation system. Amity Global Bus. Rev. **7** (2012)

46. Wen, H.: Development of personalized online systems for web search, recommendations, and e-commerce. Development **1**, 1–2011 (2011)
47. Hong, J.: A simple discussion about user-model technique in individuation service of digital library. J. Libr. Sci. Jiangxi **4** (2006)
48. Zheng, J., Guo, S., Gao, L., Xue, D., Zhao, N., Ma, H.: Inferring gender of micro-blog users based on multi-classifiers fusion. International Journal of Performability Engineering **14**, 349 (2018)
49. Meenakshi, S., Agarwal, G., Bakshi, S., Bhatter, S., Sivakumar, P.: Cognitive agents for web based search engines: a review. In: 2017 Second International Conference on Recent Trends and Challenges in Computational Models (ICRTCCM), pp. 201–206 (2017)
50. Fun Li, K., Wang, Y., Yu, W.: Personalised search engine evaluation: methodologies and metrics. In: Web Search Engine Research, pp. 163–202. Emerald Group Publishing Limited (2012)
51. Ignatova, T., Ivichev, V.: Technologies for streaming processing of news for personalized search engine results for news content. Probl. Econ. Transit. **57**, 39–45 (2015)
52. Esbitan, S.M.Y.: A Personalized Context-Dependent Web Search Engine Using Word Net (Sama Search Engine). A Personalized Context-Dependent Web Search Engine Using Word Net (Sama Search Engine) (2012)
53. Choochaiwattana, W.: An architecture of an academic search engine with personalized search result ranking mechanism. In: Proceedings of the Fifth International Conference on Network, Communication and Computing, pp. 161–165 (2016)
54. Ai, Q., Zhang, Y., Bi, K., Chen, X., Croft, W.B.: Learning a hierarchical embedding model for personalized product search. In: Proceedings of the 40th International ACM SIGIR Conference on Research and Development in Information Retrieval, pp. 645–654 (2017)
55. Babekr, S., Fouad, K.M., Arshad, N.: Personalized semantic retrieval and summarization of web based documents. Int. J. Adv. Comput. Sci. Appl. (IJACSA) **4** (2013)
56. Mahdi, M.N., Ahmad, A.R., Ismail, R.: Intelligent metadata web search engines: a brief review of literature on intelligent metadata based search engines. In: 2014 International Conference on Information Technology and Multimedia (ICIMU), pp. 255–258 (2014)

System Integration for Large-Scale Software Projects: Models, Approaches, and Challenges

Mohamed A. Shibl(✉), Iman M. A. Helal, and Sherif A. Mazen

Faculty of Computers and Artificial Intelligence, Cairo University, Giza, Egypt
msheb1@pg.cu.edu.eg, {i.helal,s.mazen}@fci-cu.edu.eg

Abstract. Moving forward in digital transformation programs in many governments around the world created the need for integrating large organizations. Large software projects have become increasingly complex. They are often formed by integrating independent software systems. As a result, system integration has gained the attention of the information and communications technology (ICT) community in complex organizations. The purpose of this research is to analyze the challenges facing the existing system integration research, especially in Large-scale Software Projects, during the last ten years. We begin by discussing large-scale software projects challenges continued by a review on system integration issues and challenges facing existing approaches. Systems interoperability and the complexity of system architecture are on top of the integration challenges. We found that most studies describe individuals and teams who have worked in isolation to develop solutions to certain problems without widespread adoption of a form of integration. Thus, more integration issues were created. Lastly, we mentioned from the currently available research, the success and failure factors influencing system integration.

Keywords: Large-scale software projects · Systems integration · Interoperability

1 Introduction

The organizations are working on several software projects. A large-scale project is defined as a project with a planned budget of over $100M. It has an impact on either the size of the application domain (e.g., development scope includes several applications of the enterprise application domain), the size of the project organization (e.g., large development organization for product line), or the time scale of the project (projects span over years). A taxonomy was proposed with three levels: small-scale (i.e. single team), medium (2–9 teams), and large-scale (10+ teams) development projects. They are defined based on the number of teams and their impact on the coordination approaches required. The main characteristic of large-scale software projects is that the included systems arise from the interoperability between their components. From a technical perspective, large-scale projects are characterized by a large number of integrated systems [1]. From an organizational perspective, it involves the number of project members and stakeholders often globally distributed processes and organizational structures.

M. Al-Emran et al. (Eds.): ICETIS 2021, LNNS 299, pp. 99–113, 2022.
https://doi.org/10.1007/978-3-030-82616-1_10

System integration is a critical process, which can reduce complexity, risks, and costs of product management, as well as improve the system quality and user satisfaction. However, it is still an open research area for software system integration for large-scale projects. It has many gaps for interdisciplinary research in various application areas, such as software engineering, enterprise architecture, management information systems, and project management. Systems Integration can occur inside a single organization or it can cross the organization boundaries. The latter is considered external system integration [2]. We aim to investigate the barriers preventing a successful system integration and which factors affect the integration process [4].

The rest of this paper is structured as follows: In Sect. 2, we introduce (from literature) large-scale software projects challenges. Then, we present in Sect. 3 various aspects of system integration, challenges, and factors of influence. Finally, in Sect. 4, we conclude the paper and highlight possible future work.

2 Large-Scale Software Projects Challenges

Large-scale software system projects, also known as Ultra Large-Scale Systems, are a novel generation of the worldwide distribution software system and they are made up of various changing, lacking consistency, or even contradictory constituents that are distributed in a wide-ranging domain. Large-scale projects face many challenges due to the nature of these integrated projects and the interoperability between their components. These challenges can be categorized based on:

1. Project/organization size
   - The severity of the identified barriers in the integration process increase in large-size projects [3].
   - The complexity of IS architecture and software as a product and unpredictability of the development process [4, 5].
   - Interoperability of large-scale systems component systems [2, 6].
   - Sharing the same understanding in the same large-scale project across large organizations [2].
2. Multiple/Distributed teams [7]
   - Appointing responsibilities and roles through multiple teams.
   - Decision-making (technical/business) through multiple teams.
   - Communication through distributed/multiple teams.
   - Coordination and dependence management throughout multiple/distributed teams.
   - Dealing with a large number of customers/stakeholders.
3. Software Engineering and Management process [2]
   - Dealing with changes (in requirements, business goals, and priorities) and unpredictability of software engineering problems [8].

- Dealing with the maintenance of parallel software (or other organizational processes).
- Human resources, personnel changes, and sourcing decisions e.g. outsourcing/offshoring.

3 System Integration

It is important to know that system integration is one of the main activities of the project/product creation process. The integration of software components is the most challenging. It is a vital phase in all types of software projects, whether small or large, and whether the component is developed in-house or outsourced/purchased from the market as an off-the-shelf component [3]. Information exchange and integration are important within and across various areas as well as domains including industrial automation [9]. This research gives an overview of typical challenges of system integration in complex technological projects in many areas in enterprise architecture, system analysis, and design, development, testing, and deployment.

3.1 Existing Approaches and Models

The whole integration issue cannot be considered only from a technical viewpoint but it has to take into consideration other aspects such as service organization, deployment scenario, educational and business context, and resource sharing with other services [2]. The perspectives of practitioners had been elicited and analyzed from three organizational groups. They are project management, business analysis/development, and testing/deployment. Perspectives may differ between groups, but on some issues, there is an agreement across all groups. The elicited information may be employed to plan process development for integration projects in the organization, and in building a general ontology for integration challenges and their solutions [10]. Many approaches are oriented towards architecting new IT systems in Fig. 1.

Level of Abstraction Perspective: It has several approaches:

File Transfer [11]*:* It is applicable when the system requires:

- No real-time synchronization needs.
- Batch jobs for data transfer.

Shared Data Repository [11]*:* It is applicable when the system requires:

- Common data format.
- Real-time data synchronization.

Remote Procedure Call [11]*:* It is applicable when the system requires:

- Real-time data synchronization.
- Integration engineers have gained access to functional layer and source code.

Messaging [12]*:* It is applicable when the system requires:

- Loose coupling.
- Online transfer and fast message passing.

Software services [13]*:* It is applicable when the system requires:

- Real-time data synchronization.
- Loose coupling.
- Real-time data synchronization.
- Wrapping and legacy systems.

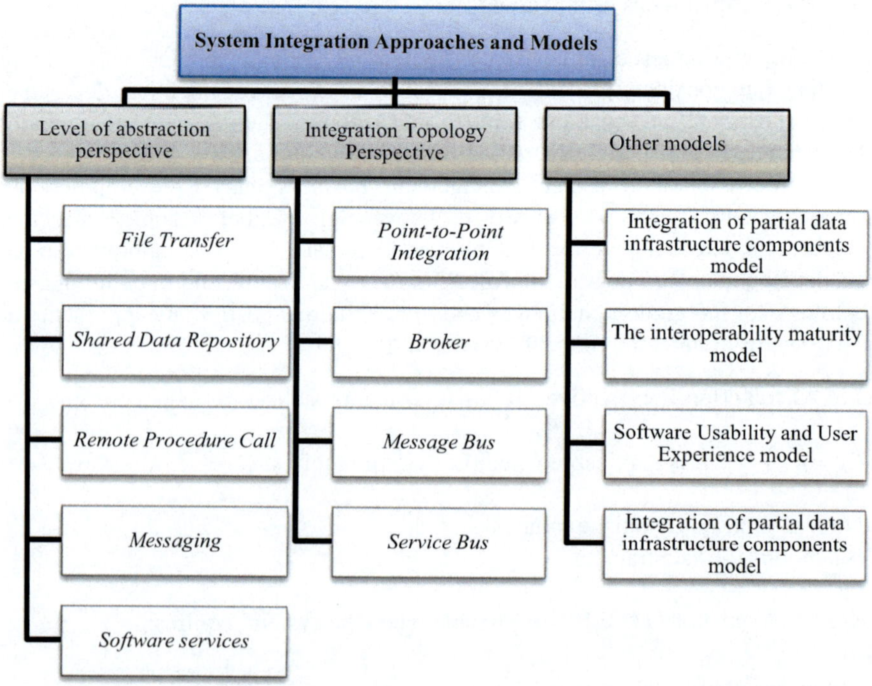

Fig. 1. System integration approaches/models

Integration Topology Perspective: It has several approaches:

- *Point-to-Point Integration* [11]*:* It is applicable for environments with a minor number of systems with no explicit demand for low cohesion.
- *Broker (Huband-Spokoe)* [11]*:* It is applicable for environments, that need a lower level of cohesion and in an environment where a clustered or redundant broker can be utilized.
- *Message Bus* [11]*:* It is applicable for environments that prerequisite a lower level of cohesion and it is useful for environments that require real-time data transfer.
- *Service Bus* [14]*:* It is applicable for heterogeneous environments, that utilize a broad number of technologies and protocols. Besides, it is used for environments, that contain legacy systems, to be wrapped as services.

Other Models

- *Integration by application interoperability model* [15]: It is used for the development and alteration of applications for gathering and the retrieval of data to provide the collection of data from various databases. The procedure of gathering and retrieval of data can be conducted via web services and similar applications, and services based on CSV or XML protocols for the transformation of numerous record types from distributed databases into a content string or some other data structure needed by the primarily user applications.
- *The interoperability maturity model* [16]*:* It covers all types of interoperability. The required interoperability types for systems are technical, syntactic, semantic, and organizational interoperability. The proposed model for ultra largescale systems should support and consider the interoperability domains [6].
- *Software Usability and User Experience model* [17]*:* A user-focused approach to software development has some conceptual overlap with agile software development practices, such as iterative and quick (user) testing.
- *Integration of partial data infrastructure components model* [15]: It redefines the belief of mutually unrelated and independent databases into a logically integrated set of components. This proposes that the current databases included in existing legacy systems require to be reconstructed in such a way that some database components must be shared as public registries which are to be treated as reference data for precise organizational or business activities.

3.2 System Integration Challenges Facing Existing Approaches

Several challenges face system integration. These challenges are presented in Fig. 2. These challenges are categorized into 5 main perspectives: (1) system design, (2) Environment culture and restrictions, (3) software development life cycle (SDLC), (4) global software development projects, and (5) continuous integration challenges.

1. **System design**

1.1 *Interoperability among systems:* Data interoperability and portability are critical aspects in system integration [18]. For example, interoperability in healthcare can be the capacity of different software applications and information technology systems to communicate, to utilize the information that has been exchanged, and to exchange data accurately [19].

1.2 *System Architecture complexity:* The complexity of the IT architecture and data transfer between several different systems like in ERP systems raises several challenges [11]. It is one of the biggest challenges in achieving interoperation and cooperation among systems via some sort of system integration [5, 20].

1.3 *Time to select/use Integration Approaches:* There are trials to integrate subsystems and functions as early as possible to find unforeseen problems early [21]. This implies that integration takes place, whereas most of the new components, functions, and subsystems are being developed. Modified existing systems or normal partial systems are used within the early phases of integration as a substitute for the not yet available parts. This causes unforeseen problems [19].

1.4 *Incompatible data models:* schema mapping problem characterized by differences in structure and semantics. It is difficult is to define correspondences between schema elements (e.g., attributes) and determine how they relate to each other. Most business data is currently stored in relational databases [2]. However, many formats for information exchange deal with the structure of given messages, not with their meaning. This needs further investigation to get better results [9].

2. **Environment culture/restrictions**

2.1 *Company and Agile culture challenges:* Agile principles such as quick and iterative work are well suited to be linked with user integration. In a company, multiple people with multiple perspectives need to have voices in the process and should be able to challenge decision-making processes without retribution. Space and opportunity for informal exchanges and the grapevine are vital. In addition to challenges related to differentiating roles, channels, and tools [10].

2.2 *Industry 4.0 challenges:* Four critical domains of synergy challenge were found, (1) Man-to-Man interaction, (2) Man-to-Machine interaction, (3) Machine-to-Man interaction, and finally (4) Machine-to-Machine interaction [15, 22] Recently, IoT technology faces several challenges such as security [11, 23, 24].

2.3 *Engineer-to-order business:* In several case studies, system integration is often mistaken as a simple plug-and-play principle. However, technology and engineering projects are becoming increasingly complex. This is due to the increasing size of technical systems, globally distributed organizations, stricter environmental legislation, and increasing technological standards. Challenges were found at the time of integration, and object integration was grouped in system, organization, or process elements [4].

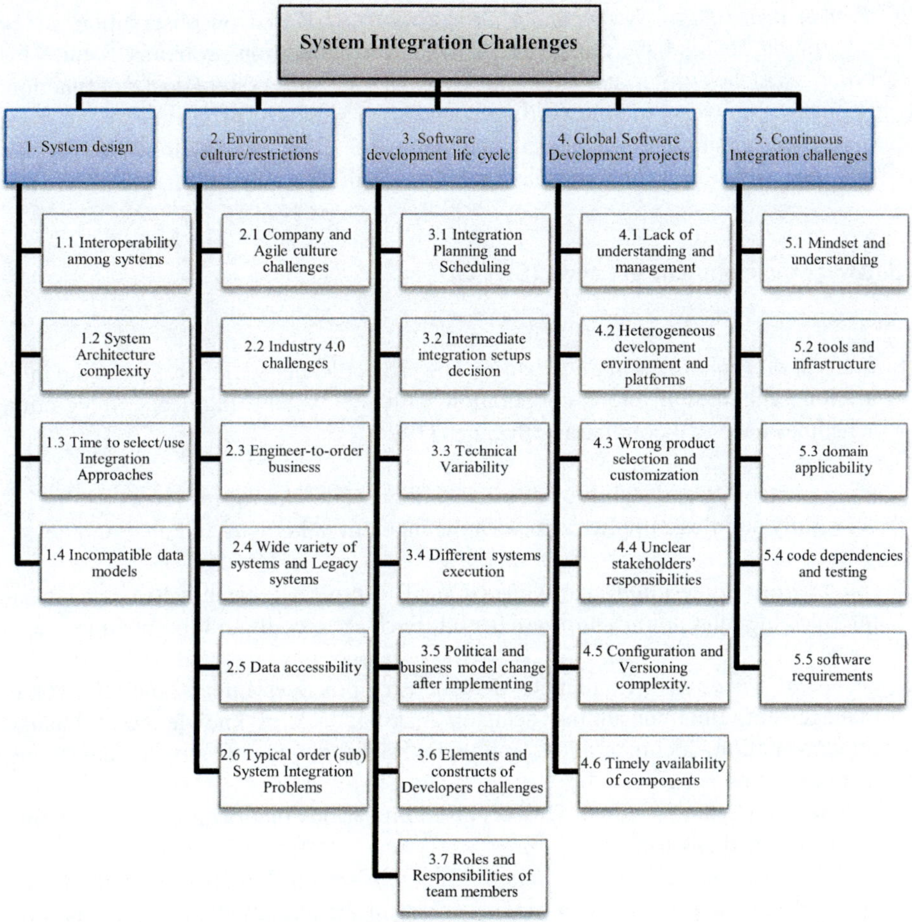

Fig. 2. System integration challenges

2.4 *Wide variety of systems and Legacy systems:* This is due to lack of common data format, integration and autonomy, global scale integration and transparency, partner systems, system coupling level, methodological and management issues. System of System (SoS) integration is part of the whole system improvement life cycle. There is increased complexity in the case of integrating legacy systems. Existing methods, processes, and tools are inadequate for integrating legacy systems [21].

2.5 *Data accessibility:* has challenges related to confidentiality, integrity, triple-A (Authentication, Authorization, and Accounting), availability of data anywhere at any time, data fidelity, and privacy [18, 20].

2.6 *Typical order (sub) System Integration Problems:* Based on observation, many integration phases have typical order whenever integration problems occur. This order is: (a) the (sub) system does not build, (b) the (sub) system does not function, (c) interface errors, (d) the (sub) system is too slow, (e) problems by the main performance parameter, such as image quality, (f) the (sub) system is not reliable, and finally (g) the system does not operate well in its context [19, 21].

3. Software development life cycle (SDLC)

3.1 *Integration Planning and Scheduling:* The project leader will therefore make integration schedules in close cooperation with the system engineers. Integration schedules have two conflicting attributes [19]:

- Predictability and stability to ensure timely availability of resources.
- Agility and flexibility to cope with the inherent unknowns and uncertainties.

3.2 *Intermediate integration setups decision:* The project team has to decide what intermediate integration setups are useful. Each extra configuration includes more costs[19].

3.3 *Technical Variability:* Challenges and industry practices impact analysis, status management, time-consuming decision making, lack of knowledge to change implementation, documentation, testing, tool support, and defect management and documentation activities [25]

3.4 *Different systems execution:* This occurred on various platforms, communicating in a distributed environment [11].

3.5 *Political and business model change after implementation:* It is integrating the environment, detects the defect management in the already deployed integration middleware, different vendors and implementations of integration technologies, maintainability of the large-scale integration solutions, etc. [11, 18].

3.6 *Elements and constructs of developers' challenges:* Missing flow of information developers in the integration project work on their areas and do not discuss their solutions with each other. Elements such as Information spread, Documentation, Information about the system, Errors in external components, and Lack of knowledge [10].

3.7 *Rolesand Responsibilities of team members:* During integration, many project team members are involved with different roles and responsibilities: Project leader - System architect/engineer/integrator - System tester - logistics and administrative support personnel – engineers - machine owner. In practice, projects face many problems because of decomposition phases. Every time an activity is fragmented, the fragmented activities run well, nevertheless, crosscutting functionality and qualities suffer from the decomposition. Lack of attention, lack of ownership, and absence of communication across organizational boundaries are root causes for these problems [5, 19].

4. Global Software Development projects

Researchers have ranked the most of barriers/challenges as critical barriers/challenges in Global Software Development (GSD) [3].

4.1 *Lack of understanding and management:* This includes lack of communication, proper documentation, compatibility, architecture mismatch, integration planning and management, proper component interfaces, common understanding of requirements, and common development process [8].
4.2 *Heterogeneous development environment and platforms* [26].
4.3 *Wrong product selection and customization.*
4.4 *Unclear stakeholders' responsibilities.*
4.5 *Configuration and Versioning complexity.*
4.6 *Timely availability of components.*

5. Continuous Integration challenges

While developing fully functional applications, web services, and online applications is taken for granted, there are many other reasons for poor indicators for availability and usage of services. Some of these reasons are cultural, technical, legal, and organizational issues [15, 27].

5.1 *Mind-set and understanding:* This addresses the change of old habits, exposes work intention, understands unclear goals, and increased pressure on different interpretations. It uses a button-up approach.
5.2 *Tools, infrastructure: Code review, maturity, regression feedback time, integration queue, test automation,*
5.3 *Domain application:* It addresses the process suitability and product complexity.
5.4 *Code dependencies and testing:* This is related to integration coordination, dead code, unstable test cases, too many manual testing, implementation, and test dependencies preserving quality.
5.5 *Software requirements:* It is related to the consequences of requirements breakdown and the integration of big-impact changes [8].

All these challenges are mapped to existing approaches in Table 1. This table illustrates each approach's advantages and disadvantages.

Table 1. System integration approaches and challenges

Approach	Challenges	Advantages	Disadvantages
File Transfer [11]	1.1, 1.2, 1.4, 2.2, 2.3, 2.4, 2.5, 3.3, 3.4	- Loose coupling - Simplicity: No additional tools are required	- Transfer of large data Volumes - No common data format - Concerns about time slot of file export and transfer

(*continued*)

Table 1. (*continued*)

Approach	Challenges	Advantages	Disadvantages
Shared Data Repository [11]	1.1, 1.2, 1.4, 2.4, 3.5, 4.5,	- Easy to implement, straightforward - All systems have a common data model, no risk of semantic dissonance	- The lower performance of business-critical applications - Higher complexity in data access concurrency management
Remote Procedure Call [11]	1.1, 1.2 2.2, 2.4 3.4, 3.5 4	- Easy implementation, portability, and good code readability - Data type checking and control - Escaping semantic dissonance with different data views with getter methods	- Harder support of doubled data methods - Low performance of remote calls - A lot of knowledge about the business layers of the system, which leads to tight coupling
Messaging [12]	1.1, 1.2 2.2, 2.4 3.3, 3.4, 5	- Loose coupling due to a central mediator element - Asynchronous calls are not blocking, so the calling application can continue work - Fast transfers and real-time actual data	- Harder and complex - Management of central element activities like addressing, routing, etc.
Software Service [13]	1.2, 1.4, 2.2, 2.3, 2.4, 3.1, 3.3, 3.4, 4, 5	- Control over granularity, business process optimization - Loose coupling between systems - Usage of open standards & wide adopted protocols - Service registry and discovery mechanisms - Interface self-definition - Wrapping legacy systems	- Risk of network or communication issues - Security risks, the higher need for security control

(*continued*)

Table 1. *(continued)*

Approach	Challenges	Advantages	Disadvantages
Point-to-Point Integration [11]	1.1, 1.2, 1.4, 2.4, 2.5, 3.1, 4	- A straightforward approach, easy to implement for a small number of systems	- Tight coupling - The dependency of the adapters on every system - Low scalability - Hard to manage and. maintain for a larger number of systems
Broker (Huband-Spokoe) [11]	1.1, 1.2, 1.4, 2.2, 2.4, 3.4, 4	- A lower level of cohesion - A lower number of communication channels and adapters - Easier to maintain - Better scalability	- The central broker is a single point of failure
Message Bus [11]	1.1, 1.2, 2.2, 2.4, 3.3, 3.5, 4, 5	- A lower level of cohesion - A lower number of communication channels and adapters - Easier addition of new systems - High performance and reliability	- The bus is a single point of failure - Harder to install and configure due to the wide range of out-of-the-box features
Service Bus [14]	1.1, 1.2, 2.2, 2.3, 2.4, 3.3, 3.5, 4, 5	- A lower level of cohesion - A lower number of communication channels - Better scalability and extensibility, easy addition of new systems - Adoption of open, widely used standards - Better applicability for heterogeneous environments	- The bus is a single point of failure - Harder to install and configure due to the wide range of out-of-the-box features

(continued)

Table 1. (*continued*)

Approach	Challenges	Advantages	Disadvantages
Integration by application interoperability [15]	1.1, 1.3, 1.4, 2.1, 3.1, 3.2, 3.7, 5	- Better for gathering and data retrieval from various databases - Use protocols for the transformation of various record types from distributed databases	- Certain risks in mapping data
The interoperability maturity model [16]	1.3, 2.1, 2.6, 3.1, 3.6, 3.7, 5	- Covers all types of interoperability (technical, syntactic, semantic, and organizational)	- It should support interoperability domains - Factors comprise four attributes: procedures, applications, infrastructure, and data
Software usability & User experience model [17]	1.3, 2.1, 2.6, 3.1, 3.2, 3.6, 3.7, 5	- Focused on software development - Lead to Usability and User Experience	- Overlap with agile software development practices, as iterative and quick (user) testing
Integration of partial data infrastructure components model [15]	1.3, 2.1, 2.5, 3.1, 3.2, 3.6, 3.7, 5	- Integration of partial data infrastructure components into one public data infrastructure - Redefines the concept of unrelated databases into a logically integrated set of contents	- Databases included in present legacy systems need to be reconstructed in a way that some database contents must be published as public registries

3.3 Factors Influencing System Integration

It is important to explore which barriers stand in the way of successful integration and what factors ease and speed the integration process [3]. We conclude and list success and failure factors that affect system integration including experts' participants.

Success Factors:

- Enterprise architecture and complexity of the IT architecture [2, 5].
- Standardized organizational process framework e.g. CMMI [6].
- Cross-functional teams for a successful collaboration of distributed teams [7].
- Frequent communication between developers at remote sites.
- Specified and detailed requirements and design documentation [8].

- Stakeholders' involvement in requirements analysis.
- Choice of a suitable process model for development.
- Early user participation and regular feedback.
- Executing and planning integration activities at development early stages.
- Rich communication during requirements clarification (i.e., video-conference) [8].
- Integrating (end) users in software development processes such as in agile software development processes [17].
- Software usability and user experience [17].
- Established communication plan [28].
- Quality assurance procedures and plans should be discussed early in projects.
- Frequent reviews are significant for receiving a timely response from the stakeholders.

Failure Factors:

- The communication gap among remote locations because of inadequate language skills arises despite using English as a shared language [7].
- Long-term awareness lack about activities at remote sites.
- Integration strategy/plan issues related to responsibilities assignment.
- Lack of detailed requirements specification documentation for global teams [7].
- Small-time zone overlaps among the locations involved in software development.
- Poorly communicated module objectives or requirements [8].
- Coordination issues among remote locations lead to many technical issues including ill-tested constituents and redundant code.
- Deficient understanding of interface issues and requirements.
- Lack of informal external communication.
- Diversity in processes, techniques, and standards at remote locations.
- Fast-changing unforeseen technical interdependencies and requirements [8].

4 Conclusion and Future Work

Digital transformation programs include many Large-Scale Software Projects. System integration is a critical activity and requires to be appropriately planned in Large-scale Software Projects. It is a special type of software evolution that has become more and more important in recent years, which brings new challenges and complexities. In this paper, we report the results from existed researches in Large-scale Software Projects and System Integration. We provide an overview of large-scale challenges and system integration issues. Moreover, we discovered the factors influencing system integration. This field is maturing at a slower pace than other software engineering fields. Therefore, there is a need for additional research to understand and raise its maturity.

In most research, individuals and teams are working in isolation while developing techniques that solve system integration problems. There is a need to develop generic procedures, techniques, and tools to design, analyze, evaluate, and evolve system integration. In this paper, we aim for providing researchers and professionals with a piece of comprehensive knowledge on the existed system integration approaches and the challenges facing them while using a single framework. Furthermore, this study can help technical project managers and system architects in selecting the right approach that fits their needs. Based on this research, we found that there is no comprehensive criteria, approach, or framework that help professionals in integrating a Large-scale project efficiently and aid in decision making.

As future work, we target to implement system integration that mitigates interoperability and architecture challenges in Large-scale Software Projects efficiently with overcoming its challenges. Also, we aim for utilizing system integration models for proposing an architecture framework for integrating e-government services to reduce the gap between e-government infrastructure to the public sector.

References

1. Aintila, K.: Agile Methodologies in Large Scale Information Systems Project Context – A Literature Review and Reflections. University of Helsinki (2016)
2. Kähkönen, T., Maglyas, A., Smolander, K.: ERP system integration: an inter-organizational challenge in the dynamic business environment. In: Cordeiro, J., Hammoudi, S., Maciaszek, L., Camp, O., Filipe, J. (eds.) Enterprise Information Systems, pp. 39–56. Springer, Cham (2015). https://doi.org/10.1007/978-3-319-22348-3_3
3. Ilyas, M., Khan, S.U.: Software integration in global software development: challenges for GSD vendors. J. Softw. Evol. Process. **29**, 1–17 (2017). https://doi.org/10.1002/smr.1875
4. Foehr, M., Gepp, M., Vollmar, J.: Challenges of system integration in the engineer-to-order business. In: IECON 2015 - 41st Annual Conference of the IEEE Industrial Electronics Society, pp. 73–79 (2015). https://doi.org/10.1109/IECON.2015.7392078
5. Sunyaev, A.: Design of good information systems architectures. In: Sunyaev, A. (ed.) Internet Computing, pp. 51–81. Springer, Cham (2020). https://doi.org/10.1007/978-3-030-34957-8_3
6. Patnaik, P., Pattnaik, S., Singh, P.: A simulation model for quality assurance of e-governance projects. In: Rautaray, S.S., Eichler, G., Erfurth, C., Fahrnberger, G. (eds.) I4CS 2020. CCIS, vol. 1139, pp. 309–325. Springer, Cham (2020). https://doi.org/10.1007/978-3-030-37484-6_18
7. Martini, A., Stray, V., Moe, N.B.: Technical-, social- and process debt in large-scale agile: an exploratory case-study. In: Hoda, R. (ed.) XP 2019. LNBIP, vol. 364, pp. 112–119. Springer, Cham (2019). https://doi.org/10.1007/978-3-030-30126-2_14
8. Koski, A., Mikkonen, T.: What we say we want and what we really need: experiences on the barriers to communicate information system needs. In: Ramachandran, M., Mahmood, Z. (eds.) Requirements Engineering for Service and Cloud Computing, pp. 3–21. Springer, Cham (2017). https://doi.org/10.1007/978-3-319-51310-2_1
9. Jirkovský, V., Obitko, M., Kadera, P.: Information exchange and integration within industrial automation domain. In: Mařík, V., et al. (eds.) Industrial Applications of Holonic and Multi-Agent Systems: 9th International Conference, HoloMAS 2019, Linz, Austria, August 26–29, 2019, Proceedings, pp. 159–170. Springer International Publishing, Cham (2019). https://doi.org/10.1007/978-3-030-27878-6_13
10. Rognerud, H.J., Hannay, J.E.: Challenges in enterprise software integration: an industrial study using repertory grids. In: 2009 3rd International Symposium on Empirical Software Engineering and Measurement, ESEM 2009, pp. 11–22 (2009). https://doi.org/10.1109/ESEM.2009.5316045

11. Kouzmanov, S.: Analysis and classification of business software systems integration approaches. Econ. Altern. **2**, 77–94 (2015)
12. Trowbridge, D., Roxburgh, U., Hohpe, G., Manolescu, D., Nadhan, E.G.: microsoft Libr (2014)
13. Menascé, D.A., Casalicchio, E., Dubey, V.: On optimal service selection in service oriented architectures. Perform. Eval. **67**(8), 659–675 (2010)
14. Rademakers, T., Dirksen, J.: Open Source ESBs in Action. Manning Publications Co. (2009)
15. Tomicic-Pupek, K., Dobrovic, Z., Furjan, M.T.: Strategies for information systems integration. In: Proceedings of International Conference on Information Technology Interfaces, ITI, pp. 311–316 (2012). https://doi.org/10.2498/iti.2012.0459
16. Rezaei, R., Chiew, T.K., Lee, S.P.: An interoperability model for ultra large scale systems. Adv. Eng. Softw. **67**, 22–46 (2014). https://doi.org/10.1016/j.advengsoft.2013.07.003
17. Stickel, O., Ogonowski, C., Jakobi, T., Stevens, G., Pipek, V., Wulf, V.: User integration in agile software development processes: practices and challenges in small and medium sized enterprises. In: Cockton, G., Lárusdóttir, M., Gregory, P., Cajander, Å. (eds.) Integrating User-Centred Design in Agile Development. HIS, pp. 49–76. Springer, Cham (2016). https://doi.org/10.1007/978-3-319-32165-3_2
18. Maharatna, K., Bonfiglio, S. (eds.): Systems Design for Remote Healthcare. Springer, New York (2014). https://doi.org/10.1007/978-1-4614-8842-2
19. Muller, G.: Coping with system integration challenges in large complex environments. In: 17th Annual International Symposium of the International Council on Systems Engineering, INCOSE 2007 - Systems Engineering: Key to Intelligent Enterprises, vol. 2, pp. 1373–1387 (2007). https://doi.org/10.1002/j.2334-5837.2007.tb02930.x
20. Kazman, R., Nielsen, C., Nielsen, C.: Understanding patterns for system-of-systems integration. Res Showcase@CMU (2013)
21. Madni, A.M., Sievers, M.: System of systems integration: key considerations and challenges. Syst Eng **17**, 330–347 (2014). https://doi.org/10.1002/sys.21272
22. Kaivo-oja, J., Kuusi, O., Knudsen, M.S., Lauraeus, T.: Digital twins approach and future knowledge management challenges: where we shall need system integration, synergy analyses and synergy measurements? In: Uden, L., I-Hsien Ting, Juan Manuel Corchado, (eds.) KMO 2019. CCIS, vol. 1027, pp. 271–281. Springer, Cham (2019). https://doi.org/10.1007/978-3-030-21451-7_23
23. Abbou, A.N., Baddi, Y., Hasbi, A.: Software defined networks in internet of things integration security: challenges and solutions. In: Proceedings - 2018 International Conference on Wireless Networks and Mobile Communications, WINCOM 2018, pp. 1–6 (2019). https://doi.org/10.1109/WINCOM.2018.8629722
24. Bures, M., et al.: Interoperability and integration testing methods for IoT systems: a systematic mapping study. In: de Boer, F., Cerone, A. (eds.) SEFM 2020. LNCS, vol. 12310, pp. 93–112. Springer, Cham (2020). https://doi.org/10.1007/978-3-030-58768-0_6
25. Ihme, T., Pikkarainen, M., Teppola, S., Kääriäinen, J., Biot, O.: Challenges and industry practices for managing software variability in small and medium sized enterprises. Empir. Softw. Eng. **19**(4), 1144–1168 (2013). https://doi.org/10.1007/s10664-013-9253-0
26. Baresi, L., Garriga, M.: Microservices: the evolution and extinction of web services? In: Bucchiarone, A., et al. (eds.) Microservices, pp. 3–28. Springer, Cham (2020). https://doi.org/10.1007/978-3-030-31646-4_1
27. Couto, L.D., Tran-Jørgensen, P.W.V., Nilsson, R.S., Larsen, P.G.: Enabling continuous integration in a formal methods setting. Int. J. Softw. Tools Technol. Transf. **22**(6), 667–683 (2019)
28. Zafar, A., Ali, S., Shahzad, R.K.: Investigating integration challenges and solutions in global software development. In: Proceedings - 2011 9th International Conference on Frontiers of Information Technology, FIT 2011, pp. 291–297 (2011). https://doi.org/10.1109/FIT.2011.61

Comparison of Performances of Jaya Algorithm and Cuckoo Search Algorithm Using Benchmark Functions

Mashuk Ahmed[1], Abdullah B. Nasser[1](✉), Kamal Z. Zamli[1], and Sulistyo Heripracoyo[2]

[1] Universiti Malaysia Pahang, 26300 Kuantan, Pahang, Malaysia
MCS20004@student.ump.edu.my, {abdullahnasser,kamalz}@ump.edu.my
[2] Bina Nusantara University, Jakarta, Indonesia
hpracoyo@binas.edu

Abstract. Nowadays, selecting the best possible solution among several solutions becomes an important skill for engineering and research. Therefore, engineers are turning to optimization methods as a complementary alternative strategy of exhaustive searching. Metaheuristic algorithms have been used successfully for solving different optimization problems. To help engineers select the best metaheuristic algorithms for their problems, there is a need to evaluate the performance of different metaheuristic algorithms against each other using common case studies. This paper aims to compare the performance of two metaheuristic algorithms which are Jaya Algorithm (JA) and Cuckoo Search (CS) using some common benchmark functions. CS and JA have implemented in the same platform (Intellij IDEA Community Edition 2020.2.3) using the same language (Java). The experimental results show that JA has better and consistent performance as compared to CS in most cases in terms of execution time and test suite size; however, the performance of JA is still within acceptable ranges.

Keywords: Jaya Algorithm · Cuckoo Search Algorithm · Metaheuristic algorithm · Optimization · Execution time

1 Introduction

Besides the analysis, optimization has been indicated as one of the vital stages of engineering design [1]. Optimization is a vital part of software engineering which allows the engineer to find the optimum solution in the presence of design constraints and criteria. Whereas the analysis stage of the engineering design is the use of the mathematic model to predict the design results optimization method such as metaheuristic algorithms have been used successfully for solving different optimization problems such as robotic path planning, optimization of neural networks, minimizing weights of Truss Structures, task scheduling over cloudlets, optimization of integrated process planning, dynamic vehicle routing, susceptibility mapping of forest fire, etc. [2] Engineers and researchers have been working with the implementation of metaheuristic algorithms for a long time. As

M. Al-Emran et al. (Eds.): ICETIS 2021, LNNS 299, pp. 114–125, 2022.
https://doi.org/10.1007/978-3-030-82616-1_11

a result, metaheuristic algorithms such as Genetic Algorithm (GA), Flower Pollination Algorithm (FPA), Particle Swarm Optimization (PSO), Jaya Algorithm (JA), Hill Climbing (HC), Cuckoo Search (CS), have become very popular in the field of optimization. Metaheuristic algorithm based optimizations are also being mentioned as metaheuristic optimization now [3]. Metaheuristic based optimizations possess stochastic characteristics [4]. These algorithms have a wide range of applications. Every metaheuristic algorithm has unique characteristics. While performing an optimization, there may be a need for an algorithm with particular characteristics to solve the optimization problem. So, a specific algorithm with some characteristics may not be enough to solve a specific optimization problem.

Selecting the most suitable metaheuristic algorithm is vital in optimization. Also, it's quite a challenging task, especially when the optimization problem is complex [5]. So, it's necessary to have ideas about the comparison of different metaheuristic algorithms. The comparisons of the algorithms help to find out their unknown characteristics. For this reason, a comparison between two of the metaheuristic algorithms have been done, which are Jaya Algorithm (JA) and Cuckoo Search (CS) Algorithm. There has been a minimization process of some functions. The coding of the algorithms has been done in Java using Intellij IDEA Community Edition 2020.2.3. The result of the comparison will be helpful to determine which one is more suitable for a specific optimization problem.

The next sections will have a detailed discussion of the algorithms and experiments. In Sect. 2, there will be an overview of JA and CS. Section 3 will discuss the related works and applications of JA and CS. Section 4 will discuss the methodologies. Section 5 will describe the tests and results, and related data. Section 6 will analyze the 2 algorithms based on the test results. Finally, Sect. 7 will conclude the overall discussions with future works.

2 Background

Jaya Algorithm (JA) and Cuckoo Search (CS) Algorithm, both are efficient in generating test suite, trying to reach the best solutions and avoid worse solutions.

2.1 Jaya Algorithm

JA is a global algorithm in the field of optimization. It's one of the recent metaheuristic algorithms. The algorithm is powerful and simple to implement. JA has some advantages. The potential solution for JA is based on finding the most optimum solution and avoiding the worst solution. JA doesn't require tuning [6]. It's because, mainly the common controlling parameters are required for JA. There's no need for control parameters that are specific to the algorithm.

2.2 Cuckoo Search Algorithm

CS is an evolutionary optimization algorithm [7]; it's based on swarm intelligence. Yang and Deb, being inspired by cuckoos' natural behaviors, developed this algorithm. The algorithm mimics the behavior of some cuckoo species' obligate brood parasitism where they lay eggs in other host birds' nests [8]. Obligate brood parasitism is a special characteristics of cuckoos. As a bird, cuckoo's behavior and activity are fascinating. It's not only because of their beautiful sounds, but also due to their aggression in reproduction strategy. Mature cuckoos lay their eggs in the nests of other host birds. Basically, this algorithm is based on how cuckoos lay eggs and breed. If the host birds find out that the eggs aren't theirs, they will either throw away these unknown eggs or leave their nests and build up new nests somewhere else.

The CS follows 3 important rules:

1. Each of the cuckoos lays an egg at a time and drops the egg in a nest chosen randomly.
2. The nest having high-quality eggs is the best one. This nest carries over to the next stage or generation.
3. There's a fixed number of available nests of host birds. The probability Pa ε [0, 1] determines whether the host bird will discover the cuckoo's laid egg or not [9].

3 Related Works and Applications

This section discusses the existing works of JA and CS. JA and CS have been implemented in various fields and applications.

3.1 Related Works and Applications of Jaya Algorithm

Jaya algorithm (JA) has a wide range of applications. Researchers have implemented the algorithm for adaptive control for the problem of surge tank indirectly. They have scrutinized the adaptive controls based on JA for the surge tanks' nonlinear models and components having nonlinearities. Usually, the formulations of the controllers are single objective optimizing problems with different controlling variables to find out the optimal solutions for satisfying JA's different constraints. This approach ensures improvement in randomness and performance of systems [10].

JA is an effective option to extract different PV model parameters [11]. This process controls and tracks maximum power point on photovoltaic systems, and also identifies reliable and accurate model parameters of PV modules and cells. This process uses an improved JA that quantifies the individual functioning in the population. Each individual selects various evolution strategies depending on probability. The strategies are designed for exploitation abilities and balancing exploration for the search process. The quantified performance is for constructing the searching direction by selecting the exemplar. This process ensures an improved population for exploring better solutions; for this, around the present best solution, it introduces a perturbation mechanism, which is self-adaptive and chaotic.

JA can be implemented for load balancing in the cloud. Cloud computing has various challenges like automated resource provisioning, server consolidation, event content dissemination, security, virtual machine migration, etc. For load balancing in the cloud, the challenges are decreasing response time, decreasing service request time of data center, improving the system's overall performance, etc. JA uses less controlling parameters and ensures a very good optimized result [12]. This approach has been proved to have great efficiency while comparing with other approaches.

Researchers have used the JA to perform fuzzy analysis using approaches based on pressure on benchmark networks [13]. Using fuzzy analysis, we can understand the uncertainty in different independent parameters of the network of water distribution. The parameters include pipe roughness values, nodal demands, pipe diameters, reservoir heads, etc. Obtaining the dependent parameters' membership functions are based on considering uncertain independent parameters' membership functions. According to the method of Impact Table, there's supposed to be a repetitive analysis due to the monotonous relationship between independent and dependent parameters. Methods based on optimization are more useful for fuzzy analysis when there's a non-monotonous relationship between independent and dependent parameters. JA has been found to be an efficient option for optimization. The analysis can be done by using a hydraulic model in EPANET, linked up with MATLAB for optimization.

JA can be implemented for tuning PID controllers for DC servo motor's position control [14]. Here, the unit step input's integral of squared error or ISE is the performance index. Using JA, ISE is minimized for obtaining the controller settings. Tuning based on JA ensures satisfactory response.

3.2 Related Works and Applications of Cuckoo Search Algorithm

Cuckoo Search (CS) based applications have shown very good efficiency in solving optimization problems. This algorithm provides better solutions as compared to many other algorithms.

It's possible to implement enhanced scatter search algorithms using CS [15]. An example is the problem of traveling salesman using improved and original scatter search. The improved edition of the scatter search algorithm performs better than the original one.

CS is an efficient algorithm for solving nurse schedule problems. Lim Huai Tein used this algorithm for nurse scheduling, which is very useful in healthcare institutions. Also, with the CS, we can solve problems in manufacturing optimization. CS has been effective in optimizing machine parameters in operations, and has provided better solutions as compared to other algorithms [16].

Quantum Inspired CS is an improved CS developed by A. Layeb [17]. This method is based on CS and Principles of Quantum Computing. The process includes defining the algorithm with a proper representation scheme, allowing the application of the algorithm on combinatorial optimizing tasks and some principles of quantum computing, such as measurement, qubit state superposition, interference, representation, etc. Representing the quantum solutions with a probability, it's possible to code the solutions. This improved algorithm ensures efficient and optimal solutions with the number of iterations and population size.

Another application of the CS is a flow of OP-AMP optimization assisted by a polynomial metamodel of 3 steps [18]. This improved algorithm provides solutions to the issues of inefficient system performance of optimized OP-AMPs. The CS provides the desired optimization results, ensuring a design flow for OP-AMP optimization. The process estimates the performance of OP-AMP; for this, it generates extremely accurate and ultra-fast polynomiameta models and facilitates quick time-domain simulating system of a metamacromodel of OP-AMP. These are integrated into a module of Verilog-AMS.

The particle approach based on CS helps to achieve energy efficiency in multimodal objective functions and wireless sensor networks [19]. This approach formulates network optimization. This process randomly deploys the nodes, and organizes those as static clusters using CS. The collection and aggregation of information are done after the selection of cluster heads. Then by the generalized algorithm of particle approach, this process forwards the information to the base station. CS helps to select cluster heads and form clusters among sensor nodes. This approach provides comparable results as compared to simulation results of LEACH protocols. Sensor network's longevity increases due to this protocol. Also, complications in chain formation reduces due to this approach.

4 Methodologies

4.1 Implementing Jaya Algorithm

While implementing the JA, the process is based on searching for the best solution and avoiding the worst solution for a certain problem. In the process, there are basic parameters, such as the size of the population, termination condition, number of design variables, etc. The termination condition is usually the maximum number of iteration. Maximizing or minimizing an objective function f(x) is the primary objective of this algorithm [20]. Suppose we have m design variables at i^{th} iteration. Also, there are n candidate solutions. The best-obtained candidate solution for f(x) from all candidate solutions is represented as f(x)best. Similarly, we represent the worst candidate solution as f(x)worst. We use $X_{j,k,i}$ to represent the jth designing variable for k^{th} candidate at iteration number i. The modification of $X_{j,k,i}$ is $X'_{j,k,i}$. We do this modification by the following equation:

$$X'_{j,k,i} = X_{j,k,i} + r_{1,j,i}(X_{j,best,i} - |X_{j,k,i}|) - r_{2,j,i}(X_{j,worst,i} - |X_{j,k,i}|) \tag{1}$$

In the equation, $X_{j,best,i}$ represents j variable's value for the best candidate, and $X_{j,worst,i}$ represents j variable's value for the worst candidate. $r_{1j,i}$ and $r_{2j,i}$ are two random numbers ranging from 0 to 1 [21]. With this equation, the search moves towards the best solution. Depending on which one is bigger between $X'_{j,k,i}$ and $X_{j,k,i}$, we update the solution. The process continues as a loop. As the iteration keeps moving forward, the solution becomes more and more optimum.

4.2 Implementing Cuckoo Search Algorithm

The CS has some stages. Usually, we begin with an objective function f(x). Then initial population is generated having n host nests xi (i = 1,2, …., n). The search continues till the maximum generation. In each generation, a cuckoo is selected randomly by lévy flight. We evaluate the fitness or quality of the cuckoo. We choose a nest randomly. Then we compare the fitnesses of cuckoo and nest and replace the one having lower fitness [22].

While generating new solution $x^{(t+1)}$ for, suppose cuckoo i, we can use the following equation,

$$x^{(t+1)} = x^{(t)} + \alpha \oplus \text{Lévy}(\beta) \quad (2)$$

We use the product $\oplus$ meaning entry-wise walk during multiplications. There's essentially a random walk due to the Lévy flights [23]. We draw random steps of Lévy flights from a distribution of Lévy for large steps.

$$\text{Lévy} \sim u = t^{-1-\beta}(0 < \beta \leq 2) \quad (3)$$

This has an infinite variance, and with it, there's an infinite mean. The consecutive steps or jumps of a cuckoo develop a random walk process following a power-law and step-length distribution having a heavy tail. We abandon the worst nest's fraction Pa to help to build up new nests at new locations by randomly walking and mixing. The eggs or solutions get mixed by the random permutation depending on the difference or similarity to the host eggs. The step size sample generation isn't trivial using the Lévy flights. A simple scheme can be shown as:

$$x_i^{(t+1)} = x_i^{(t)} + \alpha \oplus \text{Lévy}(\beta) \sim 0.01(u/|v|)\left(x_i^{(t)} - x_b^{(t)}\right) \quad (4)$$

Here, we have drawn u and v from normal distribution.

$$u = N\left(0, \sigma_u^2\right) \quad (5)$$

$$v = N\left(0, \sigma_v^2\right) \quad (6)$$

$$\sigma_u = \left\{ \frac{\Gamma(1+\beta)\sin(\pi\beta/2)}{\frac{\Gamma\frac{[1+\beta]}{2}\beta 2(\beta-1)}{2}} \right\} 1/\beta \quad (7)$$

$$\sigma_v = 1 \quad (8)$$

Here, we have used Γ as the standard Gamma function.

5 Tests and Results

All tests of the functions have been done on Intellij IDEA Community Edition 2020.2.3. The tests were performed on Intel(R) Core(TM) i5-8250U (1.60 GHz, 3.4 GHz) with 8 GB of DDR4 RAM on Windows 10 operating system. We have done the minimization of the functions and updated the minimized values and time performances in milliseconds. Each test has been done with 1000 iterations.

For JA, there was no need for any fixed parameter values. After taking the initial population, the process had moved towards the next iterations, where the population was updated using Eq. (1). For CS, there were some fixed parameter values. Probability, Pa = 0.25. σ_u = 0.6969 was used based on previous study[24]. The parameters varied in the iterations.

We have used 11 functions for the tests as shown in Table 1.

Table 1. Benchmark functions

No	Functions	Conditions
1.	$f_1(x) = x_1{}^2 - x_1x_2 + x_2{}^2 + 2x_1 + 4x_2 + 3$	$-100 \le x_1, x_2 \le 100$
2.	$f_2(x) = x_1{}^2 + 2x_2{}^2 - 0.3\cos(3\pi x_1) - 0.4\cos(4\pi x_2) + 0.7$	$-100 \le x_1, x_2 \le 100$
3.	$f_3(x) = (x_1{}^2 + x_2 - 11)^2 + (x_1 + x_2{}^2 - 7)^2$	$-5 \le x_1, x_2 \le 5$
4.	$f_4(x) = \lvert x_1{}^2 + x_2{}^2 - 2x_1x_2\rvert + \lvert\sin x_1\rvert + \lvert\cos x_2\rvert$	$-500 \le x_1, x_2 \le 500$
5.	$f_5(x) = 10x^4 - 8x^2 + 12x + 16$	$-100 \le x \le 100$
6.	$f_6(x) = 17x^5 - 11x^2 + 11x + 6$	$-100 \le x \le 100$
7.	$f_7(x) = 3x_1{}^2 + 5x_2{}^2 - 0.6\cos(3\pi x_1 + 4\pi x_2) + 4$	$-5 \le x_1, x_2 \le 5$
8.	$f_8(x) = 3x_1{}^2 + 7x_2{}^2 + 15(\sin^2 x_1 + \sin^2 x_2)$	$-100 \le x_1, x_2 \le 100$
9.	$f_9(x) = (2.7 - x_1 + x_1x_2)^2 + (1.85 - x_1 + x_1x_2{}^2)^2 + (3.1 - x_1 + x_1x_2{}^3)^2$	$-100 \le x_1, x_2 \le 100$
10.	$f_{10}(x) = (x_1{}^2 + x_2 - 19)^2 + (3x_1 + x_2{}^2 - 16)^2$	$-100 \le x_1, x_2 \le 100$
11.	$f_{11}(x) = 10^4 x_1{}^2 + x_2{}^2 - (x_1{}^2 + x_2{}^2) + \{10^{-4}(x_1{}^2 + x_2{}^2)\}$	$-100 \le x_1, x_2 \le 100$

Table 2. Results for Jaya Algorithm for minimized problems

No	Best results	Worst	Mean	Time (ms)
f_1	−6.33	33.2	−2.934	53.6
f_2	0	0	0	59
f_3	0	4.3	0.891	56.3
f_4	1.00	1.41	1.116	64
f_5	5.23	5.23	5.23	54.3
f_6	-1.699×10^{11}	-1.677×10^{11}	-1.692×10^{11}	74.9
f_7	3.4	3.45	3.407	63.46
f_8	0	0.01	0.001	65.55
f_9	0.68	2.88	2.265	83.67
f_{10}	0	1.02	0.104	59.1
f_{11}	0	0.84	0.107	44.1

Table 2 and Table 3 show the test obtained results. Each test was run 20 times and the best-minimized value, the worst minimized value, mean value, and the execution time were reported. The execution time of the two algorithms are shown in the tables. Also, Fig. 1 and Fig. 2 show the comparison of the two algorithms in terms of mean values and execution times for the functions.

Table 3. Results for Cuckoo Search Algorithm for minimized problems

No	Best results	Worst	Mean	Time (ms)
f_1	−6.33	−6.33	−6.33	33.5
f_2	0	0	0	35.1
f_3	0	0	0	32.9
f_4	1.00	1.15	1.055	151.5
f_5	5.23	5.23	5.23	29.6
f_6	-1.7×10^{11}	-1.7×10^{11}	-1.7×10^{11}	672391
f_7	3.4	3.4	3.4	40.85
f_8	0	0	0	37.5
f_9	0	0.68	0.612	99.3
f_{10}	0	0	0	40.5
f_{11}	0	0	0	26.5

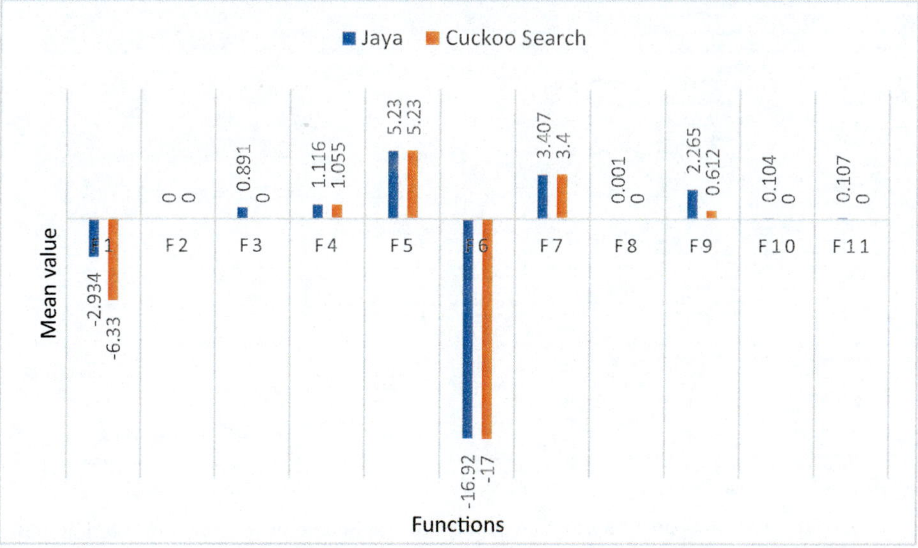

Fig. 1. Performance Comparison between Jaya algorithm and Cuckoo Search in term of test size

Figure 3 shows the screenshot of a Cuckoo Search program done in Intellij IDEA Community Edition 2020.2.3. The coding of the algorithms has been done in Java.

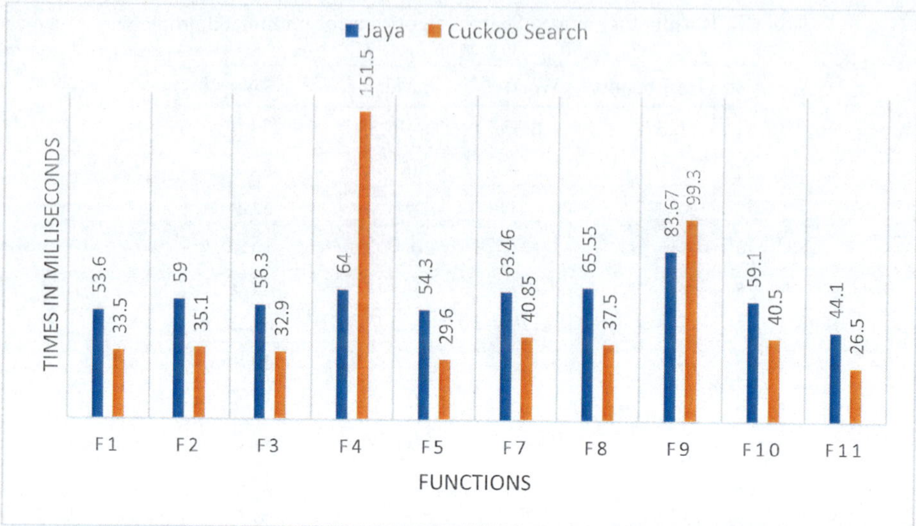

Fig. 2. Execution time comparison between Jaya algorithm and Cuckoo Search.

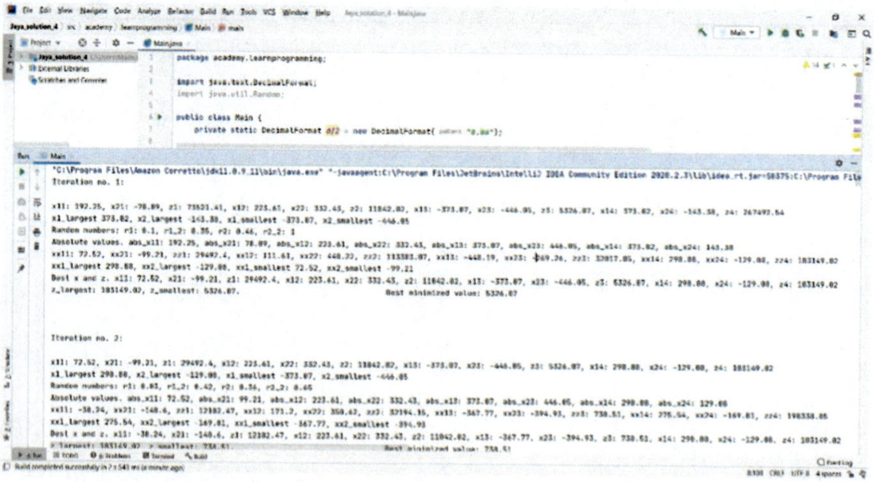

Fig. 3. Screenshot of the part of a program of the Cuckoo Search based function solution with output in Intellij IDEA Community Edition 2020.2.3

6 Discussions

As observed from the tests, JA has performed well consistently for all the 11 functions. CS hasn't been consistent like JA, but the mean values determined from the tests of the CS were nearer to the best-minimized values as compared to JA for many functions. As shown in Table 2 and 3, for 8 of the 11 functions, CS required less execution time as compared to that of JA but the differences weren't much. For 3 functions, which are function 4 (f_4), function 6 (f_6), and function 9 (f_9), JA took less execution time as compared to that of CS. But in these cases, the time differences were larger as compared to those of other functions. In the case of function 6 (f_6), the average execution time of JA was 74.9 ms, but the average execution time of CS algorithm was 672391 ms. Here, the difference is huge. For function 4 (f4), CS takes 151.5 ms, where JA takes only 64 ms. But when JA takes larger execution time as compared to CS, the difference isn't much. There has been no case in the tests where JA took equal to or more than twice the execution time of CS, according to Fig. 2. So, it indicates that CS may take much larger execution time for some functions as compared to JA. Considering this case, JA is more consistent.

CS showed better performance for 8 of the 11 functions in terms of mean values and execution times as shown in Table 2 and 3, but lacked consistency as compared to JA. For function 9 (f_9), CS was better than JA, but the execution time was a little higher. Regarding execution time, in Fig. 2, CS's performance isn't always efficient for all functions. There are some functions, where CS performs at a much slower speed. For some functions, CS can take extremely large execution time, as shown in the case of function 6 (f_6). Regarding these issues, JA is a simpler and more consistent algorithm, providing solutions to most functions with an acceptable execution time. But regarding mean values, JA performs well but is less optimum as compared to CS. From the overall scenario, JA is more consistent due to its performances in acceptable ranges for most functions.

7 Conclusion and Future Works

This paper evaluated the performances of CS and JA using the benchmark functions in terms of mean values and execution times. Each of the algorithms has shown impressive results in many cases. There has also been complexities for both algorithms regarding execution time and mean values in some cases. In the overall scenario, JA has turned out to be a more consistent algorithm. The analysis of these 2 algorithms will help engineers to select the right algorithm for specific problems.

As for the future works, we can make a detail analysis explaining for which functions CS works well with acceptable execution times and for which functions, CS struggles to maintain a faster execution time. Also, there are works to be done with the functions, for which, CS takes much longer execution time.

Acknowledgments. This research is funded by Universiti Malaysia Pahang (UMP) under grant: "Prioritized T-way Test Suite Generation Strategy Based on Chaotic Flower Pollination Algorithm", Grant no: RDU190372. We thank UMP for the contribution and supports.

References

1. Kelley, T.R.: Optimization, an important stage of engineering design. Technol. Teach. **69**(5), 18–23 (2010)
2. Mansour, N.: Search Algorithms and Applications (2011)

3. Yang, X.-S.: Metaheuristic optimization: algorithm analysis and open problems. In: Pardalos, P.M., Rebennack, S. (eds.) Experimental Algorithms. LNCS, vol. 6630, pp. 21–32. Springer, Berlin (2011). https://doi.org/10.1007/978-3-642-20662-7_2
4. Wong, W.K., Ming, C.I.: A review on metaheuristic algorithms: recent trends, benchmarking and applications. Paper presented at the 2019 7th International Conference on Smart Computing & Communications (ICSCC) (2019)
5. Alsariera, Y.A., Alamri, H.S., Nasser, A.M., Majid, M.A., Zamli, K.Z.: Comparative performance analysis of bat algorithm and bacterial foraging optimization algorithm using standard benchmark functions. Paper presented at the 2014 8th Malaysian Software Engineering Conference (MySEC), Langkawi, Malaysia (2014)
6. Syafruddin, W.A., Köppen, M., Benaissa, B.: Does the jaya algorithm really need no parameters?. Paper presented at the 10th International Joint Conference on Computational Intelligence
7. Pandey, H.M.: Jaya a novel optimization algorithm: what, how and why? Paper presented at the 2016 6th International Conference - Cloud System and Big Data Engineering (Confluence) (2016)
8. Sharma, S., Bhushan, B.: Adaptive control using jaya algorithm. In: AIP Conference Proceedings, vol. 2136, p. 020001 (2019)
9. Yu, B.Q., Yue, C., Ge, S., Chen, X., Liang, J.: A performance-guided JAYA algorithm for parameters identification of photovoltaic cell and module. Appl. Energy **237**, 241–257 (2019)
10. Mohanty, S., Patra, P.K., Ray, M., Mohapatra, S.: An approach for load balancing in cloud computing using JAYA algorithm. Int. J. Inf. Technol. Web Eng. **14**(1), 27–41 (2019)
11. Subrahmanian, S., Gupta, R.: Fuzzy node flow analysis of water distribution networks using Jaya algorithm. Paper presented at the IOP Conference Series Earth and Environmental Science, vol. 491, p. 012010
12. Sahu, A.K., Bokam, J.K., Singh, V.P., Singh, S.P.: JAYA algorithm based tuning of PID controller. Int. J. Ind. Electron. Electr. Eng. **4**(12) (2016)
13. Mohamad, A., Zain, A.M., Bazin, N.E.N., Udin, A.: Cuckoo search algorithm for optimization problems - a literature review. Appl. Mech. Mater. **421**, 502–506 (2013)
14. Fister Jr., I., Fister, D., Fister, I.: A comprehensive review of cuckoo search: variants and hybrids. Int. J. Math. Model. Numer. Optim. **4**(4), 387–409 (2013)
15. Mohamad, A.B., Zain, A.M., Bazin, N.E.N.: Cuckoo search algorithm for optimization problems—a literature review and its applications. Appl. Artif. Intell. Int. J. **28**(5), 419–448 (2014)
16. Al-Obaidi, A.T.S.: Improved scatter search using cuckoo search. Int. J. Adv. Res. Artif. Intell. **2**(2), 61–67 (2013)
17. Lim, H.T., Ramli, R.: Recent advancements of nurse scheduling models and a potential path. In: Proceedings of the 6th IMT-GT Conference on Mathematics, Statistics and its Applications (ICMSA2010), Kedah, Malaysia (2010)
18. Yildiz, A.R.: Cuckoo search algorithm for the selection of optimal machine parameters in milling operations. Int. J. Adv. Manuf. Technol. **64**, 55–61 (2012)
19. Layeb, A.: A novel quantum-inspired cuckoo search for Knapsack problems. Int. J. Bio-Inspired Comput. **3**(5), 297–305 (2011)
20. Zheng, G., Mohanty, S.P., Kougianos, E.: Metamodel-assisted fast and accurate optimization of an OP-AMP for biomedical applications. Paper presented at the 2012 IEEE Computer Society Annual Symposium on VLSI, Amherst, MA, USA (2012)
21. Salih, A.-H.A., Ali, A.H., Hashim, N.Y.: Jaya: an evolutionary optimization technique for obtaining the optimal Dthr value of evolving clustering method (ECM). Int. J. Eng. Res. Technol. **11**(12), 1901–1912 (2018)

22. Dhivya, M., Sundarambal, M.: Energy Efficient Computation of data fusion in wireless sensor networks using cuckoo based particle approach (CBPA). Int. J. Commun. Netw. Syst. Sci. **4**(4), 249–255 (2011)
23. Zheng, H., Zhou, Y.: A novel cuckoo search optimization algorithm base on gauss distribution. J. Comput. Inf. Syst. **8**(10), 4193–4200 (2012)
24. Nasser, A.B., Alsewari, A.R.A., Zamli, K.Z.: Tuning of cuckoo search based strategy for t-way testing. In: International Conference on Electrical and Electronic Engineering, vol. 9, p. 8948 (2015). Journal of Engineering and Applied Sciences

Urban Road Marker Classification Using Histogram of Oriented Gradient and Local Binary Pattern with Artificial Neural Network

Zamani Md Sani[1](✉), Liang Jin Chuan[1], Tarmizi Ahmad Izzudin[1], Hadhrami Abd Ghani[2], and Aerun Martin[3]

[1] Universiti Teknikal Malaysia Melaka, Malacca, Malaysia
zamanisani@utem.edu.my
[2] Universiti Malaysia Kelantan, Pengkalan Chepa, Malaysia
[3] Universiti Multimedia, Cyberjaya, Malaysia

Abstract. Road safety is paramount and depends on different factors, one of which is road makers. To regulate the road traffic while ensuring its safety, several common types of road markers are typically used such as the dashed, double solid and solid-dashed markers. This paper proposed a vision-based system to classify three types of markers using Artificial Neural Network (ANN) and image processing approaches. The accuracy was compared between two types of feature extraction techniques, namely the vector on Histogram of Oriented Gradient (HOG) and Local Binary Pattern (LBP), with the latter is lower at 98.3% vs 98.9% . The accuracy is further improved at 99.4% with both HOG and LBP features were combined and used as the input vectors for the classification process.

Keywords: Road marker · Image processing · Feature HOG and LBP · Artificial neural network and classification

1 Introduction

Road accident is an alarming issue in many countries including Malaysia. It is one of the top five principal causes of death in Malaysia between 2017 and 2018 [1] as shown as in Fig. 1. As clearly seen in the figure, the number of road deaths increases from 2017 to 2018, which proves the significance of this problem [2].

Road markers are used as an indicator for the drivers and the road users to make the right decisions at important moments while they are on the wheel. These decisions include the lane changing, overtaking, parking and so on. Hence, the road markers are essential especially when the traffic is heavy such as in the urban roads because the number of road users is higher in the urban areas. Furthermore, the markers are also employed on roadways and highways to guide the road users which will typically move faster than those in the urban areas. Different markers indicate distinct kinds of information to alert the drivers to take the necessary decision while driving. For example, a double solid marker indicates that vehicle overtaking is prohibited to avoid

M. Al-Emran et al. (Eds.): ICETIS 2021, LNNS 299, pp. 126–135, 2022.
https://doi.org/10.1007/978-3-030-82616-1_12

accidents, which are caused by various reasons including the road surface conditions, human errors and vehicle conditions [3, 4]. To improve the situation, researchers have been researching autonomous car driving. This paper presents three types of markers, namely, double dashed, dashed and double solid normally found in the urban area are being classified by using different features (HOG and LBP) and artificial neural network (ANN) for the classifier.

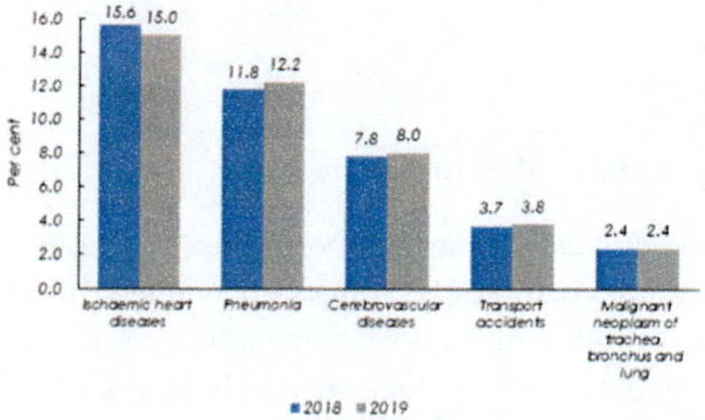

Fig. 1. Top five principal causes of death in Malaysia between 2018 and 2019

1.1 Previous Research

In [5], a road marker classification mechanism has been proposed to classify up to five distinct types of markers. In the first stage of classification, three road markers are classified using Bayesian approach. In the second stage, two more road markers, which are solid dashed and dashed solid are classified. However, the classification's results were found to be prone to abrupt changes in each frame, causing inconsistent results while driving. Furthermore, zooming into the confusion matrix table results indicated that the classes for solid-dashed and dashed-solid had lower accuracy than the rest of the classes.

Two types of lane marker had been classified by Toan et al. [6], which are the dashed and solid lanes under various environmental condition. A threshold is identified to measure the distance in the static images before detecting the right and left boundaries of the road, by using a perspective camera modelling technique. In [7], Zhang et al. had used the Stage Feature Extraction with You Only Look Once (YOLO) for a robust lane detection which improves lane markers' detection. However, the classification of types of markers was not performed in this work.

Interestingly, Tang et al. [8] and Ali et al. [9] worked on 6 and 5 types of lane marker classification, which similar to Paula et al. work and based on the numbers of markers being classified. In [8], the markers' semantic information is extracted, and the decision trees model is used. Whereas in [9], the seed fill algorithm was used as a classifier for the road markers. Nevertheless, both works did not comprehensively compare their marker classification methods with [5] in terms of intensive accuracy using the same videos provided by Paula.

In the continuous research, Zamani et al. [10, 11] worked on road marker classification, initially with only two types of markers using features and Artificial Neural

Networks, and later expanding to five types of marker classification using the features extracted from customised Region of Interest.

Throughout the review, the classification types of markers and not widely being studied compared to lane detection. Furthermore, it is hard to find a standard database to classify the markers' types due to the different types available in different countries. In this paper, only 3 types of markers normally found in the urban area are being classified using the neural network.

2 Methodology

2.1 Hardware Preparation and Setup

Camera positioning is crucial as an incorrect position of camera mounting will result in recording failure. The camera is mounted at the same angle as the x-axis. There may be some illumination issue on the video frame when strong sunlight is projected into the camera view. The best angle is to mount the camera slightly lower, about 5 to 10 degrees with respect to the x-axis as shown in Fig. 2. The image acquisition tool in MATLAB is used with Logitech C310. The image is acquired in a RGB format, with a resolution of 1021 × 720 pixels, and capturing at 30 frames per second.

Fig. 2. Camera position at the car in the box

2.2 Preprocessing Module

There are 5 stages of processing on the frames which are grayscale conversion, image sharpening, noise removal, cropping and resizing, such as shown in Fig. 3. This preprocessing module is created to make sure the process of feature extraction smooth and easier.

The first step to process the frame is to convert the frame colour from RGB format to grayscale format. The step is known as a grayscale conversion and often widely used in image processing. Next, all the image samples are sharpened by using the unsharp masking techniques by subtracting a blurred (unsharp) version of the image from itself. Its principle is to extract the high-frequency components by subtracting the blurred version from the original [12]. Noise removing, using a median filter is applied to the images to remove any unwanted noise from the image, and once all the process had been completed, it will crop to the region of interest for the feature extraction.

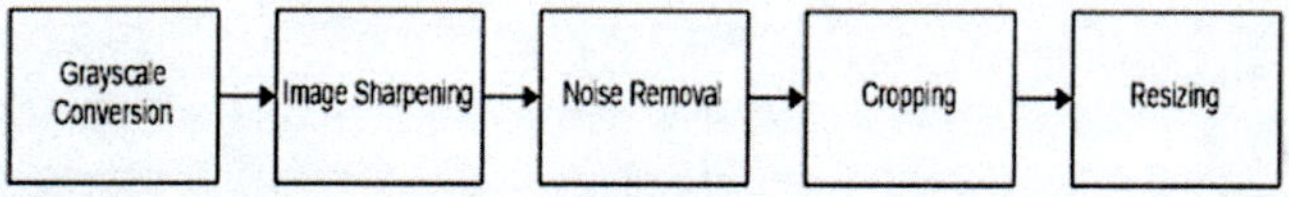

Fig. 3. Stages of the pre-processing methods

The cropping process is a manual procedure that is performed using the Matlab function to remove the unwanted information in the images. All crop images must have a similar dimension to avoid high error and low accuracy when going through the feature extraction. The results for the preprocessing is shown as in Fig. 4(a). After the grayscale conversion (b) After the sharpen process (c) After the noise removal (d) The cropped and resized images.

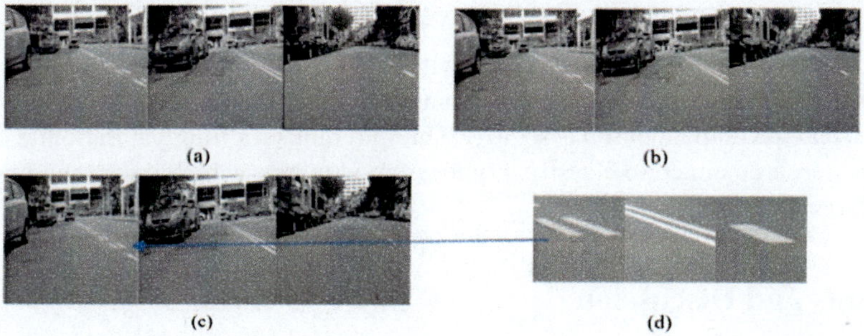

Fig. 4. Results from the pre-processing module (a) Grayscale conversion (b) Sharpen process (c) Noise removal (d) Cropped images

2.3 Feature Extraction

After the sample images going through the preprocessing module, the output image samples are ready for the feature extraction. In this research, two types of feature extraction have been used, Histogram of Oriented Gradient (HOG), a feature descriptor, and Local Binary Pattern (LBP), a type of visual descriptor. HOG is that local object appearance and shape in an image can be described by the distribution of intensity gradients or edge directions, whereas LBP is a useful feature for texture classification [13]. These two types of feature extraction are a common method to extract the feature from the image, and the feature data are in the form of feature vector. The only difference between these two methods is the value in the feature vector. For HOG, the feature vector's value is float number, whereas the value in the feature vector for LBP is scalar. Example of the sample frames on HOG visualisation are shown in Fig. 5.

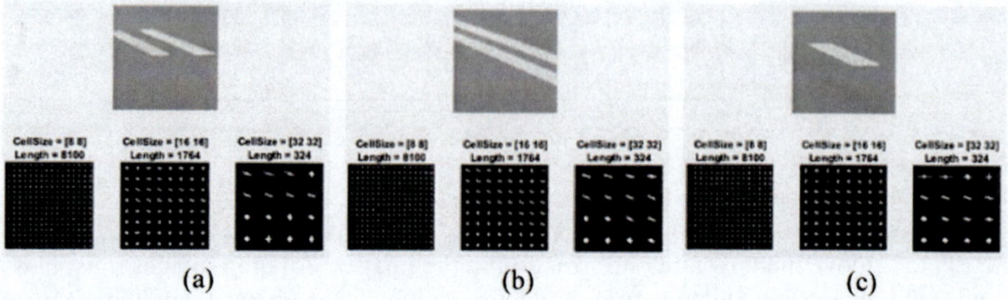

Fig. 5. HOG visualization for different markers (a) Double dashed (b) Double solid (c) Single dashed

2.4 Training and Classification Module

Features extracted are used to build models for accurate classification. Two Layer Feed Forward Artificial Neural-Network (ANN) with a sigmoid transfer function in the output layer was used to classify the type of road lane markers. The number of hidden neuron is set to 10 and the output neuron is set to 3. The split ratios for train/validation/test were 7:2:1. This research aimed to classify 3 types of road marker which meant there were 3 types of target data.

3 Results and Discussion

3.1 Training and Testing Set Images

In this research, a total of 180 image frames are selected with 60 images per road markers. These 180 selected image frames are stored in a database to be used during the training and classification process. A total of 180 samples for 3 classes of road markers were divided into 60 samples in each class to use the HOG for the feature's selection. The input data set which are the total of 180-by-1764 features vector extracted were used for the ANN for training process, testing and validating the network. The neural network's performance, training state, error histogram, and overall confusion matrix using the HOG feature are shown in Fig. 6(a).

From the performance plot, the cross-entropy error is maximum at the beginning of training. The best validation performance can be seen at epoch 18, with the cross-entropy error found is very close to zero. For the training state plot, maximum validation checks 6 can be found at epoch 24 and at this point, the network halts the training process for the best performance. The error histogram which showed the error for the system is very close to zero. The combination of the overall matrix (training, validation and testing) resulted in 98.9% correct classification for the three classes, such as shown in Fig. 6(a). The true positive rate versus the false-positive rate at various threshold settings of the network, the Receiver Operating Characteristic (ROC) curve of the network, is shown in Fig. 6(b). The area under the curve (AUC) shows a slightly result for this proposed system. To conclude for the training, test and validation, the network performs 98.9% correct in classifying three classes of road markers.

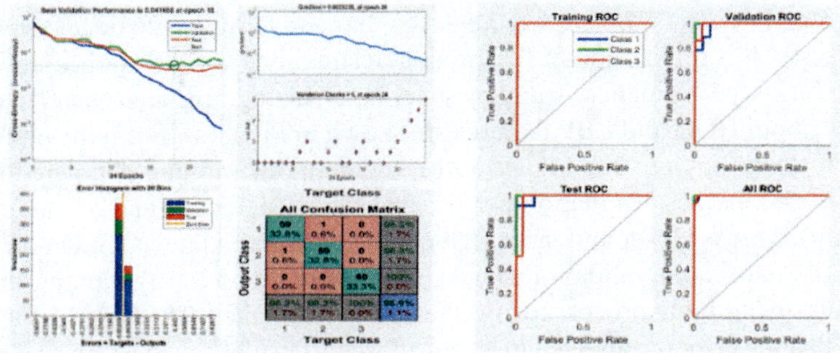

(a) (b)

Fig. 6. (a) Neural network (HOG): performance, training state, error histogram, and overall confusion matrix (b) HOG network Receiver Operating Characteristic (ROC)

For LBP feature, the input data set which are a total of 180-by-256 features vector extracted were used for the ANN for training, testing and validating the network. The neural network's performance, training state, error histogram, and overall confusion matrix using the LBP features are shown in Fig. 7(a). It can be seen from the performance plot, the cross-entropy error is higher at the beginning of training compared to HOG. The best validation performance can be seen at epoch 31, and at this point, the cross-entropy error is closer and more stable to zero compared to HOG. For the training state plot, maximum validation checks 6 can be found at epoch 37 and at this point, the network halts the training process for the best performance. The error histogram which showed the error for the system is very close to zero. The combination of the overall matrix (training, validation and testing) for the three classes results in an overall 98.3% correct classification, shown in Fig. 7(a). The true positive rate versus false-positive rate at various threshold settings of the network, the Receiver Operating Characteristic (ROC) curve of the network, is shown in Fig. 7(b). The area under the curve (AUC) shows a slightly result for this proposed system. To conclude for the training, test and validation, the network performs 98.3% correct in classifying three classes of road markers.

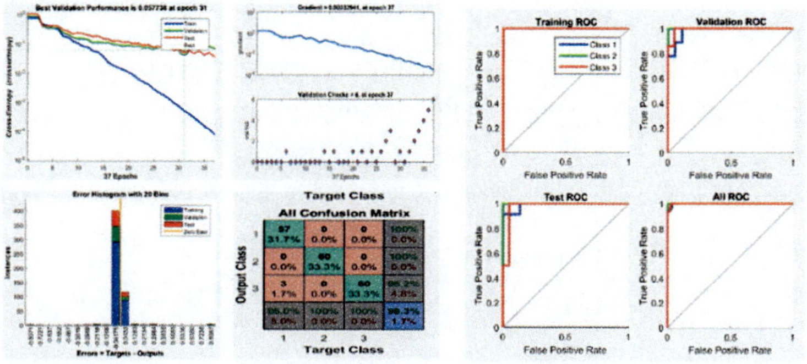

(a) (b)

Fig. 7. Neural network (HOG): performance, training state, error histogram, and overall confusion matrix (b) HOG network Receiver Operating Characteristic (ROC)

For using HOG and LBP as a feature, the extracted 180-by-2020 features vectors are used as an ANN input data set for training, testing, and validating the network. The neural network's performance, training state, error histogram, and overall confusion matrix using both HOG and LBP features are shown in Fig. 8(a). From the performance plot, the cross-entropy error is higher at the beginning of training compared to HOG. For this system, the best validation performance is at epoch 23, and at this point, the cross-entropy error is closer and more stable to zero compared to HOG. On the training state plot, the maximum validation checks 6 at epoch 29 and at this point, the neural network halts the training process to give the best performance. The error histogram plot represents that the error of this system is very close to zero. The overall confusion matrix plot shows 99.4% correct classification for this system. The true positive rate versus false-positive rate at various threshold settings of the network, the Receiver Operating Characteristic (ROC) curve of the network, is shown in Fig. 8(b). The area under the curve (AUC) shows a slightly result for this proposed system. The area under the curve (AUC) shows a slightly result for this proposed system. To conclude the training, test and validation, the network performs 99.4% correct in classifying three classes of road markers.

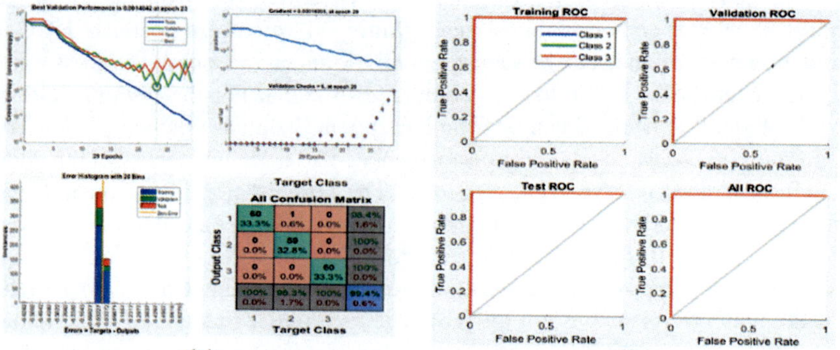

(a) (b)

Fig. 8. (a) Neural network (HOG): performance, training state, error histogram, and overall confusion matrix (b) HOG network Receiver Operating Characteristic (ROC)

Based on Table 1, it is clearly that the combination of HOG and LBP feature extraction method will result in higher accuracy in the classification process.

Table 1. Accuracy of the classification determined by feature extraction method

Feature extracted	Accuracy (%)
HOG	98.9
LBP	98.3
HOG and LBP	99.4

3.2 Video Testing

After getting the trained network, the system was tested with new video input on the real time situation. Different types of road markers situation were tested with the trained network. The accuracy and result were shown in Table 2 and Table 3.

Table 2. Classification accuracy using real-time video

Experiment	Type	Frame	Accuracy (%)
1	Double dashed	91	41.7
2	Double solid	89	100
3	Single dashed	180	68.3

Table 3. Confusion matrix of the overall accuracy on the video

	Predicted class			
	Double dashed	Double solid	Single dashed	
Double dashed	38 (10.56%)	35 (9.72%)	18 (5%)	**41.76%**
Double solid	0	89 (24.72%)	0	**100.00%**
Single dashed	48 (13.33%)	9 (2.5%)	123 (34.17%)	**68.33%**
	44.19%	**66.92%**	**87.23%**	**69.44%**

Based on Table 2, three experiments were carried out on a real-time video. Each experiment was tested with the trained network on each type of road marker. In experiment 1, the road marker type was double dashed, and 91 samples of frames were tested and classified. The accuracy for experiment 1 was 41.76%. In experiment 2, the road marker type was double solid, and 89 samples of frames were tested and classified. The accuracy of the classification in for experiment 2 was 100.00%. In experiment 3, the road marker type was single dashed, and 180 samples of frames were tested and classified. The classification accuracy for experiment 3 was 68.33%.

In Table 3, the first three diagonal cells showed the number and percentage of correct classifications by the trained network. For example, 38 samples were correctly classified as Class 1, a double dashed road marker. This corresponded to 10.56% of all 360 samples. Secondly, 89 samples were correctly classified as a double solid road marker—this led to 24.72% of all 360 samples. Similarly, 123 samples were correctly classified as a single dashed road marker. This corresponded to 34.17% of all 360 samples. 35 samples of the double dashed markers were incorrectly classified as a double solid marker. This corresponded to 9.72% of all 360 samples in the data. 18 samples of the double dashed markers were wrongly classified as a single-dashed marker. This corresponded to 5% of all data. At the same time, 48 samples of single dashed markers were wrongly classified as double dashed markers, which corresponded to 13.33% of all data. Other 9 samples of single dashed markers were incorrectly classified as double solid markers, which corresponded to 2.5% of all the samples.

Out of 91 double dashed marker predictions, 41.76% were correct, and 58.24% were wrong. Out of 89 double solid marker predictions, all 89 samples were predicted correctly, indicating 100% correct. Out of 180 single dashed marker predictions, 68.33% were correct, and 31.67% were wrong.

On the other hand, out of 86 double dashed marker cases, 44.19% were correctly predicted as a double dashed marker, and 55.84% were wrongly predicted. Out of 133 double solid marker cases, 66.92% were correctly classified as double solid markers, and 33.08% were classified as other types of road markers. Out of 141 single-dashed marker cases, 87.23% were correctly classified as a single-dashed marker, and 12.77% were classified as other types of road marker. Overall, 69.44% of the predictions were correct and 30.56% were wrong.

In addition, there are also some errors. Examples of these errors are shown in Fig. 9(b), where the classifier incorrectly classified unmarked road as a double solid marker. This is because the trained network had only 3 classes of target markers: double dashed marker, double solid marker, and single dashed marker, as shown in Fig. 9(a). The road without marker is not included in the classification process. Due to this reason, the feature extracted from the image in Fig. 9(b) may have almost the same value as in feature vector in double solid class, so it is classified in the double solid class even though it is not, which is not accurate.

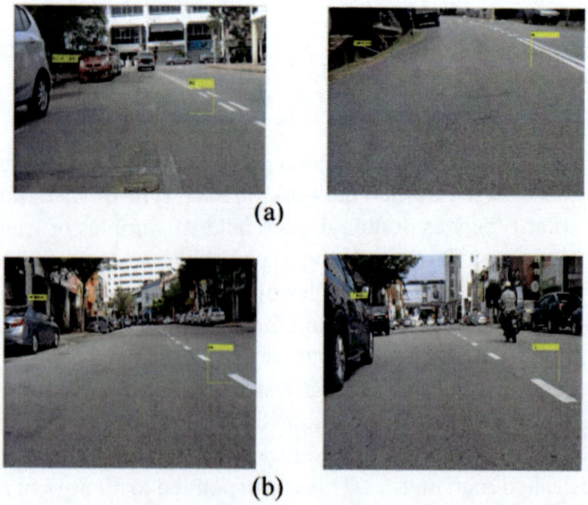

Fig. 9. Testing on Video and Error on real-time video (a) With correct results (b) With wrong results

4 Conclusion

Among all the feature extraction methods, the combination of HOG and LBP as the feature extraction method showed the highest accuracy, 99.4%. The testing was also carried out using a trained network against the real-time video. The classification accuracy results on the road marker of double dashed, double solid and single dashed markers are 41.76%, 100% and 68.33%, respectively. The overall accuracy of the classification on the road marker is 69.44%. However, there are some errors in the video, which is due to the region selection. The errors can be solved by implementing a detection system on the road marker in the future.

References

1. Department of Statistics Malaysia Official Portal. Social Statistic Bulletin, 2019, 26 November 2019. https://www.dosm.gov.my/
2. ASEAN Road Safety Centre, General Road Accident Data in Malaysia (1997 – 2016), Malaysian Institute of Road Safety Research (MIROS) official website (2017)
3. Subramaniam, S.K., Ganapathy, V.R., Subramonian, S., Hamidon, A.H.: Automated traffic light system for road user's safety in two lane road construction sites. WSEAS Trans. Circuits Syst. **2**(9), 71–80 (2010)
4. Gannapathy, V.R., Subramaniam, S.K., Diah, A.M., Suaidi, M.K., Hamidon, A.H.: Risk factors in a road construction site. Proc. World Acad. Sci. Eng. Technol. **46**, 640–643 (2008)
5. de Paula, M.B., Jung, C.R.: Automatic detection and classification of road lane markings using onboard vehicular cameras. IEEE Trans. Intell. Transp. Syst. **16**(6), 3160–3169 (2015). https://doi.org/10.1109/TITS.2015.2438714
6. Hoang, T.M., Hong, H.G., Vokhidov, H., Park, K.R.: Road lane detection by discriminating dashed and solid road lanes using a visible light camera sensor. Sensors (Switzerland) **16**(8) (2016). https://doi.org/10.3390/s16081313
7. Zhang, X., Yang, W., Tang, X., Liu, J.: A fast-learning method for accurate and robust lane detection using two-stage feature extraction with YOLO V3. Sensors (Switzerland) **18**(12) (2018). https://doi.org/10.3390/s18124308
8. Tang, L., Deng, T., Ren, C.: Semantic information extraction of lanes based on onboard camera videos. Int. Arch. Photogramm. Remote Sens. Spat. Inf. Sci. ISPRS Arch. **42**(3), 1643–1649 (2018). https://doi.org/10.5194/isprs-archives-XLII-3-1643-2018
9. Ali, A.A., Hussein, H.A.: Real-time lane markings recognition based on seed-fill algorithm. In: ACM International Conference Proceeding Series, pp. 190–95 (2019). https://doi.org/10.1145/3321289.3321306
10. Sani, Z.M., Ghani, H.A., Besar, R., Sen, L.W.: Real-time daytime road marker recognition using features vectors and neural network. In: IEEE Conference on Sustainable Utilization and Development in Engineering and Technology (CSUDET), pp. 1–6 (2015). https://doi.org/10.1109/CSUDET.2015.744622
11. Sani, Z.M., Ghani, H.A., Besar, R., Azizan, A., Abas, H.: Real-time video processing using contour numbers and angles for non-urban road marker classification. Int. J. Electr. Comput. Eng. (IJECE) **4**(8), 2540–2548 (2018). https://doi.org/10.11591/ijece.v8i4
12. Krasula, L.: Quality assessment of sharpened images. Czech Technical University in Prague, Faculty of Electrical Engineering, Department of Radio Engineering (2013)
13. Dalal, N., Triggs, B.: Histograms of oriented gradients for human detection. In: IEEE Computer Society Conference on Computer Vision and Pattern Recognition, vol. 1, pp. 886–893, June 2005

Rice Leaf Disease Classification Using Deep Learning and Target for Mobile Devices

Nguyen Tien Su[1], Phan Duy Hung[1], Bui Trong Vinh[2], and Vu Thu Diep[3](✉)

[1] FPT University, Hanoi, Vietnam
su18mse13007@fsb.edu.vn, hungpd2@fe.edu.vn
[2] Hanoi Procuratorate University, Hanoi, Vietnam
vinhbt@tks.edu.vn
[3] Hanoi University of Science and Technology, Hanoi, Vietnam
diep.vuthu@hust.edu.vn

Abstract. Rice is the main staple food in many Asian countries and the quality of these crops plays an important part in security. Therefore, the ability to timely identify and accurately diagnose common rice diseases such as bacterial leaf blight, rice brown spot and leaf smut is highly desirable. This paper compares automatic plant disease identification methods using deep convolutional neural networks and targeting for mobile platforms. A dataset of 120 images of the three common diseases samples was collected, and the model was trained to identify them on rice paddy. As a result, it can effectively detect and recognize those classes of rice diseases with an accuracy of 81.87% on training data and 81.25% on the validation data.

Keywords: Rice leaf disease classification · Convolutional neural network · Mobilenet · NASNetMobile · DenseNet

1 Introduction

Rice is that the most generally consumed staple food within the world, and particularly in Asia. It's tormented by a large type of diseases, and it can place a significant constraint on the assembly and be a significant threat to food security. Hence, early and accurate identification of said diseases is extremely desirable.

In recent years, the amount of diseases on rice plants and their impact have increased because of the styles of pathogens, changes in cultivation methods and inadequate plant protection techniques.

Leaf diseases in rice have numerous characteristics and so their type and severity may be difficult to spot by farmers without expert knowledge. And even for people with experience, the diagnosis process are often time-consuming and inaccurate.

There are various studies on automatic disease classification using different algorithms to boost the speed and accuracy of disease identification.

The authors in [1] used the NN method on a dataset with three classes: Rice Blast, Sheath Blight, Brown Spot. Despite the quantity of images in each class being 50, their

M. Al-Emran et al. (Eds.): ICETIS 2021, LNNS 299, pp. 136–148, 2022.
https://doi.org/10.1007/978-3-030-82616-1_13

accuracy was 85%. With a similarly small dataset of three classes, the researchers in [2, 3] used SVM classifier and obtained accuracies up to 97.2% and 88.87% respectively.

A Bayes classifier was applied on a bigger dataset (with two classes being Rice Blast and Brown Spot) and led to 68.1% of accuracy [4].

With another dataset of three sorts of diseases (Bacterial blight, Brown Spot, Rice Blast), the researchers in [5] had applied a deep learning network to supply a formidable result with 100% accuracy.

Deep learning techniques have shown great potential in image classification. Besides rice, they need been also wont to classify traffic signs [6], diabetic retinopathy [7], surfaces of damaged concrete [8], diseases in mango [9]. In most cases, images of leaves or fruits were accustomed detect the diseases.

The contribution of this work is that the comparison and evaluation of deep learning architectures targeting mobile devices like MobileNet-v1 [10], MobileNet-v2 [11], DenseNet121 [12] and NasNetMobile [13] in terms of their accuracy and sizes of the models. From there, the most effective model for classification rice plant disease which might run on mobile devices is proposed.

The content of the study is arranged as follows: Sect. 2 mentions the methodology. Next, Sect. 3 analyzes the experiment and evaluation. Finally, conclusions and perspective are showed in Sect. 4.

2 Methodology

2.1 Overview

This paper sets resolute compare four deep learning models in classifying images of rice diseases. The models are MobileNetV1, MobileNetV2, NasNetMobile, and DenseNet121, all of which have a tiny low size suitable for running on mobile platforms.

The proposed method includes 3 stages: (1) data preparation, (2) implementation of the models using the transfer learning technique and (3) experimentation and assessment (Fig. 1).

2.2 Data Preparation

This project uses data available on UCI (Table 1), which was created by manually separating infected leaves into different disease classes [14]. This dataset has also been used by some researchers, for example [15, 16] to classify rice leaf diseases. The authors had consulted with farmers and asked them to provide the names of diseases for sample leaves. The farmers provided names in their native languages (Gujarati) and the author verified their English names by consulting agricultural experts.

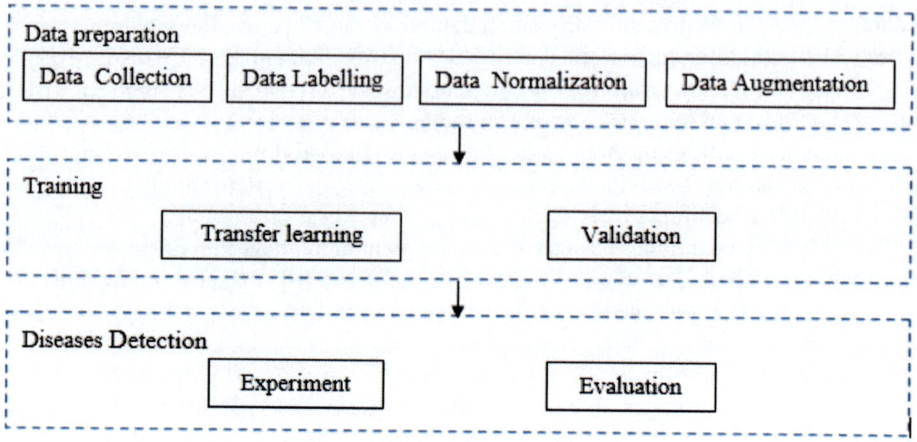

Fig. 1. Proposed method for rice disease detection.

Table 1. The numerical distribution of the images for each class.

Species	Number of instance
Bacterial leaf blight	40
Brown spot	40
Leaf smut	40

Rice bacterial blight, shown in Fig. 2, also called bacterial blight of rice, a deadly bacterial disease that's among the foremost destructive afflictions of rice (Oryza sativa and O. glaberrima). In severe epidemics, crop loss is also as high as 75%, and voluminous hectares of rice are infected annually. The disease was first observed in 1884–85 in Japan, and also the causal agency, the bacterium Xanthomonas oryzae pathovar oryzae, was identified in 1911, at that point having been named Bacillus oryzae.

Fig. 2. Bacterial blight.

Brown spot, shown in Fig. 3, is that the most typical rice disease and causes the foremost damage to the standard and quantity of production. It's a mycosis that affects the leaves and leaf sheathes. The foliar lesions may appear as dark brown or black spots measuring 0.5–2.0 mm in size. it should be circular or oval and appears sort of a flavorer. This infection affects the plant the least bit stages from seedling to maturity.

Fig. 3. Brown spot.

Rice leaf smut, shown in Fig. 4, is additionally as referred to as Entyloma oryzae. It's widespread where rice is grown, but this sort of infection doesn't often cause serious damage. However, leaf smut can make your rice prone to other diseases, and ultimately causes a yield reduction.

Fig. 4. Leaf smut.

To prepare for the training of the classification models, the dataset is split into a training set and a validation set with a ratio of 80% and 20% respectively. Then, data augmentation techniques are applied to increase the size of the dataset.

2.3 Transfer Learning Implementation

Convolutional neural networks are a specific type of neural networks that are generally composed of the at lease layers: Convolution (CONV), Pooling (POOL), Fully Connected (FC). Each of layer carry out different tasks. The convolution layer (CONV) uses filters to perform convolution operations by scanning through the input data in its dimensions. Hyperparameters, for example the filter size F and stride S can be tuned. The output O is named feature map or activation map. The pooling layer (POOL) is a downsampling operation, which applied after a convolution layer. There are two kind of pooling: max and average pooling. In the same current view, max pooling selects the maximum value, and average pooling averages the values. Finally, the fully connected layer (FC) run on a flattened input where each input is connected to all neurons. In almost case, FC layers are usually present at the end of CNN architectures and can be used to calculate score for each class.

The creating a deep learning model for solving rice leaf disease classification problem is the most important part of the present research. However, with a tiny low dataset, accuracy is low and unstable. A good thanks to rectify this problem is transfer learning, where a model developed for a previous task is reused for a model on a second task. It's a well-liked technique in Deep Learning because it makes the training a deep learning model available with quite a bit of data.

In this paper, four different pre-trained CNN models including MobileNetV1, MobileNetV2, DenseNet121 and NasNetMobile are experimented with to gauge their performance.

The main idea behind MobileNetV1 is that the CONV layers, which are essential to computer vision jobs but quite expensive to compute, could also be replaced with the so called depthwise separable convolutions (Fig. 5).

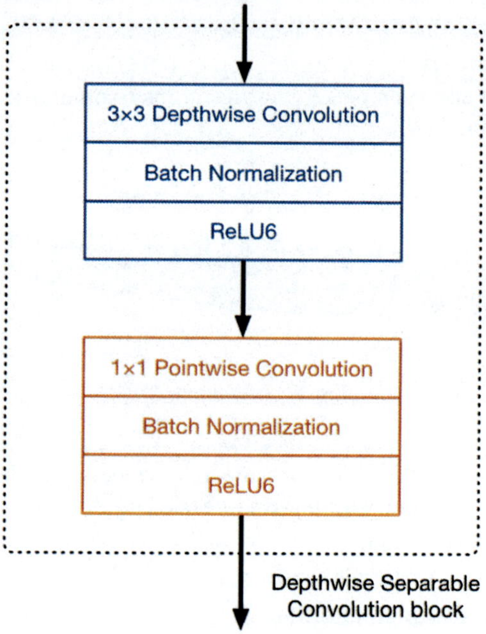

Fig. 5. A MobileNet model.

In MobileNetV1 model, there are two subtasks which performed by two blocks. Firstly, the input is filtered by a depthwise convolution layer, and then these filtered values are transformed to form new features by 1 × 1 convolutional layer. The combining the depthwise and pointwise convolutions togeather create a "depthwise separable" convolutional block, which is similar to the standard convolution but will be done faster.

In terms of overall architecture of MobileNetV1 (Fig. 6), there is a 3 × 3 convolution at the first layer, and the 13 times of the above building block follow it. POOL layers does not exist in between these depthwise separable blocks. Instead, some the depthwise layers with a stride of two in value to scale back the spatial dimensions of the info. Parallel with that, the corresponding pointwise layer also doubles the quantity of output channels. If the input image have dimention 224 × 224 × 3 then the network will have ouput may be a 7 × 7 × 1024 feature map.

In a classifier using Mobilenet architecture, average pooling layer offten appears at the end, which is followed by a totally classification layer or constant 1 × 1 convolution, and a softmax [18].

Type / Stride	Filter Shape	Input Size
Conv / s2	$3 \times 3 \times 3 \times 32$	$224 \times 224 \times 3$
Conv dw / s1	$3 \times 3 \times 32$ dw	$112 \times 112 \times 32$
Conv / s1	$1 \times 1 \times 32 \times 64$	$112 \times 112 \times 32$
Conv dw / s2	$3 \times 3 \times 64$ dw	$112 \times 112 \times 64$
Conv / s1	$1 \times 1 \times 64 \times 128$	$56 \times 56 \times 64$
Conv dw / s1	$3 \times 3 \times 128$ dw	$56 \times 56 \times 128$
Conv / s1	$1 \times 1 \times 128 \times 128$	$56 \times 56 \times 128$
Conv dw / s2	$3 \times 3 \times 128$ dw	$56 \times 56 \times 128$
Conv / s1	$1 \times 1 \times 128 \times 256$	$28 \times 28 \times 128$
Conv dw / s1	$3 \times 3 \times 256$ dw	$28 \times 28 \times 256$
Conv / s1	$1 \times 1 \times 256 \times 256$	$28 \times 28 \times 256$
Conv dw / s2	$3 \times 3 \times 256$ dw	$28 \times 28 \times 256$
Conv / s1	$1 \times 1 \times 256 \times 512$	$14 \times 14 \times 256$
5× Conv dw / s1	$3 \times 3 \times 512$ dw	$14 \times 14 \times 512$
5× Conv / s1	$1 \times 1 \times 512 \times 512$	$14 \times 14 \times 512$
Conv dw / s2	$3 \times 3 \times 512$ dw	$14 \times 14 \times 512$
Conv / s1	$1 \times 1 \times 512 \times 1024$	$7 \times 7 \times 512$
Conv dw / s2	$3 \times 3 \times 1024$ dw	$7 \times 7 \times 1024$
Conv / s1	$1 \times 1 \times 1024 \times 1024$	$7 \times 7 \times 1024$
Avg Pool / s1	Pool 7×7	$7 \times 7 \times 1024$
FC / s1	1024×1000	$1 \times 1 \times 1024$
Softmax / s1	Classifier	$1 \times 1 \times 1000$

Fig. 6. MobileNet Model architecture.

The depthwise separable convolutions are also used in MobileNetV2 architecture as shown in Fig. 7. In this architecure, there are three convolutional layers where the last two are similar MobileNetV1: a depthwise CONV and a 1×1 pointwise CONV layer. If depthwise convolution layer perform same task as MobileNetV1, the 1×1 layer now attachs with a new job, opposite to before. It makes the quantity of channels smaller.

The DenseNet architecture could be a new (Fig. 8a), logical extension of ResNet which simplifies the connectivity pattern between layers introduced within the other architectures: High-way Networks [19], Residual Networks [20], Fractal Networks [21]. DenseNet has several compelling advantages: DenseNet needs less parameters than Resnet but will have same accuracy; DenseNet is very effective against overfitting; the ability to reuse features more efficiently, etc.

Unlike other models, NasNetMobile (Fig. 8b) wasn't been designed manually. Instead, a reinforcement learning technique called AutoML [22] was wont to generate this model, specifically to perform well on the ImageNet [23] dataset. This technique searches for the most effective CONV layer (cell) on alittle dataset and so transfers the block to a bigger dataset. By changing the quantity of CONV cells and also the number of filters in CONV cells, different versions of NASNet were developed. during

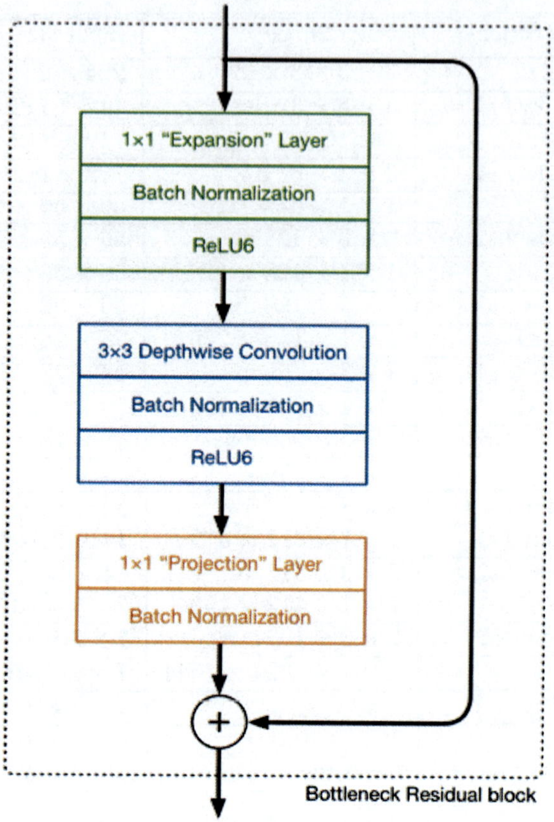

Fig. 7. A MobileNetV2 model.

this paper, we considered the tiniest of all the NASNets, called NASNetMobile, which targets mobile devices.

To use transfer learning method, the convolutional layers which stands before the FC layers in model architecture are initiated. Then it's run on the training and validation data to export the output features. When finishing, new FC layers (Fig. 9) are trained on top of those output features. The output is a vector that represents the probability distributions of three classes: Brown Spot, Bacterial blight, Leaf Smut. The training task is performed using the Keras framework [26], with loss function is Categorical Cross-entropy and optimizer algorithm is Adam [24] or RAdam [25].

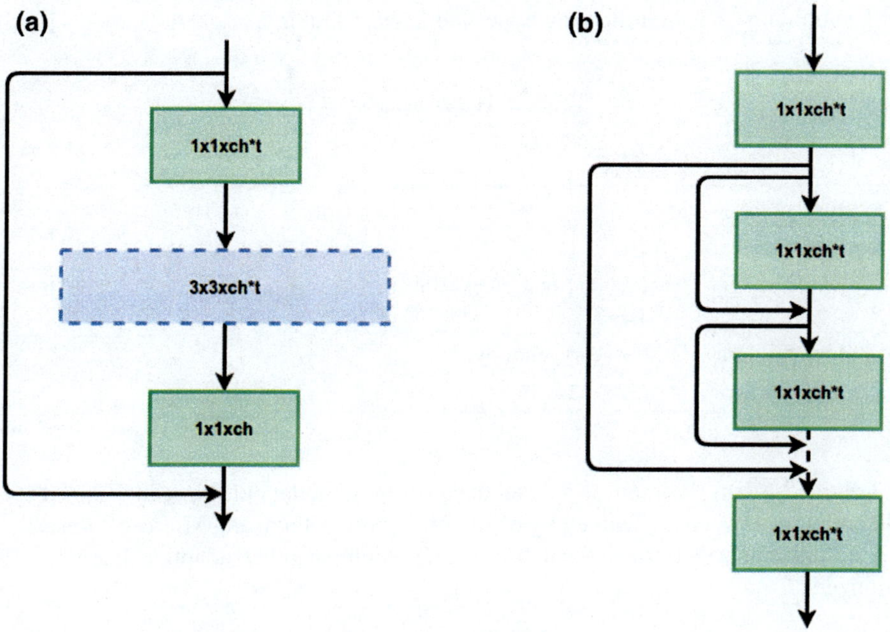

Fig. 8. a) DenseNet model; b) NasNetMobile model.

Layer (type)	Output Shape	Param #
global_average_pooling2d_4 (	(None, 1024)	0
dropout_4 (Dropout)	(None, 1024)	0
batch_normalization_3 (Batch	(None, 1024)	4096
dense_3 (Dense)	(None, 1024)	1049600
dropout_5 (Dropout)	(None, 1024)	0
batch_normalization_4 (Batch	(None, 1024)	4096
dense_4 (Dense)	(None, 3)	3075

Fig. 9. Customized FC layers architecture.

3 Experiment and Evaluation

3.1 Environment and Hyperparameters Setup

This work uses a computer with the configurations: Intel® Core™ i7-7700HQ CPU, 2.80 GHz, 16 GB of RAM and a GeForce GTX 1050 GPU with 4 GB of VRAM.

The training uses the following hyper-parameters (Table 2):

Table 2. Hyper-parameters

No	Hyper-parameter	Value
1	Batch size	4
2	Epoch	20
3	Optimizer	Radam (total_steps = 5000, warmup_proportion = 0.1, min_lr = 1e−5)
4	Loss function	Category_entropy
5	Image dim	(224, 224, 3)

Rectified Adam (RAdam) is a novel deep learning model optimizer introduced by a collaboration between the University of Illinois, Georgia Tech, and Microsoft Research. This optimizer corrects the variance issues apparent in other adaptive learning rate optimizers.

3.2 Results

The model's size, the training loss, the validation loss, the training accuracy and the validation accuracy are shows as in bellow Table 3 and graphs (Figs. 10, 11, 12, 13):

Table 3. The models' size after training

No.	Model	Size
1	MobileNet	37.748 MB
2	MobileNetV2	32.401 MB
3	NASNetMobile	49.307 MB
4	DenseNet121	68.242 MB

From the above graphs, it is observed that:

- MobileNetV1: The accuracy is quite high with 81.87% for training and 81.25% for validation. However, it also shows a little overfitting.
- MobileNetV2: The accuracy is lower, (training: 82.39%, validation: 67.97%), and there is overfitting, too.
- DenseNet121: The accuracy is highest among the models (training: 82.09%, validation: 77.00%). Overfitting appears quite much
- NasNetMobile: The accuracy is lowest (training: 80.56%, validation: 42,36%). Validation loss oscillates and have great value.

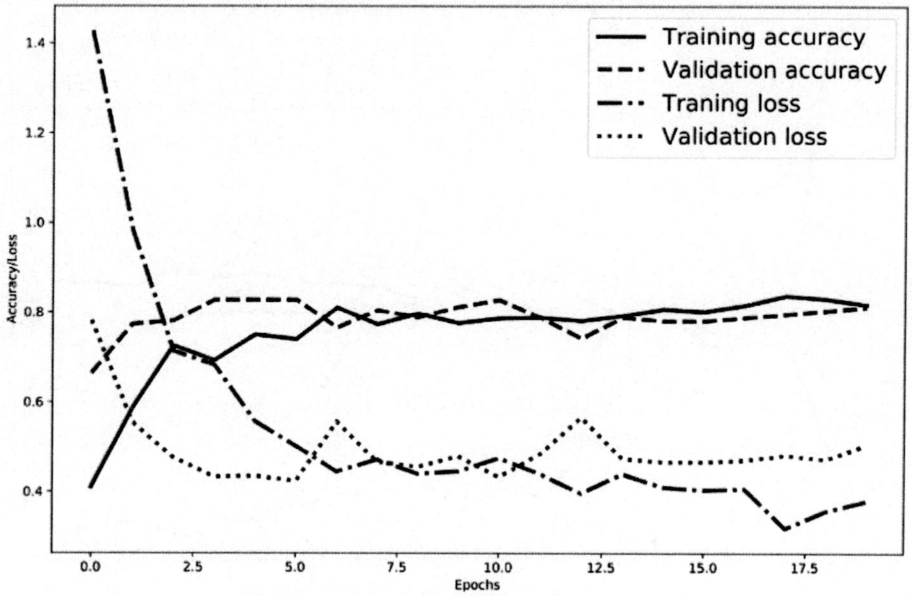

Fig. 10. Results of the MobileNetV1 model.

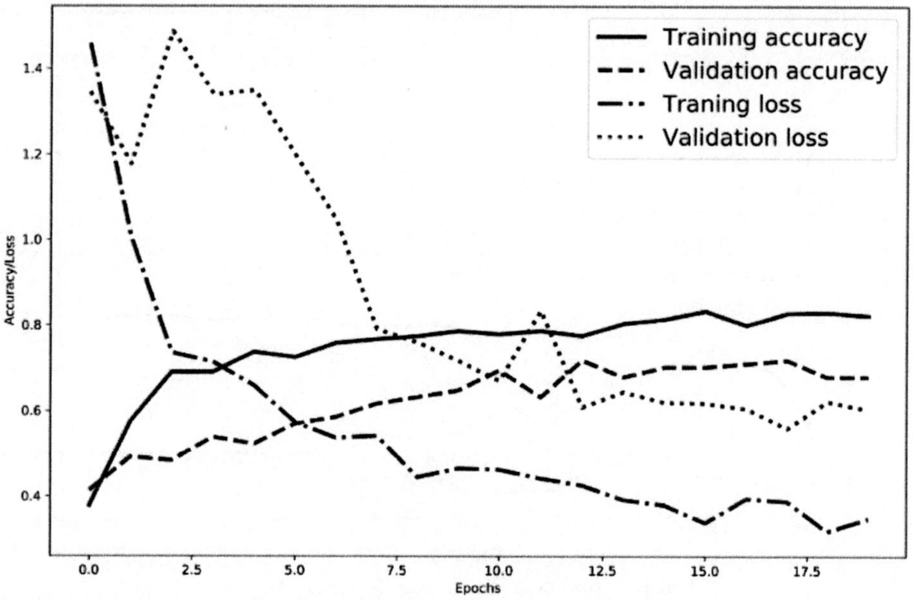

Fig. 11. Results of the MobileNetV2 model.

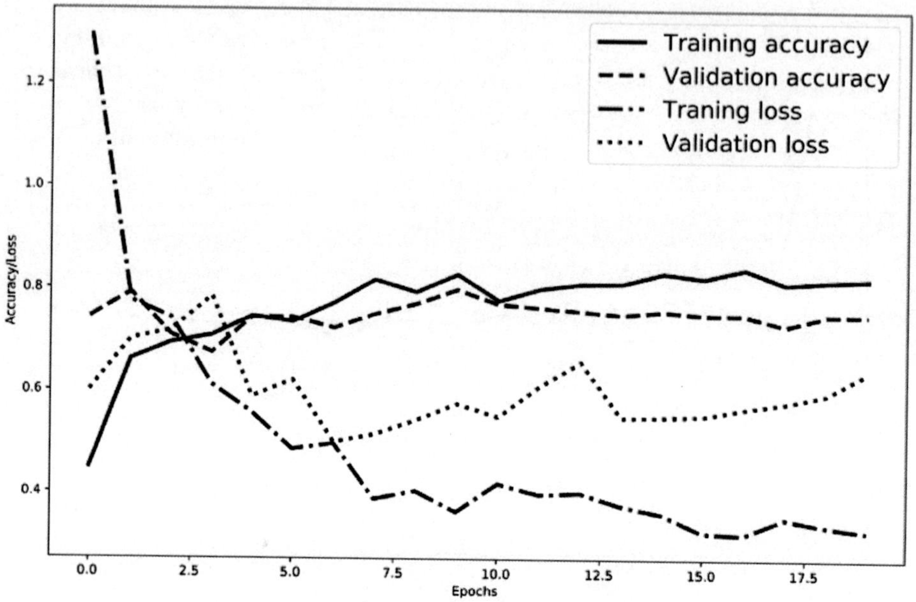

Fig. 12. Results of the Dense-Net121 model.

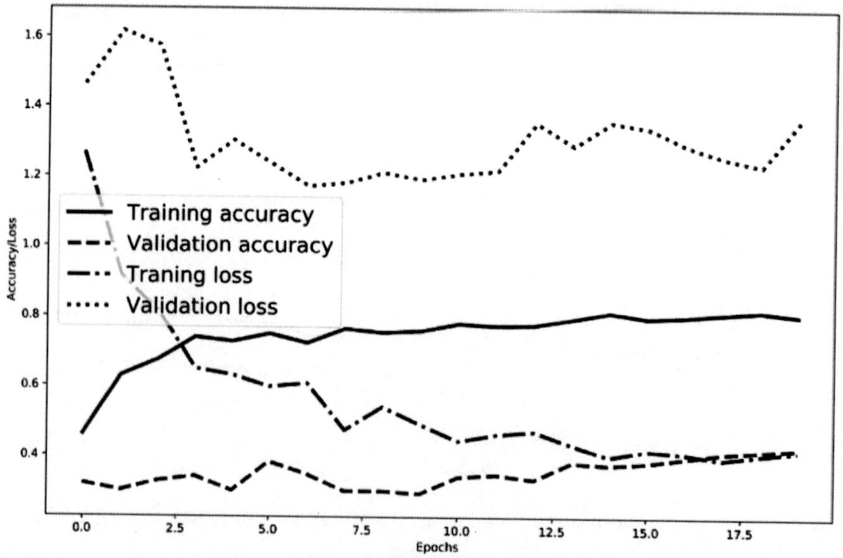

Fig. 13. Results of the NasNetMobile model.

4 Conclusion and Perspective

This paper compares the accuracy of models of deep learning targeting mobile devices on a rice diseases leaf dataset.

Experimental results have shown that the MobileNetV1 model size isn't much bigger than MobileNetV2's smallest size, but its accuracy is that the highest among the models. This model is capable of being employed on many platforms like the online, mobile, and is that the easiest to place into practical applications.

This work can be used as reference for the matter of Pattern Recognition [26–28]. It may use for several domains in Deep learning, for instance, CNN, Computer Vision, Parameterized Learning in Deep learning, etc.

References

1. Anthonys, G., Wickramarachchi, N.: An image recognition system for crop disease identification of paddy fields in Sri Lanka. In: Proceedings of the International Conference on Industrial and Information Systems (ICIIS), pp. 403–407. IEEE (2009)
2. Yao, Q., Guan, Z., Zhou, Y., Tang, J., et al.: Application of support vector machine for detecting rice diseases using shape and color texture features. In: Proceedings of the International Conference on Engineering Computation, pp. 79–83 (2009)
3. Maharjan, G., Takahashi, T., Zhang, S.H.: Classification methods based on pattern discrimination models for web-based diagnosis of rice diseases. J. Agric. Sci. Technol. **1**(1), 48–56 (2011)
4. Phadikar, S., Sil, J., Das, A.K.: Classification of rice leaf diseases based on morphological changes. Int. J. Inf. Electron. Eng. **2**(3), 460–463 (2012)
5. Orillo, J.W., Cruz, J.D., Agapito, L., et al.: Identification of diseases in rice plant (oryza sativa) using back propagation Artificial Neural Network. In: Proceedings of the International Conference on Humanoid, Nanotechnology, Information Technology, Communication and Control, Environment and Management (HNICEM), pp. 1–6. IEEE (2014)
6. Staravoitau, A.: Traffic sign classification with a convolutional network. Pattern Recognit. Image Anal. **28,** 155 (2018). https://doi.org/10.1134/S1054661818010182
7. Raja, C., Balaji, L.: An automatic detection of blood vessel in retinal images using convolution neural network for diabetic retinopathy detection. Pattern Recognit. Image Anal. **29,** 533 (2019). https://doi.org/10.1134/S1054661819030180
8. Hung, P.D., Su, N.T., Diep, V.T.: Surface classification of damaged concrete using deep convolutional neural network. Pattern Recognit. Image Anal. **4**, 676–687 (2019)
9. Arivazhagan, S., Ligi, S.V.: Mango leaf diseases identification using convolutional neural network. Int. J. Pure Appl. Math. **120**(6), 11067–11079 (2018)
10. Howard, A.G., Zhu, M., Chen, B., et al.: MobileNets: efficient convolutional neural networks for mobile vision applications. arXiv preprint arXiv:1704.04861v1 (2017)
11. Sandler, M., Howard, A., Zhu, M., et al.: MobileNetV2: inverted residuals and linear bottlenecks. arXiv preprint arXiv:1801.04381 (2018)
12. Huang, G., Liu, Z., Maaten, L., et al.: Densely connected convolutional networks. arXiv preprint arXiv:1608.06993 (2016)
13. Zoph, B., Vasudevan, V., Shiens, J., et al.: Learning transferable architectures for scalable image recognition. arXiv preprint arXiv:1707.07012 (2017)
14. UCI: Machine Learning Repository. https://archive.ics.uci.edu/ml/datasets/Rice+Leaf+Diseases. Accessed 20 Oct 2020
15. Prajapati, H.B., Shah, J.P., Dabhi, V.K.: Detection and classification of rice plant diseases. Intell. Deci. Technol. **11**(3), 357–373 (2017). https://doi.org/10.3233/IDT-170301(2017)
16. Shah, J.P., Prajapati, H.B., Dabhi, V.K.: A survey on detection and classification of rice plant diseases. In: Proceedings of the Current Trends in Advanced Computing (ICCTAC), pp. 1–8 (2016)

17. Uniqtech: Understand the Softmax Function in Minutes. https://medium.com/data-science-bootcamp/understand-the-softmax-function-in-minutes-f3a59641e86d. Accessed 20 Oct 2020
18. Kumar, R., Gredd, K., Schmidhuber, J.: Highway networks. arXiv preprint arXiv:1505.00387 (2015)
19. He, K., Zhang, X., Ren, S., et al.: Deep residual learning for image recognition. arXiv preprint arXiv:1512.03385 (2015)
20. Larsson, G., Maire, M., Shakhnarovich, G.: FractalNet: ultra-deep neural networks without residuals. arXiv preprint arXiv:1605.07648 (2017)
21. He, X., Zhao, K., Chu, X.: AutoML: a survey of the state-of-the-art. arXiv preprint arXiv:1908.00709 (2019)
22. Deng, J., Dong, W., Socher, R., et al.: ImageNet: a large-scale hierarchical image database. In: Proceedings of the IEEE Conference on Computer Vision and Pattern Recognition, Miami, FL, USA (2009)
23. Kingma, D.P., Ba, J.: Adam: a method for stochastic optimization. arXiv preprint arXiv:1412.6980 (2017)
24. Liu, L., Jiang, H., He, P., et al.: On the variance of the adaptive learning rate and beyond. arXiv preprint arXiv:1908.03265v1 (2019)
25. Keras-team: Keras: the Python deep learning library. https://keras.io/. Accessed 20 Oct 2020
26. Hung, P.D., Su, N.T.: Incremental learning for classifying vietnamese herbal plant. In: Dang, T.K., Küng, J., Takizawa, M., Chung, T.M. (eds.) FDSE 2020. CCIS, vol. 1306, pp. 434–442. Springer, Singapore (2020). https://doi.org/10.1007/978-981-33-4370-2_31
27. Hung, P.D.: Counting people using images from two low cost webcams. In: Dang, T.K., Küng, J., Takizawa, M., Bui, S.H. (eds.) Future Data and Security Engineering. LNCS, vol. 11814, pp. 688–695. Springer, Cham (2019). https://doi.org/10.1007/978-3-030-35653-8_48
28. Hung, P.D., Loan, B.T.: Automatic vietnamese passport recognition on android phones. In: Dang, T.K., Küng, J., Takizawa, M., Chung, T.M. (eds.) Future Data and Security Engineering. Big Data, Security and Privacy, Smart City and Industry 4.0 Applications. CCIS, vol. 1306, pp. 476–485. Springer, Singapore (2020). https://doi.org/10.1007/978-981-33-4370-2_36

Robotic Hand Movement Using If-Statement Technology

Ansam A. Alwan[1] (✉), Zinah T. Nayyef[1], and Mohammed S. Fadose[2]

[1] Dijlah University College, Baghdad, Iraq
{ansam.ahmed,Zeena.tariq}@duc.edu.iq
[2] Baghdad, Iraq

Abstract. Robotic hand systems fall under the areas of artificial intelligence (AI) and machine learning (ML), and are an important field nowadays. Today's technology shows a great interest in robotic systems, evident in the number of studies being conducted in this area. There are numerous studies in various fields on robotic systems, with the aim to facilitate human life. In this work, a robot hand was designed that is capable of replicating finger movements by relying on flexible sensors attached to a wearable glove. The sensor is able to measure finger angle with high accuracy, albeit inexpensively. Thanks to the manual robotic prototype created in this study using different materials and mechanisms, it can be adopted in areas that are unsafe for human health. Through the gloves, the angle data on the sensors and servo motor position information are transmitted to the robotic handheld system through a digital AI software. Determinations related to the robot hand control are carried out with an Arduino card. Five servo motors, one for each finger, constrain the mechanical hand. These servo engines can be moved if necessary. The Main Control Unit (MCU) is attached to these servo engines. Arduino is used in its MCU. We used the IF statement algorithm to allow the robotic hand's fingers to move smoothly in a variety of angles, despite the device's resistance, and at a low cost, so that it could be made available to anyone.

Keywords: Artificial Intelligent · Main Control Unit (MCU) · Machine learning (ML) · Robotic hand · Arduino · Robotic Arm · Five Fingered Robotic Hand (FFRH)

1 Introduction

Arduino is an easy-to-use assistive robot in medical procedures and hazardous conditions, with the aim to improve the independence of people with physical disabilities. The motor vehicle and arm movement are controlled by hand movements with an Arduino device. A robotic arm displacement is monitored so that it can be used to locate and position objects from any source to any destination. This device enables physically disabled individuals to communicate with a robot by recognizing hand movements [7].

For over three decades, the concept of robotic hands has been an active field of study. In the literature, over 70 hand designs can be found, and there is a comparable number of designs in the patent literature. Descriptions can be found in the vast body

M. Al-Emran et al. (Eds.): ICETIS 2021, LNNS 299, pp. 149–159, 2022.
https://doi.org/10.1007/978-3-030-82616-1_14

of design research that has accrued over the decades. Among the earliest were some of the most popular designs, and a minimalistic non-anthropomorphic hand was the most prominent. Several hand designs were published every year, not long afterwards. The number of new hand designs peaked in 2001, with nine new designs published that year, and a steady stream of new designs have been produced since then. Many of these hands are designed to imitate the human hand's ability by an having identical degree of freedom (DOF), range of motion, and tie size. Despite the large number of robot hand prototypes, there are very few commercially available robot hands in the market. This is partly because a robot's hand, to be considered commercially viable, must excel in several performance metrics, including handling capacity, grasp capacity, sensing ability, modularity, durability, size, mass, speed, and cost [8].

To enhance the regulation of robotic hands and object grasping, neural networks and deep learning models have been successfully implemented. A data-driven strategy for controlling robotic fingers was proposed to assist users in manipulating bi-hand objects. The trained neural network is used to regulate the robotic movement of the finger on ten bi-manual activities, such as running a tablet, holding a large object, gripping a bottle, and opening a bottle cap [9].

The main purpose of this work is to develop and implement a Five Fingered Robotic Hand (FFRH) that can be used for a variety of objects, and provides a simple reflective understanding. Additionally, the technical design allows for moving the fingers in various angles to make the hand's movement more flexible and easier to capture objects.

2 Literature Review

The study by Gosselin et. al. (2008) presents the development and experimental validation of a 15-degree-of-freedom stereoscopic robotic hand with a single actuator. The design is based on fixed fingers reinforced with a series of differential mechanisms to treat lack of activity between the fingers [1]. Another assessment composed by F. Lotti et. al. (2002) focused on finding solutions for the mechanical structure of fingers in robotic hands, and a different design approach was proposed to consider finger structures composed of rigid bonds connected by flexion hinges. The folds within each finger can be guided. A simplified model has been presented for one of these structures, and the feasibility of this concept was assessed through preliminary simulation [2]. O'Toole et al. (2007) proposed the mechanical design of synthetic SMA beam actuators in the signage area, with a 12° range of freedom of movement. The design focuses on ease of assembly and replacement of SMA wires with a modular approach. The prosthesis can either be used as a separate prosthesis, or it can be paired with standard wrist and forearm prostheses to form a complete lower arm prosthesis [3]. Raghavan et. al. (2016) introduced a robotic hand that is distinguished by the way the system works, and depends on the behavior of photoelectric signals (EMG) taken from the human body. EMG signal capture ability has been added to power DC motors. The EMG signal was then filtered and calculated with MATLAB to produce a value that reflects the EMG signal's amplitude, which was then used to control the motor's orientation [4]. J. Wang et. al. (2013) discusses the issues related to surface design EMG signal amplification and filter circuits. The proposal circuit was merged using a PC-based EMG data acquisition system through three electrodes. This research focused on amplifying and designing a filter circuit to process surface EMG signals, and aims to install amplifiers and filters inside a robotic hand with a limited area for command and control robot hand movement [5]. This work describes the AR III side, which is based on a passive mechanism, as well as the associated EMG monitoring system. Only three motors are required to move the 5-finger, 15-joint hand. Without the use of several complicated motors and control methods, this function helps the hand to adapt to a variety of grips [6].

3 Arduino and Servos Devices

The Arduino is a low-cost, open-source microcontroller that can be programmed, disabled, and reprogrammed with ease. It is an open source computing platform focused on simple microcontroller boards for designing and programming electronic devices. It can act as a mini computer, accepting inputs and adjusting outputs for a wide range of electronic devices, much like other microcontrollers. With the aid of various Arduino shields, it is also capable of receiving and transmitting data over the internet. Arduino generates code with the Arduino IDE using the hardware and software known as the Arduino development board (Integrated Development Environment). These microcontrollers are equipped with Atmel's 8-bit AVR or 32-bit ARM microcontrollers, and can be easily programmed using the Arduino IDE's C or C++ language. Electronics-related people are now increasingly emerging and embracing the position of Arduino for their own ventures. By connecting it to a computer via USB, this development board can also be used to upload (burn) new code. The Arduino IDE is a simplified integrated framework that allows users to write Arduino programs in C, C++, Python, or other languages on standard computers. The decision to develop a particular Arduino board requires a series of surveys to assess its specifications and capabilities, especially with many Arduino boards in the market that can be used to implement a project based on their respective applications [12].

Servomotors are small, powerful motors used in a variety of products, including robots and helicopters. An electric motor, a feedback voltage meter connected to an output shaft, and a joystick are the three primary components of a servo. By keeping track of its current angular position, the servo rotates to odd angles. This allows it to rotate to specific angles by tracking its current angular location, and since the servo could rotate to a specific position rather than a specific speed, it is very useful in prosthetic devices because it could remove time variables by very precise movements [11].

4 Arduino Programming

The Arduino was programmed on the Arduino website using the development environment available. The main part of the project code was to locate the servo by which to determine the movement of the fingers. The servo library provides good control over the positioning of the servo by containing the functions of connecting the servos with the pins and setting their angles. The *write ()* function sets the angular position of the servo to a value between 0 and 180°, and the *attach ()* function connects each servo to a specific pin on the Arduino. The amount of finger flexion is determined by the servo position; a lower angle relaxes the finger, while a higher angle bends it toward the palm. Movements are made that correspond to the number of times the finger is bent. When the hand makes this movement, this is done by adjusting each servo to a specific angle [11].

5 Robotic Hand

A robotic hand is a complex electromechanical device with several moving parts. The electrical components and the mechanical structure that allows it to move are the two main components of the device. The human hand is one of the most complicated body parts, second only to the brain. We can understand why its activity has piqued the interest of previous researchers. In recent decades, the focus of research has not only been performed in the medical field, but also in engineering. The robot hand has a movement structure similar to a human hand, and each finger's joints can be regulated individually. Antagonistic tendons that run to a motor for each finger joint actuate the fingers. The problem with these adaptable mechanisms is that each finger's joint must be measured for a particular position or torque. To gain power, the five-fingered hand, which has three joints on each finger, requires twelve desired positions. This flexibility is required for manipulation, but gripping does not require the robot to control all finger joints independently [10].

Given the challenge of controlling a robot's hand from a distance with precision and versatility when a wide range of degrees of freedom is necessary, a "Artificial Human Hand Model" using hand gloves was developed that can precisely follow the movements of a human hand. Trajectories for different types of articulated movements were developed, and a learning model was proposed. Wearing gloves decreases sensor noise and allows for a more flexible response. Data was measured directly with hand and/or arm joint angles and spatial position, rather than using gesture-based interaction, which is transferred to the robotic arm via the Arduino. This ensures simple and fast processing in real time, without any delay of execution. The robotic hand was able to go up and down vertically, thanks to a movable foundation under the hand and a rack-and-pinion mechanism. These characteristics can be very effective in preventing danger to human users in geographical, radioactive, poisonous, and chemically hazardous areas [13].

6 Application of Robot Hands

In addition to humanoid robots, humanoid robot hands are used in various applications such as multi-fingered haptic interfaces, electromyogram prosthetic hands, hand rehabilitation support systems, robotic breast engorgement measuring systems, breast cancer palpation training systems, and hand rehabilitation training systems [14].

7 Proposed System Architecture

In this work, we designed a robotic arm using different parts of hardware. These pieces connected with each other in order to make the movement of the arm fingers flexible using different angels, and enable a stronger grip. We used a micro metal servo SG90, which has the role of controlling the rotation angle, as shown in Fig. 1. We also utilized a breadboard medium for building and testing circuits rapidly, prior to settling any circuit plan. The breadboard has various openings into which circuit portions such as ICs and resistors can be inserted, as exhibited in Fig. 2. We used Arduino Nano for a number of purposes, and the findings are as follows: The Arduino Nano has a 32 Kb stripe memory. It comes with a preinstalled bootloader that uses 2 kb of shoot memory. This microcontroller board's SRAM memory is 8 kb, as shown in Fig. 3. We also used a square, which is used to reduce power flow, adjusts signal levels, segregates voltage, contains propensity amazing bits, and terminates transmission lines, among other functions, as shown in Fig. 4. Finally, we used a flex sensor that is 11.4 cm in size, which is used in wide locales of assessment from PC interfaces, reclamation, security systems and even music interfaces. It is notable among understudies and hobbyists, as exhibited in Fig. 5.

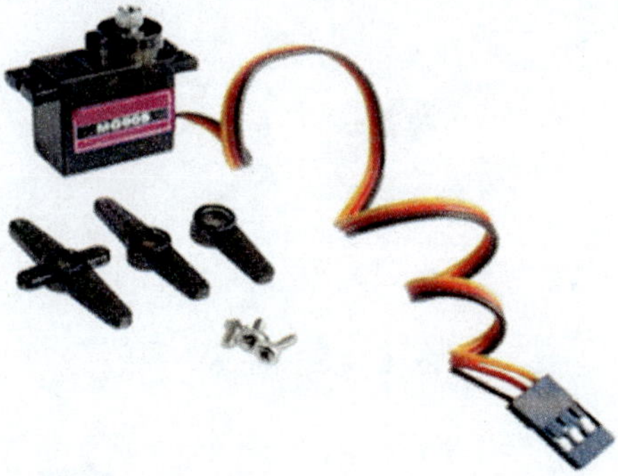

Fig. 1. Micro metal servo.

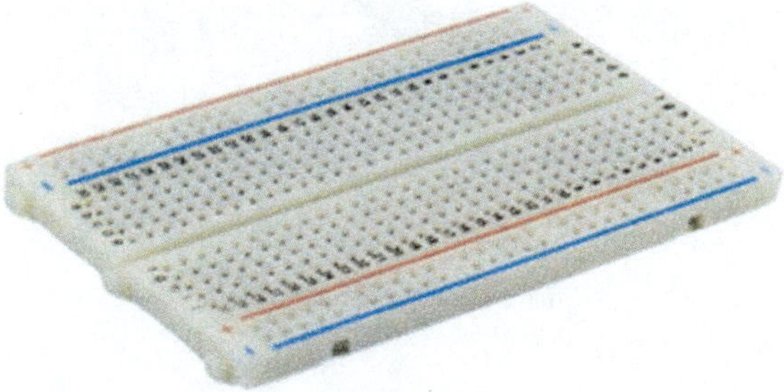

Fig. 2. Bread board medium.

Fig. 3. Arduino nano.

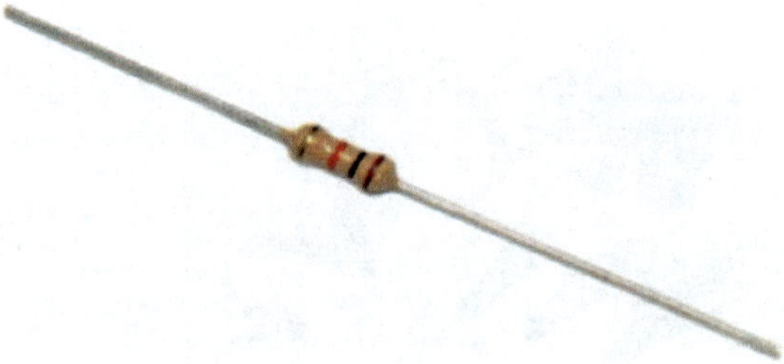

Fig. 4. Resistance.

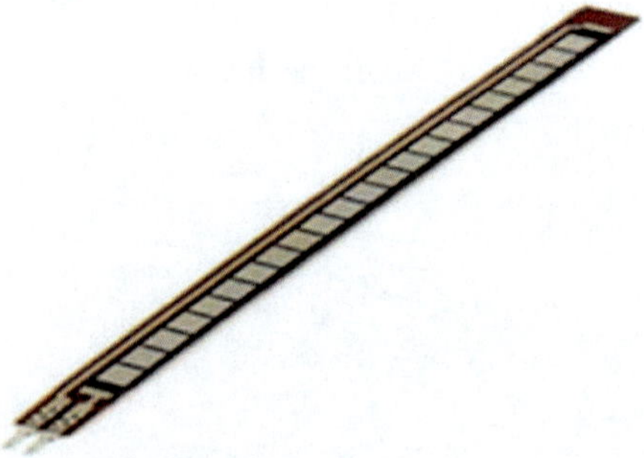

Fig. 5. Flex sensor.

8 Methodology and Results

The IF-statement technology was used to obtain the best movement of the fingers in order to overcome friction and noise. We used two strategies. One of them is for hand movement, as shown in Algorithm (1), and the other is for the test sensor, as shown in Algorithm (2).

Hand_New Algorithm
Step 1: We used five servos for each finger (from 1 to 5).
Step 2: Configuring analog inputs for the sensors (flexpin "1, 2, 3, 4, 5").
Step 3: Configuring PWM pins for servos where: servo1.attach(3)for pinky finger servo2.attach(5) for ring figure servo3.attach(6) for middle finger servo4.attach(9) for index finger and last servo5.attach (11) for thumb finger.
Step 4: Declaring servo position and flex position.
Step 5: Flex position is equivalent to the sensor esteem .
Step 6: Mapping the sensor incentive to servo position that is compose
step 7: Servo position is an incentive between max (servo) and min (servo)
Step 8: Printing the servo position.

Test Sensor Algorithm
Step 1: flexpin1 = 0.
Step 2: Serial begins with (9600). The angle instructions from computer are read.
Step 3: Print the flex position angle.

Figure 6 shows the connection of hardware devices (flex sensor, bread board medium and Arduino nano) to form a single sensor and determine the length of its range. In addition, Fig. 7shows the connection of the servo, which connects between the Arduino and the servo.

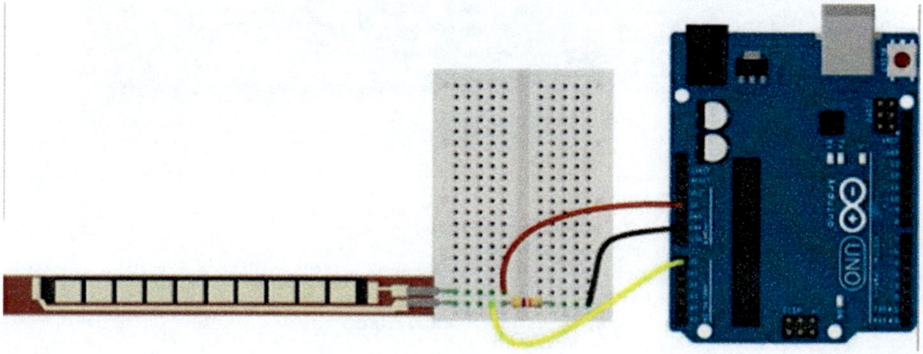

Fig. 6. Single sensor connection.

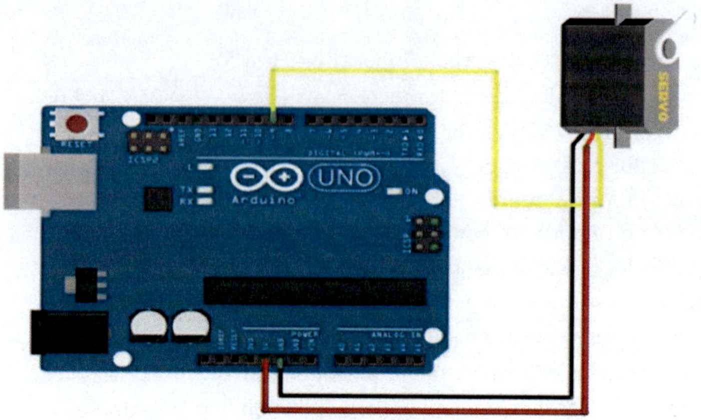

Fig. 7. Servo connection.

In the proposed arm design, plastic pieces made by a 3D printer were used for the hand, arm and fingers. The arm was attached with the hand by using a diameter bolt. Aluminum wires were used to attach each joint in all fingers. Plastic wires were used to attach the fingers with the servo motors that have already been placed in the arm. Rubber wires were used to help each finger to return to its original state after moving (Fig. 8).

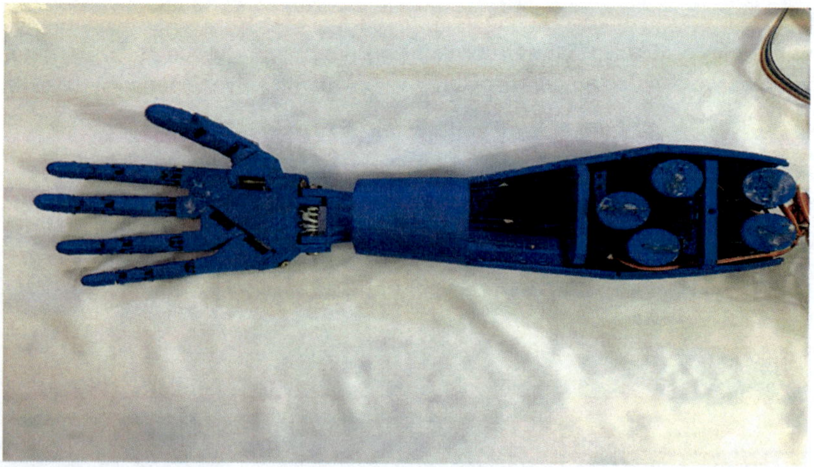

Fig. 8. Arm robotic design.

Each movement in the hand of the robotic arm was controlled by a person' hand that dressed palm, where each movement in the person's fingers made changes in the robotic arm. Each finger in the palm is connected with wires that are connected to the software using the C++ language. The connection of the palm is shown in Fig. 9.

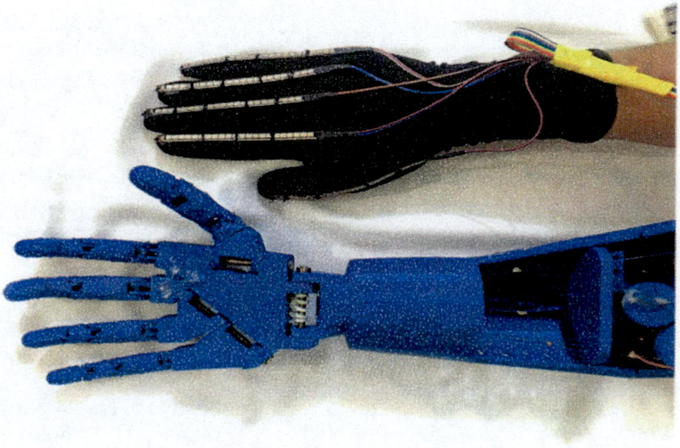

Fig. 9. The connection of the palm.

There are several movements of the fingers that are shown in the figure below. When the palm moves the thumb figure, the thumb finger in the robotic arm changes with angels, depending on the movement of the palm, as shown in Fig. 10.

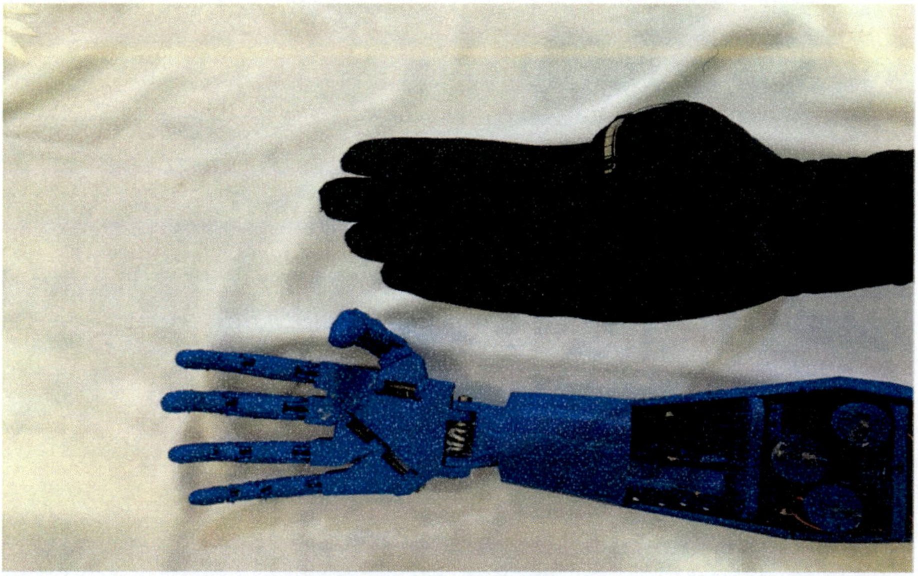

Fig. 10. Thumb finger movement.

When we move two fingers in the palm, then the two fingers in the robotic arm move with the same movement of the palm, as shown in Fig. 11.

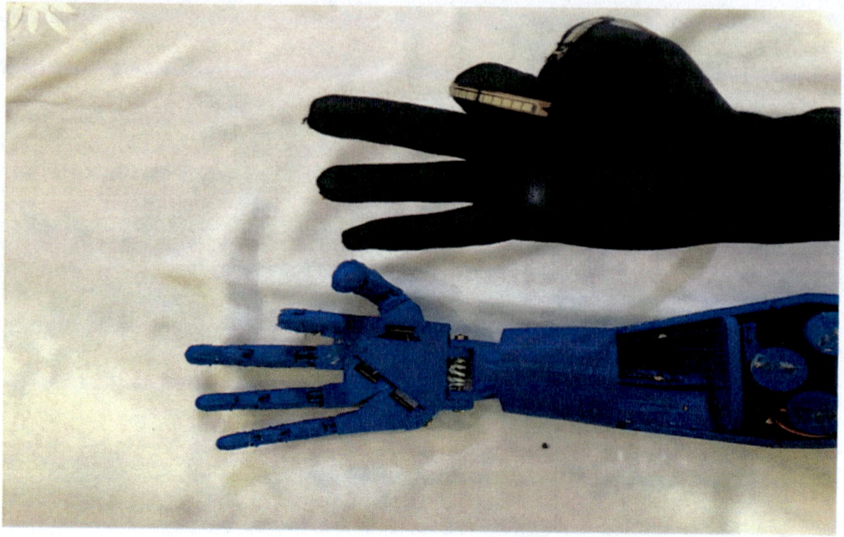

Fig. 11. Thumb and the index fingers movement.

In Fig. 12, all fingers were moved as the same movement of the palm of the person used it. Any changes in the palm make changes in the robotic hand.

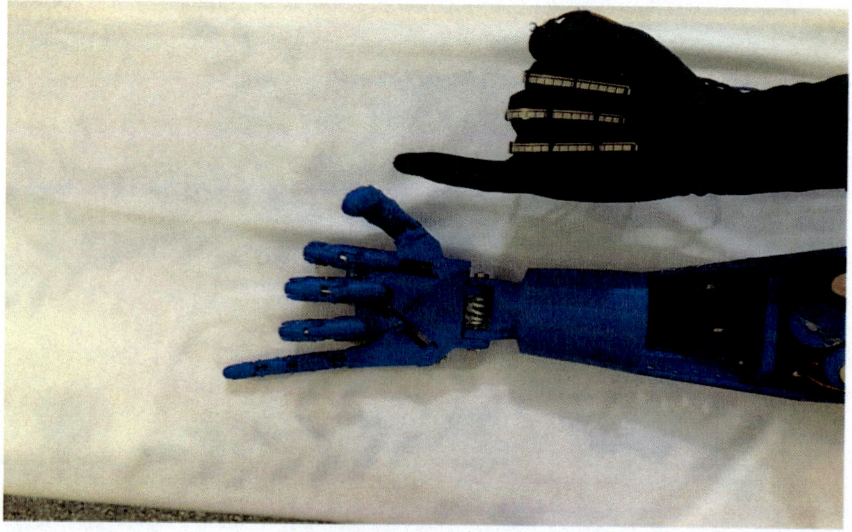

Fig. 12. Five finger movement.

9 Conclusion

A robotic hand with five fingers was developed, constructed, and tested. The hand was designed to be inexpensive, mechanically simple, and easy to control using a simple algorithm (IF-statement method). The proposed system allows the hand's differential mechanism to adjust to finger movements, allowing for the use of a clear and simple control algorithm. The goal of this hand design is to show that inexpensive mechanisms can achieve reliable hand movements and grasping.

References

1. Gosselin, C., Pelletier, F., Laliberte, T.: An anthropomorphic underactuated robotic hand with 15 dofs and a single actuator. In: IEEE International Conference on, pp. 749–754. IEEE (2008)
2. Lotti, F., Vassura, G.: A novel approach to mechanical design of articulated fingers for robotic hands. In: IEEE/RSJ International Conference on, vol. 2, pp. 1687–1692. IEEE (2002)
3. O'toole, K.T., McGrath, M.M.: Mechanical design and theoretical analysis of a four fingered prosthetic hand incorporating embedded SMA bundle actuators. Eng. Technol. (2007)
4. Raghavan, A., Joseph, S.: EMG analysis and control of artificial arm. Int. J. Cybern. Inf. (IJCI) **5**(2) (2016)
5. Wang, J., Tang, L., Bronlund, J.E.: Surface EMG signal amplification and filtering. Int. J. Comput. Appl. (2013)
6. Yang, D., et al.: An anthropomorphic robot hand developed based on underactuated mechanism and controlled by EMG signals. J. Bionic Eng. (2009)
7. Meenatchi Sundaram, N., Shiva Rama Krishnan, R., Suraj Kamal, T.K., Rajeswari, J.: Robotic vehicle movement and arm control through hand gestures using Arduino. Int. Res. J. Eng. Technol. (IRJET) **05**(04) (2018)
8. Quigley, M., Salisbury, C., Ng, A.Y., Kenneth Salisbury, J.: Mechatronic design of an integrated robotic hand. Int. J. Rob. Res. **33**(5), 706–720 (2014)
9. Devaraja, R.R., Maskeliú, R., Damaševičius, R.: Design and evaluation of anthropomorphic robotic hand for object grasping and shape recognition. MDPI an Open Access Journal (2021)
10. Ramaiah, P.S., Venkateswara Rao, M., Satyanarayana, G.V.: A microcontroller based four fingered robotic hand. Int. J. Artif. Intell. Appl. (IJAIA) **2**(2) (2011)
11. Bonini, N., Iyer, N., Kim, D., Mathison, K., Wellons, L.: Robotic hand in motion using arduino controlled servos. New Jersey Governor's School of Engineering and Technology (2014)
12. Louis, L.: Working principle of arduino and using it as a tool for study and research. Int. J. Control Autom. Commun. Syst. (IJCACS) **1**(2) (2016)
13. Noor, S., Shohan, J.A., Waresi, M.S.: Real time hand movement controlled robotic arm for risk prevention. In: IEEE Region 10 Humanitarian Technology Conference (R10-HTC), Dhaka, Bangladesh, 21–23 December 2017
14. Yanagido, G.: Humanoid robot hand and its applied research. J. Rob. Mech. **31**(1) (2019)

Check for updates

African Buffalo Optimization Algorithm Based T-Way Test Suite Generation Strategy for Electronic-Payment Transactions

Julius B. Odili[1](✉), Abdullah B. Nasser[2], A. Noraziah[2,3], Mohd Helmy Abd Wahab[4], and Mashuk Ahmed[2]

[1] Faculty of Science and Science Education, Anchor University Lagos, Lagos, Nigeria
[2] Faculty of Computing, Universiti Malaysia Pahang, 26300 Kuantan, Malaysia
{abdullahnasser,noraziah}@ump.edu.my
[3] Centre for Software Development and Integrated Computing, Pekan, Malaysia
[4] Advanced Communication Research Center, Universiti Tun Hussein Onn Malaysia, Parit Raja, Malaysia
MCS20004@sudent.ump.edu.my

Abstract. The use of meta-heuristics in Combinatorial Interaction Testing (CIT) is becoming more and more popular due to their effectiveness and efficiency over the traditional methods especially in authenticating electronic payment (e-payment) transactions. Concomitantly, over the past two decades, there has been a rise both in the development of metaheuristics and their application to diverse theoretical and practical areas including CIT in e-payments. In the implementation of t-way strategies (the *t* is used to represent the interaction strength), mixed results have been reported; some very exciting but, in other cases, the performance of metaheuristics has been, to say the least, below par. This mixed trend has led many researchers to explore alternate ways of improving the effectiveness and efficiency of metaheuristics in CIT, hence this study. It must be emphasized, however, that available literature indicates that no particular metaheuristic testing strategy has had consistent superior performance over the others in diverse testing environments and configurations. The need for effectiveness, therefore, necessitates the need for algorithm hybridization to deploy only the component parts of algorithms that have been proven to enhance overall search capabilities while at the same time eliminating the demerits of particular algorithms in the hybridization procedure. In this paper, therefore, a hybrid variant of the African Buffalo Optimization (ABO) algorithm is proposed for CIT. Four hybrid variants of the ABO are proposed through a deliberate improvement of the ABO with four algorithmic components. Experimental procedures indicate that the hybridization of the ABO with these algorithmic components led to faster convergence and greater effectiveness superior to the outcomes of existing techniques, thereby placing the algorithm among the best when compared with other methods/techniques.

Keywords: African Buffalo Optimization · e-payment system · T-way testing · Software testing · Combinatorial problem · Optimization techniques

M. Al-Emran et al. (Eds.): ICETIS 2021, LNNS 299, pp. 160–174, 2022.
https://doi.org/10.1007/978-3-030-82616-1_15

1 Introduction

With the massive deployment of electronic payment systems, especially during this period of Lockdown across many countries of the world as a result of the New Corona Virus, otherwise called COVID-19, the need to examine the software used in e-payments have not been greater in human history than at present. Different money-transaction organizations, the world over have developed several payment platforms/software, many of these software have, to say the least, been quite successful. However, the need for continuous improvements cannot be over-emphasized. The need to ensure foolproof transactions in our monetary systems has necessitated several research investigations to ensure stability, reliability and security of financial transactions in our financial institutions. This has led to the involvement of artificial intelligence techniques to assist is ascertaining the reliability of existing financial software. One of the areas attracting researcher's attention in this regard is the use of optimization algorithms is software testing.

Optimization has gradually been entrenched in virtually all aspects of Software Engineering ranging from software requirements engineering to software management, software testing and software refactoring respectively. Optimization, fundamentally, is sometimes defined as the application of economic principles to science and engineering which basically deals with generating the highest amount of output with as little input as possible [1]. In other words, optimization is the process of discovering the best alternative out of a given number of alternatives utilizing limited inputs as much as possible [2]. In the past few decades, there has been a marriage of convenience between optimization and metaheuristic algorithms. Some of the popular metaheuristic algorithms include: Simulated Annealing (SA) [3]; Tabu Search (TS) [5], African Buffalo Optimization [6], Great Deluge (GD) [4];; Ant Colony Optimization (ACO) [3]; and Particle Swarm Optimization (PSO) [7], etc.

Optimization becomes very important in Combinatorial Interaction Testing (CIT) in order to identify the optimal cases from a large range of combinatorial values based on the specified interaction strength (t) since the major task of CIT is to determine the optimal test cases from among a large test suite. Since it may not be possible or efficient to test every possible test combination in a large test suite, there is the need for combinatorial optimization [8]. Combinatorial optimization is relevant in a situation where an exhaustive search is practically impossible or totally inefficient when a choice has to be made, thereby emphasizing the need for optimization search algorithms [9].

In most testing situations (similar to our focus in this study), CIT is a (NP-hard (Non-Polynomial-hard) problem. The term NP-hard problem refers to classes of problems that there exists no algorithm that provides a solution to the problem as well as the verification of such solutions in a polynomial time. Polynomial time is simply a linear time. This further underscore the need for metaheuristics. In response, today, the research community has designed several metaheuristics which have been applied to several NP-hard problems such as the travelling salesman's problem [10]; N-Queen problems; numerical function optimization [11]; and software testing [12] etc. With particular reference to software testing, the PSO, SA, and (HS) [13] have proven to be quite successful. It must, however, be emphasized that in spite of the huge contributions of the above algorithms, there still exists the need for further improvements, hence this paper.

The remainder of this paper is structured thus: section two gives a general overview of t-way testing in addition to discussing its theoretical background; section three presents a review of some existing t-way strategies; section four examines the hybridization of variants of the ABO algorithm for an appropriate t-way test suite generation of an e-payment system; section five is concerned with the experimental investigation of the ABO algorithm for a t-way test suite of an e-payment system and discusses the experimental outcomes while section six presents the conclusions drawn from the study.

2 Explanation of Background and History

2.1 The Need for T-Way Test Suite Generation for E-payment Platforms

Of the several CIT techniques in use, the t-way testing has proven to be one of the most effective combinatorial techniques that can dramatically reduce the size of the test parameters without compromising effectiveness. T-way testing, basically, is a sampling mechanism used to generate test samples with a clear focus on the attitude of interaction system's component parts. To further elaborate on the need for t-way testing in test suite reduction, consider the online payment systems such as MasterCard, Visa etc. Online Payment systems enable customers to carry out banking transactions electronically.

To carry out an electronic payment transaction, a customer is required to simply fill out the online Payment Details form and then submit to the merchant's website. Usually, the transaction form consists of six different inputs, namely: The Payment System (Visa, MasterCard, PayPal, American Express etc.); Name on Card; Expiration Date (MM and YY); Card Number; Card and Card's CVV. In the Transaction Method, a customer is required to choose the particular Payment method from among a number of choices including: "Discover"; "Visa Card"; "American Express"; "MasterCard"; and "PayPal". The next input data is the Name on Card, then the Card Number which takes numeric string values. Following this is the Card Expiration Date, which requires further numeric inputs (i.e. MM takes values 1 ... 12, and YY takes values 16 ... 31). The last compulsory input is the Card CVV which requires another numeric value. An example of the online payment system is presented in Fig. 1.

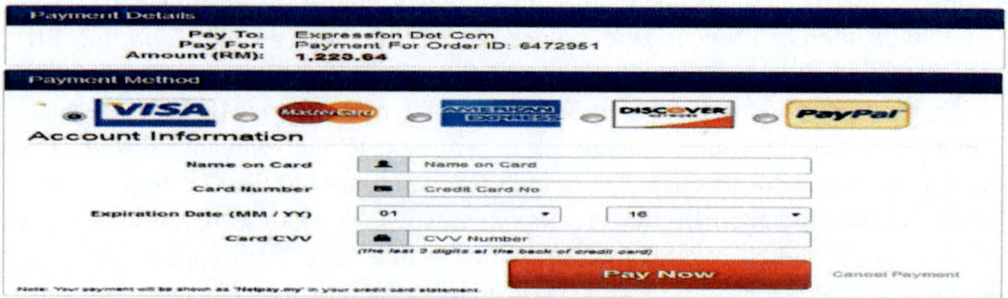

Fig. 1. Electronic payment system

To exhaustively test the electronic payment system, 900 test cases are required [14]. Nevertheless, utilizing a two-way test suite reduces it to only 180 test cases [15]. In general, the test suite T is $n \times m$ array which consists of n rows of the generated test cases and m columns in such a way that each test case becomes a combination of m input values. Thus, whereas the t-way test suite T1 covers all valid pair values of input parameters, one test case incorporates many pairs of input parameter values. Therefore, a good software tester would need to discover an effective test suite T1 which is a subset of T that should have the smallest number of rows and columns with the embedded capacity of unravelling a defect wherever one exists.

In scientific literature, there exist many test suite generation strategies such as: partition testing [16]; random testing [17]; combinatorial interaction testing [19]; etc.

2.2 Theoretical Foundation of T-Way Test Suite Generation

To formulate CIT, many researchers adopt Covering Array (CA). Basically, from literature, any system under test (SUT) is made up of a number of component parts which interact with each other parts /parameters cooperatively with each component making available its values to other components in the system. Please note that throughout this paper, the symbols p, t, and v refer to number of parameters, interaction strength and values, respectively [21, 22].

Similarly, it is important to emphasize that whenever the number of values (v) is the same with the number of parameters (p), the covering array is assumed to be a uniform Covering Array, CA (N, t, v^p). For instance, anytime a covering array of CA (6; 2, 2^4) is made up of six (6) rows of the test cases being generated from the four (4) columns of the appropriate parameters with only two (2) values, then it is a case of a uniform Covering Array. On the other hand, anytime the number of parameters being investigated does not correlate (that is, when. each parameter under consideration generates a different number of values), then such covering array is a the Coverage Array symbol of MCA (N, t, $v_1{}^{p1}$ $v_2{}^{p2}$ $v_3{}^{p3}$.....$v_j{}^{pj}$).

3 A Detailed Explanation of Recent Findings

3.1 Hybrid African Buffalo Optimization for T-Way Test Suite Generation

The primary objective of this section is to present an efficient, effective cum fast t-way strategy for a good test suite generation based on a meta-heuristic algorithm for the authentication of an electronic payment system. The need for hybridization emanates from the effectiveness of hybrid algorithm over the classical algorithms because hybrid algorithm draws from the strength of each of the algorithms in the hybrid [23]. Hybridization is a very successful concept in the development and use of meta heuristic algorithms. In literature, many hybridization methods have been proposed by simply selecting two algorithms to a new one [24]. Even though the exceptional performance of a new hybrid is not guaranteed, some hybridization has been very successful [24].

Therefore, finding an effective way for algorithm-hybridization is still an open issue for researchers [25]. To design an effective hybrid algorithm, there is the need for a thorough grasp of the algorithms' components and employ the concept of component grafting [26]. In this study, first, a new t-way suite test generation strategy is proposed; called Buffalo Strategy (BS) based on the ABO algorithm [27, 28]. The ABO has been improved to obtain several hybridization variants of the original ABO. Lastly, the hybrid variants of BS have been evaluated and their performance compared with existing strategies.

The ABO is one of the 21st century swarm intelligence-based algorithm. The ABO was inspired by the unique intelligence displayed by migrant buffalos searching for food in the African continent. ABO being is a swarm-based technique where simulates the aggregate intelligence of individual buffalo as they optimize their search for food sources [29]. ABO attempts to model three qualities of the African Buffalo as follows:

A. Extensive memory capacity of African Buffalos. These animals keep track of their migrant paths for hundreds of miles;
B. Democratic nature: The major decisions of the herd are determined by "election";
C. Cooperative behavior of buffalos: African Buffalos have two sounds: the alert '/waaa/' sound prompting the herd to move since their present because location is dangerous or unfavorable for some reasons; while the sound "/maaa/" requests other Buffalos to stay in the same location [30].

The ABO algorithm starts by initializing a number of buffalos and then updates each buffalo's fitness using Eq. 1. At this point, the algorithm saves the fitness of each buffalo *bp* and the global fitness overall of the herd b*g*. *lpr1* and *lpr2* are learning factors. The three parts of Eq. 1 (i.e. w^t, lpr1 $\times$ $(bg - w^t)$, and lpr2 $\times$ $(bg - w^t)$ represent: memory capacity; the cooperative behavior of buffalos; and the intelligence part of the buffalos, respectively. Equation 2, basically, is used to update the buffalo's locations. Please note that λ is a random number. The ABO algorithm searching strategy is governed by the equations [31]:

$$m^{(t+1)} = w^{(t)} + lpr1 * (bp - w^{(t)}) + lpr2 * (bg^{(t)} - w^{(t)}) \tag{1}$$

and

$$\mathrm{W}_{\mathrm{k}'} = (\mathrm{W}_{\mathrm{k}} + \mathrm{m}_{\mathrm{k}})/\lambda \tag{2}$$

3.2 Buffalo Strategy T-Way Test Suite Generation for E-payment

In this section, our concern with the presentation of the Buffalo Strategy (BS) for the design and implementation of a tway test suite generator [32]. BS uses the original ABO in generating optimized test suite through a diligent search for particular test cases which incorporates the maximum number of t-combinations of inputs at least once. It treats a buffalo (that is, a feasible solution and interaction elements) within the search space. Basically, BS starts by first generating all interaction elements that are within the present search space into a list and representing them as component units of a population of buffalos.

At the evaluation loop, which is the next step, the buffalos are repeatedly subjected to a cycle of ABO search procedures in order to generate the expected final optimized test suite. In general, the BS comprises of two steps:

- Produce interaction test elements;

- Produce T-way Test Suite by diligent search procedures of the ABO to discover the optimal test cases. Please see Fig. 2.

```
1. Producing Interaction Element Step:
      Let IE be a set of Interaction Elements
      B = generate the smallest binary number contain t ones.
      Until all t binary number generated
          Generate all possible interaction elements for B
       Add all possible interaction elements to
   IE  B  =  Generate_next_binary_number(B).
   End loop
2. Generating T-way Test Suite Step:
     While IE is not empty do
 //Perform search using ABO while <Max
   Generation  or stop criterion do
        For i =0 to buffalo_size loop
          Update buffalo's fitness:
          m(t+1) = w(t) + lpr1 * (gbest − w(t)) + lpr2 * (bg_best(t) − w(t))
          Update buffalos location
             wk'= (wk + mk) / λ
          Find the best buffalo
        End for
     End while
     Add the best buffalo into TS.
     Remove covered interactions elements from IE.
End while
```

Fig. 2. BS for T-way test suite generation

3.3 Parameter Adjustment of BS

The behavior of BS is determined by the population size *buffalo_size*, learning factors *lp1* and *lp2* as well as the iteration number *n*. As a result, these parameters need to be properly tuned. To do this, one of the well-known covering arrays CA (N; 2, 10^5) is used. To ensure systematic tuning, the values of two parameters are fixed as we investigate the best value of the other parameter. For instance, the value of buffalo size is iteration fixed (i.e. buffalo size = 10, and iteration = 30) and those of *lp1* and *lp2* (0.1, 0.2, 0.3… 0.6) are investigated (See Tables 1 and Fig. 3). Moreover, the reverse process is performed for each (See Table 3, and Table 4). To ensure reliability, BS is executed 20 times with every parameter value, and the mean is recorded (See Table 1).

Table 1. Mean Test Suite for CA (N; 2, 10^5) with varied value for Iteration and pollen size

Buffalo Size	Iteration										
	5	10	20	30	40	50	100	200	300	500	700
10	35.65	32.15	32.95	30.30	29.95	29.15	28.55	27.65	26.40	25.65	25.25
20	32.20	29.65	28.80	28.45	27.55	27.40	26.95	25.10	24.95	24.80	24.50
30	30.95	28.65	28.00	26.80	26.75	25.50	25.35	24.65	24.50	24.20	24.00
50	29.05	26.65	27.20	25.95	25.95	25.15	25.55	24.10	24.00	23.00	23.40
100	27.15	26.10	25.65	25.25	24.65	24.70	24.55	23.85	24.10	23.90	23.50
200	26.25	25.15	24.80	24.65	24.50	24.35	24.00	23.40	23.50	23.90	23.75
300	25.90	24.90	24.55	24.05	24.45	24.15	23.95	23.80	23.70	23.45	24.00
500	24.80	24.50	24.15	24.05	23.70	23.80	24.10	23.35	23.55	23.80	23.65

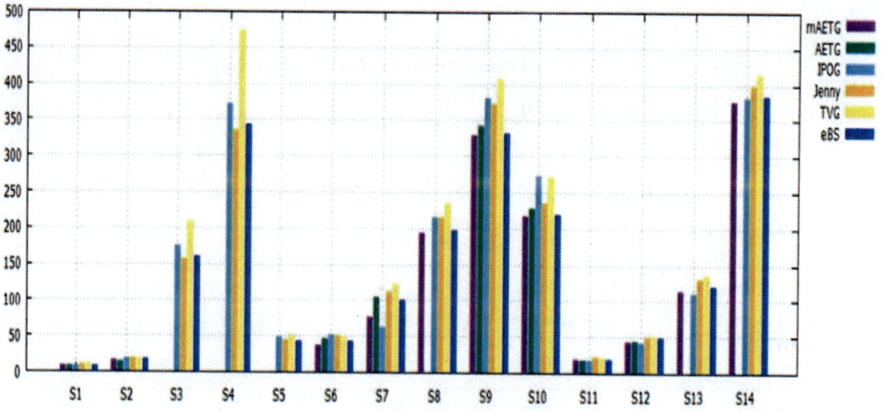

Fig. 3. Comparison of eBS with Computational-based Strategies

Table 1 clearly shows that using a large buffalo population leads to better results and vice-versa. So using a large population of up to 30 buffalos, the performance of BS improved significantly. Nonetheless, increasing the buffalo population to as many as 500 did not necessarily give better results. The best outcomes were realized when the buffalo population size ranges from 50 to 100 buffalos. Also, as the iteration value increased, the results improved (See Table 1). The best outcome was obtained when the iteration number was between 300 and 500.

With regards to the learning factors *lp1* and *lp2* (See Table 2), results show that using a higher value of *lp1* and *lp2* led to better results. Nevertheless, the best results were obtained when *lp1* and *lp2* was between 0.4 and 0.6. To summarize, BS obtained the optimal test suite when the buffalo population was between 50 and 100, iteration was between 300 and 500 and learning parameters between 0.4 and 0.6.

3.4 ABO-Based Strategies for Test Generation

From our experiments, the obtained results of BS using original ABO were very competitive when compared with those from other strategies such as: Simulated Annealing (SA); Harmony Search (HS); Particle Swarm Optimization (PSO) and Genetic Algorithm (GA) One of the observed drawbacks of ABO-based strategy is low randomization of the population.

Therefore, to enhance ABO's performance, we added a few components consistent with other efficient algorithms to the ABO algorithm. The additional components required to enhance local intensification and global diversification are discussed below.

- *Elitism Technique:* Elitism Technique is simply a way to make the randomization more efficient by replacing a part of the population to ensure that the solution quality will not be degraded in the next iteration [33];
- *Mutation operator:* This helps the algorithms to ensure the diversification of solutions of the population from one generation to the next. In mutation, one or more solution values are changed. There are different types of mutation including: Flip bit; Boundary; Bit string mutation; Gaussian mutation [34]; etc.
- *Local Search*: This is a straight-forward and highly effective technique for finding a local optimum solution. Local search deliberately moves from the current location to one of the neighboring states while seeking a local optimum. That is to say, local search emphasizes diligent exploitation of the search area that has a likelihood of good solutions [35].

The hybridization of ABO with the above components promotes greater search effectiveness (See Figs. 3 and 5). In light of the greater effectiveness, this paper proposes four different hybrid variants of BS: mutation Buffalo Strategy (mBS); local-search Buffalo Strategy (lBS); elitism Buffalo Strategy (eBS); and elitism local-search Buffalo Strategy (elBS). The eBS variant uses the elitism technique to fine-tune the buffalo population randomly replacing the unprofitable buffalos with new ones. The mBS variant, on the other hand, uses mutation operator to include diversity in the population of buffalos. On its part, the lBS uses intensive local search to improve local intensification while the elBS injects elements of both elitism technique and local search into BS.

3.5 Evaluation of Hybrid Variants of ABO

The four hybrid variants of BS were tested so as to select the best hybrid variant which can compete with existing test generation strategies techniques (See Fig. 3). Determining the optimal parameters of the various variants of BS parameters is challenging but necessary. To achieve this, a detailed parametric study was conducted for BS parameters as can be seen in Sect. 3.2. To do this, we deliberately chose the mutation rate to be 0.03 and elitism probability to be 0.25 as published in [36, 37], respectively. We applied the Hybrid Algorithms for three well-known covering array problems such as CA(N; 2, 10^5), and CA(N; 2, 4^6) (See Fig. 4). The result is presented in Table 2.

Table 2. Evolution of hybrid variants of ABO

Hybridization	CA(N; 2, 10^5)			CA(N; 2, 4^6)			CA(N; 3, 5^6)		
	Avg	Best	Time(s)	Avg	Best	Time(s)	Avg	Best	Time(s)
BS	142.0	141	9.42	25.7	25	1.03	221.4	220	108.89
eBS	122.4	120	7.89	23.5	22	1.09	198.2	197	155.35
lBS	127.10	126	35.34	23.9	23	5.01	204.8	202	11467.32
mBS	140.4	139	11.566	25.8	25	1.64	215	219.4	205.69
elBS	123.3	121	34.35	23.7	22	7.93	198.4	197	11472.83

The results in Table 2 show that all hybrid algorithm variants of BS outperformed the original BS in terms of average test suite size and best test suite size. Similarly, the results show that ABO-Elitism (eBS) produced better results than the other variants of BS with average (122.4) and Best (120). Moreover, the performance of elBS with average (123.3) and Best (121) is next to that of eBS. The next best performer is the lBS with average (127.1) and Best (126). The least efficient performer is the mBS with average (140.4) and Best (139).

In terms of time taken to achieve results, the eBS still maintained its superiority with 7.89 s, followed by BS with 9.42 s; mBS 11.566 s; elBS, 34.35 s and lBS, 35.34 s. From this analysis, it is evident that, although the BS did not produce very good results, it is an efficient algorithm since the CPU time correlates with less use of computer resources to produce results [38]. Nevertheless, kudos to eBS for producing the best results in the shortest possible time.

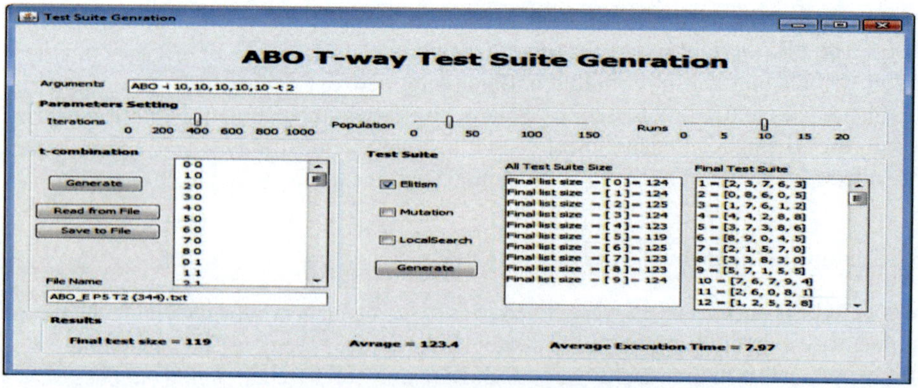

Fig. 4. Implementation of hybrid ABO-based Strategies

Next, we investigated the convergence behavior of hybrid ABO-based variants since convergence behavior is an important issue in performance evaluation of any hybridization (see Fig. 3). To evaluate the convergence rate of these hybrid variants of BS, we executed the hybrid variants of BS 20 times with different test suite sizes and iteration limits (5, 10, 20, 30, 40, 50, 100, 200, 300, 500, and 1000). The mean value of the 20 runs for the two well-known covering arrays CA CA (N; 2, 10^5) and (N; 2, 4^6) are taken (See Fig. 5).

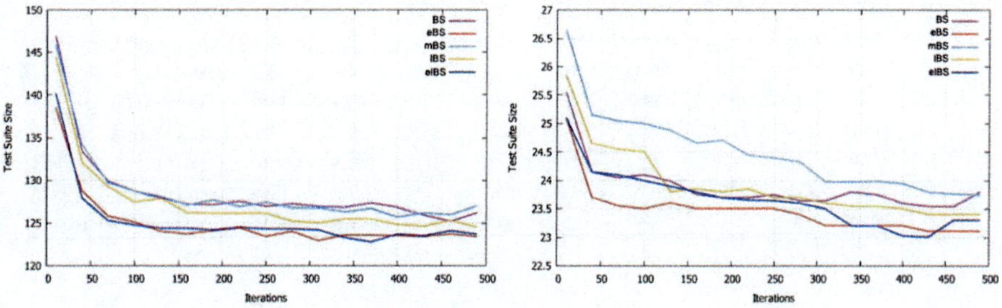

(a) (b)

Fig. 5. The Hybrid Variants of Buffalo Strategy's Convergence Rates for CA (N; 2, 10^5) & (N; 2, 4^6)

As can be seen in Fig. 4, hybridizing ABO improved its convergence capacity; the eBS and elBS converged faster than the other variants. This result shows that the addition of an elitism component improved ABO's result, speed and convergence capacity. This is because the elitism mechanism ensures the retention of good solutions for the next generation/iteration of the algorithm. Again, the elitism component makes for efficient randomizations through the replacement of poorer quality solutions with new random solutions.

3.6 More Experimental Evaluations and Comparative Analysis of Results

To further validate our earlier findings (see Section four), we investigated the efficiency cum effectiveness of the selected strategy (the eBS algorithm) by comparing its results with the existing t-way strategies in terms of test suite size. The outcome of this investigation is displayed in the tables and graphs below. Please note that all the experiments in this study were performed on Core i7–3770 CPU@ 3.40 GHz–3.40 GHz, Windows 7 Operating System. Again, we used several benchmark problems popular with many researchers [39–41]. The eBS parameters are: lpr1 = 0.5, lpr2 = 0.5, maximum iteration = 400, buffalo size = 30 and elitism factor = 0.25. For convenience and reliability, the experiments were divided into four configurations:

1. Comparing proposed strategy with results obtained earlier strategies [13, 40, 41] for the different configuration.
2. Comparing proposed strategy with popular existing strategies for CA (N; t, 2^{10}), where *t* varies from 2 to 10.
3. Comparing proposed strategy with preexisting popular strategies for CA (N; 4, 5^P), where *p* varies from 5 to 10
4. Comparing proposed strategy with existing popular strategies for CA (N; 4, v^{10}), *where* v varies from 2 to 7 The experimental outcomes are presented in Tables 3, 4 and Fig. 5.

Table 3. Comparison with existing strategies (N; t, 2^{10}), CA (N; 4, 5^{P}), P varied from 5 to 10

p	Computational-based Strategies									Meta-heuristic-based Strategies			
	IPOG	ITCH	Jenny	PICT	TConfig	TVG	GTWay	MIPOG	CTEXL	PSO	HSS	CS	eBS
5	908	837	810	773	849	731	625	779	NS	779	751	776	782
6	1239	1074	1072	1092	1128	1027	625	1001	NS	1001	990	991	985
7	1349	1248	1279	1320	1384	1216	1125	1209	NS	1209	1186	1200	1162
8	1792	1424	1468	1532	1595	1443	1384	1417	NS	1417	1358	1415	1329*
9	1793	1578	1643	1724	1795	1579	1543	1570	NS	1570	1530	1562	1486*
10	1965	1791	1812	1878	1971	1714	1643	1716	NS	1716	1624	1731	1622*
11	2091	1839	1957	2038	2122	1852	1722	1902	NS	1902	1860	2062	1787
12	2285	1964	2103	NA	2268	2022	1837	2015	NS	2015	2022	2223	1946

Table 4. Comparison with existing strategies CA (N; t, 2^{10}), CA(N; 4, v^{10}) with v varied from 2 to 7

V	Computational-based Strategies									Meta-heuristic-based Strategies			
	IPOG	ITCH	Jenny	PICT	TConfig	TVG	GTWay	MIPOG	CTEXL	PSO	HSS	CS	eBS
2	49	58	39	43	45	40	46	43	NA	34	37	28	28*
3	241	336	221	231	235	228	224	217	NA	213	211	211	208*
4	707	704	703	742	718	782	621	637	NA	685	691	698	666
5	1965	1750	1719	1812	1878	1917	1714	1643	NA	1716	1624	1731	1622*
6	3935	NA	3519	3735	NA	4159	3514	3657	NA	3880	3475	3894	3329*
7	7061	NA	6462	NA	NA	7854	6459	5927	NA	NA	6398	NA	6261*

In Table 3, different systems configurations were used by simply varying the number of parameters (p), levels of interaction strength (t) and number of value (v) to compare the effectiveness of eBS strategy with existing strategies. We were obliged to use N/A in instances where comparative results were unavailable (such as the results for some tools like the mAETG, AETG and CS).

The results presented in Table 3 show the minimum test suite size obtained by existing strategies for different wellknown problems. All the systems addressed only small measures of interaction strength (i.e. $t = 2$ and $t = 3$) because most computational-based strategies are limited in terms of range of interaction strength. In this experiment, eBS produced the smallest test suite size in two cases, CA (N; 2, 3^{4}), and CA (N; 2, 5^{10}). Even though the eBS was unable to produce the smallest size for all cases (See Fig. 5), the novel strategy outperformed popular strategies such as PSO, ACO, CS and HSS in many cases. In general, it is safe to say that the eBS produced very competitive results since it matched the performance of GA, outperformed HSS, CS, ACO and PSO but was only outperformed by SA.

Furthermore, Table 3 shows that, in most cases, computational-based strategies (such as mAETG, AETG, IPOG, TVG and Jenny) performed worse (with just two optimal results) than the meta-heuristic-based strategies (with 12 optimal solutions): ACO, GA,

PSO, eBS etc. Aside from the meta-heuristic-based strategies, the mAETG strategy obtained close to optimum results in a few cases; while TVG had the worst results. The table also shows SA, ACO and GA (see Problems 81, 86, 88-s14) generated some good results in most cases whereas PSO, HSS, CS, and eBS (See Problems no 81, 83, 85) generated some good results too. In fact, eBS had the optimum result in two instances (see the shaded portions).

In addition, Tables 4 show the capacity of the eBS to address the problems of software testing with a high configuration setting. In these experiments, we increased the three parameter values of covering array v, *t,* and *p.* As can be seen in the tables, most of the existing strategies were unable to produce good results beyond $t > 6$ due to their heavy computation costs (as in the case of GA, ACO GA and PSO). Referring to the results, eBS produced the optimum results in four cases (i.e. CA (N; 10, 2^{10}), CA (N; 3, 2^{10}), CA (N; 4, 2^{10}), CA (N; 7, 2^{10})). The eBS generated the best results in three out of the eight instances. There were another three instances (see the shaded portions) where the eBS produced the best results among other comparative metaheuristics.

Finally, Table 4 reports results for CA (N; 4, 5^P) where P is varied from 5 to 10. The comparative results indicate that eBS outperformed the other strategies in five out of the six cases, when p is equal to 8, 9 and 10 respectively (See CA (N; 4, 5^8), CA (N; 4, 5^9) and CA (N; 4, 5^{10}). In addition, the comparative performance of the eBS with existing strategies, when *v* is varied from 2 to 7 (Table 4), shows that eBS produced satisfactory test sizes for all cases. In all, it is safe to conclude that the proposed strategy outperformed the existing metaheuristic strategies or at least produced competitive results equal to the best results generated by other strategies.

4 Conclusion

In this paper, we proposed and evaluated a new t-way test suite strategy based on African Buffalo Optimization, called the Buffalo Strategy (BS). Next, we investigated four hybridization variants of the BS. The hybridization variants were obtained by employing the concept of component-grafting such as Elitism Technique, Mutation operator and Local Search into the BS. Experimental outcomes indicated that the Elitism-BS (eBS) outperformed the other variants. The experimental output of the eBS was compared with those from the existing strategies in the context of t-way test suite generation. The results of eBS were very competitive. In some cases, the eBS outperformed the other strategies; while, in a few other cases, the proposed strategy failed to produce optimum results. Despite this, the results of eBS were still within an acceptable range. In the future, we recommend further enhancement to improve the efficiency and effectiveness of eBS not just by incorporating algorithm components but full hybridization with other algorithms, such PSO, GA, ACO etc.

Acknowledgment. The authors appreciate the Research Fund by Universiti Tun Hussein Onn, Malaysia and the UMP Short Term Grant RDU1903372 for additional supporting under Fundamental Research Grant Scheme, RDU190185 with Reference no: FRGS/1/2018/ICT03/UMP/02/3, and UMP Short Term Grant RDU1903122.

References

1. Odili, J.B., Mohmad , M.N.: Solving the traveling salesman's problem using the African buffalo optimization. Comput. Intell. Neurosci. **2016**, 1–12 (2016)
2. Odili, J.B., Mohmad Kahar, M.N.: African buffalo optimization approach to the design of PID controller in automatic voltage regulator system. In: National Conference for Postgraduate Research, Universiti Malaysia Pahang, pp. 641–648 (2016)
3. Stardom, J.: Metaheuristics and The Search For Covering and Packing Arrays. Trent University (2001)
4. Özcan, E., Mısır, M., Ochoa, G., Burke, E.K.: A Reinforcement Learning: Great-Deluge Hyper-Heuristic, Modeling, Analysis, and Applications in Metaheuristic Computing: Advancements and Trends: Advancements and Trends, p. 34 (2012)
5. Nurmela, K.J.: Upper bounds for covering arrays by Tabu search. Discrete Appl. Math. **138**, 143–152 (2004)
6. Odili, J.B., Fatokun, J.O.: The mathematical model, implementation and the parameter-tuning of the African buffalo optimization algorithm. In: 2020 International Conference in Mathematics Computer Engineering and Computer Science (ICMCECS), pp. 1–8 (2020)
7. Ahmed, B.S., Zamli, K.Z., Lim, C.P.: Constructing A T-way interaction test suite using the particle swarm optimization approach. Int. J. Innov. Comput. Inf. Control **8**, 431–452 (2012)
8. Grötschel, M., Lovász, L., Schrijver, A.: Geometric algorithms and combinatorial optimization vol. 2: Springer Science & Business Media (2012)
9. Ausiello, G., Crescenzi, P., Gambosi, G., Kann, V., A., Protasi, M.: Complexity and Approximation: Combinatorial Optimization Problems and Their Approximability Properties. Springer Science & Business Media, Heidelberg (2012). https://doi.org/10.1007/978-3-642-58412-1
10. Odili, J.B., Kahar, M.N.M., Anwar, S., Azrag, M.A.K.: A comparative study of African buffalo optimization and randomized insertion algorithm for asymmetric travelling salesman's problem. In: 2015 4th International Conference on Software Engineering and Computer Systems (ICSECS), pp. 90–95 (2015)
11. Odili, J.B., Kahar, M.N.M.: Numerical function optimization solutions using the African buffalo optimization algorithm (ABO). Br. J. Math. Comput. Sci. **10**, 1–12 (2015)
12. Myers, G.J., Sandler, C., Badgett, T.: The art of software testing. Wiley, New York (2011)
13. Alsewari, A.R.A., Zamli, K.Z.: Design and implementation of a harmony-search-based variable-strength<i> t</i>-way testing strategy with constraints support. Inf. Softw. Technol. **54**, 553–568 (2012)
14. Song, P.H., A., Robbins, J., McCullough, J.S.: Exploring the business case for ambulatory electronic health record system adoption. J. Healthcare Manage. **56**, 169 (2011)
15. Garvin, B.J., Cohen, M.B., Dwyer, M.B.: Evaluating improvements to a meta-heuristic search for constrained interaction testing. Empir. Softw. Eng. **16**, 61–102 (2011). https://doi.org/10.1007/s10664-010-9135-7
16. Sugitani, T., Hamasaki, T., Hamada, C.: Partition testing in confirmatory adaptive designs with structured objectives. Biometrical J. **55**, 341–359 (2013)
17. Arcuri, A., Iqbal, M.Z., Briand, L.: Random testing: theoretical results and practical implications. IEEE Trans. Softw. Eng. **38**, 258–277 (2012)
18. Lee, C.-C., Friedman, J.: Requirements modeling and automated requirements-based test generation. SAE Int. J. Aerosp. **6**, 607–615 (2013)
19. Petke, J., Yoo, S., Cohen, M.B., Harman, M.: Efficiency and early fault detection with lower and higher strength combinatorial interaction testing. In: Proceedings of the 2013 9th Joint Meeting on Foundations of Software Engineering, pp. 26–36 (2013)

20. Gargantini, A., Vavassori, P.: Citlab: a laboratory for combinatorial interaction testing. In: 2012 IEEE Fifth International Conference on Software Testing, Verification and Validation, pp. 559–568 (2012)
21. Burr, K., Young, W.: Combinatorial test techniques: table-based automation, test generation and code coverage. In: Proceedings of the International Conference on Software Testing Analysis and Review (1998)
22. Yilmaz, C., Cohen, M.B., Porter, A.A.: Covering arrays for efficient fault characterization in complex configuration spaces. IEEE Trans. Softw. Eng. **32**, 20–34 (2006)
23. Odili, J.B.: Implementation analysis of Cuckoo search for the benchmark Rosenbrock and Levy test functions. J. Inf. Commun. Technol. **17**, 17–32 (2020)
24. Neshat, M., Sepidnam, G., Sargolzaei, M., Toosi, A.N.: Artificial fish swarm algorithm: a survey of the state-of-the-art, hybridization, combinatorial and indicative applications. Artif. Intell. Rev. **42**(4), 965–997 (2012). https://doi.org/10.1007/s10462-012-9342-2
25. Yang, X.-S., Press, L.: Nature-Inspired Metaheuristic Algorithms, 2nd edn. Luniver Press, London (2010)
26. Qiu, X., Lau, H.Y.: An AIS-based hybrid algorithm for static job shop scheduling problem. J. Intell. Manuf. **25**, 489–503 (2014). https://doi.org/10.1007/s10845-012-0701-2
27. Odili, J.B., M.N., Noraziah, A., Abiodun, O.E.: African buffalo optimization and the randomized insertion algorithm for the asymmetric travelling salesman's problems. J. Theor. Appl. Inf. Technol. **87**, 356–364 (2016)
28. Odili, J.B., M.N.: African buffalo optimization. Int. J. Softw. Eng. Comput. Syst. **2**, 28–50 (2016)
29. Odili, J.B., Kahar, M.N.M., Anwar, S.: African buffalo optimization: a swarm-intelligence technique. Procedia Comput. Sci. **76**, 443–448 (2015)
30. Odili, J.B., Noraziah, A., Ambar, R., Abd Wahab, M.H.: Implementation strategies for the Cuckoo search and the African buffalo optimization for the benchmark Rosenbrock function. In: The Eurasia Proceedings of Science Technology Engineering and Mathematics, pp. 395–402
31. Odili, J.B., Kahar, M.N.M.: African buffalo optimization. Int. J. Softw. Eng. Comput. Syst. **2**, 28–50 (2016)
32. Odili, J.B., Kahar, M.N.M.: African buffalo optimization (ABO): a new meta-heuristic algorithm. J. Adv. Appl. Sci. **3**, 101–106 (2015)
33. Malik, M.S., Wadhwa, M.S.: Preventing premature convergence in genetic algorithm using DGCA and elitist technique. Int. J. **4** (2014)
34. Gong, W., Cai, Z.: Differential evolution with ranking-based mutation operators. IEEE Trans. Cybern. **43**, 2066–2081 (2013)
35. Rodrigues, E., Gaspar, A.R., Gomes, Á.: An evolutionary strategy enhanced with a local search technique for the space allocation problem in architecture, Part 1: Methodology. Comput. Aided Des. **45**, 887–897 (2013)
36. Shiba, T., Tsuchiya, T., Kikuno, T.: Using artificial life techniques to generate test cases for combinatorial testing. In: International Computer Software and Applications Conference, pp. 72–77 (2004)
37. Nasser, A.B., Alsewari, A.R.A., Zamli, K.Z.: Tuning of cuckoo search based strategy for T-way testing. In: International Conference on Electrical and Electronic Engineering (2015)
38. Khompatraporn, C., Pintér, J.D., Zabinsky, Z.B.: Comparative assessment of algorithms and software for global optimization. J. Global Optim. **31**, 613–633 (2005). https://doi.org/10.1007/s10898-004-9971-3
39. Alsewari, A.R.A., Zamli, K.Z.: Design and implementation of a harmony-search-based variable-strength t-way testing strategy with constraints support. Inf. Softw. Technol. **54**, 553–568 (2012)

40. Ahmed, B.S., Abdulsamad, T.S., Potrus, M.Y.: Achievement of minimized combinatorial test suite for configuration aware software functional testing using the cuckoo search algorithm. Inf. Softw. Technol. **66**, 13–29 (2015)
41. Cohen, M.B.: Designing Test Suites for Software Interaction Testing. Citeseer (2004)

ICT in Education

Jom eLearning Unavoidable Acceptance

S. Y. AliciaLim and P. S. JosephNg(✉)

Institute of Computer Science and Digital Innovation, UCSI University, UCSI Heights, Cheras, 56000 Kuala Lumpur, Malaysia
josephng@ucsiuniversity.edu.my

Abstract. Technology is an important component in higher education institutions in the form of LMS and e-learning to enhance the learning experience of students. The goal of the implementation of e-learning in higher educational institutions is to help save cost, improve efficiency and branding for both students and institutions. Every student has their study methods and a specific e-learning platform cannot fulfil students' different study methods. Other than functions, user behaviour can also affect the effectiveness of e-learning. In this paper, data is being collected from students by survey and interviews from private universities about their views on e-learning. Students can freely use the time in between classes and able to view recorded classes to boost their knowledge. Students also have complaints about communication problems, easily distracted during e-learning, not motivated to study, and often encounter technical problems. Managerial and practical implication is discussed.

Keywords: e-learning · Student · Student's behaviour · Lecturer · University · College · Higher education · Malaysia · Cost-effective · Efficiency · Branding

1 The Usage of Technology

In this generation, technology has advanced in a way that people can never imagine in the past. Technology brings improvement and convenience to our daily life. Currently, many people in Malaysia have internet access whether it is a public or private internet. In Malaysia, there are more than 20.1 million users that are active on the internet [1]. People use the internet and technology in their work, business, school, house, transportation, and entertainment. The internet and technology help access information in a blink of an eye when others may need to go through books and experience to get the same piece of information. It can also help in communication between distanced people. It can help in the working environment to save records in the database. Technology can also help simplifies the tasks of the data recording because it can be done through technology and not through paper or physical recording [2].

The education industry adopts technology to improve the educational system. In the last decade, many studies have been done about e-learning [3]. E-learning has been a helpful tool for lecturers and students in higher education [4]. Other than replacing face-to-face teaching, e-learning has complemented traditional face-to-face teaching in many learning environments. Lecturers use this system to upload syllabus, assignments,

M. Al-Emran et al. (Eds.): ICETIS 2021, LNNS 299, pp. 177–186, 2022.
https://doi.org/10.1007/978-3-030-82616-1_16

information, and also perform discussions with students. Students can 24 h access to updated materials, learn at their own pace and it is cost-effective [5]. Other than that, LMS also use to provide live class [6]. Studies show that live classes can improve the learning experience and help students have more feeling of engagement with content because students can take control as well, not only the lecturers [7].

Traditional face-to-face teaching requires classes to be held at a specific time and a specific area to have the class. It depends on the lecturer's arranged time and classroom to perform the class. There is also a time limit and the number of students is limited due to the space constraints of the classroom. E-learning is a computing environment and students can join e-learning anytime, anywhere. In e-learning, the lecturer and student can also set a specific time to have class but do not need to have fix place [8–11]. Every student is an individual and they may have a preferred way of learning. Students who do not perform well in classroom teaching may perform well in e-learning. It will be hard for the lecturer to tailor their teaching method to suit each of the students. By understanding the different learning preferences, e-learning can provide different methods of learning of the same content for different students [12]. The online learning environment can add different activities to meet students' needs such as animations, games, and visual objects [13].

2 Pass Study and Comparison of the e-learning Platform

E-learning is used to test the student's acceptance and effectiveness. Self-assessments such as assignments and quizzes are important to test the students' understanding of the course. Blogging activity, experiments, and classes are also effective. It makes studying more interesting and may stimulate students to learn. Practical Assignment is important because it evaluates the knowledge of the students and can they use this knowledge in practical ways [14]. Student's behaviour or characteristic can affect the effectiveness of e-learning such as time management, discipline, motivation, individual experience, and technology dependence [15]. Research shows that motivated students actively engage with the activity, enjoy and adopt e-learning better which enhances study performance [16].

User Interface is also very important. Some students may access e-learning through their smartphone, computer, laptop, or tablet. Due to the different screen sizes, the learning material will not fit nicely into the screen. Mobile phones are smaller in screen size and have limited input capacity [17]. Having a more colourful and well-made user interface can motivate the student to access e-learning. Institutions need to know how teachings are delivered and how students and lecturers interact in the institution to know which e-learning platform is most suitable for the institution. It may be a well-developed e-learning platform or a brand new platform [18].

3 Problem Statement

Regardless of the satisfaction and the advantage of online programs, there are students in universities or colleges that are uncomfortable using e-learning [21]. They will face problems such as confusion about the usage of technologies, hard to interpret their thoughts to the lecturer, and lack of manual guidance. This software provides LMS and e-learning for universities and colleges. But these software systems are made to generally fit all institutions. They may not fulfil the need of the student as to traditional face-to-face teaching. Despite the benefits that come with e-learning, one of the most obvious disadvantages is a student might experience social isolation because e-learning is not as interactive as face-to-face teaching. This may deteriorate the student's communication skills [22]. From the objectives derived, these are hypotheses for each point of it and summarized in Fig. 1 and Table 1.

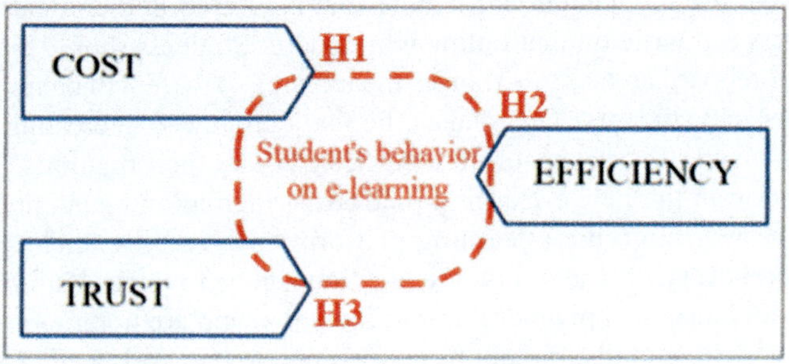

Fig. 1. Hypothesis research model

H1: E-learning will help reduce costs compared to face-to-face classes. Students do not need to travel to the institution which helps save transportation, petrol, and parking fees. It also saves printing fees as students do not need to print notes and food costs are reduced when eating from home.

H2: E-learning is more efficient because students can view the class anytime and anywhere. Students only need a smart device and internet connection to access e-learning. They can attend classes, viewing the recording of classes, get study material, and updated information from the lecturer anytime.

H3: As was mentioned, some students are not comfortable with e-learning and it affects the performance of the students. The new e-learning system will be developed with the student's preferences and opinions in mind. The new e-learning system will help increase the students' acceptance.

Table 1. Research hypothesis.

H1: E-learning is more cost-effective than a physical classroom because it helps reduce travel and material costs. (COST)
H2: E-learning is more efficient because students can access anytime and anywhere. (EFFICIENT)
H3: The acceptance of e-learning will improve using our e-learning. (TRUST)

4 Value Creations

The final product is to provide students with an e-learning system that have a wider range of function in comparison to the other university e-learning platform. There will be different functions to fulfil the different needs of the students. With the new features, students can monitor their study progress, manage time, and improve their study experience. Collecting badges can also help lecturers to evaluate the students' performance and encourage students to keep doing better each time [23]. During live class, lecturers have the option to use the whiteboard feature that is offered in the e-learning platform but only lecturers can write on that online whiteboard. Students should be able to write on that same whiteboard at the same time as the lecturer so there will be more interaction between lecturer and students. This makes the live class more interesting and students can be more involved. It is easier for both parties to share their thoughts. There will be a resources section in the new e-learning platform which allows students to save their different form of work into the e-learning platform such as files, slides, pictures, and videos. Therefore, everything about their education can be found and store in one place. Lastly, there would also be a planner feature. Students who are not good at planning or lack of discipline will lead insufficient time to complete the work well. This will affect the physical and mental health due to stress and lack of rest. It will also affect the final result of the student and their motivation to do well on that subject. By having a planner feature, the e-learning platform will provide a brief planner for the students according to their daily class schedule and task due date. Students then can follow the planner or customize the planner. The system will observe the student's progress and adapts according to the specific student's needs. The goal is to help students to reach their goals under complex information space [24, 25].

5 Methodology

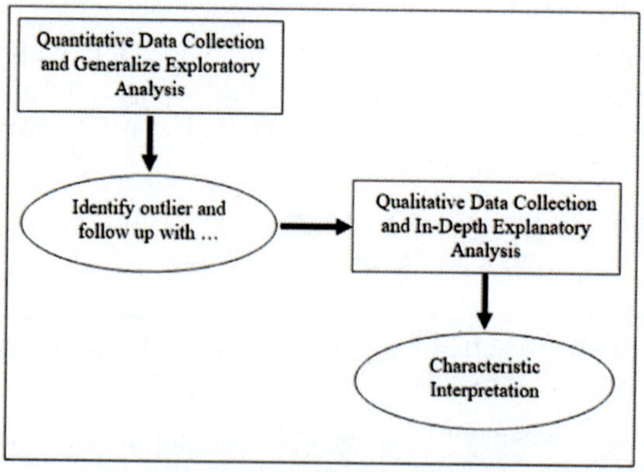

Fig. 2. Sequential design [26–39]

This research will apply the mixed method of data collection to get more accurate findings. A survey and interview will occur during the data collection. The research dimension will be explained in sequential design which will explain each step of data collection. The mixed-mode is applied by doing both surveys and interviews as its primary data collection. To elaborate on the steps of data collection, the sequential design will illustrate as shown in Fig. 2. The survey is conducted in various private and public universities. The responses are from students of different courses and ages. Quantitative data collection will be done by using Google form. Quantitative is more general while the qualitative method will be done through interviews by using Google form because it is more efficient. More in-depth information and personal opinion about e-learning can be taken. Finally, all information will sum up to be an impactful summary of higher education students' opinions about e-learning.

6 Results and Findings

From Fig. 3, 70.2% of respondents vote that their experience with e-learning is good. They agree that e-learning brings positive impacts to their studies. However, 19.3% of respondents rated their e-learning experience as bad and 8.8% of respondents rated their e-learning experience as terrible. Some students have difficulty adapting to the e-learning way of studying and students in online courses perform worse than traditional face-to-face courses [25]. They encounter difficulties such as technical problems, unable to focus in the live class, difficulty in asking questions, and many more.

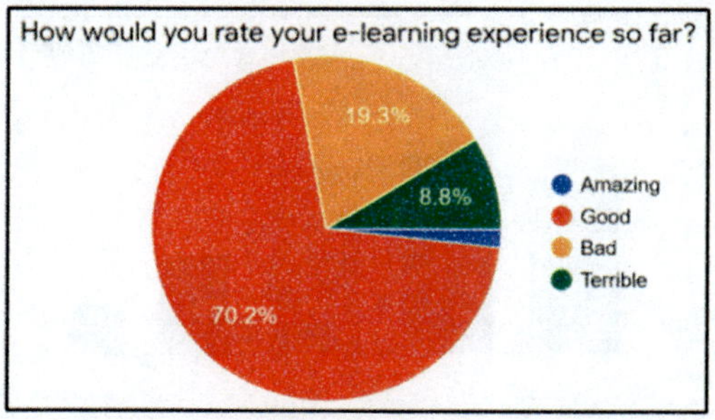

Fig. 3. Rating of E-learning experience

Figure 4 shows what respondents think about e-learning from the perspective of cost. 33.3% of respondents are neutral about whether e-learning is more cost-effecting than traditional learning. From the interview, the respondents are neutral because institutions are saving costs such as electricity cost and internet fees but also need to invest in equipment and hired technical people to manage and maintain the e-learning platform. Some respondents also stated that they save printing and transportation fee but the electrical bill for house use also increase significantly. 24.6% of respondents agree and 14% of respondents strongly agree that e-learning is cost-effective because respondents stated that e-learning reduces the cost of transportation, parking, printing, textbooks, and eating fees. 52.6% of respondents strongly agree and 31.6% of respondents agree that e-learning help saves transportation fees. 40.4% strongly agree and 19.3% agree that e-learning help saves reduce paper and printing costs. 17.5% of respondents disagree and 10.5% of respondents strongly disagree that e-learning is cost-effective because the students need to spend more money to maintain their own devices such as laptops and desktops, higher electrical cost, and overall cost of living. The most mentioned problem is that the overall fee of the e-learning is about the same as face-to-face learning and inconvenient to use institutions' facilities even though they already pay for the facilities. 8.8% of respondents show a negative response that e-learning help saves transportation fees. 22.8% of respondents show negative response e-learning help saves reduce paper and printing costs.

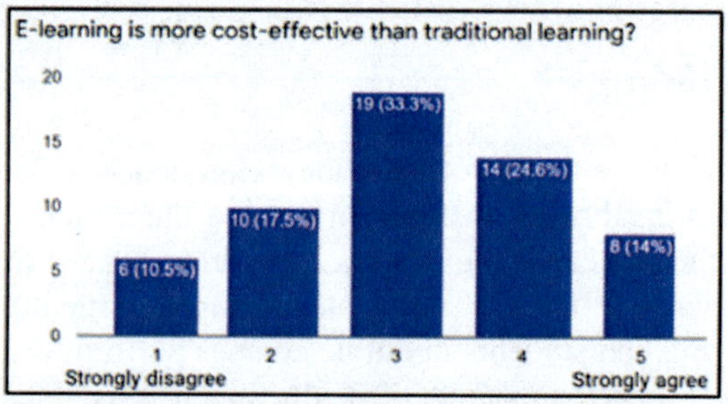

Fig. 4. E-learning in the perspective of cost

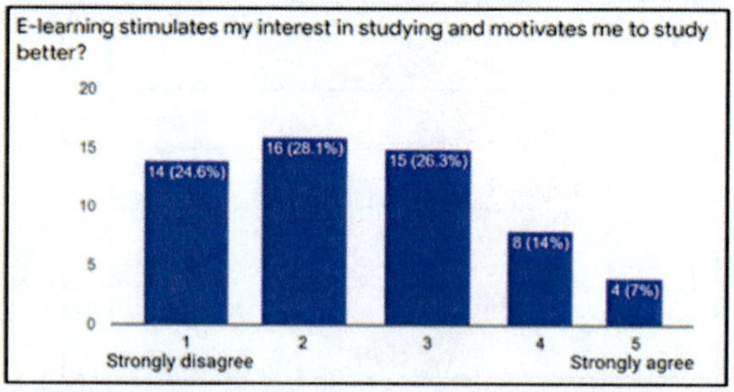

Fig. 5. E-learning in the perspective of efficiency

In the perspective of efficiency in Fig. 5, 26.3% of respondents are neutral because, in the interview, the respondents stated that it depends on the attitude of the students. If someone is willing to learn, e-learning will be effective. If not, it's the same result as one who refuses to work hard. Some state that it is effective as long as students can learn at their own paced. Some state that not having to travel from class to class will not interrupt the study flow but students will be tempted to go to sleep during the evening time. There are 14% agree and 7% of respondents strongly agree and stated students have to focus to know what the lecturer is teaching in live classes. Students can view recorded classes to help revise and improve their knowledge of the subject. When students get better results, it makes students want to study more. Students can take important notes that they have missed during the live class. There are 28.1% disagree respondents and 24.6% of respondents strongly disagree. From the interview, the students responded that e-learning causes a lack of interactions. Students wish to ask question face to face with lecturers and face to face can allow students to portrait their idea better. Some stated that they didn't pay attention due to a lack of interest. There are also a lot of distracting factors in the surroundings that cause less concentration in class. The whiteboard feature in the new e-learning platform can increase interaction and help students be more involved in the class.

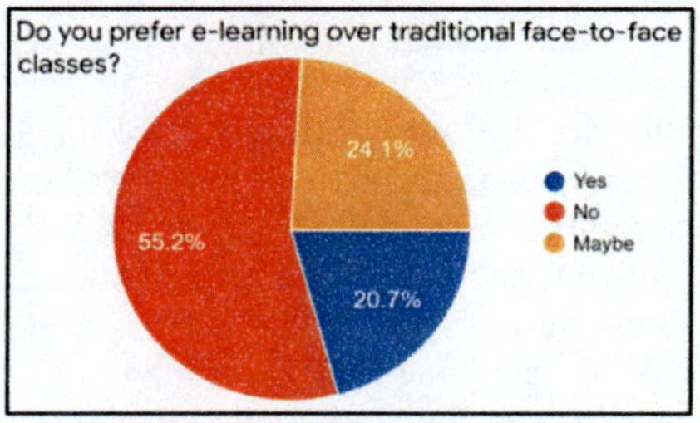

Fig. 6. Preference for e-learning over traditional classes

20.7% of respondents voted they prefer e-learning and 24.1% voted on maybe as shown in Fig. 6. In the interview, respondents stated things that they like best about e-learning. Respondents stated that the most are how e-learning saves time and bring convenience to the students. E-learning is accessible within and outside of campus which is convenient for students and lecturers [26]. Students get easy access to lecture and assignment materials, updated information, and recording classes that students would need to make their study life better. Students can freely use the time between classes. They can do revision, sport, rest, hobby, and even part-time work. The next most mention benefit is the recording function during the live class. It can help students who can't catch up in class, help students to know better, revise before an examination, and study at their own pace. The next benefit is sufficient rest. Students can have more rest before class by not having to prepare to go to a physical class and move from one class to another. Sufficient rest will help in improved students' performances and attendance [40]. 55.2% of respondents voted that they do not prefer e-learning over the traditional class. In the interview, the respondents mention the problems that they face in e-learning. The problem that mentioned the most is technical problems such as weak internet connection, system problem, run out of battery, poor quality of audio and video.

7 Conclusion, Limitation and Future Works

In conclusion, e-learning is an unavoidable platform for the future of educational industries. It is a quick and easy way for lecturers and students to transfer knowledge. It is accessible within and outside of campus which is convenient for users. The goal of e-learning is to help save costs, improve efficiency, and branding for both students and institutions. Although e-learning brings benefits, problems in e-learning that are mentioned by students also need to be improved. Other than technical aspects, actions should also be taken in the learning aspect. Institutions can take action to help solve the problem that students mentioned which is the tuition fee of the e-learning class should be cheaper than traditional classes. Students can learn to have better time management, a hard-working attitude, discipline to fight distraction and a mindset to accept new technology. Lecturers can improve and try to use different methods of teaching to stir up the interest of the students in studying.

References

1. Kunachagaran, K., Govindaraju, V., Ramakrishnan, S., Hisham, S.S.: End users' perception on public WiFi system—a study conducted in Kuala Lumpur, Malaysia. Adv. Sci. Lett. **23**(9), 9265–9270 (2017)
2. Gomez, J.F.: Higher Education Faculty Use of a Learning Management System in Face-To-Face Classes, California State University Stanislaus Scholar Works, pp. 1–116 (2015)
3. Ouadoud, M., Chkouri, M.Y.: Studying and comparing free e-learning platforms. In: International Colloquium on Information Science and Technology, pp. 581–586 (2016)
4. Ouadoud, M., Nejjari, A., Chkouri, M.Y., Kadiri, K.E.E.: Educational modelling of a learning management system. In: 2017 International Conference on Electrical and Information Technologies (ICEIT), pp. 1–6 (2017)
5. Mohd Salleh, M.F.: E-learning Issues in Malaysian Higher Education, Universiti Teknologi Malaysia International Campus, pp. 1–150 (2008)
6. Carliner, S., Shank, P.: The e-learning Handbook: Past Promises, Present Challenges, USA, Pfeiffer, pp. 1–356 (2016)
7. Scagnoli, N., Choo, J., Tian, J.: Students' insights on the use of video lectures in online classes. Br. J. Edu. Technol. **50**(1), 399–414 (2017)
8. Kang, C.M., et al.: A study on integrating penetration testing into the information security framework for Malaysian higher education institutions. In: International Symposium on Mathematical Science and Computing Research, Malaysia, pp.156–161 (2015)
9. AdrianChin, Y.K., et al.: JomDataMining: learning behaviour affecting their academic performance, really? In: IEEE 6th International Conference on Engineering Technologies and Applied Science, Malaysia, pp. 1–4 (2020)
10. AdrianChin, Y.K., et al.: JomDataMining: Academic Performance and Learning Behavior Dubious Relationship, International Journal of Business Information Systems (2020, Forthcoming article)
11. Waweru, B.W. et al.: Gamesy: how videogames serve as a better replacement for school? In: IEEE Student Conference on Research and Development, pp. 10–15 (2020)
12. Jung, S., Huh, J.H.: Efficient LMS platform test bed. Electronic **8**(2), 1–154 (2019)
13. Truong, H.M.: Integrating learning styles and e-learning system: current development, problem and opportunities. Comput. Hum. Behav. **55**, 1185–1193 (2016)
14. Basarmak, U., Mahiroglu, A.: The effect of online learning environment based on caricature animation used in science and technology course on success and attitude of the student for humor. Turk. J. Educ. Technol. **15**(4), 107–118 (2016)
15. Goosen, L., van Heerden, D.: E-learning environment tools to address online and open distance education context challenges. In: ICEL2016-Proceedings of the 11th International Conference on e-Learning, pp. 275–284 (2015)
16. Nakayama, M., Mutsuura, K., Yamamoto, H.: Student's reflections on their learning and note-taking activities in a blended learning course. Electron. J. e-Learning **14**(1), 43–53 (2016)

17. Hartnett, H.: Motivation in Online Education. Springer, Singapore (2016). https://doi.org/10.1007/978-981-10-0700-2
18. Sabharwal, R., Chugh, R., Hossain, M.R., Wells, M.: Learning management systems in the workplace: a literature review. In: 2018 IEEE International Conference on Teaching, Assessment, and Learning for Engineering (TALE), pp. 387–393 (2018)
19. Ananga, P., Kofi , I.: Comparing face-to-face and online teaching and learning in higher education. J. Educ. Stud. Pract. **7**(2), 165–179 (2017)
20. Al-, W.M., Alias, N., Alzahrani, A.I., Alfarraj, O., Saged, A.A., Rahman, N.S.A.: Use of E-learning by university students in malaysian higher educational institutions: a case in Universiti Teknologi Malaysia. IEEE Access **6**, 14268–14276 (2018)
21. Borboa, D., Joseph, M., Spake, D., Yazdanparast, A.: Perceptions and use of learning management system tools and other technologies in higher education: a preliminary analysis. J. Learn. High. Educ. **10**(2), 17–23 (2017)
22. Ahmad, N., Chua, L.: Technology and higher education: using an E-learning tutorial as a pedagogy for innovation and flexible learning. Malaysian J. Distance Educ. **17**(1), 21–31 (2015)
23. Beluce, A.C., Oliveira, K.L.D.: Students' motivation for learning in virtual learning environments. Paidéia (Ribeirão Preto) **25**(60), 105–113 (2015)
24. Imran, H.: Evaluation of awarding badges on student's engagement in Gamified e-learning systems. Smart Learn. Environ. **6**(1), 1–12 (2019). https://doi.org/10.1186/s40561-019-0093-2
25. Morales, L., Garrido, A., Serina, I.: Planning and execution in a personalised E-learning setting. In: Lozano, J.A., Gámez, J.A., Moreno, J.A. (eds.) CAEPIA 2011. LNCS (LNAI), vol. 7023, pp. 233–242. Springer, Heidelberg (2011). https://doi.org/10.1007/978-3-642-25274-7_24
26. JosephNg, P.S., Eaw, H.C.: Making financial sense from EaaS for MSE during economic uncertainty. In: Arai, K. (eds.) FICC 2021. AISC, vol. 1363, pp. 976–989. Springer, Cham (2021). https://doi.org/10.1007/978-3-030-73100-7_67
27. JosephNg, P.S., Loh, Y.F., Eaw, H.C.: Grid Computing for MSE during volatile economy. In: International Conference on Control, Automation and Systems, pp. 709–714. IEEE Explore, Busan (2020)
28. JosephNg, P.S., et al.: EaaS: available yet hidden infrastructure inside MSE. In: 5th International Conference on Network, Communication and Computing, ACM International Conference Proceeding Series, Kyoto, Japan, pp. 17–20 (2016)
29. JosephNg, P.S., Kang, C.M., Choo, P.Y., Wong, S.W., Phan, K.Y., Lim, E.H.: Beyond cloud infrastructure services in medium size manufacturing. In: International Symposium on Mathematical Sciences and Domputing Research, pp. 150–155. IEEE Explore, Ipoh (2015)
30. Joseph Ng, P.S., Choo, P.Y., Wong, S.W., Phan, K.Y., Lim, E.H.: Malaysia SME ICT during economic turbulence. In: International Conference on Information and Computer Network, Singapore, pp. 67–71 (2012)
31. Joseph Ng, P.S., Yin, C.P., Wan, WS., Nazmudeen, M.S.H.: Energizing ICT infrastructure for Malaysia SME during economic turbulence. In: Student Conference on Research and Development, pp. 328–322. IEEE Explore, Cyberjaya (2011)
32. JosephNg, P.S.: EaaS infrastructure Ddisruptor for MSE. Int. J. Bus. Inf. Syst. **30**(3), 373–385 (2019)
33. JosephNg, P.S.: EaaS optimization: available yet hidden information technology infrastructure inside medium size enterprises. J. Technol. Forecast. Soc. Change **132**(July), 165–173 (2018)
34. Joseph Ng, P.S., Kang, C.M., Choo, P.Y., Wong, S.W., Phan, K.Y., Lim, E.H.: Exostructure services for infrastructure resources optimization. J. Telecommun. Electron. Comput. Eng. **8**(4), 65–69 (2016)

35. JosephNg, P.S., Kang, C.M.: Beyond barebone cloud infrastructure services: stumbling competitiveness during economic turbulence. J. Sci. Technol. **24**(1), 101–121 (2016)
36. Joseph, N.P.S., Mahmood, A.K., Choo, P.Y., Wong, S.W., Phan, K.Y., Lim, E.H.: Barebone cloud IaaS: revitalization disruptive technology. Int. J. Bus. Inf. Syst. **18**(1), 107–126 (2015)
37. Joseph, N.P.S., Mahmood, A.K., Choo, P.Y., Wong, S.W., Phan, K.Y., Lim, E.H.: IaaS cloud optimization during economic turbulence for Malaysia small and medium enterprise. Int. J. Bus. Inf. Syst. **16**(2), 196–208 (2014)
38. Joseph, N.P.S., Mahmood, A.K., Choo, P.Y., Wong, S.W., Phan, K.Y., Lim, E.H.: Battles in volatile information and communication technology landscape: the Malaysia small and medium enterprise case. Int. J. Bus. Inf. Syst. **13**(2), 217–234 (2013)
39. Joseph Ng, P.S., Choo, P.Y., Wong, S.W., Phan, K.Y., Lim, E.H.: Hibernating ICT infrastructure during rainy days. J. Emerg. Trends Comput. Inf. Sci. **3**(1), 112–116 (2012)
40. Bettinger, E., Fox, L., Loeb, S., Taylor, E.: Virtual classrooms: how online college courses affect student success. Am. Econ. Rev. **107**(9), 2855–2875 (2017)
41. Ahmad, N.A., Chua, L.N.: Technology and higher education: using an E-learning tutorial as a pedagogy for innovation and flexible learning. Malays. J. Distance Educ. **17**(1), 21–31 (2015)

Behavioral Intention to Use E-Learning Among Academic Staff During COVID-19 Pandemic Based on UTAUT Model

Alaa S. Jameel[1,2](✉), Mohammed A. Karem[1], and Abd Rahman Ahmad[2]

[1] Department of Public Administration, Cihan University-Erbil, Kurdistan Region Erbil, Iraq
{alaa.salam,mohammad.abdulkarem}@cihanuniversity.edu.iq
[2] Faculty of Technology Management and Business, Universiti Tun Hussein Onn, Parit Raja, Malaysia
arahman@uthm.edu.my

Abstract. This study investigates the impact of technology acceptance on behavioral intention (BI) to use e-learning among academics at three private universities in Iraq that implemented the e-learning. The study used the stratified sampling technique to ensure equal representation of the population. The data were collected by using questionnaires sent via google form to academics due to the movement control order during the COVID-19 pandemic. The study employed the Unified Technology Acceptance and Use Technology (UTAUT) which consisted of Performance Expectancy (PE), Effort Expectancy (EE), Social Influences (SI), Facilitating Conditions (FC) and a new element, the Personal Innovativeness (PI). 187 valid responses were obtained and the data were analyzed by using the structural equation model. The results indicated the PE, EE and FC increased the willingness to use e-learning among academics and had a positive and significant impact on BI, while SI and PI had an insignificant impact on BI to use e-learning. The findings implied that decision-makers at universities should enhance the technological infrastructure at the universities and provide sufficient training to academics to encourage them to e-learning.

Keywords: UTAUT · e-learning · Behavioral intention · Academic staff

1 Introduction

Information technology (IT) had changed in many ways since the 1990s when scholars started researching electronics and exploring transformative options of remote technology learning made possible through the World Wide Web. Meanwhile, IT has a groundbreaking effect on almost all areas of education; thus, education instructors have to master the technologies available to achieve education excellence [1–3].

Recently, the pandemic of COVID-19 demanded education institutions, including universities, to swiftly adopt distance learning and online learning, to maintain the continuity of the teaching and learning process. In response, the number of academics who use e-learning to enrich their technical skills or obtain specific qualifications had increased [1].

M. Al-Emran et al. (Eds.): ICETIS 2021, LNNS 299, pp. 187–196, 2022.
https://doi.org/10.1007/978-3-030-82616-1_17

Online learning is not new to most universities, especially in developed countries. However, the universities in developing countries are facing several challenges to adopting e-learning, particularly the universities in Iraq. The Iraqi universities are facing several issues related to the IT infrastructure, the awareness of e-learning usage for both academics and students and the lack of e-learning training [5, 6]. The closure of universities and schools has many detrimental effects on students, such as disrupted learning, which means students have been deprived of growth and development opportunities (UNESCO,2020). E-learning is an essential source of information because of its omnipresence, low costs and interactive nature [7].

The e-learning system, like Moodle, has many useful features to be utilized for distance learning. These features appear to be handy for learning in the period of a pandemic. For example, students can text or participate in learning activities with teachers using laptops or mobile devices from their homes through the e-learning system. Students can also quickly access learning material on their mobile devices or laptops to be linked to mobile networks or wireless networks locally. Learning management system (LMS) is considered as one of the e-learning methods [8]. E-learning is thus about delivering, arranging, and handling e-learning activities through a framework such as students' enrolment, tests, assignments, course details, curricula, introductory courses, etc. [9]. By switching from conventional learning to online learning, the student can access e-learning platforms, such as Moodle 24 h a day, which provides many advantages, including enhanced communication with teachers, and better access to teaching materials [10].

In Iraq, there is a lack of interest in using e-learning among academics [11]. Therefore, the primary purpose of this study is to identify the elements that hinder university academics from using e-learning. Not all academics are ready to participate in e-learning and consequently become less interested in using e-learning applications. It is assumed that every academic can actively use e-learning in the subjects they taught; however, they do not use e-learning in teaching and learning. The purpose of this study is to investigate the impact of BI to use e-learning based on the UTAUT model among academics in Iraqi higher education context.

2 Literature Review

E-Learning platforms are a particular category of a software system designed for open education and learning forum, including online access to learning materials and online learning support. E-learning provides students with engaging, time- and location-appropriate knowledge and learning opportunities. The implementation of e-learning is based on a participant's decision to use the technology as a behavioral function [12–14]. The technology acceptance model (TAM) describes behavioral intentions as a result of the user's behavioral towards technical features: perceived usefulness and perceived ease of use [12, 15]. Venkatesh et al. [5] revamped the UTAUT model, which expanded the explanatory power of the original TAM [16]. The theoretic foundation for UTAUT was developed by Venkatesh et al. [16]. UTAUT focuses on evaluating the efficacy of theoretical paradigms in various information technology industries [16]. UTAUT has a clear theoretical basis and is highly detailed [17]. The UTAUT model synthesizes key factors of the models previously described such as the theory of reasoned action (TRA) [18], the technology acceptance model (TAM) [19], and the theory of planned behaviour (TPB) [20]. UTAUT consists of four main determinants, namely PE, EE, SI and FC [16]. This study extended the UTAUT model by adding the Personal Innovativeness (PI) factor. The PI could play an essential role in increasing the BI to use e-learning among academic staff. Previously, limited studies examined the impact of PI on BI to use e-learning, particularly among academics [21, 22].

2.1 Performance Expectancy

PE refers to the level a user feels will improve his work efficiency through the use of a technology [1]. Performance expectancy can be defined as the state of belief that e-learning can help academics accomplish their work more efficiently and effectively [10]. The success expectancy reflects on users' perception that e-learning is of great benefit to e-learners. It enables learners to conduct learning tasks immediately or enhance their education and success [23].

Xian [1] reported that PE was able to increase the BI to use e-learning among university academics. Similarly, PE increased the BI to use e-learning and significantly impacted academics in Turkish higher education context [4]. Similarly, BI to use e-learning was significantly impacted by PE among academic staff in Indonesian higher education context [24]. PE has significant effect on BI to use e-learning in several earlier studies [23, 25–27].

H1: PE increases BI to use e-learning among academic staff.

2.2 Effort Expectancy

UTAUT has been used to investigate the adoption of e-learning and academic staff acceptance of vocational e-learning in the higher education context. EE is defined as the level of ease associated with operation or system use [16, 28]. EE refers to the perception of academics that e-learning tools are simple and easy to use [21]. Since e-learning is in its initial phases, the presumption of effort is considered an important component of the behavioral purpose of e-learning [23]. Individual e-learning acceptance is hypothesized to be depended on the ease and accessibility of e-learning, thus the effect of EE on behavioral purpose. Empirically, EE significantly affected BI to use e-learning among academics in Indonesian higher education context [24]. In further support, EE had a significant effect on BI to use e-learning [25, 26]. The similar result was also reported by Uğur and Turan [4] among academics in the higher education context. On the other hand, EE had an insignificant impact on BI to use e-learning among academics [1]. EE also had an insignificant impact on BI in several earlier studies [23, 27].

H2: EE increases the BI to use e-learning among academic staff.

2.3 Social Influences

SI is defined as the level to which a staff perceives that significant others agree that the new system should be used [16]. SI refers to the extent a person feels that society requires him to use technology [29]. It refers to external pressure (for example, peer pressure or supervisory pressure, faculty motivation, etc.) that affects one's impression of e-learning [21]. SI is an important determinant of BI to use e-learning [23].

SI describes the pressure of subjective norm rooted from the impact of a group on a person's decision [25]. SI refers to the extent a person recognizes the value of "other individual's" thought that they should use a technology [7]. People may be affected by others and be involved in certain conduct even though they do not realize the direct effect of social impact on the purpose of behaviour [26]. SI had a significant impact on behavioural intention and played an important role to increase the behavioural intention [23, 26]. On the other hand, SI had an insignificant impact on BI to use e-learning among academics in the education context [1]. Similarly, SI has an insignificant impact on BI to use e-learning [25, 27].

H3: SI increases the BI to use e-learning among academic staff.

2.4 Facilitating Conditions

The environmental factors or the physical actions that convince a consumer to perform tasks are called the "Facilitating Conditions". FC is a significant determinant that influences the application of information systems based on the UTAUT model [16]. FC can be explained by the perception of using organizational and technological infrastructure to promote the use of new systems [7]. FC are measured by users' perception of their ability to access the resources required and support necessary for e-learning services [26]. FC describes how staff consider the technological and organizational infrastructure can be improved and enhance that will aspire teachers to use e-learning platforms [12]. FC including support services comprise of individual counselling, training, tools available to develop skills and awareness and the accessibility of the e-learning system [23].

FC was found to increase the BI to use e-learning among academic staff [1]. In line with this finding, Gunasinghe [21] reported a significant increase of BI to use e-learning due to FC. Nevertheless, FC also showed an insignificant impact on BI to use e-leaning in a different study [26].

H4: FC increases the BI to use e-learning among academic staff.

2.5 Personal Innovativeness

Employees with higher PI levels are more likely to believe in new information system creativity than those with lower levels. PI appears to be related to adventurous and audacious characteristics. People with high PI are more likely to embrace new technology innovation even in highly volatile new IT adoption situation [30]. Farooq [22] recognized

PI as a sustainable attribute that makes people want to try new technology. The PI of users will influence their decision to involve in a complete e-learning environment [31]. When individuals adjust to a new system, they reveal their utility and ease of use more efficiently [32]. The more innovative the learner is, the more likely it is to demonstrate the utility of total e-learning and plans to use it.

Empirically, PI enhanced the BI to use e-learning and able to improve it [31]. In parallel, Liu et al. [30] reported the PI had a significant effect on BI. In addition, PI has positive and significant impact on BI [22]. However, PI also had an insignificant impact on BI to use e-learning among academic staff [21].

H5: PI increases the BI to use e-learning among academic staff.

3 Methodology

This study was conducted among academic staff in three private universities located in Erbil, Kurdistan Region, Iraq that had been implementing e-learning. The study employed the stratified sampling technique to ensure equal representation of the sample from the three universities due to different academic staff number in each university. The data were collected by using questionnaires sent to academic staff e-mail via google forms due to closure of the campus during the COVID-19 pandemic.

283 questionnaires were distributed among academic staff, but only 196 returned. The data were analyzed by two softwires i.e. SPSS for the primary analysis, such as demographic descriptive analysis, missing values, outliers, and Cronbach alpha, while the second software, AMOS was utilized for the hypotheses testing. After treating the missing values and outliers, only 187 valid questionnaires were obtained and analyzed by AMOS. The instrument was adapted from prior studies related to UTAUT proposed by Venkatesh et al.

The BI consisted of 5 items, PE consisted of 4 items, EE 5 items and SI 6 items adopted from [16, 26, 27].However, FC consisted of 5 items adopted form [16, 22, 26] and PI consisted of 3 items adopted form [22].

4 Results

4.1 Demographic Analysis

the majority of respondents were male (56%). Most of the participants were be-tween 30 to 40 years old (41%), and followed by 41 to 50 years old (39%). Respondents aged between 51 to 65 were only 13%. Respondents younger than 29 years old and older than 66 years old were 5% and 3% respectively.

5 Measurement Model

The purpose of this step was to find the validity and reliability of the model. Hair [33] indicated the factor loading for each item should be 0.6 and above. Based on Table 1, the factor loading of items in the model exceeded the cut-off level of 0.6. On a different note, the reliability was measured by composite reliability (CR) and according to Hair [33], CR should be 0.7 or above to ensure good internal consistency. Table 1 depicts that all the constructs had high internal consistency and exceeded the cut-off level of 0.7. Lastly, the average variance extracted (AVE) should be higher than 0.5, and all the constructs in this study exceeded the minimum cut-off value. However, the model fit has been achieved after several modifications. According to Hair [33], the CFI, IFI and TLI should be higher than 0.90. The model of this study achieved all the requirements with CFI = .928, IFI = .929 and TLI = .920. RMSEA should be <0.08 [33] (see Fig. 1) and the present model achieved the cut-off value with .062. By fulfilling the test for validity, reliability and model fit, the next step was to examine the hypotheses.

Table 1. Validity and reliability

Construct	Items	Loading	CR > 0.7	AVE > 0.5
BI	BI1	0.67	0.753	0.615
	BI2	0.84		
	BI3	0.84		
	BI4	0.83		
	BI5	0.72		
PE	PE1	0.85	0.865	0.725
	PE2	0.92		
	PE3	0.95		
	PE4	0.66		
EE	EE1	0.71	0.824	0.530
	EE2	0.86		
	EE3	0.61		
	EE5	0.71		
SI	SI1	0.82	0.773	0.680
	SI3	0.83		
	SI4	0.83		
	SI5	0.84		
	SI6	0.81		
FC	FC1	0.83	0.765	0.681
	FC2	0.85		
	FC3	0.77		
	FC4	0.82		
	FC5	0.76		
PI	PI1	0.71	0.768	0.533
	PI2	0.80		
	PI3	0.66		

5.1 Structural Model

The purpose of this step was to measure the proposed hypotheses. Based on Table 2 and Fig. 1 the BI to use e-learning was positively impacted by PE, EE and FC with p-value 0.02, 0.00 and 0.000 (<0.05) respectively. Thus, H1, H2 and H4 were Supported. Meanwhile, SI (p-value = 0.159) and PI (p-value = 0.367) had an insignificant impact on BI to use e-learning at $\alpha = 0.05$. Thus, H3, H5 were Not Supported.

Table 2. Hypotheses test

Hypotheses		Estimate	S.E	C.R	P	Label
H_1	BI < --- PE	.170	.057	2.977	.002	Supported
H_2	BI < --- EE	.293	.081	3.616	***	Supported
H_3	BI < --- SI	.086	.061	1.408	.159	Not supported
H_4	BI < --- FC	.310	.086	3.586	***	Supported
H_5	BI < --- PI	.157	.174	.901	.367	Not supported

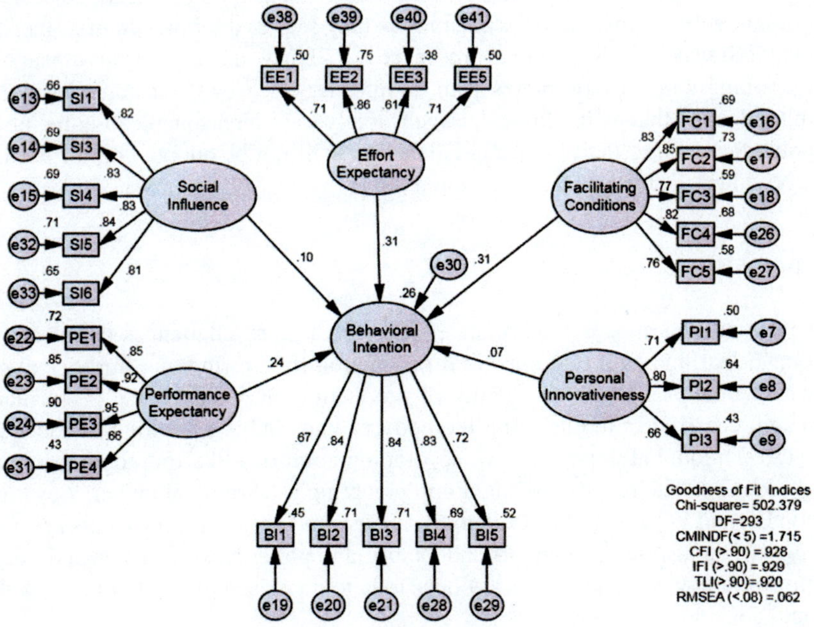

Fig. 1. Structural model

6 Discussion

The purpose of this study is to examine the impact of BI to use of e-learning among academics. The research objective has been achieved based on the mentioned results.

The result showed that PE increased the BI to use e-learning and this finding is supported by prior studies [1, 4, 25]. PE was a significant determinant of the decision to use e-learning. Academics who perceived e-learning as helpful would be more likely to adopt the system. In such event, the e-learning will drive the academics to enhance the quality of learning material by offering adequate and updated material needed to meet students' needs, thus attracting more e-learner users. EE increased the BI to use E-learning, and this result was supported by previous findings [24–26]. It means that, if the e-learning system is user-friendly and less difficult, the academic staff will have a higher tendency to use it. Therefore, it is important to implement frequent device tests, user-friendly configurations and swift support options. Moreover, this result explains the academics' belief that if e-learning is considered more useful than conventional education, the use of e-learning for teaching will expand. FC increased the BI to use e-learning among the academics, which in line with earlier findings [1]. This finding indicates that university administrators must provide appropriate facilities (e.g. computers, Wi-Fi), support resources and rapid e-learning access in growing the number of e-learning users. BI to use e-learning was insignificantly impacted by SI among the academics, and this result is supported by previous findings [1, 25]. The academics in Iraqi higher education institutions have the liberty to use e-learning or other methods of teaching during the closure of university due to COVID-19 outbreak. Without any compulsory requirement to adopt e-learning, and lesser interaction between the academics as they stay in their own homes, therefore the impact of SI was less visible in the present study. PI had an insignificant impact on BI to use e-learning which corroborates with findings reported by Gunasinghe et al. [21]. The result indicated that technological usage is likely to be a reasonable judgment based on available facts rather than intuition. E-learning will not be adopted out of curiosity or courage unless the criteria for assessment are met.

7 Conclusion

Effective use of e-learning can help universities solve many challenges that occur in a classroom. E-learning supports improved production monitoring and improvement of skills, which finally enhance the quality of production and institutional performance. The research contributes to enriching the body of knowledge by testing the adequacy of the UTAUT model at universities in developing countries like Iraq to understand e-learning adoption. The results indicated implementing e-learning at universities highly depend on EE and FC among the academics rather than SI and PI. The results are useful for university management, educators and faculty, and policymakers to plan and execute their online strategy and take necessary decisions to increase the number of academics who would consider e-learning.

References

1. Xian, X.: Empirical investigation of e-learning adoption of university teachers: a PLS-SEM approach. In: Cheung, S.K.S., Jiao, J., Lee, L.-K., Zhang, X., Li, K.C., Zhan, Z. (eds.) ICTE 2019. CCIS, vol. 1048, pp. 169–178. Springer, Singapore (2019). https://doi.org/10.1007/978-981-13-9895-7_15

2. Al-Emran, M., Teo, T.: Do knowledge acquisition and knowledge sharing really affect e-learning adoption? An empirical study. Educ. Inf. Technol. **25**(3), 1983–1998 (2019). https://doi.org/10.1007/s10639-019-10062-w
3. Jameel, A.S., Ahmad, A.R.: The mediating role of job satisfaction between leadership style and performance of academic staff. Int. J. Psychosoc. Rehabil. **24**(04), 2399–2414 (2020)
4. Uğur, N.G., Turan, A.H.: E-learning adoption of academicians: a proposal for an extended model. Behav. Inf. Technol. **37**(4), 393–405 (2018)
5. Jameel, A.S., Ahmad, A.R.: Leadership and performance of academic staff in developing countries. In: Proceedings of the 33rd International Business Information Management Association Conference, IBIMA 2019: Education Excellence and Innovation Management through Vision 2020, pp. 6101–6106 (2019)
6. Karem, M.A., Mahmood, Y.N., Jameel, A.S., Ahmad, A.R.: The effect of job satisfaction and organizational commitment on nurses' performance. Humanit. Soc. Sci. Rev. **7**(6), 332–339 (2019)
7. Jameel, A.S., Moshfeghyan, M.: Factors Affecting E-Commerce By E-Learning. LAP LAMBERT Academic Publishing, Germany (2017)
8. Jameel, A.S., Ahmad, M.A.: Determine some factors that affect to adoption of e-commerce among small and medium enterprises in Erbil. Polytech. J. **8**(1) (2018)
9. Jameel, A.S., Hamdi, S.S., Karem, M.A., Raewf, M.B., Ahmad, A.R.: E-Satisfaction based on E-service quality among university students. J. Phys. Conf. Ser. **1804**(1), 12039 (2021)
10. Mousa, T.S., Jameel, A.S., Ahmad, A.R.: The impact of attitude, subjective norm and information communications technology on knowledge sharing among academic staff. Int. J. Psychosoc. Rehabil. **23**(02), 704–717 (2019)
11. Jameel, A.S.: Challenges facing students toward ICT library adoption. In: International Conference on Accounting, Business, Economics and Politics, pp. 231–237 (2018)
12. Davis, F.D.: Perceived usefulness, perceived ease of use, and user acceptance of information technology. MIS Q. **13**, 319–340 (1989)
13. Al-Emran, M., Mezhuyev, V., Kamaludin, A.: Towards a conceptual model for examining the impact of knowledge management factors on mobile learning acceptance. Technol. Soc. **61**, 101247 (2020)
14. Al-Emran, M., Arpaci, I., Salloum, S.A.: An empirical examination of continuous intention to use m-learning: an integrated model. Educ. Inf. Technol. **25**(4), 2899–2918 (2020). https://doi.org/10.1007/s10639-019-10094-2
15. Salloum, S.A., Alhamad, A.Q.M., Al-Emran, M., Monem, A.A., Shaalan, K.: Exploring students' acceptance of e-learning through the development of a comprehensive technology acceptance model. IEEE Access **7**, 128445–128462 (2019)
16. Venkatesh, V., Morris, M.G., Davis, G.B., Davis, F.D.: User acceptance of information technology: toward a unified view. MIS Q. **27**, 425–478 (2003)
17. Al-Emran, M., Granić, A., Al-Sharafi, M.A., Ameen, N., Sarrab, M.: Examining the roles of students' beliefs and security concerns for using smartwatches in higher education. J. Enterp. Inf. Manag. ahead-of-print(ahead-of-print) (2020)
18. Fischbein, M., Ajzen, I.: Belief, Attitude, Intention and Behavior. Addison-Wesley, Reading (1975)
19. Davis, F.D.: A Technology Acceptance Model for Empirically Testing New End-user Information Systems: Theory and Results. Massachusetts Institute of Technology (1985)
20. Ajzen, I.: The theory of planned behavior. Organ. Behav. Hum. Decis. Process. **50**(2), 179–211 (1991)
21. Gunasinghe, A., Hamid, J.A., Khatibi, A., Azam, S.M.F.: The adequacy of UTAUT-3 in interpreting academician's adoption to e-Learning in higher education environments. Interact. Technol. Smart Educ. **17**(1), 86–106 (2019)

22. Farooq, M.S., et al.: Acceptance and use of lecture capture system (LCS) in executive business studies. Interact. Technol. Smart Educ. **14**(4), 329–348 (2017)
23. Salloum, S.A., Shaalan, K.: Factors affecting students' acceptance of e-learning system in higher education using UTAUT and structural equation modeling approaches. In: International Conference on Advanced Intelligent Systems and Informatics, pp. 469–480 (2018)
24. Putri, V.Q., Shihab, M.R., Hidayanto, A.N.: Does inertia effect e-learning system acceptance among university lecturers? Insights from Sriwijaya University. In: International Conference on Advanced Computer Science and information Systems (ICACSIS) 2019, pp. 307–312 (2019)
25. Mehta, A., Morris, N.P., Swinnerton, B., Homer, M.: The influence of values on e-learning adoption. Comput. Educ. **141**, 103617 (2019)
26. Tarhini, A., Masa'deh, R., Al-Busaidi, K.A., Mohammed, A.B., Maqableh, M.: Factors influencing students' adoption of e-learning: a structural equation modeling approach. J. Int. Educ. Bus. **10**(2), 164–182 (2017)
27. Mikalef, P., Pappas, I.O., Giannakos, M.: An integrative adoption model of video-based learning. Int. J. Inf. Learn. Technol. **33**(4), 219–235 (2016)
28. Al-Emran, M., Al-Maroof, R., Al-Sharafi, M.A., Arpaci, I.: What impacts learning with wearables? An integrated theoretical model. Interact. Learn. Environ. 1–21 (2020)
29. Venkatesh, V., Zhang, X.: Unified theory of acceptance and use of technology: US vs. China. J. Glob. Inf. Technol. Manag. **13**(1), 5–27 (2010)
30. Liu, Y., Li, H., Carlsson, C.: Factors driving the adoption of m-learning: An empirical study. Comput. Educ. **55**(3), 1211–1219 (2010)
31. Al-Busaidi, K.A.: An empirical investigation linking learners' adoption of blended learning to their intention of full e-learning. Behav. Inf. Technol. **32**(11), 1168–1176 (2013)
32. Alshurideh, M., Al Kurdi, B., Salloum, S.A., Arpaci, I., Al-Emran, M.: Predicting the actual use of m-learning systems: a comparative approach using PLS-SEM and machine learning algorithms. Interact. Learn. Environ. 1–15 (2020)
33. Hair, J.F., Black, W.C., Babin, B.J., Anderson, R.E.: Multivariate Data Analysis, 7th edn. Prentice Hall, New Jersey (2010)

Factors Affecting the Personal Development and Online Education of the Computer Science Students During Covid-19 Pandemic Lockdown

Ochilbek Rakhmanov(✉) and Abzalbek Ulasbekov

Nile University of Nigeria, Plot 681, Cadastral Zone C-OO, Research and Institution Area, Jabi, Abuja, Nigeria
{ochilbek.rakhmanov,abzalbek.u}@nileuniversity.edu.ng

Abstract. In this study, the factors affecting personal development and online education of undergraduate computer science students during pandemic lockdown have been investigated. Due to sudden outbreak of Covid-19, undergraduate students were forced to shift to home-schooling, an experience they never tried before. The results of the conducted survey and some academic performance features were combined to identify the factors affecting personal development and online education. Online lecture attendance and having personal timetable were found to be effective factors on personal development. We observe that final year students are more likely to take part in personal development activities. The availability of the technological facilities has direct effect on online course attendance and satisfaction, moreover, there was strong correlation between online education satisfaction and personal development.

Keywords: Education · Personal development · Distance education · Technology in education

1 Introduction

The quote from Stanford University senior "It is funny that we are talking about things outside the classroom because I feel like that is the place that I have done my most growing" can give a broad view on personal development and its effect on student's academic and work life performance [1]. Personal development can be summarized as intentional involvement in activities that improve awareness, develop talents, boost human potential and all of these factors contribute to the realization of goals and dreams [2]. There are many researches supporting the fact that what happens outside the classroom can contribute to valued outcomes of college, but not all the students benefit equally from the same experience [1]. The intention of student for personal growth can be measured [2], or it can be set as a goal by developing a Personal Development Plan [3]. Goal-oriented programs for struggling students can have a positive effect on academic performance.

Personal development (PD) is extremely important in computer science education [3]. As today's technology is improving rapidly, the need to follow up on PD plan and

M. Al-Emran et al. (Eds.): ICETIS 2021, LNNS 299, pp. 197–206, 2022.
https://doi.org/10.1007/978-3-030-82616-1_18

to stay continuously updated about recent software and IT solutions is highly important for students studying Computer Science (CS). Some common activities for PD for CS students include but not limited to reading new CS related books, taking educative online courses about recent developments in the field, participate in some CS related projects during out-of-classroom time to improve skills [1, 3].

An important aspect of today's education is of course online education and its effect on student's development. Computer-based education allows students to evolve into active learners and offers more freedom [4]. As online education has become a mainstream tool in modern education [5], it brings its own challenges as well. Two main challenges are retention and satisfaction of the student [6]. There are many studies that proposed some factors, which affect the retention and satisfaction of the students during online education. Such factors include: availability (internet, laptops, etc.) and the use of the technology, instructor's performance, instructor-student and student–student interaction level and ease of use of course materials and practically useful website [4, 6–8].

While long-term planned online education may show a good level of satisfaction from student's aspect, that may not be the same case for the universities forced to shift to online education suddenly due to unexpected reasons [9, 10]. This research specifically focused on exploring the outcomes of one of such situations, factors affecting the personal growth and online education of the computer science students during Covid-19 pandemic lockdown. The study concentrates on students from private university in Abuja, Nile University of Nigeria.

Unfortunately, the technological facilities in most of the Nigerian educational institutions didn't allow them to proceed to online education during pandemic lockdown period and this was presented in several previous researches as well [12–16]. Only few private educational institutions were able to continue online education and Nile University of Nigeria was one of them, within 2-week time frame, the management and students managed to shift to online education, which they never experienced before.

The lack of technological facilities is not familiar to educational institutions only, households also suffer same obstacles, like poor access and interruption of the electricity or expensive internet providers with restricted coverage area [12–15, 17].

Covid-19 forced many schools in the world to shift to homeschooling creating tremendous challenges [18]; students social life was affected, education moved to uncertain method with many trials and errors, evaluation methods changed and most importantly, the spread of Covid-19 created serious anxiety level in students [18]. In one of such recent studies, Rakhmanov and Dane stated that more than 70% of the Nigerian students participated in their study were experiencing some level of anxiety, while almost one-fourth of them felt into the group with severe anxiety [11, 19, 20]. But there is one aspect which can be classified as opportunity during pandemic period, students were forced to stay in their houses for more than 45 days during lockdown (in Nigeria from 21st of March to 5th of May, 2020), and such a long time was ideal for CS students to develop a PD plan and invest on their abilities and skills. One such supportive study presented by Gonzalez et al. where the authors derived that Covid-19 confinement actually showed positive effect on students' performance, stating that students learning strategies became more continuous, in comparison to campus education [21].

With no serious previous experience and governmental support, private universities in Nigeria were forced to quick adoption of distance learning to prevent learning disruption during Covid-19. This was to prevent postponement of education until post-pandemic period like federal and state universities did. Thus, it is highly essential for private universities to understand the behavior of the students during suddenly scheduled online classes; how to keep them engaged, what to advise and how to keep them on track without losing focus.

As the Covid-19 pandemic eruption caught the education field without any preparation in the world, it is not only important to understand how the students may behave during online classes, but also there is a need to understand how the students may behave in total lockdown. Researchers should look into the link on whether a student in total lockdown is expected behave in same way like ordinary students who are not in lockdown. In other words, if the daily activity and engaging them in online classes would keep the students out of pandemic stress and make them concentrate on their own personal development.

1.1 Purpose of the Study

In this study, the factors affecting the personal development and online education during the pandemic lockdown were investigated. The relation between study level, gender, academic performance, online education attendance and satisfaction, self-control through using personal daily timetable, availability of the facilities and personal development were investigated.

2 Review of Related Literature

Several previous studies helped to develop the methodology during this research. Online education, factors on online education satisfaction and factors effecting personal development are the main fields investigated during literature review.

Bolliger stated that instructor, technology and interactivity are discovered to be the main components out of selected seven components effecting the satisfaction of online education [7]. Same components are also discovered as important factors in some other researches, while student characteristics, course and website content quality, students support and other circumstantial variables are found as effective factors too [4, 6, 8]. Several comparative studies have been conducted to see the effect of online education and its satisfaction between different groups. Cole et al. couldn't detect any statistically significant difference in online education satisfaction, when they compared different groups like gender, age or level of study [24]. Kuo et al. stated that self-regulation has importance on online education performance and satisfaction [8]. The same claim was supported by Herbert, deriving from results the conclusion that students whose attendance was continuous and completed the course materials were more satisfied in comparison to those who didn't [6].

The effect of personal development and usage of personal development plan on student's life were addressed by several researchers. Patanapu et al. used Personal Growth Initiative (PGI) to conclude that students with better academic performance have higher average of intentional behavior for personal growth, while final year students' readiness for the pandemic change resulted in higher average score when compared to lower levels [2]. Irons strongly supported that personal development plan should be used during education of computer science students, as this would result with development of creativity and computing skills [3]. Kari and Savolainen looked into personal development from internet usage aspect [25]. They established 11 different relationships between internet searching and the goals of personal development.

The aforementioned studies show that retention and satisfaction of students may depend on several factors like gender, level of study, the course website content, attendance and self-regulation. Personal development was also investigated and several factors like academic performance, year of study and ability of search were significant factors effecting the development. Distance learning is the field where all factors like satisfaction, retention and personal development are coming together to produce an efficient factor to evaluate student's development as an online learner.

3 Methodology

3.1 Ethics and Standards

The experimental protocol was by following international ethical standards. The study was performed per under the Helsinki Declaration (1975, revised in 1996–2013). It was a descriptive cross-sectional study. The aims and objectives of the study were explicitly explained to the participants before the commencement of the study. All participants voluntarily gave written informed consent to participate in the study. All the names were removed.

3.2 Participants

All students of Computer Science Department, Nile University of Nigeria, were personally invited to participate in the survey. An online questionnaire was created and shared with students who volunteered to participate. The survey was conducted during pandemic lockdown, when all students were in self-isolation and continued homeschooling. To keep privacy, the secured internet link was produced for online survey (protected with password). A total of 107 students (53 male (49.5%) and 54 female (50.5%) students) participated in this survey, which is almost 40% of total student population in the department. We have 31 students (29.0%) from 100 level (1st year of study), 36 students (33.8%) from 200 level (2nd year of study), 31 students (29%) from 300 level (3rd year of study) and 9 students (8.5%) from 400 level (final year of study). While number of students from 100,200 and 300 level were statistically sufficient, the number of 400 levels was low, but still it represented 35% of total number of 400 level students. CGPA scores were used during the study, and permission of the students and the management were obtained. At Nile University of Nigeria, the CGPA score is continuous variable, ranging from 0.0 to 5.0. At the same time, the CGPA scores are divided into 3 categories; First class (scores between 5.00–4.00), Second class (3.99–3.00) and finally Third class (2.99 and below).

3.3 Data Collection

A survey questionnaire consisted of 5 parts. Each subsection of the questionnaire was designed with respect to existing literature. To be able to do comparative study, the total scores for OES, AF, PD were needed, thus, Likert scale was the preferred measurement.

Part 1. In this subsection, demographic information of the students, study level and gender, were asked. CGPA score (and CGPA class) of the student was also included later after survey conclusion.

Part 2. Online education satisfaction (OES) was measured in this subsection. Students were asked to rate four items on 5-point Likert-scale; performance of the instructors, content of the course, the technology used during online class and interactivity level [7]. An OES index was subsequently built by adding the value of four items, where minimum score was 4 and maximum was 20 points.

Part 3. Availability of facility (AF) was measured in this subsection. Students were asked to rate four items on 3-point Likert-scale (Unlimited/Nonstop, Manageable, Serious challenges); the availability of internet, the availability of continuous electricity, the availability of personal electronics (laptops, tablet, etc.) and lastly availability of personal space in the house for self-study. Questions were selected referring some previous literatures [4, 13, 15]. An AF index was subsequently built by adding the value of four items, where minimum score was 4 points and maximum score was 12 points.

Part 4. Personal development (PD) was measured in this subsection. Students were asked to rate four items on 3-point Likert-scale (Two or more, One, None); during the 45-day lockdown, how many CS related books they have finished, how many online courses they have finished, how many computing and coding related projects they have started or were involved in, and lastly if they tried to learn new professional skills [3, 26, 27]. A PD index was subsequently built by adding the value of four items, where minimum score was 4 points and maximum score was 12 points.

Part 5. In this part students were asked to answer if they have daily timetable to follow. Also, the attendance to online lectures of the students were obtained and classified into two classes; students who participated to at least 75% of the lectures and those whose participation rate was lower than 75%. While Kuo et al. stated that self-regulation has importance on online education [8], Herbert stated that students whose attendance was continuous were more satisfied with courses [6]. Thus, personal timetable and attendance were used during the research. During statistical analysis, students were separated into two groups in terms of timetable usage and online class attendance. While students were grouped as High (attendance is higher than 75%) and Low (attendance is lower than 75%) in terms of online class attendance, they were also grouped as Yes (those who used daily timetable) and No (those who didn't use daily timetable) in terms of timetable usage.

Cronbach's alphas for the subsections OES ($\alpha = 0.85$), AF ($\alpha = 0.75$) and PDS ($\alpha = 0.74$) were calculated and all found to have good internal consistency.

3.4 Statistical Analysis

Measured values are given as a mean±standard deviation (std.dev.). Statistical analysis was performed using the software SPSS for Windows, version 26. To compare the online education satisfaction and personal development scores in the participants, the Student's t-test was used to get the gender, level, class attendance and timetable differences. One-way ANOVA (post-Hoc Tukey test) was used to get study level and CGPA class differences. A p-value of less than 0.05 was considered statistically significant. The Pearson correlation was employed to establish correlation between subsection measures.

4 Results

4.1 Factors Effecting Personal Development

Student's t-test results showed that there was significant difference between online lectures attendance and timetable groups, while no such significance was found for gender. Personal development score was significantly higher ($t = 2.93$, $p = 0.004$) for students whose online lecture attendance was high in comparison to those who was low. In the same manner, personal development score of the students who have daily timetable was higher and it was statistically significant ($t = 3.83$, $p < 0.001$), in comparison to those who didn't have one. Table 1 presents the results for t-test for both online lectures attendance and timetable. It can be derived from the results of the Table 1 that students with high level of lecture attendance and those who use daily timetable tend to have higher personal development score when compared to other groups, respectively. So, attendance and timetable usage are not only correlated to retention and satisfaction of students, but they are also effective on personal development of the students.

Table 1. Personal development scores for different groups in attendance and timetable

Test variable	Groups	Mean ± Std.dev.	T value	p value
Attendance	High (N = 63)	7.43 ± 2.21	2.93	0.004
	Low (N = 44)	6.27 ± 1.66		
Timetable	Yes (N = 32)	8.06 ± 1.99	3.83	0.000
	No (N = 75)	6.48 ± 1.93		

The difference between personal development scores was compared for different study levels and CGPA classes using one-way ANOVA test. 400 level students had statistically significant higher personal development score ($F(3,103) = 3.37$, $p = 0.02$) in comparison to other study levels; 100 level (difference = 0.61), 200 level (difference = 0.92) and 300 level (difference = 2.06). Table 2 shows descriptive statistics for each level. At this point, it is important to mention again that number of samples for 400 levels were relatively small. Unlike the study level, there was no statistically significant difference for CGPA classes. Thus, it can be concluded that students of final year study are likely to have higher personal development score comparing to other levels, and it doesn't matter if the education is on-campus or online, final year students try to pay attention to personal development to get ready for the industry.

Table 2. Personal development scores for all study levels

Level	Mean ± Std.dev.	N
100	7.39 ± 2.18	31
200	7.08 ± 2.32	36
300	6.06 ± 1.41	31
400	8.00 ± 1.73	9

There was statistically significant correlation between online education satisfaction and personal development ($r = 0.36$, $p < 0.001$), which supports the results of Student's t-test on online lecture attendance and timetable. So, if the student is satisfied with online education, he/she is likely to have higher personal development score as well.

4.2 Other Important Findings

It turns out that online lecture attendance not only affects personal development, but also has strong relation with online education satisfaction and availability of the facilities. While there is statistically significant difference ($t = 3.66$, $p < 0.001$) in terms of online education satisfaction score for students with high and low attendance, it can be observed that lack of availability of facilities is strongly related with the attendance. Table 3 is summary for these t-test calculations which concludes that high rate of attendance to online lectures trigger the satisfaction as well, or vice versa, while the availability of facilities has serious effect on attendance.

Table 3. The satisfaction and availability of the facilities scores for attendance groups

Test variable	Attendance	Mean ± st.dev.	T value	p value
Online education satisfaction	High (N = 63)	13.17 ± 3.23	3.66	0.000
	Low (N = 44)	10.7 ± 3.7		
Availability of the facilities	High (N = 63)	9.97 ± 1.7	4.00	0.000
	Low (N = 44)	8.34 ± 2.5		

There is a strong statistically significant correlation ($r = 0.50$, $p < 0.001$) between online lecture satisfaction and availability of the facilities score, which supports the fact that higher attendance triggers higher satisfaction score and high attendance depends on availability of these facilities.

5 Discussion of the Results

In this paper, several factors and their relations between online education and personal development were investigated. In this section, we compare our findings with existing literature and further discussion are presented.

The results in Sect. 4 show that having a daily timetable and good online course attendance have strong correlation with personal development and online education satisfaction. This supports the claim from research of Kuo et al., that self-regulation has importance on online education [8], and from Herbert, who stated that students whose attendance was continuous were more satisfied with courses [6]. Thus, as Herbert stated, retention problem of the students in online education is creating a great weakness and affecting further personal development. The results also supports the claim of Irons, who underlined that having personal development plan is highly important for computer science students [3]. Students characteristics was listed as important measurement for evaluation of the online education satisfaction by Volery [4]. Indeed, self-control factors like having daily timetable and online lecture attendance affect not only online education satisfaction, but also personal development.

The personal development score of the final year students found to be higher than other study levels and this finding supports the claim of Patanapu et al., that the factor "Readiness for change and improvement" was higher for 4th-year students (final year). But the results of this study conflicts with Patanapu et al. finding about academic performance and personal growth. The calculations during this study didn't present any significant fact that students who have higher CGPA or fall into higher CGPA class have higher personal development score.

Access to internet and technological facilities were mentioned by Volery and Herbert as important factors for online education, while Oyelaran-Oyeyinka and Fawehinmi stated that lack of technological infrastructures puts serious challenges for education in Nigeria [4, 6, 12, 15]. The results in Sect. 4 strongly support these claims, as lack of facilities (internet and electricity) and electronic devices affect the online education participation and satisfaction.

The results of this study also highlight that irrespective of whether a student is forced to learn under a pandemic lockdown, the factors affecting his/her satisfaction and personal development are almost same with a student who is not, as the results found during this study supports the existing literature. Thus, if an educational institution is forced for sudden (like Covid-19 lockdown) online education, they are advised to follow up previous researches on how to keep students engaged during online classes.

To summarize the results of the study, having daily attendance and keeping high level of online class attendance are some of the key factors to foster student's personal development. Having a better facility is one other key factor that facilitates online-class satisfaction and personal development.

6 Conclusion

This research was conducted in the capital city of Nigeria, Abuja, which has better technological infrastructure when compared to other regions of Nigeria. But even in the capital territory, the factors discussed in this paper seriously affected undergraduate students' online education and personal development during the pandemic lockdown. For institutions who have never experienced homeschooling before, and are planning to shift to online education, we recommend the following; keeping attendance on high level, ensuring that students follow the daily timetable, and discussion with parents for provision of best possible facility options for the students during online education. These listed factors directly affect personal development as well.

On the other hand, the engagement of the students during online classes appears to be highly important to keep student away from pandemic stress. Setting goals for students' personal development may boost this. Thus, promoting the setting goals for personal development would help the educational institutions to keep students focused on their studies without losing focus.

References

1. Kuh, G.D.: The other curriculum: out-of-class experiences associated with student learning and personal development. J. High. Educ. **66**, 123–155 (1995)
2. Patanapu, S.K., Doshi, D., Kulkarni, S., Reddy, P., Adepu, S., Reddy, S.: Does academic performance influence personal growth initiative? An institutional-based study among undergraduate dental students. J. Educ. Health Promot. **7** (2018)
3. Irons, A.: Using Personal Development Plans (PDPS) to facilitate learning for computing students (2003)
4. Volery, T., Lord, D.: Critical success factors in online education. Int. J. Educ. Manage. (2000)
5. Ferguson, J.M., DeFelice, A.E.: Length of online course and student satisfaction, perceived learning, and academic performance. Int. Rev. Res. Open Distrib. Learn. **11**, 73–84 (2010)
6. Herbert, M.: Staying the course: a study in online student satisfaction and retention. Online J. Distance Learn. Adm. **9**, 300–317 (2006)
7. Bolliger, D.U.: Key factors for determining student satisfaction in online courses. Int. J. E-learn. **3**, 61–67 (2004)
8. Kuo, Y.-C., Walker, A.E., Belland, B.R., Schroder, K.E.: A predictive study of student satisfaction in online education programs. Int. Rev. Res. Open Distrib. Learn. **14**, 16–39 (2013)
9. Rakhmanov, O.: A novel algorithm to classify hand drawn sketches with respect to content quality. In: Gervasi, Osvaldo, Murgante, Beniamino, Misra, Sanjay, Garau, Chiara, Blečić, Ivan, Taniar, David, Apduhan, Bernady O., Rocha, Ana Maria A C., Tarantino, Eufemia, Torre, Carmelo Maria, Karaca, Yeliz (eds.) ICCSA 2020. LNCS, vol. 12252, pp. 179–193. Springer, Cham (2020). https://doi.org/10.1007/978-3-030-58811-3_13
10. Rakhmanov, O.: On validity of sentiment analysis scores and development of classification model for student-lecturer comments using weight-based approach and deep learning. In: Proceedings of the 21st Annual Conference on Information Technology Education, pp. 174–179 (2020)
11. Rakhmanov, O., Dane, S.: Knowledge and anxiety levels of African university students against COVID-19 during the pandemic outbreak by an online survey. J. Res. Med. Dent. Sci. **8**, 53–56 (2020)
12. Fawehinmi, A.S., Oyerinde, O.V.: Household energy in Nigeria: the challenge of pricing and poverty in fuel-switching. J. Energy Dev. **27**, 277–284 (2002)
13. Monyei, C.G., Adewumi, A.O., Obolo, M.O., Sajou, B.: Nigeria's energy poverty: Insights and implications for smart policies and framework towards a smart Nigeria electricity network. Renew. Sustain. Energy Rev. **81**, 1582–1601 (2018)
14. Osang, F.: Internet access in Nigeria: perception of National Open University of Nigeria (NOUN) students. Int. J. Emer. Technol. Adv. Eng. **2**, 492–497 (2012)
15. Oyelaran-Oyeyinka, B., Adeya, C.N.: Internet access in Africa: empirical evidence from Kenya and Nigeria. Telematics Inform. **21**, 67–81 (2004)

16. Al-Emran, M., Granić, A., Al-Sharafi, M.A., Ameen, N., Sarrab, M.: Examining the roles of students' beliefs and security concerns for using smartwatches in higher education. J. Enterp. Inf. Manage. (2020)
17. Arpaci, I., Karataş, K., Baloğlu, M.: The development and initial tests for the psychometric properties of the COVID-19 Phobia Scale (C19P-S). Pers. Individ. Differ. **164**, 110108 (2020)
18. Burgess, S., Sievertsen, H.H.: Schools, skills, and learning: the impact of COVID-19 on education. VoxEu.org. **1** (2020)
19. Rakhmanov, O., Demir, A., Dane, S.: A brief communication: anxiety and depression levels in the staff of a nigerian private university during COVID 19 pandemic outbreak. J. Res. Med. Dent. Sci. **8**, 118–122 (2020)
20. Rakhmanov, O., Shaimerdenov, Y., Dane, S.: The effects of COVID-19 pandemic on anxiety in secondary school students. J. Res. Med. Dental Sci. **8**, 186–190 (2020)
21. Gonzalez, T., et al.: Influence of COVID-19 confinement in students performance in higher education (2020). arXiv preprint arXiv:2004.09545
22. Okpaga, A.: Distance learning, sandwich and open education system in Nigeria: challenges and prospects for educational development. Int. J. Acad. **1**, 343–352 (2016)
23. ASUU gives reasons why online learning cannot commence in Nigeria amidst coronavirus lock down. https://www.kemifilani.ng/2020/04/asuu-gives-reasons-why-online-learning-can-not-commence-in-nigeria-amidst-coronavirus-lock-down.html. Accessed 08 Nov 2020
24. Cole, M.T., Shelley, D.J., Swartz, L.B.: Online instruction, e-learning, and student satisfaction: a three year study. Int. Rev. Res. Open Distrib. Learn. **15**, 111–131 (2014)
25. Kari, J., Savolainen, R.: Relationships between information seeking and context: a qualitative study of Internet searching and the goals of personal development. Lib. Inf. Sci. Res. **29**, 47–69 (2007)
26. Cegielski, C.G., Hall, D.J.: What makes a good programmer? Commun. ACM **49**, 73–75 (2006)
27. Goodliffe, P.: Becoming a Better Programmer: A Handbook for People Who Care About Code. O'Reilly Media, Inc. (2014)

JomNFCSignIN: Seamlessly Register Your Class Attendance

A. T. I. Mazumdar[1], P. S. JosephNg[1(✉)], H. C. Eaw[2], K. Y. Phan[3], and Y. P. Yeck[3]

[1] Institution of Computer Science and Digital Innovation, UCSI University, UCSI Heights 56000 Cheras, Kuala Lumpur, Malaysia
josephng@ucsiuniversity.edu.my

[2] Faculty of Business and Management, UCSI University, UCSI Heights 56000 Cheras, Kuala Lumpur, Malaysia
eawhc@ucsiuniversity.edu.my

[3] Faculty of Information and Communication Technology, Universiti Tunku Abdul Rahman, Kampar, Perak, Malaysia
{kyphan,yeckyp}@utar.edu.my

Abstract. Attendance in educational institutions worldwide has become a major requirement. The evidence is showing that the traditional pen-paper method lost a tremendous amount of time and numerous forgeries occur. Smartphones are increasingly widespread with technology and Near Field Communication is a crucial factor for the payment and other contactless methods of transferring information. This allowed faculty members to reduce their workload and to focus more on teaching. This proposed study will explain how the application will help the students participate in the class with ease and efficiency. Besides that, it will also allow the lecturers to have more time to spend. The application will track student attendance, follow the rules of the university, and provide a significantly cheaper operating and set-up infrastructure. The data are gathered by interview sessions and surveys from various government and private universities using a mixed-mode method and with very optimistic results.

Keywords: Attendance system · Near Field Communication · Automated · NFC mobile application · Wireless attendance system

1 Introduction

Monitoring the attendance of every student at every educational institution is a major factor because it is related directly to their academic success [1]. In addition to that, every student has a certain attendance rate. It will be calculated according to the average percentage for each of the courses they enrolled in. Research has proven that there is a compelling bond between students' academic level and the level of attendance [2]. Sometimes the student decides not to attend the class and that gives the lecturer a tough time to complete their tasks as per their schedule [3]. Therefore, students in Malaysia must have their attendance percentage of more than 80% every semester. This will allow them to proceed forward with their exam at the end of their semester [4]. Besides that,

M. Al-Emran et al. (Eds.): ICETIS 2021, LNNS 299, pp. 207–216, 2022.
https://doi.org/10.1007/978-3-030-82616-1_19

the traditional way of recording students' attendance has a couple of drawbacks. For instance, the lecturer calling out the names of the student, sign on a paper which is being reused for the whole semester that sometimes gets ruined due to continuous usage [1]. Other students sign for his or her friend if they are absent and if the lecture class has more than 100 students [5] then either it will take a long time or there will be a long queue to sign up after the class which disrupts the next class [6].

To correct the flaw, we can overcome the new technical system and reduce the time and hassle for the attendance registration by implementing technology like Near Field Communication (NFC) [7, 8]. The world is relying more on technologies and the development of the technology will for sure increase the reliability [9] and That will save the lecturer a precious amount of time [10]. With the help of the technology, the University be leading technology out from the other competitors Not to mention, this method will also reduce the cost of the university plus help the community to transition towards going green. With the expansion of online payment systems such as Samsung Pay and Google Pay, NFC is a leading wireless technology. Particularly when it comes to high-end devices and even many mid-range devices. But what exactly is NFC? This paper will explain, how it works and how is it being implemented in the attendance system. NFC is called "Near Field Communication," enabling short-range communication among compatible devices, as its name implies. At least one transmitter, and another to receive the signal. Several devices can be considered passive/active by the NFC standard. Tags and other small transmitters are part of passive NFC devices that can send data to other NFC devices without their power source. However, the information transmitted from other sources is not processed and cannot communicate with other passive components.

1.1 Literature Review

NFC is a type of communication protocol that allows you to exchange data wirelessly in a short distance which is approximately 5 to 10 cm [11]. It is and an upgraded version of the previous RFID technology which adds up both reader and smart card interface into a single standalone device. This can be used as a contactless method for identifying, authenticating, and transfer information. Plus, the technology which it packs gives more flexibility in data integration and much higher security compared to the traditional method [11]. That kind of secure communications in NFC is made possible by the implementation of cryptographic algorithms. Which urged many tech giants like Apple, Samsung & Google to endorse NFC with their payment system [12]. Therefore, it proves that there will be no compromise in terms of security while using NFC while signing in for the lass. It does not need any sort of initial setup as it uses magnetic field induction to start up its communication between the devices [13]. The radiofrequency of NFC is 13.56 MHz which is operated within the global and unlicensed radio band [14]. NFC is designed to operate as a short-range wireless technology which has been explained more in detail with comparison in Table 1. There are three different modes for NFC to operate, which is Card emulator mode, reader/writer mode, and peer-to-peer mode [15]. In Card emulator mode, Technology will the NFC-enabled device to work as an actual card where it accepts the user interaction. Next, we talk about, Read/Write mode. It allows the user to store a piece of short information in the device which will be used later. Lastly, we talk about Peer to peer mode, this allows the user to exchange information in between NFC enabled devices [16, 17] as illustrated in Fig. 1.

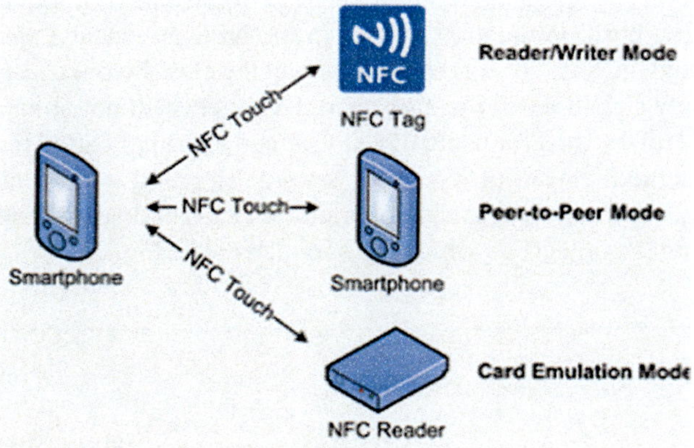

Fig. 1. Modes of NFC to operate [18]

1.2 Problem Statement, Question and Objective

Using the traditional method can lead to many security vulnerabilities [19, 20] as mentioned earlier, which can lead to affect the students post-university real-life scenario as they have missed some vital pieces of information during the time of the class and that will eventually lead to ruin the University's reputation and cause blame to their education system. With all features developed for the app JomSignIN, the students can easily and securely register for their class. check their live attendance percentage and keep track of whether they got barred from their class or not. Apart from that, this method can be used during the final exam to reduce the long queue during the entrance of the hall and reduce the possibility of fraud students attending the exam. Thus, having a better examination standard. In addition to that, every student's parents can also download the application and log in as guardians and keep up to track their child's attendance record. The purpose of this study was to answer the question of research in Table 1.

Table 1. Research objective

R01: To motivate University to use NFC based attendance system application and save costs on printing the papers
R02: To support the students to build a better effective academic rate by notifying the students with a push notification service in the NFC attendance application
R03: To build up the trust for using the NFC attendance application rather than using the traditional method

R01: To encourage the University to use NFC based attendance system application and save costs on printing the papers. By using the application, in the long run, will help the university to reduce paper waste and be more sustainable to the environment.

R02: To support the students to build a better effective academic rate by notifying the students with a push notification service in the NFC attendance application, if by any chance the student is late or forgetting to attend the class by oversleeping. But if the student has already signed in for the class then the system will not notify the student.

R03: To build up the trust for using the NFC attendance application rather than using the traditional method instead of using conventional method will eventually take the initial step to digitalize a system to the community [21]. These are hypotheses for each point derived from the objectives which are summarized in Fig. 2.

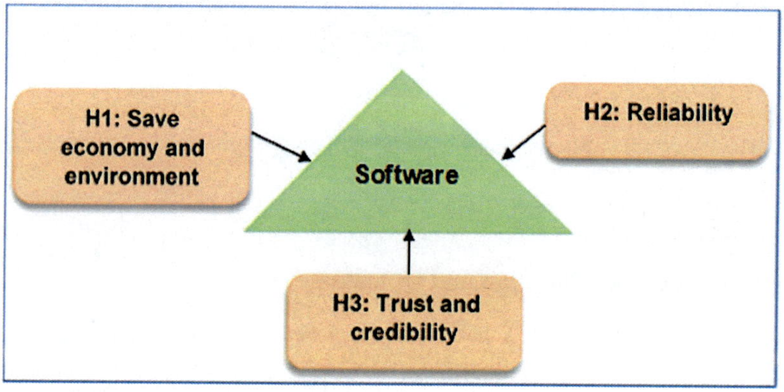

Fig. 2. Hypothesis research model

H1: University operation cost will be reduced using NFC based mobile application.

As the world is more focused on transitioning towards eco saving or going green. Using papers for printing the student record is a very aged method. Besides that using the same paper for the whole semester is a bit concern to the hygiene as the sheet undergoes many students every lecture class. The development will not only clear out the hygiene, it will save a huge amount of resources which can be used elsewhere for a better purpose.

H2: Student attendance quality will be improved using NFC based mobile application. By allowing the students to see their live absence of the course in which they have selected the course to see how many classes they have missed. In addition to that, if the student is reaching the level to which they will get barred. The lecturer will be notified and an automated message will be sent to the student for counselling to check on the student's whereabouts. Besides that, if the student has valid reasoning then the administrator can take the necessary steps to help the student to deal with the immigration officer for the international students.

H3: University technology leadership branding will be improved using NFC based mobile application. By using this application, the students will sign-in securely to the system so that no one will be able to edit or alter the student's attendance information as shown in Table 2. User interaction will make the attendance system more convenient plus, by using this app the student can also have secure access to the library and other specific places that have been approved by the administration like if a scholarship student is having a part-time job then this app can be used to access the authorized places. Not only that, but they also track down the last known location of the student when they use this application [22–26].

Table 2. Research hypothesis

H1: University operation cost will be reduced using NFC based mobile application
H2: Student attendance quality will be improved using NFC based mobile application
H3: University technology leadership branding will be improved using NFC based mobile application

The features which have been developed in the app will allow the student to record the attendance and check the live record of their attendance percentage so that at the end of the semester they will be assured to sit for the exam. Besides that, the app allows the student to see the future schedule and create his/her tasks which will give them better time management and engage in more social activity in the community. The students will also be notified with a push notification service if they have not attended the class. For those who have missed classes and their percentage level has been decreased from the minimum required percentage, then the student will get an automated message from the lecturer's class and the faculty will be notified as well and if there is any sort of confusion then the student will be able to be in touch with the lecturer or course coordinator so that there will not be any sort of colossal damage in the future.

2 Methodology

This study would use a hybrid research approach, composed of both quantitative and qualitative research techniques [27–30]. To do so, By making surveys and interviews which will occur during the data collection, we will collect data more efficiently and easily. Figure 3 provides a sequential design for the analysis dimension which clarifies the different stages of the data collection. A random survey and interviews are the primary data collection tool which is implemented in the mixed-mode or mixed-method. The sequential design is shown in Fig. 3 to expand the steps to collect data.

The survey will be conducted by a randomly selected student at a few of the private and public Universities from different countries like Malaysia, the data collection be conducted at both public and private universities.

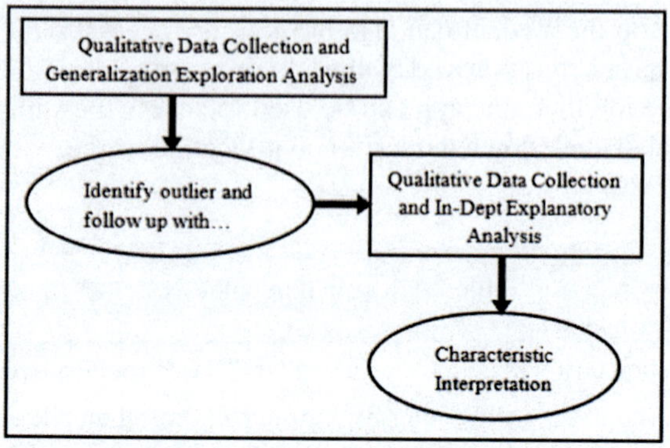

Fig. 3. Mix mode sequential design [31–37]

3 Results and Findings

According to the research that has found out that the preference of using the application is more favourable over the traditional method in both the preliminary and technical data collection which has been shown in below Fig. 4 of the responses in percentage form. Merely 84.6% of the respondents agree to use the application. Apart from that, 29.9% highly recommend using the app to sign in. This suggests that the application should be implemented in the university's attendance record system.

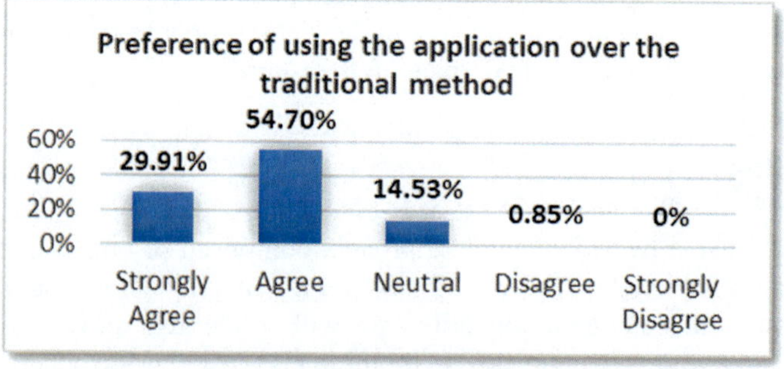

Fig. 4. Post-test results of preference method.

54.70% of the data shown in Fig. 4 voted for the application with the implementation of the push notification service application for their courses, which will raise their awareness. If the last-minute changes are not discovered, then the application will alert the students. It is important to keep students informed. Moreover, it will also warn the student if during the time he or she did not sign up. Besides, during the interview, many positive responses were received, as most of them emphasized that some students can typically be forgetful as shown in Fig. 6. A notification is also always a saver and decreases the risk of missing schools (Fig. 5).

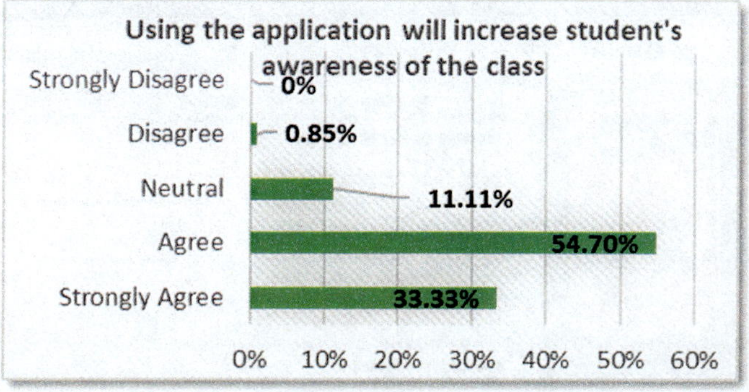

Fig. 5. Post-test result of using the application will increase student's awareness of the class

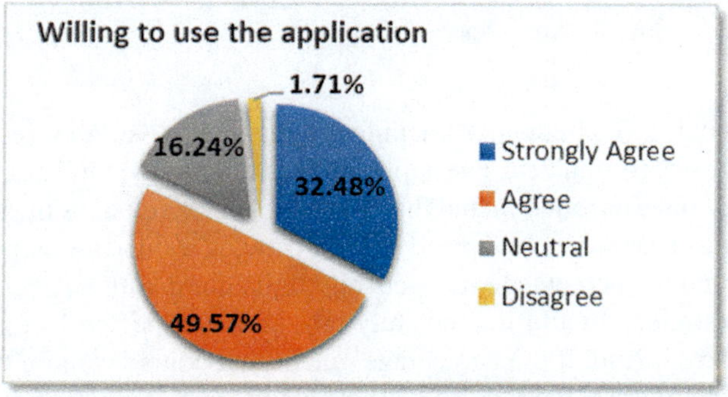

Fig. 6. Post-test result of students willing to use the application.

The data in Fig. 6 shows that it is much safer and easier to use their smartphone application to sign-in rather than to wait for an attendance form or wait for the teacher to call out the student's names. This will encourage the teachers to spend more time and have less distraction in the classroom. Qualitative research showed that 84% agreed with the NFC approach, that the university would benefit the economy, as close to 85% of worldwide mobile users have incorporated NFC into their phones [38]. That's why the application should be used at the university. Moreover, 91% of respondents believe that printing is just a waste of paper. And 78% of respondents reacted positively to have better time management, thus managing their schedule effectively without conflict at other activities. The respondents from the interview gave positive feedback on the university's brand image as they shift in the digital era. In addition to enhancing the attendance system, this would also send a good image to the community in which the university is more sustainable and affect other society in the progress of digital technology.

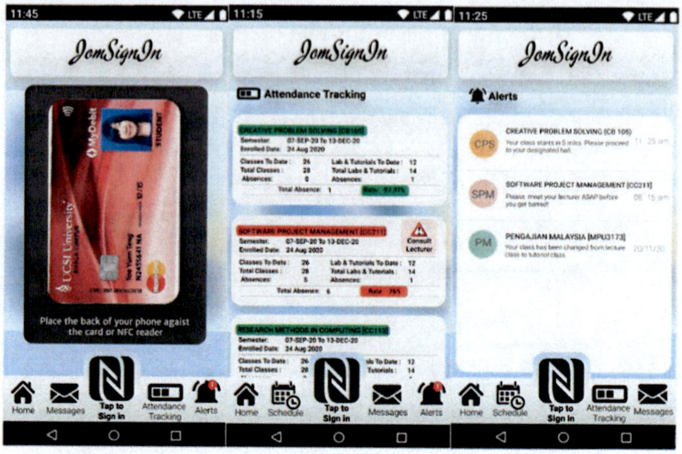

Fig. 7. Attendance tracking and alert on the app

As illustrated in Fig. 7, attendance tracking is completely live. As soon as the student signs in, the database gets updated. The application calculates the total number of lecture classes which they have enrolled in and then looks at how many days they have missed, then shows the result as a percentage. If the percentage value is less than 80% then the course name and the percentage rate gets highlighted red with an exclamation mark showing that the student should immediately talk to the respected lecturer before the student getting barred. And, if the percentage rate of that course is more than 80% then in green colour. The figure at the right shows how alert the student is being alerted before 5 min of each class and of any change of events. Figure 7 shows how the student can easily tap and sign in with their phone's NFC. The user taps to sign in, and then the student's ID will show as a verification of the user so that other students will not be able to use it and alter their attendance. After a few users tested the app, with all the features included while development, the process of signing in for class has never been easier. And also as per the interviews would like to use this app in the nearby future. Without fault, all features have been effectively built.

4 Conclusion and Future Works

To conclude, JomNFCSignIn will drastically help to boost the student's life experience in the university and have established a stronger relationship with the lecturers with the help of live tracking of attendance and direct communication with them. There was some complication while doing the data collection and analysis due to time constrain and limited resources. Nevertheless, this application has shown a lot of improvement and has the potential to grow which will lead other future users to modify and optimize as time passes by. Some of the future possible improvements could be having a stronger verification login so that the user's identification cannot be stolen and have much more seamless sign inexperience.

References

1. Qureshi, R.: The proposed implementation of RFID based attendance system. Int. J. Softw. Eng. Appl. **11**(3), 59–69 (2020)
2. Jacksi, K., Ibrahim, F., Ali, S.: Student attendance management system. Scholars J. Eng. Technol. (SJET) **6**(2), 49–53 (2018)
3. Mohandes, M.A.: Class attendance management system using NFC mobile devices. Intell. Autom. Soft Comput. **23**(2), 251–259 (2017)
4. Hooi, Y.K., Tachmammedov, S.: Multi-factor attendance authentication system. J. Softw. Eng. Comput. Syst. **4**(2), 2–79 (2018)
5. Daramola, C.Y., Folorunsho, O., Ayogu, B.A., Adewole, L.: Near field communication (NFC) based lecture attendance management system on android mobile platform. In: International Science Conference, Nigeria, vol. 32(7), pp. 34–38 (2019)
6. Mwangi, W.P.: Class attendance monitoring system using NFC. J. Adv. Res. Comput. Commun. Eng. **4**(2), 350–353 (2017)
7. Chew, C.B., Mahinderjit-Singh, M., Wei, K.C., Sheng, T.W., Husin, M.H., Hassain, M., Nurul, H.A.: Sensors-enabled smart attendance systems using NFC and RFID technologies. Int. J. New Comput. Architect. Appl. (IJNCAA) **5**(1), 19–28 (2015)
8. Hamzah, M.L., Desnelita, Y., Purwati Sastri, A., Rusilawati, E., Kasman, R., Rizal, F.: A review of near field communication technology in several areas. Rev. Espacios **32**(40), 2–19 (2019)
9. Liow, K.M., et al.: JomDesignLab: bringing artwork design nearer. In: IEEE International Conference on Control Systems, Computing & Engineering, Penang, Malaysia, pp. 1–6 (2020)
10. Kang, C.M., et al.: A study on integrating penetration testing into the information security framework for Malaysian higher education institutions. In: International Symposium on Mathematical Science & Computing Research, Malaysia, pp. 156–161 (2015)
11. bin Mohd Nasir, M.A.H., bin Asmuni, M.H., Salleh, N., Misra, S.: A review of student attendance system using near-field communication (NFC) technology. In: Gervasi, O., et al. (eds.) ICCSA 2015. LNCS, vol. 9158, pp. 738–749. Springer, Cham (2015). https://doi.org/10.1007/978-3-319-21410-8_56
12. Widiyanto, A., Nuryanto: Prototype of NFC reader as an attendance sign at the presence system. IOP Conf. Ser. J. Phys. **1196**(1), 6–11 (2019)
13. Cao, Z., Chen, P., Ma, Z., Li, S., Gao, X., Wu, R.-X., Pan, L.: Near-field communication. Sensors **19**(1), 39–47 (2019)
14. Liébana-Cabanillas, F., Ruiz, M.: To use or not to use, that is the question: analysis of the determining factors for using NFC mobile payment systems in public transportation. Technol. Forecast. Soc. Change **139**(July), 266–276 (2019)
15. Ooi, K.-B., Tan, W.-H.: Mobile technology acceptance model: an investigation using mobile users to explore smartphone credit card. Expert Syst. Appl. **59**(2016), 33–46 (2016)
16. Kommey, B., Amuzu, W.E.: SwyftTapp: an NFC based attendance system using fingerprint. J. Eng. Sci. Technol. **10**(1), 23–39 (2018)
17. Maity, A.S., Patil, M.A., Jadhav, R.S., Shinde, R.P.: Near field communication (NFC) based employee tracking system (ETS). Int. Res. J. Eng. Technol. **05**(03), 496–499 (2018)
18. Lamberti, F.: The survey on NFC. Sensors **15**(6), 13348–13405 (2015)
19. Li, B., Tian, M., Zheng, S., Jiang, S.: Design of access control system based on the near field. IMMAEE **452**(1), 1–7 (2018)
20. Coskun, V., Ozdenizci, B., Ok, K.: The survey on near field communication. Sensors **15**(6), 13348–13405 (2015)
21. Lai, P.C., Scheela, W.: Convergence of technology in the E-commerce world and venture capital landscape in South East Asia. In: Global Entrepreneurship and New Venture Creation in the Sharing Economy, vol. 6(2), pp. 149–168. IGI Global, Hershey, PA (2017)

22. Opoku, S.K.: An automated biometric attendance management. Int. J. Comput. Sci. Mobile Comput. **2**(3), 18–25 (2018)
23. Esfahani, S.S., Bulent Ozturk, A.: The influence of individual differences on NFC-based mobile payment adoption in the restaurant industry. J. Hosp. Tour. Technol. **10**(2), 219–232 (2019)
24. Dutot, V.: Factors influencing near field communication adoption: an extended TAM approach. J. High Technol. Manage. Res. **26**(1), 45–57 (2015)
25. Tan, W.-H., Ooi, K.-B., Chong, S.-C., Hew, T.-S.: NFC mobile credit card: the next frontier of mobile payment? Telematics Inform. **31**, 292–307 (2014)
26. Mane, S., Tikka, K., Deshpande, G., Toke, S.: Smart attendance and location tracking using IoT. Int. J. Eng. Res. Technol. (IJERT) **11**(8), 681–683 (2019)
27. JosephNg, P.S., Eaw, H.C.: Still technology acceptance model? Reborn: exostructure as a service model. Int. J. Bus. Inf. Syst. (2021, Forthcoming)
28. JosephNg, P.S., Eaw, H.C.: Making financial sense from EaaS for MSE during economic uncertainty. In: Springer Advances in Intelligent Systems and Computing (2021, Forthcoming)
29. JosephNg, P.S., Loh, Y.F., Eaw, H.C.: Grid computing for MSE during volatile economy. In: International Conference on Control, Automation and Systems, IEEE Explore, Busan, Korea, pp. 709–714 (2020)
30. JosephNg, P.S., et al.: EaaS: available yet hidden infrastructure inside MSE. In: 5th International Conference on Network, Communication and Computing, ACM International Conference Proceeding Series, Kyoto, Japan, pp. 17–20 (2016)
31. JosephNg, P.S.: EaaS infrastructure disruptor for MSE. Int. J. Bus. Inf. Syst. **30**(3), 373–385 (2019)
32. JosephNg, P.S.: EaaS optimization: available yet hidden information technology infrastructure inside medium size enterprises. J. Technol. Forecast. Soc. Change **132**(July), 165–173 (2018)
33. JosephNg, P.S., et al.: Exostructure services for infrastructure resources optimization. J. Telecommun. Electron. Comput. Eng. **8**(4), 65–69 (2016)
34. JosephNg, P.S., Kang, C.M.: Beyond barebone cloud infrastructure services: stumbling competitiveness during economic turbulence. J. Sci. Technol. **24**(1), 101–121 (2016)
35. Joseph, N.P.S., Mahmood, A.K., Choo, P.Y., Wong, S.W., Phan, K.Y., Lim, E.H.: Barebone cloud IaaS: revitalization disruptive technology. Int. J. Bus. Inf. Syst. **18**(1), 107–126 (2015)
36. Joseph, N.P.S., Mahmood, A.K., Choo, P.Y., Wong, S.W., Phan, K.Y., Lim, E.H.: IaaS cloud optimization during economic turbulence for Malaysia small and medium enterprise. Int. J. Bus. Inf. Syst. **16**(2), 196–208 (2014)
37. Joseph, N.P.S., Mahmood, A.K., Choo, P.Y., Wong, S.W., Phan, K.Y., Lim, E.H.: Battles in volatile information and communication technology landscape: the Malaysia small and medium enterprise case. Int. J. Bus. Inf. Syst. **13**(2), 217–234 (2013)
38. Pu, X., Chan, F.T.S., Chong, A.Y.-L., Niu, B.: The adoption of NFC-based mobile payment services: an empirical analysis of Apple Pay in China. Int. J. Mobile Commun. **18**(3), 343–371 (2020)

Motivation Factors Attributed to Engaging in Online Studies Amongst Public University Students

Balan Rathakrishnan[1(✉)], Soon Singh Bikar Singh[1], Mohd Fahmi Ghazali[2], Azizi Yahaya[1], Noor Hassline Mohammed[1], and Mohammad Rahim Kamaluddin[3]

[1] Faculty of Psychology and Education, Universiti Malaysia Sabah, 88400 Kota Kinabalu, Sabah, Malaysia
{rbhalan,soonbs,azizi.yahaya,hassline}@ums.edu.my

[2] Centre of Global Business and Digital Economy, Faculty of Economics and Management, Universiti Kebangsaan Malaysia, 43600 Bangi, Selangor, Malaysia
fahmi@ukm.edu.my

[3] Centre for Research in Psychology and Human Well-Being, Faculty of Social Sciences and Humanities, Universiti Kebangsaan Malaysia, 43600 Bangi, Selangor, Malaysia
rahimk@ukm.edu.my

Abstract. This research study investigated the engagement and motivation levels amongst young adults of different genders from a public university, with regards to engaging in online studies. The data was obtained through the use of a questionnaire, where a total of 190 public university students in Sabah participated. This study revealed three major outcomes attributed to the motivation levels to engage in online studies. Firstly, there were no significant differences in online learning factors between both male and female students. Secondly, there was a significant relationship between the motivation and engagement factors attributed to online learning amongst the young adults in the public university. Thirdly, it was found that motivation helps predict the engagement outcomes in online learning amongst these young adults. This research contributed to the understanding of online studies, and how it helps students be well prepared in future digital learning environments.

Keywords: Engagement online study · Motivation · Adolescent

1 Introduction

Over the past decade, there have been many topics on the motivation and engagement factors associated with online studies. The internet has been a powerful global tool, which has become an essential aspect of life for many people. It is a mechanism that initiates, directs, and preserves goal-oriented activities along with the principles of motivation. If having a plate of food to alleviate hunger, or reading a book to acquire wisdom, it essentially represents what prompts you to act in that particular situation. Motivation involves biological, emotional, social, and cognitive forces, that activate a particular behaviour.

M. Al-Emran et al. (Eds.): ICETIS 2021, LNNS 299, pp. 217–226, 2022.
https://doi.org/10.1007/978-3-030-82616-1_20

The internet has provided better access to learning, convenience, and flexibility, to help support educational opportunities. Technology has enabled learning, which can be done from any part of the world. Amidst the COVID-19 pandemic, Malaysians, alongside the global community, also struggled in the fight against the deadly virus. The pandemic has indeed highlighted the importance of online learning platform to provide uninterrupted access to study materials.

However, online learning is challenging, as not everyone will agree on this option over a long period. Recently, there was a case in Sabah which caught Malaysians by surprise, when a young student had to climb trees to get online access for her e-learning class. After a few days, the internet service provider arrived at the subject's house to connect the internet. As we can see, students' motivation to gain access to knowledge is high, and it can be correlated to online learning platforms.

Online learning has been introduced for some time now, and it is now being used worldwide, especially in e-learning education platforms. The problem to highlight here is the motivation and engagement relationship in the e-learning process. There is a high correlation toward motivation, be it intrinsic or extrinsic, with regards to the student's engagement in online learning. Since the events that occurred during COVID-19, the world has changed, including Malaysia's learning system. When online learning is available to students at home, it is a challenge for them to gain access to the internet, as certain students have limited access to it.

Online educators and teachers have been involved in the effort to promote the efficiency of interactions and communications, in order to resolve the issue. However, it can be challenging and stressful to keep students involved in interactions and communication, without motivating them, especially in learning environments where physical contact and natural communication are missing [1]. Therefore, this is a very challenging and demanding task for the student, as not everyone is born with a silver spoon. Therefore, this study aims to determine the levels of motivation and engagement in online learning amongst the students. Furthermore, it is also intended to determine if there is a significant difference between genders with regards to online learning, as well as motivation and engagement factors amongst them. Lastly, this study aims to determine if the motivation factor can also predict public university students' engagement in online learning.

2 Literature Review

There have been various studies associated with gender differences with regards to technology use, which caters to the fact that men are more likely to use technology and new media than women [2]. This shows that male students are more likely to engage in online learning than face-to-face classes. So far, there is no empirical evidence for the presence and impact of online learning, and the few available academic findings are often quite contradictory. One position claimed that there were gender-specific patterns of actions, which may contribute to sexism against women, for using e-learning platforms. Another argued that e-learning especially benefits women through its versatile and immersive learning approach. There is evidence that men and women show differing degrees of anxiety, acceptance, and interest over time in emerging technologies [2]. Since online learning is still relatively new to some students, both females and males are anxious and worried about their progress and learning regarding the tasks and assignments. Men tend to give longer and more frequent statements, while women show more openness and willingness to cooperate in online learning. Even though it is not easy to adapt to technology for some people, both genders are seen to be motivated to try new things, especially for online learning.

The relationship between motivation and engagement is often reported, as this relationship significantly influences learning. Young people with adaptive motivational attitudes towards education, are inspired to learn and master new content, have a strong sense of interaction with their school or classroom, and seem to be much more academically capable [3]. This is supported by [4], who stated that both engagement and motivational attitudes in schools are related to academic achievement, attendance, classroom behaviour, and even graduation rates. Students' motivation to learn, successfully participate in learning activities, and persist in challenging circumstances when learning independently and/or being in a class are subjects that have kept researchers in academia and educators busy over the past decade [5]. There is evidence in terms of internal variables, which show a strong relationship between teachers' learning dedication and their associated motivational factors [6]. They also found that theories of motivation achievement have been commonly used to describe if learners participate in learning tasks, or are disengaged from them. The most often used theory is the self-determination theory, to help explain the learner's motivational regulation. For the online learning approach, [7] supported it by mentioning that individuals' intrinsic motivation positively affects the gamification of objects and their involvement. Hence, this increased the student gamer's motivation and helped engage them much more toward online learning courses. In the research by [8], the research outcome affirmed the importance of motivational regulation in the online learning environment. The findings of their study showed the correlation between motivational factors, and online learning participation. Nevertheless, there is still a lack of research and study on the content of online learning. Thus, much more research is still needed on the motivation and engagement factors that affect students in online learning environments.

3 Methodology

In this study, a quantitative method was used to obtain information on the topic, through the use of an online survey questionnaire. The questionnaire was developed using Google Forms. It is an easily accessible tool that comes with any Google account, which most students have. Thus, it is convenient for students to access. The subject of this research was focused on public university students in Sabah. A total of 190 respondents participated in the research, and it was conducted via convenient sampling.

A sum of 51 items questionnaires was used for this research, which included three sections. The first section addressed the respondent's demographics, which consisted of four questions pertaining to their name, age, gender, and educational level. The second section was made up of a Revised Study Process Questionnaire Two-Factor (R-SPQ-2F) developed by [5]. It had 20 questions in total, which addressed the motivation levels. The R-SPQ-2F scale was developed to measure the process of learning and motivation, and the ensuing approach towards learning. This initial evaluation of the R-SPQ-2F indicated that the questionnaire had an acceptable internal consistency. It is a promising short instrument for diagnosing student and teachers in-depth surface approach to learning [9]. The total number of questions were then computed and recorded into a new datasheet for convenience. The third section consisted of the Engagement Scale (ES), developed by [1], where only the engagement part of the questionnaire was used. This scale consisted of three parts of engagement, i.e., behavioural, cognitive, and voice, of which there were a total of 28 questions on online learning engagement.

4 Data Analysis

The descriptive data in Table 1 showed that out of 190 respondents, 67 (35.3%) identified as male, while another 123 (64.7%) identified as female. 50 respondents were aged below 20 years old (26.31%), 134 respondents were between 21–25 years old (70.53%), 5 respondents were between 26–30 years old (2.63%), and while 1 respondent was above 36 years old (0.53%). As for the respondent's education level, 60 respondents were diploma students (31.6%), where the remaining 130 respondents were degree level students (68.4%).

Table 1. Descriptive data by gender, age, and education level

Items		Frequency (f)	Percentage (%)
Gender	Male	67	35.3
	Female	123	64.7
Age	Below 20	50	26.31
	21–25	134	70.53
	26–30	5	2.63
	Above 36	1	0.53
Education level	Diploma	60	31.6
	Degree	130	68.4

To evaluate the differences between males and females in online learning, the data was tested using an independent t-test. The independent sample t-test was conducted to compare the male and female level in motivation and engagement with regards to online learning. The test results with conditions of $t(188) = 1.03$, $\alpha > .05$, $p = .30$ revealed that for gender and engagement, males had a value of M = 101.61, SD = 11.65, while and females had a value of M = 99.54, SD = 14.06. Therefore, there was no significant difference between genders and their engagement in online learning.

The evaluation of gender and motivation levels with conditions of $t(188) = 1.09$, $\alpha > .05$, $p = .28$, showed that males (M = 66.15, SD = 8.77) and females (M = 64.52, SD = 10.30), had no significant difference between them, with regards to gender and their motivation for online learning. Hence, the null hypothesis failed to be rejected, and it was accepted that there was no significant difference between the genders and their motivation and engagement for online learning (Table 2).

Table 2. Comparison of an independent *t*-test between gender engagement and motivation

Variable	Engagement	Mean	SD	*t*-value	df	*p*-value
Gender	Male	101.61	11.65	1.03	188	0.30
	Female	99.54	14.06	1.09	158.32	0.28
Variable	Motivation	Mean	SD	*t*-value	df	*p*-value
Gender	Male	66.15	8.77	1.09	188	0.28
	Female	64.52	10.30	1.15	155.07	0.25

Table 3 shows that $r = .56$, thus it is a positive and strong relationship. The $\alpha = 0.05$, $p < .05$, $p = .00$. Since $p < 0.05$, therefore, the null hypothesis was rejected. Hence, there is a moderately significant relationship between the motivation and engagement factors for online learning amongst public university students.

Table 3. Pearson's correlation between motivation and engagement factors

		Engagement	Motivation
Engagement	Pearson Correlation	1	$.562^{**}$
	Sig. (2-tailed)		.000
Motivation	Pearson Correlation	$.562^{**}$	1
	Sig. (2-tailed)	.000	

$^{**}p < 0.05$

To determine the prediction between motivation and engagement, a simple linear regression was plotted on the collected data (Table 4). The prediction equation used was $Y = 50.75 + 0.76X$. The significant value of $\alpha = 0.05$, $p = 0.00$, deemed that the null hypothesis was rejected. $F = 86.71$, and the $R = 0.56$, which indicated a good correlation. The $R^2 = 0.32$, which means that a 32% variance is fairly explainable with regards to the engagement, which is predicted by motivation factors. Therefore, the regression model fits with the data, and the null hypothesis is rejected. Hence, the motivation significantly predicted the engagement in online learning amongst public university students.

Table 4. Simple linear regression for motivation and engagement

Model	*R*	*R* square	Adjusted *R* square	Std. error of the estimate
1	.562[a]	.316	.312	11.00061

a. Predictors: (Constant), motivation
b. Dependent Variable: engagement

5 Discussion

5.1 Gender and Online Learning

The findings showed that male and female students who attend online lectures have the same motivation and engagement levels. Since this research was done during the Movement Control Order (MCO) in Malaysia, it can be seen that both males and females were equally concerned with their learning progress. It is safe to say that both genders were aware of the stay at home order and on the national news. Thus, gender did not affect their urge to study online, even though classes were cancelled almost immediately. The findings were aligned with the research, and showed no considerable evidence that there is a difference between female and male students in terms of their quality of motivation [10]. Also, since gender differences were not significant on this scale of results, the present questionnaire might need to be considered in future works, if there is a need to include and indicate the respondent's gender in the future. [10] also mentioned that the results revealed no main effects of the genders across the two cognitive variables, or interactive effects amongst the motivational variables.

The discussion can be synced with literature reviews, as it supported the fact that there is little evidence to support gender differences in online learning environments. This may happen because there are still not many students exploring online learning to access their subjects, which may have increased online learning's daily use amongst students and lecturers. The students in this research were motivated and engaged in online learning, with the intent to increase their knowledge, as opposed to just wasting their time on the internet. In this sense, regardless of the university's origin, there is no significant effect on the perceived effectiveness of this e-learning activity on the student's social life. Both genders perceived it as a source of motivation, to easily engage in online learning. If the female or male participant was comfortable studying in isolation, then their engagement and motivation tended to increase in online learning. However, in this case, there was no significant difference to support previous research outcomes, as this study occurred during a pandemic. Online learning is a platform for students to gain knowledge, and to associate themselves with tasks. The biggest detractor of gender-separate education sees separation as discrimination based on sex, thus setting up inequality between boys' and girls' education [11]. This can be called discrimination towards genders, as the learning only involves either females or males, and they are not treated equally. However, the conclusions of the research on the subject have produced no definitive results.

The study provided evidence that female students demonstrated greater self-efficacy for self-regulated learning, with regards to gender disparities [12]. Female students are more confident in their learning, and perform tasks much more completely than male students. Nevertheless, in this study, the result was inversed, where both females and males were content and confident in their learning, and performed tasks well during the online courses. This may also have occurred because, during the online classes, both the male and female students were required to focus much more on the gadget used, as they were worried they might lose their internet connection during the lecture. Therefore, this applies only to online classes and works across both genders. Thus, there was no significant difference between both male and female students in online learning.

5.2 Relationship Between Motivation and Engagement in Online Learning

A positive correlation was found from the data analysis, showing a positive relationship between motivation and online learning engagement. Online learning is convenient and easy to use, and is applicable despite hectic schedules, as it can be accessed conveniently anywhere. Hence, this extrinsic and intrinsic motivational factor boosts the engagement of students in online learning. Students themselves need to identify their self-motivation drive, to engage in online lectures positively. Engagement and motivation have a high correlation, as it helps students in their online learning environment. [13] found that if students assumed that online learning activities were related to their learning practices and career development, students were more likely to participate in them. The perceived role was valued as a predictive variable of human learning participation and academic success, which was consistent with the stipulation of the expectation of the value theory [14].

Despite the fact that there was a high correlation between motivation and engagement in online learning, there were also disadvantages. Online learning is complex, as learners can get distracted, and fail to self-regulate themselves during classes. In this research, many other factors could distract the public university student's focus from staying engaged in class. Staying at home during the MCO is hard for many, since some students do not have stable Internet connections. Some may be working part-time, or even helping their parents with house chores. With the intention to improve motivation and engagement during classes, students need to take good care of their mental health [15]. As for online discussions, teachers need to support them by offering critical guidance, to improve the quality of the posted messages in the learning management systems (LMS), the student's interactions, and the student's understanding of learning resources. However, in this study, the online discussion environment is limited to the ability to provide for multiple learning activities and teaching strategies. The approach of traditional face-to-face learning is entirely different from the online learning approach.

On the contrary, [1] mentioned that students are isolated from social interaction in the online class environment, helping them concentrate on the course material. Students must initiate such participation in online classes. The interaction is usually carried out in written form, takes much more time, and is much more commitment than a verbal conversation. The statements were found to be true. These were noted to be disadvantages associated with engaging in online learning, especially during a pandemic, where students cannot seek help from their teachers for knowledge. This research has significant results for the relationship between motivation and engagement in online learning, as both are highly correlated. The findings of this discussion are that the higher the student's motivation, the higher the engagement involved in their online learning. Thus, motivation helps a lot in online learning engagement, besides other external factors. As the saying goes, when there is a will, there will always be a way. Therefore, there is a significant correlation between motivation and engagement in online learning amongst public university students.

5.3 Prediction of Motivation in the Engagement in Online Learning

The results showed that motivation fits the prediction of engagement in learning. The relationship can be seen as a means for driving the students' motivation at home, or when the students are stuck in their colleges, which can highly predict their engagement for learning virtually in front of a computer, or mobile phone. In another research by [16], it was stated that students' autonomous self-regulation predicted their dedication, academic success, and happiness, while controlled self-regulation did not demonstrate any correlation with these effects. However, recent researchers have found that high technical capabilities, or interactions in online learning environments, do not necessarily predict the learners' level of activity. Neither profound learning methods, nor anxiety, forecasted the real level of success in the online learning environment. It can be justified that research on the motivation in online learning could cause lower engagement, as certain students do not have the ability to afford expensive and high technology personal computers, or even consistently pay for their mobile data. This can be proven in work by [1], who stated that motivation, or self-determination, also failed to measure learning results, such as the study hours a week, number of hits, or even predicting the 128 grades in the online courses. This means that some researchers do not find any correlation in motivation on engagement, as the students might feel that other factors affect the student's engagement in online learning.

These findings support the literature review, where most of the students in this research found that motivation does exist in the prediction of engagement in online learning. A typical student tends to show a stronger preference for face-to-face learning, than for online learning. Students who preferred face-to-face learning struggled with adapting to online learning [17]. Thus, the more motivated the students became, the better their cognitive involvement was. Again, although this preference is similar to findings in the previous literature works, the experience cannot be completely compared due to the current pandemic circumstances. Students who have not had prior experience with online learning may think that online learning is not suitable. Consistent with the existing literature works [18], it was confirmed that when students are not motivated, their level of cognitive engagement is lower, and vice versa [19]. [20] reported that intrinsic motivation helped predict engagement, while extrinsic motivation predicted online learning usage. Importantly, engagement was predictive of grades, whereas the usage of online learning was not. As for this statement, motivation does help predict students' engagement in their learning, but not in the context of online learning. This was because the respondents in work by [21] were not familiar with the use of the internet, especially for online learning.

6 Conclusion

The results revealed three major outcomes; there is no significant difference in online learning between both male and female students; there is a significant relationship between motivation and engagement in online learning amongst public university students; motivation does help predict online learning engagement amongst public university students. This study proved that both genders helped in their online learning development, but the implications cannot be identified clearly, as not every student shared the same experience. This study's findings are useful for understanding the connection between the students' motivational values, the management of motivation, and their online learning involvement. Therefore, further studies on the correlation between the students' involvement in learning, and their motivational factors, are needed. Every student needs to adapt to this new era, as online classes and assignments are available on the internet. Teachers and students can easily interact with each other, even though they are miles apart. However, the students' motivation and engagement need to be improvised and practised to be more comfortable and approachable in an online learning environment.

Acknowledgement. This work is a part of a project submitted to Universiti Malaysia Sabah.

References

1. Russell, J.L.: Supporting students' motivation in college online courses. Doctoral thesis, University of Iowa (2013)
2. Cuadrado-García, M., Ruiz-Molina, M., Montoro-Pons, J.D.: Are there gender differences in e-learning use and assessment? Evidence from an interuniversity online project in Europe. Procedia Soc. Behav. Sci. **2**, 367–371 (2010)
3. Heatly, M.C., Votruba-Drzal, E.: Developmental precursors of engagement and motivation in fifth grade: linkages with parent- and teacher-child relationships. J. Appl. Dev. Psychol. **60**, 144–156 (2019)
4. Alexander, K., Entwisle, D., Horsey, C.: From first grade forward: early foundations of high school dropout. Sociol. Educ. **70**(2), 87–107 (1997)
5. Kubischta, F.: Engagement and motivation: questioning students on study motivation, engagement and study strategies. Bachelor thesis, Haaga-Helia University of Applied Sciences (2014)
6. Zhang, S., Liu, Q.: Investigating the relationships among teachers' motivational beliefs, motivational regulation, and their learning engagement in online professional learning communities. Comput. Educ. **134**, 145–155 (2019)
7. Feng, Y., Ye, H.J., Yu, Y., Yang, C., Cui, T.: Gamification artifacts and crowdsourcing participation: examining the mediating role of intrinsic motivations. Comput. Hum. Behav. **81**, 124–136 (2018)
8. Uzun, A.M., Kilis, S.: Does persistent involvement in media and technology lead to lower academic performance? Evaluating media and technology use in relation to multitasking, self-regulation and academic performance. Comput. Hum. Behav. **90**, 196–203 (2019)
9. Martinelli, V., Raykov, M.: Evaluation of the revised two-factor study process questionnaire (R-SPQ-2F) for student teacher approaches to learning. J. Educ. Soc. Res. **7**(2), 9–13 (2017)

10. Kusurkar, R.A., Cate, T.J.T., Asperen, M.A., Croiset, G.: Motivation as an independent and a dependent variable in medical education: a review of the literature. Med. Teach. **33**, 242–262 (2011)
11. Schacter, D.L., Gilbert, D.T., Wegner, D.M., Nock, M.K.: Psychology, 3rd edn. Worth Publishers, New York (2014)
12. Alghamdi, A., Karpinski, A.C., Lepp, A., Barkley, J.: Online and face-to-face classroom multitasking and academic performance: moderated mediation with self-efficacy for self-regulated learning and gender. Comput. Hum. Behav. **102**, 214–222 (2020)
13. Wigfield, A., Eccles, J.S.: Expectancy–value theory of achievement motivation. Contemp. Educ. Psychol. **25**(1), 68–81 (2000)
14. Rathakrishnan, B., Chan, A.: Stress and mental health problem on secondary school student in Sabah. Malaysia J. Youth Stud. **8**, 167–176 (2013)
15. Rathakrishnan, B., Molugulu, N., Parasuraman, B., Narasappa, K.: The relationship of stress, alcoholism and sexual behavior with mental health among secondary school students: a study in Sabah. Malaysia. Euro. J. Soc. Sci. **31**(3), 376–383 (2012)
16. Black, A.E., Deci, E.L.: The effects of instructors' autonomy support and students' autonomous motivation on learning organic chemistry: a self-determination theory perspective. Sci. Educ. **84**(6), 740–756 (2000)
17. Patricia Aguilera-Hermida, A.: College students' use and acceptance of emergency online learning due to COVID-19. Int. J. Educ. Res. Open **1**, 100011 (2020). https://doi.org/10.1016/j.ijedro.2020.100011
18. Sun, J.C., Lin, C., Chou, C.: Applying learning analytics to explore the effects of motivation on online students' reading behavioral patterns. Int. Rev. Res. Open Distrib. Learn. **19**(2), 209–227 (2018)
19. Rathakrishnan, B., Samsuddin, A.R., Singh, S., Juliana, J.: Job preferences among marginalised and non-marginalised youths: A multi-ethnic study in Sabah. Pertanika J. Soc. Sci. Hum. **25**, 55–66 (2017)
20. Dunn, T.J., Kennedy, M.: Technology enhanced learning in higher education; motivations, engagement and academic achievement. Comput. Educ. **137**, 104–113 (2019)
21. Parasuraman, B., Satrya, A., Rathakrishnan, B., Muniapan, B.: Analysing the relationship between unions and joint consultation committee: case studies of Malaysian and Indonesian postal industries. Int. J. Bus. Soc. **10**(1), 41–58 (2009)

Factors that Accelerate the Rise of Acceptance of Big Data Platforms for Academic Teaching: Personal Innovativeness as Moderating Variable

Adnan Aldholay[1](✉), Osama Isaac[2], Abdullah Nabeel Jalal[1], Farah Akmar Anor[1], and Ahmed M. Mutahar[1]

[1] Faculty of Business Management and Professional Studies, Management and Science University, Shah Alam, Malaysia
adnan_aldholay@msu.edu.my
[2] Guangdong University of Finance, Guangdong, China

Abstract. Unless higher education institutions react to technological changes and the unprecedented interruption that resulted from the COIVD-19 pandemic, it will face great difficulty to remain relevant. Among the new technological applications that enrich teaching activities is the utilization of big data platforms in teaching. This study investigates the factors that would influence the acceptance of such platforms by academicians in higher education institutions in Malaysia. 143 respondents participated to examine the effect of interactivity on system quality, knowledge quality, service quality, and perceived enjoyment which in turn influence acceptance of big data platforms for academic teaching. Besides, examining the moderating role of personal innovativeness. The result shows that system quality has an insignificant effect on the acceptance with all remaining direct relationships significant. Further, interactivity significantly influences all but system quality. Lastly, personal innovativeness moderated only the relationship between perceived enjoyment and acceptance of big data platforms.

Keywords: Big data · Personal innovativeness · Interactivity · Higher education

1 Introduction

Technology is today's world denotation; it has altered all aspects of human life [1]. Internet as a technology application has allowed information to be accessible at any time and place, as we are approaching the internet of things (IoT) era where billions of devices will be connected and produce a huge amount of data which will be processed and analyzed to extract patterns that present valuable information in every industry, and that is the essence of big data. Nowadays, big data has become a buzzword that immediately draws attention and can be considered the future platform for innovation and productivity [2]. In the context of higher education, big data is expected to enhance this vital sector in the foreseen future, through the function of e-learning, by boosting the students' lecturer interaction, or by facilitating the realization of individual goals and requirements of students [3]. Further, the covid-19 pandemic has accelerated the

M. Al-Emran et al. (Eds.): ICETIS 2021, LNNS 299, pp. 227–243, 2022.
https://doi.org/10.1007/978-3-030-82616-1_21

full deployment of technology in the higher education institution as it became the viable solution to continue in the teaching and learning process during lockdown [4].

Many studies have examined the use of big data in different industries including the education industry which is the focus of this study. Some researchers examined the impact of big data on the scientific research process [5], others looked at its impact on students' learning experience and students' behavior [6–8], and how it presents a powerful tool for higher education institutions' to make quality decisions [9]. Nevertheless, this study focuses on the use of big data platforms in the teaching process by examining technology and personal factors that may motivate academic staff to utilize big data platforms. The information system success model by Delone and Mclean is distinguished in technology adoption and use which has been identified as efficacious in investigating various technology devices. [10] found that system, information, and service qualities have a significant impact on online learning usage. However, researchers have constantly examined different types of factors along those of technology nature to try to have a more comprehensive understanding of technology usage. For instance, interactivity is seen as a detrimental characteristic that makes any internet technology application more attractive and likely to be used quite often [11]. On the other hand, [12] described how significant is the perceived enjoyment of the technology to increase and sustain its use. Further, the personal trait of innovation has been widely linked to play a role in determining an individual's use of new technologies [13–15].

In the current era of COVID-19, higher education institutions have been forced to dramatically accelerate their efforts and make the leap to the digitalization of their teaching and learning. This study is tapping into this opportunity to promote the use of big data platforms in teaching activities by examining the factors that encourage its use in the context of higher education institutions in Malaysia.

2 Literature Review

2.1 Knowledge Quality (KQ)

Knowledge is important for all sectors, especially to the academic world. The enforcement of gaining new knowledge is rather imperative and significant in the teaching and learning process. Thus, securing good quality new knowledge is highly encouraged and the procedures have long been discussed among academicians. In meeting the objectives, big data platforms in academic teaching have been introduced and emphasized to every educator in sharing the knowledge details.

The fundamental components of knowledge quality in teaching and delivery gained via big data platforms not only depend on the capacity of the information received but rather the current data collection across the globe. Hence, the potential of big data platforms to significantly impact academic teaching is attributed to its size and complexity. It is beyond the concept of traditional datasets in which the comprehensive system provided by big data platforms is proven to provide a decent quality of eLearning system in the education setting [16].

In the face of the Covid-19 outbreak recently, the educational sectors in many countries have taken a novel inclination towards online teaching and delivery. This aligns with the growth in information technology which witnesses the innovativeness of electronic-learning (e-Learning) to be incorporated into teaching using technology [17]. Since eLearning is rather imperative to the academic world recently, the services provided via big data simplify educators in securing additional information. A similar view by [18] described that big data platforms support users in gaining diverse types of knowledge from online content especially when most current and comprehensive information is in demand.

Since the need to enhance virtual education platforms received high demand in the current immense globalization development [19], all contents in the big data platforms must be highly trustworthy, dependable, and reliable. In complying with the procedures, all details received are thoroughly screened and inspected. [20] have amplified the accuracy provided in the big data platform, whereby the correctness of data used must be screened and vetted before accepting the details. This acknowledges that errorless, correct, and accurate content on big data platforms is well sheltered to ensure the quality of knowledge is protected. Therefore, all details provided above help to accelerate the acceptance of big data platforms to academic teaching. Therefore, this study postulates the following hypothesis:

H1: Knowledge Quality has a positive effect on Acceptance of Big Data Platforms for Academic Teaching.

2.2 Service Quality (SEQ)

Service quality is hailed as an integral part of the information system field, Delone and Mclean have emphasized its significance in their well-known model [21]. It refers to the characteristics of the value of the process offered by big data platforms in terms of responsiveness, functionality, and reliability [22, 23]. [24] examined the service quality effect on student's acceptance of online learning and found that a significant effect exists between the two variables. Moreover, in the same context of online learning, [10] proved the significance of service quality on performance impact among higher education institutions in Yemen. Further, a study on the end-user acceptance of telemedicine technology in the Netherlands has concluded that the quality of the service is vital when it comes to users' acceptance of this new technology [25]. Therefore, this study aims to test the following hypothesis:

H2: Service Quality has a positive effect on the Acceptance of Big Data Platforms for Academic Teaching.

2.3 System Quality (SYQ)

System quality is a desirable characteristic of an information system. System quality is an influential element in determining the outcome of an overall system. System quality is viewed as a critical component for users, developers, and project supervisors. System quality refers to the criteria that the industry believes should be added to improve the success of a product [26]. Furthermore, the information system model stipulates that system quality is an influential factor in affecting the intention to use and user satisfaction. The quality of an information system is usually evaluated based on the perceived easiness to learn the system. Moreover, system quality is observed in the flexibility of system usage since users prefer system features that are convenient, intuitive, sophisticated, and minimum response time [27, 28]. The amount of time needed for a system to respond to instruction is referred to as response time. This is vital as a longer response time contributes to reduced user satisfaction. Likewise, quality is presented in the system reliability, the degree to which users can trust the system. Therefore, system quality is crucial in the acceptance of big data platforms by higher learning institutions. Hence, this study postulates the following hypothesis:

H3: System Quality has a positive effect on the Acceptance of Big Data Platforms for Academic Teaching.

2.4 Perceived Enjoyment (PE)

Perceived enjoyment is defined as the users' perception of the enjoyability of activity, other than any expected performance outcomes [29]. The literature concluded that perceived enjoyment is crucial in user technology acceptance and has many practical implications, specifically among big data platforms. Nevertheless, the perceived enjoyment of big data platforms should be more flexible, cost-efficient, and interactive than traditional platforms for individuals who are well-versed in technology. Big data orientation must not underestimate the significance of enjoyment either in user satisfaction from improved accessibility, easier delivery, and enhanced technology management [30]. Convenience and perceived enjoyment are identified as significant factors in big data platforms' satisfaction and participation. Satisfaction can be realized as a "feeling of joy, elation, and pleasure" that correlates with an individual's engagement with a system. Perceived enjoyment has a crucial impact on the acceptance of big data platforms. Consequently, the following hypothesis is proposed:

H4: Perceived Enjoyment has a positive effect on Acceptance of Big Data Platforms for Academic Teaching.

2.5 Interactivity (INT)

Technology usage in education is rapid at all levels. This suggests that the utilization of systems and platforms requires interactivity to attract more users. Interactivity has emerged as a key player to be considered carefully under the empirical lenses. Interactivity is characterized as response performance underwent by users towards a platform or a system [31]. A quick response by a system encourages smooth communication between users and the system. In this study interactivity is defined as how interactive the big data platforms in terms of retrieving data, illustrating information when are used.

Interactive has been widely researched in many settings such as online shopping which identified interactivity is capable of boosting cognitive and affective participation of online shoppers, and promotes positive perception and behavioral intention among online shoppers [32–35]. Although interactivity is deemed as an essential component of web design for e-commerce [36], minimal studies have investigated the merge between background music and interactivity in encouraging more active customer experience. Website interactivity offers users the technical competence to communicate with an interface and the opportunity to alter content instantaneously [37–39]. Improving the authenticity and control of user experience, interactivity increases user engagement and preference for the website [40, 41]. Website interactivity was identified to boost users' utilitarian principles [42] and positively impact user engagement [33].

A big data platform creates a domain that offers fun activities that improve user engagement and learning motivation [43, 44]. Many users find multimedia inclusion interesting. This paints the impact of the multimedia environment in establishing meaningful learning. The distinction between the influence of different interactivity modalities on user experience has prompted the current study to investigate the influence of interaction on service quality, system quality, service quality, perceived enjoyment, and knowledge quality of big data that can be employed in the teaching and learning process. Consequently, the following hypotheses are proposed:

H5: Interactivity significantly influences Knowledge Quality.

H6: Interactivity significantly influences Service Quality.

H7: Interactivity significantly influences System Quality.

H8: Interactivity significantly influences Perceived Enjoyment.

2.6 Personal Innovativeness (PI)

[45] initiated the concept of personal innovativeness as an element in technology usage in information technology literature. Rogers' Theory stipulates that personal innovativeness in information technology refers to an individual's readiness to study new technology. PI refers to an individual's innovative personality characteristic that is associated with new technologies [46, 47], and an individual's innovative personality is generally observed as the essence of risk-taking inclination [46, 47]. Individuals who have a higher innovative personality tend to adapt easily or innovate [48–50]. Therefore, individuals possessing active personal innovativeness are predicted to grow positive perspectives on new technology [51].

[45] indicated that personal innovativeness is critical to comprehend new information technology and the acceptance of technology usage. Research on personal innovativeness in technology has been performed in different settings which include online shopping [52], cyberlearning [53], blog [54], and wireless service provider [55, 56].

Individuals who have a stronger innovative personality are more inclined to find exploring new technologies and solving the problem more enjoyable which contributes to a higher acceptance of new technological innovations [57, 58]. Hence, this study aims to examine personal innovativeness as a potential moderator on the relationship between perceived enjoyment and acceptance of big data platforms in the teaching and learning process. Furthermore, this study assesses the moderating effect of personal innovation on the relationship among service quality, knowledge quality, system quality and big data platform acceptance for academic teaching. Consequently, the following hypotheses are proposed:

H9: Personal Innovativeness strengthen the positive effect of Knowledge Quality on Acceptance of Big Data Platforms for Academic Teaching.

H10: Personal Innovativeness strengthen the positive effect of Service Quality on Acceptance of Big Data Platforms for Academic Teaching.

H11: Personal Innovativeness strengthen the positive effect of System Quality on Acceptance of Big Data Platforms for Academic Teaching.

H12: Personal Innovativeness strengthen the positive effect of Perceived Enjoyment on Acceptance of Big Data Platforms for Academic Teaching.

2.7 Acceptance of Big Data Platforms for Academic Teaching (ACC)

The logical flow of actions when studying the use of any new technology application is to identify the factors that will influence, motivate, and prompt an individual to have an intention to use or accept that particular technology which will eventually lead us to study the consequences of that use [59]. Numerous theories in the information system literature have explored this idea such as the theory of reasoned action (TRA) developed by [60], the unified theory of acceptance and use of technology (UTAUT) designed by [61], and technology acceptance model (TAM) constructed by [62]. In this study, acceptance of big data platforms for academic teaching is the main construct of investigation. It refers to the positive perception of academic staff in higher education institutions in Malaysia towards the use of such platforms in their teaching activities. This has become especially important in the era of Covid-19 which accelerated the incorporation of technology in all aspects of education [59, 63].

3 Research Method

3.1 Overview of the Proposed Conceptual Framework

The framework below is created based on several concepts and models on the field of information systems that are derived from literature. The information system success model by Delone and Mclean is built by three factors which are service quality, knowledge quality, and system quality [64]. Besides, interactivity has been building steam in the field as a crucial factor in the technology usage and acceptance research effort as in [65] and [66]. Moreover, perceived enjoyment and personal innovativeness were derived from the technology acceptance model 3 (TAM3) and technology readiness index (TRI) respectively [67, 68]. The model comprises 12 hypotheses to be examined (Fig. 1).

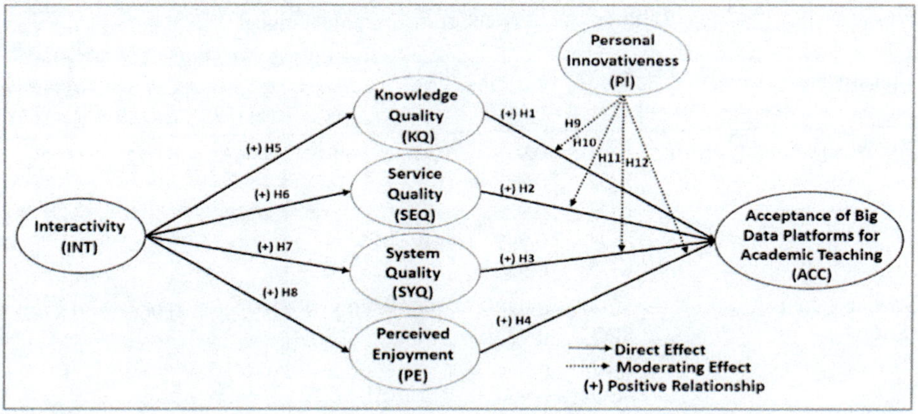

Fig. 1. Study framework

3.2 Instrument Development and Collection of Data

The study utilizes a sample of 143 academic staff in higher education institutions in Malaysia, using an online-based survey during the period of June-July 2020. The questionnaire adopted a total of 20 items from the literature and incorporated a five-point Likert scale where 1 indicates strongly disagree and 5 indicates strongly agree (refer Appendix A). Recommendations by [69] were followed in determining the number of items to measure each construct to minimize overlapping.

4 Analysing the Data and Results

SEM-VB via SmartPLS 3.0 [70] was employed for data analysis and hypothesis testing as suggested by previous research [71].

4.1 Assessment of Measurement Model

The composite reliability (CR), Cronbach's alpha, average variance extracted (AVE), and the factor loadings fulfil the requirements [72, 73] as illustrated in Table 1.

Table 1. Assessment of measurement model

Constructs	Item	Loading (> 0.7)	M	SD	α (> 0.7)	CR (> 0.7)	AVE (> 0.5)
Knowledge Quality (KQ)	KQ1	0.916	3.604	0.814	0.963	0.972	0.872
	KQ2	0.928					
	KQ3	0.926					
	KQ4	0.949					
	KQ5	0.950					
Service Quality (SEQ)	SEQ1	0.928	3.602	0.810	0.954	0.966	0.878
	SEQ2	0.927					
	SEQ3	0.938					
	SEQ4	0.954					
System Quality (SYQ)	SYQ1	0.942	2.891	0.938	0.853	0.931	0.872
	SYQ2	0.925					
Perceived Enjoyment (PE)	PE1	0.973	3.447	0.920	0.944	0.973	0.947
	PE2	0.973					
Interactivity (INT)	INT1	0.971	3.542	0.789	0.938	0.970	0.941
	INT2	0.969					

Fornell-Larcker was used to test the discriminant validity, Table 2 shows that all constructs of the model fulfilled satisfactorily [74, 75].

Table 2. Fornell-Larcker criterion

	ACC	*INT*	*KQ*	*PE*	*PI*	*SEQ*	*SYQ*
ACC	**0.959**						
INT	0.638	**0.970**					
KQ	0.841	0.613	**0.934**				
PE	0.814	0.680	0.750	**0.973**			
PI	0.689	0.481	0.556	0.521	**0.996**		
SEQ	0.860	0.660	0.861	0.774	0.579	**0.937**	
SYQ	0.222	0.183	0.179	0.219	0.134	0.220	**0.934**

4.2 Assessment of Structural Model

According to [75] the structural model can be tested by bootstrapping procedure with a resample of 5,000, see Fig. 2.

Direct Hypothesis Results Table 3 and Fig. 2 detail the hypothesis testing results of the structural model. Knowledge quality, perceived enjoyment, and service quality were

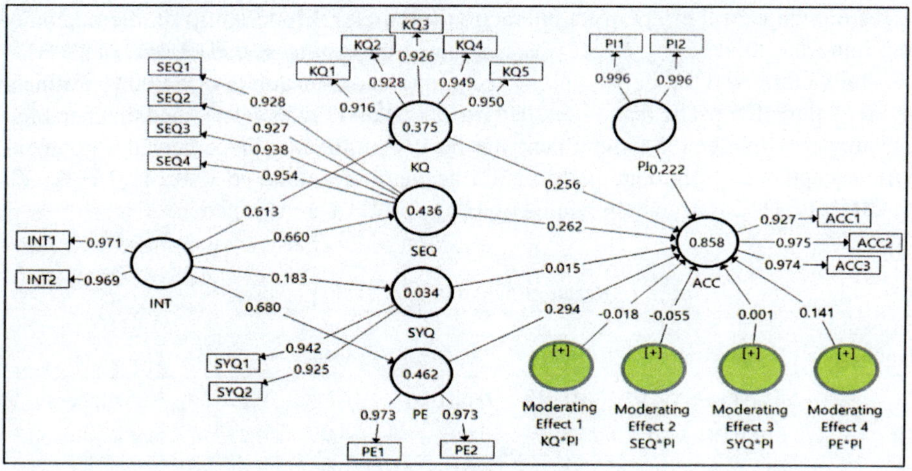

Fig. 2. PLS algorithm results

identified to positively influence the acceptance of big data platforms in academic teaching. Hence, H1, H2, and H4 are accepted with ($\beta = 0.256$, t = 2.774, $p < 0.001$), ($\beta = 0.262$, t = 2.156, $p < 0.05$), and ($\beta = 0.294$, t = 4.193, $p < 0.001$) respectively. Interactivity positively influences knowledge quality, service quality, and perceived enjoyment. Hence, H5, H6, and H8 are accepted with ($\beta = 0.613$, t = 8.254, $p < 0.001$), ($\beta = 0.660$, t = 9.948, $p < 0.001$) and ($\beta = 0.680$, t = 10.774, $p < 0.001$) respectively, while H3 and H7 are rejected. According to [76] and [77] the value of 0.86 of $R2$ indicating the explanatory power as an acceptable level.

Table 3. Direct hypothesis results

Hypothesis	Relationship	Beta	Std Error	t-value	p-value	Decision	R^2
H1	KQ → ACC	0.256	0.092	2.774	0.006	Supported	0.86
H2	SEQ → ACC	0.262	0.121	2.156	0.031	Supported	
H3	SYQ → ACC	0.015	0.038	0.399	0.690	Not supported	
H4	PE → ACC	0.294	0.070	4.193	0.000	Supported	
H5	INT → KQ	0.613	0.074	8.254	0.000	Supported	0.38
H6	INT → SEQ	0.660	0.066	9.948	0.000	Supported	0.44
H7	INT → SYQ	0.183	0.101	1.821	0.069	Not supported	0.03
H8	INT → PE	0.680	0.063	10.774	0.000	Supported	0.46

Moderating Effect Hypotheses Figure 2 and Table 4 demonstrate that three sub-hypotheses were tested for four main hypotheses which are: (1) Testing the causal effect of a predictor on an outcome, (2) Testing the causal effect of a moderator on an outcome,

(3) Testing the causal effect of an interaction (predictor*Moderating) on the outcome. Examining the moderating effect was conducted by testing H9, H10, H11, and H12. This study employed the bootstrapping technique and a resample of 5,000 to evaluate the interaction effect. The results detailed in Table 4 and Fig. 3 determined that personal innovativeness moderates (strengthen) the positive influence of perceived enjoyment on the acceptance of big data platforms for academic teaching ($\beta = 0.141$, $t = 2.130$, $p < 0.05$), so, H12 is accepted, while H9, H10, and H11 are rejected.

Table 4. Result of moderating effect hypotheses

Hypothesis	Relationship	Beta	Std. error	t-value	p-value	Decision
H9	KQ*PI → ACC	-0.018	0.081	0.221	0.825	Not supported
H10	SEQ*PI → ACC	-0.055	0.098	0.564	0.573	Not supported
H11	SYQ*PI → ACC	0.001	0.033	0.037	0.971	Not supported
H12	PE*PI → ACC	0.141	0.066	2.130	0.033	Supported

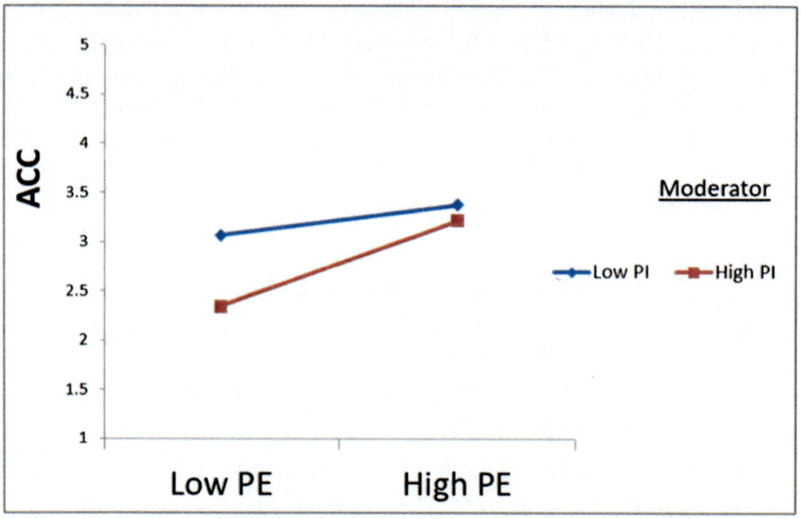

Fig. 3. Moderating effects result

5 Discussion

The study examined the impact of interactivity on each of system quality, knowledge quality, service quality, and perceived enjoyment which in turn the four constructs influence acceptance of big data platforms for academic teaching within higher education institutions in Malaysia. The study found that both knowledge quality and service quality had a significant influence on the acceptance of big data platforms, that is, the respondents tend to accept such platforms increasingly as long as they provide up-to-date, relevant, comprehensive, and accurate data. Besides, big data platforms provide modern interfaces with quick responses. This result fits well with previous studies that proved the significance of knowledge and service quality in numerous contexts [24, 71, 78, 79]. On the other hand, system quality was not found to have a significant impact on the acceptance of big data platforms, this is in contrast to some available studies [80–82], and could be due to the advancement of technology whereby the quality of the system became competitive and less of a concern for users. Perceived enjoyment was identified to exert a significantly positive effect on the acceptance of big data platforms which aligns with findings reported in the literature [9, 12], as it is attributed by the fact that the acceptance of new technology applications triggers excitement in exploring the new application.

Moreover, the study outcome shows that interactivity significantly influences all related variables in the model but one which is system quality. Many studies in the literature have extensively examined the role played by interactivity as a new determinant factor in technology acceptance [65, 83, 84]. Those relationships of interactivity in the model can be explained by the fact that having a new technology application that provides an interactive environment and communication will enhance the knowledge relevance and make it comprehensive, provide a modern interface and quick responses, besides, to make the whole experience more fun and pleasant.

Lastly, the study results show that the moderation role of personal innovativeness was not found to statistically exist except on the relationship between perceived enjoyment and acceptance of big data platform, this can be seen as an expected outcome as both personal innovativeness and perceived enjoyment are individual factors and goes in line with many studies in the literature [13, 85], that is the more time to be spent by the user in exploring new technologies in teaching and always strive to be the first among peers to use them, the stronger is the effect of perceived enjoyment on the acceptance of big data platforms in terms of using and recommending them to other. Nevertheless, the unfound moderating role of personal innovativeness in the relationships between technology variables and acceptance of big data platforms may be due to the incapability of individual characteristics to alter the course of relationships which include variables of technological nature.

6 Implications

With the emergence of Big Data platforms, most educational institutions seem to have a big challenge in motivating their staff to adopt those platforms. It is of utmost importance that the educational institutions look into the factors and the performance outcomes while adopting Big Data tools. The study has provided a grounded framework that can be considered in many different educational institutions. The theoretical implications of this study are one of the few types of research that provide in-depth knowledge about how Big Data can be adopted and utilized emphasizing the educational sector in the context of Malaysia. On the other hand, the practical implications of this study guide educational providers during their efforts to develop effective Big Data platforms, and advance the benefits, particularly in the educational profession.

7 Conclusion

Many sectors have been significantly influenced by Big Data and the educational sector is one of the sectors that has incorporated this concept. The educational field is one of the biggest creators of digital data. Big data technologies have created new opportunities and benefits for students, educational units, and higher learning institutions. The adoption of suitable big data technology and structure must be meticulously examined to fully reap the benefits. Although big data technology is unarguably essential for institutions of higher learning, more development is needed to maximize its potential. The conventional educational structure has not fully adjusted to this significant change in technology. This study examined technology and individual characteristics' role in the acceptance of big data platforms in academic teaching among academicians in higher education institutions in Malaysia. The result shows that system quality has an insignificant effect on the acceptance with all remaining direct relationships significant. Further, interactivity significantly influences all but system quality. Lastly, personal innovativeness moderated only the relationship between perceived enjoyment and acceptance of big data platforms.

Acknowledgment. This study was sponsored by the Management and Science University under seed grant number **(SG-20190527017-FBMP)** for the year 2020.

References

1. Kumar, T.A., Ferdous, A.S.M., Nabeel, J.A.: Operational or strategic benefits: Empirical investigation of internet adoption in supply chain management. Manag. Res. Rev. **40**(1), 28–52 (2017). https://doi.org/10.1108/MRR-10-2015-0225
2. Sedkaoui, S., Khelfaoui, M.: Understand, develop and enhance the learning process with big data. Inf. Discov. Deliv. **47**(1), 2–16 (2019). https://doi.org/10.1108/IDD-09-2018-0043
3. Md Talib, Z., Mosbah, A., Al-Jubari, I.: Intention to pursue postgraduate studies in Malaysian Universities. Int. J. Innov. Creat. Chang. **5**(2), 591–609 (2019)
4. Drayton-Brooks, S.M., Gray, P.A., Turner, N.P., Newland, J.A.: The use of big data and data mining in nurse practitioner clinical education. J. Prof. Nurs. **36**, 1–5 (2020). https://doi.org/10.1016/j.profnurs.2020.03.012
5. Jackson, N.C.: Managing for competency with innovation change in higher education: Examining the pitfalls and pivots of digital transformation. Bus. Horiz. **62**(6), 761–772 (2019). https://doi.org/10.1016/j.bushor.2019.08.002
6. Cantabella, M., Martínez-España, R., Ayuso, B., Yáñez, J.A., Muñoz, A.: Analysis of student behavior in learning management systems through a Big Data framework. Futur. Gener. Comput. Syst. **90**, 262–272 (2019). https://doi.org/10.1016/j.future.2018.08.003
7. Mosbah, A., Razak, A., Al-Jubari, I.: Students and cultural issues in international contexts: the case of Malaysian international businesses students in UK and China. Int. J. Innov. Creativ. Chang. **5**, 573–590 (2019)
8. Al-Jubari, I.: College students' entrepreneurial intention: testing an integrated model of SDT and TPB. SAGE Open **9**(2), 1–15 (2019). https://doi.org/10.1177/2158244019853467

9. Verma, S., Bhattacharyya, S.S., Kumar, S.: An extension of the technology acceptance model in the big data analytics system implementation environment. Inf. Process. Manag. **54**(5), 791–806 (2018). https://doi.org/10.1016/j.ipm.2018.01.004
10. Aldholay, A., Isaac, O., Abdullah, Z., Ramayah, T.: The role of transformational leadership as a mediating variable in DeLone and McLean information system success model: the context of online learning usage in Yemen. Telemat. Inf. **35**(5), 1421–1437 (2018). https://doi.org/10.1016/j.tele.2018.03.012
11. Jalal, A.N., Bahari, M., Tarofder, A.K., Musa, W.M.N.M.W.: Factors influencing customer social relationship management implementation and its benefits in healthcare industry. Polish J. Manag. Stud. **19**(2), 196–205 (2019). https://doi.org/10.17512/pjms.2019.19.2.16
12. Gursoy, D., Chi, O.H., Lu, L., Nunkoo, R.: Consumers acceptance of artificially intelligent (AI) device use in service delivery. Int. J. Inf. Manage. **49**(April), 157–169 (2019). https://doi.org/10.1016/j.ijinfomgt.2019.03.008
13. Ng, H.S., Kee, D.M.H., Ramayah, T.: Examining the mediating role of innovativeness in the link between core competencies and SME performance. J. Small Bus. Enterp. Dev. **27**(1), 103–129 (2019). https://doi.org/10.1108/JSBED-12-2018-0379
14. Salem, S.F., Alanadoly, A.B.: Personality traits and social media as drivers of word-of-mouth towards sustainable fashion. J. Fash. Mark. Manag. **25**, 24–44 (2020). https://doi.org/10.1108/JFMM-08-2019-0162
15. Purwanto, E., Deviny, J., Mutahar, A.M.: The mediating role of trust in the relationship between corporate image, security, word of mouth and loyalty in m-banking using among the millennial generation in Indonesia. Manag. Mark. **15**(2), 255–274 (2020). https://doi.org/10.2478/mmcks-2020-0016
16. Masoumi, D., Lindström, B.: Quality in e-learning: a framework for promoting and assuring quality in virtual institutions. J. Comp. Assist. Learn. **28**, 27–41 (2012). https://doi.org/10.1111/j.1365-2729.2011.00440.x
17. Bhuasiri, W., Xaymoungkhoun, O., Zo, H., Rho, J.J., Ciganek, A.P.: Critical success factors for e-learning in developing countries: a comparative analysis between ICT experts and faculty. Comput. Educ. **58**(2), 843–855 (2012). https://doi.org/10.1016/j.compedu.2011.10.010
18. Qwaider, W.Q., Arabia, S.: Integrated of Knowledge Management and E- Learning System (2011)
19. Yoo, D.K.: Perceived knowledge quality: a sensemaking perspective. In: 18th American Conference Information and System. AMCIS 2012, Vol. 4, pp. 2572–2579 (2012)
20. Rao, L., Osei-Bryson, K.-M.: Towards defining dimensions of knowledge systems quality. Expert Syst. Appl. **33**(2), 368–378 (2007). https://doi.org/10.1016/j.eswa.2006.05.003
21. Jeyaraj, A.: DeLone & McLean models of information system success: critical meta-review and research directions. Int. J. Inf. Manage. **54**(April), 102139 (2020). https://doi.org/10.1016/j.ijinfomgt.2020.102139
22. DeLone, W.H., Mclean, E.R.: The DeLone and McLean model of information systems success: a ten-year update. J. Manag. Inf. Syst./Spring **19**(4), 9–30 (2003). https://doi.org/10.1073/pnas.0914199107
23. Pituch, K.A., Lee, Y.: The influence of system characteristics on e-learning use. Comput. Educ. **47**(1), 222–244 (2006)
24. Lee, J.W.: Online support service quality, online learning acceptance, and student satisfaction. Internet High. Educ. **13**(4), 277–283 (2010). https://doi.org/10.1016/j.iheduc.2010.08.002
25. Jansen-Kosterink, S., Dekker-van Weering, M., van Velsen, L.: Patient acceptance of a telemedicine service for rehabilitation care: a focus group study. Int. J. Med. Inform. **125**(Jan. 2018), 22–29 (2019). https://doi.org/10.1016/j.ijmedinf.2019.01.011
26. Gorla, N., Somers, T.M., Wong, B.: Organizational impact of system quality, information quality, and service quality. J. Strateg. Inf. Syst. **19**(3), 207–228 (2010). https://doi.org/10.1016/j.jsis.2010.05.001

27. Dubey, S., Ghosh, S., Rana, A.: Comparison of Software Quality Models: An Analytical Approach (2012)
28. Jalal, A.N., Bahari, M., Musa, M.W., Sherief, S.R.: Social media and traditional customer relationship management: an overview of theoretical challenges and future directions in healthcare context. Int. J. Adv. Trends Comput. Sci. Eng. **8**(1.6), 76–82 (2019). https://doi.org/10.30534/ijatcse/2019/1281.62019
29. Ha, S., Stoel, L.: Consumer e-shopping acceptance: Antecedents in a technology acceptance model. J. Bus. Res. **62**(5), 565–571 (2009). https://doi.org/10.1016/j.jbusres.2008.06.016
30. Shin, D.: Determinants of customer acceptance of multi-service network: an implication for IP-based technologies. Inf. Manag. **46**, 16–22 (2009). https://doi.org/10.1016/j.im.2008.05.004
31. Merikivi, J., Nguyen, D., Tuunainen, V.K.: Understanding percieved enjoyment in mobile game context. In: 2016 49th Hawaii International Conference on System Sciences (2016)
32. O'Brien, H.L., Toms, E.G.: What is user engagement? A conceptual framework for defining user engagement with technology. J. Am. Soc. Inf. Sci. Technol. **59**(6), 938–955 (2008). https://doi.org/10.1002/asi.20801
33. Xu, Q., Sundar, S.S.: Lights, camera, music, interaction! interactive persuasion in e-commerce. Commun. Res. **41**(2), 282–308 (2014). https://doi.org/10.1177/0093650212439062
34. Blazquez Cano, M., Perry, P., Ashman, R., Waite, K.: The influence of image interactivity upon user engagement when using mobile touch screens. Comput. Human. Behav. **77**, 406–412 (2017). https://doi.org/10.1016/j.chb.2017.03.042
35. Hwang, A.H.C., Oh, J.: Interacting with background music engages E-Customers more: the impact of interactive music on consumer perception and behavioral intention. J. Retail. Consum. Serv. **54**(June 2019), 101928 (2020). https://doi.org/10.1016/j.jretconser.2019.101928
36. Shweta, P., Deepak, C.: Online customer experience (OCE) in clothing e-retail: exploring OCE dimensions and their impact on satisfaction and loyalty – Does gender matter? Int. J. Retail Distrib. Manag. **46**(3), 323–346 (2018). https://doi.org/10.1108/IJRDM-01-2017-0005
37. Sundar, S.S., Jia, H., Waddell, T.F., Huang, Y.: Toward a theory of interactive media effects (TIME). In: The Handbook of the Psychology of Communication Technology, pp. 47–86, January 2015, Doi:https://doi.org/10.1002/9781118426456.ch3
38. Liu, Y., Shrum, L.J.: A dual-process model of interactivity effects. J. Advert. **38**(2), 53–68 (2009). https://doi.org/10.2753/JOA0091-3367380204
39. Steuer, J.: Defining virtual reality: dimensions determining telepresence. J. Commun. **42**(4), 73–93 (2006). https://doi.org/10.1111/j.1460-2466.1992.tb00812.x
40. Van Noort, G., Voorveld, H.A., Van Reijmersdal, E.A.: Interactivity in brand web sites: cognitive, affective, and behavioral responses explained by consumers' online flow experience. J. Interact. Mark. **26**(4), 223–234 (2012)
41. Oh, J., Sundar, S.S.: How does interactivity persuade? An experimental test of interactivity on cognitive absorption, elaboration, and attitudes. J. Commun. **65**(2), 213–236 (2015). https://doi.org/10.1111/jcom.12147
42. Yoo, W.-S., Lee, Y., Park, J.: The role of interactivity in e-tailing: creating value and increasing satisfaction. J. Retail. Consum. Serv. **17**(2), 89–96 (2010). https://doi.org/10.1016/j.jretconser.2009.10.003
43. Shamir, A., Korat, O.: Educational electronic books for supporting emergent literacy of kindergarteners at-risk for reading difficulties—what do we know so far? Comput. Sch. **32**(2), 105–121 (2015). https://doi.org/10.1080/07380569.2015.1027868
44. Weng, C., Otanga, S., Weng, A., Cox, J.: Effects of interactivity in e-textbooks on 7th graders science learning and cognitive load. Comput. Educ. **120**(February), 172–184 (2018). https://doi.org/10.1016/j.compedu.2018.02.008

45. Agrawal, R., Sambamurthy, V., Stair, R.M.: Research report the evolving relationship between general and specific computer self-efficacy--an empirical assessment..pdf. (2000)
46. Agarwal, R., Prasad, J.: A conceptual and operational definition of personal innovativeness in the domain of information technology. Inf. Syst. Res. **9**(2), 204–215 (1998)
47. Lee, H.Y., Qu, H., Kim, Y.S.: A study of the impact of personal innovativeness on online travel shopping behavior—a case study of Korean travelers. Tour. Manag. **28**(3), 886–897 (2007). https://doi.org/10.1016/j.tourman.2006.04.013
48. Jeong, N., Yoo, Y., Heo, T.: Technological forecasting & social change moderating effect of personal innovativeness on mobile-RFID services : based on Warshaw ' s purchase intention model. Technol. Forecast. Soc. Chang. **76**(1), 154–164 (2009). https://doi.org/10.1016/j.techfore.2008.08.007
49. Zhang, Z., Hou, Y.: The effect of perceived risk on information search for innovative products and services : the moderating role of innate consumer innovativeness. J. Consum. Mark. **34**, 241–254 (2017). https://doi.org/10.1108/JCM-04-2016-1768
50. Cheng, Y.-M.: Exploring the intention to use mobile learning: the moderating role of personal innovativeness. J. Syst. Inf. Technol. **16**, 40–61 (2014). https://doi.org/10.1108/JSIT-05-2013-0012
51. López-Nicolás, C., Molina-Castillo, F.J., Bouwman, H.: An assessment of advanced mobile services acceptance: contributions from TAM and diffusion theory models. Inf. Manag. **45**(6), 359–364 (2008). https://doi.org/10.1016/j.im.2008.05.001
52. Ruiz-Mafé, C., Bigné-Alcañiz, E., Sanz-Blas, S., Aldás-Manzano, J.: Influence of online shopping information dependency and innovativeness on internet shopping adoption. Online Inf. Rev. **32**(5), 648–667 (2008). https://doi.org/10.1108/14684520810914025
53. van Raaij, E.M., Schepers, J.J.L.: The acceptance and use of a virtual learning environment in China. Comput. Educ. **50**(3), 838–852 (2008). https://doi.org/10.1016/j.compedu.2006.09.001
54. Wang, C.-Y., Chou, S., Chang, H.-C.: Exploring An Individual's Intention to Use Blogs: The Roles of Social, Motivational and Individual Factors (2010)
55. Lu, X., Viehland, D.: Factors influencing the adoption of mobile learning factors influencing the adoption of mobile learning. In: ACIS 2008 Proceedings (2008)
56. Lu, J., Yao, J.E., Yu, C.-S.: Personal innovativeness, social influences and adoption of wireless Internet services via mobile technology. J. Strateg. Inf. Syst. **14**(3), 245–268 (2005)
57. Mutahar, A.M., Mohd Daud, N., Ramayah, T., O. Isaac, T. and I. Alrajawy, "Integration of Innovation Diffusion Theory (IDT) and Technology Acceptance Model (TAM) to Understand Mobile Banking Acceptance in Yemen: The Moderating Effect of Income," *Int. J. Soft Comput.*, vol. 12, no. 3, pp. 164–177, 2017, doi: https://doi.org/10.3923/ijscomp.2017.164.177.
58. Salem, S.F., Salem, S.O.: Effects of social media marketing and selected marketing constructs on stages of brand loyalty. Glob. Bus. Rev. **22**, 650–673 (2019). https://doi.org/10.1177/0972150919830863
59. Cabrera-Sánchez, J.P., Villarejo-Ramos, Á.F.: Acceptance and use of big data techniques in services companies. J. Retail. Consum. Serv. **52**, 101888 (2020). https://doi.org/10.1016/j.jretconser.2019.101888
60. Fishbein, M.: Reasoned action, theory of. The International Encyclopedia of Communication (2008). Doi:https://doi.org/10.1002/9781405186407.wbiecr017
61. Venkatesh, V., Thong, J.Y.L., Chan, F.K.Y., Hu, P.J.-H., Brown, S.A.: Extending the two-stage information systems continuance model: incorporating UTAUT predictors and the role of context. Inf. Syst. J. **21**(6), 527–555 (2011). https://doi.org/10.1111/j.1365-2575.2011.00373.x
62. Davis, F.D.: Perceived usefulness, perceived ease of use, and user acceptance. MIS Q. **13**(3), 319–339 (1989). https://doi.org/10.2307/249008

63. Purwanto, E., Mutahar, A.M.: Examine the technology of acceptance model among mobile banking users in Indonesia. Technol. Reports Kansai Univ. **62**(07), 3969–3979 (2020)
64. Delone, W.H., Mclean, E.R.: The DeLone and McLean model of information systems success : a ten-year update. J. Manag. Inf. Syst. **19**, 9–31 (2003)
65. Cheng, Y.M.: Roles of interactivity and usage experience in e-learning acceptance: a longitudinal study. Int. J. Web Inf. Syst. **10**(1), 2–23 (2014). https://doi.org/10.1108/IJWIS-05-2013-0015
66. Lee, J.H., Lee, C.F.: Extension of TAM by perceived interactivity to understand usage behaviors on ACG social media sites. Sustain. **11**(20), 1–19 (2019). https://doi.org/10.3390/su11205723
67. Venkatesh, V., Bala, H.: Technology acceptance model 3 and a research agenda on interventions. Decis. Sci. **39**(2), 273–315 (2008). https://doi.org/10.1111/j.1540-5915.2008.00192.x
68. Parasuraman, A.: Technology readiness index (TRI): a multiple-item scale to measure readiness to embrace new technologies. J. Serv. Res. **2**(4), 307–320 (2000). https://doi.org/10.1177/109467050024001
69. Hayduk, L.A., Littvay, L.: Should researchers use single indicators, best indicators, or multiple indicators in structural equation models? BMC Med. Res. Methodol. **12**(1), 159 (2012)
70. Ringle, C.M., Wende, S., Becker, J.-M.: SmartPLS 3. Bonningstedt: SmartPLS. (2015)
71. Aldholay, A., Abdullah, Z., Isaac, O., Mutahar, A.M.: compatibility perspective of Yemeni students on use of online learning extending the information systems. Inf. Technol. People **33**(1), 106–128 (2019). https://doi.org/10.1108/ITP-02-2018-0095
72. Kline, R.B.: Principles and practice of structural equation modeling, 3rd edn. The Guilford Press, New York (2010)
73. Hair, J.F., Black, W.C., Babin, B.J., Anderson, R.E.: Multivariate Data Analysis, 7th edn. Pearson, New Jersey (2010)
74. Fornell, C., Larcker, D.F.: Evaluating structural equation models with unobservable variables and measurement error. J. Mark. Res. **18**(1), 39–50 (1981)
75. Hair, J.F., Hult, G.T.M., Ringle, C., Sarstedt, M.: A Primer on Partial Least Squares Structural Equation Modeling (PLS-SEM), 2nd edn. SAGE, Thousand Oaks (2017)
76. Chin, W.W.: Issues and opinion on structural equation modeling. MIS Q. **22**(1), 7–16 (1998)
77. Cohen, J.: Statistical Power Analysis for the Behavioral Sciences, 2nd edn. Routledge, New York (1988)
78. Jang, H.Y., Noh, M.J.: Customer acceptance of IPTV service quality. Int. J. Inf. Manage. **31**(6), 582–592 (2011). https://doi.org/10.1016/j.ijinfomgt.2011.03.003
79. Aldholay, A., Abdullah, Z., Isaac, O., Mutahar, A.M.: Perspective of Yemeni students on use of online learning: Extending the information systems success model with transformational leadership and compatibility. Inf. Technol. People **33**(1), 106–128 (2019). https://doi.org/10.1108/ITP-02-2018-0095
80. Wang, Y.S., Liao, Y.W.: Assessing eGovernment systems success: a validation of the DeLone and McLean model of information systems success. Gov. Inf. Q. **25**(4), 717–733 (2008). https://doi.org/10.1016/j.giq.2007.06.002
81. Chang, C.: Exploring the determinants of e-learning systems continuance intention in academic libraries. Libr. Manag. **34**(1/2), 40–55 (2013). https://doi.org/10.1108/01435121311298261
82. Almaiah, M.A., Jalil, M.A., Man, M.: Empirical investigation to explore factors that achieve high quality of mobile learning system based on students' perspectives. Eng. Sci. Technol. Int. J. **19**(1), 1314–1320 (2016). https://doi.org/10.1016/j.jestch.2016.03.004
83. Chan, S.C.H., Wan, J.C.L., Ko, S.: Interactivity, active collaborative learning, and learning performance: the moderating role of perceived fun by using personal response systems. Int. J. Manag. Educ. **17**(1), 94–102 (2019). https://doi.org/10.1016/j.ijme.2018.12.004

84. Obeidat, Z.M., AlGharabat, R.S., Alalwan, A.A., Xiao, S.H., Dwivedi, Y.K., Rana, N.P.: Narcissism, interactivity, community, and online revenge behavior: The moderating role of social presence among Jordanian consumers. Comput. Hum. Behav. **104**, 106170 (2020). https://doi.org/10.1016/j.chb.2019.106170
85. Parveen, F., Sulaiman, A.: Technology complexity, personal innovativeness and intention to use wireless internet using mobile devices in Malaysia. Int. Rev. Bus. **4**(5), 1–10 (2008)

A Framework to Implement Blockchain in Higher Education Institutions

Suaad Al Mansoori(✉) and Piyush Maheshwari

The British University in Dubai (BUiD), Dubai, UAE
20170871@student.buid.ac.ae, piyush.maheshwari@buid.ac.ae

Abstract. This paper presents a framework to implement business solutions based on Blockchain technology (BCT) for the higher education institutions (HEIs). The first part of the paper provides an overview of the Blockchain technology, its implementation in the sector, and its advantages. The second part discusses the common challenges of implementing technology and then identifies higher education stakeholders' insights. Getting stakeholders' insights is the root of creating a comprehensive framework to help higher education institutions successfully implement the solutions based on BCT.

Keywords: Blockchain · Centralization · Decentralization · Stakeholders · Framework

1 Introduction

Traditional systems used to manage students' credentials are no longer as beneficial as they were. With the current technological wave set by smart technologies, such as the Internet of Things, Smart Cities, Artificial Intelligence, and Blockchain, centralized learning management systems might be considered drawbacks for Higher Education Institutions (HEIs). Since the last century, the traditional systems are exposed to the central point of failure that might cause losing data if not timely backed up at another sites. They are also facing issues of security, data duplication, and redundancy. There are no clear criteria on what data to be entered or who is authorized to access systems. There is also the difficulty of sharing data and information between HEIs and stakeholders and extracting reports for data analysis and management purposes. Such issues make it valid for other smarter systems that can easily manage the data transparent yet securely. Blockchain Technology (BCT), a decentralized and distributed public ledger, has received extensive attention as it applies a linked block structure to verify and store data and applies the trusted consensus mechanism to synchronize changes, which helps to create a tamper-proof digital platform for storing and sharing data [1]. BCT is argued to be the new wave that will challenge several organizations and institutions in society, bringing wide-scale social, cultural, economic, and political transformations rendering the once-powerful nation-state somewhat powerless [2]. Building a smart platform is no issue. There are expertise in the market who are eager to follow the technological trends

M. Al-Emran et al. (Eds.): ICETIS 2021, LNNS 299, pp. 244–254, 2022.
https://doi.org/10.1007/978-3-030-82616-1_22

and offer the updated intelligent systems and applications as customized as HEIs need. There are also some HEIs willing to build platforms as they have the required resources.

The issue relies on preparing the environment to embrace the new technology and take away the traditional historical systems operating for ages. HEIs have a history of being slow and ineffective in adopting new and smart technologies [2], which appears in having most HEIs not embracing BCT yet [3]. Another critical point is that HEIs are gradually changing to be more of businesses enterprises in its organizational structures and missions, which can confuse the stakeholders of educational background who need to shift their intentions to the business side; and vice versa. Stakeholders who are making decisions to embrace the BCT in HEIs need a detailed, friendly framework or roadmap to help those making informed decisions and systematically implement the new technology. This research answers two main questions: what are the challenges faced by the HEIs when embracing BCT? How can HEIs reform their environments in the light of BCT?

2 Overview of Blockchain in Education

2.1 What is Blockchain?

Blockchain is a public ledger, in which all transactions are stored in a chain of blocks. This chain continuously grows when new blocks are appended to it. BCT has key characteristics, such as decentralization, persistency, anonymity and auditability [4]. BCT works in decentralized environments, which integrates several core technologies such as cryptographic hash, digital signature, asymmetric cryptography; and distributed consensus [4]. BCT emerged after emerging Bitcoin in 2009 by Satoshi Nakamoto [5]. Bitcoin was the first implementation of peer-to-peer trustless electronic cash. Subsequently, many other forms of electronic cash or cryptocurrencies have been created using similar structures [6]. BCT allows the decentralized transactional data sharing across a large network of participants [7]. In comparison to other technologies, BCT gives the strengths of compatibility, sustainability, data sharing, and interconnectivity [9]. BCT has three broad classifications: fully private, public, and consortium [11]. Private BCT has the write permissions kept to one organization, while read permissions may be available or restricted to specific participants. In public BCT, all nodes are potentially deciders and can create new blocks [10]. Consortium BCT is a blend of the public and private BCT. It conserves the distributed structure and assets security because only the participants verified by the preset nodes could trade in the BCT. In the consortium BCT, a node needs permission before exercising its rights [12].

2.2 Characteristics of BCT

Yang *et al.* (2019) highlighted the significant characteristics of BCT [13], which are: (1) Decentralized: all nodes are equal in the network as there is no central node, allowing the existence of malicious nodes, and the high robustness of BCT makes the network run stably, (2) Trustless: data can be exchanged between nodes in a trusted environment and the operation behavior is transparent and open, (3) Reliable Database: every node has all the data exchange databases. Even if there is data destruction of the wrong node, the correct data can be recovered quickly, (4) Collective Maintain: all nodes participate in maintaining the blocks in the chain. To explain the collective maintain point, BCT can reach a consensus regarding the state of data and agree on the network rules without a central governing entity [14]. Accordingly, any user can propose an improvement to the system. Still, the update will be implemented only after the acceptance of all parties involved in the network, which helps in enhancing transparency and trust [14].

BCT promises a significant security level. To transcend a target system's hashing power, the attacker needs to compromise 51% of the system [15]. One computer is not enough to hack a BCT. Changing one small data on the BCT takes a large amount of computing power to access each participant's instance, so bigger networks have a higher level of security [16]. Those advantages ensure information credibility and traceability, which traditional technologies lack [17].

2.3 Implementation of BCT in Education

Williams (2018) contributed to the literature with a research about the implementation of BCT in the education field [18]. The paper provided a timeline of using this disruptive technology as following: (1) Blockcerts, was developed at the Massachusetts Institute of Technology by MIT Media Lab and the Learning Machine company as an open standard for creating library tools, (2) Sony Global Education, worked in the maintenance of 'Next Generation of School ICT Environment' for the Ministry of Internal Affairs and Communications in Japan, (3) several universities started to offer modules in BC, (4) in March 2018 plans were announced for the world's first Blockchain-powered university which is Woolf University; formed by a small group of Humanities academics at Oxford University. The University aims to be an online university that allows academics to publicize their knowledge and students to choose courses as credits towards undergraduate degrees. Turkanovic *et al.* (2018) proposed EduCTX, a global higher education credit BCT-based platform [19]. This platform is based on the concept of the European Credit Transfer and Accumulation System (ECTS). It provides a globally trusted and decentralized higher education credit, creates a grading system that can offer a globally unified viewpoint for students and higher education institutions and their stakeholders. The authors conducted a proof of concept by a prototype implementation of the environment. The authors mentioned that EduCTX would process, manage, and control ECTX tokens, representing credits that students achieve for fulfilled modules [19]. Ocheja *et al.* (2019) proposed a BCT-based approach for connecting learning data across different Learning Management Systems (LMS), Learning Record Stores (LRS), institutions, and organizations [20].

2.4 Advantages of BCT in Education

BCT's decentralized nature can ensure so many advantages to surpass the weaknesses of the traditional centralized systems. Ralston [3] highlighted the applications of BCT on higher education as: managing students' academic degrees, evaluating the learning outcomes, accessing and storing certifications and degree records, reinforcing the security of stored information, and reducing the costs of administrative work such as admissions and registration. The transparency of transaction recording and the complexity of coding in BCT eliminate the possibility of copying transactions, which makes the information safe [2]. Sharples & Domingue discussed the implementations of BCT in higher education as the following [21]: the educational records are distributed over a large network so they are resilient to the risk of infrastructure loss, the identity confirmation of adding records, and no modification or alteration of records even by original authors. It is also apparent that BCT contributes heavily to the reputation of the HEIs [21] with the vast level of competition between institutions to gain students' trust and academics. Adding such innovative technology to the institute environment can reform its image and identity if planned well. While a huge project like BCT-based systems can cost a fortune to start, it will help cut costs in the long run when it comes to economics and transactions [22], mainly including administration work costs.

3 Challenges of BCT in Education

The literature highlighted some common challenges that all industries and organizations might face when adopting BCT:

1) *Institution's Theatres:* While BCT promises to dramatically change the education field similar to any other new technology [21], many of its advantages are exaggerated due to clever marketing strategies [3]. Some organizations might implement BCT to keep up with the innovation waves but not getting the main benefits out it, which may lead to huge costs in finance, human resources, and confusion in the institution's policies and regulations. Tterm "Theatre" was created by Steve Blank in a Harvard Business Review article of 2019 [3]. The report explained the term as Organizational Theatre: shorts-term gains resulted from hiring consultants to advise in re-organizing the institution, Innovation Theatre: innovation activities such as workshops and design-thinking classes to try changing the culture of the institution; and Process Theatre: creating more processes and operations to conduct work in the institution. Maloney & Kim (2020) mentioned in an online article regarding "Innovation Theatre" that investment for innovation will be ineffective if the learning structure is unaltered [24]. This indicates the importance of setting a reliable investment purpose rather than only following technological trends. Promises of BCT cannot always justify its adoption in HEIs, especially when there is no regulations, ready environment, and primary goals that drive adopting the disruptive technology [3]. Implementing BCT should be preceded by a precise setting for reliable short and long-term goals to not fall into one of the theatre's traps. Generally, innovation is a challenge in organizations because it involves high uncertainty and unpredictability [23].
2) *Choosing the technology:* BCT platforms differ in many aspects, requiring different types of setting and readiness in HEIs environment and infrastructures. The Blockchain Council; which is an authoritative team of experts in Blockchain Research and Development; highlighted in the council's official website that BCT platforms differ in type of Network, Popularity, Language, Pricing and the Consensus Mechanism [25].

3) *Privacy and Security:* BCT has three broad classifications: fully private, public, and consortium [11]. Private blockchain has the write permissions kept to one organization while read permissions may be public or restricted to certain participants. In public blockchain, all nodes are potentially deciders and can participate in the creation of new blocks [10]. Consortium blockchain is a blend of the public blockchain and private blockchain. It conserves the distributed structure as well as assets security because only the participants verified by the preset nodes could trade in the blockchain. In the consortium blockchain, a node needs permission before it exercises its rights [12]. Blockchain contains immutable information, and at the same time, users' privacy is highly protected. Public blockchain offer pseudonymity to its users so each user is able to interact with the ledger through a newly created address without revealing their real identity [26].
4) *Users:* Hawlitschek *et al.* discussed users' effect, especially regarding their trust [27]. The authors highlighted the definition of trust as the intention to accept vulnerability based upon positive expectations. They also highlighted user behavior, especially when shifting from current or old systems to the new blockchain-based systems. It is more of a challenge for HEIs to build trust and amend the user's behavior towards a new system because of different users such as students, faculty, staff, managers, etc. in HEIs. Each of those users has different expectations of behavior when using systems. Most importantly, there should be solutions for the resistance in HEIs environments that are unwilling to have a room for trial in technology adoption and manage external stakeholders' trust and interests [2].
5) *Resilience:* Merriam-Webster defines resilience as an ability to recover from or adjust easily to misfortune or change. There is a need to understand how an HEI can evolve and adopt external demands and trends [2] with minimal level of confusion in its environment. BCT might shift the focus of many HEIs from creating knowledge to research and inventions [21].
6) *Finance:* What are the financial values gained by implementing innovative ideas like BCT [21]? There are so many financial challenges associated with huge projects. Examples of those challenges are: policy support, the fund-raising model, project development approach and the financial structure of the organization itself [28].
7) *Infrastructure:* The infrastructure layer provides the environment to deploy the BCT-based system with the support of other facilities such as the hardware and network resources as well as other storage and computing resources [29]. HEIs are hoping to have a more powerful infrastructure that can ease the institution's operational procedures alongside the powerful features of decentralization and higher security levels. Software programs, standards, dashboards, and visual analytics technologies that constitute the infrastructure [30] have a radical impact on BCT's implementation success.

4 Methodology

BCT's complexity and its associated challenges raise the need for a comprehensive framework or a guideline that can help successfully implement the technology. There are several frameworks available that can help any organization [23, 31–33]. However, higher education institutions' unique environment requires a unique comprehensive framework that addresses the different challenges of implementing BCT. In order to achieve this goal, qualitative and quantitative methods were used to get insights from higher education stakeholders. The main quantitative method was an online survey questionnaire, in which 250 stakeholders of HEIs successfully participated. The main qualitative methods were structured and unstructured interviews in which 6 interviewers provided valuable inputs based on their experiences in BCT as well as their experiences in implementing smart systems. This mixed approach helped to provide insights on the most important challenges facing HEIs when implementing solutions using BCT. Participants were students, admin staff, faculty members, and IT managers. The collected data was used to draw meaningful conclusions about the most important challenges that must be considered in the new proposed framework.

4.1 Results

Participants in the online survey questionnaire had to provide their opinion on the most significant challenge that HEIs face when implementing BCT. Those answers would help in proposing a new framework. The answers fall into similar categories, as displayed in Fig. 1.

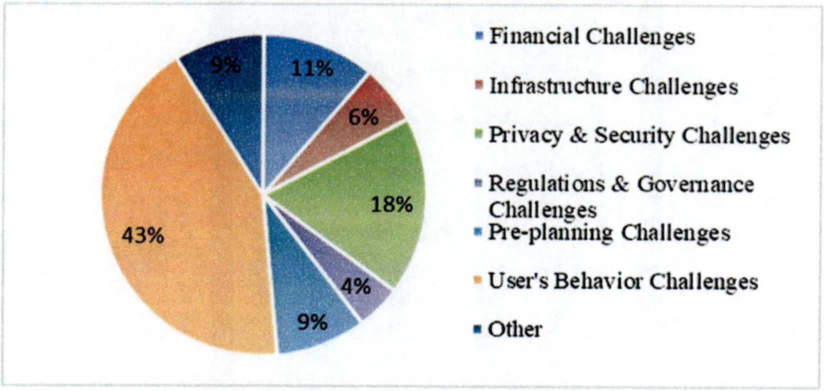

Fig. 1. Challenges of implementing BCT in HEIs

The majority of the participants (43%) believed that the HEIs' most significant challenge is the users' behavior. 18% indicated challenges related to privacy and security, usually the case of any new technology. Less significant, yet effective, challenges referred to pre-planning, regulations, and financial challenges. While the majority believed that user's behavior is the most significant challenge, 52% mentioned training and organizing workshops to overcome the obstacles facing HEIs when implementing Blockchain.

In addition to the questionnaire results, three interviewees from senior levels with computer science and business backgrounds provided their opinion on the challenges of implementing BCT in HEIs. The emphasis was mainly given on the following points:

- How to change the tradition of centralization in students and staff mentality. Conducting some studies before allowing the project, for example, analyzing Return on Investment and Feasibility Study. Such tools must be accomplished appropriately to deliver real results.
- Involving users (e.g., students and staff) in the process and provide comprehensive training to get to know the new system and reduce the change resistance level.

The challenges reported by the participants became the necessary requirements that this research weighed in proposing a new framework.

5 A New Proposed Framework

The new framework resulted from the available frameworks in addition to the gathered insights from the stakeholders of HEIs (see Fig. 2). Each component of the framework includes sub-components and items that affect the success of implementing BCT in HEIs. Below we provide an overview of each component.

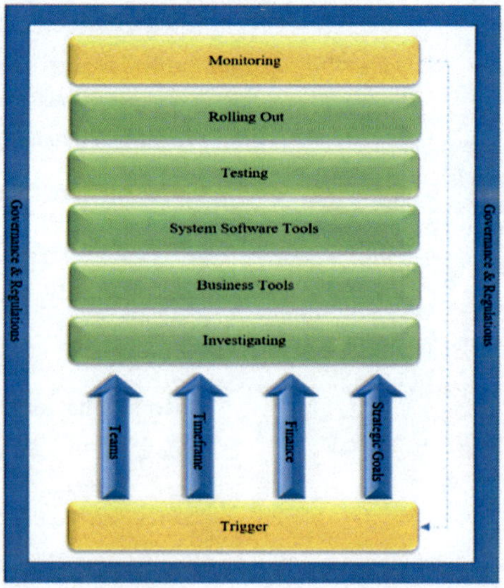

Fig. 2. New framework to implement BCT in HEIs

1) *Trigger:* Some researches refer to triggering implementing BCT as "Organizational Case" that motivates decision-makers to adopt the new technology. In this new proposed framework, the trigger is mainly the need to manage students' credentials as the current practices are not fully efficient. Another important trigger can be governmental instructions or initiatives from the Ministry of Higher Education.
2) *Strategic Goals:* None of the existing frameworks mention the overall strategies of HEIs as important elements to implement BCT. Projects in any organization should

be parallel with the overall strategies that are decided by upper management. The board of trustees and other stakeholders at the top of the decision-making pyramid are usually the ones who decide the HEI's strategic goals, which usually last for more than a year.

3) *Finance:* There should be a clear decision on how much financial resources would be available for the project of implementing the new technology or system (e.g., limited budget, open-budget, etc.). The existing researches and frameworks mentioned finance as part of the feasibility study that can be conducted to minimize the risk of adopting BCT [31, 32].
4) *Timeframe:* None of the existing frameworks mention the time as an element in the process of implementing BCT; and that it probably due to the complexity of such project. While time is always a subject of changing, it is crucial to know the deadlines and milestones of the project of implementing the new system.
5) *Teams:* None of the existing frameworks mention the importance of identifying the members that are going to work in implementing BCT. Such project needs a diverse team members who have certain skills and capabilities; not only in the project management, but also in BCT itself. The initial team can identify and nominate the personnel that will work in each phase of the project. A certain member may be involved in each team and phase. It is also possible to outsource some members if the institution's current human resources lack the required skills.
6) *Investigating:* Existing frameworks refer to investigating as an organization case, feasibility study, effective factors, discovering, etc. [23, 31, 32]. However, in this new framework, the initial investigation works as an overall look for the current practice's entire situation and the expectations for the new one.
7) *Business Tools:* Most of the available Blockchain frameworks start from the business tools' point [23, 31–33] and miss out the previous six points. There is no limit in the business tools that can help in implementing BCT. SWOT (strengths, weaknesses, opportunities, and threats) for the current and new practices, ROI (Return on Investment) for the new practice to be presented for the decision makers, a feasibility study will also be crucial to identify the availability of resources (including human resources) that will carry out the technical work such as programmers, software engineers and system developers [32].
8) *System Software Tools:* The available frameworks put emphasis on business tools but miss out the importance of system software tools. The deliverables of the business tools will help in identifying if the new system can be built in-house, purchased, hosted in cloud, etc. If the institution has the capabilities and resources to build the new system, the classic SDLC (system development life cycle) process will be carried out, software engineering and other computer science related tasks that can help identify the technical aspects.
9) *Testing:* This testing phase is separate from the testing phase carried out in SDLC. Representatives from all end-user categories (e.g., students, admin staff, management, etc.) must be involved in this phase. Several test types can help ensure the success of the new system. For example, using dummy data allows real end-users to use the new system in a simulation environment. More testing helps in identifying new insights that can be suggested back to the technical team.

10) *Rolling out:* While the new framework replaces the current traditional systems of managing students' credentials with BCT, there are three options for the rolling out phase [32]: running the existing system and keeping the Blockchain based system as a backup one, running the new Blockchain based system and keep the old system as a backup one, or using the new Blockchain system as a standalone system.
11) *Monitoring:* The project ends during the rolling out. However, there will be continuous monitoring roles assigned to all end-users responsible for providing feedback and suggestions. There should be an official channel to receive comments and suggestions. Input in this phase can be a trigger for another upgrade or replacement.
12) *Governance and Regulations:* Available frameworks emphasize the importance of laws, policies, regulations and rules that can manage and organize the process of implementing BCT [23, 31–33].

6 Conclusion

Blockchain technology can help to mitigate the issues of the current practices in HEIs. HEIs need a framework or a guideline to adopt the new smart technology successfully. We presented a framework that is derived from the strengths of some available frameworks and insights collected from the stakeholders of HEIs. It consists of a unified framework of a few components and processes to guarantee success of blockchain projects. A future project can be figuring out how to link all HEIs and other related organizations under one blockchain platform.

References

1. Feng, Q., He, D., Zeadally, S., Khan, M., Kumar, N.: A survey on privacy protection in blockchain system. J. Netw. Comput. App. **126**, 45–58 (2019)
2. Haugsbakken, H., Langseth, I.: The blockchain challenge for higher education institutions. Eur. J. Educ. **2**(3), 41 (2019)
3. Ralston, S.: Postdigital Prospects for Blockchain-Disrupted Higher Education: Beyond the Theater, Memes and Marketing Hype. Postdigital Sci. Educ. **2**(2), 280–288 (2019)
4. Zheng, Z., Xie, S., Dai, H., Chen, X., Wang, H.: Blockchain challenges and opportunities: a survey. Int. J. Web Grid Serv. **14**(4), 352 (2018)
5. Chen, G., Xu, B., Lu, M., Chen, N.: Exploring BCT and its potential applications for education. Smart Learn. Environ. **5**(1) (2018)
6. Novo, O.: Blockchain Meets IoT: An Architecture for Scalable Access Management in IoT. IEEE Internet of Things J. **5**(2), 1184–1195 (2018)
7. Xu, X., Pautasso, C., Zhu, L., Gramoli, V., Ponomarev, A., Tran, A., Chen, S.: The blockchain as a software connector. In: 2016 13th Working IEEE/IFIP Conference on Software Architecture (WICSA) (2016). Accessed 13 Nov 2019
8. Yli-Huumo, J., Ko, D., Choi, S., Park, S., Smolander, K.: Where is current research on BCT?—a systematic review. PLOS ONE **11**(10), e0163477 (2016)
9. Yu, S., Lv, K., Shao, Z., Guo, Y., Zou, J., Zhang, B.: A high performance blockchain platform for intelligent devices. In: 2018 1st IEEE International Conference on Hot Information-Centric Networking (HotICN) (2018). Accessed 27 Dec 2019

10. Vizier, G., Gramoli, V.: ComChain: bridging the gap between public and consortium blockchains. In: 2018 IEEE International Conference on Internet of Things (iThings) and IEEE Green Computing and Communications (GreenCom) and IEEE Cyber, Physical and Social Computing (CPSCom) and IEEE Smart Data (SmartData) (2018). Accessed 27 Dec 2019
11. Sankar, L., Sindhu, M., Sethumadhavan, M.: Survey of consensus protocols on blockchain applications. In: 2017 4th International Conference on Advanced Computing and Communication Systems (ICACCS) (2017). Accessed 27 Dec 2019
12. Wang, X., Feng, Q., Chai, J.: The research of consortium blockchain dynamic consensus based on data transaction evaluation. In: 2018 11th International Symposium on Computational Intelligence and Design (ISCID) (2018). Accessed 27 Dec 2019
13. Yang, A., Li, Y., Liu, C., Li, J., Zhang, Y., Wang, J.: Research on logistics supply chain of iron and steel enterprises based on block chain technology. Futur. Gener. Comput. Syst. **101**, 635–645 (2019)
14. Abeyratne, S.A., Monfared, P.R.: Blockchain ready manufacturing supply chain using distributed ledger. Int. J. Res. Eng. Technol. **05**(09), 1–10 (2016)
15. Biswas, K., Muthukkumarasamy, V.: Securing smart cities using BCT. In: 2016 IEEE 18th International Conference on High Performance Computing and Communications; IEEE 14th International Conference on Smart City; IEEE 2nd International Conference on Data Science and Systems (HPCC/SmartCity/DSS) (2016). Accessed 27 Dec 2019
16. Gemeliarana, I., Sari, R.: Evaluation of Proof of Work (POW) blockchains security network on selfish mining. In: 2018 International Seminar on Research of Information Technology and Intelligent Systems (ISRITI) (2018). Accessed 27 Dec 2019
17. Wang, R., He, J., Liu, C., Li, Q., Tsai, W., Deng, E.: A privacy-aware PKI system based on permissioned blockchains. In: 2018 IEEE 9th International Conference on Software Engineering and Service Science (ICSESS) (2018). Accessed 27 Dec 2019
18. Williams, P.: Does competency-based education with blockchain signal a new mission for universities? J. Higher Educ. Policy. Manage. **41**(1), 104–117 (2018)
19. Turkanovic, M., Holbl, M., Kosic, K., Hericko, M., Kamisalic, A.: EduCTX: a blockchain-based higher education credit platform. IEEE Access **6**, 5112–5127 (2018)
20. Ocheja, P., Flanagan, B., Ueda, H., Ogata, H.: Managing lifelong learning records through Blockchain. Res. Pract. Technol. Enhanc. Learn. 14(1) (2019)
21. Sharples, M., Domingue, J.: The blockchain and kudos: a distributed system for educational record, reputation and reward. Adaptive and Adaptable Learning, pp. 490–496 (2016). Accessed 27 Sep 2020
22. Ahluwalia, S., Mahto, R., Guerrero, M.: BCT and startup financing: A transaction cost economics perspective. Technol. Forecast. Soc. Chang. **151**, 1198–54 (2020)
23. Beck, R., Müller-Bloch, C.: Blockchain as radical innovation: a framework for engaging with distributed ledgers as incumbent organization. In: Proceedings of the 50th Hawaii International Conference on System Sciences (2017). Accessed 2 Oct 2020
24. Maloney, E., Kim, J.: How Universities Can Avoid Learning Innovation Theater|Inside Higher Ed. Insidehighered.com (2020). https://www.insidehighered.com/digital-learning/blogs/technology-and-learning/how-universities-can-avoid-learning-innovation. Accessed 27 Sep 2020
25. Sharma, T.: Top 10 Blockchain Platforms You Need To Know About. Blockchain-council.org (2020). https://www.blockchain-council.org/blockchain/top-10-blockchain-platforms-you-need-to-know-about/. Accessed 27 Sep 2020
26. Litke, A., Anagnostopoulos, D., Varvarigou, T.: Blockchains for supply chain management: architectural elements and challenges towards a global scale deployment. Logistics **3**(1), 5 (2019)

27. Hawlitschek, F., Notheisen, B., Teubner, T.: perspective on the limits of trust-free systems: a literature review on BCT and trust in the sharing economy. Electron. Comm. Res. App. **40**, 1009–35 (2020)
28. Thanh Truong, T., Friedrich, H., Charoenngam, C.: Success factors for financial sustainability of toll road projects: empirical evidence from China. Transp. Res. Procedia. **48**, 1848–1860 (2020)
29. Wang, Q., Su, M.: Integrating BCT into the energy sector—from theory of Blockchain to research and application of energy blockchain. Comput. Sci. Rev. **37**, 1002–75 (2020)
30. Williamson, B.: The hidden architecture of higher education: building a big data infrastructure for the 'smarter university.' Int. J. Educ. Technol. Higher Educ. **15**(1), 12 (2018)
31. Wibowo, S., Hw, E.: Blockchain implementation assessment framework, case study of IoT LPWA licensing in Indonesia. In: 2018 International Conference on ICT for Smart Society (ICISS) (2018). Accessed 27 Dec 2019
32. Wang, H., Chen, K., Xu, D.: A maturity model for blockchain adoption. Finan. Innov. **2**(1), 1–5 (2016)
33. Janssen, M., Weerakkody, V., Ismagilova, E., Sivarajah, U., Irani, Z.: A framework for analysing BCT adoption: integrating institutional, market and technical factors. Int. J. Inf. Manage. **50**, 302–309 (2020)

The Acceptance of E-Learning Service in a Higher Education Context

Alaa S. Jameel[1,3](✉), Mohammed A. Karem[1], Saeed Hameed Aldulaimi[2], Ahmed Kh. Muttar[2], and Abd Rahman Ahmad[3]

[1] Department of Public Administration, Cihan University-Erbil, Kurdistan Region, Erbil, Iraq
{alaa.salam,mohammad.abdulkarem}@cihanuniversity.edu.iq

[2] College of Administrative Sciences, Applied Science University, Al Eker, Kingdom of Bahrain
{saeed.aldulaimi,ahmed.almuhamadi}@asu.edu.bh

[3] Faculty of Technology Management and Business, Universiti Tun Hussein Onn Malaysia, Parit Raja, Malaysia
arahman@uthm.edu.my

Abstract. The purpose of this study is to determine the impact of information quality (IQ) and computer self-efficacy (CSE) on perceived ease of use (PEOU) and perceived usefulness (PU), which were hypothesized to impact behavioral intention (BI) to use E-learning. Based on the Technology Acceptance Model, this study was conducted among undergraduate students of the College of Administrative and Financial Sciences, Cihan University-Erbil. The students participated in E-learning that supported the traditional teaching method via Moodle platform during COVID-19 pandemic. 209 valid questionnaires were collected and analyzed using Structural Equation Modelling to examine seven proposed hypotheses. This study found that CSE had a positive and significant impact on PEOU but an insignificant impact on PU. Moreover, IQ positively and significantly impacted PU and PEOU. PEOU and PU significantly impacted BI to use E-learning among undergraduate students. PEOU had a significant impact on PU in the context of the education system. The implication of this study is universities and administrators of E-learning programs must focus on CSE, IQ, PEOU and PU that play a significant role in increasing E-learning adoption by students, and consequently influence the success of the educational process and productivity in the time of COVID-19 outbreak.

Keywords: E-Learning · Computer self-efficacy · Behavioral intention · Information quality · TAM

1 Introduction

Electronic learning (E-learning) programs, like online learning tools and Learning Management System (LMS), have been widely embraced by universities due to the pandemic of COVID-19. E-learning is known as a service for the delivery of information through Internet technologies [1]. Meanwhile, E-learning platforms are popular innovations for cyberspace that support distance learning, face-to-face and combined (blended) teaching/learning processes [1].

M. Al-Emran et al. (Eds.): ICETIS 2021, LNNS 299, pp. 255–264, 2022.
https://doi.org/10.1007/978-3-030-82616-1_23

The programs provide teaching and learning support resources and functions such as the supervision of classes, online community interactions and debates, related content (i.e. lesson content and tasks), lecture slides and evaluations of courses [2]. E-learning systems are increasingly being incorporated into higher education institutions worldwide, particularly during the COVID-19 pandemic. It can be used as a substitute or supplement to the conventional mode of learning even after this pandemic.

Technology Acceptance Model (TAM) describes whether a user embraces or refuses information technology based on their perceived usefulness (PU) and perceived ease of use (PEOU) [3]. Centered on the principle of rational intervention, TAM focuses on the behavioral intention (BI) to use the technology and assumes two main determinants: PU and PEOU [4]. However, limited studies about TAM were conducted in developing countries [5, 6], thus, the findings of studies performed in the developed countries were transferred directly to the developing ones [6] without considering country-specific factors such as the computer self-efficacy and information quality. There is a lack of understanding about BI to use E-learning in the Iraqi higher education system [1]. Iraqi universities face several challenges particularly during the pandemic of COVID-19, such as poor IT infrastructure [7–10] and the lack of IT awareness among students and academic staff [11]. In response, the purpose of this study is to investigate the impact of computer self-efficacy and information quality as external factors on the TAM among undergraduate students in the Iraqi higher education context.

2 Literature Review

2.1 E-learning

E-learning is described as the teaching and learning approach that completely or partially represents the educational paradigm focusing on using electronic media and computers to increase the availability of instruction, communication and interaction [1]. The use of technological means is an integral aspect of E-learning because E-learning is presently described as learning through numerous computational devices such as laptops, cell phones, tablets and virtual worlds [1].

Students should have good ability to control their time and keep updated about the new programming skills and ICT knowledge to succeed in the current E-learning environment caused by COVID-19 pandemic. Students' experience with technology use and E-learning behaviors are essential for the successful implementation of E-learning nowadays. The productive use of IT in E-learning is vital in terms of students' acceptance and effective E-learning. The university facilities should also be sufficient, secure, and able to provide students with material as expeditiously as possible.

Similar to any other learning platform, the E-learning has its strengths and weaknesses. Regarding the strengths, E-learning can create a cooperative spirit between students, promote individual students, build strong relationships and enhance problem-solving skills between students and teachers. As far as the shortcomings are concerned, E-learning is less trustworthy in peer reviews and collaborative tasks than traditional learning [5, 12].

2.2 Theory and Research Hypotheses

Technology Acceptance Model (TAM)

TAM is considered one of the most common and effective theories among the mentioned theories. TAM consisted of six variables including PU, PEOU, attitude, BI and actual use. BI to use technology is regarded as the most significant element in classical behavioral theories of TRA [13] and TPB [14]. It was defined as the motives that drive a person to use technology. PU and PEOU are considered the major variables in the model that impact on BI to use technology and actual use [3, 15–18]. PU is to the degree that an individual assumes that applying a certain application or framework enhances his work efficiency. In contrast, PEOU is the level to which the user assumes it is effortless to use a specific technology [3].

TAM clarifies and predicts how users adopt and use information technologies, and the factors influencing these decisions [15, 19, 20]. TAM has been applied in many prior studies ranging from distance learning (e.g. measuring technology adoption by students), to marketing (e.g. online or internet banking) [21]. For example, a study compared TAM and TPB to identify the theory that leads to accurate prediction of IT use. The result showed that TAM predicted IT use and acceptance better than TPB [22]. Meanwhile, earlier researchers also used the foundation of TRA to develop TAM in explaining the connections between PEOU, PU, attitudes and BI in the direction of the receiver system [3]. The model described the adoption mechanisms and the reasons behind the acceptance of technologies. On a different note, TAM was proposed to explore user awareness of IT at work [3, 16], and the result confirmed user's technology adoption actions directly from the perspective of information systems [3, 16, 23]. In many pertinent experiments, the TAM model was used and thoroughly tested and generally accepted because of its simplicity [21, 24, 25].

However, studies that used TAM as the underlying theory in determining the impact of BI to use technology in the Iraqi higher education system during COVID-19 pandemic were limited, although studies on TAM and E-learning were generally numerous. For instance, a study was conducted among Greek university students about the adoption of e-Class based on TAM, and the results indicated the PEOU had an insignificant impact on BI to use e-Class [21]. On the other hands, PU had a positive and significant impact on BI among university students [26]. Al Adwan [27] reported the PU and PEOU had a positive and significant impact on BI and PU had more impact on BI than PEOU among students. Additionally, Al- Adwan [27] indicated PEOU had a positive and significant impact on PU. Several other studies confirmed the impact of PEOU on PE [2, 5, 26–28] and the impact of PU and PEOU on BI [2, 5, 27–30]. Thus, the following hypotheses were developed.

H1: PU has a positive and significant impact on BI to use e-learning among students.
H2: PEOU has a positive and significant impact on BI to use e-learning among students.
H3: PEOU has a positive and significant impact on PU among students.

Computer Self-Efficacy (CSE)

CSE refers to the degree of trust that people have in their ability to cope effectively with a particular challenge [27, 31]. CSE is not seen as an attribute of individuals' computer capabilities but reveals what people hope they will achieve in the future with computing skills and abilities. Self-efficacy refers to users' trust in their abilities to take action to cope with potential circumstances [5]. It is connected to users' trust in their capacity to use the E-learning method in computer systems. Self-efficacy considers one of the essential elements that enhance the E-learning system users' satisfaction. Students with high trust in their computer abilities can perform in E-learning, and these students do not quickly get discouraged in the face of technological difficulties.

Past researchers reviewed 107 articles and identified 152 external variables used to extend the TAM and recognized self-efficacy as a variable extensively used to extend the TAM [15]. In further support, self-efficacy was considered the most commonly used external factor for TAM [5]. Several prior studies had used CSE as an external variable when using TAM [5]. PEOU positively and significantly impacted self-efficacy in a study conducted in Greece among university students [21]. A study in Jordan found that CSE positively and significantly impacted PU and PEOU among university students [27]. Several other studies confirmed the impact of self-efficacy on PU [6, 27], and PEOU [5, 6, 18, 27]. On the other hand, Salloum et al. [5] reported the CSE has an insignificant impact on PU.

H4: CSE has a positive and significant impact on PU of the E-learning system.
H5: CSE has a positive and significant impact on PEOU of the E-learning system.

Information Quality (IQ)

IQ is considered as one of the main elements of the Information System (IS) [32, 33]. IQ relates to producing accurate and appropriate information for learning purposes by a system [29]. IQ refers to the use of E-learning for knowledge that may be essential for learning and modified in order to allow the learner to understand it [5, 30]. IQ often relates to the users' opinion about the quality of a website [5].

IQ able to enhance PEOU and PU [20]. IQ performance is recognized as one of the most widely used and accepted measures for investigating the user acceptance of various IS [20]. According to several studies, IQ is considered an essential external element of the TAM [20, 34]. A study that empirically examined the impact of IQ on PEOU and PU confirmed that IQ positively and significantly impacts PEOU and PU [5, 20, 29]. Previous researchers confirmed that IQ significantly impacts PU when the individuals receive the information [34, 36]. Conversely, a study reported that IQ has a significant impact on PEOU but not on PU among bachelor degree students [30].

H6: IQ has a positive and significant impact on PU of the E-learning system.
H7: IQ has a positive and significant impact on PEOU of the E-learning system.

3 Methodology

The quantitative method has been used to test the seven proposed hypotheses of this study. The population of this study is undergraduate students at Cihan University-Erbil.

Cihan University was selected because it implemented the E-learning since three years ago via Moodle platform to support the traditional teaching process at the campus. 300 questionnaires were randomly distributed among students at the College of Administrative and Financial Sciences, where only 224 questionnaires were returned, and 209 were valid for analysis. The questionnaire was close-ended and consisted of two main parts. The first part contained the demographic questions, and the second part comprised 24 questions representing the five variables. The questionnaire was measured by a five-point Likert scale ranging from (1) Strongly disagree, to (5) Strongly agree. All the items were adopted from previous studies. PU, CSE and IQ consisted of 5 items adopted several studies [5, 20, 27]. Additionally, BI and PEOU consisted of 6 items [5, 20, 27]. The data were analyzed using two software. The preliminary analytical analysis was conducted via SPSS version 21 for data coding, data entry, outlier values, missing values and descriptive demographic analysis. The main analysis was conducted via AMOS version 24 for hypothesis testing.

4 Findings

4.1 Confirmatory Factor Analysis

The analysis began with the Confirmatory Factor Analysis (CFA) to assess the validity and reliability of the proposed model. The CFA indices consist of Composite Reliability (CR), Factor Loadings and Average Variance Extracted (AVE) [35]. The factor loading should exceed the cut-off value of 0.60 [35]. The reliability was measured by Cronbach's Alpha (CA) and CR to find the internal consistency, and the cut-off level for both criteria were 0.70 [35]. All the CR and CA values in this study exceeded 0.70. The AVE should be higher than 0.5 [35]. To achieve the model fit a minimum three fit indices should be achieved [35], and the suggested fit indices should achieve RMSEA and Relative Chi-Square as a primary requirement and at least one of the other indices such as GFI, CFI, TLI and IFI.

4.2 Measurement Model

All the item loadings exceeded the recommended value of 0.60, and the loadings showed between .625 (CSE3) and .969 (PEOU3), except for item CSE4 which had poor loading and removed to enhance the model fit. However, the model reliability examined by CR and CA indicated that all the constructs were reliable and achieved the reliability level equal to or higher than 0.7. Table 1 depicted the results of validity and reliability for all the items and constructs. The AVE showed to be greater than 0.5 for all the constructs (see Table 1).

In the current model, the value of RMSEA is .063, which is less than 0.08 and the Relative Chi-Square value is 1.816, which is less than the cut-off level of 5. In contrast, other indices, namely CFI (.949), IFI (.949) and TLI (.940) achieved and exceeded the recommended cut-off value of 0.90 [34], but the GFI (.857) were less than .90. However, if RMSEA, Relative Chi-Square and any other indices exceeded the recommended cut-off value, the analysis can be continued to the next step [35]. Figure 1 illustrated the model fit indices, which were generated by AMOS. After ensuring the model fit, the next step was analyzing the structural model.

Table 1. Factor loadings, AVE, and construct reliability of the variables

Constructs	Items	Factor loading	CA	CR	AVE > 0.5
BI	BI1	.681	.895	0.760	0.638
	BI2	.859			
	BI3	.841			
	BI4	.836			
	BI5	.758			
PU	PU1	.610	.881	0.746	0.602
	PU2	.886			
	PU3	.904			
	PU4	.788			
	PU5	.639			
PEOU	PEOU4	.975	.944	0.877	0.806
	PEOU3	.969			
	PEOU2	.896			
	PEOU1	.728			
IQ	IQ1	.855	.896	0.765	0.652
	IQ2	.851			
	IQ3	.751			
	IQ4	.812			
	IQ5	.759			
CSE	CSE1	.720	.830	0.824	0.532
	CSE2	.878			
	CSE3	.625			
	CSE5	.669			

4.3 Structural Model

The purpose of this step is to determine the influence of independent variables on dependent variables. The current study proposed seven hypotheses, and the results indicated as following;

The first hypothesis indicated the PU significantly impacted BI among undergraduate students with p-value 0.00 < 0.05 and the CR (t-value) > 1.96. Thus, H1 is supported. The second and third hypotheses indicated the PEOU had a positive and significant impact on BI (t-value = 3.712 > 1.96) and PU (t-value = 3.931 > 1.96) with p-value 0.00 < 0.05. Thus, H2 and H3 supported.

The fourth and fifth hypotheses showed the CSE had an insignificant impact on PU because the p-value 0.270 > 0.05 and t-value 1.102 < 1.96. Besides, the CSE showed a significant impact on PEOU with p-value 0.001 < 0.05 and the t-value 3.202 > 1.96. Thus, H4 was not supported, and H5 was supported.

IQ significantly impacted PU and PEOU with p-value 0.00 and t-value = 0.034 < 0.05 respectively and t-value 7.498 and 2.110 > 1.96 respectively. Thus, H6 and H7 were supported. Table 2 and Fig. 1 depicted the results of the seven proposed hypotheses.

Table 2. Hypotheses test

Label	Path	Estimate	SE.	CR.	P-value	Label
H_1	BI ← PU	.330	.072	4.569	***	Supported
H_2	BI ← PEOU	.164	.044	3.712	***	Supported
H_3	PU ← PEOU	.161	.041	3.931	***	Supported
H_4	PU ← CSE	.076	.069	1.102	.270	Not supported
H_5	PEOU ← CSE	.389	.121	3.202	.001	Supported
H_6	PU ← IQ	.521	.069	7.498	***	Supported
H_7	PEOU ← IQ	.228	.108	2.110	.034	Supported

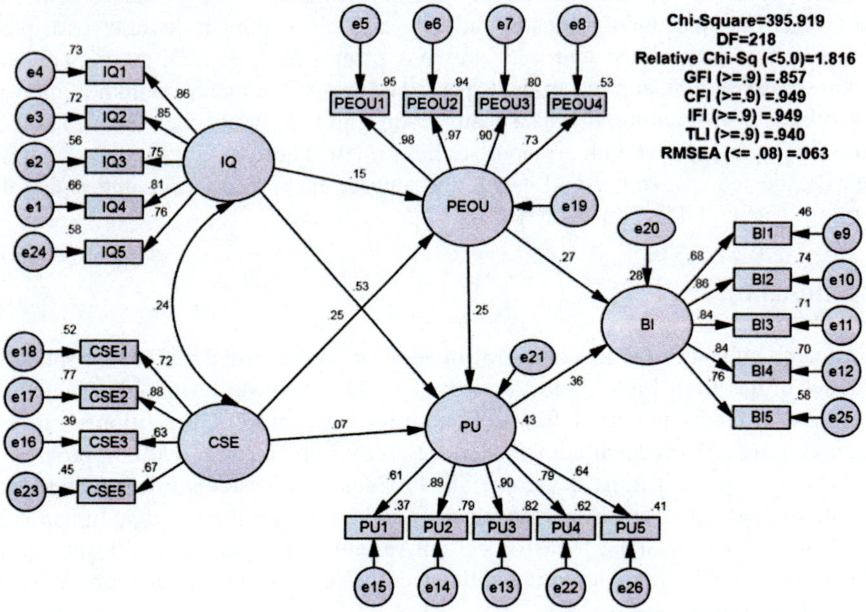

Fig. 1. Structural model

5 Discussion

The results indicated that PU and PEOU have a positive and significant impact on BI to use E-learning. This results are in line with previous findings [2, 5, 27, 28]. The simpler and more useful E-learning tools like Moodle are viewed, the more confident students will use them. These results indicated that students' adoption of E-learning was regarded as practical behavior. Students chose to embrace emerging technology such as Moodle because they thought they could produce great results with fewer efforts. The people involved in this study were undergraduate students. They feel that their BI will benefit from simple use and efficient accessibility of good quality content by E-learning tools such as the Moodle. The results indicated the PU had more impact on BI to use E-learning than PEOU. Although Moodle has essential functions, students are more likely to take up the difficulty with E-learning. PEOU has a significant impact on PU, and this result confirmed by previous studies [2, 5, 26, 27]. The research respondents assumed that E-learning by Moodle is intuitive to use and thus considered useful for learning. This finding indicates that the more quickly the usability of Moodle is perceived, and the less work they need to do than conventional approaches, the more useful Moodle is perceived. CSE has a significant impact on PEOU, and this result is consistent with previous studies [5, 6, 18, 27]. Students with a high degree of belief in their programming abilities or computer skills are better at using E-learning. These students are not easily discouraged in the face of technical difficulties but will often attempt to overcome problems or challenges. They understand that the use of E-learning for self-learning is simple and easy to use. A high CSE helps students explore the different features and choices needed to execute their assignments and communicate successfully with e-learning. CSE has an insignificant impact on PU among undergraduate students, and this finding is parallel with prior finding [5], but also inconsistent with previous studies [6, 27]. CSE among students cannot increase the PU, and the students perceived that self-efficacy could not improve the usefulness of E-learning. IQ has a positive and significant impact on PEOU and PU, and this finding is parallel with previous studies [5, 29]. These findings indicate that the students' expected ease of use will inevitably improve as the knowledge underlying the e-learning system and its consistency became deeply rooted.

6 Conclusion

This research enhances the use of TAM, in addition to the added variables applied to the model in the Iraqi higher education context. The analysis results also include a clearer view of environmental influences and offer policymakers, practitioners, developers, and designers practical advice to successfully implement E-learning programs. University management must develop suitable E-learning systems infrastructure and assess student readiness for E-learning systems. The policymakers and administrators of E-learning programs must focus on certain variables that play a significant role in increasing adoption by students, which influences the success of the educational process and the productivity of the students.

References

1. Jameel, A.S., Karem, M.A., Mahmood, N.Z.: A review of the impact of ICT on business firms. Int. J. Latest Eng. Manag. Res. **2**(1), 15–19 (2017)
2. Fathema, N., Shannon, D., Ross, M.: Expanding the Technology Acceptance Model (TAM) to examine faculty use of Learning Management Systems (LMSs) in higher education institutions. J. Online Learn. Teach. **11**(2), 210–232 (2015)
3. Davis, F.D., Bagozzi, R.P., Warshaw, P.R.: User acceptance of computer technology: a comparison of two theoretical models. Manag. Sci. **35**(8), 982–1003 (1989)
4. Ahmad, A.R., Keerio, N., Jameel, A.S., Karem, M.A.: The relationship between national culture and succession planning in Malaysian public universities. J. Educ. e-Learn. Res. **7**(3), 242–249 (2020)

5. Salloum, S.A., Alhamad, A.Q.M., Al-Emran, M., Monem, A.A., Shaalan, K.: Exploring students' acceptance of e-learning through the development of a comprehensive technology acceptance model. IEEE Access **7**, 128445–128462 (2019)
6. Valencia-Arias, A., Chalela-Naffah, S., Bermúdez-Hernández, J.: A proposed model of e-learning tools acceptance among university students in developing countries. Educ. Inf. Technol. **24**(2), 1057–1071 (2018). https://doi.org/10.1007/s10639-018-9815-2
7. Jameel, A.S., Ahmad, A.R.: The mediating role of job satisfaction between leadership style and performance of academic staff. Int. J. Psychosoc. Rehabil. **24**(04), 2399–2414 (2020)
8. Jameel, A.S., Ahmad, A.R.: Factors impacting research productivity of academic staff at the Iraqi higher education system. Int. Bus. Educ. J. **13**(1), 108–126 (2020)
9. Jameel, A.S., Moshfeghyan, M.: Factors Affecting E-Commerce By E-Learning. LAP LAMBERT Academic Publishing, Germany (2017)
10. Jameel, A.S.: Challenges facing students toward ICT library adoption. In: International Conference on Accounting, Business, Economics and Politics, pp. 231–237 (2018)
11. Mousa, T.S., Jameel, A.S., Ahmad, A.R.: The impact of attitude, subjective norm and information communications technology on knowledge sharing among academic staff. Int. J. Psychosoc. Rehabil. **23**(02), 704–717 (2019)
12. Jameel, A.S., Ahmad, A.R.: Leadership and performance of academic staff in developing countries. In: Proceedings of the 33rd International Business Information Management Association Conference, IBIMA 2019: Education Excellence and Innovation Management through Vision 2020, pp. 6101–6106 (2019)
13. Fishbein, M., Ajzen, I.: Belief, Attitude, Intention, and Behavior: an Introduction to Theory and Research. Addison-Wesley, Read (1975)
14. Ajzen, I.: The theory of planned behavior. Organ. Behav. Hum. Decis. Process. **50**(20), 107–114 (1991)
15. Abdullah, F., Ward, R.: Developing a General Extended Technology Acceptance Model for E-Learning (GETAMEL) by analysing commonly used external factors. Comput. Hum. Behav. **56**, 238–256 (2016)
16. Davis, F.D.: Perceived usefulness, perceived ease of use, and user acceptance of information technology. MIS Q. **13**, 319–340 (1989)
17. Jameel, A.S., Hamdi, S.S., Karem, M.A., Raewf, M.B., Ahmad, A.R.: E-Satisfaction based on E-service Quality among university students. J. Phys. Conf. Ser. **1804**(1), 12039 (2021)
18. Li, H., Yu, J.: Learners' continuance participation intention of collaborative group project in virtual learning environment: an extended TAM perspective. J. Data Inf. Manag. **2**(1), 39–53 (2019). https://doi.org/10.1007/s42488-019-00017-8
19. Ifinedo, P.: Examining students' intention to continue using blogs for learning: perspectives from technology acceptance, motivational, and social-cognitive frameworks. Comput. Hum. Behav. **72**, 189–199 (2017)
20. Tsai, Y.-Y., Chao, C.-M., Lin, H.-M., Cheng, B.-W.: Nursing staff intentions to continuously use a blended e-learning system from an integrative perspective. Qual. Quant. **52**(6), 2495–2513 (2017). https://doi.org/10.1007/s11135-017-0540-5
21. Revythi, A., Tselios, N.: Extension of technology acceptance model by using system usability scale to assess behavioral intention to use e-learning. Educ. Inf. Technol. **24**(4), 2341–2355 (2019)
22. Taylor, S., Todd, P.A.: Understanding information technology usage: a test of competing models. Inf. Syst. Res. **6**(2), 144–176 (1995)
23. Venkatesh, V., Morris, M.G., Davis, G.B., Davis, F.D.: User acceptance of information technology: toward a unified view. MIS Q. **27**, 425–478 (2003)
24. Al-Emran, M., Arpaci, I., Salloum, S.A.: An empirical examination of continuous intention to use m-learning: an integrated model. Educ. Inf. Technol. **25**(4), 2899–2918 (2020). https://doi.org/10.1007/s10639-019-10094-2

25. Hsu, H., Chang, Y.: Extended TAM model: impacts of convenience on acceptance and use of Moodle. Online Submiss. **3**(4), 211–218 (2013)
26. Al-Emran, M., Mezhuyev, V., Kamaludin, A.: Towards a conceptual model for examining the impact of knowledge management factors on mobile learning acceptance. Technol. Soc. **61**, 101247 (2020)
27. Al-Adwan, A.S.: Investigating the drivers and barriers to MOOCs adoption: the perspective of TAM. Educ. Inf. Technol. **25**(6), 5771–5795 (2020). https://doi.org/10.1007/s10639-020-10250-z
28. Al-Emran, M., Teo, T.: Do knowledge acquisition and knowledge sharing really affect e-learning adoption? An empirical study. Educ. Inf. Technol. **25**(3), 1983–1998 (2019). https://doi.org/10.1007/s10639-019-10062-w
29. Rui-Hsin, K., Lin, C.-T.: The usage intention of e-learning for police education and training. Polic. Int. J. **41**, 98–112 (2018)
30. Jasim, Y.A., Raewf, M.B.: Information technology's impact on the accounting system. Cihan Univ. J. Humanit. Soc. Sci. **4**(1), 50–57 (2020)
31. John, S.P.: Influence of computer self-efficacy on information technology adoption. Int. J. Inf. Technol. **19**(1), 1–13 (2013)
32. Delone, W.H., McLean, E.R.: The DeLone and McLean model of information systems success: a ten-year update. J. Manag. Inf. Syst. **19**(4), 9–30 (2003)
33. DeLone, W.H., McLean, E.R.: Information systems success: the quest for the dependent variable. Inf. Syst. Res. **3**(1), 60–95 (1992)
34. Chen, J.V., Jubilado, R.J.M., Capistrano, E.P.S., Yen, D.C.: Factors affecting online tax filing–an application of the IS success model and trust theory. Comput. Hum. Behav. **43**, 251–262 (2015)
35. Hair Jr, J.F., Hult, G.T.M., Ringle, C.M., Sarstedt, M.: A Primer on Partial Least Squares Structural Equation Modeling (PLS-SEM), 2nd ed. Sage, Thousand Oaks (2017)
36. Arpaci, I.: The role of self-efficacy in predicting use of distance education tools and learning management systems. Turk. Online J. Distance Educ. **18**(1), 52–62 (2017)

The Effects of Serious Games and Blended Learning on Students' Achievement in Reading

Anas Hanandeh[1](✉), Zaleha Abdullah[2], and Jamalludin Harun[2]

[1] Universiti Teknologi Malaysia, Johor Bahru, Malaysia
mkhahanas2@graduate.utm.my

[2] Faculty of Social Sciences and Humanities, School of Education, Universiti Teknologi Malaysia, Johor Bahru, Malaysia

Abstract. Reading is considered one of the main skills that English as a foreign language learner needs to acquire successfully in their studies and lives. Yet, Jordanian learners suffer from a deficiency in reading and their achievement scores show weaknesses in reading. One of the suggested solutions to improve their reading is integrating serious games in the educational process since that serious games have been successfully applied in many disciplines. Still, there is a lack of research that explores the best practice of using serious games in classrooms. Hence, this paper aims to examine the effects of a serious game and blended learning model called lab rotation on students' achievement in reading. The research design used in this study was the pre-experimental design that was conducted among fifty Jordanian undergraduate students who registered in the "Effective communication skills" subject for semester II, 2017 at Yarmouk University, Jordan. Moreover, a combination of quantitative and qualitative methods was employed in the data collection and analysis processes. The study's results showed that utilizing a serious game and lab rotation model can improve students' achievement in reading. Moreover, the study explored how students' achievement in reading was improved after integrating the serious game and lab rotation model.

Keywords: Serious game · Blended learning · Achievement · Reading

1 Introduction

The video games market is growing rapidly and is considered one of the main largest domains of the IT industry [1, 2]. What is more, Newzoo, which is a global provider of games and esports analytics, expects that the global games market will grow to reach $180.1 billion by 2021 [3]. Millions of people are spending many hours playing games on their devices [4]. For instance, young American adults spend 10,000 h playing games by the age of 21 [5]. Moreover, video games are considered attractive and engaging since they provide players with various elements and features [6] such as interactive stories, visual elements, freedom to fail feature, progression and feedback. As well, many types of purposeful games have been designed and developed to meet the demands of the new generation learners such as serious games (SGs). SGs are defined as "Games primarily

M. Al-Emran et al. (Eds.): ICETIS 2021, LNNS 299, pp. 265–275, 2022.
https://doi.org/10.1007/978-3-030-82616-1_24

focused on education rather than Entertainment" [7] and as "Games used for purposes other than mere entertainment" [8]. As well, Serious games have considerable success in various fields and contexts. Therefore, researchers have explored how serious games can be applied to improve language learning process.

2 Literature Review

2.1 Serious Games and Reading

SGs have been widely used in various application contexts including health care, education, military and cultural heritage. What is more, the main factors that helped the wide and rapid usage of SGs around the world are the new models of learning that have been developed and created focusing on changing the learner's role to be more active and in the center of the learning process. As well, serious game designers have employed new interactive technologies to assist students in their learning and attract their concentration [9]. Also, serious games can provide learners with interactive learning experiences that can improve their motivation and achievement in language learning process. For example, Guillén-Nieto and Aleson-Carbonell [10], Johnson, Vilhjálmsson [11] stated that applying video games in language learning is considered more engaging than the traditional methods since that video games can provide learners with various interactive environments that can attract and sustain their attention towards language learning. Hence, serious game designers have developed new techniques to promote learners' motivation and engagement to cater English language learners' new demands [11]. And since it has been noticed that such games have shown a remarkable success [12–14], researchers have explored how serious games can improve reading as shown in a study by Gaytán-Lugo, Santana-Mancilla [15] which investigated the effectiveness of using serious games in developing learners' reading abilities in English where serious games showed promising results in developing such skills.

As well, a study by Cauchi-Saunders [16] indicated that using serious games has significantly improved students' reading comprehension skills in Tasmanian classrooms where an experimental design was employed by conducting pre and posttests. Furthermore, it was found that providing students with more gameful versions of serious games can help to enhance students' comprehension skills. Likewise, a successful serious game called "Mingoville", that focuses on improving learners' skills such as reading and spelling, was translated into 31 languages to be applied in various countries [17, 18]. As well, Tsihouridis and Batsila [19] examined the effectiveness of 'Silang', which is a serious game, on students' listening and reading skills. The findings of this study revealed that 'Silang' successfully enhanced students' listening and reading skills and motivated them towards the learning process. Building and designing serious games is a complex process as well as developing such games is a costly process that requires a lot of efforts, time and budget. So, researchers have tried to examine all optimal solutions to benefit from the existed serious games in the efficient ways [20]. Hence, there is a need to develop and apply tools and strategies that integrate suitable pedagogical approaches to use serious games effectively in teaching and learning processes [21]. One of the suggested solution for enhancing the practical usage of serious games in the classroom is by applying a suitable pedagogical approach with serious games such as blended learning that can enhance the learning experience and provides learners with an enjoyable environment that support SGs in achieving learning goals successfully [12]. Blended learning is "a mixture of face-to-face and online learning or computer-mediated instruction, combining instructional modalities (or delivery media) and instructional methods" [22, 23]. And according to Graham [24], There are three main reasons to adopt a blended learning approach in the educational process which are: it improves the learning effectiveness, increases the accessibility and convenience and saves the money of the institutions.

2.2 Serious Games and Blended Learning

SGs are considered more motivating and engaging comparing to the traditional teaching methods [25] and many studies showed that serious games have a positive impact on the learning process [10, 11, 26]. Yet, there is a need to have new models that help to create a balance between learning and entertainment in serious games since that the designers lack the appropriate skills of the instructional design [27]. From this point, researchers have focused on the stages before and after developing the serious games. For example, a project called Serious Games–Engaging Training Solutions (SG-ETS) was conducted to enhance SGs effectiveness by grouping game developers and pedagogical experts. Moreover, one of the goals of this project was to apply blended learning to enhance SGs effectiveness and support and satisfy learners with a suitable learning environment by analyzing learners and identifying the best learning approaches based on interviews and surveys that were given to the learners [28].

By applying serious games in the educational process, many recommendations of replacing the teachers' roles in the classroom with serious games were suggested. Still, researchers have realized that it is difficult to replace teachers' roles in the classroom with serious games alone [29]. Hence, they started to examine the best practice of SGs inside classes to affirm that learning and teaching objectives will be achieved successfully. Yet, there is a lack of research that explores the best practice of using serious games inside classrooms. Therefore, researchers have investigated the integration of serious games and blended learning models and founded that SGs are more effective when combined with blended learning since that BL offers an appropriate scaffolding for learners to meet the desired learning goals. In many cases it is essential to provide serious games with a blended learning approach and suitable scaffolding to help them achieve complex cognitive and behavioral aims. For example, in a stroke rehabilitation case [30], it was very difficult for learners to achieve complex aims alone without a suitable support. Likewise, Kirriemuir and Mcfarlane [31] mentioned that the use of games in the schools is rare for many reasons, one of them is that teachers don't have the appropriate time to practice how to apply the games successfully inside the classroom. Thus, applying one of blended learning models seems to be one of the suggested solutions to enhance the practical usage of serious games in the classrooms since they can improve learning and help learners to succeed [32, 33].

In this study, a popular model called lab rotation model is used. Lab-rotation model has been applied successfully in many contexts and it can help in achieving a great financial sustainability [34, 35]. Moreover, the students rotated on a fixed schedule between teacher-led instructions, played a serious game and involved in different learning activities and tasks such as group discussions, quizzes and worksheets. And to construct the learning activities in a pedagogically well-planned manner, a model named Attention, Relevance, Confidence and Satisfaction model (ARCS) was applied to design the learning tasks since that ARCS helps to attract learners' attention, relate the content to their lives, improve their confidence and satisfy their needs.

2.3 Attention, Relevance, Confidence and Satisfaction Model (ARCS)

ARCS model was introduced by John Keller in 1984 in order to enhance learners' motivation towards the instructional materials [36]. And in order to promote motivation towards the instructional material, Keller listed four steps as follows:

- Attention: stimulating learners' knowledge seeking and grabbing their interests towards the subject taught can attract their attention. Also, this can be achieved by posing problems to be solved, asking challenging questions and varying instructional methods to sustain learners' attention.
- Relevance: explaining that mastering a specific skill can help learners to achieve their prospective goals. Moreover, providing them with real life examples is considered important to relate the subject to their future goals.
- Confidence: confidence can influence learners' accomplishment and success that can be gained by employing suitable strategies and techniques that provide them with appropriate feedback and offer personal control.
- Satisfaction: learners' satisfaction can be achieved through their participation in various learning activities which can motivate them towards the newly gained skills where they can practice them in their daily lives and careers.

3 Methodology

3.1 Participants

A purposive non-random sampling technique was employed in this study to select 50 first-year undergraduate students from Yarmouk University, Jordan. Furthermore, the selected student registered in "Effective communication skills" subject for semester II, 2017. Moreover, a purposive sampling was used to select six students to be interviewed after playing the serious game and completing the learning activities and tasks.

3.2 Instruments

One group pre and post-test design of pre experimental design was conducted in this study. Furthermore, pre and post reading tests were carried out in order to check if there are any significant differences between students' achievements in reading before and after playing the serious game and completing the learning tasks. Also, the pre and post tests were designed based on bloom's taxonomy categories which are: remembering, understanding, applying, analyzing, evaluating and creating. Moreover, a serious game called 'Silang' was used in this study to improve students' reading. As well, through playing 'Silang', the learners are supposed to interact with various characters in different settings through dialogues which can help to improve their reading. At the same time, 'Silang' provides learners with challenging activities and tasks to meet their skills and keep them motivated in playing the serious game till they master the required skills.

3.3 Procedures

The study lasted for a period of ten weeks and conducted in four phases based on Analysis, Design, Develop, Implement and Evaluate (ADDIE) model as follows:

- Analysis Phase: The purpose of this phase was to identify a convenient SG and learning tasks to help learners improve their reading. Furthermore, the lecturers and students were interviewed to find the convenient learning tasks that will be employed in the educational process.
- Design and Development Phases: The aims of these phases were to set the objective and goals of the learning process and design the learning tasks. Furthermore, ARCS model was employed to design the learning tasks such as the worksheets and presentations in order to attract students' attention towards the taught subject, relate the content to their daily lives, improve their confidence and satisfy their needs. Moreover, the SG and learning tasks were organized, planned and developed for the following phase which is the implementation phase.
- Implementation Phase: The lab rotation model which is a popular blended learning model was applied in this study. Furthermore, lab rotation model is considered suitable to be applied since that most Jordanian public universities' classes lack technological infrastructure [37, 38]. The lab rotation model has two elements which are the classroom instruction and the computer lab [35] where the students rotated on a fixed schedule between playing the 'Silang' serious game and completing learning activities and tasks. At first, a pre-test was carried out among the students before playing the serious game and completing the learning tasks. Following that, a face-to-face lecture was given by the instructor discussing the benefits of mastering reading and how to improve it by employing various strategies. As well, the students rotated on a fixed schedule between lecturer-led instruction, playing 'Silang', completing learning tasks which were group discussions, worksheets, presentations and quizzes.
- Evaluation Phase: A post test was carried out to measure students' achievements in reading after playing the SG and completing the learning tasks. Moreover, the duration between the pre and post-test was four weeks. Following, the collected data from the pre- and post-tests were analyzed, and a paired sample t-test was conducted between the scores of the pre and post-tests. Also, six students were interviewed to get in depth information about the effects of the serious game and learning activities and tasks on their achievement in reading. Additionally, Nvivo software (version 11) was used to code the transcribed data by employing descriptive coding method to identify the major themes to explore how serious games and lab rotation model improved students' achievements in reading.

4 Results

A pre and post-tests were employed to measure the students' achievements before and after the applying the serious game and lab rotation model. Figure 1 shows the students' scores in the pre and post-tests. Based on Fig. 1 below, the students' scores are higher in the post test compared to the pretest.

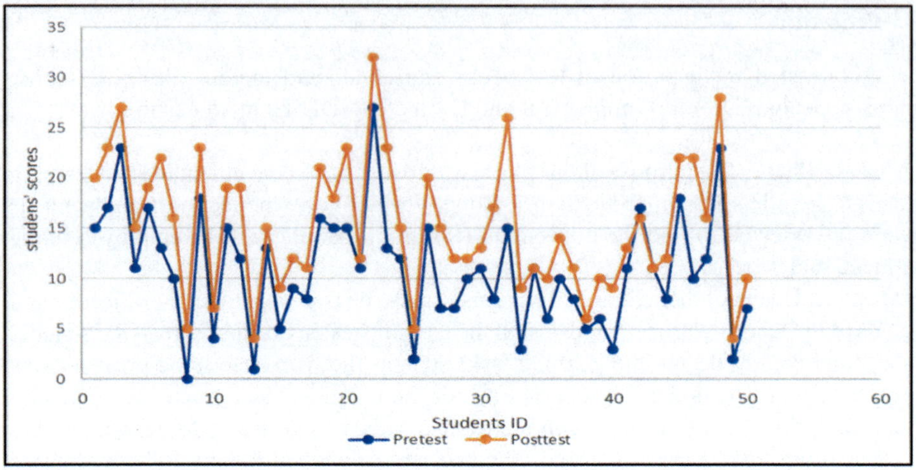

Fig. 1. Students' scores in the pre and post-tests

Furthermore, a paired sample t-test was employed between both tests scores as shown in Table 1 below.

As shown in Table 1, the (P-vale) is 0.00 that is less than 0.05, which indicates that There is a statistically significant difference between the pre and post-tests' scores. Hence, the integration of the SG and lab rotation model improved students' achievements in reading.

Table 1. Paired sample T-test

		Paired devices					t	df	Sig. (2-tailed)
		Mean	Std. deviation	Std. error mean	95% confidence interval of the difference				
					Lower	Upper			
Pair 1	Pretest - postest	−4.36000	2.73123	.38625	−5.13621	−3.58379	−11.288	49	.000

Furthermore, six students were interviewed to answer four questions which are:

1. Did the SG and learning activities/tasks help you to improve your reading? If yes, how is that so?
2. Did you make any mistakes while playing the serious game and completing the learning activities/tasks? What did you do after that mistake?
3. When given a task in the serious game, describe the process you follow to execute the tasks? For example, describe the steps you follow to ensure it is completed.

4. What do you do if you get stuck while playing the serious game?

The students' answers on the first interview question "Did the SG and learning activities/tasks help you to improve your reading? If yes, how is that so?" showed that the serious game and learning activities helped them to improve their achievement. What is more, the weighted percentage, which represents the frequency of the word relative to the total counted codes, were analyzed to identify the codes and the major themes which are: simplicity, challenging, exploration, interactivity, flow experience and organization. Moreover, the weighted percentage of the first three main themes; simplicity, challenging, exploration was (21.42%) for each. (14.28%) for the interactivity, and the same percentage (14.28%) for the flow experience, and (7.18%) for the organization.

As well, the students' answers on the second interview question "Did you make any mistakes while playing the serious game and completing the learning activities/tasks? What did you do after that mistake?" showed that the six students made some mistakes while playing the serious game, "I made some mistakes, after that I tried to read again to find the correct answer" and "Yes, I made some mistakes such as in the beginning of the game where I had to fill the brackets with the suitable words". As well, all six students indicated that they kept trying to correct their mistakes and find the right answers to execute the tasks as they mentioned. "I had to read the sentences multiple times to figure the right answer(s)".

The themes and their weighted percentage that were identified from the third question "When given a task in the serious game, describe the process you follow to execute the tasks? For example, describe the steps you follow to ensure it is completed" are: following instructions (36.3%), repeating to learn from mistakes (27.3%), paying attention (27.3%). and planning (9.1%). The students start to read the instructions and follow them to execute the serious game tasks "I try to read and follow the instructions provided word by word". Following, if they cannot finish the tasks, they keep trying till they learn from their mistakes. Besides, they pay attention to the characters and their body language to help them proceed in the serious game's levels as indicated by the students "I can figure that I finished the level successfully by paying attention to the characters' actions and body language". As well, the students mentioned that they follow an organized plan to finish the game "I try to read the instruction carefully and try to have an organized plan to make sure that the mission or the task will be successfully completed".

The identified themes and their weighted percentages from the fourth interview question "What do you do if you get stuck while playing the serious game?" were: trying multiple times then ask for help (50%), asking instructor (37.5%) and colleagues (12.5%) for help. The students stated that they try multiple times to proceed through the tasks until they find a way to finish them "I try to answer more than one since I like challenge" and if they cannot find a way to finish the task they ask for help from the instructor or colleagues "having an instructor while playing will help us more in our study or when we get stuck" and "The instructor can help me in the game activity where he/she can present the game objectives and also can help me if I get stuck while playing the game" and "I will try to ask a friend, and if he does not know the answer, I will ask the instructor".

5 Discussion

Based on the results of the pre and post-tests shown in Table 1, there is a statistically significant difference between the students' scores in the pre and post-tests. Thus, the integration of the SG and lab rotation model improved students' achievements in reading. Moreover, the findings of this study are in line with studies that found applying technologies improves students' achievement in English [39–44]. Similarly, the results agree with a study by Tsihouridis and Batsila [19] which showed that the scores of the reading and listening post-test were higher than the scores of the pre-test after employing 'Silang' serious game. Hence, the serious game was found to have improved students' reading skills.

Applying a suitable learning environment that includes different activities improves learners' reading as well as their confidence in learning English [45]. Furthermore, students' interviews showed that simplicity, challenging activities, exploration, interactivity, flow experience and organization helped them to improve their reading. The simplicity of the learning activities and tasks aims to lower the students' cognitive load by employing easy words, tasks, rules and tasks [46] which helped to improve students confidence throughout the learning process. And since that fear of failure in learning "may shut down students' ability to think creatively" [47], serious games provide learners with freedom to fail feature that encourages them to play the serious game without fear of losing which helps to improve their confidence and build their skills. And based on the answers, their confidence improved because of the challenging tasks they participated in since that Silang game and the learning tasks provided them with challenging tasks to meet their skills and keep them motivated in playing the game till the end.

Providing learners with serious games and learning activities through employing lab rotation model has its own advantages. For instance, the serious games offer a friendly environment where learners can explore, play, interact, try, make mistakes and learn from their choices and actions through the game [48–50]. Also, serious games can enhance learning process by providing a flexible learning environment that traditional classroom activities cannot offer. Furthermore, challenging, interesting and encouraging tasks helped to stimulate students' curiosity towards exploring Silang game as stated by the students. As well, the interactivity feature helped them to communicate and interact where they can participate in dialogues, make decision and solve puzzles. Besides, the students indicated that the game helped them to focus more through their playing experience which indicates that the there was a balance between students' skills and the learning tasks. Also, the students' involvement and immersion in Silang game indicates that the "Flow experience" state was achieved [51]. What is more, the organization and arrangement of the learning activities and tasks -where they rotated among various stations based on lab rotation model- helped them to think and learn in a planned way as shown in their answers to the interview questions.

Students' answers on the interview questions showed that they followed a pattern in executing learning tasks and activities by following instructions, repeating to learn from their mistakes, paying attention to the game's elements and planning to solve such tasks. What is interesting about these results is that it is similar to what persistent students do to execute tasks. They exert more efforts to accomplish tasks, follow instructions, take actions to understand the task, pay attention and focus more and identify the best strategy or plan to finish the task successfully [52]. As well, students' persistence is considered one of the behavioural dimensions of engagement [53], which suggests that the integration of the SG and lab rotation model actively engaged the students' and helped to improve their achievements in reading.

6 Conclusion

A SG and lab rotation model were employed to enhance students' achievement in reading. The results of this study showed that there is a significant improvement of the students' achievement in reading. Moreover, the results of the interview questions explored in-depth information on how utilizing serious games and lab rotation model helped the students in their learning process. At the same time, it is suggested to conduct more studies that investigate the effectiveness of applying serious games with different blended learning models to improve students' skills and particularly in the language learning contexts [26].

References

1. Choi, H.S., et al.: The effect of intrinsic and extrinsic quality cues of digital video games on sales: an empirical investigation. Decis. Support Syst. **106**, 86–96 (2018)
2. Szopik-Depczyńska, K., Kędzierska-Szczepaniak, A., Szczepaniak, K.: Application of crowdfunding to video game projects financing. Procedia Comput. Sci. **176**, 2714–2724 (2020)
3. Wijman, T.: Mobile revenues account for more than 50% of the global games market as it reaches $137.9 billion in 2018. Acesso em, 2018. 23
4. Furness, R., Wardle, D.: Developing innovative, sustainable and effective educational and training programs for the Asia Pacific Region: a case study. Int. Hydrogr. Rev. **2016**(15) (2016)
5. Nguyen, T.: The gamification of education: a case study in using positive psychology and game design to increase student engagement. In: Blended Learning in Engineering Education, pp. 109–124. CRC Press, Boca Raton (2018)
6. Arpaci, I.: Design and development of educational multimedia: the software development process for mobile learning. In: Handbook of Research on Wearable and Mobile Technologies in Education, pp. 272–285. IGI Global, Hershey (2016)
7. Miller, L.M., et al.: Learning and motivational impacts of a multimedia science game. Comput. Educ. **57**(1), 1425–1433 (2011)
8. Susi, T., Johannesson, M., Backlund, P.: Serious Games: An Overview. University of Skövde, Skövde (2007)
9. Garris, R., Ahlers, R., Driskell, J.E.: Games, motivation, and learning: a research and practice model. Simul. Gaming **33**(4), 441–467 (2002)
10. Guillén-Nieto, V., Aleson-Carbonell, M.: Serious games and learning effectiveness: the case of It'sa Deal! Comput. Educ. **58**(1), 435–448 (2012)
11. Johnson, W.L., Vilhjálmsson, H.H., Marsella, S.: Serious games for language learning: how much game, how much AI? In: AIED (2005)
12. Breuer, J.S., Bente, G.: Why so serious? On the relation of serious games and learning. Eludamos. J. Comput. Game Cult. **4**(1), 7–24 (2010)

13. Eichenberg, C., Grabmayer, G., Green, N.: Acceptance of serious games in psychotherapy: an inquiry into the stance of therapists and patients. Telemed. e-Health **22**(11), 945–951 (2016)
14. Abidin, S.R.Z., Noor, S.F.M., Sahari, N.: Low-fidelity prototype design for serious game for slow-reading students. Int. J. Adv. Comput. Sci. Appl. **10**(3), 270–276 (2019)
15. Gaytán-Lugo, L.S., et al.: Developing a serious game to improve reading comprehension skills in third graders. Avances en Interacción Humano-Computadora 71 (2015)
16. Cauchi-Saunders, A.: Mininauts: a serious game platform for enhanced instruction in reading comprehension for primary school students. University of Tasmania (2019)
17. Sørensen, B.H., Meyer, B.: Serious games in language learning and teaching–a theoretical perspective. In: Proceedings of the 3rd International Conference of the Digital Games Research Association (2007)
18. Anyaegbu, R., Ting, W., Li, Y.: Serious game motivation in an EFL classroom in Chinese primary school. TOJET: Turkish Online J. Educ. Technol. **11**(1) (2012)
19. Tsihouridis, C., Batsila, M.: The silang simulation tool for business secondary technical education ESP students. Int. J. Emerg. Technol. Learn. **11**(4) (2016)
20. Stapleton, A.J.: Serious games: serious opportunities. In: Australian Game Developers Conference, Academic Summit, Melbourne (2004)
21. Hendrix, M., et al.: Integrating serious games in adaptive hypermedia applications for personalised learning experiences. In: 5th International Conference on Mobile, Hybrid, and On-Line Learning (2013). eLmL
22. Watson, J.: Promising Practices in Online Learning: Blended Learning: The Convergence of Online and Face-to-Face Education. North American Council for Online Learning, Vienna (2008)
23. Allan, B.: Blended Learning: Tools for Teaching and Training. Facet Publishing, Abingdon (2007)
24. Graham, C.: Blended learning system: definition, current trends, and future directions. In: The Handbook of Blended Learning. Wiley, San Francisco (2006)
25. Girard, C., Ecalle, J., Magnan, A.: Serious games as new educational tools: how effective are they? A metaanalysis of recent studies. J. Comput. Assist. Learn. **29**(3), 207–219 (2013)
26. Backlund, P., Hendrix, M.: Educational games-are they worth the effort? A literature survey of the effectiveness of serious games. In: 2013 5th International Conference on Games and Virtual Worlds for Serious Applications (VS-GAMES). IEEE (2013)
27. Younis, B., Loh, C.: Integrating serious games in higher education programs. In: Proceedings of the Academic Colloquium 2010: Building Partnership in Teaching Excellence (2010)
28. De Freitas, S., Jarvis, S.: Serious games-engaging training solutions: a research and development project for supporting training needs. Br. J. Edu. Technol. **38**(3), 523 (2007)
29. Alklind Taylor, A.-S.: The active instructor: benefits and barriers to instructor-led serious gaming. In: 2015 7th International Conference on Games and Virtual Worlds for Serious Applications (VS-Games) (2015). IEEE.
30. Burke, J.W., et al.: Optimising engagement for stroke rehabilitation using serious games. Vis. Comput. **25**(12), 1085–1099 (2009)
31. Kirriemuir, J., Mcfarlane, A.: Literature review in games and learning (2004)
32. Murphy, R., et al.: Blended Learning Report. Michael & Susan Dell Foundation (2014)
33. Prescott, J.E., et al.: Elementary school–wide implementation of a blended learning program for reading intervention. J. Educ. Res. **111**(4), 497–506 (2018)
34. Vander Ark, T.: Making math work: K-8 blended learning (2013)
35. Staker, H., Horn, M.B.: Classifying K-12 Blended Learning. Innosight Institute (2012)
36. Keller, J.M.: Development and use of the ARCS model of instructional design. J. Instr. Dev. **10**(3), 2–10 (1987)
37. Alassaf, H.A.: Effective strategies on using ICT for teaching and learning undergraduate level at Jordanian universities. Significance **5**(3) (2014)

38. Mohammad Al-Omoush, M.: Blended e-learning constraints from the viewpoint of faculty members. Am. Acad. Scholar. Res. J. **8**(4) (2016)
39. Hussain, M.A., et al.: Technology based learning environment and student achievement in English as a foreign language in Pakistan. Int. J. Acad. Res. **2**(5), 398–402 (2010)
40. Suleman, Q., et al.: Effectiveness of educational technology in strengthening students' academic achievement in English at secondary school level in Kohat Division, Pakistan (2013)
41. Basheer, H., Ibrahim, A.: The effect of technology-based English language teaching on university students' achievement (a case study of the faculty of medicine and health sciences Nile Valley university). Doctoral dissertation (2016)
42. Ngwako, A.: Technology integration in teaching, student motivation, and reading achievement. Adv. Soc. Sci. Res. J. **3**(12) (2016)
43. Taj, I.H., et al.: Effect of technology enhanced language learning on EFL reading comprehension at tertiary level. Arab World Eng. J. **8**(1) (2017)
44. Rahmani, M.: The effect of CALL on EFL learners' reading comprehension and attitude towards the use of technology. J. Teach. Eng. Lang. Stud. **1**(3), 39–56 (2013)
45. Nezami, S.R.A.: A critical study of comprehension strategies and general problems in reading skill faced by Arab EFL learners with special reference to Najran university in Saudi Arabia. Int. J. Sci. Educ. **2**(3), 306–316 (2012)
46. Chesham, A., et al.: What older people like to play: genre preferences and acceptance of casual games. JMIR Serious Games **5**(2), e8 (2017)
47. Cheng, W.K.R.: Relationship between visual attention and flow experience in a serious educational game: an eye tracking analysis. Doctoral dissertation. George Mason University (2014)
48. Burgos, D., Tattersall, C., Koper, R.: Re-purposing existing generic games and simulations for e-learning. Comput. Hum. Behav. **23**(6), 2656–2667 (2007)
49. Anolli, L., et al.: Emotions in serious games: from experience to assessment. Int. J. Emerg. Technol. Learn. (iJET) **5**(2010) (2010)
50. Van Eck, R.: Building artificially intelligent learning games. In: Games and Simulations in Online Learning: Research and Development Frameworks, pp. 271–307. IGI Global, Hershey (2007)
51. Faiola, A., et al.: Correlating the effects of flow and telepresence in virtual worlds: enhancing our understanding of user behavior in game-based learning. Comput. Hum. Behav. **29**(3), 1113–1121 (2013)
52. Dumdumaya, C.E., Banawan, M.P., Rodrigo, M.M.T.: Identifying students' persistence profiles in problem solving task. In: Adjunct Publication of the 26th Conference on User Modeling, Adaptation and Personalization (2018)
53. Skinner, E., et al.: Engagement and disaffection in the classroom: part of a larger motivational dynamic? J. Educ. Psychol. **100**(4), 765 (2008)

Covid-19 and Online Learning Engagement: Effects of Internal Crisis Communication, Technology and Social Responsibility

Suha Fouad Salem(✉), Ibrahim Al-Jubari, Adnan Aldholay, Abdullah Nabeel Jalal, and Ahmed M. Mutahar

Faculty of Business Management and Professional Studies, Management and Science University, Shah Alam, Malaysia
Suha_fksalem@msu.edu.my

Abstract. Online education has been around for many years and it has been growing rapidly in many parts of the world. However, many educational service providers were reluctant to utilise it fully till Covid-19 hit the world. As a result, almost most of these educational institutions have been pushed to adopt this method of teaching as an alternative mode to the conventional approach. They adapted to this new norm very quickly. However, with this new norm there arise inevitable challenges, one of which is the students' engagement in the learning process, where in normal times it has its challenge. Therefore, the current study was planned to uncover the effects of an organisational factor (e.g. internal crises communication), technological factors (e.g. knowledge quality, interactivity, complexity), and personal factor (e.g. social responsibility) on students' online learning engagement. Data have been obtained from a convenience sample of 365 business students from a private university in Selangor state in Malaysia. The results of the partial least square structural equation modelling have revealed that the students online learning engagement is positively and directly predicted by internal crises communication, interactivity and social responsibility as well as negatively predicted by complexity. Findings also show that knowledge quality, internal crises communication, and complexity have influenced online learning engagement indirectly through interactivity. Theoretical and practical implications have also been discussed.

Keywords: Covid 19 · Online learning engagement · Internal crises communication · Social responsibility · Knowledge quality · Interactivity · Complexity

1 Introduction

Computer-based learning can be traced back to the early 1960s [1] and it has been evolving ever since. During the last couple of decades institutions of higher learning around the world started to promote and exploit this new technology [2]. Although it remained popular mostly among students who are tech-savvy, now it is being seen as the future

M. Al-Emran et al. (Eds.): ICETIS 2021, LNNS 299, pp. 276–289, 2022.
https://doi.org/10.1007/978-3-030-82616-1_25

of higher education. The United Nations described online learning as a reliable solution in disruptive times such as wars and natural disasters. Consequently, the COVID-19 pandemic is set to make a profound impact on humanity, socially, economically, and politically [3]. Day by day, the magnitude of this pandemic is ravelling. The disruption and damage that have been inflicted on our lifestyle are severe specifically in the education sector [4]. Globally, countries have instilled strict measures by locking down cities and restricting the movement of the people with stay at home orders for prolonged periods of time, where an end is yet to be seen [5]. Online learning platforms are currently the champions of education delivery with almost all higher education institutions, both public and private have been utilising in these difficult times. However, many students are having their first online learning experience, which brings the notion of learners' engagement into the spotlight from multiple aspects including behavior [6], emotion, and cognitive. In Malaysia, most universities have stepped up efforts to develop their own online platforms and enhance their staffs' skills. However, despite all the research that has been conducted to examine student's engagement toward online platform, the real effects of Covid-19 pandemic on online students' engagement still remain unclear. To address this gap, in this study, the effect of crisis factor on online learning engagement was investigated. To the best of our knowledge, this is the first study to explore the interaction between internal crisis communication, social responsibility and online learning engagement in this context. Therefore, this study aims to investigate the role of internal communication and social responsibility on online learning engagement. The results of this study hope to benefit relevant universities and stakeholders in many ways. Universities mangers within this field would be able to understand students learning engagement and their behaviors towards online learning during the pandemic crisis to develop various yet precise learning and teaching strategies to be executed accordingly.

2 Theoretical Background

2.1 Crisis Theory

A crisis develops when a person or system is incapable of dealing efficiently with a situation or a problem Dixon [7]. According to Hussain [8], a personal crisis can be defined as having the feeling of being overwhelmed due to problems in which the person becomes unable to function normally. Crisis management is the practices that are conducted by the organisation to have a pre-crisis preventing management or to have a post-crisis management strategy to deal with the crisis that is on-going or has already happened. There are executive and also practical sides of crisis, therefore, it comprises practices of avoiding practical shortage and enriching of organisation communication and public relation strategy for post-crisis management. We need to consider the essential principles as well as concepts of a crisis when applying the crisis theory to university's management of the epidemic. The vital factors of crisis management are about assessing danger and risks within a short time, handling feelings as well as drama related to a crisis, calculated preparation and reconstruction of the crisis, establishing communication channels and also education as well as mobilisation of resources to meet the difficulty.

2.2 Conceptual Model and Research Hypotheses

Based on the crisis theory and other significant articles, the authors can propose a revised conceptual model. This study focused on the crisis factor, technological factors, and social responsibility on the online learning engagement during the Corona virus pandemic. Figure 1 shows the cause-effect relational model of this particular study.

2.3 Online Learning Engagement

Academic engagement generally refers to the active involvement students would have during any learning activity [9]. It has been conceptualised differently in educational research. However, a common agreement is that engagements feature three interrelated but distinct constructs including behavioural, emotional and cognitive engagement. Other dimensions were also considered as important facets of academic engagement such as agentic engagement, and recently social engagement. Another conceptualisation of students' engagement was borrowed from work engagement, which reflect three distinct dimensions of engagement such as dedication, absorption and vigor. In the current study, we will follow the four dimensionalities of engagement including behavioural, emotional, cognitive and agentic engagement. Students' engagement can lead to positive learning outcome such as retention of the students and academic achievement. As has been reported in previous research by Paulsen [10], online learning has positively correlated with other benefits as compared to the face to face learning mode including satisfaction, better habits of learning, learning gains and perceived academic challenge. On the other hand, benefits such as a collaborative environment for academic interaction and support were more apparent in the face-to-face mode of learning.

2.4 Knowledge Quality

Knowledge quality is defined as the gained knowledge from the interactive content that was accessible through online learning platforms [11]. It has been the focus of research in the IS arena all over the world since its inception in [12] and its update in 2003. Extensive studies have highlighted the important role that knowledge quality plays in determining online learning usage as well as the user engagement [13]. For instance, a recent study by [14] in Turkey to investigate the effect of information quality on system usage has found that there is a significant relationship between the aforementioned constructs. Accordingly, it can be assumed that:

- H1a. knowledge quality has a positive relationship with students' engagement during a pandemic crisis.
- H1b. knowledge quality has a positive relationship with Interactivity during a pandemic crisis.

2.5 Internal Crisis Communication

Internal crisis communication is defined as the helpful and useful collaboration between managers and any internal stakeholders (such as employees) during any stage of the crisis (prior, during, or post) and in any type of organisations (private, public, or NGOs). According to Adamu [15], ICC can be defined as the organisation's condition of communication during an unfortunate situation which can develop a relationship based on trust and allows a two-sided communication with all of the internal stakeholders. In the situation of crisis, many studies on the internal stakeholders have described that students must be observed as human capital and are to be considered as the most important asset of the higher education institutions [16]. In the academic context, we can say that effective internal communication is when students receive updated information massages about how the university is going to handle these crises and how it will continue to create positive learning experiences during a crisis. Therefore, this form of systematic communication can lead to individual engagement and positive communication behaviours. Accordingly, it can be assumed that:

- H2a. Internal communication has a positive relationship with students' online engagement during a pandemic crisis.
- H2b. Internal communication has a positive relationship with Interactivity during a pandemic crisis.

2.6 Complexity

Complexity is defined as the level of difficulty that is encountered by the users of a technology or a system and this difficulty could be in understanding or the use matters [17]. Form the individual's perspective, it denotes the difficulties and obstacles that are experienced by the user who is trying to adopt a new technology or innovation. Many studies have validated the significance of the negative impact of complexity on the adaption or use of different innovations. Parveen [18] have examined the influence of complexity on the intention to use the wireless internet using mobile devices and found that it has a significant impact: the more difficult the technology the lesser the intention the user will have towards using that technology. The same conclusion was found by Mndzebele [19] on their studies in e-commerce and e-learning, respectively. This study examined the same negative impact of complexity on the level of learning engagement while using online learning:

- H3a. Complexity has a negative relationship with online learning engagement.
- H3b. Complexity has a negative relationship with interactivity during a pandemic crisis.

2.7 Interactivity

Interactivity can be defined as a collection of system permissions which allow the users through a system- depending on the most central elements for the transmission of information in the traditional models of communication, to edit medium, source and the message of their communications. Studies have identified e-learners' interactivity as a critical success factor in virtual learning environments, since e-learners do not relate face to face with instructors, university staff and classmates [20]. Essentially, online platform increases the learners' engagement and motivation to learn by bringing in an environment with enjoyable activities. The addition of multimedia are most interesting to young students whose attention and focus increase as they listen to the story, thereby improving their motivation to re-read the stories [21]. This illustrates why and how multimedia environments promote active learning [22]. The current study explores the impact of interaction techniques during interacting with blended learning at the time of the COVID-19, and this is the result of the differences regarding the effects of different interactivity modalities on the user's experience, which are discussed in the previous literature. Consequently, the following hypothesis is proposed:

- H4. Interactivity significantly has a positive impact on online learning engagement

2.8 Social Responsibility

Social Responsibility is defined as the ability of an individual to understand his or her responsibilities on one hand and the legal, ethical responsibilities on the other hand [23]. Social responsibility initiatives refer to individual attitudes and responses. The positive responses and perception of individuals appear to increase the benefits towards better online learning engagement. Therefore, studying individual perception can improve our understanding of how people perceive social responsibility activities, and how that further increases their intention to further engage in online platforms. The individual perception and awareness reflect the social responsibility and its impact on the online learning engagement. Thus, societies with higher levels of awareness or concern of social responsibility are more likely to have positive attitudes in responding and reacting to situations such as Covid 19. Therefore, this is deemed to be a reasonable hypothesis:

- H5. A higher level of individual awareness of social responsibility leads to a higher level of online learning engagement.

2.9 Interactivity as a Mediator

Nowadays, keeping students engaged in online learning during a pandemic crisis is very challenging [24]. According to [24], knowledge quality and complexity have proven to be significant variables that improve the students' engagement. However, the current situation in which the world is under lockdown because of the coronavirus pandemic has a negative influence on the engagement of students online [25]. Blasco-Arcas argue that students will be more engaged if they have a system that allows them to interact with it in an effective way. Interactivity could improve the students' perception towards the quality of information, communication and perceived complexity of the online learning system. Therefore, it is proposed that:

- H6a: online learning engagement is indirectly impacted by knowledge quality through interactivity during a pandemic crisis.
- H6b: online learning engagement is indirectly impacted by internal communication through interactivity during a pandemic crisis.
- H6c: online learning engagement is indirectly impacted by complexity through interactivity during a pandemic crisis.

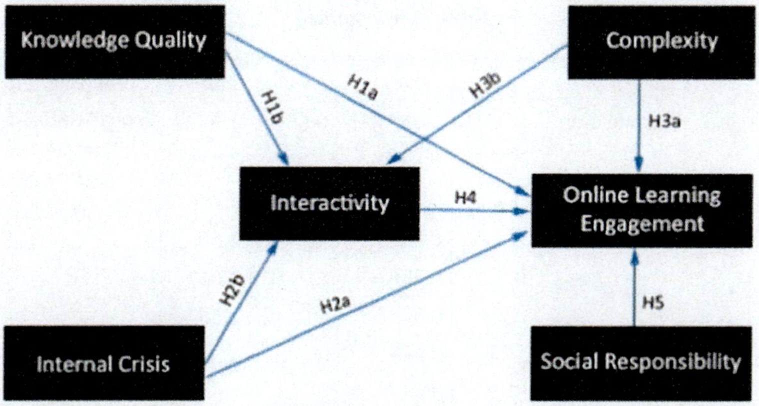

Fig. 1. Conceptual model

3 Methodology

3.1 Validity and Reliability of the Measurement Model

Internal consistency of the items-variable association is an essential reliability analysis, which can be estimated by the measure of composite reliability or Cronbach's alpha [27]. The rule of thumb is clearly provided by many scholars including [28]; he has stated that any composite reliability score above 0.7 is acceptable, but any score above 0.96 must be avoided because of the high potential similarity between items. In addition, convergent validity is achieved by assuring that the average variance extracted (AVE) score is above the threshold of 0.5. And finally, outer loading of items into its variable is in the acceptable level above 0.708. All results are evaluated based on rule of thumbs by Hair. Another examination of validity is the discriminant validity between variables of the designed model. The results of the heterotrait-monotrait correlations (HTMT) ratio, in which the results are acceptable because it is below the cut off value of 1.0 as the rule of thumb proposed by Henseler [29] and that the discriminant validity is achieved as stated in Table 2. Table 1 and Fig. 2 show the discussed results. Table 3, shows the results of another complementary examination that is known as Fornell Larcker criterion matrix. For the results to be acceptable, the scores at the diagonal cell must be higher than any other results in the same row and column (Byrne 2013); the results in the table shows an acceptable level and that the discriminant validity is achieved.

Table 1. Construct reliability and validity

Construct	Item	Loading	AVE	VIF	Composite reliability
Knowledge quality (KQ)	KQ1	0.949	0.878	3.033	0.956
	KQ2	0.917		4.729	
	KQ3	0.945		4.552	

(*continued*)

Table 1. (*continued*)

Construct	Item	Loading	AVE	VIF	Composite reliability
Internal crisis communication (ICC)	ICC1	0.735	0.662	1.8	0.921
	ICC2	0.837		2.485	
	ICC3	0.827		2.394	
	ICC4	0.855		2.56	
	ICC5	0.796		2.067	
	ICC6	0.825		2.367	
Social responsibility (SR)	SR1	0.847	0.696	1.865	0.92
	SR2	0.844		2.649	
	SR3	0.81		2.286	
	SR4	0.837		2.235	
	SR5	0.831		2.391	
Interactivity (IY)	IY1	0.895	0.8	2.552	0.923
	IY2	0.88		2.068	
	IY3	0.909		2.76	
Complexity (CY)	CY1	0.861	0.78	2.717	0.934
	CY2	0.899		3.306	
	CY3	0.777		1.863	
	CY4	0.875		2.531	
Behavioural engagement (BE1)	BE1	0.887	0.748	3.711	0.922
	BE2	0.916		4.394	
	BE3	0.798		1.752	
	BE4	0.854		2.217	
Emotional engagement (EE)	EE1	0.873	0.765	2.713	0.928
	EE2	0.919		4.093	
	EE3	0.926		4.13	
	EE4	0.771		1.64	
Cognitive engagement (CE)	CE1	0.898	0.805	3.124	0.943
	CE2	0.916		3.514	
	CE3	0.922		3.689	
	CE4	0.851		2.309	
Affective engagement (AE)	AE1	0.802	0.75	1.77	0.923

(*continued*)

Table 1. (*continued*)

Construct	Item	Loading	AVE	VIF	Composite reliability
	AE2	0.887		2.971	
	AE3	0.888		3.022	
	AE4	0.885		2.598	

Table 2. Discriminant validity – heterotrait- monotrait ratio

Construct	AE	BE	CE	CY	EE	IY	ICC	KQ
BE	0.59							
CE	0.579	0.738						
CY	0.517	0.554	0.614					
EE	0.611	0.661	0.801	0.796				
IY	0.553	0.558	0.682	0.758	0.796			
ICC	0.342	0.524	0.57	0.536	0.614	0.711		
KQ	0.411	0.493	0.601	0.643	0.691	0.761	0.66	
SR	0.112	0.353	0.305	0.095	0.112	0.158	0.294	0.326

Notes: KQ, Knowledge quality; ICC, Internal crises communication; CY, Complexity; IY, Interactivity; SR, Social responsibility; OLE Convenience.

Table 3. Fornell-Larcker criterion matrix (discriminant validity)

Construct	AE	BE	CE	CY	EE	IY	ICC	KQ	SR
AE	0.866								
BE	0.521	0.865							
CE	0.524	0.666	0.897						
CY	0.465	0.497	0.561	0.883					
EE	0.544	0.59	0.727	0.718	0.874				
IY	0.49	0.492	0.612	0.677	0.708	0.895			
ICC	0.311	0.472	0.521	0.485	0.553	0.632	0.814		
KQ	0.375	0.448	0.557	0.59	0.631	0.687	0.604	0.937	
SR	0.1	0.328	0.292	0.094	0.11	0.154	0.281	0.308	0.834

Notes: KQ, Knowledge quality; ICC, Internal crises communication; CY, Complexity; IY, Interactivity; SR, Social responsibility; OLE Convenience

3.2 Structural Model

For evaluation of the prediction power of the proposed design, R2 is used to reveal the ration of variance that can be explained [30–32]. The results show 62.9% of the interactivity variance can be explained by internal crises management, knowledge quality, and complexity. In addition, 59.2% of online learning engagement can be explained by internal crises management, knowledge quality, complexity, social responsibility, and interactivity. The predicting power is considered at a moderate level. Hypotheses testing. As shown in Table 4 and Fig. 3, H1a examined the positive relationships between knowledge quality and online learning engagement with a result of T-statistics = 1.359. Nonetheless, knowledge quality showed a positive relationship with interactivity with T-statistics = 5.897. Therefore, the first hypotheses (H1a) was rejected while the second one (H1b) was supported. There was, however, a powerful connection between the internal crises communication and interactivity (H2b) with T-statistics = 6.046. Similarly, the connection between internal crises communication and online learning communication (H2a) was supported with T-statistics = 1.954. Complexity demonstrated a strong connection with interactivity and online learning engagement (H3a, H3b) with the value of T-statistics at 5.44 and 7.328, respectively. The fourth hypothesis (H4), as depicted in Fig. 2 and Table 4, proved that there was a significant relationship between interactivity and online learning engagement with the T-statistic value at 5.589. Lastly, the fifth hypothesis (H5) in this study can be accepted as there was a significant relationship between social responsibility with the t-value above the threshold of 1.96 (t -value = 2.799).

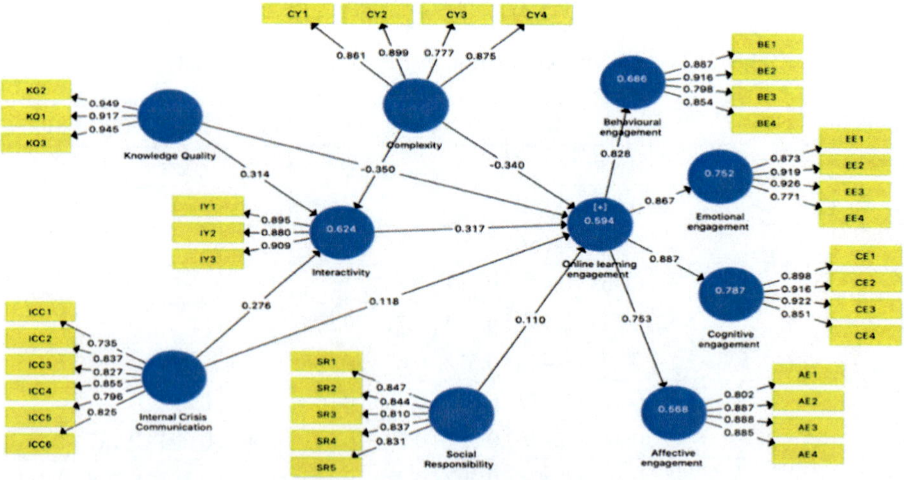

Fig. 2. PLS algorithm path model

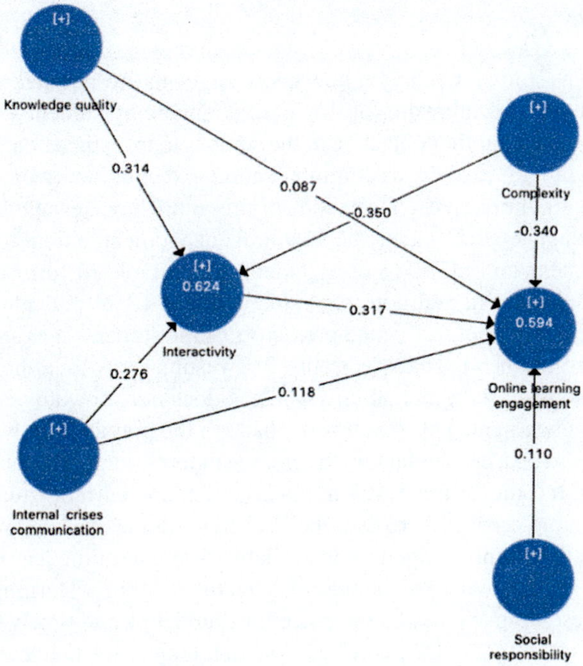

Fig. 3. Bootstrapping results

Table 4. Hypothesis testing

Hypothesis	Path	Path coefficient (β)	SE	T-statistics	Decision
H1a	KQ → OLE	0.087	1.359	1.359	Not supported
H1b	KQ → IY	0.314	0.053	5.897	Supported
H2a	ICC → OLE	0.118	0.06	1.954	Supported
H2b	ICC → IY	0.276	0.046	6.046	Supported
H3a	CY → OLE	−0.34	0.063	5.44	Supported
H3b	CY → IY	−0.35	0.048	7.328	Supported
H4	IY → OLE	0.317	0.057	5.589	Supported
H5	SR → OLE	0.11	0.039	2.799	Supported
H6a	KQ → IY → OLE	0.099	0.025	4.006	Supported
H6b	ICC → IY → OLE	0.087	0.022	3.933	Supported
H6c	CY → IY → OLE	−0.111	0.025	4.434	Supported

Notes: KQ, Knowledge quality; ICC, Internal crises communication; CY, Complexity; IY, Interactivity; SR, Social responsibility; OLE Convenience

4 Discussion

The main aim of this study is to analyse how knowledge quality, internal crisis communication, complexity and social responsibility predict university student's online learning engagement. Moreover, another objective of the study is to investigate the indirect impact of knowledge quality, internal crises communication and complexity on online learning engagement through interactivity. The results of this study have revealed that all dimensions of learning engagement, namely behavioural engagement, emotional engagement, cognitive engagement and affective engagement are suitable to form online learning engagement in Malaysia. In addition, the value of R2 (59.4%) indicated that the constructs of this research model were good predictors of the students' engagement towards online learning during the coronavirus status. The results show that knowledge quality has a strong relationship with interactivity while it does not show any connection with online learning engagement. This result is opposite to the previous study that has found knowledge quality to be a crucial factor to increase students' engagement towards online learning [33]. The reason for this result might be due to the current situation that is the coronavirus pandemic crisis, where everybody is living under lockdown and isolation. However, the quality of knowledge that is available through online learning is found to be a main factor in improving the interactivity feature within a learning system. This implies that by presenting and maintaining accurate, updated, and timely information on the online system can improve the perceived system interactivity, that lead to an increase in the students' engagement with the online system [34]. In investigating how internal communication crises are directly connected to online learning engagement and interactivity, it is found that both have strong connections, especially on interactivity. The effective and professional communication with students during a crisis might enhance students' interactivity with the system, classmates, and lecturers. Students will be more sharing of their learning experience if there are effective communication strategies and platforms that connect them with all parties of the learning system [35]. The results of this study also show that there is a negative relationship between complexity and interactivity as well as online learning engagement. The findings were in line with the outcomes of earlier studies, which had suggested that complexity was a critical factor of learning engagement and interactivity. Moreover, results from this study have also shown positive relationships between social reasonability and online learning engagement. The crisis of coronavirus shows that the more students are aware of their responsibility towards the society, the more they become motivated in their learning process. Finally, the results of this study indicate that knowledge quality, internal crises communication and complexity have indirect relationship with online students' engagement through interactivity.

5 Research Implications

The theoretical implications and contributions of this study mainly develop an online learning engagement framework. Which will fill one of the most important research gaps in online learning. In view of the increased need for the guidelines, it is obvious that most educational organisations are having serious challenges in understanding the online learning engagement mechanism. The possible results of this research react to the questions about the engagement in online learning in the educational context during the outbreak. The results of this study would help universities, institutions and education providers to consider this online learning engagement framework. Therefore, this research provides them with the standard guidelines to follow in addressing the factors that lead to more affective engagement.

The findings of the current study have implication for education stakeholders including students, lecturers and researchers, and universities as well. It is evident that online education is becoming the new norm of learning and thus achieving learning outcomes should be ensured. First, students may be trained and educated on the technical and technological aspects of the learning process. They should be aware of the importance of online learning and its impact on their overall performance and thus they should be involved in the learning process and their feedback should be taken into consideration. Second, lecturers play a central role in enhancing the students learning engagement either in the conventional mode of teaching or the online mode and may exploit the innovativeness in the various learning platforms which make the interaction with students more meaningful. Lecturers should support students in learning motivational tendencies by giving them a voice and choice, making them feel involved and related and giving tasks that are challenging but accomplishable.

6 Limitations and Future Research

The study is an empirical examination based on 256 participants of one academic institute in a cross-sectional time span by using self-directed questionnaire. Future studies can replicate the same design but with a higher number of participants to enrich the generalisation of the design. In addition, the study can be conducted in a regional context that has many academic institutes. Doing the same study in different environments, countries, and contexts can provide a better understanding for the research topic. Other future studies direction can consider doing a longitudinal empirical methodology.

The model can explain 0.629 of the interactivity variances, and 0.569 for online learning engagement variance. There are more than 35% of non-explained variance that needs further investigations for the inclusion of other factors of interaction and online learning. Future research can explore a wider span of concepts.

References

1. Rodríguez-Ardura, I., Meseguer-Artola, A.: E-learning continuance: the impact of interactivity and the mediating role of imagery, presence and flow. Inf. Manag. **53**(4), 504–516 (2016). https://doi.org/10.1016/j.im.2015.11.005
2. Alharbi, J., Almahdi, H., Mosbah, A.: The impact of entrepreneurship education programmes (EEPs) on the entrepreneurial attitudes among higher education students in Saudi Arabia. Int. J. Manag. Econ. Soc. Sci. **7**(3), 245-271 (2018)
3. Chick, R.C., et al.: Using technology to maintain the education of residents during the COVID-19 pandemic. J. Surg. Educ. (2020). https://doi.org/10.1016/j.jsurg.2020.03.018
4. Chiodini, J.: Online learning in the time of COVID-19. Travel Med. Infect. Dis. 101669 (2020). https://doi.org/10.1016/j.tmaid.2020.101669.

5. Masri, R., et al.: The management dilemma on the delivery of tacit-explicit knowledge in the diffusion of innovation theory. Turkish Online J. Des. Art Commun. **8**(March), 686–692 (2018)
6. Ibrahim, Z., Rahman, N.R.A., Johar, M.G.M.: A job satisfaction of emotional intelligence, leadership, employee performance with information technology. Int. J. Recent Technol. Eng. **8**(2), 712–718 (2019). Special Issue 9
7. Dixon, S.L.: Working with People in Crisis: Theory and Practice. CV Mosby, Maryland Heights (1979)
8. Hussain, S.B.: Crisis communication at higher education institutions in South Africa: a public relations perspective. J. Econ. Behav. Stud. **6**(2), 144–151 (2014). https://doi.org/10.22610/jebs.v6i2.477
9. Reeve, J.: A self-determination theory perspective on student engagement. In: Christenson, S.L., Reschly, A.L., Wylie, C. (eds.) Handbook of Research on Student Engagement, pp. 149–172. Springer US, Boston (2012). https://doi.org/10.1007/978-1-4614-2018-7_7
10. Paulsen, J., McCormick, A.C.: Reassessing disparities in online learner student engagement in higher education. Educ. Res. **49**(1), 20–29 (2020). https://doi.org/10.3102/0013189X19898690
11. Aldholay, A., Isaac, O., Abdullah, Z., Ramayah, T.: The role of transformational leadership as a mediating variable in DeLone and McLean information system success model: the context of online learning usage in Yemen. Telemat. Inform. **35**(5), 1421–1437 (2018). https://doi.org/10.1016/j.tele.2018.03.012
12. DeLone, W.H., McLean, E.R.: Information systems success: the quest for the dependent variable. Inf. Syst. Res. **3**(1), 60–95 (1992)
13. Isaac, O., Aldholay, A., Abdullah, Z., Ramayah, T.: Online learning usage within Yemeni higher education: the role of compatibility and task-technology fit as mediating variables in the IS success model. Comput. Educ. **136**(February), 113–129 (2019). https://doi.org/10.1016/j.compedu.2019.02.012
14. Dorobăţ, I., Corbea, A.M.I., Muntean, M.: Integrating student trust in a conceptual model for assessing learning management system success in higher education: an empirical analysis. IEEE Access **7**, 69202–69214 (2019)
15. Adamu, A.A., Mohamad, B.: A reliable and valid measurement scale for assessing internal crisis communication. J. Commun. Manag. **23**(2), 90–108 (2019). https://doi.org/10.1108/JCOM-07-2018-0068
16. Jamal, J., Abu Bakar, H.: The mediating role of charismatic leadership communication in a crisis: a Malaysian example. Int. J. Bus. Commun. **54**(4), 369–393 (2017)
17. Rogers, E.M.: Diffusion of Innovations. Free Press, New York (1995)
18. Kim, Y.: Enhancing employee communication behaviors for sensemaking and sensegiving in crisis situations. J. Commun. Manag. **22**, 451-475 (2018)
19. Mndzebele, N.: The effects of relative advantage, compatibility and complexity in the adoption of EC in the hotel industry. Int. J. Comput. Commun. Eng. **2**(4), 473–476 (2013). https://doi.org/10.7763/ijcce.2013.v2.229
20. Peltier, J.W., Schibrowsky, J.A., Drago, W.: The interdependence of the factors influencing the perceived quality of the online learning experience: a causal model. J. Mark. Educ. **29**(2), 140–153 (2007)
21. Verhallen, M., Bus, A.: Video storybook reading as a remedy for vocabulary deficits: outcomes and processes. J. Educ. Res. Online **1**, 172-196 (2009)
22. Arpaci, I.: The role of self-efficacy in predicting use of distance education tools and learning management systems. Turkish Online J. Distance Educ. **18**(1), 52–62 (2017)
23. Öberseder, M., Schlegelmilch, B.B., Murphy, P.E.: CSR practices and consumer perceptions. J. Bus. Res. **66**(10), 1839–1851 (2013)

24. Liaw, S.-S., Hatala, M., Huang, H.-M.: Investigating acceptance toward mobile learning to assist individual knowledge management: based on activity theory approach. Comput. Educ. **54**(2), 446–454 (2010)
25. Jafari, S.M., Salem, S.F., Moaddab, M.S., Salem, S.O.: Learning management system (LMS) success: an investigation among the university students. In: 2015 IEEE Conference on e-Learning, e-Management e-Services, IC3e 2015, pp. 64–69 (2016). https://doi.org/10.1109/IC3e.2015.7403488
26. Blasco-Arcas, L., Buil, I., Hernández-Ortega, B., Sese, F.J.: Using clickers in class. The role of interactivity, active collaborative learning and engagement in learning performance. Comput. Educ. **62**, 102–110 (2013)
27. Salem, S.F., Salem, S.O.: Self-identity and social identity as drivers of consumers' purchase intention towards luxury fashion goods and willingness to pay premium price. Asian Acad. Manag. J. **23**(2), 161–184 (2018)
28. Hair, J.F., Jr., Hult, G.T.M., Ringle, C., Sarstedt, M.: A primer on partial least squares structural equation modeling (PLS-SEM). Sage Publications, London (2017)
29. Henseler, J., et al.: Common beliefs and reality about PLS: comments on Rönkkö and Evermann (2013). Organ. Res. Methods **17**(2), 182–209 (2014)
30. Chin, W.W.: The partial least squares approach to structural equation modeling. Mod. Methods Bus. Res. **295**(2), 295–336 (1998)
31. Salem, S.F., Salem, S.O.: Effects of social media marketing and selected marketing constructs on stages of brand loyalty. Glob. Bus. Rev. **22**, 650-673 (2019). 0972150919830863
32. Ferdous Azam, S.M., Tham, J., Ahmad, A.: Psycho-social perspectives of knowledge sharing and job performance in Malaysia: conceptual articulation. Int. J. Sci. Technol. Res. **9**(4), 3500–3509 (2020)
33. Bradford, G., Wyatt, S.: Online learning and student satisfaction: academic standing, ethnicity and their influence on facilitated learning, engagement, and information fluency. Internet High. Educ. **13**(3), 108–114 (2010). https://doi.org/10.1016/j.iheduc.2010.02.005
34. Sims, R.: Promises of interactivity: aligning learner perceptions and expectations with strategies for flexible and online learning. Int. J. Phytoremediation **24**(1), 87–103 (2003). https://doi.org/10.1080/01587910303050
35. Al-Samarraie, H., Teng, B.K., Alzahrani, A.I., Alalwan, N.: E-learning continuance satisfaction in higher education: a unified perspective from instructors and students. Stud. High. Educ. **43**(11), 2003–2019 (2018)

Fostering Students' Attitude Towards Online Learning: The Mediation Effect of Satisfaction and Perceived Performance

Abdul Hafaz Ngah[1], Rosyati Abdul Rashid[2], Nur Asma Ariffin[3,4], Fauzayani Ibrahim[4], Noor Azuan Abu Osman[5], Nurul Izni Kamalrulzaman[4], Mohamad Firdaus Halimi Mohamad[1], and Nor Omaima Harun[6,7](✉)

[1] Faculty of Business, Economics and Social Development, Universiti Malaysia Terengganu, Kuala Nerus, Malaysia
[2] Centre for Foundation and Continuing Education, Universiti Malaysia Terengganu, Kuala Nerus, Malaysia
[3] Faculty of Fisheries and Food Science, Universiti Malaysia Terengganu, Kuala Nerus, Malaysia
[4] Centre for Academic Management and Quality, Universiti Malaysia Terengganu, Kuala Nerus, Malaysia
[5] Faculty of Engineering, University of Malaya, Kuala Lumpur, Malaysia
[6] Faculty of Science and Marine Environment, Universiti Malaysia Terengganu, Kuala Nerus, Malaysia
omaima@umt.edu.my
[7] Data & Digital Development Centre, Universiti Malaysia Terengganu, Kuala Nerus, Malaysia

Abstract. The COVID-19 pandemic outbreak has caused Malaysian education institutions to embrace online learning due to the Movement Control Order (MCO) and closure of universities. By adopting the Stimulus-Organism-Response (S-O-R) theory, this study aims to identify the factors influencing students' attitudes to continue online learning. Since attitude can influence behaviours, the lecturers must stimulate a positive attitude among students during the online learning sessions. The study also looked into the potential of student satisfaction and performance as a mediating variable. A purposive sampling method was employed in the study, and 1349 data were gathered using an online survey. The data were analysed using a structural equation modelling with Smart Partial Least Squares. The results indicate that perceived ease to use positively affected satisfaction and performance; satisfaction and performance were confirmed to positively related to attitude. Satisfaction and performance explain 69.8% of the variance of attitude. Data analysis also revealed that performance and satisfaction mediated the relationship between perceived ease to use and attitude. These findings suggest that the university's top management must foster a positive attitude among students to continue to use online learning to prevent COVID-19 from spreading on campus. The present study limited the variables studied to ease of use, satisfaction, and performance of science students at Universiti Malaysia Terengganu. Additionally, the present study significantly contributes to the body of knowledge about student's attitudes in online learning amidst the Covid-19 pandemic. From a practical perspective, it is crucial for the top management of higher institutions in Malaysia

M. Al-Emran et al. (Eds.): ICETIS 2021, LNNS 299, pp. 290–302, 2022.
https://doi.org/10.1007/978-3-030-82616-1_26

and around the globe to provide online learning systems and platforms which are easy to use.

Keywords: Online learning during COVID-19 · Stimulus-Organism-Response Theory · Attitude

1 Introduction

The COVID-19 pandemic has shaped the year 2020 by halting all service provision industries and economy; agriculture, manufacturing, and education. This pandemic caused severe economic and social disruption and has changed all human activities. The pandemic has led to the drastic loss of human life and presents an unprecedented challenge to public health, food systems, and the whole working world.

The biggest challenge posed by the COVID-19 pandemic for academic institutions, from lower to tertiary levels, is making sure the teaching and learning process can be resumed and replaced with online classes from the traditional physical classrooms. In response to the COVID-19 pandemic, even without proper physical and mental preparation for students and lecturers, many institutions immediately shifted all courses to being entirely online. Therefore, most education institutions are undertaking tremendous ongoing efforts to provide the necessary support to their faculties, instructors, and students. Several studies have proposed that online learning and blended instruction can be as effective as traditional classroom models in focusing on learners' satisfaction with online method, especially during the transition period from a traditional approach to an online learning approach [1].

Even e-learning and its multiple blended approaches have become a choice in this pandemic, re-designing a curriculum to meet the traditional components is not the best move. The literature confirms the role of attitude in predicting students' behaviour. Most of the studies on e-learning in the COVID-19 pandemic setting were concerned with performance [2], satisfaction [3], and continuance of online learning [4, 5]. However, in this case, COVID-19 forces us to apply social distancing, hence continuing learning via online is not a choice, but compulsory for students all around the globe. Thus, all higher institutions started to enforce online learning even without proper information on students' readiness in terms of facilities and attitude. Nevertheless, learning must be via online. On top of that, the factors influencing attitude among students in this setting are still scarce in the available literature. Owing to the established literature, it was confirmed that attitude has a great influence on behaviours; it is crucial for the lecturers and higher institutions to identify what are the factors to stimulate a positive attitude among students during online learning sessions in the COVID-19 pandemic. Therefore, to fill the literature gaps, the present research aims to unearth factors influencing the students' attitude to continue learning online in this critical situation of the COVID-19 pandemic.

The investigation focuses on online learning during the COVID-19 pandemic and uses individual (student) as a unit of analysis. Besides revealing the factors that influence students' attitudes, the study also aims to discover the mediation effects of performance and satisfaction. The findings from this study will no doubt benefit lecturers in designing their online learning sessions and the management of higher institutions to craft a better policy to escalate a positive attitude towards online learning among students who mostly miss being present physically on campus.

The following sections of this article discuss the literature review, research methodology, data analysis, and discussion and conclusion. The literature review section focuses on research framework development while in the methodology, this section explains the methodological approach and data collection procedure. Additionally, the analysis section explains the measurement and structural model analysis. The last section of the current study explains the findings of the study and the implications.

2 Literature Review

2.1 Stimulus-Organism-Response Theory

Stimulus-Organism-Response (S-O-R) model was introduced in 1974 by Mehrabian and Russell. This theory proposes that stimuli (S) from an individual's environment can elicit the cognitive and affective response (O), which leads to a behavioural response (R) [6]. The stimuli are the external elements of the physical atmosphere to a person, while the organism is referred to as a person's internal processes and structures occurring between stimuli and responses [7]. The theory is appropriate for the context of this study since it has been used in researching users' behaviour in online learning [8]. In the current study, the perceived ease of use represents stimuli since it comes from students' surrounding environment using the system. The stimuli can influence students' internal response, which is their satisfaction and performance; satisfied students who use the online learning system may display a positive behavioural response reflected by their positive attitude towards online learning. Figure 1 demonstrates the research model of the study.

2.2 Attitude

Attitude is an individual level of approval toward the behaviour in question [9]. This notion indicates that attitude predicts an individual behavioural intention and actual behaviour. The present study proposes that satisfaction and performance will influence students' attitudes. Besides being a construct commonly studied in marketing studies, attitude is also used in educational studies to determine consumer behaviour [10]. Past literature has recognized the relationship between attitude and satisfaction [11], and the relationship between attitude and performance [12]. However, studies on the relationship between satisfaction and attitude and performance and attitude are still scarce. The current study uses satisfaction and perceived performance to identify the students' attitudes toward online learning. It is important to study students' attitudes since attitude can influence the students' behaviour in an online learning setting. Maintaining a positive attitude could enhance online learning continuance.

2.3 Perceived Ease of Use (PEOU)

Perceived ease of use is an individual perception of applying particular system which free of effort [13]. Similarly, perceived ease of use refers to the assumption of whether performing a task would command effort from the users' part [14]. Users might adapt swiftly to the latest innovation if they perceive it to be easy [15]. In the current study, PEOU is defined as the student's degree of effort required to conduct online learning sessions.

Past studies revealed that PEOU has positive effect on satisfaction. A study by Al-Rahmi et al. [16] regarding online learning confirms the presence of such a relationship. On the effect of ease of use and performance, Davis and Wiedenbeck [17] found that PEOU has a positive effect on computer usage performance. Moreover, the ease of using a system will influence higher usage, leading to better performance. Hence, the present study proposes the following hypotheses:

H1: PEOU is positively influence student's satisfaction.
H2: PEOU is positively influence student's performance.

2.4 Satisfaction

Satisfaction is the students' perception of the learning experience and learning environment and how they aid their academic success [18]. According to Efiloğlu Kurt [19], satisfaction is the level to which the user believes the system meets their initial purpose. Having a successful interaction with the online learning system will lead to high satisfaction while failing to have a good interaction with the system will lead to dissatisfaction. For this study, students' satisfaction with prior experience is argued to influence their attitude towards the learning system. Past study reveals that satisfaction positively influences attitude in the internet usage context [20]. Therefore, following hypothesis is proposed:

H3: Satisfaction positively influences students' attitude.

2.5 Performance

Students' performance is the students' ability to complete tasks and respond to the learning environment [21]. Factors that may influence learning performance are students' ability, motivation, and desire for knowledge [22]. In the current study, it is argued that a positive performance of using an online learning system will develop a positive attitude towards the learning system. The literature shows that performance has positive effect on attitude in the financial service context [23]. Hence, the current study postulates the following hypothesis:

H4: Performance positively influence an attitude.

2.6 Mediation

Mediation analysis enhances the model and theoretical advancement [24]. Therefore, this study aims to improve the model's predictive power by adding satisfaction and performance as mediators for the relationship between PEOU and attitude. Studies in the literature showed that perceived ease of use positively correlates with satisfaction and performance, and satisfaction and performance positively influence attitude. Besides that, past work revealed that PEOU positively influences attitude in the education context [25]. Hence, this study theorized that satisfaction and performance are suitable for mediation.

H5: Satisfaction mediates the relationship between PEOU and attitude.
H6: Performance mediates the relationship between PEOU and attitude.

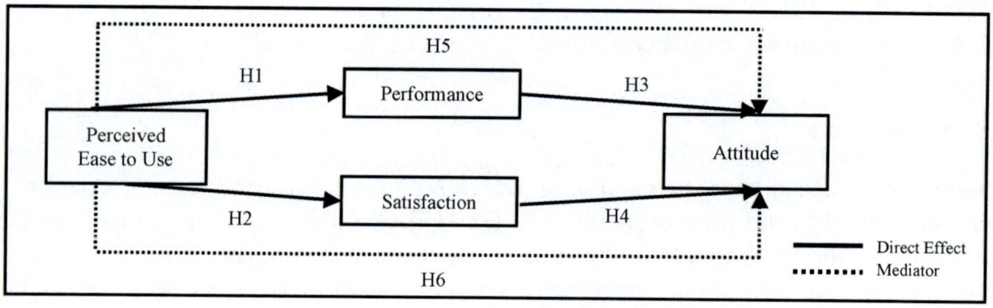

Fig. 1. Research framework

3 Research Methodology

3.1 Instrument Development

All the items were adopted from previous established studies. Items of PEOU were adopted from Abdullah et al. [26], satisfaction adopted from Isaac et al. [27], while the items of performance were adopted from Damnjanovic et al. [28], and items measuring attitude were from Salloum et al. [29].

3.2 Sampling Method

Since the study only focused on UMT's Science degree students attending courses using online learning during the second semester of the 2019/2020 session, the purposive sampling method was used. Since the present study only looked at the theoretical effect of variables in the research framework, convenience sampling would suffice for this purpose [30]. The survey for the study was made using Google Forms and distributed online via the UMT Official Facebook, from 15 July 2020 to 11 August 2020, which was over one month. Following Hair et al. [31] recommendation, the research determined the sufficient sample size using the power of analysis, which is based model's complexity. Gefen et al. [32] suggested that with the minimum of 80% power, medium effect size, and $p = 0.05$, the minimum sample size needed is 84. The sample size was not an issue for this study as a total of 1349 completed questionnaires were usable for the study.

A total of 77.7% of the overall respondents in the study were female students, while the remaining 22.3% were males. Second-year students made up 37.1% of the respondents, 33% were first-year students, 27.1% were third-year students, and 2.9% were fourth-year students. With regards to electronic device usage, 80.1% of the respondents had more than one electronic device, 5.4% had a laptop, 4.0% had a smartphone, and 0.5% had a desktop. A total of 85.6% of the respondents had internet access at home, while 14.4% used mobile data.

4 Data Analysis

The Smart Partial Least Squares (PLS) software was used to identify the relationship between variables in the research framework due to the nature of study is concern on the predictive purposes. As mentioned by Hair et al. [33] the PLS-SEM software is suitable for the study, as the study is focused on predictive purposes.

4.1 Common Method Bias (CMB)

The study relied on a single source data, where the exogenous and endogenous variables data were derived from the same respondents simultaneously. To address this issue, as recommended by [35, 36], the study applied the procedural and statistical methods to overcome the CMB issue of the study. As for the procedural, the study engaged different anchor scales to measure the exogenous variables (1–5) and endogenous variables (1–7) [34, 36] and the marker variable technique for the statistical method [37]. The unmeasured marker variables were used as an exogenous variable predicting every endogenous variable in the model. The analysis found that all the supported hypothesis with marker variable remained supported even the model without the marker variable. Hence, we can conclude that the CMB was not a severe issue of the study.

4.2 Measurement Model

The present study employed a two-stage approach proposed by Hair et al. [33]. The approach consists of a measurement model and a structural model. According to Ngah et al. [38], two categories of validities must be confirmed while establishing the outer model, namely convergent validity (CV) and discriminant validity (DV). The CV can be achieved if the loading is ≥ 0.708, the average variance explained (AVE) ≥ 0.5, and the Composite Reliability (CR) is ≥ 0.7 [33]. Table 1 illustrates the results of the CV test. All the loadings, AVE, and CR were superior than the minimum values, thus confirming that the CV is established.

Table 1. Convergent validity

Variable	Item	Loading	CR	AVE
Perceived ease to use	PEU1	0.913	0.935	0.826
	PEU2	0.923		
	PEU3	0.891		
Satisfaction	S1	0.950	0.966	0.905
	S2	0.950		
	S3	0.955		
Performance	PER1	0.947	0.945	0.852
	PER2	0.957		
	PER3	0.862		
Attitude	ATT1	0.913	0.929	0.814
	ATT2	0.911		
	ATT3	0.883		

Note: ATT4 was deleted due to discriminant validity

Next, the DV was measured by the heterotrait-monotrait (HTMT). Table 2 depicts the values of HTMT are lower compared to the required threshold value of 0.9 as suggested by Franke and Sarstedt [39]. Hence, the discriminant validity was confirmed.

Table 2. Discriminant validity: HTMT ratio

	1	2	3	4
Perceived ease to use (1)				
Satisfaction (2)	0.829			
Performance (3)	0.777	0.817		
Attitude (4)	0.865	0.893	0.813	

Once the measurement model was examined and established, the next step was to evaluate the structural model. The process involved evaluating the relationships between the variables and the structural model's predictive abilities. Hair et al. [33] suggested that before assessing the structural model, collinearity should be examined to confirm that collinearity will not affecting the results of the analysis. The results show that collinearity was not critical since all the VIF values are less than 5 for all variables included in the study [31].

The hypotheses were tested by running a bootstrapping procedure with resampling of 5000. The results show that for the main hypotheses, all of the propose hypotheses are supported. With PEU -> SAT ($\beta = 0.763$, $p < 0.001$) and PEU -> PER ($\beta = 0.702$, $p < 0.001$), thus indicating the positive effects of PEU, SAT and PER. Thus, supporting the H1 and H2 of the study. For H3 and H4, the finding shows that there are positive effects for the relationship between SAT and PER with ATT ($\beta = 0.621$, $p < 0.001$) and ($\beta = 0.260$, $p < 0.001$), hence supporting H3 and H4. All the results of the main hypotheses are illustrated in Fig. 2 and Table 3.

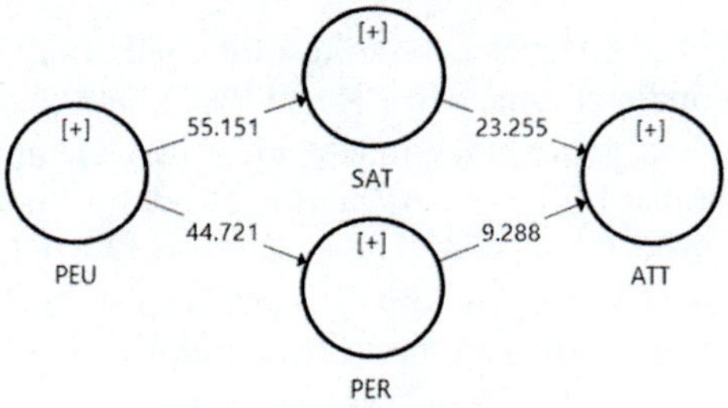

Fig. 2. Structural model

Table 3. Hypothesis testing

Hypothesis	Path co-efficient	Beta	Std error	T value	P value	LL	UL	Decision	VIF
H1	PEU -> SAT	0.763	0.014	55.151	0.000	0.739	0.785	Supported	1.000
H2	PEU -> PER	0.702	0.016	44.721	0.000	0.674	0.726	Supported	1.000
H3	SAT -> ATT	0.621	0.027	23.255	0.000	0.577	0.665	Supported	2.369
H4	PER -> ATT	0.260	0.028	9.288	0.000	0.213	0.305	Supported	2.369

As recommended by Hair et al. [33], the study also looked at the assessment of the coefficient of determination (R^2), predictive relevance (Q^2), and effect size (f^2). The R^2 value of 0.583 for SAT, 0.492 for PER, and 0.698 for ATT indicate that PEU explains 58.3% of the variance of the SAT, and PEU explains 49.2% of the PER variance. As for SAT and PER explain 69.8% of the variance of ATT. As for predictive relevance, using the blindfolding technique, a value of $Q^2 \geq 0$, shows that the model of the study has a meaningful predictive relevance [31]. The analysis indicated that Q^2 is 0.524, 0.417, and 0.565 for the SAT, PER, and ATT, respectively, showing that the model has good predictive power.

Lastly, the f^2 was assessed. According to Cohen [40], an effect size of 0.35, is considered as large 0.15, is medium and 0.02 is small effect size. PEU was found to have a large effect size towards the SAT and PER with a value of 1.396 and 0.970, respectively. The SAT has a large effect size (with a value of 0.539,) on ATT. On the contrary, the PER has a small effect (0.094) on ATT. Table 4 illustrates the results for R^2, Q^2, and f^2 of the study.

Table 4. Effect size

Relationship	R^2	Q^2	f^2	Decision
PEU -> SAT	0.583	0.524	1.396	Large
PEU -> PER	0.492	0.417	0.970	Large
SAT -> ATT	0.698	0.65	0.539	Large
PER -> ATT			0.094	Small

Aligned with Preacher and Hayes [41], to test the mediation effect, the study used the bootstrapping of the indirect effect. If, LL and UL do not straddle a 0 in between, indicating a mediation effect between the independent variable and dependent variable [41]. The results indicate that PER has a mediation effect for the relationship between PEU -> PER -> ATT ($\beta = 0.182$, $t = 8.764$: LL = 0.143, UL 0.224, $p < 0.001$), thus supporting H5. As for H6 (PEU -> SAT -> ATT) ($\beta = 0.474$, $t = 20.754$: LL = 0.428, UL 0.518, $p < 0.001$), indicated that satisfaction has a mediation effect for the relationship between PEU and ATT. Table 5 shows the results of the mediation effect of the study.

Table 5. Mediation analysis

Hypothesis	Relationship	Beta	SE	T value	P value	LL	UL	Decision
H5	PEU -> PER -> ATT	0.182	0.021	8.764	0.000	0.143	0.224	Supported
H6	PEU -> SAT -> ATT	0.474	0.023	20.754	0.000	0.428	0.518	Supported

5 Discussion and Conclusion

The study revealed that perceived ease to use positively influences students' satisfaction (H1). The finding confirms the results reported by Al-Rahmi et al. [16]. The study also found that perceived ease to use positively impacts students' performance (H2). This result supports the finding by Davis and Wiedenbeck [17]. Thus, the result indicates that when an online learning system is easy to use, students' satisfaction level and performance will improve.

As for H3 and H4, students' satisfaction and performance were found to influence students' attitudes towards online learning. These findings align with those found by Chang et al. [20] where students' satisfaction significantly affects students' attitude in terms of internet usage and online learning. The present study also seconds Akhgari et al. [23], who highlight that satisfaction positively affects attitude. Hence, lecturers and university management need to create an excellent online learning system or environment to ensure that students are satisfied with the system; students' satisfaction with the system could increase their performance and positive attitude towards online learning.

Lastly, the results showed that students' performance and satisfaction mediated the relationship between perceived ease to use and students' attitude (H5 and H6). Students' positive attitude is crucial in determining whether students intend to continue using the online learning system. Therefore, the availability of satisfaction and the improvement of students' performance while using the online learning system can positively affect online learning. Besides that, students' satisfaction can be further enhanced with the improvement of the online learning system usability features.

The current study provides a new SOR model to explain the learning environment effect on students' attitude using online learning in the COVID-19 pandemic setting. By introducing new combination factors of the SOR model, the study enriched the capability of the SOR model to explain students' attitudes using the online learning system. As online learning has become a new norm in Malaysia's higher education because of the COVID-19 pandemic, understanding the factors that can create student's positive attitudes towards online learning is crucial. Hence, the present study significantly contributes to the body of knowledge about students' attitudes in online learning amidst the COVID-19 pandemic. From a practical perspective, it is crucial for the top management of higher institutions in Malaysia and around the globe to provide online learning systems and platforms which are easy to use. Ease of use can positively impact learners' satisfaction and performance. Learners' increased satisfaction and performance will, in turn, enhance the formation of a positive attitude towards online learning among the learners.

The present study limited the variables studied to ease of use, satisfaction, and performance of science students studying at Universiti Malaysia Terengganu. Future studies should explore other variables which may influence students' attitude towards online learning and involve students from other universities, especially the higher-ranking universities in Malaysia.

Acknowledgements. The authors would like to thank Mohd Hafriz Nural Azhan for his support and encouragement throughout this research. The authors also would like to thank the top management, all staff, and students of Universiti Malaysia Terengganu for their full support and contribution to this research. This research was supported by Centre for Talent Development and Innovation, and Centre for Research and Innovation Management, Universiti Malaysia Terengganu (Vot No. 53372).

References

1. Smart, K.L., Cappell, J.J.: Students' perceptions of online learning: a comparative study. J. Inf. Technol. Educ. **5**, 201–219 (2006)
2. Noor Azizah, Y., Khairul Rijal, M., Nuriyatur Rohmah, U., Adymas Pranajaya, S., Ngiu, Z., Mufid, A., HaliahMàu, D.: Transformational or transactional leadership style: which affects work satisfaction and performance of islamic university lecturers during COVID-19 pandemic? Syst. Rev. Pharm. **11**(7), 577–588 (2020)
3. Basuony, M.A.K., EmadEldeen, R., Farghaly, M., El-Bassiouny, N., Mohamed, E.K.A.: The factors affecting student satisfaction with online education during the COVID-19 pandemic: an empirical study of an emerging Muslim country. J. Islamic Market. (2020). https://doi.org/10.1108/JIMA-09-2020-0301

4. Al-Emran, M., Arpaci, I., Salloum, S.A.: An empirical examination of continuous intention to use m-learning: an integrated model. Educ. Inf. Technol. **25**(4), 2899–2918 (2020). https://doi.org/10.1007/s10639-019-10094-2
5. Alshurideh, M., Al Kurdi, B., Salloum, S.A., Arpaci, I., Al-Emran, M.: Predicting the actual use of m-learning systems: a comparative approach using PLS-SEM and machine learning algorithms. Interact. Learn. Environ. 1–152020https://doi.org/10.1080/10494820.2020.1826982
6. Mehrabian, A., Russell, J.A.: An Approach to Environmental Psychology. MIT Press, Cambridge (1974)
7. Bagozzi, R.P.: Principles of Marketing Management. Science Research Associates, New York City (1986)
8. Yang, S., Zhou, S., Cheng, X.: Why do college students continue to use mobile learning? Learning involvement and self-determination theory. Br. J. Edu. Technol. **50**(2), 626–637 (2019). https://doi.org/10.1111/bjet.12634
9. Ajzen, I.: The theory of planned behavior. Organ. Behav. Hum. Decis. Process. 179–2111991). https://doi.org/10.4135/9781446249215.n22
10. Zhu, Y., Zhang, J. H., Au, W., Yates, G.: University students' online learning attitudes and continuous intention to undertake online courses: a self-regulated learning perspective. Educ. Technol. Res. Dev. **68**, 1485–1519 (2020). https://doi.org/10.1007/s11423-020-09753-w
11. Taghizadeh, M., Hajhosseini, F: Investigating a blended learning environment: contribution of attitude, interaction, and quality of teaching to satisfaction of graduate students of TEFL Asia-Pac. Educ. Res. (2020).https://doi.org/10.1007/s40299-020-00531-z
12. Laguador, J.M., Dotong, C.I.: Engineering students' challenging learning experiences and their changing attitude towards academic performance. Eur. J. Educ. Res. **9**(3), 1127–1140 (2020). https://doi.org/10.12973/EU-JER.9.3.1127
13. Davis, F.D.: Perceived usefulness, perceived ease of use, and user acceptance of information technology. MIS Q.: Manag. Inf. Syst. **13**(3), 319–339 (1989). https://doi.org/10.2307/249008
14. Rouibah, K., Abbas, H., Rouibah, S.: Factors affecting camera mobile phone adoption before e-shopping in the Arab world. Technol. Soc. **33**(3–4), 271–283 (2011). https://doi.org/10.1016/j.techsoc.2011.10.001
15. Morosan, C.: Theoretical and empirical considerations of guests' perceptions of biometric systems in hotels: extending the technology acceptance model. J. Hosp. Tour. Res. **36**(1), 52–84 (2012). https://doi.org/10.1177/1096348010380601
16. Al-Rahmi, W.M., Alias, N., Othman, M.S., Marin, V.I., Tur, G.: A model of factors affecting learning performance through the use of social media in Malaysian higher education. Comput. Educ. **121**(February), 59–72 (2018). https://doi.org/10.1016/j.compedu.2018.02.010
17. Davis, S., Wiedenbeck, S.: The mediating effects of intrinsic motivation, ease of use and usefulness perceptions on performance in first-time and subsequent computer users. Interact. Comput. **13**(5), 549–580 (2001). https://doi.org/10.1016/S0953-5438(01)00034-0
18. Lo, C.C.: How student satisfaction factors affect perceived learning. J. Scholarsh. Teach. Learn. **10**(1), 47–54 (2010)
19. Efiloğlu Kurt, Ö.: Examining an e-learning system through the lens of the information systems success model: empirical evidence from Italy. Educ. Inf. Technol. **24**(2), 1173–1184 (2019). https://doi.org/10.1007/s10639-018-9821-4
20. Chang, Y.C., Tsai, C.L., Chiu, W.Y.: The influence of life satisfaction and well- being on attitude toward the internet, motivation for internet usage and internet usage behavior. J. Interdiscip. Math. **18**(6), 927–946 (2015). https://doi.org/10.1080/09720502.2015.1108111
21. Chang, C.S., Liu, E.Z.F., Sung, H.Y., Lin, C.H., Chen, N.S., Cheng, S.S.: Effects of online college student's Internet self-efficacy on learning motivation and performance. Innov. Educ. Teach. Int. **51**(4), 366–377 (2014). https://doi.org/10.1080/14703297.2013.771429

22. Bandura, A.: Self-Efficacy in Changing Societies. Cambridge University Press, New York (1997)
23. Akhgari, M., Bruning, E.R., Finlay, J., Bruning, N.S.: Image, performance, attitudes, trust, and loyalty in financial services. Int. J. Bank Mark. **36**(4), 744–763 (2018). https://doi.org/10.1108/IJBM-06-2017-0118
24. Ngah, A.H., Gabarre, S., Eneizan, B., Asri, N.: Mediated and moderated model of the willingness to pay for halal transportation. J. Islamic Mark. (2020). https://doi.org/10.1108/JIMA-10-2019-0199
25. Weng, F., Yang, R.-J., Ho, H.-J., Su, H.-M.: A TAM-based study of the attitude towards use intention of multimedia among school teachers. Appl. Syst. Innov. **1**(3), 36 (2018). https://doi.org/10.3390/asi1030036
26. Abdullah, F., Ward, R., Ahmed, E.: Investigating the influence of the most commonly used external variables of TAM on students' Perceived Ease of Use (PEOU) and Perceived Usefulness (PU) of e-portfolios. Comput. Hum. Behav. **63**, 75–90 (2016)
27. Isaac, O., Aldholay, A., Abdullah, Z., Ramayah, T.: Online learning usage within Yemeni higher education: the role of compatibility and task-technology fit as mediating variables in the IS success model. Comput. Educ. **136**(2019), 113–129 (2019)
28. Damnjanovic, V., Jednak, S., Mijatovic, I.: Factors affecting the effectiveness and use of Moodle: students' perception. Interact. Learn. Environ. **23**(4), 496–514 (2013)
29. Salloum, S.A., Alhamad, A.Q.M., Al-Emran, M., Monem, A.A., Shaalan, K.: exploring students' acceptance of e-learning through the development of a comprehensive technology acceptance model. IEEE Access **7**, 128445–128462 (2019)
30. Hulland, J., Baumgartner, H., Smith, K.M.: Marketing survey research best practices: evidence and recommendations from a review of JAMS articles J. Acad. Mark. Sci. 1–17 (2017). https://doi.org/10.1007/s11747-017-0532-y
31. Hair, J.C.L., Hollingsworth, A.B., Randolph, A.C.: An updated and expended assessment of PLS SEM in information systems research. Ind. Manag. Data Syst. **117**(3), 1–41 (2017)
32. Gefen, D., Rigdon, E., Straub, D.: An update and extension to SEM guidelines for administrative and social science research. MIS Q. **35**(2). iii-A72011https://doi.org/10.1016/j.lrp.2013.01.001
33. Hair, J.F., Risher, J.J., Sarstedt, M., Ringle, C.M.: When to use and how to report the results of PLS-SEM. Eur. Bus. Rev. **31**(1), 2–24 (2019). https://doi.org/10.1108/EBR-11-2018-0203
34. MacKenzie, S.B., Podsakoff, P.M.: Common method bias in marketing: causes, mechanisms, and procedural remedies. J. Retail. **88**(4), 542–555 (2012). https://doi.org/10.1016/j.jretai.2012.08.001
35. Ngah, A.H., Ramayah, T., Ali, M.H., Khan, M.I.: Halal transportation adoption among pharmaceuticals and comestics manufacturers. J. Islamic Mark. **11**(6), 1619–1639 (2019). https://doi.org/10.1108/JIMA-10-2018-0193
36. Ngah, A.H., Thurasamy, R., Aziz, N.A., Ali, M.H., Khan, M.I.: Modelling the adoption of halal warehousing services among halal pharmaceutical and cosmetic manufacturers. J. Sustain. Sci. Manage. **14**(6), 103–116 (2019)
37. Rönkkö, M., Ylitalo, J.: PLS marker variable approach to diagnosing and controlling for method variance. In: Proceedings of International Conference on Information Systems Shanghai, China (2011)
38. Ngah, A.H., Zainuddin, Y., Thurasamy, R.: Contributing factors of Halal warehouse adoption. In: Management and Technology in Knowledge, Service, Tourism & Hospitality, pp. 89–94 (2014)
39. Franke, G., Sarstedt, M.: Heuristics versus statistics in discriminant validity testing: a comparison of four procedures Internet Res. 431–447 (2019). https://doi.org/10.1108/IntR-12-2017-0515

40. Cohen, J.: Statistical Power Analysis for the Behavioral Sciences. Laurence Erlbaum and Associates, Hillsdale (1988). https://doi.org/10.4324/9780203771587
41. Preacher, K.J., Hayes, A.F.: Asymptotic and resampling strategies for assessing and comparing indirect effects in multiple mediator models. Behav. Res. Methods **40**(3), 879–891 (2008). https://doi.org/10.3758/BRM.40.3.879

Blockchain Adoption Intention in Higher Education: Role of Trust, Perceived Security and Privacy in Technology Adoption Model

Nishant Kumar[1](✉), Mitushi Singh[2], Kamal Upreti[3], and Divya Mohan[4]

[1] Amity School of Business, Amity University, Noida, India
nkumar23@amity.edu
[2] Jaypee Business School, Jaypee Institute of Information Technology, Noida, India
mitushi.singh@jiit.ac.in
[3] Dr. Akhilesh Das Gupta Institute of Technology & Management, Shastri Park, New Delhi 110053, India
kamal.upreti@adgitmdelhi.ac.in
[4] Delhi Institute of Advanced Studies, Rohini, Delhi, India
divyasahni2002@dias.ac.in

Abstract. Blockchain technology with the potential to disrupt established business processes has emerged with a different student and institution-centric applications for higher education. This study aims to investigate educated individual beliefs and behavior toward blockchain adoption in higher education. The study proposes to extend Technology Adoption Model (TAM) with the inclusion of additional variables: perceived security and privacy and trust. Scale validated from the literature has been adopted for the study. Based on the survey of 208 respondents variance-based partial least square structural equation modeling was employed to examine the hypothesized relationship in the model. Findings of the study revealed that the factors incorporated in the model have a significant positive influence on individual adoption intention for blockchain in higher education. Perceived security and privacy has emerged as powerful factor affecting perceived usefulness, ease of use, and trust towards blockchain technology. Study respondents believe that they feel secured and confident with the blockchain implication in higher education.

Keywords: Blockchain · Higher education · Technology adoption model · Perceived security and privacy · Trust · Structure equation modeling

1 Introduction

The advent of technology and the increase in usage of electronic gadgets across the world has changed human lives. Technology advancement has developed the entire

The original version of this chapter was revised: the removal of old affiliation of the author Kamal Upreti. The correction to this chapter is available at https://doi.org/10.1007/978-3-030-82616-1_58

M. Al-Emran et al. (Eds.): ICETIS 2021, LNNS 299, pp. 303–313, 2022.
https://doi.org/10.1007/978-3-030-82616-1_27

market space and significantly influenced the economy as well. Many technology applications evolved frequently but very few of them get survived because of their future implications. Over the years blockchain has emerged as a technology for the future. The applications of blockchain technology have been exploited in various fields such as financial services, manufacturing, food, agriculture, pharmaceutical, hotels, airline, healthcare, and government, including the supply chain. Blockchain promotes traceability, visibility, authenticity, legitimacy, aggregation, automation, resiliency, accurate and timely decisions, reduced cost, improving supply chain performance, and customer satisfaction [1]. In the light of the progressive technological environment, higher education has adopted new paradigms like flipped classroom teaching, blended learning, and other advancements to enrich the teaching-learning experience. The dynamic nature of blockchain provides its assistance to higher education institutions in terms of certification, authentication of records, recruitment, maintenance, and accessibility of records [2]. Furthermore, it opens the door for addressing some current issues in higher education like counterfeiting, security, and trust. The dispersion of blockchain technology is still in the initial stage so a structured review of recent rationale on technological innovation, its potential, and efficacy are likely to support both academics and industry people. Blockchain provides the infrastructure for creating own learning pathways and support to the education field. Studies are highlighting the techniques to revolutionaries the education field but limited literature showcases the relevance of blockchain adoption intention of individuals. Individual actions toward such advanced technology are imperative to be observed to evaluate its scalability. The paucity of research studies focusing on individual adoption intention provides a strong foundation for this research. To bridge the literature gap, this study attempts to investigate educated individual adoption behavior towards blockchain technology through the Technology Adoption Model (TAM).

This paper is organized into six sections. Section one covers the introduction of the study. The second section reviews the blockchain application in higher education followed by the conceptual framework developed in section three. The survey instrument and study methodology are described in the fourth section. Data analysis and empirical findings of the study are covered under section five and the study is concluded with future research directions under section six.

2 Blockchain in Higher Education

Blockchain has evolved as a disruptive technology with the potential to transform the established business processes. Blockchain is showing steady progress in integrating itself with various domains like healthcare, logistics, banking, energy, manufacturing, retail, life science, etc. The field of higher education is also not untouched by the application of blockchain. It has been highlighted that blockchain has the likelihood to revolutionaries' higher education through the authenticity of documents, transparency, immutability, and trust [3]. Some of the potential application areas of blockchain in higher education extracted from literature are detailed in Table 1.

Table 1. Blockchain application

Blockchain implication	Framework
Transcripts & record	Counterfeiting protection as well as secure access and management of certificates through bitcoin chain [4] The blockchain-based solution to save the academic record and automatically issue degree after course completion [5] Digital certificates based on blockchain to avoid the problem of counterfeiting certificates [6] Blockchain is a safe, fast, and cost-efficient way to structure record and data [7]
Transaction	Preventing fraud by introducing high data transparency using blockchain [8] Blockchain-based assessment of learning outcomes [9] Blockchain may be used to achieve sustainability and transparency [10]
Libraries	Using blockchain to create a time-stamped and verifiable version of journal article [11] Blockchain to provide a unique identity and ease the transferability of digital documents [12]
Human resource	Student achievement can be streamlined as per the interest of the employer [13] Blockchain-based solution for Internationalization of Higher Education Institutions [14] Data encryption through blockchain and accessibility only to authorized personals to prevent data leakage [15]
Digital badges	Application of blockchain to earn badges after reaching certain academic milestone [16] Digital badge system with blockchain support [17] Learners possess badge and achievement visualization space called passport to represent their skills in front of the employer [16]

3 Conceptual Framework

Examining consumer adoption intention towards blockchain in higher education is the concern of the study. Technology Adoption Model (TAM) has been deployed in study context to measure the same. TAM was derived from the Theory of Reasoned Action (TRA) that explains "people behavior is specified by their intention to carry out their behavior" [18, 19]. Perceived usefulness and perceived ease of use were introduced by Davis in TAM. Perceived usefulness was explained as "the degree to which a person believes that using a particular system would enhance his or her job performance" [20]. Furthermore, it was also entrusted that decision to use new technology also depends on the extent to which individuals believe that it would enhance the effectiveness in work [20]. Perceived ease of use was defined as "the degree to which individual believes that the use of new technology requires no effort" [20]. Easy-to-use technologies are less threatening for individuals and develop a favorable attitude towards their use [20, 21]. The attitude was explained as "positive or negative feeling towards the use of technology" [20]. Individual's attitude was proved as a precursor for actual behavior and implementation of new technology is self-sufficient for change in individual attitude toward its use, which could further lead to change an entire system [22].

TAM was proved as a good model to predict adoption intention for new technologies but in some environmental contexts, it needs to be customized as well [23]. In continuation to the same, TAM was extended through perceived security and privacy, and trust. The importance of security and privacy has been widely discussed across studies and proved as a barrier to the adoption of new technologies [24]. Security was considered as a threat to the choice or implementation of new technology. Perceived security and privacy explain the individual believes that the access, use, and disclosure of personal information will be maintained confidentially [25]. This develops a belief and trust in the system and prior studies confirm significant positive influence of Perceived security and privacy on trust [26]. Trust was also found as a strong predictor for individual attitude [27].

Based on the above discussion below mentioned framework was proposed to test the hypothesized relationship (Fig. 1):

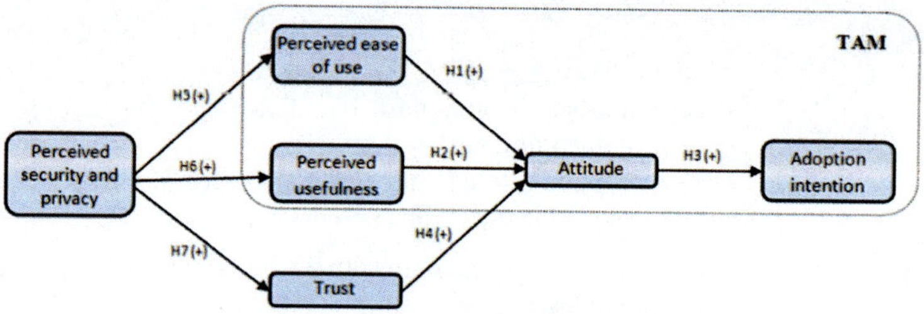

Fig. 1. Conceptual framework

H1: PEU has a significant positive influence on individual ATT towards the adoption of blockchain technology in higher education.
H2: PU of use has a significant positive influence on individual ATT towards the adoption of blockchain technology in higher education.
H3: ATT has a significant positive influence on AI towards blockchain technology in higher education.
H4: TRT has a significant positive influence on individual ATT towards the adoption of blockchain technology in higher education.
H5: PSP has a significant positive influence on PEU for blockchain technology in higher education.
H6: PSP has a significant positive influence on PU for blockchain technology in higher education.
H7: PSP has a significant positive influence on TRT.

4 Methodology

The scale validated from the literature has been adopted for the study. The latent construct perceived security and privacy (PSP) and trust (TRT) has been used in addition to the existing TAM constructs; perceived ease of use (PEU), perceived usefulness (PU), attitude (ATT), and adoption intention (AI) towards blockchain to extend the existing TAM model in the study context. Scale items for attitude were measured on a semantic differential scale (extremely bad-1 and extremely good-7) and other scale items were anchored on a seven-point Likert scale (1-strongly agree to 7-strongly disagree) to record the responses. Descriptions of latent constructs are detailed in Table 2.

A questionnaire was designed as a survey instrument to target respondents within the age group of 18 years and above. Students and professionals were targeted because of the technical nature of the research study. The educated individual can simplify and better comprehend the technical information [36]. The questionnaire administered in the Delhi-National Capital Region (India) consisted of 24 statements. Respondents were assured that their response is going to be used for academic research. Convenience sampling was used to select the sample from the population. The research instrument was circulated through various social media platforms to reach respondents. A total of 410 questionnaires were circulated and 208 complete in all aspects yielding a response 51% rate, was retained for analysis. As per the demographic profile of respondents represented in Table 3 out of the total respondents, 122 (58.65%) were male and 86(41.34%) were female. The majority of respondents were postgraduates (43%) followed by graduates (37%) and others. Data descriptive also confirms that 55% of respondents were students and 45% were professionals i.e., academicians at various levels and IT personals. The common method of bias was addressed through a full collinearity test and VIF for all latent construct was found to be below 5 [37].

5 Result and Discussion

Data analysis was performed using SPSS 21.0 and Smart-PLS 2.0. Data descriptive was ascertained through SPSS 21.0. Smart-PLS 2.0 was employed for structural equation modeling in two-step: measurement model assessment, and structural model assessment [38].

5.1 Measurement Model Assessment

The reliability and validity of each item should be measured before examining the structural relationship among constructs. Internal consistency among items was measured through composite reliability (CR). As represented in Table 4, the identified CR value for AI, ATT, PEU, PSP, PU, and TRT was found to be above 0.7 and considered reliable [39]. The AVE for all latent constructs AI, ATT, PEU, PSP, PU, and TRT was well above the defined threshold of 0.5 and $\sqrt{AVE}$ represented in the diagonal exceeds the off-diagonal values in the correlation matrix provides evidence for convergent and discriminant validity [40, 41]. As represented in Fig. 2 all item loadings were more than 0.7 and establish indicator reliability [42].

Table 2. Description of latent construct

Construct	Scale	Reference
Attitude (ATT)	ATT1:Using blockchain technology is extremely bad (1) / extremely good (7); ATT2:Using blockchain technology is extremely undesirable (1) / extremely desirable (7); ATT3:Using blockchain technology is extremely unenjoyable (1) / extremely enjoyable (7); ATT4:Using blockchain technology is extremely unfavorable (1) / extremely favorable (7)	[28]
Perceived ease of use (PEU)	PEU1: Learning a blockchain-based system would be easy for me; PEU2: I feel blockchain based system would be easy to implement; PEU3: It is easy to be skillful in blockchain-based system; PEU4: Blockchain based system is easy to use	[29]
Perceived usefulness (PU)	PU1: Blockchain-based system would enhance my performance; PU2: Blockchain-based system would increase my productivity; PU3: Blockchain-based system would enhance my effectiveness; PU4: Blockchain-based system would reduce my processing time	[29, 30]
Adoption intention (AI)	AI1: I intend to use a blockchain-based system; AI2:I would use a blockchain-based system; AI3: I would plan to use a blockchain system; AI4: I would put the required effort to use a blockchain system	[31, 32]
Perceived security and privacy (PSP)	PSP1: Use of the blockchain-based system is technically secure; PSP2: Use of the blockchain-based system is robust in terms of application; PSP3: I have confidence in the system; PSP4: I believe the system for the confidentiality of data	[33]
Trust (TRT)	TRT1:It is trustworthy; TRT2: It gives an impression of promise and commitment; TRT3: It keeps my interest in consideration; TRT4: This system can be used in long run	[34, 35]

Table 3. Demographic profile

	Number	Percentage (%)
Gender		
Male	122	58.65%
Female	86	41.34%
Education Qualification		
Graduate	77	37.01%
Post Graduate	89	42.78%
Others	42	20.19%
Occupation		
Student	114	54.80%
Academicians	48	23.07%
IT Professionals	46	22.11%
TOTAL	**208**	**100%**

Table 4. Reliability and validity result

	AI	ATT	PEU	PSP	PU	TRT
AI	***0.891***					
ATT	0.505	***0.839***				
PEU	0.498	0.490	***0.733***			
PSP	0.480	0.375	0.470	***0.795***		
PU	0.468	0.603	0.490	0.338	***0.752***	
TRT	0.534	0.595	0.495	0.582	0.583	***0.838***
CR	*0.920*	*0.876*	*0.777*	*0.838*	*0.796*	*0.904*
AVE	*0.795*	*0.704*	*0.538*	*0.633*	*0.566*	*0.703*

5.2 Structural Model Assessment

Non-parametric bootstrapping was employed to examine the hypothesized relationship and the result for the same is represented in Fig. 2. A significant path coefficient is one of the primary measures to evaluate the causal relationship in the structural model. The finding of the study provides sufficient empirical support for the proposed relationship. The core TAM variable perceived ease of use and perceived usefulness significantly influence individual attitude towards blockchain technology and support H1 ($\beta = 0.168$, $t = 5.325$) and H2 ($\beta = 0.337$, $t = 9.778$). The attitude was found to be a strong predictor for individual adoption intention for blockchain in higher education; H3 ($\beta = 0.505$, $t =$

18.269). Additionally, trust and perceived security and privacy were included to extend and better predict individual adoption intention for blockchain in the higher education system. Trust has a significant positive influence on individual attitude; H4 ($\beta = 0.316$, $t = 11.333$). Moreover, perceived security and privacy have a strong positive influence on trust followed by perceived ease of use and perceived usefulness; H7 ($\beta = 0.582$, $t = 22.839$), H5 ($\beta = 0.470$, $t = 13.473$) and H6 ($\beta = 0.338$, $t = 8.295$). R^2 explains the predictive strength of the model and measures the amount of variance in the dependent variable by the independent variable.

R^2 was assessed against the suggested range of 0.19, 0.33, and 0.67 indicating weak, moderate, and substantial effects, respectively [43]. Predictors included in the structural model explain a moderate influence on individual attitude ($R^2 = 47.3\%$) and blockchain adoption intention in higher education ($R^2 = 35.5\%$). Furthermore, the support of the hypothesis indicates that a significant effect of predictor variables on the outcome variable. The three independent variables PEU, PU, and TRT are positively related to consumer attitude toward blockchain application in higher education and represent user aspect and belief. Consumer attitude is one of the key behavioral determinants to influence behavioral action. The three predictors vary in terms of their predictive strength for attitude. With high trust and faith, convenience and usability, the innovation can develop the individual assumption to act emphatically toward blockchain-based solutions [44]. The contrary relationship between trust and risk highlights that the perceived risk for the individual is increased when trust is decreased. Trust is one of the key elements to support the confidence in the innovation to believe how proficiently this innovation can be utilized with less exertion. PSP is the primary component in this model to portray the intricacy level of blockchain innovation and application.

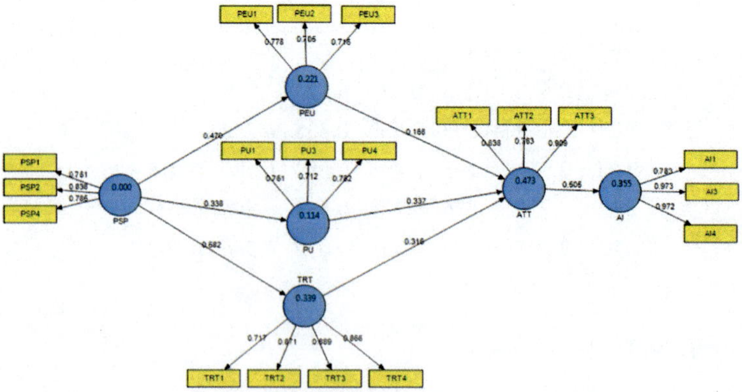

Fig. 2. Structural model

6 Conclusion

The technology Adoption Model (TAM) has been differently applied in many studies to examine individual adoption intention towards new technologies [45–48]. As of now, blockchain has shown a rapid expansion across the globe with a variety of options for a secure and safe operational environment. However, the technology was not welcomed as per the expectations and the real use is still extremely low. Sensing the limited application of blockchain this paper aimed to explore its adoption intention in higher education. Considering the ascent of innovation the proposed extended model opens a viewpoint on blockchain innovation reception. Findings of the study revealed that the factors incorporated in the model have a significant positive influence on individual adoption intention for blockchain in higher education. Perceived security and privacy have emerged as powerful factors affecting perceived usefulness, ease of use, and trust towards blockchain technology. Study respondents believe that they feel secured and confident with the blockchain implication in higher education. Moreover, the upgraded security with blockchain is basic for the achievement, urging the establishment to put resources into creating a safe climate, and reliable frameworks and stages that shield clients' delicate information from unauthorized modification, deletion, or fabrication. The adoption of blockchain in education is possible with great management support. This can further enhance the collaboration among the institutional members. Education institutions should frame guidelines for the effective implementation of innovative technologies. The government should frame regulations to promote and aware citizens regarding the benefit of blockchain implementation in the education domain. Technology implementation would diffuse as their awareness and its effectiveness through implementation gets increased among people.

This research makes a few key contributions. Firstly, the study extends TAM to examine the attitude-behavior gap while opting for blockchain-based applications in higher education. Secondly, TAM has been validated in past to understand the technology adoption behavior in different environmental contexts [45], but the study highlights the significant role of perceived security privacy and trust for blockchain adoption in higher education. Furthermore, the research can be extended by examining the relationship among personality traits, social, economic, cultural factors, and blockchain adoption.

References

1. Babich, V., Hilary, G.: Blockchain and Other Distributed ledger technologies in operations. SSRN 3232977, 19 Nov 2018
2. Rooksby, J., Dimitrov, K.: Trustless education? A blockchain system for university grades1. Ubiquity: J. Pervasive Media **6**(1), 83–8, 1 Nov 2019
3. Grech, A., Camilleri, A.F.: Blockchain in Education. JRC Science for Policy Report, European Commission (2017)
4. Gräther, W., Kolvenbach, S., Ruland, R., Schütte, J., Torres, C., Wendland, F.: Blockchain for education: lifelong learning passport. In: Proceedings of 1st ERCIM Blockchain Workshop 2018. European Society for Socially Embedded Technologies (EUSSET).
5. Palma, L.M., Vigil, M.A., Pereira, F.L., Martina, J.E.: Blockchain and smart contracts for higher education registry in Brazil. Int. J. Netw. Manage. **29**(3), e2061, May 2019
6. Cheng, J.C., Lee, N.Y., Chi, C., Chen, Y.H.: Blockchain and smart contract for digital certificate. In: 2018 IEEE International Conference on Applied System Invention (ICASI), pp. 1046–1051. IEEE, 13 Apr 2018
7. Hameed, B., et al.: A Review of Blockchain based Educational Projects. Framework **28**, 30 (2019)

8. Mauri, R.: Blockchain for fraud prevention: Industry use cases (2017)
9. Duan, B., Zhong, Y., Liu, D.: Education application of blockchain technology: learning outcome and meta-diploma. In: 2017 IEEE 23rd International Conference on Parallel and Distributed Systems (ICPADS), pp. 814–817. IEEE, 15 Dec 2017
10. Thomason, J., et al.: Blockchain—powering and empowering the poor in developing countries. In: Transforming Climate Finance and Green Investment with Blockchains, pp. 137–152. Academic Press, 1 Jan 2018
11. Irving, G., Holden, J.: How blockchain-timestamped protocols could improve the trustworthiness of medical science. F1000Research, 5 (2016)
12. Griffey, J.: Blockchain & Intellectual Property—Internet Librarian 2016, 18 Oct 2016
13. Thayer, T.L.: Promising and Ambitious Blockchain Initiatives for Higher Education (2018)
14. Kamišalić, A., Turkanović, M., Mrdović, S., Heričko, M.: A preliminary review of blockchain-based solutions in higher education. In: Uden, L., Liberona, D., Sanchez, G., Rodríguez-González, S. (eds.) LTEC 2019. CCIS, vol. 1011, pp. 114–124. Springer, Cham (2019). https://doi.org/10.1007/978-3-030-20798-4_11
15. Mori, K., Miwa, H.: Digital University Admission Application System with Study Documents Using Smart Contracts on Blockchain. In: Barolli, L., Nishino, H., Miwa, H. (eds.) INCoS 2019. AISC, vol. 1035, pp. 172–180. Springer, Cham (2020). https://doi.org/10.1007/978-3-030-29035-1_17
16. Mikroyannidis, A., Domingue, J., Bachler, M., Quick, K.: A learner-centred approach for lifelong learning powered by the blockchain. In: EdMedia+ Innovate Learning, pp. 1388–1393. Association for the Advancement of Computing in Education (AACE), 25 Jun 2018
17. Choi, M., Kiran, S.R., Oh, S.C., Kwon, O.Y.: Blockchain-based badge award with existence proof. Appl. Sci. **9**(12), 2473 (2019)
18. Davis, F.D., Bagozzi, R.P., Warshaw, P.R.: User acceptance of computer technology: a comparison of two theoretical models. Manage. Sci. **35**(8), 982–1003 (1989)
19. Azjen, I.: Understanding Attitudes and Predicting Social Behavior. Englewood Cliffs (1980)
20. Davis, F.D.: Perceived usefulness, perceived ease of use, and user acceptance of information technology. MIS. Q. **1**, 319–340 (1989)
21. Moon, J.W., Kim, Y.G.: Extending the TAM for a World-Wide-Web context. Inf. Manage. **38**(4), 217–230 (2001)
22. Rice, R.E., Aydin, C.: Attitudes toward new organizational technology: network proximity as a mechanism for social information processing. Adm. Sci. Q. **1**, 219–244 (1991)
23. Mezhuyev, V., Al-Emran, M., Fatehah, M., Hong, N.C.: Factors affecting the meta modelling acceptance: a case study from software development companies in Malaysia. IEEE Access. **10**(6), 49476–49485 (2018)
24. Aladwani, A.M.: Online banking: a field study of drivers, development challenges, and expectations. Int. J. Inf. Manage. **21**(3), 213–225 (2001)
25. Amin, H., Ramayah, T.: SMS banking: explaining the effects of attitude, social norms and perceived security and privacy. Electron. J. Inf. Syst. Developing Countries **41**(1), 1–5 (2010)
26. Roca, J.C., García, J.J., De La Vega, J.J.: The importance of perceived trust, security and privacy in online trading systems. Inf. Manage. Comput. Secur. 5 Jun 2009
27. Wu, L., Chen, J.L.: An extension of trust and TAM model with TPB in the initial adoption of on-line tax: an empirical study. Int. J. Hum. Comput. Stud. **62**(6), 784–808 (2005)
28. Kim, Y., Han, H.: Intention to pay conventional-hotel prices at a green hotel–a modification of the theory of planned behavior. J. Sustain. Tour. **18**(8), 997–1014 (2010)
29. Venkatesh, V., Thong, J.Y., Xu, X.: Consumer acceptance and use of information technology: extending the unified theory of acceptance and use of technology. MIS. Q. **1**, 157–178 (2012)
30. Taylor, S., Todd, P.A.: Understanding information technology usage: a test of competing models. Inf. Syst. Res. **6**(2), 144–176 (1995)

31. Chow, W.S., Chen, Y.: Intended belief and actual behavior in green computing in Hong Kong. J. Comput. Inf. Syst. **50**(2), 136–141 (2009)
32. Paul, J., Modi, A., Patel, J.: Predicting green product consumption using theory of planned behavior and reasoned action. J. Retail. Consum. Serv. **1**(29), 123–134 (2016)
33. Pikkarainen, T., Pikkarainen, K., Karjaluoto, H., Pahnila, S.: Consumer acceptance of online banking: an extension of the technology acceptance model. Internet Res. 1 Jul 2004
34. McCloskey, D.W.: The importance of ease of use, usefulness, and trust to online consumers: an examination of the technology acceptance model with older customers. J. Organ. End User Comput. (JOEUC) **18**(3), 47–65 (2006)
35. Jarvenpaa, S.L., Tractinsky, N., Vitale, M.: Consumer trust in an Internet store. Inf. Technol. Manage. **1**(1–2), 45–71 (2000)
36. Chan, R.Y.: Determinants of Chinese consumers' green purchase behavior. Psychol. Mark. **18**(4), 389–413 (2001)
37. Kock, N., Lynn, G.: Lateral collinearity and misleading results in variance-based SEM: an illustration and recommendations. J. Assoc. Inf. Syst. **13**(7), 26 Sep 2012
38. Anderson, J.C., Gerbing, D.W.: Structural equation modeling in practice: a review and recommended two-step approach. Psychol. Bull. **103**(3), 411 (1988)
39. Nunnally, J.C.: Psychometric theory 3E. Tata McGraw-hill education (1994)
40. Fornell, C., Larcker, D.F.: Evaluating structural equation models with unobservable variables and measurement error. J. Mark. Res. **18**(1), 39–50 (1981)
41. Chiu, C.M., Wang, E.T.: Understanding Web-based learning continuance intention: the role of subjective task value. Inf. Manage. **45**(3), 194–201 (2008)
42. Hair, J.F., Black, W.C., Babin, B.J., Anderson, R.E., Tatham, R.L.: Multivariate data analysis. Upper Saddle River, NJ, Prentice hall, 23 Mar 1998
43. Chin, W.W.: The partial least squares approach to structural equation modeling. Mod. Methods Bus. Res. **295**(2), 295–336 (1998)
44. Albayati, H., Kim, S.K., Rho, J.J.: Accepting financial transactions using blockchain technology and cryptocurrency: a customer perspective approach. Technol. Soc. **62**, 101320, 1 Aug 2020
45. Arpaci, I.: Understanding and predicting students' intention to use mobile cloud storage services. Comput. Hum. Behav. **1**(58), 150–157 (2016)
46. Arpaci, I., Yardimci Cetin, Y., Turetken, O.: Impact of perceived security on organizational adoption of smartphones. Cyberpsychol. Behav. Soc. Netw. **18**(10), 602–608 (2015)
47. Arpaci, I., Kilicer, K., Bardakci, S.: Effects of security and privacy concerns on educational use of cloud services. Comput. Hum. Behav. **1**(45), 93–98 (2015)
48. Venkatesh, V., Davis, F.D.: A theoretical extension of the technology acceptance model: four longitudinal field studies. Manage. Sci. **46**(2), 186–204 (2000)

Perceptions About English as Second Language Teachers' Technology Based English Language Teaching in Pakistan: Attitudes, Uses of Technology and Challenges

Wajahat Taj Abbasi(✉), Ainol Haryati Ibrahim, and Fatima Binti Ali

Faculty of Modern Language Centre, Universiti Malaysia Pahang, 26000 Pekan, Pahang, Malaysia
{ainol,tym}@ump.edu.my

Abstract. This study aims to find out the perceptions of English as a second language (henceforth, ESL) teachers' attitudes towards the use of technology for teaching of English to the undergraduate students in Pakistan. The study exposes the uses of information and communication technologies (henceforth, ICT) in teaching, and the difficulties faced in integrating technology for teaching of English. Data is collected through survey questionnaire and interview protocol from ESL teachers teaching at college level in Abbottabad, Pakistan. Survey data is analyzed through SPSS 22 while interview questions have a thematic analysis. Findings of survey and interview protocol used in the study reveal positive attitudes on the part of teachers towards the use of technology. No significant difference is found among teachers' use of technology due to gender or academic qualification. However, the difference between teachers' technology use can be ascribed to their competence. The use of technology is beneficial for teaching and learning environment, but its' use is not optimal in educational environment in Pakistan. The findings highlight limited use of technology due to shortage of resources, unavailability of tools and lack of training. The study puts forward some recommendations for the effective use of technology in language teaching and the removal of the potential barriers in its use by teachers in Pakistan.

Keywords: Attitudes · ICT · Barriers

1 Introduction

Changes in contemporary teaching and its' methodologies urge teachers to realize the importance of integration of technology in educational environment in Pakistan. Integration of technology is also a much focused area because of the needs of students at present [1]. It is very important to have an acquaintance with the use of technology for teaching and learning environments because of the demands of the contemporary era [2]. Modern technologies are a part of the norms of developed countries yet they are still a neglected area by teachers in Pakistan due to some factors [3].

M. Al-Emran et al. (Eds.): ICETIS 2021, LNNS 299, pp. 314–325, 2022.
https://doi.org/10.1007/978-3-030-82616-1_28

Use of technology has affected every field of life including language education [4]. Studies show that the use of technology in teaching has played an important role in the development of students' autonomy, motivation and language proficiency [4, 5]. It is not only helpful for teaching and learning inside the classrooms, but it also assists language learners to learn language outside the classroom. Web technology provides material such as videos and audios in this respect without any charges.

Teachers' attitudes are very significant in integrating educational technology [6]. The basic agent for technology enhanced teaching is teachers. Teachers are likely to control what comes inside a classroom [7]. Technology cannot be successfully integrated without the cooperation of teachers [8]. Technology can be applied in teaching environments only when teachers will be willing to transform teaching practices. It is how teachers play their role in the integration of technology. However, Chohan et al. [3] pointed some other factors such as unavailability of computers and lack of teachers' training pointed in relation to limited use of technology in teaching environment.

Teachers' attitude towards technology integration is one out of many variables which affect the use of technology for teaching [9]. Therefore, the present study focuses on the perceptions of college teachers' attitudes towards the use of technology for teaching English in Pakistan. A number of researches focus on factors affecting teaching of English [10] yet no research is found in the context of teachers' attitude towards integration of technology in conventional and traditional setting of English teaching in Pakistan. The study will also highlight the potential obstacles faced by teachers in their incorporation of technology in perspectives of technology based ESL teaching. This point necessitates the existence of such a research that addresses the following questions aiming at the objectives of the study.

Q1. What are the general perceptions about English teachers' attitudes towards the use of technology?
Q2.Which different challenges are faced by ESL teachers in using technology for teaching English?
Q3. Which different technology applications are used by ESL teachers for teaching English?

2 Literature Review

2.1 Teachers' Attitudes in Using Technology for Language Teaching

ICT is mainly concerned with the introduction of overhead projectors, multimedia projectors, smart computer labs, computer assisted language learning lab and free access to wifi in teaching and learning in Pakistan [11]. A lot of educational programs throughout the world implement different ICT tools like interactive boards, laptops, tablets, projectors, high level access software and internet infrastructure to ensure technology integration [12, 13] in teaching and learning environments. Many educational programs are running with the help of computer applications, software and programs [14] to make English language teaching interesting, easy and effective [15].

A number of studies discussed the perceptions about the role of ESL teachers' attitudes towards the implementation of technology for education [16, 17] and teaching of English language [18]. A study conducted by scholars [19] in Spain opined that there is a relationship between teachers' attitudes and technology integration. The researchers also claimed that teachers with positive attitudes are found to be more likely to use technology whereas the use of technology will be hardly optimal if teachers have negative attitudes about its incorporation.

Teachers' positive attitude is however not sufficient in itself for the successful implementation of technology. There are other factors that are also responsible for the use of technology beside teachers' attitudes. Training, knowledge and availability of digital tools are among these factors according to the scholars [20] that are necessary for the incorporation of technology for teaching and learning of English. It is due to this fact that the present study aims at perceptions about teachers' attitudes, uses and challenges in integrating technology for teaching and learning of English. Additionally, the study will also put forth some suggestions and recommendations of what should be done to make technology a part of teaching and learning in educational ESL contexts.

2.2 Previous Research

This section sheds light on the past literature to reveal the relation of attitudes and the uses of technology. The part of study deals with advantages of using technology in teaching English, and the challenges that teachers face in the implementation of technology for teaching English in ESL classes.

Some scholars [21] discussed the benefits of multimedia computers in English language teaching. The researchers proposed the use of multimedia to provide fun and amusement in teaching environments. They also expressed the idea that the use of multimedia-assisted language teaching could motivate students learning of English, enhance their enthusiasm for communication and improve their listening and speaking proficiency.

A study about the use technology [22] was conducted in Punjab, Pakistan. The data of the study was collected with the help of a survey questionnaire and interviews. The study reveals positive perception of secondary teachers, ICT coordinators, administrators and students of the school regarding integration of technology in teaching and learning. The findings of the study also highlight some barriers such as unavailability of computers and lack of computer skills which obstruct the implementation of technology due to social economic conditions of Pakistan.

Attitudes of Chinese teachers towards implementing technology in teaching English at university level were focused in the study [23]. The researchers collected data of their study through observation, questionnaire and semi-structured interviews of teachers and students. The findings of the study exposed the positive attitude of teachers towards technology integrated language teaching. Different barriers to the integration of technology were introduced such as limited ICT skills, inadequate support from school administrators and domination of traditional English language teaching through the findings of the study.

Some researchers [24] shed light on Iranian English as Foreign Language (EFL) teachers' perception of the use of computer tools for teaching of English. Survey questionnaire is used by the researcher to collect data from 100 male and female EFL teachers. Teachers have positive attitudes towards the use of technology for teaching English according to the findings of the study. The findings of the research support the use of ICT tools for enhancing students learning beyond time and space through reducing cultural barriers with the provision of authentic material. The correlational analysis of the data shows that teachers' positive attitudes towards computers can enhance technology integration in their teaching practices.

A study [25] was conducted in Khartoum, Sudan. The researcher of the study collected the data from state teachers with a survey questionnaire. Findings of his study highlighted positive attitude of teachers towards the use of technology in the teaching of EFL teachers in Khartoum. The study revealed that scarcity of equipment on the part of teachers is a problem in using technology for teaching of English. The findings of the study also reflected on incompetency of Sudanese teachers due to the lack of training to integrate technology into their teaching practices. The researcher recommended acquisition of computer skills by EFL teachers for using technology in their teaching practices.

The present study is based on the theory of Technology Acceptance Model (TAM). The theory [26] was originally propounded to offers insight into the use and acceptance of technology both on individual and organizational level. TAM is a broadly used theory to comprehend individuals' belief about the use and the acceptance of technology because it focuses on the use of information system, and is based on the theory of social psychology.

TAM is an adaptation of the Theory of Reasoned Action (TRA), and is intended to focus the acceptance of technology. This theory reveals that the use of technology is determined by behavioral intention of an individual which is influenced by attitudes, perceived usefulness (PU) and perceived ease of using technology (PEOU). Attitudes determine intention and they (attitudes) are in turn shaped by PU and PEOU. TAM is also used in combination with other theories for extension and specification to include more social factors which impact the intention of individuals in their use of technology.

3 Methodology

The present study uses mixed research methods to address the research questions. The research design applies both quantitative and qualitative strands [27] of data collection namely interview protocol and survey questionnaire. Quantitative tool of the present study was built through the review of the previous studies while qualitative instrument was based on the literature review and the questionnaire used in the present study. The triangulation of the data was deliberate to achieve a rich data [28]. The aim of the present study was to give both quantitative and qualitative interpretations of the data on teachers' attitude towards technology based English Language Teaching in Pakistan and challenges faced in the incorporation of technology in ESL classes.

3.1 Instruments

Two types of tools are used in the study for the collection of data: survey questionnaire and interview protocol.

Questionnaire
The questionnaire was adapted from pre-existing research [29] and modified to suit the context of the study. The content of the questionnaire was validated by seven experts in the field of the study. Their recommendations were incorporated to upload the questionnaire onto Google form which allowed the researchers to include the responses of informants without any limitation. The questionnaire is comprised of 26 items and has three sections. The first section is about attitudes, the second section consists of hurdles and the third section deals with the use of technology applications. The questionnaire has five point likert scale items ranging from Strongly Agree with a score of 5 to Strongly Disagree with a score of 1. The questionnaire was sent to the participants through social media apps like WhatsApp, and was edited in SPSS 22 for analysis. The reliability of the three sections of the questionnaire was (.86; .76 and.81) on Cronbach's Alpha.

Interviews
Interview is the second tool used in the study for the collection of data. The questions of the interviews were based on the items of the questionnaire. Different data source for the triangulation [30] of findings. Three interview questions were sent to ten interviewees. The responses of the interviews were coded and major themes were discussed in the qualitative analysis.

3.2 Participants

The participants of the study were 34 lecturers and professors of English (14 male and 20 female) working at different colleges of Abbottabad District. All the respondents had relevant experience. A total of 7 ESL teacher were PhD holders, 8 of them had MPhil and the rest of 19 possessed BS in English (Table 1).

3.3 Data Analysis Procedure

Both quantitative and qualitative data tools were sent to the participants of the study through email and also by social media apps such as WhatsApp through gatekeepers. The participants were then given a period of one week to complete the questionnaire and answer the interview questions. The data was then received through social media apps and through personal email account of the researcher. Descriptive statistics were used for the analysis of quantitative data whereas codes were generated and themes of the qualitative data were discussed in the study.

Table 1. A summary of the demographic of respondent

Gender	Number	Total
Male	14	34
Female	20	
Qualification	Number	Percentage
BA	0	0%
BS/MA	19	55.9%
MS/Mphil	8	23.5%
PhD.	7	20.6%
Participant with computer	Yes/No	Percentage
33	Yes	97.6%
1	No	2.94
Years of experience	Number	Percentage
20 and above	4	11.8
16–20 years	2	5.9
11–15 years	11	32.4
6–10 years	8	23.5
1–5 years	9	26.5

4 Results

4.1 What are the General Perceptions About English Teachers' Attitudes Towards the use of Technology?

Perception about teachers' attitudes towards the use of technology covers opinions, thinking and feelings for the use of technology in their teaching practices. A high overall mean score of 4.3382 reflects on teachers attitudes towards the utility of technology for the teaching of English. The positive feeling of teachers towards the use of technology can facilitate English teaching, saves teachers time and energy. Also, the use of technology helps teachers to bring authentic material and culture into the classrooms.

Interview questions were also answered by 10 interviewees who participated in the research survey of the study. They were asked to specify their perceptions regarding teachers' attitudes towards technology, hurdles in the use of technology and technology applications that were used in teaching and learning of English in ESL classes. The findings of the study revealed that majority of teachers had positive (60%) attitude towards the use of technology. The respondents favored the use of technology because it created interest in the learners to learn English, and it also helped them to improve their abilities to compete with the rest of the world. Another 20% of the interviewees perceived the use of technology as the need of hour whereas 10% of teachers thought the

Table 2. A summary of the attitudes of teachers towards the use of technology

No	Item	Mean	SD
6	Using technology facilitates English Language teaching	.4118	.49995
7	Technology in teaching motivates learners' to learn English	.3529	.64584
8	Using Technology in teaching saves teachers' time and energy	.1471	.70205
9	Using Technology helps teachers to bring authentic English language culture to their classrooms	.3824	.77907
10	Technology should be used in teaching English	.4706	.61437
11	I like using ICT in teaching English	.2647	.75111

use of technology as a time saving practice. There is also a 10% of teachers who endorses the use of something complementary for both teaching and learning of a language.

4.2 Which Different Challenges are Faced by ESL Teachers in Using Technology for Teaching English?

Table 3 illustrates the results of survey questionnaire of different challenges faced by ESL teachers in the integration of technology for teaching and learning English in ESL teaching practices. An overall mean of 4.2843 reflects on the problems that teachers confronted in their teaching practices in ESL classes. These challenges consisted of low speed of internet, limited internet connectivity, power cuts, unavailability of computer gadgets, lack of training, and limited knowledge about the use of technology.

Table 3. A summary of teachers' perception about challenges faced in technology use

No	Item	Mean	Sd. Deviation
12	Low speed of internet makes the technology use difficult	4.4118	.60891
13	Electric power cuts obstructs the use of technology	4.1765	.86936
14	Unavailability of computer gadgets can impede the use of technology	4.1765	.79661
15	I will be more willing to use technology if have training about it	4.4706	.61473
16	Limited knowledge about the use of technology can hitch its integration in teaching	4.2941	.57889
17	Limited availability of internet makes the use of technology impossible	4.1765	.79661

In response to the hurdles faced by teachers in their use of technology, four out of ten interviewees (40%) thought lack of resources as a major inhibitor in the use of technology. The lack of resources was comprised of unavailability of internet, un-affordability of technology tools and frequent power breakdowns. Another three out of ten interviewees (30%) pointed out poor signals, internet connectivity and shortage of language labs for integrating technology in teaching and learning of English. Also, ten percent of interviewees called lack of training as a problem in incorporating technology for teaching and learning of English. There were also 10% of interviewees who considered lack of knowledge as an impediment in the use of technology whereas another 10 percent deemed lack of class interaction as a problem for the incorporation of technology into teaching and learning of English.

4.3 Which Different Technology Applications are Used by ESL Teachers for Teaching English?

Table 4 sheds light on the uses of technology in teaching of English. An overall mean score of 3.3921 expresses the interest of teachers in using technology for teaching of English. Technology is basically used by teachers for teaching of language skills. Teachers also use applications such as WhatsApp, Youtube, Facebook and other educational websites to teach English in their classrooms. The difference between overall mean score of the participants' attitudes towards technology as 4.3382 given in Table 2, and the overall mean score of the participants uses of technology out of five items as 3.24 expresses that use of technology is not up to the mark in teaching practices of teachers in Pakistan.

Table 4. A summary of teachers' perception about the uses of technology

No	Item	Mean	Sd. Deviation
18	How often do you use ICT in your language teaching	3.5294	.86112
19	How often do you use ICT in teaching spoken English	3.3824	.77907
20	How often do you use ICT in teaching written English	3.0882	.90009
21	How often do you use technology in teaching English grammar	3.2059	.94643
22	I use power point in my classes to present teaching material	3.0000	.88763
23	I use internet in preparing my lessons to achieve teaching objectives	4.0588	.81431
24	I use social media such as WhatsApp and Facebook for communication with learners and development of their language skills	3.9412	.81431
25	I display native English speakers movies and dialogues in my classes for better language learning opportunities	3.0882	.93315
26	I can design technology based teaching activities for my students	3.2353	.74096

Figure 1 highlights teachers' overall mean scores in relation to their use of technology in their teaching practices. Male mean score is slightly higher than the female mean score. However, there is not a big difference in the use of computer based on gender.

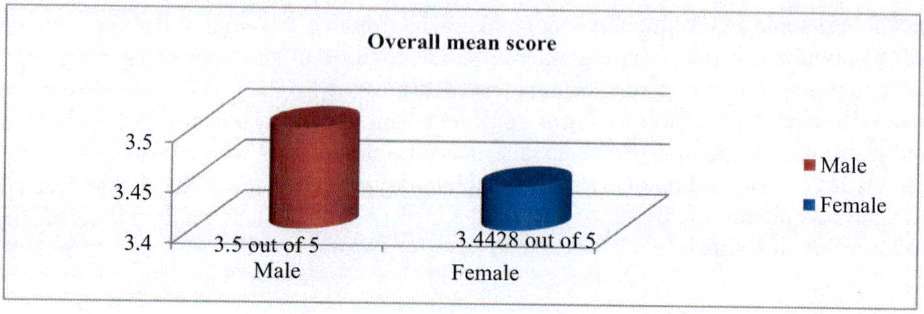

Fig. 1. Male and female mean score

Figure 2 shows a significant difference in the use of technology because of teachers' competency. This difference is due to the correlation that is found between the level of competency and the use of technology on the part of teachers.

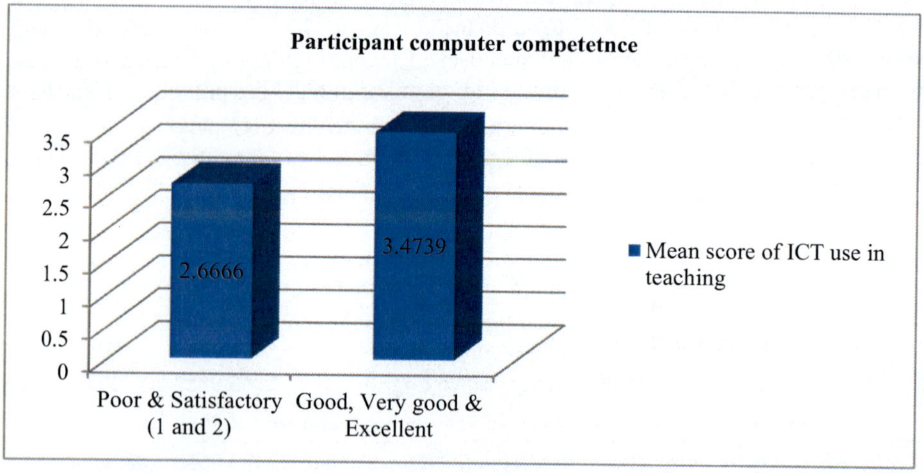

Fig. 2. Difference between teachers' use of computer on the basis of gender

In response to the third question of the qualitative tool, four interviewees from a total of ten (40%) suggested Google meets, Google classrooms, computer labs, zoom, power point program, Grammarly, Whatsapp and MS word programs to be educational apps used in teaching and learning of English. There were also another 5 interviewees (50%) who considered audio/video players, British council apps, Phonetic chart Apps, YouTube video and educational movies as suitable audio and visual material for teaching and learning English inside classrooms. Another 10% of the interviewees favored the use of dictionaries as online dictionary apps for teaching and learning of English.

5 Discussion

The findings of the study given in Table 2 of the survey question number one, and interview question based on the items used in the Table 2 of the survey reveal positive attitudes and inclinations of ESL teachers towards the use of technology. The teachers believe in the potentials of technology to help students in their process of learning. Technology is also helpful for teachers themselves since it saves their time and energy. The integration of technology also helps teachers to bring authentic teaching material in their classrooms. The use of technology also creates interest among learners, and it also helps them to improve their abilities to compete with the world. The findings of the study are in line with the study [25] who favors teachers' attitudes as a good predictor for the incorporation of technology if other influential factors are properly addressed.

Technology integration is not only possible through teachers' attitudes because there are some hurdles which also contribute in the successful incorporation of technology in ESL classes. These hurdles and challenges are low speed of internet, limited internet connectivity, power cuts, unavailability of computer gadgets, lack of training and limited knowledge and skills of teachers to use technology in their teaching. Qualitative findings of the study reveal other challenges like unaffordability of technology tools, shortage of language labs and lack of class interaction as problems in the use of technology for teaching of English. A significant correlation has been found between teachers' computer competence and the practical use of technology on their part. The study has connection with previous research [14] in regard to the correlation which is found between competence and the use of technology for teaching of English in ESL classes.

The study also highlights the use of technology in teaching of ESL teachers. Different applications are used by teachers for teaching such as Whatsapp, Zoom, Facebook and some educational websites. However, the use of technology is not up to the mark in the educational environments. There is a contrast between the teachers' attitudes and their practical use of technology due to multitude factors. It has also been found in the research [9]. Lack of technology integration can be resulted due to shortage of technology tools, low speed of internet and internet connectivity problems. Some other hurdles in this regard are competence rather than gender difference according to the findings of the study.

6 Conclusion and Recommendation

The present study probes into ESL teachers' attitudes towards the use of technology in teaching and learning of English, uses of ICT, and hurdles which affect the incorporation of technology. Findings of the study show positive attitudes on the part of teachers towards integration of technology. However, there is no significant influence of teachers' attitudes on their uses of technology. The use of technology in teaching practice in colleges in Pakistan is not up to the mark because of other different factors. It is limited to the use of internet, MS Word, MS Power point and other educational websites. Significant difference in the use of technology is found because teachers' competence instead of the difference of gender.

The use of technology is very low because of different challenges which teachers face in teaching English in ESL classes. Therefore, the study recommends sound internet connections for English Departments and language classrooms. The present study also suggests establishment of language labs in the colleges for the teaching of English. Training sessions are also recommended for teachers and learners to make the use of technology in English language teaching classes. Pedagogical trainings are also recommended for teachers to learn the designing of technology based teaching and learning activities. Further empirical researches should be done in different contexts to make teachers ready for the incorporation of technology in teaching ESL.

References

1. DeCoito, I.: Teachers and technology: present practice and future directions. Contemp. Issues Technol. Teac. Educ. **18**(2), 362–378 (2018)
2. Amin, M.: The role of educational technology in the ESL classroom. Glob. J. Archaeol. Anthropol. **11**(1) (2019)
3. Chohan, M.N., Abbas, F., Saleem, M.: CALL as a tool in teaching EFL in Pakistani religious institutes (Madaris): a survey of issues and challenges. Al Qalam., 31–44 (2018)
4. Ali, J.K.M., Bin-Hady, W.R.A.: A study of EFL students' attitudes, motivation and anxiety towards whatsapp as a language learning tool. Arab World English J. (AWEJ) Spec. Issue CALL (5), 289–298 (2019)
5. Lai, C., Yeung, Y., Hu, J.: University student and teacher perceptions of teacher roles in promoting autonomous language learning with technology outside the classroom. Comput. Assist. Lang. Learn. **29**(4), 703–723 (2016)
6. Sabzian, F., Gilakjani, A.P.: Teachers attitudes about computer technology training, professional development, integration, experience, anxiety, and literacy in english language teaching and learning. Int. J. Appl. Sci. Technol. **3**(1), 67–75 (2013)
7. Oda, K.: Post-Secondary Foreign Language Teachers' Belief Systems about Language Teaching/Learning and Technology/Teaching with Technology. Unpublished dissertation, University of Minnesota. ProQuest LLC (2011)
8. Coklar, A.N., Yurdakul, I.K.: Technology integration experiences of teachers. Discourse Commun. Sustain. Educ. **8**(1), 19–31 (2017)
9. Semerci, A., Aydin, M.K.: Examining high school teachers' attitudes towards ICT use in education. Int. J. Progress. Educ. **14**(2), 93–105 (2018)
10. Asif, S., Bashir, R., Zafar, S.: What are the factors affecting the use of English language in English-only classrooms: student's perspectives in Pakistan. Engl. Lang. Teach. **11**(6), 67–79 (2018)
11. Mahmood, A.S., Khattak, N., Haq, N., Umair, S.: Technology integration and upgradation of higher secondary education: need of the hour in Pakistan. In: Handbook of Research on Mobile Devices and Smart Gadgets in K-12 Education, pp. 115–133 (2012)
12. Arpaci, I., Basol, G.: The impact of pre service teachers' cognitive and technological perceptions on their continuous intention to use flipped classroom. Educ. Inf. Technol. (2020)
13. Falloon, G.: What's the difference? Learning collaboratively using iPads in conventional classrooms. Comput. Educ. **84**, 62–77 (2015)
14. Ahmed, S.T.S., Qasem, B.T.A., Pawar, S.V.: Computer-assisted language instruction in South Yemeni context: a study of teachers' attitudes, ICT uses and challenges. Int. J. Lang. Educ. **4**(1), 59–73 (2020)
15. Al-Emran, M., Arpaci, I., Salloum, S.A.: An empirical examination of continuous intention to use m-learning: an integrated model. Educ. Inf. Technol., 1–20 (2020)

16. Gilakjani, A.P., Leong, L.M.: EFL teachers attitudes toward using computer technology in English language teaching. Theory Pract. Langu. Stud. **2**(3) (2012)
17. Qasem, A.A.A., Viswanathappa, G.: Teacher perceptions towards ICT integration: professional development through blended learning. J. Inf. Technol. Educ. Res. **15**, 561–575 (2016)
18. Mahmood, Q., Iqbal, M.J., Nadeem, M.W., Javed, M.A., Hassan, M.U.: Perceptions of English teachers towards teaching English through information and communication system in Pakistan. Bull. Bus. Econ. **3**(3), 150–154 (2014)
19. Domingo, M.G., Garganté, A.B.: Exploring the use of educational technology in primary education: teachers' perception of mobile technology learning impacts and applications' use in the classroom. Comput. Hum. Behav. **56**, 21–28 (2016)
20. Ertmer, P.A., Ottenbreit-Leftwich, A.T., Sadik, O., Sendurur, E., Sendurur, P.: Teacher beliefs and technology integration practices: a Critical relationship. Comput. Educ. **59**(2), 423–435 (2012)
21. Guan, N., Song, J., Li, D.: On the advantages of computer multimedia aided English teaching. Procedia Comput. Sci. **131**, 727–732 (2018)
22. Hassan, T.U., Sajid, A.R.: ICTs in learning: problems faced by Pakistan. J. Res. Reflect. Educ. **7**(1), 52–64 (2013)
23. Hu, Z., McGrath, I.: Innovation in higher education in China: are teachers ready to integrate ICT in English language teaching? Technology Pedagogy Educ. **20**(1), 41–59 (2011)
24. Sadeghi, B., Rahmany, R., Doosti, E.: L2 teachers' reasons and perceptions for using or not using computer mediated communication tools in their classroom. J. Lang. Teach. Res. **5**(3), 663–674 (2014)
25. Saeed, Y.B.M.: The effect of using computer technology on English language teachers' performance. SUST J. Hum. **1**, 64–79 (2015)
26. Davis, F.D.: Perceived usefulness, perceived ease of use, and user acceptance of information technology. MIS Q. **13**(3), 319–340 (1989)
27. Creswell, J.W., Plano Clark, V.L.: Designing and conducting mixed methods research (3rd edition). Sage, Thousand Oaks (2018)
28. Leung, L.: Validity, reliability, and generalizability in qualitative research. J. Fam. Med. Primary Care **4**(3), 324 (2015)
29. Khan, R.M.I., Radzuan, N.R.M., Shahbaz, M., Ibrahim, A.H.: EFL instructors' perceptions on the integration and implementation of MALL in EFL classes. Int. J. Lang. Educ. Appl. Linguist. **8**(2), 39–50 (2018)
30. Creswell, J.W.: Research Design: Qualitative, Quantitative, and Mixed Methods Approaches, 4th edn. Sage, Thousand Oaks (2014)

Pre-service Teachers' Readiness and Engagement for Online Learning During the Covid-19 Pandemic: A Rasch Analysis

Pasttita Ayu Laksmiwati[1](✉), Donnie Adams[2], and Eka Sulistyawati[3]

[1] SEAMEO Regional Centre for QITEP in Mathematics, Yogyakarta 55281, Indonesia
pasttita.laksmi@staff.qitepinmath.org

[2] Faculty of Education, University of Malaya, 50603 Kuala Lumpur, Malaysia

[3] Institut Agama Islam Negeri (IAIN) Kediri, Kediri 64127, Indonesia

Abstract. The widespread of the COVID-19 pandemic has challenged training institutions to rethink the execution of their training programs. New approaches to teaching and learning are now needed due to the forced shift to e-learning modes. However, questions remain about how ready pre-service teachers are and their level of engagement in an online learning model of instruction. The purpose of this study was to investigate the readiness and engagement of pre-service teachers for online learning in Indonesia. This study used a non-experimental quantitative research design. Data were gathered from a sample of 285 pre-service teachers using a questionnaire. Data were analysed using WINSTEPS Rasch model measurement software version 4.30. Descriptive statistical analysis was used to examine pre-service teachers' readiness and engagement of pre-service teachers for an online learning model of instruction, and a Differential Item Functioning (DIF) analysis was carried out to specifically assess their readiness and engagement for online learning based on their gender and age. Findings indicate pre-service teachers were not ready for online learning during the Covid-19 Pandemic in Indonesia. However, pre-service teachers were found to be actively engaged in online learning. Further analysis indicated there were differences in pre-service teachers' readiness and engagement in online learning based on their gender and age. This study provides insight into pre-service teachers towards an online learning model of instruction in Indonesia, discusses implications, and offers recommendations for future research.

Keywords: Online learning · Readiness · Engagement · Pre-service teachers · Rasch model

1 Introduction

The COVID-19 pandemic has impacted many sectors around the world. One of the main sectors affected is higher education, where universities are closed in Indonesia. A study reporting the impact on learning and its management in 424 universities worldwide states that 59% of universities are closed and more than 80% of universities are doing online learning [1]. Likewise, all universities in Indonesia have been implementing online

M. Al-Emran et al. (Eds.): ICETIS 2021, LNNS 299, pp. 326–340, 2022.
https://doi.org/10.1007/978-3-030-82616-1_29

learning since March 2020. The pandemic has caused a shift from a formal (face-to-face) approach to online learning [2]. As a result, teachers have to face real challenges, because these conditions can have a negative effect on students [3], where learning during the COVID-19 pandemic is known to affect student's learning outcomes [4].

The implementation of online learning requires adaptations made by teachers to utilize technology, for example, the use of online learning software and tools [5, 6]. However, there are still many challenges that occur, for example, the ability of teachers to use technology [7], especially in Indonesia [8, 9, 10]. The use of online learning is still very limited [5], and this situation also happens in Indonesia [8, 9, 10]. Therefore, the readiness and engagement for online learning among pre-service teachers in Indonesia is questionable. Based on the aforementioned conditions, this study focused on investigating the readiness and engagement of pre-service teachers for online learning. The following are the research questions of the study:

1. What are pre-service teachers' readiness and engagement for online learning in Indonesia?
2. Are there any significant differences in pre-service teachers' readiness and engagement for online learning based on their gender and age?

2 Literature Review

2.1 Online Learning in Higher Education

Online learning refers to learning that is more student-centred through synchronous or asynchronous learning experiences using several devices such as cell phones, laptops, and others, and this learning is mediated by the use of the internet [3, 11, 12]. Through online learning, teachers' and students' interactions can be done more easily from the side of students who can easily learn, save costs, and help organize tasks properly according to the deadlines given by the teacher [13, 14]. Synchronous online learning is carried out with real-time interactions between teachers and students which are structured in the form of live lectures, where feedback can be given directly [3]. On the other hand, online learning with an asynchronous system is not structured in real-time but learning materials can be provided through indirect systems such as the use of a learning management system [2].

Online learning provides easy access and flexibility and provides a sense of psychological security to learning communities affected by COVID-19 [3, 15, 16]. Several studies have also found that students have positive perceptions of online learning, and it can increase students' engagement in learning [17–20], as well as during a pandemic [2]. Furthermore, knowledge management practices (access, storage, and application) have a significant and positive effect on the perception of online learning usefulness [21]. In addition, it is also known that different perceptions of online learning occur in male and female students [6]. Several factors affect the success of technology-based learning, for example, the availability of supporting human resources, the availability of facilities, and continuous professional development to ensure the continued effectiveness [22].

Given the many advantages that online learning can provide, it does not rule out that online learning also has drawbacks. The drawbacks are, for example, that students will find it difficult to get direct two-way interaction if they experience difficulties in learning; learning is also difficult to develop students' potential if students do not learn independently; often online learning provides a lot of theories and lack of practice, lack of collaboration between students; and even students also have the possibility of having difficulty in understanding the learning objectives [3].

Based on the challenges faced by online learning, several solutions are offered by several studies. This challenge can be solved, one of which is by making a recording of the lesson so that it makes it easier for students to relearn while taking notes [2, 3]. Furthermore, there is a need for proper evaluation to ensure that online learning is carried out well in terms of student learning and interaction [23].

2.2 Pre-service Teachers' Readiness and Engagement for Online Learning

The essential factor in online education is students' engagement [24]. Several indicators influence students' engagement in online learning, such as self-regulation, reasoning skills, and activities that improve students' thinking [25]. In addition, environmental conditions also affect students' engagement, such as intrinsic motivation and attitudes, interactions between students, interactions between teachers and students, and teachers' efficacy [26–29].

Likewise, pre-service teachers' readiness is also crucial for online learning, for example, how technology is used and the perceptions of online learning [19, 30]. Pre-service teachers' readiness in learning is influenced by several components such as motivation, individual characteristics, attitudes, access to technology, and technical skills [31, 32]. In addition, several dimensions affect pre-service teachers' readiness for online learning, such as a positive view of technology and belief about its benefits, the use of technology, attitudes, self-regulated learning skills, and collaborative and interactive skills [33, 34]. In the same vein, Karataş and Arpaci [35] describe that self-directed learning, metacognitive awareness, and 21st-century skills may also enhance students' readiness for online learning.

Given the importance of readiness and engagement in online learning, this study aims to investigate pre-service teachers' readiness and engagement for online learning in higher education institutions in Indonesia.

3 Methodology

This study used a cross-sectional quantitative survey method. The data were collected from 285 pre-service teachers using a questionnaire. The participants consisted of 234 pre-service teachers from public universities and 51 pre-service teachers from private universities. A convenience sampling technique was employed. The questionnaire used in this research consists of three sections: Demographic, pre-service teachers' readiness, and engagement for online learning. There are 41 items for the readiness scale consisting of six dimensions such as technology skills (11 items), technology usage (8 items), technology availability (4 items), self-directed learning (6 items), computer and internet efficacy (8 items), and attitude (4 items). The pre-service teachers' engagement scale consists of three dimensions covering Cognitive Engagement (4 items), Emotional

Engagement (5 items), and Behavioural Engagement (7 items). The questionnaires use a 4-point Likert scale which allows respondents to choose from strongly disagree (1) to strongly agree (4).

This study employed the Rasch model in which raw data (Likert-type data) is converted to a logit scale; where the higher the logit score, the better the readiness and engagement of the pre-service teachers [36]. The data obtained in this study were further analysed using the Rasch modelling software WINSTEPS version 4.30. The software mathematically converts raw data into a logit scale based on the probability of frequency response using algorithmic functions [36, 37]. In addition, a descriptive statistical analysis was carried out which included means and standard deviations. The data analysis was conducted to answer the first and second research questions. Furthermore, Differential Item Functioning (DIF) analysis was carried out to identify responses based on the demographic data profile of the pre-service teachers, namely age and gender. DIF analysis provides an overview of the different types of responses from respondents based on their demographic data.

Table 1. Reliability index values.

Index of reliability	Value
Person	0.93
Item	0.99
Cronbach's Alpha	0.94

Table 1 indicates the Person Reliability index (0.93) and Item Reliability index (0.99), indicating that the consistency of person and item responses was 'very good' [38] and that the reliability of both item and person is very high. In addition, the Cronbach Alpha coefficient value is 0.94, suggesting the questionnaire has 'very good' internal consistency and a highly reliable instrument [39]. Further analysis shows the questionnaire has good unidimensionality (see Appendix 1) which indicates that the raw variance index (46.1% for readiness and 41.1% for engagement) is above the standard of 40% [40].

4 Result

4.1 Pre-service Teachers' Readiness for Online Learning

Firstly, pre-service teachers' readiness for online learning was analysed. The level of pre-service teachers' readiness was classified based on the mean logit = 1.26 and SD = 1.25 and into 5 levels.

Based on Fig. 1, many pre-service teachers were not ready for online learning during the Covid-19 Pandemic as majority of the respondents were below the average (M). Pre-service teacher's readiness at Level 1 indicates teachers are unprepared for online learning. Level 5 indicates teachers are very ready for online learning. Results in Table 2 indicate more than half (50%) of pre-service teachers felt unprepared for online learning. Only 18% of the pre-service teachers were ready for online learning.

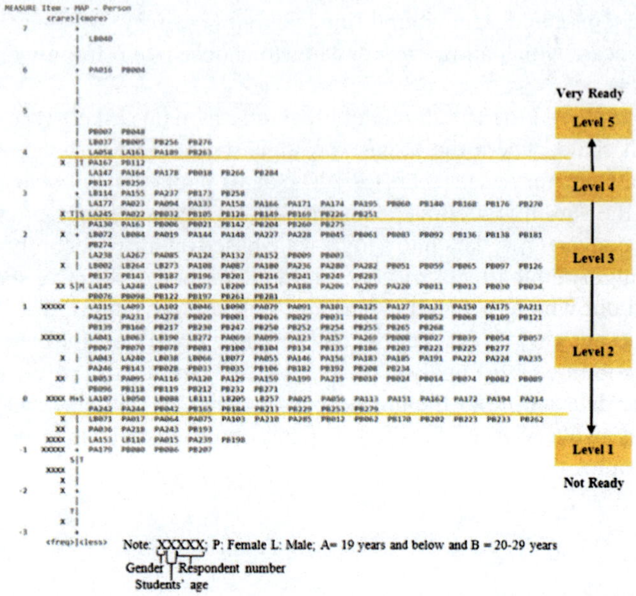

Fig. 1. Rasch Wright person logit map of online learning for readiness

Table 2. Level of pre-service teachers' readiness (x = MEASURE).

Criteria	Level of perception	Number of respondents	Percentage
$x > 3.77$	Level 5	13	5%
$2.51 < x \leq 3.77$	Level 4	38	13%
$1.26 < x \leq 2.51$	Level 3	74	26%
$0.01 < x \leq 1.26$	Level 2	133	47%
$-1.24 < x \leq 0.01$	Level 1	27	9%

4.2 Pre-service Teachers' Engagement for Online Learning

Secondly, pre-service teachers' engagement in online learning was analysed. Based on Fig. 2, most pre-service teachers are actively engaged in online learning during the Covid-19 Pandemic as most respondents are above the average (M).

Then, the level of the pre-service teachers' engagement was classified into 6 levels (see Table 3). Level 1 indicates that teachers have no engagement during online learning, while Level 6 indicates that teachers have very high engagement during online learning. The level classifications are made based on the mean logit = 1.03 and SD = 1.5.

Based on Table 3, 14% of pre-service teachers have little to no engagement during online learning. 70% of pre-service teachers were moderately engaged. Meanwhile, only 12% of pre-service teachers were highly engaged in online learning.

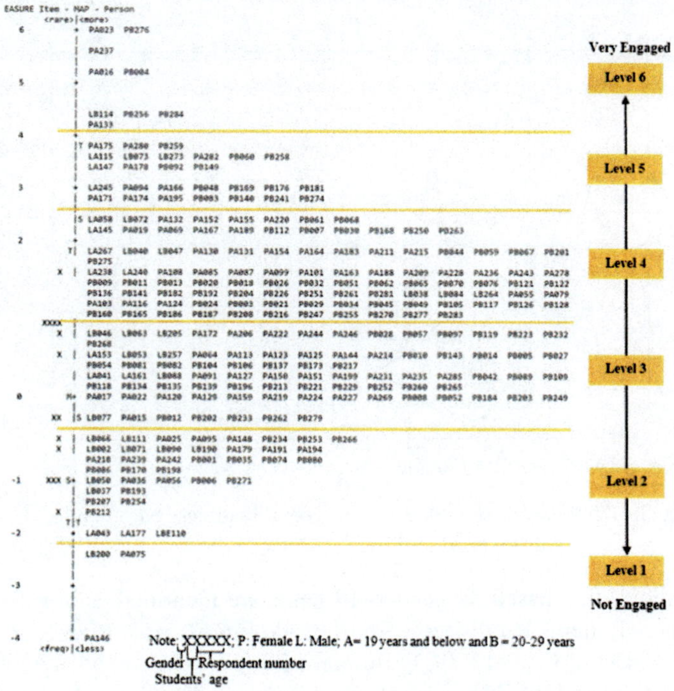

Fig. 2. Rasch Wright person logit map of online learning for engagement

Table 3. Level of pre-service teachers' engagement (x = MEASURE).

Criteria	Level of perceptions	Number of respondents	Percentage
x > 4.03	Level 6	7	2%
2.53 < x ≤ 4.03	Level 5	27	9%
1.03 < x ≤ 2.53	Level 4	127	45%
− 0.47 < x ≤ 1.03	Level 3	83	29%
− 1.97 < x ≤ −0.47	Level 2	38	13%
x ≤ −1.97	Level 1	3	1%

4.3 Differences in Pre-service Teachers' Readiness for Online Learning Based on Gender and Age

The next analysis stage was figuring out the differences between the respondents' gender and age and their readiness for online learning. These were analysed using Differential Item Functioning (DIF) analysis, which suggests that respondents of separate subgroups respond differently to some items [41], thus measuring distinctive readiness at the item level.

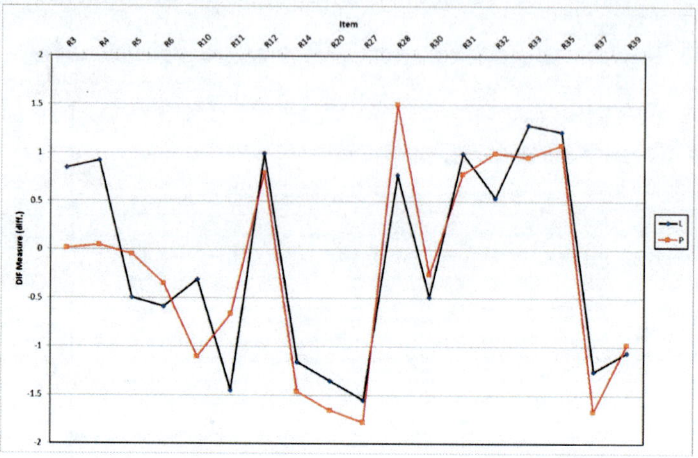

Note: L= Male and P = Female

Fig. 3. DIF analysis of pre-service teachers' readiness based on gender

Figure 3 shows that based on gender, 18 items are identified as having significant differences, namely items R3, R4, R5, R6, R10, R11, R12, R14, R20, R27, R28, R30, R31, R32, R33, R35, R37, and R39. From items R3 (*I have a computer/laptop with an internet connection*) and R4 (*I am a highly independent learner*), it is known that female pre-service teachers had more access to computer or laptop as well as internet connection and were more independent learners than male pre-service teachers. However, items R5 (*I feel confident in using online tools (e.g. e-mail, internet chat, instant messenger to communicate effectively with others*), R6 (*I think online learning is useful for universities*) and R11 (*I feel confident in expressing myself (e.g. emotions and humour) in my university's Learning Management System (e.g. Blackboard, Moodle)*), indicate that male pre-service teachers felt more confident in communicating online and expressing themselves in the learning management system than female pre-service teachers and felt online learning was useful. As from items R10 (*I am able to learn new technologies*), R12 (*I am interested to participate in Online Learning activities*), R14 (*I often use office software (e.g. M.S. Word, PowerPoint, Excel)*), R20 (*I often use social networking sites to share information (e.g. Facebook, Twitter, Instagram, Snapchat)*), R35 (*I know how to use word processing software (e.g. M.S. Word)*), and R37 (*I know how to use presentation software (e.g. M.S. PowerPoint)*), female pre-service teachers were found to be more confident in expressing themselves in the learning of new technologies. They were also more familiar with office software and often used social networking sites. Similarly, the results of items R27 (*I am not distracted by other online activities when learning online (e.g. Facebook, Gaming, Internet surfing)*), R31 (*I know how to access an online library or database*), and R33 (*I can read the online instructional materials on the basis of my needs*) indicate that female pre-service teachers were more focused when learning online, they were also more familiar with the access of online library and the use of web browser particularly to download files, and how to deal with the online instructional materials than male pre-service teachers. Meanwhile, results of items R28 (*I feel confident in performing the basic functions of presentation software (e.g. M.S. PowerPoint)*), R30 (*I often use cloud-based file hosting services to store or share documents (e.g. Google Drive, Dropbox, OneDrive)*), and R32 (*I often use Learning Management System (e.g. Blackboard, Moodle)*) indicate that male pre-service teachers used hosting services and learning management systems as well as mobile technologies more often than female pre-service teachers. Interestingly, male pre-service teachers also were more familiar with spreadsheet software than female pre-service teachers, as indicated in R39 (*I know how to use spreadsheet software (e.g. M.S. Excel)*).

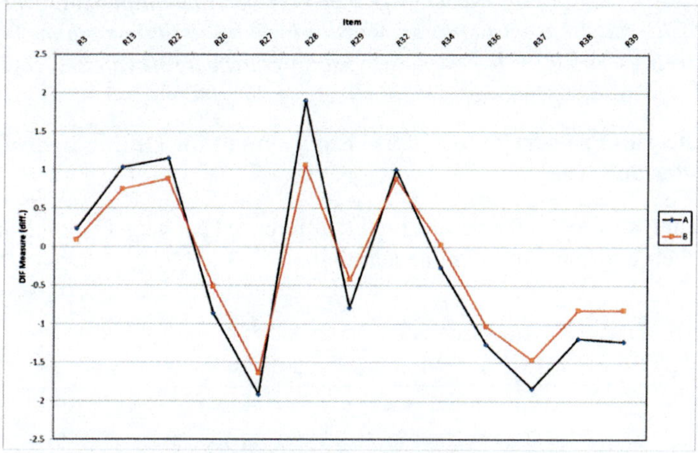

Note: A= 19 years and below and B = 20-29 years

Fig. 4. DIF analysis of pre-service teachers' readiness based on age

Figure 4 shows pre-service teachers' readiness based on age groups. Results indicate 12 items having significant differences, namely items R3, R13, R22, R26, R27, R28, R29, R32, R34, R36, R37, R38, and R39. Items R3 (*I have a computer/laptop with an internet connection*) and R13 (*I know how to open and send an email with file attachments*) indicate that pre-service teachers aged 20–29 years had stable access to the internet, and they were also more familiar with email functionalities than pre-service teachers aged 19 years and below. Similarly, items R22 (*I would describe myself as a self-starter in learning using technology*), R28 (*I feel confident in performing the basic functions of presentation software (e.g., M.S. PowerPoint)*) and R32 (*I often use Learning Management System (e.g., Blackboard, Moodle)*) indicate that pre-service teachers aged 20–29 years were more confident in performing the basic functions of presentation software and often used learning management systems. They described themselves as a self-starter in learning using technologies, more than pre-service teachers aged 19 years and below did. However, based on items R26 (*I often use instant messaging (e.g. WhatsApp, Viber, WeChat, Line, Telegram)*), R27 (*I am not distracted by other online activities when learning online (e.g. Facebook, Gaming, Internet surfing)*), and R29 (*I know how to download files using web browsers (e.g. Google Chrome, Internet Explorer, Mozilla Firefox) and view them*), it is found that pre-service teachers aged 19 years and below often used instant messaging and were familiar with web browsers usage as compared to pre-service teachers aged 20–29 years. Interestingly, for items related to the use of a spreadsheet, pre-service teachers aged 19 years and below were confident in using this software, as it is indicated by item R34 (*I feel confident in performing the basic functions of spreadsheet (e.g., M.S. Excel)*). Moreover, these age group teachers often used mobile technologies to communicate and felt more confident in using browsers, knew how to use presentation software, and knew how to use spreadsheet as compared to pre-service teachers aged 20–29 years, as indicated by items R36 (*I often use mobile technologies (e.g. Smartphone, Tablet) to communicate*), R37 (*I know how to use presentation software (e.g. M.S. PowerPoint)*), R38 (*I feel confident in using web browsers (e.g. Google Chrome, Internet Explorer, Mozilla Firefox) to find or gather information for online learning*), and R39 (*I know how to use spreadsheet software (e.g. M.S. Excel)*).

4.4 Differences in Pre-service Teachers' Engagement for Online Learning Based on Gender and Age

This part of analysis reports the results on the differences between respondents' gender and age and their engagement in online learning.

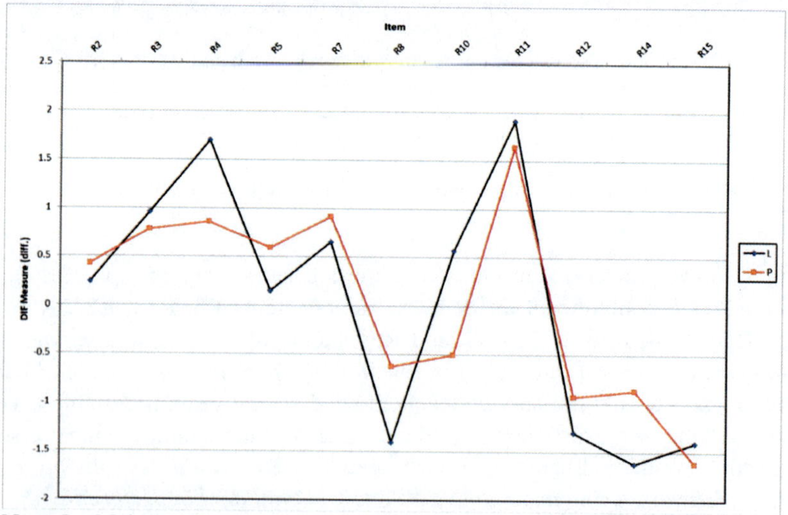

Note: L= Male and P = Female

Fig. 5. DIF analysis of preservice teachers' engagement based on gender

Figure 5 shows eleven items, namely, items R2, R3, R4, R5, R7, R8, R10, R11, R12, R14, and R15 recorded a DIF score. Item R3 (*I participate actively in online activities*) and item R4 (*I am able to discuss on assignments with my classmates using my university's Learning Management System (e.g., Blackboard, Moodle)*) indicates that female pre-service teachers participated more actively in the online learning process than male pre-service teachers.

Similarly, female pre-service teachers were more interested and made more effort in the online learning process than male pre-service teachers, as indicated by items R10 (*I spend enough time and effort to learn online*), R11 (*I feel my classes are more interesting when performing task online with my classmates*), and R15 (*I try to do my best in online group work*). Interestingly, male pre-service teachers were more motivated enough to complete online tasks and to collaborate with their peers, as shown by item R2 (*I am able to do my best to complete online tasks*), and item R5 (*I give importance to studying together with my classmates in a group online*). In addition, items R7 (*I prepare myself in reading materials online before attending classes*) and R8 (*I feel my classmates respect my thoughts and views during online discussion*) indicate male preservice teachers were also more ready for online classes, and they also felt that their peers respect them during the discussion as compared to female pre-service teachers. Moreover, the results indicate that male pre-service teachers were more punctual in submitting the tasks, and they were more careful to read other teachers' views during the discussion, as found in items R12 (*I do my assignments and submit it on time online*) and R14 (*I carefully read other students views during online discussion*).

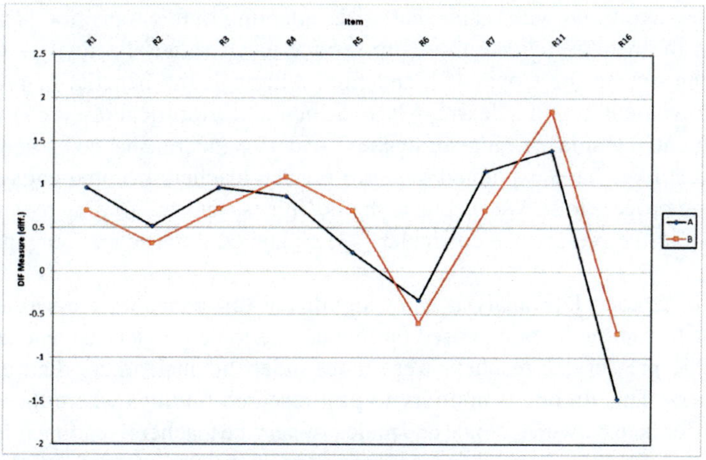

Note: A= 19 years and below and B = 20-29 years

Fig. 6. DIF analysis of preservice teachers' engagement based on age

Figure 6 illustrates significant differences in nine items based on pre-service teachers age and their engagement in online learning, namely items R1, R2, R3, R4, R5, R6, R7, R11, and R16. Items R1 (*I am able to do my best to complete online tasks*), R2 (*I am able to motivate myself to learn when performing online tasks*), and R3 (*I participate actively in online activities*) indicate preservice teachers aged 19 years and below were motivated and participated more during online learning than pre-service teachers aged 20–29 years. The findings also show that teachers aged 19 years and below were fairer in completing tasks with their classmates, had more preparation before joining classes, and felt more interested in the online learning process, as shown in the DIF results for items R6 (*I behave fairly to all my classmates when performing task online*), R7 (*I prepare myself in reading materials online before attending classes*), and R11 (*I feel my classes are more interesting when performing task online with my classmates*). Finally, pre-service teachers aged 20–29 years were more active participants in learning management system and online discussion as compared to teachers aged 19 years and below, as indicated by items R4 (*I am able to discuss on assignments with my classmates using my university's Learning Management System (e.g. Blackboard, Moodle)*), R5 (*I give importance to studying together with my classmates in a group online*), and R16 (*I share information with my classmates during online discussion*).

5 Discussion

This study aimed to investigate pre-service teachers' readiness and engagement for online learning in Indonesia and the significant differences based on their gender and age. Firstly, the findings of this study indicate many pre-service teachers were not ready for online learning during the Covid-19 Pandemic with more than half of the teachers unprepared for online learning. This finding is hardly surprising as the COVID-19 pandemic caused a sudden shift from a formal (face-to-face) approach to online learning [2] and teachers are faced with many challenges adapting to this transition [4].

Secondly, the results indicate most pre-service teachers were actively engaged in online learning during the Covid-19 Pandemic. Although the benefits of pre-service teachers' involvement in online learning have been noted, empirical research evaluating their engagement in online learning during the Covid-19 pandemic has so far been limited [42]. Limited studies have explained how pre-service teachers become engaged in an online learning process [42]. We speculate the skilfulness and scaffolding techniques by the instructors in the pre-service curricular [43] enhanced the teachers' engagement in this study.

In terms of gender, DIF analysis show significant difference on pre-service teachers' readiness for online learning based on female and male pre-service teachers. Most notably, female pre-service teachers were more independent learners than male pre-service teachers. This finding is contrary to past research findings where there was no significant difference between female and male pre-service teachers' readiness for online learning [32, 44]. Female pre-service teachers were also regular users of social networking sites, mobile technologies, and were more familiar with office software as compared to male pre-service teachers. This is consistent with past findings where females were found to be active users of social networking sites and mobiles, more often than males and that females perceived themselves to be more skilful in using Office software more often than males [40, 45]. Interestingly, male pre-service teachers in this study also found online learning very useful as compared to female pre-service teachers, consistent with the findings of Naresh, Reddy and Pricilda's study [46]. Additionally, male pre-service teachers were more focused when learning online and were more confident in communicating and dealing with online instructional materials, similar to reports from past studies [47].

DIF analysis for teachers' readiness based on age groups show significant difference based on teachers aged 19 years and below, and those aged 20–29 years. Pre-service teachers aged 20–29 years tended to use e-mail more often and were more independent learners in using technologies as compared to teachers aged 19 years and below. These findings are consistent with Adams et al.'s study [40]. Generally, younger teachers are more comfortable with using mobile technologies and instant messaging to communicate and are more familiar with web browsing as compared to older teachers, as they are from a technology savvy generation [40].

As for pre-service teachers' engagement in online learning based on gender, DIF analysis showed significant difference based on female and male pre-service teachers. Notably, female pre-service teachers participated more actively and were more excited to participate in online learning process compared to male pre-service teachers. These findings are consistent with other studies (e.g. [47, 48]). Interestingly, male pre-service teachers were more motivated to complete the tasks, collaborate with their peers, and were more punctual in submitting the tasks online. Additionally, male pre-service teachers were also more ready for online classes as compared to female pre-service teachers [46].

Finally, the findings of pre-service teachers age and their engagement in online learning show significant differences based on teachers aged 19 years and below, and those aged 20–29 years. Younger teachers were motivated and participated more than older pre-service teachers. This finding is contrary to other studies (e.g. [40, 42, 48]). The findings of this study is also contrary to Adams, et al. [48] and Alwafi, Downey and Kinchin's studies [42] in which older teachers were more active participants in learning management system and online discussion as compared to younger teachers. Interestingly, in this study, older teachers were more interested in online learning and were better prepared before joining the online classes.

6 Conclusion

The widespread of the COVID-19 pandemic has challenged training institutions to rethink the execution of their training programs. New approaches to teaching and learning is now needed due to the forced shift to e-learning modes. Pre-service teachers need to adapt to this change. Therefore, knowing pre-service teachers' readiness and current engagement in the "new normal" is a must. Overall findings of this study revealed many pre-service teachers were not ready for online learning during the Covid-19 Pandemic in Indonesia. However, pre-service teachers were found to be actively engaged in online learning. Further analysis indicated there were differences in pre-service teachers' readiness and engagement in online learning based on their gender and age. Therefore, future pre-service teachers' online trainings, class activities, and its facilitation need to be devised according to pre-service teachers' interest, competencies, and skills.

This study has few limitations of its own. Firstly, the study concentrated only on a small sample of pre-service teachers from public and private universities in Indonesia. Therefore, the generalisability of the results is limited. Future studies could be expanded to more public and private higher education institutions in Indonesia involving a larger sample of pre-service teachers to ensure that the data are representative and generalisable. Secondly, the research design of this study is limited to a cross-sectional quantitative survey method. Thus, a mixed-methods research design with the addition of qualitative methodology such as interviews is highly recommended to further explore and explain the quantitative findings.

We conclude that as the thrust towards implementing online learning in courses gathers pace and continues to grow in times of this global pandemic, training programmes and support systems must be developed and put in place to ease the adaption process for all learners. Granted, pre-service teachers' engagement and experience with online learning classes will only increase their readiness to embrace online learning (Rapanta et al., 2020) as we head towards the new normal in the foreseeable future.

References

1. Marinoni, G., Land, H.V., Jensen, T.: The impact of COVID-19 on higher education around the world IAU global survey report. France: InternationalAssociation of Universities (2020). https://www.iauaiu.net/IMG/pdf. Accessed 21 Dec 2021
2. Mishra, L., Gupta, T., Shree, A.: Online teaching-learning in higher education during lockdown period of COVID-19 pandemic. Int. J. Educ. Res. Open **1**, 100012 (2020). https://doi.org/10.1016/j.ijedro.2020.100012
3. Dhawan, S.: Online learning: a panacea in the time of COVID-19 crisis. J. Educ. Technol. Syst. **49**(1), 5–22 (2020). https://doi.org/10.1177/0047239520934018
4. Gonzalez, T., et al.: Influence of COVID-19 confinement on students' performance inhigher education. PLoS ONE **15**(10), 1–23 (2020). https://doi.org/10.1371/journal.pone.0239490
5. König, J., Jäger-Biela, D.J., Glutsch, N.: Adapting to online teaching during covid 19 school closure: teacher education and teacher competence effects among early career teachers in Germany. Eur. J. Teach. Educ. **43**(4), 608–622 (2020). https://doi.org/10.1080/02619768.2020.1809650
6. Shahzad, A., Hassan, R., Aremu, A.Y., Hussain, A., Lodhi, R.N.: Effects of COVID 19 in e-learning on higher education institution students: the group comparison between male and female. Quality and Quantity (2020). https://doi.org/10.1007/s1113502001028-z
7. Mahona, J.P., Mkulu, D.G.: Online classes during COVID-19 pandemic in higher learning institutions in Africa. Glob. Res. High. Educ. **3**(3) (2020). https://doi.org/10.22158/grhe.v3n3p1
8. Febrianto, P.T., Mas'udah, S., Megasari, L.A.: Implementation of online learningduring the COVID-19 pandemic on Madura island, Indonesia. Int. J. Learn. Teach. Educ. Res. **19**(8), 233–254 (2020). https://doi.org/10.26803/ijlter.19.8.13
9. Giatman, M., Siswati, S., Basri, I.Y.: Online learning quality control in thepandemic COVID-19 era in Indonesia. J. Nonformal Educ. **6**(2), 168–175 (2020)
10. Irfan, M., Kusumaningrum, B., Yulia, Y., Widodo, S.A.: Challenges during thepandemic: Use of e-learning in mathematics learning in higher education. Infinity J. **9**(2), 147 (2020). https://doi.org/10.22460/infinity.v9i2.p147-158
11. Rapanta, C., Botturi, L., Goodyear, P., Guàrdia, L., Koole, M.: Online university teaching during and after the covid-19 crisis: refocusing teacher presence and learning activity. Postdigit. Sci. Educ. **2**(3), 923–945 (2020). https://doi.org/10.1007/s42438-020-00155-y
12. Zawacki-Richter, O.: The current state and impact of COVID-19 on digital higher education in Germany. Hum. Behav. Emerg. Technol. (2020). https://doi.org/10.1002/hbe2.238
13. Benta, D., Bologa, G., Dzitac, I.: E-learning platforms in higher education-Case study. Procedia Soc. Behav. Sci. **31**, 1170–1176 (2014). https://doi.org/10.1016/j.procs.2014.05.373
14. Nguyen, Q.L.H.T.T., Nguyen, P.T.: Roles of e-learning in higher education. J. Crit. Rev. **6**(4), 7–13 (2019). https://doi.org/10.22159/jcr.06.04.02
15. Ali, W.: Online and remote learning in higher education institutes: a necessity in light of COVID-19 pandemic. High. Educ. Stud. **10**(3), 16 (2020). https://doi.org/10.5539/hes.v10n3p16

16. Teräs, M., Suoranta, J., Teräs, H., Curcher, M.: Post-covid-19 education and education technology 'solutionism': a seller's market. Postdigit. Sci. Educ. **2**(3), 863–878 (2020). https://doi.org/10.1007/s42438-020-00164-x
17. Adams, D., Mabel, H.J.T., Sumintono, B.: Students' Readiness for Blended Learning in a Leading Malaysian Private Higher Education Institution. Interactive Technology and Smart Education (2020)
18. Dumford, A.D., Miller, A.L.: Online learning in higher education: exploring advantages and disadvantages for engagement. J. Comput. High. Educ. **30**(3), 452–465 (2018). https://doi.org/10.1007/s12528-018-9179-z
19. Ngampornchai, A., Adams, J.: Students' acceptance and readiness for E-learning in Northeastern Thailand. Int. J. Educ. Technol. High. Educ. **13**(1), 1–13 (2016). https://doi.org/10.1186/s41239-016-0034-x
20. Pham, L., Limbu, Y.B., Bui, T.K., Nguyen, H.T., Pham, H.T.: Does e-learning service quality influence e-learning student satisfaction and loyalty? Evidence from Vietnam. Int. J. Educ. Technol. High. Educ. **16**(1), 1–26 (2019). https://doi.org/10.1186/s41239-019-0136-3
21. Arpaci, I., Al-Emran, Al-Sharafi, M.A.: The impact of knowledge management practices on the acceptance of Massive Open Online Courses (MOOCs) by engineering students: a cross-cultural comparison. Telemat. Inform. **54**, 101468 (2020). https://doi.org/10.1016/j.tele.2020.101468
22. Turugare, M., Rudhumbu, N.: Integrating technology in teaching and learning in universities in Lesotho: opportunities and challenges. Educ. Inf. Technol. **25**(5), 3593–3612 (2019). https://doi.org/10.1007/s10639-019-10093-3
23. Balula, A., Moreira, A.: Evaluation of Online Higher Education: Learning, Interaction and Technology. Springer , New York (2014)
24. Kim, H.J., Hong, A.J., Song, H.-D.: The roles of academic engagement and digital readiness in students' achievements in university e-learning environments. Int. J. Educ. Technol. High. Educ. **16**(1), 1–18 (2019). https://doi.org/10.1186/s41239-019-0152-3
25. Reschly, A.L., Christenson, S.L.: Jingle, Jangle, and Conceptual Haziness: Evolution and future directions of the engagement construct. In: Christenson, S.L., Wylie, C., Reschly, A.L. (eds.) Handbook of Research on Student Engagement, pp. 3–20 (2012). https://doi.org/10.1007/978-1-4614-2018-7
26. Bakhshaee, F., Hejazi, E.: Student's academic engagement: the relation between teacher's academic optimism and female student's perception of school climate. Int. J. Mental Health Addict. **15**(3), 646–651 (2016). https://doi.org/10.1007/s11469-016-9674-2
27. Godec, S., King, H., Archer, L., Dawson, E., Seakins, A.: Examining student engagement with science through a Bourdieusian notion of field. Sci. Educ. **27**(5–6), 501–521 (2018). https://doi.org/10.1007/s11191-018-9988-5
28. Groves, M., Sellars, C., Smith, J., Barber, A.: Factors affecting student engagement: a case study examining two cohorts of students attending a post-1992 university in the United Kingdom. Int. J. High. Educ. **4**(2), 27–37 (2015). https://doi.org/10.5430/ijhe.v4n2p27
29. Skilling, K., Bobis, J., Martin, A.J.: The "ins and outs" of student engagement in mathematics: shifts in engagement factors among high and low achievers. Math. Educ. Res. J. **1**, 25 (2020). https://doi.org/10.1007/s13394-020-00313-2
30. Atman, C.J., Adams, R.S., Cardella, M.E., Turns, J., Mosborg, S., Saleem, J.: Engineering design processes: a comparison of students and expert practitioners. J. Eng. Educ. **96**(4), 359–379 (2007). https://doi.org/10.1002/j.21689830.2007.tb00945.x
31. Baiseitova, Z.B., Amangeldinovna, N.B., Ahatayeva, G.B.: Diagnosis of readiness of the future teachers to work with students of deviant behavior. Procedia Soc. Behav. Sci. **190**, 180–185 (2015). https://doi.org/10.1016/j.sbspro.2015.04.933

32. Yağcı, M., Sırakaya, D.A., Özüdoğru, G.: The investigation of attitude and readiness of information and communication technologies pre-service teacherstoward web-based learning. Procedia Soc. Behav. Sci. **174**, 1099–1106 (2015). https://doi.org/10.1016/j.sbspro.2015.01.800
33. Elliott, K.M., Hall, M.C., Meng (Gloria), J.: Student technology readiness and itsimpact on cultural competency. Coll. Teach. Methods Styles J. (CTMS) **4**(6), 11 (2011). https://doi.org/10.19030/ctms.v4i6.5555
34. Forson, I.K., Vuopala, E.: Online learning readiness: perspective of students enrolled in distance education in Ghana. Online J. Dist. Educ. E-Learn. **7**(4), 277–294 (2019)
35. Karataş. K., Arpaci, I.: The role of self-directed learning, metacognitive awareness, and 21st century skills and competences in predicting the readiness for online learning. Contempor. Educ. Technol. (2021)
36. Eklund, M., Erlandsson, L.K., Persson, D., Hagell, P.: Rasch analysis of an instrument for measuring occupational value: implications for theory and practice. Scand. J. Occup. Therapy **16**, 118–128 (2009). https://doi.org/10.1080/11038120802596253
37. Linacre, J.M.: A User's guide to WINSTEPS MINISTEP rasch-model computerprograms (3.91.0) (2006). http://www.winsteps.com/a/WinstepsManualPDF.zip. Accessed 21 December 2021
38. Sumintono, B., Widhiarso, W.: The application of rasch model in educational assesment [In Bahasa]. Trim Komunikata Publishing House, Cimahi (2015)
39. Bond, T.G., Fox, C.M.: Applying the Rasch Model: Fundamental Measurement in the human sciences. Routledge, New York (2015)
40. Adams, D., Sumintono, B., Mohamed, A., Mohamad Noor, N.S.: E-learning readiness among students of diverse backgrounds in a leading Malaysian higher education institution. Malaysian J. Learn. Instr. **15**(2), 227–256 (2018)
41. Boone, W.J.: Rasch analysis for instrument development: why, when, and how? CBE Life Sci. Educ. **15**(4), 1–7 (2016). https://doi.org/10.1187/cbe.16-04-0148
42. Alwafi, E.M., Downey, C., Kinchin, G.: Promoting pre-service teachers' engagement in an online professional learning community. J. Profession. Capit. Commun. **5**(2), 129–146 (2020)
43. Dorner, H., Kumar, S.: Pre- and in-service teacher satisfaction with online collaborative mentoring for technology integration: applying the Kano quality attributes. Online Learn. J. **21**(4), 283–301 (2017)
44. Hung, M.-L., Chou, C., Chen, C.H., Own, Z.-Y.: Learner readiness for online learning: scale development and student perceptions. Comput. Educ. **55**(3) (2010)
45. Ikolo, V.E., Okiy, R.B.: Gender differences in computer literacy among clinical medical students in selected Southern Nigerian Universities. Library Philos. Pract. (e-journal) **745** (2012)
46. Naresh, B., Reddy, B.S., Pricilda, U.: A study on the relationship between demographic factor and e-learning readiness among students in higher education. Sona Global Manag. Rev. **10**(4), 1–11 (2016)
47. Win, N.L., Wynn, S.D.: Introducing blended learning practices in our classrooms. J. Inst. Res. South East Asia **12**(2), 17–27 (2015)
48. Adams, D., Mabel, H.J.T., Sumintono, B., Siew, P.O.: Blended learning engagement in public and private higher education institutions: a differential item functioning analysis of students' backgrounds. Malaysian J. Learn. Instruct. **17**(1), 133–158 (2020)

Digital Oriented Learning Based on the UTAUT Model During Covid-19 Pandemic

Azliza Yacob[1], Aini Zuriyati Abd Kadir[1](✉), Nur Sukinah Aziz[1], Noor Suhana Sulaiman[1], and W. M. Amir Fazamin W. Hamzah[2]

[1] Faculty of Computer, Media and Technology Management, University College TATI (UCTATI), Teluk Kalong, 24000 Kemaman, Terengganu, Malaysia
azliza@tatiuc.edu.my, aini@uctati.edu.my

[2] Faculty Informatics and Computing, Universiti Sultan Zainal Abidin (UniSZA), Besut Campus, 22200 Besut, Terengganu, Malaysia

Abstract. The presence of the Covid-19 virus invades countries around the world, has paralyzed all human activities including education. The enforcement of the Movement Control Order (MCO), has resulted in all teaching and learning (TnL) activities being transformed into virtual or online learning forms. This scenario has changed the existing learning system, and provided a new perspective on the importance of digital technology in education. However, the implementation of online learning is not as easy as one might think, yet presents challenges for students, parents and teachers. This study was conducted to analyze parents' perceptions of digital-oriented learning during MCO based on the UTAUT Model. For this purpose, a quantitative study was conducted randomly in which a total of 103 respondents were involved in this survey. Statistical Package for the Social Sciences (SPSS) software was used to analyze the collected data. The results of this study can provide input for improvement in the national education sector in the future. Hopefully, the country's education sector in general, will be more advanced compared to the passage of time.

Keywords: Covid-19 · Digital oriented · e-Learning · Digital Malaysia · MCO · UTAUT

1 Introduction

At the end of 2019, the world was shocked by the presence of a Coronavirus (CoV) that invaded the human respiratory tract simultaneously causing death. In an effort to stop the spread of the Covid-19 network in Malaysia, Movement Control Order (MCO) is implemented depending on the level of increase in the number of positive ones periodically. For better education, online learning is implemented in compliance with Standard Procedures (SOPs), therefore the role of the latest technology becomes very important in disseminating knowledge.

Technological sophistication can show better performance and can really change the way people work. The same goes for the technology used in schools and educational

M. Al-Emran et al. (Eds.): ICETIS 2021, LNNS 299, pp. 341–350, 2022.
https://doi.org/10.1007/978-3-030-82616-1_30

institutions. It can be an important tool in implementing and ensuring the success of online learning. Besides, technology is also capable of contributing to effectiveness in classroom teaching, further leading to knowledge exploration [1]. Thus [2] argue that educational institutions should prepare students by considering the appropriate integration of technology in their curriculum. It is considered an important element for improvement and expansion and also serves as an important tool in implementing and ensuring the success of online learning. Research around the world has also shown that good technological facilities can lead to improved student learning and better pedagogical practices.

Increasing number of mobile users provided by mobile technologies has created new opportunities for learning. In addition it can be used anytime and anywhere, mobile learning is able to save costs, reduce distance and time, support distance learning, and reduce paper usage [3]. The use of mobile technology able to provide opportunities for students to learn better and more effectively [4, 5]. It have been used in many subject including mathematic, biology, Malay foreign language and programming subject [6–10].

Overall, mobile learning seeks to increase students' motivation to learn [8], however according to [11] the success of mobile learning will depend on human factors in the use of the new mobile and wireless technologies. It also limited to the hardware and software constraints of mobile devices, for example low storage, low screen resolution, compatible issues, short battery life, small screen size and many other technical problem, regarding to mobile been used [12]. It also was found that students have not fully accepted the use of tablets in learning. This is due to the lack of learning content, non-interactive learning applications and their level of readiness to use technology in learning [13].

Therefore, the role of good ICT is very important to support online learning throughout the covid-19 epidemic [14]. In addition to improve the management of online learning during pandemic, the integration of a good ICT also able to develop satisfaction among students to have better communication.

To that end, this study was conducted to discuss about parents perceptions on digital oriented learning, based on the UTAUT Model during the Covid-19 epidemic. The discussion in this study is divided into 5 sections. Section 1 for introduction; Sect. 2 on related studies; Section 3 deals with research methodologies; Sect. 4 discusses the results, and Sect. 5 discusses and discussion and conclusion.

2 Related Issues

2.1 Covid-19 Outbreak

Following this life-threatening epidemic, the first version of the MCO was issued by the Malaysian central government on 18 March 2020 as a precautionary measure. This MCO is not only issued by the Malaysian government, but most other countries in the world also implement MCO with the aims to help stop the Covid-19 network.

As a result, Movement Control Order (MCO) was issued on 18 Mac 2020 as a precautionary measure by the federal government of Malaysia. Also, many of the other country in the world were also implement MCOs too, for the same purpose. However, because of the new norm and better education, the teaching and learning process continues through the online or virtual learning. In order to realize the implementations of teaching and learning, technology become an important tool to deliver the knowledge. A good technology can perform better and also becomes an important tool in ensuring the successful of online learning. Besides that, the technologies also contribute the effectiveness at teaching in the class and encourage the exploration of knowledge in various ways [1]. According to [15], it was quite impossible to implement an online learning without appropriate equipment. Educational institutions supposed to prepare students purposely to consider ICT integration in their curriculum [2] which plays as an important tool in ensuring the successful of online leaning.

A good and updated technology has seen considerable as useful in improving the effectiveness of teaching and learning processes. As in line with Industrial Revolution 4.0 (IR 4.0) which changing the way we live and work. IR 4.0 in education need for high technology especially teaching aids, power point slides, internet access and also updated teaching and learning tools. Even the use the updated technology can improve the effectiveness and facilitate students to understand the teaching better.

2.2 Digital Oriented Learning

Digital technologies are electronic tools or systems that generate and process data. Digital technology nowadays is replacing traditional educational methods more and more each day. The inclusion of digital learning in the classrooms can vary from simply using tablets instead of paper to using elaborate software programs and equipment as opposed to the simple pen. This could entail using sites, services, programs, teaching tools, and technologies like study aids built for at-home use. Even social networks and communications platforms can be used to create and manage digital assignments and agendas. Irrespective of how much technology is integrated into the classroom, digital learning has come to play a crucial role in education. It empowers students by getting them to be more interested in learning and expanding their horizons.

Digital learning as define by [16], is an unplanned and implicit process with unpredictable results using several types of technological devices like smartphones, tablets, computers, and others. Mobile phone technology can be one of the alternative learning media in the classroom as a teacher's aid in delivering learning [17, 18]. It can happen across all curriculum learning areas including chemistry subject, programming and mathematics that will make effective use of technology. Among digital learning strategies can be or combine from blended learning, online learning, adaptive learning, mobile learning and online learning.

Online learning is a constructivism learning approach that able to improve student's learning interest, creativity and learning skills. The constructivist learning environment is greatly aided by a combination of advanced multimedia and online technologies. It was suggested that teachers or lecturers to use the most effective methods as an alternative to motivate students in learning. Furthermore, during Covid-19 outbreak, this type of learning approach not just become more important, but become compulsory for people to realize the teaching and learning process. Online learning is often related to areas that use technology to deliver learning and training programs. According to [19], online learning should basically be defined as learning and teaching facilitated online through network technology.

Since the spread of the Covid-19 pandemic worldwide, it has forced many employees to work from home and hold meetings online only. This has resulted in many online meeting or conference web applications such as Google Classroom, Cisco Webex, Zoom, Skype and Google Meet being an alternative way to stay connected. However, some students also prefer to use social networks as a medium for the learning purpose which valuable in the educational curriculum [20]. The development of online learning applications through this method must meet the needs of students, that is, through the study of student behavior [10].

2.3 UTAUT Model

In this study, we propose to study the direct implications of online learning during MCO instruction based on Unified Theory of Acceptance and Use of Technology or UTAUT. According to [21], UTAUT has four direct determinants of consumer acceptance, namely Performance Expectations (PE), Effort Expectations (EE), Social Influence (SI) and Facilitator Conditions (FC). While the other two are dependent variables; Behavioral Intention (BI) and Use Behavior (UB). Four moderators were used, namely gender, age, experience and willingness to use.

However, to achieve the objectives of this study, 4 identical independent variables are still used, but modifications are made to the dependent variables. The two dependent variables referred to are Behavioral Intention (BI) and Learning Impact (LI). All items used in this study were processed based on online learning issues during the Movement Control (MCO) period.

3 Methodology

This study is a quantitative study, namely by obtaining parents survey information in Peninsular Malaysia. The total number of respondents involved in this study is 103. These survey questions were distributed randomly to respondents regardless of gender, race, occupation and even academic qualifications.

3.1 Instruments

In this study, a set of survey questions were developed and were distributed randomly. A total of 32 items were used as the main instrument in this study to analyze parents' perceptions on digital oriented learning based on the UTAUT Model during the MCO period. These set of questionnaires consist of 3 sections, namely demographics, parent's perceptions of online learning and finally improvement suggestions.

The first part is about the demographic background of the respondents which consists of 6 items which include gender, age, race, highest academic qualifications, employment and ability in using ICT. For the second part, the 26 items produced are based on the 6 constructs found on the UTAUT model. These constructs include Effort Expectancy (EE), Performance Expectancy (PE), Facilitating Condition (FC), Social Influence (SI),

Behavioral Intention (BI) and Learning Impact (LI), whose items are processed based on deep learning issues line during the Movement Control period (MCO). This study also involved three moderators such as gender, age, level of education. The questions produced are based on the 4-Likert scale which is 4 = Strongly Disagree, 3 = Disagree, 2 = Agree and 1 = Strongly Agree. While the Third section contains 2 items related to improvement suggestions.

3.2 Analysis

The analysis carried out is descriptive analysis and inference analysis. Descriptive analysis summarizes the demographics of the respondents and the data collected from the set of questions distributed. While inference analysis will study the correlation and regression of the constructs used in the UTAUT model. Correlation analysis that is Pearson moment correlation is conducted to find out the strength of the relationship between constructs. To study the relationship between independent variables namely Effort Expectancy (EE), Performance Expectancy (PE), Facilitating Condition (FC), Social Influence (SI) and dependent variables namely Behavioral Intention (BI) and Learning Impact (LI)), regression analysis will be used. Regression analysis was used to answer the research questions: What is the Online Learning Impact (LI) during the Covid-19 outbreak.

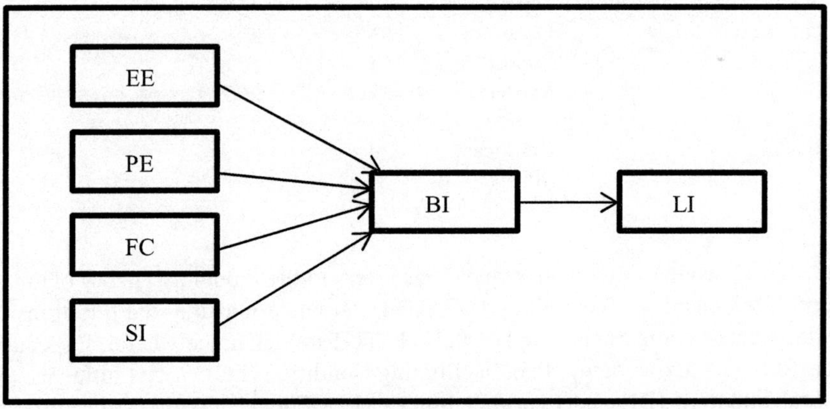

Fig. 1. UTAUT model for digital oriented learning

The models for regression analysis to be used are:

$$y = \beta_0 + \beta_1 x_1 + \beta_2 x_2 + \beta_3 x_3 + \beta_4 x_4 \tag{1}$$

$$y = \beta_0 + \beta_1(\text{EE}) + \beta_2(\text{PE}) + \beta_3(\text{FC}) + \beta_4(\text{SI}) \tag{2}$$

y is the dependent variables i.e. BI and LI. Both analyzes will be conducted using SPSS. The UTAUT model for this study is as shown in Fig. 1 below.

4 Result

4.1 Descriptive Analysis

The Table 1 below shows the results for descriptive analysis where the analysis is explained. Descriptive analysis is divided into two parts, namely the demographics of the respondents and descriptive for the questions in parts two of the set of questions. Descriptive questions are also asked to the group of parents related to personal matters such as gender, age and level of education.

Table 1. Demographic information

	Demographic Information	Frequency	Percent
Gender	Male	20	19.4
	Female	83	80.6
Age	30–39	43	41.7
	40–49	37	35.9
	50–59	19	18.4
	60 and above	4	3.9
Education level	Diploma	17	16.5
	Doctor of	8	7.8
	Philosophy	13	12.6
	Masters	27	26.2
	Degree	9	8.7
	Certificate	29	28.2
	SPM		

The second group of descriptive questions refers to the model of Theory of Acceptance and Use of United Technology (UTAUT). Parents are also asked questions that refer to the same six constructs based on the UTATUT model, namely Effort Expectancy (EE), Performance Expectancy (PE), Facilitating Condition (FC), Social Influence (SI), Behavioral Intention (BI) and Learning Impact (LI). For the first construct of Effort Expectancy (EE), the average mean is 2.65 referring to online learning is very effective. Parents also think that online learning can open the minds of students where the average mean is 2.56 expressing support related to this question. For the second construct, Performance Expectancy (PE), parents think online learning interactions are clearer and easier. In addition, they think online learning is more flexible where the average means shows support of 2.50. The third construct is Facilitating Condition (FC), where an average of 2.50 means learning is faster online.

Even with an average mean of 2.23 states more references are easy to find. Followed by the Social Influence (SI) construct where the average mean of 2.47 parents stated that friends and acquaintances always give encouragement in the learning process. Not to be missed are the teachers who give a lot of encouragement and guidance in ensuring the success of students who study online with an average mean of 2.33. Next, the Behavioral

Intention (BI) construct with an average mean of 2.59 expressed support using online methods for future learning. While the average mean of 2.56 indicates learning support is made online in the future. As for the last construct of Learning Impact (LI), the mean of 2.06 stated that the effect of Covid-19 has changed the way their children learn. In conclusion, parents support online learning and good methods should be implemented to improve the existing learning methods, especially during the Covid-19 outbreak.

4.2 Inference Analysis

Table 2 shows the correlation between the six main constructs of parents/guardians. It is found that the value of the Learning Impact (LI) construct is less than 0.5. This indicates that this construct has a weak relationship with other variables. While Table 3 shows Cronbach Alpha for the six main constructs. It shows an overall value of 0.856, where the questions used is reliable and consistent.

Table 2. Correlation between constructs

	EE	PE	FC	SI	BI	LI
EE	1					
PE	0.762	1				
FC	0.572	0.691	1			
SI	0.719	0.658	0.619	1		
BI	0.770	0.685	0.488	0.619	1	
LI	0.090	0.071	0.163	0.230	0.109	1

Table 3. Cronbach's Alpha construct

Construct	Cronbach's Alpha
Effort Expectancy (EE)	0.795
Performance Expectancy (PE)	0.804
Facilitating Condition (FC)	0.835
Social Influence (SI)	0.814
Behavioural Intention (BI)	0.819
Learning Impact (LI))	0.896
Total	0.856

Table 4 **shows the regression model is insignificant. The model equation is as follows,** $y = 1.566 + 0.244(\mathrm{SI})$. Referring to the Cronbach's Alpha construct table for the parent/guardian respondent shows the strength factors of the expected effect of Effort

Expectancy (EE), expected performance of Performance Expectancy (PE), Facilitating Condition (FC), social influence Influence (SI), intention and behavior Behavioral Intention (BI) and the learning impact of Learning Impact (LI), are predictors of online learning success. The results show that the correlation of all independent variables with dependent variables is as high.

Table 4. Regression model

Model	Unstandardized coefficients		t	Sig.
	B	Std. error		
(Constant)	1.566	.201	7.782	.000
EE	−.052	.090	−.579	.564
PE	−.102	.113	−.900	.371
FC	.110	.125	.881	.380
SI	.244	.113	2.156	.034

Thus it can be stated that the findings of this study show that the independent variable is positively related, Beta (Effort Expectancy EE = −−.052, Performance Expectancy PE =−.102, Facilitating Condition FC = .110, Social Influence SI = .244). Beta values indicate that the correlation for Social Influence (SI) condition facilities is the highest while Effort Expectancy (EE) is the lowest at −.052. This shows that Social Influence (SI) is very relevant for successful online learning. While Effort Expectancy (EE) is dependent on how students should cope with new learning phenomena and improve their performance in learning. Therefore, facilities and motivation as well as cooperation should be nurtured to further enhance positive values to ensure that student performance is not affected in the future.

5 Discussion and Conclusion

Based on the study, several conclusions can be drawn regarding parents perceptions on digital oriented learning based on the UTAUT Model during covid-19 epidemic. From the parent's perspective, it was found that 99.1% of parents understand the meaning of Covid-19 and 88% of them think that the problems from this epidemic have changed the way they learn. The study also showed that the Facilitating Condition (FC) construct showed the highest value compared to the 3 constructs of the other independent variables. When the Covid-19 epidemic hit, these constructs were a very important contributor to the success of online learning. It is also influenced by factors such as simple reference sources, in addition to the short time factor required for online learning. From a student perspective as well, the Performance Expectancy (PE) construct indicates the lowest value, and it is about how students should cope with new learning phenomena and improve their performance in learning. This includes simple and clear interactions during online learning, in addition to learning methods that are said to be simple and very flexible. Therefore, facilities and motivation as well as cooperation should be fostered to further enhance positive values to ensure that student performance is not affected in the future.

It is likely due to the current existing facilities such as unstable internet connection in certain areas especially rural areas. Therefore, to ensure effective online learning, the constraints in restricted internet access need to be addressed immediately. Internet Service Provider (ISP) companies in Malaysia including TM, MAXIS, P1, DIGI and CELCOM should take immediate steps in expanding their internet coverage to the maximum territory. This is to ensure that people, especially students from rural area can access the internet wherever and whenever they wish. It is also in line with the government's call for the implementation of Digital Malaysia in all aspects of life.

All of the mentioned above can promote an effective online learning especially during pandemic. This also in line with the Industrial Revolution 4.0 (IR 4.0) which means that technologies and trends such as the Internet of Things (IoT), robotics, virtual reality (VR) and artificial intelligence (AI) can change the way we live and work, including in the field of education. It is hoped that this study will be able to provide some input for improvement in the country's education sector in developing better education for all generation. Hopefully the country's education sector in general, will be more advanced compared to the passage of time.

Acknowledgement. We would like to extend sincere appreciation to University College TATI (UC TATI) and also for all members of Faculty Computing, Media and Management Technology, UC TATI for the supports and encouragement.

References

1. Yacob, A., Kadir, A.Z.A., Zainudin, O., Zurairah, A.: Student awareness towards e-learning in education. Procedia Soc. Behav. Sci. **67**, 93–101 (2012). https://doi.org/10.1016/j.sbspro.2012.11.310
2. Ghavifekr, S., Razak, A., Ghani, M., Ran, N., Meixi, Y., Tengyue, Z.: ICT integration in education: incorporation for teaching & learning improvement. Malaysian Online J. Educ. Technol. **2**(2), 24–45 (2014)
3. Series, C.: The effectiveness of mobile learning role play game (rpg) maker mv in improving students' critical thinking ability The effectiveness of mobile learning role play game (rpg) maker mv in improving students ' critical thinking ability (2020). https://doi.org/10.1088/1742-6596/1567/4/042088
4. Arpaci, I.: A hybrid modeling approach for predicting the educational use of mobile cloud computing services in higher education. Comput. Human Behav. **90**, 181–187 (2019). https://doi.org/10.1016/j.chb.2018.09.005
5. Al-emran, M.: An empirical examination of continuous intention to use m-learning: An integrated model (2020)
6. Farahah Abdul Halim, N., Nincarean Eh Phon, D.: Mobile learning application impact towards student performance in programming subject. IOP Conf. Ser. Mater. Sci. Eng. **769**, 012056 (2020). https://doi.org/10.1088/1757-899x/769/1/012056
7. Hanafi, H.F., Said, C.S., Wahab, M.H., Samsuddin, K.: Improving students' motivation in learning ICT course with the use of a mobile augmented reality learning environment. IOP Conf. Ser. Mater. Sci. Eng. **226**(1) (2017). https://doi.org/10.1088/1757-899X/226/1/012114

8. Hardiansyah, H., Rusmono, R., Winarsih, M.: Teaching material development of natural environment based on mobile learning on elementary school. J. Phys. Conf. Ser. **1402**(7) (2019). https://doi.org/10.1088/1742-6596/1402/7/077075
9. Hendrayana, A., Wahyudin: Mobile learning to improve mathematics teachers mathematical competencies. J. Phys. Conf. Ser. **948**(1) (2018). https://doi.org/10.1088/1742-6596/948/1/012049
10. Sung, C.M., Ahmad, M., Mansor, N.R., Rashid, R.A., Abdullah, N.A.C.: The effectiveness of mobile application in learning Malay foreign language. J. Phys. Conf. Ser. **1529**, 042064 (2020). https://doi.org/10.1088/1742-6596/1529/4/042064
11. Kukulska-Hulme, A.: Mobile usability in educational contexts: What have we learnt? Int. Rev. Res. Open Distance Learn. **8**(2) (2007). https://doi.org/10.19173/irrodl.v8i2.356
12. Maniar, N., Bennett, E., Hand, S., Allan, G.: The effect of mobile phone screen size on video based learning. J. Softw. **3**(4), 51–61 (2008). https://doi.org/10.4304/jsw.3.4.51-61
13. Hamzah, W.M.A.F.W., Yusoff, H., Yacob, A.: The behavioural intentions of secondary school students to use tablet as a mobile learning device. Int. J. Interact. Mob. Technol. **19**(1), 161–170 (2020). https://doi.org/10.3991/ijim.v14i13.13027
14. Yacob, A., Baharum, Z., Mohd, W., Fazamin, A., Hamzah, W.: ICT Integration to Support Online Learning During The Covid-19 Outbreaks, vol. 2020 (2020)
15. Oliver, S., Towers, R.: Up time: Information communication technology: Literacy and access fortertiary students in Australia. Department of Education, Training and Youth Affairs, Canberra (2000)
16. Sousa, M.J., Rocha, Á.: Digital learning: developing skills for digital transformation of organizations. Futur. Gener. Comput. Syst. **91**, 327–334 (2019). https://doi.org/10.1016/j.future.2018.08.048
17. Cahyana, U., Yusmaniar, Y., Zahari, P.A.: Development of mobile learning on acid and base to improve student performance. J. Phys. Conf. Ser. **1402**(5) (2019). https://doi.org/10.1088/1742-6596/1402/5/055062
18. Cahyana, U., Paristiowati, M., Fauziyah, S.: Development of android-based mobile learning media on atomic structure and periodic table. IOP Conf. Ser. Mater. Sci. Eng. **434**(1) (2018). https://doi.org/10.1088/1757-899X/434/1/012095
19. Hrastinski, S.: Asynchronous and Synchronous E-Learning (2008)
20. Hamzah, W.M.A.F.W., Ali, N.H., Saman, M.Y.M., Yusoff, M.H., Yacob, A.: The use of Tin Can API for web usage mining in E-learning applications on the social network. In: ISCAIE 2015 - 2015 IEEE Symposium on Computing Applications Industry and Electronics, pp. 113–118 (2015). https://doi.org/10.1109/ISCAIE.2015.7298338
21. Roslina, I., Samsudin, W., Rasimah, C.M.Y., Khalili, K., Azizah, J.: Student perceptions of educational games in higher education: an empirical study. Issues Inf. Syst. XII(1), 120–133 (2011)

Criminality Facilitated by Digital ICTs: Exploration of a General Theory of Intention and Behavior by Juxtaposing Social Learning and Deterrence

Maryam Nasser AL-Nuaimi(✉)

Al-Buraimi University College, Al-Buraimi, Sultanate of Oman
maryam@buc.edu.om

Abstract. This study aims to develop and tentatively substantiate an inclusive theoretical model that elucidates the multiple realities that account for the over-arching nature of behavioral intention and behavior connected to Information and Communication Technologies (ICT) assisted ethical behavior among Omani undergraduates. The emergent model juxtaposes ethical intention and behavior with constructs that are compatible with socio-cognitive and self- control theories. Based on a phenomenalist interpretive framework, the study applied a constructivist approach to grounded theory (GT). Data was obtained through focus group discussion sessions with undergraduates and semi-structured interviews with academics and professionals at Sultan Qaboos University (SQU). The comparative analysis generated four themes, which the participants devised to scrutinize the phenomena of ethical intent and behavior. The interviewees reflected on the central functions of moral agency, distinguished between the different patterns of ICT-assisted deviance among undergraduates, and acknowledged the different types of motives that underlie deviance. The results of the qualitative data revealed that the unique features of ICT spur and curb anomalous behavior. Additionally, social learning, perceived deterrents, and internal controls contribute to ethical behaviors.

Keywords: ICT · Differential association · Deterrents · Behavior · Intention

1 Introduction

Information and Communication Technologies (ICTs) have permanently changed human life and identity, albeit sometimes problematically and at an accelerating pace [1]. Computers, digital data, and telecommunications have changed many aspects of life, work, travel, education, entertainment, and manufacturing [2]. Networks, computers, applications, operating systems, websites, routers, and gateways are now interconnected and managed by billions of lines of code [3]. Several ethical and social issues have risen and exacerbated, such as risk, reliability, and globalization, as a result of this evolving technology [2]. Hence, exploring patterns of ICT-assisted unethical behavior and the motivations that lie beneath these behaviors have become crucial in interdisciplinary fields researching information societies.

M. Al-Emran et al. (Eds.): ICETIS 2021, LNNS 299, pp. 351–367, 2022.
https://doi.org/10.1007/978-3-030-82616-1_31

Unethical ICT-assisted activities have created a financial threat to software and ICT-related services. For example, digital piracy, peer-to-peer sharing, and illegal downloading and streaming have infringed on copyright protections [4]. Comparably, Internet piracy and e-book piracy constitute persistent problems for publishers, libraries, and authors [5]. The unprecedented penetration of the Internet and expanded bandwidth have enhanced access to pirated digitized content and invalidated the effectiveness of Digital Rights Management (DRM) systems [6]. Therefore, "free access" on the Internet to patented digitized intellectual property has become a primary challenge [7, 8].

Ethics-related ICT behavior has been continuously scrutinized by legal perspectives, particularly ones based on social contract theories. Anomalous assisted ICT behavior is associated with cyber, digital, Internet, technological, and electronic crimes. Unethical ICT-assisted behavior is defined as the illegal use of computers and digital devices to commit diverse and incessantly harmful activities by individuals, groups, or organizations who possess knowledge of ICTs [9].

The ever-increasing access and utilization of information contribute to improvements in the effectiveness and efficiency of university students' learning and future careers. Nonetheless, instances of unethical acquisition, handling, and manipulation of information have been frequently reported among university students worldwide [10]. As the world entered the power stage, a critical stage of the digital revolution, ethical and legal questions, such as privacy, data protection, and intellectual property, presented themselves on a large scale. The availability of computers, information resources, and output devices facilitated many unethical activities, such as privacy invasion and software piracy [11].

Jung [12] showed that various studies documented the sharp rise in software piracy and privacy violations among college students worldwide. In 2008, the unauthorized reproduction of computer software resulted in annual revenue losses that exceeded USD$ 50 billion [13]. Likewise, Al-Rafee and Cronan [14], Banerjee, Cronan, and Jones [15], Cronan and Al-Rafee [16] highlighted social and economic damages that occurred as a result of ICT misuse, such as estimated losses of USD$ 265 billion dollars per year in the private sector due to the theft of copyrighted information. Other forms of unethical information behavior, such as information privacy violations through social networks and the dissemination of inaccurate data, directly hurt the well-being and mental health of individuals [17].

A vast majority of related studies have tackled ICT-assisted deviance by quantitatively examining pre-existing hypotheses derived from existing criminological theories [5, 16, 18]. However, there is a dearth of qualitative studies that circumstantially scrutinize the factors that influence unethical behavior in the area of ICTs among undergraduates. This study explores the criminological constructs that emerge when considering the intersection of intention, behavior, and ICT ethics among Omani undergraduates. Through a qualitative descriptive approach, the current study poses the following research questions:

1. What is the reality of the behaviors pertinent to ICT ethics among Omani undergraduates?
2. Which behavioral intentions are connected to ICT ethics among Omani undergraduates?

3. What are the criminological constructs that most influence the behavior and intentions related to ICT ethics among Omani undergraduates?

2 Theoretical Backgrounds and Related Studies

Attempting to delve into the conceptual foundations of information and computer ethics, using environmentalist ethics, Luciano Floridi [19] defended his hypothesis from the perspective of the ecological management and well-being of the infosphere. Floridi [19] defined information ethics as an object-oriented ontology ecological approach to macro-ethics. Information ethics assess the reciprocal operations, processes, and actions of moral agents by evaluating the degree to which moral agents contribute to the well-being of informational entities in the global infosphere, provided at any given level of a network system or application.

Due to the superconductivity of digitized data, Floridi's macro-ethics theory [1, 20] conceptualized ICTs as re-ontologizing technologies that drastically transformed the ontological friction of the infosphere. Ontological friction refers to the forces that obstruct the flow of information, and therefore, to the efforts required to generate, process, and transmit information [1, 20].

Further inspection of empirical ICT-ethics literature reveals that a variety of studies relate unethical ICT-assisted behavior to socio-psychological variables that are derived from the purposive [14–16, 18, 21–24]. Nonetheless, there is an observable trend of utilizing general theories of justice to explain deviant ICT-assisted behavior. To conclude, a multitude of studies are focused on the synthesis of a unified theoretical framework of justice that is amenable for empirical testing in the field of ICT ethics and social behavior [9, 25–28].

A paradigm comprised of three distinct yet compatible criminological theoretical factors can be discerned in the empirical approaches to unethical behavior. These factors are specific deterrence, social learning, and general integrative approaches. It is important to note that these theoretical approaches originated from a universal sociological perspective on unethical behavior (e.g., social contract theory). Mehlkop and Graeff [30] contended that the aggregated consequences of violating laws constitute conscious unethical behavior in social conditions. Therefore, unethical behavior is stimulated by awareness of its costs and benefits rather than being induced by motives exclusively unique to deviant peers. Comparatively, Agnew's inductive theory of crime and delinquency argued that individuals refrain from committing a crime when they possess higher levels of internal self-control and have a fear of the corresponding prosecution risks and punishment [26, 30].

In deterrence theories, internal and external controls account for focal interest. The classical deterrence theory states that the degrees of certainty, severity, and swiftness of the detection and punishment are directly associated with increased costs connected to the perpetration of an offense [31]. Therefore, unethical behavior can be explained through cost-benefit analysis [25]. Contemporary deterrence theory expanded on its classical counterpart by including inhibition and motivation measures for committing crimes [31].

This study attempts to expound the ICT-assisted behavior of Omani undergraduates through the social learning theory of unethical behavior. The study observes how qualitative evidence verifies the consolidation of control and social learning approaches to formulate a general theory of unethical behavior pertinent to the use of ICTs among Generation Z. The *a priori* conceptual framework is demonstrated in Fig. 1.

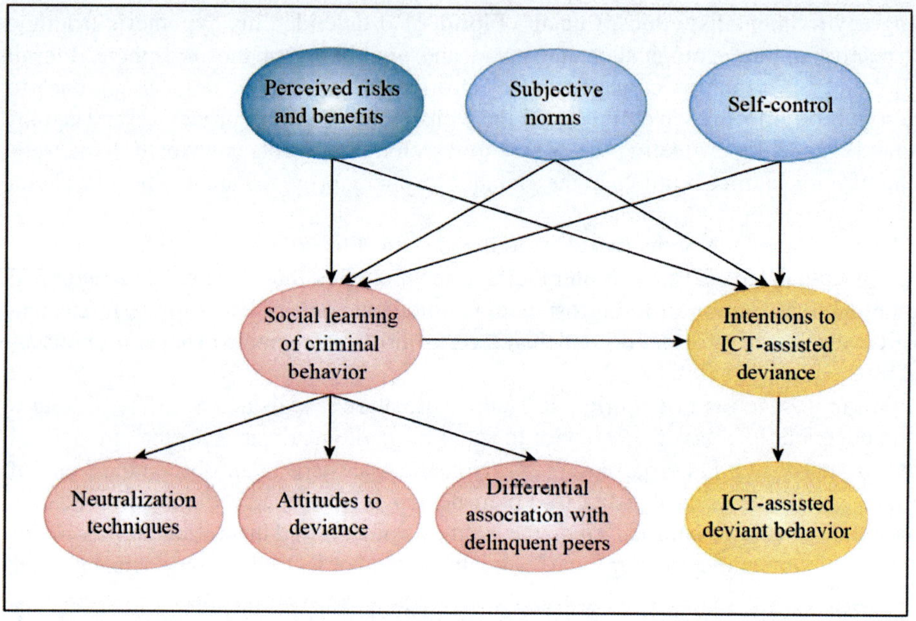

Fig. 1. A General priory conceptual model of ICT-assisted unethical behavior

Owing to the all-encompassing nature of ICT ethics, empirical studies have tackled the diverse field of problematic ICT-ethics behavior, such as academic dishonesty, clinical misconduct, digital piracy, identity theft, online offending, and the misuse of the ICT resources and computing infrastructure among undergraduates. Peace et al. [32] empirically demonstrated the viability of deterrence theory to explain variance in undergraduates' intentions to perpetrate software piracy. In contrast, Morton and Koufteros [33] demonstrated that neither the perceived certainty of punishment nor the perceived severity of punishment had a statistically significant effect on the attitudes toward music piracy.

Yu [34] attempted to test the neutralization theory as a Social Cognitive Theory (SCT) framework through the application of a sequential exploratory mixed methods design. In Yu's qualitative study, 40 preliminary semi-structured interviews were conducted on undergraduate students. The study demonstrated that students tended to countenance digital piracy behaviors. Thus, the moral judgment regarding the acceptability of such behaviors was positively correlated with its justifications among the participants [34].

Lee et al. [5] conducted a series of bivariate correlations and logistic regression models to test the hypothesized relationship between e-book piracy among undergraduate students in South Korea, the differential association with pirating peers, the perceived risks, and neutralizing techniques. Albeit at varying degrees of magnitude and significance, the differential association with delinquent peers and the perceived prosecution risks were significantly related to the propensity to be involved in illegal e-book downloading. These results were consistent with the results of previous studies conducted by Morris and Higgins [35] and Yang [36].

Through the adoption of a socio-psychological approach to e-book piracy, Camarero et al. [6] found that engagement in e-book piracy is associated with the degree to which individuals harness information technology. In addition, Fleming et al. [37] executed a meta-synthesis of illegal file-sharing related literature and asserted that perceived behavioral control and past experiences with illegal file-sharing are critical determinates of the intention to commit digital piracy.

Similar to Yu [34], Lee et al. [6] found that neutralization techniques translated into the justifications provided by participants to abolish feelings of guilt or shame through the voiding of moral judgment. Additionally, Urbonavicius et al. [38] substantiated the negative relationships between neutralization techniques, moral emotions such as shame and guilt, and attitudes on movie piracy through a sample of Generation Z participants in Lithuania. Furthermore, by utilizing a binary logistic regression, Lee et al. [6] analogously concluded that a differential association with deviant peers positively predicts the illegal downloading and uploading of copyrighted media files. In contrast, the perceived legality was not a significant predictor of engagement in digital piracy [6].

The reviewed studies emphasized the explanatory power of socio-psychological predictors on ICT-assisted deviance. A magnitude of studies demonstrated the viability of the social cognitive theory of unethical behavior alongside control theories to explain specific patterns of ICT-assisted deviance. The vast majority of the related studies adopted behavioral theories and employed structural and correlational analysis to examine the associations between socio-cognitive constructs, intent, and behavior. Few studies have sought to examine whether findings regarding the relationships between socio-cognitive constructs of unethical behavior and ICT-assisted deviance remain empirically valid against in-depth qualitative evidence.

3 Method

Reasoning in the qualitative inquiry is accompanied by the constant comparative method [39, 40], which is a commonplace occurrence among approaches to qualitative data analysis [41]. Constant comparative analysis originally advocated for the use of the grounded theory approach [42]. The constant comparative method of the grounded theory approach differs from other forms of qualitative research because constant comparison yields concepts based on which theory is constructed and gleaned from collected data [43].

Grounded theory is not committed to specific types of data or prior theoretical assumptions; it embodies distinctive methodological benchmarks that are simultaneously undertaken [44]. These methodological guidelines, including the use of constant comparative and theoretical sampling methods, in addition to the use of a coding frame of reference, are adopted to consolidate the density and saturation of the generated theory

[41–44]. Subsequently, a pivotal step between data collection and constant comparative analysis is memo-writing, which expedites the construction of theoretical categories and enhances their level of abstraction [45].

Coding the data involves a focus on patterns and insights in terms of the research objectives and questions [46]. Three types of coding are typically distinguished in methodological literature: open coding, axial coding, and selective coding [39, 42, 43, 46]. Charmaz's grounded theory [45] recognizes three types of coding: initial coding, focused coding, and theoretical coding. This paper adopts three rounds of coding from Charmaz. The emergent theory is then illustrated in a process-oriented model. ATLAS.ti software was employed throughout each of the phases to assist in qualitative data management and analysis.

Prior to being effectuated to collect data at SQU, three different qualitative interview protocols were designed, ethically reviewed, and approved by the Advisor for Academic Affairs at SQU administration. Additionally, considerable emphasis was placed on the initiation and implementation of informed consent before data collection. Three different purposeful samples were exclusively drawn from SQU. SQU undergraduates ($n = 26$) participated in nine focus group discussion sessions. Additionally, SQU faculty members ($n = 17$) and information systems professionals ($n = 2$) participated in semi-structured interviews.

4 Results

Findings from the constant comparative analysis revealed two primary theoretical constructs that were pertinent to the research questions. While exploring the data, thematic codes and conceptual categories emerged and were conducive to scrutinizing the reality of the undergraduates' intentions and behaviors in connection with ICT-ethics. Figure 2 summarize the thematic schemes that emerged from the constant analysis.

4.1 The Nature of Undergraduates' Intentions Toward ICT Ethics-Related Behavioral Intentions

4.1.1 Reflecting on the Nature and Function of Moral Agency in the Exercise of Ethical Decision Making in ICT-Ethics Pertinent Situations

The qualitative data analysis revealed that undergraduates' intentions towards ICT-ethics-related behavior are associated with the process of ethical decision making. Thus the construct of ICT-ethics-related behavioral intention is conceived as being processual in the sense that it encompasses interactive phases rather than a sequence of discrete events.

The participants linked perceived ethical self-efficacy to perceived behavioral control, and hence, they expounded ICT-assisted deviant behavior in terms of the low-self-control theory of unethical behavior [9]. Based on the processual nature of ethical decision making, the participants evoked the educational implications of ethical decision making in the field of ICT ethics.

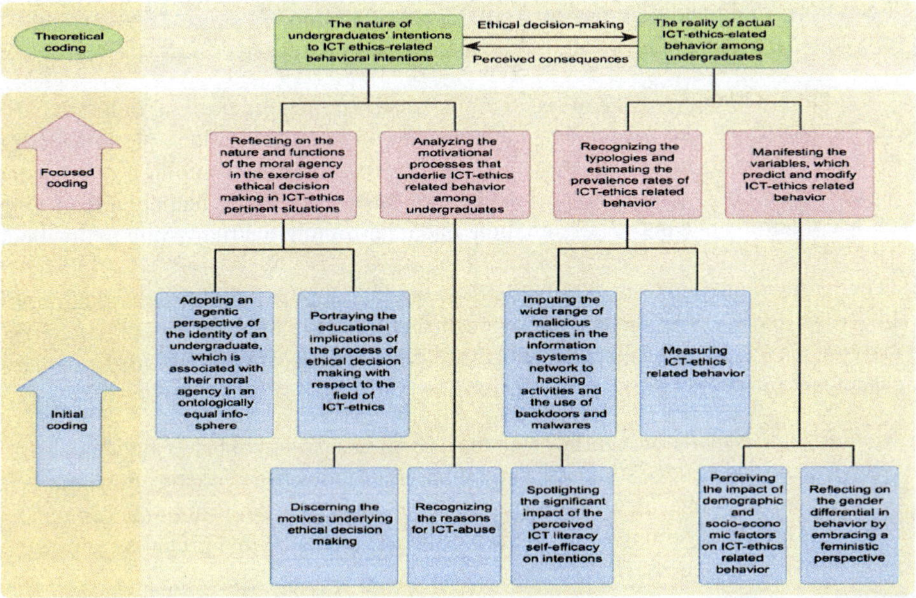

Fig. 2. The thematic scheme that emerged from the constant comparative analysis

4.1.2 Adopting an Agentic Perspective of the Identity of an Undergraduate Student, Which is Associated with Their Moral Agency in an Ontologically Equal Infosphere

The interviewees discussed the ethical implications of technological instrumentation of human actions from the perspective of the social contract theory of ethics. Although the concept of human ethical agency is at the center of the deontological theory of ethics, the ethical agency is considered a focal concept when the ethical implications of the ICT instrumentation of human actions are examined from the perspective of the social contract theory.

In this respect, the undergraduate participants in this study delineated the realm of academic integrity based on the moral philosophy of rights and duties.

4.1.3 Portraying the Educational Implications of the Process of Ethical Decision Making with Respect to the Field of ICT-Ethics

The participants emphasized the significance of cultivating self-autonomy and self-actualization skills among undergraduates to enable them to make sound ICT-related ethical decisions in dilemmatic situations. According to a participant,

> We call this in psychology 'Self-actualization.' Self-actualization refers to whether individuals realize that they are responsible for their own actions, as well as being responsible for the results of their actions without blaming other people. They have to be aware of the related laws, and they have to be responsible for every action they take. They should not claim that they did not know the law. Undergraduates

> need to be reminded that they are now responsible human beings. Their parents are no longer taking care of all their responsibilities. (P.AF.6)

To highlight the importance of solid ethical decision-making training in ICT-ethics-involving situations, the participants stressed that undergraduates lack the knowledge and skills needed to make appropriate ethical decisions. For example, a respondent maintained that undergraduates "need to be trained on two things, namely, identifying an ethical dilemma in a particular situation, and determining the right thing to do about it" (P.AF.8).

The participants highlighted the effects of the interaction between individual moderators (e.g., the locus of control and egocentrism) and situational/immediate context moderators on the relationship between the ICT-related ethical intentions and behaviors. A participant indicated that:

> Breaking into e-mail accounts has become a norm or a pattern of behavior, which is prevalent among college students. They share links to hacked accounts, download personal information, files, or images, or send messages with immoral content from hacked mail accounts. This pattern of behavior can be interpreted as a form of egoism or selfishness. (P.SN.F3)

4.2 Analyzing the Motivational Processes that Underlie ICT Ethics-Related Behavior Among Undergraduates

As a crucial predictor of overt behavior, Ajzen [47] declares that the construct of *intention* encapsulates the motivational factors that determine an individual's likeliness to perform the behavior. This is because a person's estimate of the likelihood or the subjective perceived probability of performing a given behavior is the chief underlying dimension of behavioral intentions [48].

In the context of the study, the participants examined subjectively the perceived probability of performing ICT ethics-related ethical behavior by (1) discerning the motives or types of motivational constructs underlying ethical decision making, (2) diagnosing the etiology of ICT ethics-related aberrant behavior, and (3) spotlighting the significant impact of the perceived ICT literacy self-efficacy on intentions towards ICT-ethics-related behavior.

4.2.1 Discerning the Motives or Types of Motivational Constructs Underlying Ethical Decision Making

Across the three groups of respondents, 494 quotations emphasized that having an extrinsic motivation to learn, use, and disseminate information creates a selective activation of self-sanctions, thereby leading to the moral disengagement of internal control. For example, 30 quotations indicated that undergraduates might, through ICTs, falsify empirical data for extrinsic motives such as passing exams, earning credit hours, graduating, obtaining the first degree, and so on.

Another example of extrinsic motives is hacking information systems that contain data regarding previous or upcoming exams, students' attendance records, and grades to manipulate or steal the data. Some participants traced computer-related deviances among undergraduate students to financial issues such as income instability, which leads some students to use digital piracy to access information. Other participants have pointed out that there are gratification motives that lie beneath ICT-related unethical behavior such as curiosity, challenging the forbidden, and the excitement and ecstasy that result from violating rules and regulations, especially concerning violating information privacy.

The perceived prosecution risks of unethical ICT-assisted practices are expected to be negatively associated with the construct of the moral disengagement of internal self-control and the construct of selective activation of internal self-sanctions and perceived social sanctions. However, the construct of optimism bias is anticipated to be positively related to moral disengagement, and therefore positively related to selective activation of self-sanctions and perceived social sanctions.

4.2.2 Recognizing the Reasons for ICT-Abuse

To identify the causes of ICT-ethics-allied deviance, the participants employed the ontological characteristics of cyberspace. A participant asserted that:

> *It is not unusual in social networking sites that an individual has multiple accounts. One of the accounts is under the real name of the user, while the other accounts are under pseudonyms, through which unethical information activities are carried out.* (P.SL.M2)

Additionally, when discussing the causes of hacking activities, participants tended to distinguish between hacking motivated by malicious intentions, hacking intended for benign intrusions, and hacking in response to computer intrusions [49].

4.2.3 Spotlighting the Significant Impact of the Perceived ICT Literacy Self-efficacy on Intentions Towards ICT Ethics-Related Behavior

The participants asserted that perceived ICT literacy self-efficacy had varied effects on intentions toward ICT-ethics-related behavior. However, a considerable ratio of the participants claimed that undergraduates were likely to commit unethical information practices such as hacking, digital piracy, and illegal downloading of multimedia, to prove their ICT literacy self-efficacy. For instance, a participant stated that "*the high level of IC -literacy self-efficacy leads to a high level of curiosity and thus leads to an increased probability of engagement in ICT-facilitated deviant behavior*" (P.SA.F3).

4.3 Recognizing the Typologies and Estimating the Prevalence Rates of ICT Ethics-Related Behavior

4.3.1 Imputing the Wide Range of Malicious Practices in the Information Systems Network to Hacking Activities and the Use of Backdoors and Malware

The participants were attentive to hacking activities as a significant threat to computer and network security. Some participants confined ICT-aided-deviance to hacking. Overall, the respondents considered breaching of ethical rules and ICTs in academia to access information, as a common hacking activity.

The perceived prevalence of low level of hacking activities among the undergraduates is interpreted from the stances of differential association and observational learning. For instance, a participant indicated that:

> *Students usually acquire hacking skills based on the information they communicate to their colleagues, about hacking, using websites, or via social networking sites. Usually, such information is circulated and disseminated among youths and adolescents, which facilitates the spread of hacking.* (A.PIS.1)

4.3.2 Measuring ICT Ethics-Related Behavior

Participants requested the psychometric measurement of ICT ethics-linked behavior among undergraduates; however, the students' ICT ethics-linked behavior could be assessed using several psychological measures.

4.4 Manifesting the Variables that Predict and Modify ICT-Ethics-Related Behavior

4.4.1 Underscoring the Impact of Demographic and Socio-economic Factors on ICT-Ethics-Related Behavior

The participants acknowledged the impact of demographic and socio-economic factors on behavior. However, the participants established that the impact of demographic and socio-economic factors of ICT-ethics associated behavior was not invariant. Comparably, they denied any considerable impact of age or education level on ethical behavior, though distinctions were made between students coming from urban and rural areas in terms of their ICT ethical behavior.

The participants ascribed the differences in ICT-pertinent ethical behavior to individual differences in personal, cognitive, motivational, and environmental factors. For instance, differences in ethical behavior were attributed to personal factors such as intellectual openness, analytic skills, and environmental sociocultural.

4.4.2 Reflecting on the Gender Differential in Behavior by Embracing a Feministic Perspective

The gender differential in behavior was substantiated by gender comparison. Male students who resided off-campus tended to neutralize dishonest academic behaviors based on circumstantial factors and situational philosophy of ethics, more so than female students who resided on campus, in university hostels. The participants compared male and female students in terms of digital plagiarism from a feminist perspective by posing statements or questions regarding society's male domination asymmetry and the power relations between the genders.

There were qualitative differences between genders in terms of the types of investments in which each gender invests their assets, dispositions, capacity capital, and aptitudes. To a substantial extent, the type of investments is determined by the triadic reciprocal influences among environmental, personal, and behavioral factors. Nonetheless, this typology of investments among the genders remains relative and is subject to societal shifts and contextual factors such as co-education and the diffusion of a knowledge-oriented economy.

5 Discussion

5.1 Intentions Among Undergraduates to ICT-Aided Deviant Behavior

5.1.1 Perceptions of the Risks Associated with the Illegal Usage of ICTs and the Privacy Policies Breaches

The participants in this qualitative study expressed that there is a need for disseminating intellectual property laws and counteractive laws of information technology crimes, a great deal of which are concerned with information privacy especially within the contexts of higher education institutions. Furthermore, young users of social media services like Twitter and Snapchat need to be aware the implications of privacy policies on these types of social media platforms. Among undergraduate students, the inclinations to violate information privacy can be traced to moderate levels of knowledge of the legal acts that are related to information privacy and IT-related crimes. Furthermore, these results demonstrated that ICT-ethical decision-making is influenced by the contextual affordances offered by the ICTs, such as anonymity, reproductively of information, and fluidity.

The undergraduates who expressed a willingness to expose information about other individuals online were not fully aware that such an act is considered Internet libel. The emergence of the participatory sharing culture in which users share content online with friends has created a new media model and, and simultaneously, raised ethical and legal concerns. Thus, social media posts can create legal records of communication and behavior, which can be utilized in potential lawsuit cases, especially in areas related to the invasion of information privacy and defamation [50].

The results from the qualitative data analysis expounded that the participating undergraduates tended to neutralize insecure password practices for two reasons. First, undergraduates neutralize activities that allow others to benefit from their user-assigned access privileges. Second, they underestimate or otherwise do not perceive the information security and privacy-related risks associated with sharing authentication data with other individuals, especially from outside the university community.

Based on the deterrence theory and the social cognitive theory of unethical behavior, user apathy toward information systems policies might be caused by irregular updates of staff and student manuals, as well as by the inadequate enforcement of deterrent measures [51]. Consequently, the impact of the optimism bias and differential association with deviant peers could neutralize the impact of the perceived risks of ICT-assisted deviant behavior among employees and students.

5.1.2 Stimuli that Provide the Impetus for Violating ICT Ethics Among Undergraduates

Results of the qualitative analysis demonstrated considerable intentions towards committing at least one type of academically dishonest practice among undergraduates. These results can be interpreted using a variety of reasons, such as personal and institutional determinants.

For the institutional factors, they included a lack of the use of plagiarism detection software by academics and lack of awareness-raising campaigns and enforcement of the deterrent measures stated in the academic integrity policy and academic honor codes. These reasons are consistent with the reasons identified by Yeung et al. [52], who labeled institutional determinants as academic integration.

According to the respondents (i.e., undergraduates and faculty), intrusive patterns of ICT-assisted disruptive behavior disclosed an intention among hacking undergraduates to seek thrills as these hackers already possess a basic understanding of information systems.

Comparably, the interviewed information systems administrators and academics stressed that the motives underlying hacking and cracking attempts committed by some undergraduates that were detected by the Intrusion Detection System (IDS) could not be classified as the actions of malicious insiders or cyber unethical acts. Based on interview data and existing literature, this inability to classify is because malicious insiders and cyber unethical s typically breach the security of information systems for financial gains, disrupting an organization's information systems and operations, or destroying the IT infrastructure of an institution, utilities, and emergency response units [3].

Instead, those undergraduate hackers usually aim to steal data and gain publicity among their peers but not corrupt the information systems. In their exploration of the motives that underlie the hacking behavior among undergraduates, the interviewed participants stated that academic self-concept constitutes one of the factors that predict ICT-assisted deviant behavior and intentions among undergraduates Academic self-concept comprises the attitudes, beliefs, and perceptions adopted by students regarding their academic skills and performance [53]. Areepattamannil and Freeman [53] elucidate that academic self-concept is a crucial component of academic motivation.

5.2 Prevalence of ICT-Ethics-Connected Deviant Behavior Among Undergraduates

5.2.1 Classification of ICT-Assisted Deviant Behavior

The participating undergraduates classified ICT-assisted deviance as practices that are intended to breach the privacy and security of information systems. This classification reveals how the undergraduates visualize ICT-assisted unethical behavior as deviant behavior. Ergo, the undergraduates' classification of ICT-assisted deviance is consistent with Moon et al. [9] typology of ICT-assisted crimes, including computer crime, Internet crime, and cyber-crimes.

Therefore, the data demonstrated that the participants contend that ICT-ethics-related behavior can be gauged using both quantitative and qualitative approaches, with a phenomenological inquiry. For established standardized empirically validated typologies of ICT-ethics-related behavior, the ICT-ethics-related empirical literature signaled the demand for longitudinal mixed-methods ICT-ethics studies.

5.2.2 Factors Influencing the Phenomena of Unethical ICT-Assisted Practices

The participants underscored the personal, environmental, cognitive, and behavioral variables. Based on the triadic interactive view of the social cognitive theory, the participants attributed divergences in ethical behavior among undergraduates, to individual differences in personality traits, analytic skills, motivational types, and environmental sociocultural factors. For example, the participants asserted that the personality traits of conscientiousness and intellectual openness were positively related to ICT ethics-related behavior. These assertions have been empirically substantiated by several ICT ethics-related empirical studies [54–58].

The interviewed academics arrogated the sociocultural approach to fostering academic integrity among the undergraduates by mediating and modeling compliance with ICT-ethics. According to Morris and Higgins [35], among Generation Z, parents are the primary peer group that influences ICT-ethics-related behavior. The interviewed academics refuted the effect of parental educational level on the ICT-ethics-related behavior among Generation Z, and similar to Stahl [11] asserted, that a digital divide had an impact on the differences in ICT ethics-related behavior between Generation Z and the preceding generations in terms of the accessibility to ICTs.

The interviewees also alluded to the impact of organizational culture on the ethical behavior of the undergraduates. Organizations that operate in a cloud computing environment benefit from data storage, data downloading, and data sharing services that can be accessed externally using a Web browser, which thus could provoke further information systems' security concerns [3].

6 Conclusion: A General Theory of Ethical Behavior in an IT-Configured Context

As gleaned from the qualitative data and based on the broad theory of social cognition, provisional presumptions can be induced to initiate a general criminological theory of ICT-expedited deviance. First, the perceived probability and severity of risks associated with IT-assisted deviance, especially, prosecution risks, are inversely proportional to the propensity to behave unethically using ICTs. Second, low self-control and low perceived ICT ethical self-efficacy mitigate the impact of the perceived potential external deterrents. Third, parental mediation, vicarious learning experiences, academic integration, and an information security-connected organizational culture enhance the levels of self-control and the perceived ICT-ethical self-efficacy. Fourth, the reverse is true for differential association with delinquent peers, especially in virtual environments. Differential association neutralizes the effects of self-control and perceived ICT-ethical self-efficacy. Fifth, the ramifications of differential association with deviant peers escalate the effects of neutralization techniques, and the moral disengagement mechanism on ethical intentions, which are an immediate proxy of overt behavior. These propositions are illustrated in Fig. 3.

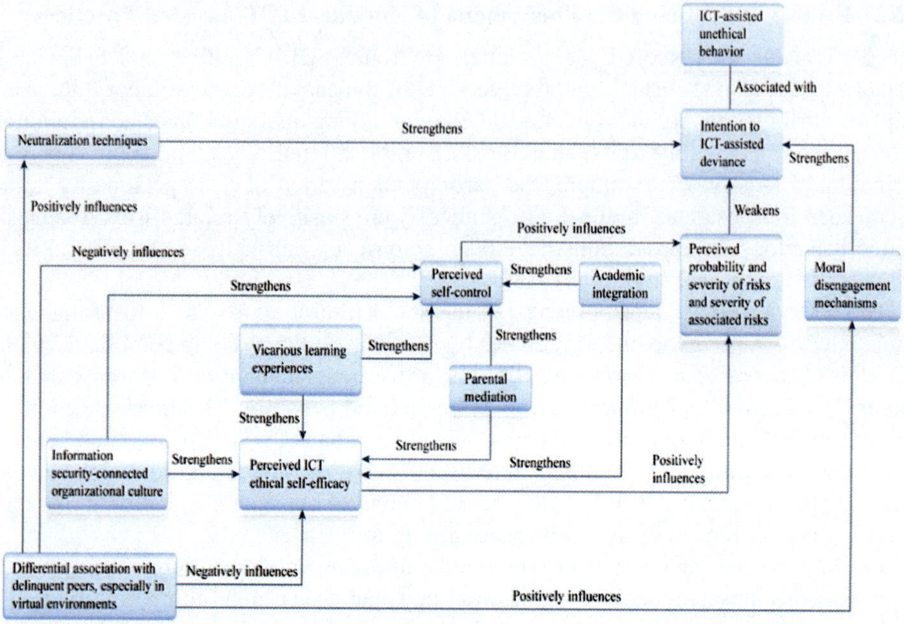

Fig. 3. A grounded theory model of unethical behavior in IT-configured contexts

6.1 Recommendations for Further Research

The participants described the unique characteristics of digital media and IT tools, in the age of information, as a source of the causes for engaging in ICT-facilitated deviance. The participants insisted on IT-based solutions as well as educative, preventive, and deterrent measures that can be effectively employed to counteract the prevalence of ICT-assisted unethical behavior among undergraduates. However, the mechanisms by which IT-based solutions, especially in higher academic institutions, can alter undergraduates' motives towards violating ICT-ethics and influence undergraduates to form unfavorable intentions to deviant ICT-linked behavior, remain equivocal. Therefore, longitudinal studies should be conducted for an in-depth exploration of these areas. Furthermore, ethnographic studies are anticipated to better understand the relationships between the information security-connected organizational culture and academic integration and ICT-ethics behavior among undergraduates.

References

1. Floridi, L.: Ethics after the information revolution. In: Floridi, L. (ed.) The Cambridge Handbook of Information and Computer Ethics, vol. 1, pp. 3–19. Cambridge University Press, New York (2010)
2. Johnson, D.G., Miller, K.W.: Ethical issues for computer scientists. In: Tucker, A.B. (ed.) Computer Science Handbook, vol. 2. Chapman & Hall/CRC, Boca Raton (2004)

3. Reynolds, G.W.: Ethics in Information Technology, 4th edn. Cengage Learning, Boston (2015)
4. Sardanelli, D., Vollero, A., Siano, A., Bottoni, G.: Lowering the pirate flag: a TPB study of the factors influencing the intention to pay for movie streaming services. Electron. Commer. Res. **19**(3), 549–574 (2019). https://doi.org/10.1007/s10660-019-09346-7
5. Lee, B., Fenoff, R., Paek, S.Y.: Correlates of participation in e-book piracy on campus. J. Acad. Librariansh. **45**(3), 299–304 (2019)
6. Camarero, C., Antón, C., Rodríguez, J.: Technological and ethical antecedents of e-book piracy and price acceptance: evidence from the Spanish case. Electron. Library **32**(4), 542–566 (2014)
7. Bynum, T.W., Rogersson, S.: Ethics in the information age. In: Bynum, T.W., Rogersson, S. (eds.) Computer Ethics and Professional Responsibility, vol. 1. Blackwell Publishing, Malden, pp. 1–13 (2004)
8. Ess, C.: Digital Media Ethics, 2nd edn. Polity Press, Malden (2014)
9. Moon, B., McCluskey, J.D., McCluskey, C.P.: A general theory of crime and computer crime: an empirical test. J. Unethic. Just. **38**(4), 767–772 (2010)
10. Lau, W.W.F., Yuen, A.H.K.: Internet ethics of adolescents: understanding demographic differences. Comput. Educ. **72**, 378–385 (2014)
11. Stahl, B.C.: Social issues in computer ethics. In: Floridi, L. (ed.) The Cambridge Handbook of Information and Computer Ethics, vol. 6, pp. 101–117. Cambridge University Press, New York (2010)
12. Jung, I.: Ethical judgments and behaviours: applying a multidimensional ethics scale to measuring ICT ethics of college students. Comput. Educ. **53**(3), 940–949 (2009)
13. Khang, H., Ki, E.-J., Park, I.-K., Baek, S.-G.: Exploring antecedents of attitude and intention toward Internet piracy among college students in South Korea. Asian J. Bus. Ethics. **1**(2), 177–194 (2012)
14. Al-Rafee, S., Cronan, T.P.: Digital piracy: factors that influence attitude toward behaviour. J. Bus. Ethics **63**(3), 237–259 (2006)
15. Banerjee, D., Cronan, T., Jones, T.: Modeling IT ethics: a study in situational ethics. MIS Q. **22**(1), 31–60 (1998)
16. Cronan, T.P., Al-Rafee, S.: Factors that influence the intention to pirate software and media. J. Bus. Ethics **78**(4), 527–545 (2008)
17. Zlatolas, L.N., Welzer, T., Heričko, M., Hölbl, M.: Privacy antecedents for SNS self-disclosure: the case of Facebook. Comput. Hum. Behav. **45**, 158–167 (2015)
18. Akbulut, Y., Dönmez, O.: Predictors of digital piracy among Turkish undergraduate students. Telematics Inform. **35**(5), 1324–1334 (2018)
19. Floridi, L.: Foundations of information ethics. In: Himma, K.E., Tavani, H.T. (eds.) The Handbook of Information and Computer Ethics, vol. 1, pp. 3–24. Wiley, Hoboken (2008)
20. Floridi, L.: The Ethics of Information. Oxford University Press, Oxford (2013)
21. Akbulut, Y.: Exploration of the antecedents of digital piracy through a structural equation model. Comput. Educ. **78**, 294–305 (2014)
22. Cilliers, L.: Evaluation of information ethical issues among undergraduate students: an exploratory study. South African J. Inf. Manag. **19**(1), 1–6 (2017)
23. Jurdi, R., Hage, H.S., Chow, H.P.H.: What behaviours do students consider academically dishonest? Findings from a survey of Canadian undergraduate students. Soc. Psychol. Educ. **15**(1), 1–23 (2012)
24. Tjiptono, F., Arli, D., Viviea, I.: Gender and digital privacy: examining determinants of attitude toward digital piracy among youths in an emerging market. Int. J. Consum. Stud. **40**, 168–178 (2016)
25. Bhattacherjee, A., Shrivastava, U.: The effects of ICT use and ICT laws on corruption: a general deterrence theory perspective. Govern. Inf. Quar. **35**(4), 703–771 (2018)

26. Ngo, F.T., Paternoster, R., Cullen, F.T., Mackenzie, D.L.: Life domains and crime: a test of Agnew's general theory of crime and delinquency. J. Unethic. Justice **39**(4), 302–311 (2011)
27. Pratt, T.C.: A self-control/life-course theory of unethical behaviour. Eur. J. Criminol. **13**(1), 129–146 (2015)
28. Scheuerman, H.L.: The relationship between injustice and crime: a general strain theory approach. J. Unethic Justice **41**(6), 375–385 (2013)
29. Mehlkop, G., Graeff, P.: Modelling a rational choice theory of unethical action: subjective expected utilities, norms, and interactions. Ration. Soc. **22**(2), 189–222 (2010)
30. Agnew, R.: Building on the foundation of general strain theory: specifying the types of strain most likely to lead to crime and delinquency. J. Res. Crime Delinquency **38**(4), 319–361 (2001)
31. Wolfe, S.E., Higgins, G.E., Marcum, C.D.: Deterrence and digital piracy: a preliminary examination of the role of viruses. Soc. Sci. Comput. Rev. **26**, 317–333 (2008)
32. Peace, A.G., Galletta, D.F., Thong, J.Y.L.: Software piracy in the workplace: a Model and empirical test. J. Manag. Inf. Syst. **20**, 153–177 (2003)
33. Morton, N.A., Koufteros, X.: Intention to commit online music piracy and its antecedents: an empirical investigation. Struct. Eq. Model. Multidisc. J. **15**, 491–512 (2008)
34. Yu, S.: College students' justification for digital piracy: a mixed methods study. J. Mixed Methods Res. **6**, 364–378 (2012)
35. Morris, R.G., Higgins, G.E.: Criminological theory in the digital age: the case of social learning theory and digital piracy. J. Unethic Justice **38**, 470–480 (2010)
36. Yang, S.C.: Attitudes and behaviours related to academic dishonesty: a survey of Taiwanese graduate students. Ethics Behav. **22**, 218–237 (2012)
37. Fleming, P., Watson, S.J., Patouris, E., Bartholomew, K.J., Zizzo, D.J.: Why do people file share unlawfully? A systematic review, meta-analysis, and panel study. Comput. Hum. Behav. **72**, 535–548 (2017)
38. Urbonavicius, S., Dikcius, V., Adomaviciute, K., Urbonavicius, I.: Movie piracy: how novelty-seeking relates to illegal downloading. EuroMed J. Bus. **14**, 21–30 (2019)
39. Boeije, H.: Analysis in Qualitative Research. SAGE Publications Inc., Thousand Oaks (2010)
40. Palmberger, M., Gingrich, A.: Qualitative comparative practices: dimensions, cases, and strategies. In: Flick, U. (ed.) The SAGE Handbook of Qualitative Data Analysis, vol. 7, pp. 94–108. SAGE Publications Inc., London, (2014)
41. Harding, J.: Qualitative Data Analysis: From Start to Finish. Sage Publications Inc, Thousand Oaks (2013)
42. Glaser, B.L., Strauss, A.L.: The Discovery of Grounded Theory: Strategies for Qualitative Research. Aldine Transaction, New Brunswick (1967)
43. Corbin, J., Strauss, A.: Basics of Qualitative Research: Techniques and Procedures for Developing Grounded Theory. SAGE Publications Inc., Thousand Oaks (2015)
44. Strauss, A.L.: Qualitative Analysis for Social Scientists. Cambridge University Press, Cambridge (1987)
45. Charmaz, K.: Constructing Grounded Theory, 2nd edn. Sage Publications Ltd., London (2014)
46. Merriam, S.B., Tisdell, E.J.: Qualitative Research: A Guide to Design and Implementation. Jossey-Bass, San Francisco (2016)
47. Ajzen, I.: The theory of planned behavior. Organ. Behav. Hum. Decis. Process. **50**, 179–211 (1991)
48. Fishbein, M., Ajzen, I.: Predicting and Changing Behaviour: The Reasoned Action Approach. Routledge, New York (2015)
49. Himma, K.E.: Ethical issues involving computer security: hacking, hacktivism, and counter-hacking. In: Himma, K.E., Tavani, H.T. (eds.) Handbook of Information and Computer Ethics, vol. 8, pp. 191–218. Wiley Inc., Hoboken (2008)

50. Lipschultz, J.H.: Social Media Communication: Concepts, Practices, Data, Law, and Ethics, 2nd edn. Routledge Taylor and Francis, New York, NY (2018)
51. Kim, D., Solomon, M.G.: Fundamentals of Information Systems Security, 3rd edn. Jones & Bartlett Learning, Burlington (2018)
52. Yeung, A.H.W., Chu, C.B.L., Chu, S.K.-W., Fung, C.K.W.: Exploring junior secondary students' plagiarism behaviour. J. Librariansh. Inf. Sci. **50**, 361–373 (2018)
53. Areepattamannil, S., Freeman, J.G.: Academic achievement, academic self-concept, and academic motivation of immigrant adolescents in the greater Toronto area secondary schools. J. Adv. Acad. **19**, 700–743 (2008)
54. Bollich, K.L., Hill, P.L., Harms, P.D., Jackson, J.J.: When friends' and society's expectations collide: a longitudinal study of moral decision-making and personality across college. PLoS ONE **11**, 1–15 (2016)
55. Bratton, V.K., Strittmatter, C.: To cheat or not to cheat? The role of personality in academic and business ethics. Ethics Behav. **23**, 427–444 (2013)
56. Brown, S.C., MacDonald, R.A.R.: Predictive factors of music piracy: an exploration of personality using the HEXACO PI-R. Musicae Scientiae **18**, 53–64 (2014)
57. Pohling, R., Bzdok, D., Eigenstetter, M., Stumpf, S., Strobel, A.: What is ethical competence? The role of empathy, personal values, and the five-factor model of personality in ethical decision-making. J. Bus. Ethics **137**, 449–474 (2016)
58. Wilks, D.C., Cruz, J.N., Sousa, P.: Personality traits and plagiarism: an empirical study with Portuguese undergraduate students. J. Acad. Ethics **14**, 231–241 (2016)

Management Information Systems

A Comparative Study of eBay and Amazon Electronic Commerce Platforms

Yousef A. Baker El-Ebiary[1], Waheed Ali H. M. Ghanem[2](✉), Syarilla Iryani A. Saany[1], Ahmad Nazari Mohd Rose[1], Julaily Aida Jusoh[1], and Hasni Hassan[1]

[1] Faculty of Informatics and Computing, UniSZA University, Kuala Terengganu, Malaysia
{yousefelebiary,syarilla,julaily,hasni}@unisza.edu.my

[2] Faculty of Ocean Engineering Technology and Informatics, Universiti Malaysia Terengganu, Kuala Terengganu, Malaysia
waheedghanem@umt.edu.my

Abstract. The key target of this documentation is to study and compare e-commerce platforms between e-Bay and Amazon. While Amazon.com gained more promotions and underlying exposure to web-based businesses, eBay has secretly built inventories that are truly Internet-friendly. The research used an analytical method to find the difference between these two platforms. Initially, Amazon was only trying to reproduce a list of action plans on the web. The innovation may be inventive, but the design of the action is certainly not. Once again, eBay has a great idea of the Internet network and has empowered supply and demand on the web with incredible outcomes. The sales position is the winner. Besides, eBay clearly shows that the benefits do not need to be in the idea of development - a dispute that Bezos never made. Amazon initially centered around BN.com as a competitor. After a while, Amazon saw eBay as a competitor. The underlying attack was a tremendous disappointment. Currently, Amazon is endeavoring to contend with eBay by driving sales and consolidating its partner programs.

Keywords: Electronic commerce · eBay · Amazon · Comparative study · E-commerce platforms

1 Introduction

An E-Commerce organization offers online sales based on items where customers try to outdo each other by placing the highest amount. It converts expenses for postings (whether the item is for sale) and imposes additional charges for Premium Auction highlights (e.g., additional images, BuyItNow, Feature It!) Or Premium Seller sign-ups [1].

E-commerce is a type of technology business that lets users buy merchandise then administration directly from merchants through the web via a website browser. Online stores inspire the physical relationship of buying or administering a block or mortar retailer; the procedure is called business-to-buyer (B2C) online shopping.

M. Al-Emran et al. (Eds.): ICETIS 2021, LNNS 299, pp. 371–380, 2022.
https://doi.org/10.1007/978-3-030-82616-1_32

For situations where a business buys from another business, this procedure is called business-to-business (B2B) online business [2].

Marketing achievements at physical stores are not the same as online stores. This is clear given the growing number of retailers now offering a web-shop interface for customers. With the expansion of internet shopping, there are many new market opportunities including openings for stores that can accurately reflect overseas market demand and administrative needs [3]. The largest web-based retail companies are Amazon.com and E-inlets. Retail achievements are not at this level about physical stores [4]. This is seen as a result of the development of retailers that now offer a web-shop interface for shoppers. With the advent of online shopping, many new market opportunities have emerged including openings for shops that meet the demands of offshore markets [5].

2 Literature Review

The start of the Amazon store was limited to a group of books and some magazines, but the fame that distinguished the store in a short time was based on the development of three basic elements:

Brand building: If the store displays a new mark on its website, a good name is important to attract attention, as the brand must reflect the product and distinguish it from its competitors in the local and international market. Also, to legally protect the name on the Internet, it is recommended that the name must be short, precise, and easy to remember. As an example, Dell is a short name that is easy to remember [6]. In a short time, he was able to announce his position among his family and friends. Although the beginning of the project was 300 thousand dollars, the signs of success were quick, and he got 20000 dollars in the first week. On this basis, the brand played an important role in raising the cost-effectiveness of the project at an amazing speed.

Growth and expansion strategy: Displaying a large assortment of products on the site contributes effectively and makes it more acceptable to visitors. Therefore, the Amazon store followed the strategy of diversification in the products presented on its website, which it offers as an effective way to gain customers, thus the store has turned into the largest retail distributor on the World level [7]. We also find the expansion strategy in terms of establishing global branches of the company in some other countries, and by using websites, it has directly and effectively contributed to creating added value for the store at the international level.

The strategy of maintaining customer loyalty: Working to satisfy the customer and gaining his trust was a key factor for the spread of the Amazon store and its electronic platform. This is based on studies that revealed that one of the most important obstacles facing electronic stores working in the field of electronic commerce is the customer's confidence and preservation of it due to competition and other factors that affect the business, directly or indirectly [8]. This situation usually affects stores offering the same goods at competitive prices and services such as multiple payment options, comparison between products and after-sales service.

Therefore, the study provides a comparison between the two platforms that are considered one of the giants of the e-commerce market, namely, Amazon and eBay.

3 Background of the Study

Electronic commerce is widely used as a way to integrate customers and suppliers into the global market economy. It is seen as an important innovation that promotes good governance, social and economic development, and poverty alleviation in certain developing countries [9]. Moreover, e-commerce mediums give Small and Medium Enterprises (SMEs) the chance to surpass their deficiency by giving them opportunities to grow up around the world, gaining new market opportunities, and attract new customers [10]. The benefits of e-commerce are the strong reasons that make e-commerce became the best medium for businesses and it is reflected in the significant growth of e-commerce users every year.

English tycoon Michael Aldrich went shopping online in 1979. His framework linked local TVs to PCs using mobile phones. He believes that videotext, a home-TV technology adapted to the direct-to-human PC-based interface, is 'new', participatory media correspondence - the first to appear on phones [11]. "This 'closed' corporate data framework was later opened to 'outside' journalists for additional processing and e-messaging and dissemination of data and distribution. It is then identified as e-business, meaning a new mass communication media by way of 'participatory' [collective, many to one], stakeholders. It is unique in its connection with the meaning of mass communication and broad communication and the introduction to long-distance informal communication on the web for twenty-five years. In March of 1980, it delivered the Redifon Office Revolution shipment, specialists, merchants, providers, and administrative organizations to connect directly to the corporate framework and progressively use the electronic business [12].

4 Data Analysis

eBay Inc. is US-predicated e-commerce company that delivers customer-to-consumer and business-to-consumer online sales platforms. Founded in 1995, eBay hires proximately 12,000 employees, with 6200 workers at the Cumulated States. Over 160 million customers buy via eBay every year and the company gains more than USD 8.6 billion only in 2015 [13].

The verbalization on eBay's purport is 'to be the world's leading destination for value and innovation'. eBay's business strategy is fixated on leveraging the benefits of the first-mover by introducing innovative accommodations and periodically integrating incipient features and capabilities to subsisting accommodations to ameliorate the utilizers' experience [14]. Senior management led by CEO and President Devin Wenig has made a strategic decision to move the business from the online auction to the most sizably voluminous online retailer, an area now dominated by Amazon [15].

Ecommerce companies have certain impotencies that include business reliance on the scope of products and accommodations performed by contestants such as Google, and the involution of the entire business model and the absence of a self-addressed distribution system for eBay competitors.

4.1 Product and Services

Started retailing collectibles and entertainment, Table 1 shows the full-fledged to contain various types of items [16].

Table 1. Items types.

Collectibles	Entertainment
	- Accessories
- Boats	- Books
- Cameras	- CD
- Cars	- Clothing
- Cell phones & PDAs	- Crafts
- Coins	- Décor
- Computers	- DVD
- Electronics	- Event Tickets
- Memorabilia	- Gardening
- Parts	- Home Improvement
- Toys & Dolls	- Jewelry
- Vehicles	- Movies, Music & Games
	- Shoes
	- VHS

4.2 e-Bay Target Segments

For online auctions and targeted shopping communities, e-Bay has a business plan for their specific segment. The main segment is “Antiques & Entertainment”. (It is the absence of an item collection organization that encouraged Pierre Omidyar to create eBay. The collection is tranquil one of the most popular in terms of clean product offerings, the delight of long-time veteran sellers.)

Motorcycle Lovers: Via the standing of Kruse Inc.’s top car collector, eBay has made a significant contribution to eBay Motors. It got one of its best parts with a $ 2,500 million Global Goods Sales at 2002. Art lovers: this platform also has several specific bomber organizations that focus on Art Collectors directly, as shown by the business processes (see Fig. 1 and 2) [17].

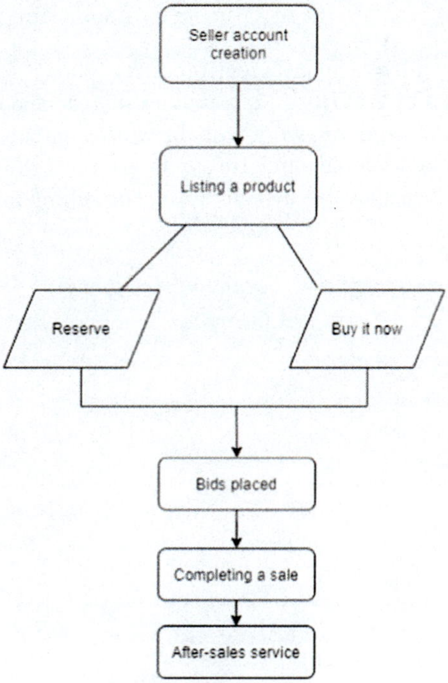

Fig. 1. Business process.

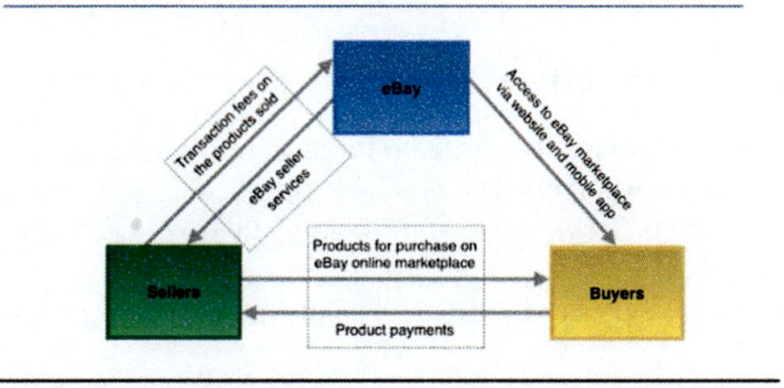

Fig. 2. e-Bay business model.

4.3 e-Bay Seller Consumer Interaction

Buyers: Get eBay listings, View product information itemized on eBay, Demonstration eBay citations on other websites, and allow comments around other customers at the end of trading transactions [18].

Retailers: Send a product used aimed toward sale on eBay, become maximum bidding data for products that sell, get an inventory of specific consumer products that are on sale, and Get an inventory of products offered by specific users.

4.4 Amazon's Vision

Amazon is an online retailer of many great products (books, toys, cooking utensils, and more) that originated by Jeff Bezos, a computer science and electrical engineering graduate. The vision of Amazon is to work with the world's greatest client-driven firm to make a platform where clients could purchase anything. Aiming to be a customer-centric company, its plan could be achievable using the Amazon selling format (see Fig. 3) [19].

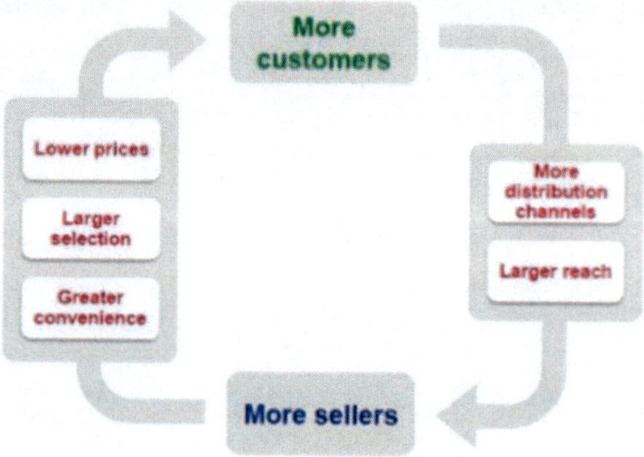

Fig. 3. Amazon selling format

4.5 Amazon Web Services (AWS)

Amazon E-Commerce Service: Exploration for inventories, recover item information, photos, and review from customers. Get preferred details, find sellers and offers
Alexa Web Information Service: Get information like page rankings and relevant websites targeted by URLs
Amazon Simple Queue Service: Distributed supply managers to store web accommodation results
Advantages of AWS: Purchase every used model, Moment viability, Very Consistent and safe, Most facilities access through basic SOAP API, and Ongoing technical backing.

The competitive tool of Amazon is the business and process model, which helps a lot the procedures between Amazon and its customers and suppliers as well, (see Fig. 4) [20].

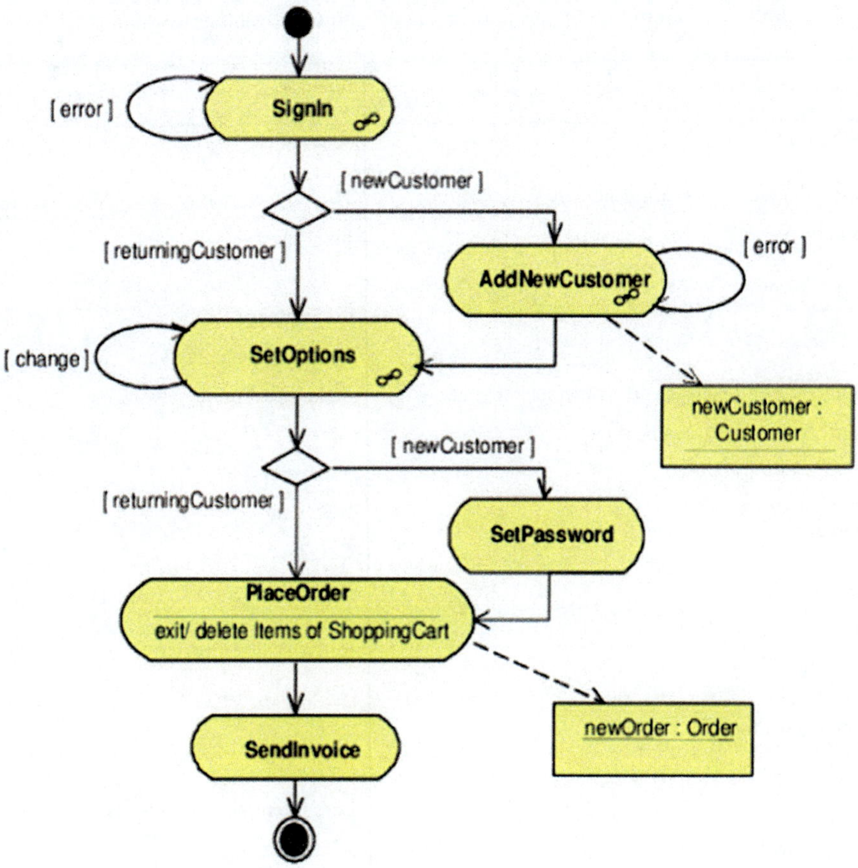

Fig. 4. Amazon process model

5 Discussions

5.1 Individual Characteristics Reasons

The main objective of this to study the differences and similarities between e-commerce platforms namely eBay and Amazon in terms of opportunities and threats as well as strengths and weaknesses. The table below shows the differences between eBay and Amazon. Table 2 shows these two online shopping platforms have shown various advantages and disadvantages also the threats experienced by these two platforms.

Table 2. Comparative in term of Advantage and Disadvantage - Summary.

eBay	Amazon
Technology Outstanding income sharing strategy with Affiliate program Strategy Empowers of all shapes and sizes traders/members to acquire at generally a same rate Business B2C lose C2C win Payment cost format Auction sites More interaction between buyers and sellers Only use PayPal for payments methods Middleman	Technology Higher technology over AWS, list and others. Strategy Prefer famous brand and agent via solid distribution network Business B2C win C2C lose Rate of sales format Fixed price Less interaction between buyers and sellers Use various kind payments method Middleman and also a retailer

6 Conclusions and Recommendations

eBay: According to this study, eBay couldn't use extra supplies for the B2C retail because markets and competitors were close to hitting market cut. eBay must surprise opponents for instance Amazon with robust B2B activities or campaigns like providing a comparison-shopping service for wholesalers. Additionally, eBay should strengthen its C2C presence through continuous appreciation for loyal customers to take care of its strengths and methods, in any case at C2C it's imperative because it can connect one another offline.

Amazon: Must contribute supplies to still dominate the B2C market by staying competitive with the worth, comfort, and consistency of its vendors and continuing to deliver in fresh retailers. What's more, Amazon should reward smaller customers because it works for network memberships too. At the end of the day, Amazon is in a better spot for its continuing B2B plan with AWS.

In summary, the marketing strategies by the Amazon store depend on some pillars, namely that are: 1) Expanding the offered products to suit all markets, 2) Using the latest technologies in designing the electronic platform interface, and 3) Using all available resources to serve the spread and ease of use of the platform.

In light of the results learned from this study, suggested recommendations include keeping pace with development and technology and what is new in the field of communications in a way that serves innovation and business development. It is also recommended that the user interface has to be changed from time to time. In terms of security, there is an urge to consider safe and consistent payment methods for each country. Finally, branding of projects is also vital in increasing the success factor for e-commerce business.

References

1. Aldrich, M.: Online shopping in the 1980s. IEEE Annals Hist. Comput. **33**(4), 57–61 (2011). ISSN 1058–6180
2. Bigne, E.: The impact of internet user shopping patterns and demographics on consumer mobile buying behavior. J. Electron. Commerce Res. (2005)
3. Snell, S.: 13 Options for Selling Digital Products on Your Website I Vandelay Design Blog. Vandelaydesign.com. Accessed 31 Dec 2012
4. King en Juhn-Shiuan Liou, S.F.: A framework for internet channel evaluation. Int. J. Inf. Manag. **24**, 473–488 (2004)
5. https://www.salehoo.com/education/selling-on-ebay-amazon#eBay-vs-Amazon-Whats-the-difference
6. Kotha, S., Basu, S.: Amazon and eBay: Online retailers as market makers. The market makers: How retailers are reshaping the global economy, pp. 155–180 (2011)
7. Ismail, C., et al.: E-payment and accounting issues in insurance companies in the use of e-commerce. JMCMS **15**(8), 660–666 (2020). https://doi.org/10.26782/jmcms.2020.08.00055
8. Bamansoor, S., Saany, S.I.A., El-Ebiary, Y.A.B.: The S-Commerce usage and acceptance modelling in Malaysia. 3C TIC. Cuadernos de desarrollo aplicados a las TIC **9**(1), 99–115 (2020). https://doi.org/10.17993/3ctic.2020.91.99-115
9. El-Ebiary, Y.A.B., Al-Sammarraie, N.A.: Mobile commerce potentials and challenges - India case study. IJRTE **8**(IC2), 1154–1157 (2019)
10. El-Ebiary, Y.A.B., Al-Sammarraie, N.A., Saany, S.I.A.: The implementation of M-commerce in supply chain management system. 3C Tecnologia, 223–240 (2019)
11. Alzubi, M.M., Alkhawlani, M.A., El-Ebiary, Y.A.B.: Investigating the factors affecting University students' e-commerce intention towards: a case study of Jordanian universities. J. Bus. Retail Manag. Res. **12**(1) (2017)
12. E-Commerce Meaning (2020). https://www.toppr.com/guides/business-environment/emerging-trends-in-business/electronic-commerce/
13. E-Commerce Usage Data in Malaysia (2020). https://www.statista.com/outlook/243/122/ecommerce/malaysia#market-paymentTypes

14. Al-Sharafi, M.A., Herzallah, F.A., Alajmi, Q., Mukhtar, M., Arshah, R.A., Eleyan, D.: Information culture effect on e-commerce adoption in small and medium enterprises: a structural equation modelling approach. Int. J. Bus. Inf. Syst. **35**(4), 415–438 (2020)
15. Al-Sharafi A.M., Arshah, R.A., Alajmi, Q., Herzallah, A.T.F., Qasem, Y.A.: The Influence of Perceived Trust on Understanding Banks' Customers behavior to Accept Internet Banking Services (2018)
16. Al-Sharafi A.M., Abdullah-Arshah, R., Alajmi, Q., Herzallah A.T.F.: Understanding Online Banking Acceptance by Jordanian Customers: The Effect of Trust Perceptions. 6th International Graduate Conference on Engineering, Science and Humanities (IGCESH). Universiti Teknologi Malaysia, Johor Bahru, Malaysia (2016)
17. Al-Saedi, K., Al-Emran, M., Ramayah, T., Abusham, E.: Developing a general extended UTAUT model for M-payment adoption. Technol. Soc. **62**, 101293 (2020)
18. Amazon, E.C.: Amazon web services, November 2012. http://aws.amazon.com/es/ec2/
19. Cloud, A.E.C.: Amazon web services (2011). Accessed 9 Nov 2011
20. Silva, A.T., Moro, S., Rita, P., Cortez, P.: Unveiling the features of successful eBay smartphone sellers. J. Retail. Consum. Serv. **43**, 311–324 (2018)

The Influence of E-Commerce in Encouraging the Tourism Industry in Malaysia

Yousef A. Baker El-Ebiary[1], Waheed Ali H. M. Ghanem[2(✉)], Syarilla Iryani A. Saany[1], Ahmad Nazari Mohd Rose[1], Julaily Aida Jusoh[1], and M. Hafiz Yusoff[3(✉)]

[1] Faculty of Informatics and Computing, UniSZA University, Kuala Terengganu, Malaysia
{yousefelebiary,syarilla,anm,julaily}@unisza.edu.my

[2] Faculty of Ocean Engineering Technology and Informatics, Universiti Malaysia Terengganu, Kuala Terengganu, Malaysia
waheedghanem@umt.edu.my

[3] Albukhary International University, Alor Setar, Malaysia
hafiz.yusoff@aiu.edu.my

Abstract. Small and Medium Enterprises (SMEs) play the significant role in country such as Malaysia, India and Indonesia (developing countries) as they build and improve the economy sector. While the e-commerce stands for electronic commerce may give a significant impact especially for the travel or tourism industry that was part of the SMEs as there were many mediums of information sharing on the Internet. The research uses an analytical method to find the enhancement ways of tourism in Malaysia. New IT customers are starting to browse items directly from the provider's website (web pages), resulting in customer-friendly pricing. As e-commerce expansion has revolutionized the tourism business, an efficient electronic commerce business framework and e-commerce strategy development for tourism business agencies management is developed. Electronic commerce for travel agencies has adopted a B2B strategic alliance to strength their competitiveness advantage on the e-commerce travel or tourism market and because it was an effective tool for getting the greater profits. However, the electronic commerce adoption in developing countries by SMEs is low because several factors such as the weakness of a suitable model for implementation of e-commerce. Objectives of the research is to investigate the reasons that affects the use of electronic commerce obviously in Malaysia (developing countries) especially in the tourism industry. Therefore, this research to provide a detail explanation of the reasons that influence e-commerce adoption on the tourism industry. In existing literature review, several research report that cover a much of model and generic framework about using the technologies and give a general understanding of technologies adoption is proposed. However, none prepare the typical contextual ideology or model of business in the particular developing countries. Therefore, the model for electronic commerce implementation in developing nations are highlighted, the Unified Theory of Acceptance and Use of Technology (UTAUT). Finally, some suggestion was highlighted that may give the researchers a more and better understanding of how to encourage the electronic commerce adoption in developing nations which they have not yet started using electronic commerce especially in the tourism or travelling industry.

M. Al-Emran et al. (Eds.): ICETIS 2021, LNNS 299, pp. 381–388, 2022.
https://doi.org/10.1007/978-3-030-82616-1_33

Keywords: E-Commerce · Tourism industry · SMEs · Conceptual model · Developing countries

1 Introduction

E-commerce was dramatically changed the promotion channel in the tourism industry at the last 4 years ago. Effect the introduction of Information Technology (IT) especially medium like e-commerce was significantly changed the role of website (traditional intermediaries) in recent years. The e-commerce became one of the most important mediums for tourism business service for entrepreneurs to provide their services and share related information to their new customers. The number of tourism business websites has grown up exponentially in the recent year and competition was becoming more vibrant with each other [1]. Tourism agencies are now looking for consumer or customer interest when using e-commerce for helping them browse the best product. Tourism agencies confident that they do not depending on intermediary way for selling the products. Using (IT) to build up and strengthen business company will reduce the marketing costs and give high quality tourism products until international market [2].

The introduction of electronic commerce was the part of (IT) has accessed using the Internet and (WWW) which is stand for World Wide Web, affecting on travel industries and also the consumer. With adoption of (IT), tourism business was growth up rapidly, entering the international and domestic tourists by biggest cost reduction and increased productivity. Besides that, consumers are depending on the Google (browser) to find tourism business information. Information technology (IT) has advanced tools such as smart phone with GPS have been purchased by users to make their travelling more conducive [3]. Furthermore, significant impact was given by using of e-commerce on the tourism sectors, specifically on the automation and promotion channels, providing related information on travel destinations, accessibility of locations, luxury hotels, attractive places and wave (map).

Recent statistics show that adoption of electronic commerce in Malaysia is significantly lower than with other developed countries around the world. Existing the serious concerns about e-commerce adoption, especially among small businesses, although their role in major economy very important for country specifically in Malaysia [4]. As described, the small businesses sector in Malaysia faces many obstacles when it comes to the e-commerce adoption and Internet. Therefore, the detail in this research explained here about the advantages of using (IT) and achievement of electronic commerce reach the purposes of small businesses, particularly in the tourism business [5].

2 Background of the Study

2.1 Electronic Commerce (E-Commerce)

Electronic commerce is widely used for a way to integrate customers and suppliers into the global market economy. It is seen as an important innovation that promotes good governance, social and the economic development and poverty alleviation in certain developing country. Moreover, e-commerce mediums give (SMEs) to surpass their deficiency by giving them opportunities to grow up around the world, gain new market opportunity and attract new customers [6]. The benefits when adopt e-commerce are the strong reasons that make e-commerce became the best medium for businesses and it is reflected in the significant growth of e-commerce users every year.

2.2 Reasons in Influencing of E-Commerce Usage or Adoption

E-commerce adoption in (SMEs) is still low. Therefore, it will be one of a various reason that will be discussed in this section or topic. Factors positively influencing the e-commerce adoption, are presented in Fig. 1, that describes of how each reason influences the other reasons, (see Fig. 1) [7].

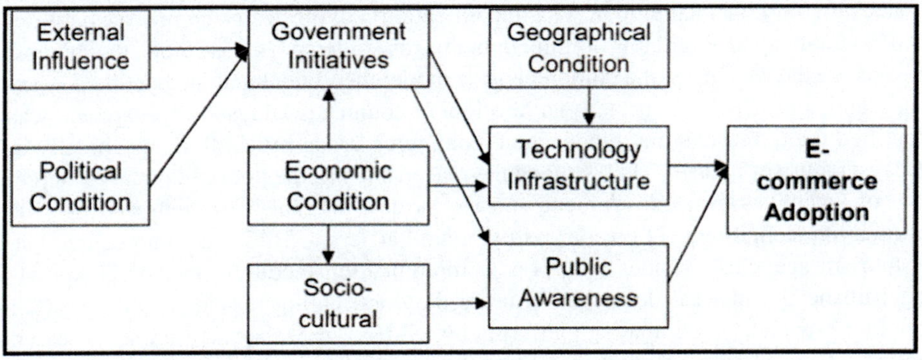

Fig. 1. Shows the reasons that affecting E-commerce adoption.

2.3 The (UTAUT) Model

(UTAUT) stand for Unified Theory of Adoption and Use of Technology was applied to study behavioral consumers when using the technology. The (UTAUT) model was the continuation of (TAM) stand for Technology Acceptance Model was introduced by Venkatesh to combine an alternative view of consumer and innovation acceptance [8]. This model contains 4 primary determinants of intent and uses which are effort expectancy, social influence, performance expectancy and facilitating conditions, as well as 4 elements that will be discussed in this research report such as age, gender, income and payment types. The Unified Theory of Acceptance and Use of Technology (UTAUT) works on the personal criterion in the use of computers in small companies. The results indicated that the expected benefit of SMEs is directly affected by using the virtual environment. As for ease of use and personal criterion, it has an indirect effect through the expected benefit from the use of technology for the employees of these companies, while the use of computers had a direct impact on the fluidity and ease of work.

Key determinants was the main reasons that directly affecting a user's behavioral intention to adopt a new technology. However, it was unclear whether (UTAUT) itself covers the contextual reasons that related with (SMEs) in Malaysia, namely the complex relationship exists between (SMEs) owners, households, consumers and workers [9].

The purpose of this research, and relying on the existing literature review, the (UTAUT) model was providing a more complete information about adoption of the technology. Furthermore, this model is related to individual characteristics, size of organizational and structure's development. Therefore, (UTAUT) capabilities, reliabilities and accessibility are basis of this study for making a report the e-commerce adoption among (SMEs), particularly in the tourism industry [10].

3 E-Commerce and Tourism Issues

As a result, our understanding of why implementation of electronic commerce between (SMEs) in Malaysia is still low or limited. In this research report, it is argued that biggest reasons are the weakness of a comprehensive model theoretical that include the contextual features of (SMEs) especially in developing country (Malaysia). Researchers was identified many reasons that hinder the e-commerce usage by (SMEs), specifically in Malaysia tourism industry [11]. Among the reason such as the poor of human resource, poor of incentives from government, security issues, internal constraints and poor of information technology (IT) resources the main barriers to B2C e-commerce that use by tourism agencies. Besides that, poor communication medium, poor of IT knowledge, financial and weak legal platform are the contributing reasons to the low (IT) adoption especially e-commerce platforms by (SMEs) in Malaysia. Another research report by Chinese researchers summarized that security privacy issues, poor of knowledgeable in understanding of e-commerce and high maintenance costs are key obstacles faces by (SMEs) on 3 developing nations (Indonesia, Malaysia and India) in adapting in e-commerce technology. In current situation need the full attention from all agencies involved, including researchers, because the (SMEs) owners played a big major in development of economy in country [12]. They contribute in this business not only popularity terms but also for job creation and reduce the poverty. The advantages of development's (SMEs) in a country will positively impact the country's economic growth, increase the GDP per capita income and vice versa [13].

4 Data Collection and Analysis

The statistics or data about electronic commerce (e-commerce) adoption in Malaysia such as age, gender, income and payment type from (statistica.com). With a market volume of US$4,179m in 2020, was generated revenue in Malaysia placed at 37th around the world, (see Figs. 2-5) [14, 15].

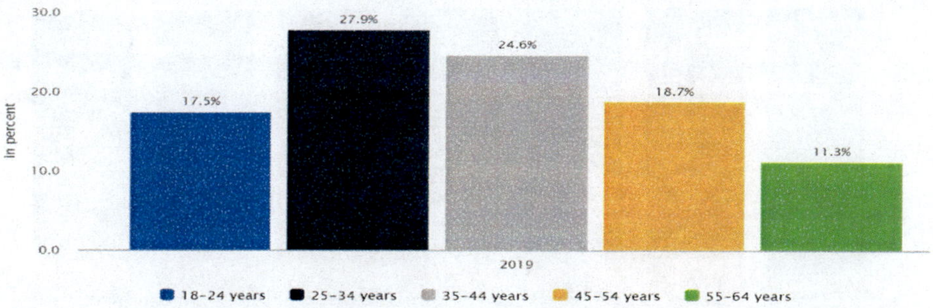

Fig. 2. Shows the electronic commerce adoption age statistics in Malaysia.

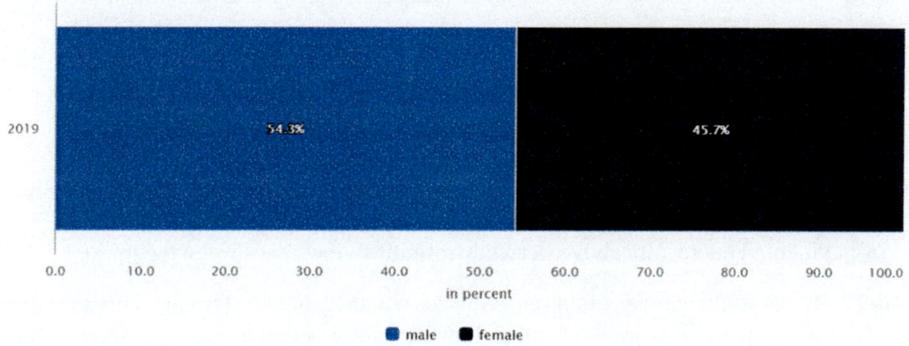

Fig. 3. Shows the comparison gender of adoption on E-commerce in Malaysia.

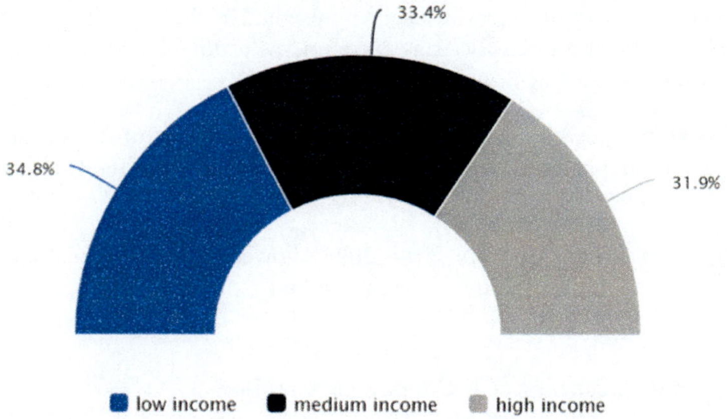

Fig. 4. Shows the comparison of E-Commerce adoption income in Malaysia.

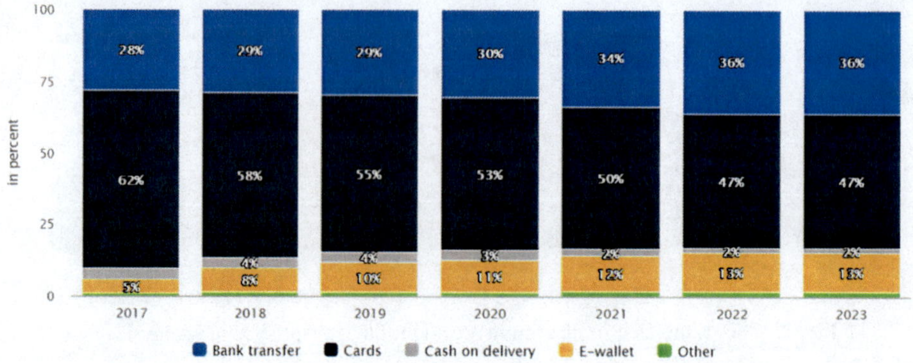

Fig. 5. Shows the comparison of E-Commerce adoption payment type in Malaysia.

5 Discussion

5.1 Individual Characteristics Reasons

For this research report, the individual characteristics reasons can influence the electronic commerce adoption according to the (UTAUT) model such as gender, age, income and type of payment. This results above derived from interview and survey techniques.

Gender: Refers to the gender of an interviewee, whether male or female. This statistic shows that male using e-commerce more than female whether to shop or travel. The result is derived from male interviews that technology is easier to use for them because they have more knowledge in handling the information technology (IT).

Age: Relates to the age of the interviewee, is represented by the years. The result shows the age from 25–34 year are mostly using e-commerce as the platform to shop or vice versa. For this research, the ages were between 18 and 64 years old are taken.

Income: This study showed that had three types or comparison of income which are low income, medium income and high income and the result show the low income are the most user.

Payment Type: The statistics from 2017–2023 showed payment types and mostly Malaysian comfort to use card as the payment method.

5.2 Reasons that Encourage of E-Commerce Adoption

The reasons have summarized below to encourage the e-commerce adoption in Malaysia tourism industry:

Performance Expectancy: Tourism agencies recognize their businesses need to adapt with new technology platforms like e-commerce, because it was a growing market especially in tourism business. It has been said this performance expectancy offer various advantages such as faster job creation, enhance productivity and time savings.

Effort Expectancy: Some businesses refer to e-commerce because it was very user friendly. Although the results show the e-commerce is user friendly, some of businesses in the tourism and accommodation sectors acknowledge that it need proper training for ease of use. Therefore, the extensive training on the system will get the more benefits from system.

Social Influence: For businesses to grow up their products in international market, it need involvement of social influence. Therefore, the influence of industry owners, competition from other businesses and the target new customers also affected. This means business decision should more innovative to gain the customers attraction.

Facilitating Conditions: A key concern regarding the impact of e-commerce adoption on the tourism business is the poor of government initiatives. Even though the evidence was found that the government is not supported (SMEs) in implementing electronic commerce applications. Governments need to set laws to regulate transactions.

National Culture: The impact of national culture is derived from interviewee' survey from indirect response and direct quotations on adoption of technology. Results shows the decision to use or adopt e-commerce was influencing by variety of cultural reasons related such as the type of businesses ownership, and an effective in strategy marketing.

6 Conclusion and Recommendations

This research shows that the e-commerce adoption by (SMEs) on the tourism business in developing countries, especially in Malaysia, is still low. Most of them are email users and websites (traditional webpages). Compared to other (SMEs) in developed nation, the level adoption of e-commerce in Malaysian (SMEs) is still far behind. It will give the big impact on the government to enhance their efforts by held the programs or campaigns and took the effective initiatives to drive faster to the e-commerce adoption by (SMEs) in Malaysia. Besides that, the tourism agencies need to be cooperate with government in giving the knowledge for the new customer of technology, especially in e-commerce. Moreover, this research showed the advantages realized by (SMEs) is increasing the e-commerce adoption. On this current situation can be highly believed for (SMEs) owners to adopt e-commerce and this it will be helpful for (SMEs) owners in making a decision whether or not to growth up to the higher level. Finally, the consumer need to follow the modern technology to catch up with the development infrastructure especially in information technology (IT) to make daily tasks easier and effective.

References

1. Abou-Shouk, M.A., Lim, W.M., Megicks, P.: Applying competing models to analyse the role of environmental aspects in e-commerce adoption by small and medium sized travel agents in the developing nations (2015)
2. Houache, H., Abdul Rahim, N.H.B., Shah, A.: Analysis using models for investigate the factors of e-commerce adoption in developing nations (2019)

3. Rahayu Bt, R., Day, J.: E-commerce usage by (SMEs) in developing nations: evidence from Indonesia (2017)
4. Mapeshoane, T.J., Pather, S.: The E-Commerce Adoption in the Lesotho Tourism Industry (2016)
5. Ismail, R.C., Majid Alawi, N.A., Mohd Hafiz, Y., Saany, S.I.A., Bamansoor, S., El-Ebiary, Y.A.B.: E-Payment and accounting issues in insurance companies in the use of E-Commerce. JMCMS **15**(8), 660–666 (2020). https://doi.org/10.26782/jmcms.2020.08.00055.
6. Bamansoor, S., Saany, S.I.A., El-Ebiary, Y.A.B.: The S-Commerce usage and acceptance modelling in Malaysia. 3C TIC: Cuadernos desarrollo aplicados a las TIC **9**(1), 99–115 (2020). https://doi.org/10.17993/3ctic.2020.91.99-115
7. El-Ebiary, Y.A.B., Al-Sammarraie, N.A.: Mobile commerce potentials and challenges - India case study. (IJRTE) **8**(IC2), 1154–1157 (2019)
8. El-Ebiary, Y.A.B., Al-Sammarraie, N.A., Saany, S.I.A.: The implementation of M-commerce in supply chain management system. 3C Tecnologia, pp. 223–240 (2019)
9. Alzubi, M.M., Alkhawlani, M.A., El-Ebiary, Y.A.B.: Investigating the factors affecting University students' e-commerce intention towards: a case study of Jordanian universities. Journal of Business and Retail Management Research, vol. 12, no. 1 (2017)
10. E-Commerce Meaning (2020). https://www.toppr.com/guides/business-environment/emerging-trends-in-business/electronic-commerce/
11. E-Commerce Usage Data in Malaysia (2020). https://www.statista.com/outlook/243/122/ecommerce/malaysia#market-paymentTypes
12. Al-Sharafi, M.A., Herzallah, F.A., Alajmi, Q., Mukhtar, M., Arshah, R.A., Eleyan, D.: Information culture effect on e-commerce adoption in small and medium enterprises: a structural equation modelling approach. Int. J. Bus. Inform. Syst. **35**(4), 415–438 (2020)
13. Al-Sharafi, A.M., Arshah, R.A., Alajmi, Q., Herzallah, A.T.F., Qasem, Y.A.: The Influence of Perceived Trust on Understanding Banks' Customers behavior to Accept Internet Banking Services (2018)
14. Al-Sharafi, A.M., Abdullah-Arshah, R., Alajmi, Q., Herzallah, A.T.F.: Understanding online banking acceptance by jordanian customers: the effect of trust perceptions. In: 6th International Graduate Conference on Engineering, Science and Humanities (IGCESH), Universiti Teknologi Malaysia, Johor Bahru, Malaysia (2016)
15. Al-Saedi, K., Al-Emran, M., Ramayah, T., Abusham, E.: Developing a general extended UTAUT model for M-payment adoption. Technol. Soc. **62**, 101293 (2020)

The Moderating Role of Perceived Risk in the Technology Acceptance Model (TAM): The Context of Mobile Banking in Developing Countries

Ahmed M. Mutahar[1](✉), Adnan Aldholay[1], Osama Isaac[2], Abdullah Nabeel Jalal[1], and Faten Elina Binti Kamaruddin[1]

[1] Faculty of Business Management and Professional Studies Management and Science, University Shah Alam, Shah Alam, Malaysia
ahmed_mutahar@msu.edu.my

[2] Guangdong University of Finance, Guangzhou, Guangdong Province, China

Abstract. The development of wireless infrastructure has prompted the amalgamation of wireless technology and mobile devices and has innovated financial services such as international mobile banking. This study aims to evaluate the moderating influence of perceived risk based on the technology acceptance model (TAM) in the mobile banking industry. The proposed model was assessed using data provided by 482 Yemini nationals through surveys. These individuals do not use mobile banking. Structural Equation Modeling (SEM) was performed by employing Smart PLS 3.0 to determine impact of connections and relationships among the assessed factors. The projected conceptual model exerted a variance of 77% for mobile service in the banking services industry. Perceived ease of use (PEOU), perceived usefulness (PU), and behavioural intention were employed to examine the influence of perceived risk (PR) as a moderator. It was determined that PR was responsible for moderation only on the association between PU and the intent to utilise mobile banking. Furthermore, this study established that PEOU applied a significantly positive effect on PU. Moreover, PEOU and PU exert a positive effect on behavioural intention in mobile banking use in Yemen. The conclusions of this study may potentially offer additional insight into the mobile banking strategy.

Keywords: Mobile banking · Perceived risk · TAM · Developing countries · Yemen · Moderating

1 Introduction

Though a majority of the Yemini banks offer mobile banking services, the clients using a mobile phone have adopted mobile banking to a significantly lesser extent than their counterparts in other Arab nations [1]. Low usage causes limited performance, productivity, and a reduced return on investment, as explained by [2–8]. Considering the massive capital invested by banks towards the development of these systems, it is imperative to

M. Al-Emran et al. (Eds.): ICETIS 2021, LNNS 299, pp. 389–403, 2022.
https://doi.org/10.1007/978-3-030-82616-1_34

ensure that customers use those services. Sadly, the factors that affect mobile banking acceptance in Yemen are not entirely understood. Strategic planning and solution finding by the banks to lead customers to use mobile services may only be done correctly when the effects of consumer perception and individual preferences are understood. A study of this kind is significantly vital for Yemen because, for this country, there is a lack of proper research on the factors affecting client behavioural intent and mobile banking service adoption.

Various theories and models were employed to assess the implementation and acceptance of specific application of information systems yet the most popular assessment is TAM [9–15]. Models such as these account for a majority of the variance, yet two essential factors, that are often overlooked or do not get due importance, are technology type and perceived risk (PR). Perceived risk is a critical factor and is modelled as a predecessor to ease of use PEOU and usefulness PU [16]. Technology type plays a crucial role in determining whether individuals adopt a technology. For instance, in marketing, it is well known that factors determining consumer decisions differ across the type of products [17–19]. Therefore, it may be reasonably presumed that PEOU and PU exert similar effects of intention for various technologies. The role of PR as a moderating aspect is evidently crucial. In TAM research, PR usually preceded PEOU, PU, or intention. However, when PR is modelled previous to PU, it is supposed that PR and PU are related, while these are independent of the other. It is, therefore, anticipated that PR moderates the effects of PEOU and PU on intention. The correlation between PU, PEOU, and intention gets weakened if users perceive more risk. This research tests perceived risk as a moderator in the TAM framework. Furthermore, this study examines the influence of its primary components of the model which are PEOU and PU.

This study aims to focus on the following research objectives: (1) to determine the influence of ease of use on usefulness, (2) To examine the influence of ease of use on intention, (3) To establish the impact of usefulness on intention, (4) To examine the influence of perceived risk as a moderator between usefulness and intention, (5) To examine the influence of perceived risk as a moderator between ease of use and intention.

2 Literature Review

2.1 Perceived Ease of Use

[20] defines PEOU as the level of the user's confidence that using a particular system will be a consequence of effort. Extensive studies were performed which indicated the existence of direct or indirect practical evidence of the substantial influence of PEOU on behavioural intention through its influence on perceived usefulness [21–31]. It is crucial to ensure that mobile banking services are convenient for users to learn and use to maximise the benefits of these services. Convenience and flexibility will improve customers' acceptance of the applications [32]. This infers that perceived ease of use is observed to positively affect customers' perceived usefulness on their communication with mobile services which directly impact customers' intention to utilise mobile banking services. Hence, the following hypotheses are postulated:

H1. Perceived ease of use has a positive effect on perceived usefulness.

H2. Perceived ease of use has a positive effect on the intention to use mobile banking services.

2.2 Perceived Usefulness

[33] outlines perceived usefulness as "the extent to which a person believes that using a particular system will enhance his or her job performance". Perceived usefulness has been thoroughly studies by IS researchers which cements the criticality of perceived usefulness towards intention to adopt [20, 34–42]. The chief reason why individuals use e-banking systems is their usefulness in bank transactions. Hence, the following hypothesis is proposed:

H3. Perceived usefulness has a positive effect on the intention to use mobile banking services.

2.3 Perceived Risk

[43] characterised perceived risk as "the potential for loss in the pursuit of the desired outcome of using an e-service". They pointed out several aspects of risk: psychological, performance, privacy, financial, time, security, social, and overall. Chen (2013) outlined that the intention to use reduces with a higher perceived risk when considering mobile banking services in Taiwan. Along the same lines, [41] were in agreement that perceived risk weighed negatively on the use of mobile banking. In other words, clients perceive the risk as a deterrent to using mobile banking as they fear to lose money, time, comfort, or information. The primary constructs of TAM (PU and PEOU) were analysed by [43] to understand the moderating influence of perceived risk. They pointed out that perceived risk modifies the effects of PEU and PU on intention [44]. Individuals who perceive greater risk in technology adoptions are affected by ease of use. Hence, perceived risk is examined as a moderator on the relationships among PU, PEOU, and intent to use in the mobile services context. Consequently, the following hypotheses are proposed:

H4. Perceived risk moderates the relation between perceived usefulness and intention.

H5. Perceived risk moderates the relation between perceived ease of use and intention.

2.4 Behavioural Intention

Intent to use is a stricture that refers to an individual's intention to attempt new technology [45]. In other word, intention can be referred as the possibility of an individuals in trying a particular system. [20, 46] conducted research using TAM and mentioned that behavioural intention leads to and affects real behaviour when interacting with a new system. A study conducted by [47] using the theory of planned behaviour (TPB) also specified that the intention to perform a specific behaviour governs actual use. Hence, the actual use of a system is influenced by the intent in using new technology. Newer studies mention an intention to use technology as a mediating attribute (e.g. [36, 48]. Nevertheless, since the focus of this study is on individuals who do not use mobile banking, the core construct remains as the intention to use mobile banking, which is gauged using three aspects, namely desire, intention to use, and preference [49–52].

Eventually, as a result for the discussion above, the following research model is proposed (Fig. 1):

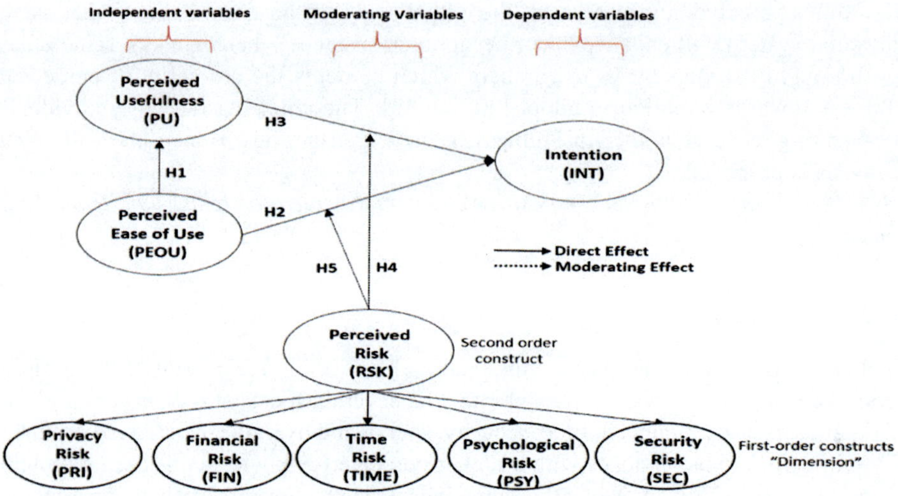

Fig. 1. Conceptual framework of the study

3 Research Method

3.1 Questionnaire Development of the Study

A survey questionnaire was used in this study to collect data designed specifically to gauge all the primary factors proposed in the study model. Survey questionnaire were administered to collect the necessary data for this study. The questionnaire contained two sections and encompassed close-ended statements that were translated to the Arabic Language to ensure a good understanding among the targeted Yemeni respondents. Strongly disagree to strongly agree are the two pillars for A 7-point Likert scale in the survey which was employed to measure respondents' agreement with the four main factors in the first section (instrument design detailed in Table 2). The second part of the questionnaire had questions pertaining to individual demographics, which were rated on an ordinal or nominal scale.

3.2 Collecting of Data

Yemini bank customers who had recently opened a bank account were chosen as the respondents. These individuals use a mobile phone but not mobile banking services. The selected respondents possess mobile phones yet do not engage in mobile banking services. This study chose a non-probability sampling strategy namely snowball sampling, in order to identify potential respondents in Sana'a, the capital of Yemen. This method is suitable when the targeted individuals are hard to reach [53]. Despite the difficulties in contacting the potential respondents and to obtain responses from them, this study successfully gathered a total of 428 complete and valid questionnaires. The first analysis employed Partial Least Squares (PLS) method was utilised using SmartPLS 3.0 software to evaluate the developed framework [54]. Also, SmartPLS 3.0 provides high accuracy in the assessment of discriminant validity of the questionnaire in comparison with exploratory factor analysis [55]. The second analysis involved SPSS V. 24.0 software for further analysis.

4 Analysis of Data and Outcomes

As per the recommendations provided by [56, 57], a two-stage analysis was performed. The initial phase assessed the reliability and validity of the measurement model and the subsequent phase was conducted to assess the postulated relationships. [58, 59] suggest these two assessment steps consisting of structural models and measurement offer benefits over a single step assessment process. [57] mentioned that the measurement model describes the measurement steps of each construct while the structural model offers evidence on the relationships among the condustructs.

4.1 Measurement Model Assessment

Cronbach's alpha coefficients were used to assess construct reliability (CR) to measure the reliability of the main constructs of the measurement model. The results demonstrated that Cronbach's alpha coefficient of each construct was between 0.855 and 0.966, exceeding the suggested value of 0.7 [60].

Moreover, construct reliability testing revealed that the values of composite reliability (CR) were within the range of 0.912 to 0.976, higher than the recommended value of 0.7 [61, 62], that shows the fulfilment of construct reliability (as shown in Table 1). Hence, the calculated Cronbach's alpha and CR values for each construct were adequately error-free.

Indicator reliability was measured using factor loading. High loading on a construct means that there is great commonality between the associated indicator, which is recorded by the construct [57]. A factor loading more than 0.5 is understood as very significant [57, 63, 64]. As shown in Table 1, the loadings for every item were higher than the recommended value of 0.5. Therefore, the loading for other items in the model has met the required criteria.

Average variance extracted (AVE) was used in the study to test convergent validity. It was found that each AVE value was in the range of 0.775 to 0.932, greater than the recommended value of 0.50 [65]. As indicated in Table 1, the convergent validity for each construct has been fulfilled, and necessary convergent validity is shown (as Table 1 shows).

Discriminant validity, which indicates the extent to which the items differentiate among constructs, or gauges distinct concepts for the measurement model, was evaluated using the Fornell-Larcker criteria. The values are shown in Table 2. The square root

Table 1. Mean, standard deviation, loading, Cronbach's Alpha, CR and AVE

First-order constructs	Second-order construct	Item	Loading (> 0.5)	M	SD	α (> 0.7)	CR (> 0.7)	AVE (> 0.5)
Perceived Ease of Use (PEOU)		PEOU1 PEOU2 PEOU3	0.890 0.929 0.877	4.94	1.76	0.882	0.926	0.808
Perceived Usefulness (PU)		PU1 PU2 PU3 PU4	0.957 0.954 0.936 0.948	4.72	1.80	0.963	0.973	0.900
Privacy Risk (PRI)		PRI1 PRI2 PRI3	0.927 0.931 0.936	4.01	1.89	0.923	0.951	0.867
Financial Risk (FIN)		FIN1 FIN2	0.951 0.953	4.06	1.85	0.896	0.951	0.906
Time Risk (TIME)		TIME1 TIME2 TIME3	0.869 0.882 0.890	3.63	1.53	0.855	0.912	0.775
Psychological Risk (PSY)		PSY1 PSY2	0.975 0.974	3.67	1.89	0.947	0.974	0.950
Security Risk (SEC)		SEC1 SEC2 SEC3	0.932 0.929 0.932	3.89	1.80	0.923	0.951	0.867
	Perceived Risk (PR)	PRI FIN TIME PSY SEC	0.923 0.890 0.886 0.890 0.934	3.85	1.79	0.966	0.958	0.819
Intention (INT)		INT1 INT2 INT3	0.961 0.972 0.964	4.77	1.88	0.964	0.976	0.932

for each AVE, shown by values on the diagonals in bold font, was higher than the correlations among constructs (corresponding row and column values). This shows that there is a strong relationship between the constructs and their indications compared to the other model constructs [66], which therefore suggest a high discriminant validity [57]. Additionally, the measured correlation among exogenous constructs is less than 0.85 (Awang, 2014). Therefore, for all constructs, discriminant validity is satisfied.

Table 2. Results of discriminant validity by Fornell-Larcker criterion

	Factors	1	2	3	4
		INT	PEOI	PR	PU
1	INT	**0.966**			
2	PEOU	0.719	**0.899**		
3	PR	−0.629	−0.486	**0.843**	
4	PU	0.838	0.681	−0.57	**0.949**

4.2 Structural Model Assessment

[57] suggest that a structural model be evaluated by measuring the R^2, beta (β), and the associated t-values via a bootstrapping technique with 5000 samples. Additionally, they recommend reporting predictive relevance (Q^2) and effect sizes (f^2). [67] emphasise that eventhough the p-values specify the influence occurs, the extent of the influence is not described.

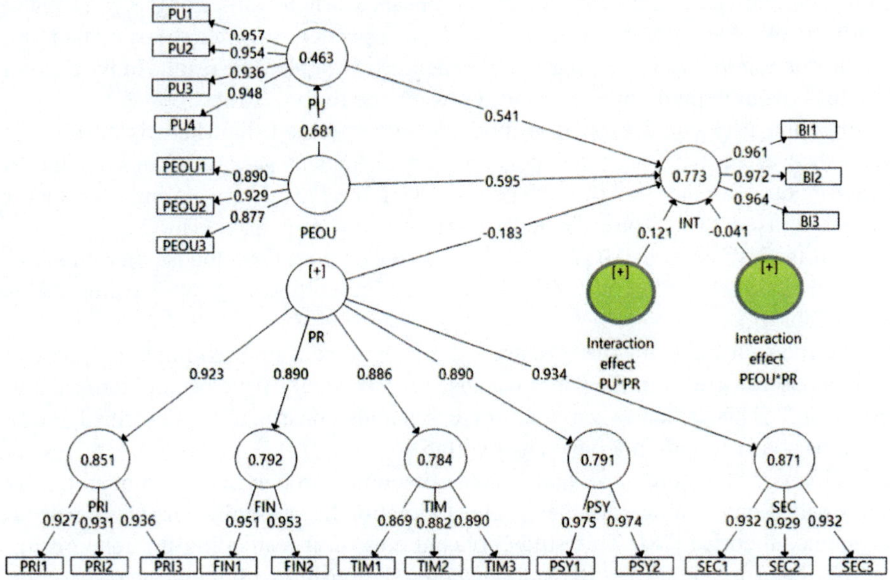

Fig. 2. PLS algorithm results

Figure 2 and Table 3 illustrate the assessment of the structural model where the hypothesis testing is detailed. PEOU predicts significantly the intention and perceived usefulness. Therefore, H1 and H2 are accepted with values ($\beta = 0.681$, t = 23.157, p < 0.001) and ($\beta = 0.227$, t = 5.723, p < 0.001), respectively. Similarly, PU is a predictor of intention. Hence, H3 is supported ($\beta = 0.541$, t = 11.476, p < 0.001). The standardised path coefficient stipulates the strength of the relationship between the independent and dependent variables. Therefore, a significantly higher direct influence was exerted by PU on intention as compared with PEOU.

Perceived ease of use (PEOU) explained a variance of 46% for perceived usefulness. A total of 77% of the variance of intention was explained by PEOU and PU. The R^2 values attained hold satisfactory value for explanatory power as stated by [68, 69], consequently signifying an effective model. The effect sizes (f^2) were also assessed in this study. Effect size, f^2, measures the significance of the influence of an independent variable on a dependent variable [70]. [57] assert that changes in the value of R^2 must also be tested. [71] suggested a standard that the values of effect size be 0.35, 0.15, and 0.02, representing large, medium, and small effects, correspondingly. As shown in Table 3, the result of f^2 indicates two relationships, one each with large and small effect size.

Moreover, the practice of the blindfolding procedure in this study measured the predictive relevance of the developed model. Hair et al. (2017) stipulated that the blindfolding process should be employed to dependent variables that possess a reflective measurement. If the value of Q^2 surpasses zero, then predictive relevance is present in the developed model for a certain endogenous construct [57, 72]. Table 3 demonstrated that the values of Q^2 were between 0.39 and 0.68, all larger than zero, displaying that the model possess sufficient predictive relevance. [57] suggest that Q^2 values should be referred as a relative measure for predictive relevance, where values of 0.35, 0.15, and 0.02 indicate that the predictive relevance of an independent variable is large, medium, and small, correspondingly, for a particular dependent variable. The results of the current study indicate a high predictive relevance between the two constructs.

Multicollinearity could exist in a model which specified that there is an overlap between the variance between independent and dependent variables, thus unable to explicate unique variance for the independent variable [73]. Variance inflation factor (VIF) was utilised to measure the level of multicollinearity in the model [73]. It is worrying if the VIF exceeds 10 [74, 75]. In accordance with [76], the multicollinearity issue is present when the highest VIF exceeds 5. Multicollinearity check using VIF is shown in Table 3.

Furthermore, it indicates the absence of any significant multicollinearity between the exogenous constructs in the study because all VIF values have been determined to be less than 5. This indicates variance for exogenous constructs, as described in the endogenous constructs, do not have any overlap.

The effects of a predictor variable on the outcome variable may be moderated by variables called moderating variables [77]. Moderation has a significant role in several social science theories [78]. This study concentrates on investigating the relationship between PEOU (predictor) and PU, and also the relationship with intention (outcome), which tested the influence of PR (moderator).

The theoretical framework of moderation highlights that a moderator variable influences the relationship between the other two. If the PR is a moderator, then the extent or direction of the association between PU and PEOU (predictor), as well as the intention (outcome) is influenced by PR [79]. The statistical moderation model forecasts the result using the predictor variable and proposed moderator along with these two considered in interaction. The interaction indicates whether there had been moderation.

However, it should consider both the predictor and moderator to have a valid interaction [80]. Researchers in numerous situations use a continuous (in contrast to a categorical) moderator variable, which they understand may increase or decrease the strength of a specific relationship among two latent variables. The direction of the associations may be changed by a moderator [76].

[77] mentions that a predictor must exert significant effects on the outcome of the introduction of a moderator variable. Hence, the introduction of a moderator (PR) into the model will contribute to varying causal effects which is attributed to the "interaction effect" between the moderator and predictor.

The moderation evaluation of this study was conducted through hypotheses H4 and H5. A resample of 5000 was chosen to perform the bootstrapping procedure to understand the interaction effect. The results, as shown in Table 3, show that perceived risk acts as a moderator on the impact of perceived usefulness on intention ($\beta = 0.121$, $t = 3.143$, $p < 0.01$), and therefore H4 is accepted. H5 was not supported because perceived risk (PR) did not exert any moderating influence on the on the impression of perceived ease of use (PEOU) on intention.

Table 3. Structural path analysis result

Hypothesis	Relationship	Std Beta	t-value	Decision	R^2	f^2	Q^2	VIF
H1	PEOU → PU	0.681	23.157***	Supported	0.46	0.865	0.39	1.945
H2	PEOU → INT	0.227	5.723***	Supported	0.77	0.118	0.68	2.298
H3	PU → IN	0.541	11.476***	Supported		0.567		
H4	PU*PR → INT	0.121	3.143**	Supported				
H5	PEOU*PR → INT	-0.041	1.185	N-supported				

***p < .001; **p < .01; *p < .05

Key: PEOU: perceived ease of use, PU: perceived usefulness, PR: perceived risk, INT: intention.

5 Discussion

The original TAM was conceptualised by [33, 46] conceived that PEOU is a predictor of PU, and both PEOU and PU function as predictors of behavioural intention in embracing new technology. In the present study, it was established that PEOU was a significant antecedent of PU having $\beta = 0.681$. Consequently, awareness of improved ease of use of mobile banking services contributes to a more positive attitude towards the usefulness of technology for financial customers. This aligns with findings from previous research such as [27, 28] in various settings and implementation, where they revealed that improved -perceived ease of use exercises a positive impact on perceived usefulness. Possible users of the Yemini banks' mobile banking systems deliberate that convenient banking systems will contribute to the perceived the usefulness of the system. This shall ultimately give a boost to customer intention to avail mobile banking facilities. It is in line with the findings of [26, 81–84]. They established empirically that PU and PEOU positively affect user intention to embrace mobile banking services. The results of this study established that H1, H2, and H3 were supported and have fulfilled the research aims. The results confirmed that perceived ease of use and usefulness are important predictors of intention to use in the initial implementation phase of mobile banking services [85–88]. This study established that perceived usefulness (PU) and perceived ease of use (PEOU) explained 77% of variable for the intention to use mobile banking services.

Regarding objectives 4 and 5, assessing perceived risk (PR) as a moderator on perceived usefulness (PU) and the intention to utilise mobile banking, it was concluded that PR moderates the relationship between PU and intention, which aligns [43]. It indicates that PR decreases the perception of usefulness for potential users. In contrast, PR did not moderate the relationship between PEOU and the intention to use mobile banking services, which contradicts [43]. While H4 was supported, H5 was not. The results inferred that PR has no influence on the relationship between PEOU and intention and the mitigated effect of PU on intention.

6 Implications

This research aligns with findings by previous literature which concluded that the likelihood of customers using mobile banking services rises if the services are beneficial and convenient. There is a robust correlation between the mobile banking platform and the technological aspects with respect to acceptance and usage of the service offering. Mobile banking platform developers, along with the banks, should invest effort and resources towards creating an effective mobile banking platform that is both useful and easy to operate because these characteristics provide for a higher probability of customer uptake towards mobile banking. The choice to engage mobile banking services is rooted on various factors such as ease of technology use, the usefulness of the service with respect to convenience, quality, and service competence.

Perceived Risk (PR) alters the effects of PU on intention. Individuals who perceive greater risk in adopting mobile banking would be influenced by the usefulness of mobile banking. This observation has many implications for the manager. When mobile banking services, understood as risky, are being set up, the usefulness of the services needs to emphasise upon.

The secondary theory in this research is (TAM) was validated in the Yemini context. The conclusions of this study may be added to the literature specific to mobile banking in Yemen. The findings of this study enrich the mobile banking literature specific to Yemen. Useful recommendation for the financial sector and government agencies to promote improved use of mobile banking services.

7 Conclusion

This study aims to validate the main constructs of TAM in the context of Yemeni banks towards the acceptance of mobile banking services. Perceived risk (PR) was also tested as a moderator on perceived ease of use (PEOU), perceived usefulness (PU), and the intention to use. The findings clarified the new parameters of the intention in using mobile banking in Yemen. In summary, it is essential for mobile banking services to provide innovative qualities to client to obtain competitive advantage in the market (e.g., e-banking). Perceived usefulness (PU) significantly influences the intention to utilise mobile banking services. This study demonstrates that clients evaluate mobile banking based on their knowledge of ease of use and usefulness. A positive perception of these predictor will lead to increased intention. The results support this concludes as this study determined that PEOU is a compulsory predictor in intention to engage in mobile banking services. Moreover, perceived risk (PR) moderates the relationship between the intention to use mobile banking and PU, no moderating effect was exerted by PR on the relationship between intention and PEOU.

References

1. E-Commerce, Mobile Banking Usage in the Middle East, Effective Measures (2013)
2. Norzaidi, M.D., Chong, S.C., Murali, R., Salwani, M.I.: Intranet usage and managers' performance in the port industry. Ind. Manag. Data Syst. **107**(8), 1227–1250 (2007)
3. Goodhue, D.L., Thompson, R.L.: Task-technology fit and individual performance. MIS Q. **19**(2), 213–236 (1995)
4. Norzaidi, M.D.: Factors determining intranet usage: an empirical study of middle managers in Malaysian port industry. Multimedia University (2008)
5. Isaac, O., Abdullah, Z., Ramayah, T., Mutahar, A.M.: Internet usage, user satisfaction, task-technology fit, and performance impact among public sector employees in Yemen. Int. J. Inf. Learn. Technol. **3**(34), 1–34 (2017)
6. Aldholay, A., Abdullah, Z., Isaac, O., Mutahar, A.M.: Perspective of yemeni students on use of online learning: extending the information systems success model with transformational leadership and compatibility. Information Technology People, vol. 33, no. 1 (2019)
7. Aldholay, A.H., Isaac, O., Abdullah, Z., Ramayah, T.: The role of transformational leadership as a mediating variable in DeLone and McLean information system success model: the context of online learning usage in Yemen. Telemat. Inform. **35**(5), 1421–1437 (2018)
8. Aldholay, A., Abdullah, Z., Isaac, O., Mutahar, A.M.: Compatibility perspective of yemeni students on use of online learning extending the information systems. Inf. Technol. People **33**(1), 106–128 (2019)
9. Oliveira, T., Martins, M.F., De Lisboa, U.N.: Literature review of information technology adoption models at firm level. Electron. J. Inform. Syst. Eval. **14**(1), 110–121 (2011)
10. Giovanis, A.N., Binioris, S., Polychronopoulos, G.: An extension of TAM model with IDT and security / privacy risk in the adoption of internet banking services in Greece. EuroMed J. Bus. **7**(1), 24–53 (2012)
11. Zhong, J., Dhir, A., Nieminen, M., Hämäläinen, M., Laine, J.: Exploring consumer adoption of mobile payments in China, pp. 318–325 (2013)
12. Hsu, C., Lin, J.C.: What drives purchase intention for paid mobile apps? – an expectation confirmation model with perceived value. Electron. Commer. Res. Appl. **14**(1), 46–57 (2015)
13. Al-Ajam, A.S., Nor, K.M.: Adoption of internet banking by yemeni consumers : an empirical investigation. Aus. J. Basic Appl. Sci. **7**(2), 182–189 (2013)
14. Mutahar, A.M., Daud, N.M., Ramayah, T., Isaac, O., Alrajawy, I.: Examing the intention to use mobile banking services in yemen: an integrated prespective of Technology Acceptance Model (TAM) with perceived risk and self-efficacy. Asian J. Inf. Technol. **2–5**(16), 298–311 (2017)

15. Kumar, T.A., Ferdous, A.S.M., Nabeel, J.A.: Operational or strategic benefits: empirical investigation of internet adoption in supply chain management. Manag. Res. Rev. **40**(1), 28–52 (2017)
16. Im, I., Kim, Y., Han, H.-J.: The effects of perceived risk and technology type on users' acceptance of technologies. Inf. Manag. **45**(1), 1–9 (2008)
17. Mandel, N.: Shifting selves and decision making: the effects of self-construal priming on consumer risk-taking. J. Consum. Res. **30**(1), 30–40 (2003)
18. Turley, L.W., LeBlanc, R.P.: An exploratory investigation of consumer decision making in the service sector. J. Serv. Mark. **7**(4), 11–18 (1993)
19. Purwanto, E., Deviny, J., Mutahar, A.M.: The mediating role of trust in the relationship between corporate image, security, word of mouth and loyalty in M-Banking using among the millennial generation in indonesia. Manag. Mark. Challenges Knowl. Soc. **15**(2), 255–274 (2020)
20. Davis, F.D.: Perceived usefulness, perceived ease of use, and user acceptence of information technology. MIS Q. **13**(3), 319–340 (1989)
21. Koenig-Lewis, N., Palmer, A., Moll, A.: Predicting young consumers' take up of mobile banking services. Int. J. Bank Mark. **28**(5), 410–432 (2010)
22. Liu, Z., Min, Q.: An Empirical Study on Mobile Banking Adoption : The Role of Trust, pp. 7–13 (2009)
23. Purwanto, E., Mutahar, A.M.: Examine the technology of acceptance model among mobile banking users in indonesia. Technol. Rep. Kansai Univ. **62**(7), 3969–3979 (2020)
24. Yu, C.: Factors affecting individuals to adopt mobile banking: Empirical evidence from the UTAUT model, pp. 104–121 (2012)
25. Akturan, U., Tezcan, N.: Mobile banking adoption of the youth market: Perceptions and intentions. Mark. Intell. Plan. **30**(4), 444–459 (2012)
26. Amin, H., Rizal, M., Hamid, A., Lada, S., Anis, Z.: The adoption of mobile banking in Malaysia: the case of Bank Islam Malaysia Berhad (BIMB). Int. J. Bus. Soc. **9**(2), 43–53 (2008)
27. Gu, J.-C., Lee, S.-C., Suh, Y.-H.: Determinants of behavioral intention to mobile banking. Expert Syst. Appl. **36**(9), 11605–11616 (2009)
28. Tan, E., Leby, J., Tan, E., Lau, J.L.: Behavioural intention to adopt mobile banking among the millennial generation. Young Consum. **17**(1), 18–31 (2016)
29. Norzaidi, M.D., et al.: Determining critical success factors of mobile banking adoption in Malaysia. Aus. J. Basic Appl. Sci. **5**(9), 252–265 (2011)
30. Hanafizadeh, P., Behboudi, M., Abedini Koshksaray, A., Jalilvand Shirkhani Tabar, M.: Mobile-banking adoption by Iranian bank clients. Telemat. Inform. **31**(1), 62–78 (2012)
31. Mawona, A., Mpogole, H.: ICT and financial inclusion : adoption of mobile phone banking among small business owners in iringa, Tanzania, pp. 1–9 (2013)
32. Moon, J.W., Kim, Y.G.: Extending the TAM for a World-Wide-Web context. Inf. Manag. **38**(6), 217–230 (2000)
33. Davis, F.: Perceived usefulness, perceived ease of use, and user acceptance of information technology. MIS Q. **13**(3), 319–340 (1989)
34. Venkatesh, V., Davis, F.D.: A model of the antecedents of perceived ease of use: development and test. Decis. Sci. **27**(3), 451–481 (1996)
35. Viswanath, V., Morris, M.G.: Why don't men ever stop to ask for directions? gender, social influence, and their role in technology acceptance and usage behavior. MIS Q. **24**(1), 115–139 (2000)
36. Venkatesh, V., Davis, F.D.: A theoretical extension of the technology acceptance model : four longitudinal field studies. Manag. Sci. **46**(2), 186–204 (2000)
37. Bhatiasevi, V., Yoopetch, C.: The determinants of intention to use electronic booking among young users in Thailand. J. Hosp. Tour. Manag. **23**, 1–11 (2015)

38. Lian, J.: Critical factors for cloud based e-invoice service adoption in Taiwan : an empirical study. Int. J. Inf. Manage. **35**(1), 98–109 (2015)
39. Raman, A., Don, Y., Khalid, R., Hussin, F., Omar, M.S., Ghani, M.: Technology acceptance on smart board among teachers in terengganu using UTAUT model. Asian Soc. Sci. **10**(11), 84–91 (2014)
40. Rana, N.P., Dwivedi, Y.K., Williams, M.D., Weerakkody, V.: Investigating success of an e-government initiative: validation of an integrated IS success model. Inf. Syst. Front. **17**(1), 127–142 (2014). https://doi.org/10.1007/s10796-014-9504-7
41. Martins, C., Oliveira, T., Popovič, A.: Understanding the internet banking adoption: a unified theory of acceptance and use of technology and perceived risk application. Int. J. Inf. Manage. **34**(1), 1–13 (2014)
42. Park, E., Kim, K.J.: An integrated adoption model of mobile cloud services: exploration of key determinants and extension of technology acceptance model. Telemat. Inform. **31**(3), 376–385 (2014)
43. Featherman, M.S., Pavlou, P.A.: Predicting e-services adoption: a perceived risk facets perspective. Int. J. Hum. Comput. Stud. **59**(4), 451–474 (2003)
44. Mutahar, A.M., Daud, N.M., Ramayah, T., Isaac, O., Aldholay, A.H.: The effect of awareness and perceived risk on the technology acceptance model (TAM): mobile banking in Yemen. Int. J. Serv. Stand. **12**(2), 180–204 (2018)
45. Seymour, L., Makanya, W., Berrangé, S.: End-Users ' Acceptance of Enterprise Resource Planning Systems : An Investigation of Antecedents (2007)
46. Davis, F.D., Bagozzi, R.P., Warshaw, P.R.: User acceptance of computer technology: a comparison of two theoretical models. Manage. Sci. **35**(8), 982–1003 (1989)
47. Aversano, N.: Technology Rejection Of Mobile Telephones (2005)
48. Yalcinkaya, R.: Police Officers' Adoption of Information Technology: A Case Study of the Turkish POLNET System. University of North Texas, Denton (2007)
49. Nasri, W., Charfeddine, L.: Factors affecting the adoption of Internet banking in Tunisia: An integration theory of acceptance model and theory of planned behavior. J. High Technol. Manag. Res. **23**(1), 1–14 (2012)
50. Venkatesh, V., Davis, F.D.: A theoretical extension of the technology acceptance model: four longitudinal field studies. Manage. Sci. **46**(2), 186–204 (2000)
51. Al-Haderi, S.: A Framework to Study Factors Influencing the Acceptance of Information Technology in Yemen Government. Universiti Utara Malaysia, Changlu (2012)
52. Md Talib, Z., Mosbah, A., Al-Jubari, I.: Intention to pursue postgraduate studies in malaysian universities. Int. J. Innov. Creat. Chang. **5**(2), 591–609 (2019)
53. Al-Qeisi, K.: Analyzing the Use of UTAUT Model in Explaining an Online Behaviour: Internet Banking Adoption (Doctoral thesis), Brunel University (2009)
54. Ringle, C.M., Wende, S., Becker, J.-M.: SmartPLS 3. Bonningstedt: SmartPLS (2015)
55. Bagozzi, R.P., Phillips, L.W.: Representing and test- ing organizational theories : a holistic construal. Adm. Sci. Q. **27**(3), 459–489 (1982)
56. Anderson, J.C., Gerbing, D.W.: Structural equation modeling in practice : a review and recommended two-step approach. Psychol. Bull. **103**(3), 411–423 (1988)
57. Hair, J.F., Hult, G.T.M., Ringle, C., Sarstedt, M.: A Primer on Partial Least Squares Structural Equation Modeling (PLS-SEM), 2nd ed. SAGE, Thousand Oaks, London (2017)
58. Schumacker, R.E., Lomax, R.G.: A Beginner's Guide to Structural Equation Modeling. Lawrence Erlbaum, New York (2004)
59. Hair, J.F., Anderson, R.E., Tatham, R.L., Black, W.C.: Multivariate Data Analysis. Seventh Edition. Prentice Hall. New Jesey, 7th ed. New Jersey: Pearson (2010)
60. Kannana, V.R., Tan, K.C.: Just in time, total quality management, and supply chain management: understanding their linkages and impact on business performance. Omega Int. J. Manag. Sci. **33**(2), 153–162 (2005)

61. Gefen, D., Straub, D., Boudreau, M.-C.: Structural equation modeling and regression: Guidelines for research practice. Commun. Assoc. Inf. Syst. **4**(1), 1–79 (2000)
62. Kline, R.B.: Principles and Practice of Structural Equation Modeling, 3rd edn. The Guilford Press, New York (2010)
63. Jalal, A.N., Bahari, M., Musa, M.W., Sherief, S.R.: Social media and traditional customer relationship management: an overview of theoretical challenges and future directions in healthcare context. Int. J. Adv. Trends Comput. Sci. Eng. **8**(16), 76–82 (2019)
64. Jalal, A.N., Bahari, M., Tarofder, A.K., Musa, W.M.N.M.W.: Factors influencing customer social relationship management implementation and its benefits in healthcare industry. Polish J. Manag. Stud. **19**(2), 196–205 (2019)
65. Hair, J.F., Black, W.C., Babin, B.J., Anderson, R.E.: Multivariate Data Analysis, 7th Editio. Prentice Hall, New York (2010)
66. Fornell, C., Larcker, D.F.: Evaluating structural equation models with unobservable variables and measurement error. J. Mark. Res. **18**(1), 39–50 (1981)
67. Sullivan, G.M., Feinn, R.: Using effect size - or why the p value is not enough. J. Grad. Med. Educ. **4**(3), 279–282 (2012)
68. Cohen, J.: Statistical Power Analysis for the Behavioral Sciences, 2nd Editio. LawreAssociatesnce Erlbaum, New Jersey (1988)
69. Chin, W.W.: The partial least squares approach to structural equation modeling. In: Marcoulides, G.A. (Ed.) Modern Methods for Business Research, pp. 295–358. Lawrence Erlbaum Associates. Mahwah, NJ: Lawrence Erlbaum, New Jersey (1998)
70. Gefen, D., Rigdon, E.E.: An update and extension to SEM guidelines for administrative and social science research. MIS Q. **35**(2), 1–7 (2011)
71. Cohen, J.: Statistical Power Analysis for the Behavioral Sciences, 2nd ed. Lawrence Erlbaum Associates, Hillsdale (1988)
72. Fornell, J., Cha, C.: Partial least squares. In: Bagozzi, R.P. (Ed.) Advanced Methods in Marketing Research, pp. 52–78. Blackwell, Cambridge (1994)
73. O'brien, R.M.: A caution regarding rules of thumb for variance inflation factors. Qual. Quant. **41**(5), 673–690 (2007)
74. Myers, R.H.: Classical and Modern Regression with Applications, 2nd ed. Duxbury, Boston, MA (1990)
75. Bowerman, B.L., O'Connell, R.: Linear Statistical Models: An Applied Approach, 2nd ed. Duxbury, Belmont, CA (1990)
76. Hair, J.F., Hult, G.T.M., Ringle, C.M., Sarstedt, M.: A Primer on Partial Least Squares Structural Equation Modeling (PLS-SEM). SAGE Publications, Thousand Oaks (2013)
77. Awang, Z.: Structural Equation Modeling Using AMOS. University Teknologi MARA Publication Center, Shah Alam. Malaysia (2014)
78. Hayes, A.F.: Introduction to Mediation, Moderation, and Conditional Process Analysis: A Regression-Based Approach, 1st ed. The Guilford Press, New York (2013)
79. Field, A.: Discovering Statistics Using IBM SPSS Statistics, 4th ed. SAGE Publications Ltd, Thousand Oaks (2013)
80. Field, A.: Discovering Statistics Using SPSS, Fourth Edi. (2013)
81. Krogstie, J.: The Adoption of Mobile Tourism Services : An Empirical Study, pp. 3–5 (2012)
82. Kim, B.G., Park, S.C., Lee, K.J.: A structural equation modeling of the Internet acceptance in Korea. Electron. Commer. Res. Appl. **6**(4), 425–432 (2007)
83. Luarn, P., Lin, H.-H.: Toward an understanding of the behavioral intention to use mobile banking. Comput. Human Behav. **21**(6), 873–891 (2005)
84. Mutahar, A.M., Daud, N.M., Ramayah, T., Putit, L., Isaac, O.: Examining the effect of subjective norms and compatibility as external variables on TAM: mobile banking acceptance in Yemen. Sci. Int. **29**(4), 769–776 (2017)

85. Riffai, M.M.M.A., Grant, K., Edgar, D.: Big TAM in oman: exploring the promise of on-line banking, its adoption by customers and the challenges of banking in Oman. Int. J. Inf. Manage. **32**(3), 239–250 (2012)
86. Zolait, A.H.S.: An examination of the factors influencing Yemeni Bank users' behavioural intention to use Internet banking services. J. Financ. Serv. Mark. **15**(1), 76–94 (2010)
87. Sathye, M.: Adoption of internet banking by Australian consumers: an empirical investigation. International Journal of Bank Marketing (1999)
88. Guriting, P., Ndubisi, N.O.: Borneo online banking: evaluating customer perceptions and behavioural intention. Manag. Res. News **29**(1/2), 6–15 (2006)

The Impact of Strategic Agility on Sustainable Competitive Advantage in the Yemeni Manufacturing Industry

Maged Barahmah[1], Mohammed Ali Al-Awlaqi[2], Ali AL-Tahitah[3](✉), Omar Habtoor[4], and Kartini Kamaruzzaman[3]

[1] Faculty of Oil and Minerals, Aden University, Aden, Yemen
[2] School of Business, Lebanon International University, Sana'a, Yemen
[3] Faculty of Leadership and Management, Universiti Sains Islam Malaysia USIM, Nilai, Malaysia
kartinikamaruzzaman@usim.edu.my
[4] Northern Border University, Arar, Northern Border province, Saudi Arabia
Omer.habtoor@nbu.edu.sa

Abstract. This study aims at examining the direct effect of strategic agility on sustainable competitive advantage in the big and medium-sized manufacturing companies operating in Yemen. A total of 227 out of 554 of companies were surveyed using a stratified random sampling technique. The questionnaire was sent to the heads of companies and their deputies. Structural equation modeling technique was used to analyze data and test the hypotheses of the study. The findings revealed that strategic agility has a significant and direct impact on sustainable competitive advantage on how the big and medium-size companies can improve their competitiveness such as by improving their chance of survival in a very resource-constrained market. This study filled the gap in the existing literature by investigating the direct role of the sustainable competitive advantage in the big and medium-sized manufacturing companies operating in Yemen.

Keywords: Strategic agility · Sustainable competitive advantage · Manufacturing industry · Yemen

1 Introduction

The world has been witnessing a dramatic development and change during the first decades of the current century which enforced business organization to operate in a very competitive and changing environment [1]. This enforces business organizations especially the one operates in least-developed countries to create and adopt new and modern concepts and systems. This would help business organizations to raise their awareness toward their current status, create more proactive strategies and reduce business surrounding threats [2, 3].

The Yemeni manufacturing companies are no different from the other companies in the least-developed countries [4, 5]. The companies encounter severe competition from

M. Al-Emran et al. (Eds.): ICETIS 2021, LNNS 299, pp. 404–418, 2022.
https://doi.org/10.1007/978-3-030-82616-1_35

foreign companies who can dominate the local market due to their superior experience and capacities. The key problem of this study is the low performance of the medium and large manufacturing companies in Yemen. The World Bank [7] emphasized that Yemeni manufacturing companies faced several problems that hinder their ability to dominate the local market and to penetrate the regional market such as poor incentives, insufficient experience, and lack of employees training. AlQershi, Abas, and Mokhtar [6] indicated that the major issue faced by large and medium-sized Yemeni manufacturing companies resulted from poor organizational capabilities and lack of resources of which contributing to dynamism.

It very important to know the key ingredient that could create sustainable competitive advantage. Previous literature had shown many proposed solutions that can improve the competitiveness of the medium and large-sized companies, for example, the theory of internal resources [7]; and dynamic capabilities [8]. Those theories concentrated on the creation of inimitable dynamic capabilities that could use to create sustainable competitive advantages [9]. Although many studies pointed at the important role of strategic agility and competitive advantage [10–12], these studies did not investigate the sustainable competitive advantage directly. Therefore, this study has been conducted to fill in this important gap by investigating the direct role of the sustainable competitive advantage in the big and medium-sized manufacturing companies operating in Yemen.

The context of large and medium-sized manufacturing companies was chosen because of several supporting factors. The industry is having problem with its market share due to intense competition from foreign companies. Moreover, big and medium-sized companies have stronger leadership that could enhance the ability to alter the strategic agility of these companies [13]. Thus, the context of big and medium-sized companies in Yemen found to be the most suitable context for this study.

The study contributes to the body of knowledge in this field by providing practical evidence on how the big and medium-sized companies can improve competitiveness improve their chance of survival in a very resource-constrained market such as Yemen.

2 Literature Review

2.1 Strategic Agility

Due to the rapid global changes, business organizations face a turbulent and uncertain working environment [4]. This has increased the competition among organizations and eroded their chance to survive. Strategic agility is also known as the process of adoption to changes which would lead to achieving greater success by exploiting the available opportunities and creating competitive advantages. Thus, Strategic agility is playing an important role for sustaining in the market when the organizations are competing in the instable business environment [14].

Accordingly, considered Strategic agility is the fundamental ability for survival in the face of serious competition [15]. Strategic agility could be applied by reconfiguring the organization and its strategies dynamically to match the changing business environment. Organizations need to continuously anticipate and interact with the new markets' trends, without compromising the organization's strategic vision [16].

Strategic agility is a multidimensional concept. Doz and Kosonen [17] outline three dimensions of strategic agility, namely; sensitivity, collective commitment, and the flow of resources. While Sharifi and Zhang [18] proposed four dimensions of strategic agility: efficiency, flexibility, speed, responsiveness. Furthermore, Dabir and Gholami [12] indicated to two dimensions of strategic agility: differentiation in innovation and differentiation in the market. Applying strategic agility supply chain practices revealed three main dimensions: customer agility, operational agility, and agility Partnership [19]. Jaworski and Kohli [20] listed three dimensions of strategic agility of sensing agility, decision-making agility. Following the nature and the objectives of, the last three dimensions of Jaworski and Kohli [20] were adopted in the current study.

2.2 Sustainable Competitive Advantage

The concept of sustainable competitive advantage has become one of the most promising concepts over the last century, particularly in the field of strategic management [4]. Porter [21], in his early writings, proposed that competitive strategies help in creating sustainable competitive advantage. Porter [22] discussed the main competitive strategies of lower-cost or product distinction strategies. However, Beal [22] argues that the traditional way of achieving sustainable competitive advantage depends on the environmental structure and overall competitive strategy. This shows that an organization must learn how to create new and innovative features that can maintain a leading competitive position [23]. This could be achieved by creating a new way of thinking and implementing a strategic intelligence that differs from the conventional way of thinking [24].

As for the sources of sustainable competitive advantage, Coyne [25] has said that sustainable competitive advantage could be achieved by exploiting potential gaps, namely: business system gaps [25]. In contrast, Harrison [26] explained that research on the potential sources of sustainable competitiveness has shifted to the importance of an effective and distinct corporate strategy. Thus, the sources of sustainable competitive advantage have become multiple and complex.

Day and Wensley [27] identified two main factors that contribute to the achievement of sustainable competitive advantage. The first factor is the existence of the employees' superior skills. The second factor is the ability of organizations to exercise those capabilities. Companies should combine their resources and skills in the core competencies to achieve sustainable competitive advantage [24, 28]. Practically, the internal resources of the organization can be used to achieve sustainable competitive advantage if they are characterized by four attributes: scarcity, value, the impossibility of imitation, and the impossibility of finding an alternative [7].

Hunt and Morgran [29] showed a range of sources of sustainable competitive advantage, namely, financial, legal, human, organizational and informational sources. Hill, Jones and Schilling [30] identified four factors in building sustainable competitive advantage: superior efficiency, superior quality, superior innovation and responding to customer needs. By drawing on those concepts, it is clear that most researchers agree that the sources of sustainable competitive advantage depend on the organization's superior skills and resources, which must be combined to achieve sustainable competitive advantage. Consequently, the factors (superior efficiency, superior quality, superior innovation and responding to customer needs) have been adopted in this study.

2.3 Strategic Agility and Sustainable Competitive Advantage

As mentioned earlier, strategic agility plays a pivotal role in identifying market movement when organizations compete in an unstable business environment [14]. Accordingly, this study comes to highlight the interaction between two effective theories, which are the RBV theory and the dynamic capabilities theory, as recommended by [31] and Chen [10]. RBV theory concentrates on the importance of managing the resources of an organization that is concerned with creating competitive advantage, but on the other hand, this theory fails to explain how to properly employ these resources. Besides, this theory does not focus much on the dynamic relationship between the surrounding environment and the resources of the organization. However, the role of dynamic capabilities theory is to fill in those gaps and serve as a complementary theory to RBV. In short, based on a study [10]. [10, 32], the theory of dynamic capabilities helps organizations improve their resources regarding the dynamic environment, where agility is an organizational resource that enhances the ability to proactively increase the capacity and productivity of companies.

Furthermore, there is a scant scholarship that discussed the relationship between strategic agility and sustainable competitive advantage. Almahamid, Awwad and McAdams [11] found a positive role of strategic agility and knowledge sharing. Besides, Chen's [9] found a mediating of strategic agility plays a significant role in the relationship between business intelligence and the flexibility of the information technology infrastructure. Chen [9] recommended studying the effect of more independent variables on the competitive advantage in the case of the existence of strategic agility. Furthermore, Dabiri and Gholami [12] indicated that there is a direct relationship between strategic agility including differentiation in innovation, market differentiation and competitive advantage. It also shows that the effect of strategic agility in terms of differentiation in innovation is more influential than differentiation in the market for achieving competitive advantage. In addition to that, Hemmati, Feiz, Jalilvand, and Kholghi [33] presented a significant relationship between a company's resources, competitive advantage, and strategic agility. Meanwhile, Baškarada and Koronios [34] study indicated that strategic agility has five dynamic capabilities: sensing, search, seizing, switching, and formulating. These capabilities help it to play a pivot role in creating a sustainable competitive advantage. Also, Al-Romeedy [35] indicated in an empirical study that strategic agility affects greatly the competitive advantage.

Based on the above argument, the researchers hypothesize the following hypotheses;

H1: There is a direct effect between strategic agility and the sustainable competitive advantage in the Yemeni large and medium manufacturing companies.

Based on the main hypothesis stated above, a total of three sub hypotheses assumed of the dimensions of the strategic agility effect (three factors) on the dependent variable "sustainable competitive advantage" which determined as ***(Ha),*** on the other hand, 12 sub-hypotheses named ***(Hb)*** assumed the effect of the all dimensions of both independent on dependent variables together, as shown below;

$Ha_{1:}$ Sensing agility has a positive and significant impact on sustainable competitive advantage in the Yemeni large and medium manufacturing companies.

$Ha_{2:}$ Decision-making agility has a positive and significant impact on sustainable competitive advantage in the Yemeni large and medium manufacturing companies.

Ha_3: Acting Agility has a positive and significant impact on sustainable competitive advantage in the Yemeni large and medium manufacturing companies.

Hb_1: Sensing agility has a positive and significant impact on superior efficiency in the Yemeni large and medium manufacturing companies.

Hb_2: Sensing agility has a positive and significant impact on superior quality in the Yemeni large and medium manufacturing companies.

Hb_3: Sensing agility has a positive and significant impact on superior innovation in the Yemeni large and medium manufacturing companies.

Hb_4: Sensing agility has a positive and significant impact on responding to customer needs in the Yemeni large and medium manufacturing companies.

Hb_5: Decision-making agility positive and has a significant impact on superior efficiency in the Yemeni large and medium manufacturing companies.

Hb_6: Decision-making agility has positive and a significant impact on superior quality in the Yemeni large and medium manufacturing companies.

Hb_7: Decision-making agility) has positive and a significant impact on superior innovation in the Yemeni large and medium manufacturing companies.

Hb_8: Decision-making agility has positive and a significant impact on responding to customer needs in the Yemeni large and medium manufacturing companies.

Hb_9: Acting Agility has a positive and significant impact on superior efficiency in the Yemeni large and medium manufacturing companies.

Hb_{10}: Acting Agility has a positive and significant impact on superior quality in the Yemeni large and medium manufacturing companies.

Hb_{11}: Acting Agility has a positive and significant impact on superior innovation in the Yemeni large and medium manufacturing companies.

Hb_{12}: Acting Agility has a positive and significant impact on responding to customer needs in the Yemeni large and medium manufacturing companies.

3 Methodology

This current study's model is a complex model which has more than one independent and dependent variables as shown in Fig. 1. Hence, simple multivariable analysis techniques are not suitable to test it. Structural equation modeling was found to be the most suitable technique that could be used test the hypotheses of this model.

Primary data was collected using a self-administered questionnaire distributed to the selected sample. The instrument was developed from previous studies where the instrument included two sections: the first section contained 19 items related to the independent variable (strategic agility). These items were adopted from Jaworski and Kohli [20]. The second section contained 16 items related to the depended variable (sustainable competitive advantage) which were adopted from Hill et al. [30].

The validity and reliability of this questionnaire were measured using Cronbach's Alpha. For the independent variable (the strategic agility) Cronbach's Alpha was .907 while the dependent variable (the sustainable competitive advantage) scored .861. Moreover, the coefficient of stability for all variables was .886. This confirmed the high reliability of the present study's instrument.

The target population was confined to large and medium manufacturing industries in the Republic of Yemen. The population of the study has comprised of 554 companies. Thompson [36] formula has been used in this study and employed a stratified sampling technique to choose a randomized sample of 227 companies. The questionnaire was sent to the heads of companies and their deputies. Despite 454 respondents were targeted, only 289 questionnaires were returned, representing a response rate of 63%. Therefore, only 271 questionnaires were analyzed, representing a response rate of (60%). The response rate was considered satisfactory and acceptable according to Sekaran and Bougie [37].

3.1 Demographic Information

The data was chosen according to the proportions for the already existed categories. A total of 227 organizations were chosen, 46% of which are from the food industry, 20% from the textile industry, 10% from the chemical industry, 15% from the cement and tiles industry, and 9% from the machine industry.

4 Findings and Discussion

The data was collected directly through site visits. To secure a high rate of response, follow-up visits were conducted. That data then tested for missing data, typos, or any outliers. The data then was tested using exploratory factors, confirmatory factors and finally, the structural equation modeling technique was used to test the research hypotheses.

4.1 Exploratory Factor Analysis

We conducted an exploratory factor analysis with factor loadings of greater than 0.40 for all variables as recommended by Hair, Hult, Ringle & Sarstedt [38]. Furthermore, Kaiser Meyer Olkin (KMO) was conducted and showed a very satisfactory result of .931. KMO values were .941 for strategic agility construct and .903 for sustainable competitive advantage construct. The Bartlett's test scores showed significant results for all the variables at a significant level of $\alpha = .05$.

4.2 CFA of the Measurement Model of the Variables

Confirmatory factor analysis was conducted to confirm the validity of the study instrument constructs. It is clear from Table 1 that the measurement model of this study showed no issues with its construct validity as all constructs' items showed very high loading scores of 0.621 [38]. Confirmatory Factor analysis indicates very promising results in its goodness a fit test as Chi-square result showed (1065.818, df = 533; $P < .000$); CIMIND 2.000; CFI .940; TLI .933; RMSEA .061; GFI .825; AGFI .793; IFI .940 and NFI .887 (Hair, et al., 2016). As shown in Table 1, the average variance extracted (AVE) is higher than the threshold ratio (0.50) which confirms the convergence validity. Moreover, the internal consistency of the variables (CR) was confirmed as the calculated values exceed the recommended value of (0.7). In addition to that, the correlation between the factors was ranging within the interval of (0.711–0.237), which is lower than the threshold value of (0.85). Based on that, we confirmed that the external correlation among the factors is less than the inner bonding value of the same factor [39]. Therefore, this asserts the discriminant validity of the study's constructs.

Table 1. Loading, Cronbach's Alpha, CR, AVE and MSV

Variables	Items	Factor loading	Squared multiple (R)	A	CR	AVE	MSV > AVE
Strategic agility				0.957			
Sensing agility				0.977	0.94	0.63	0.386
	AGI	0.96	0.906				
	AGI2I	0.99	0.941				
	AGI3I	0.944	0.894				
	AGI4I	0.919	0.857				
Decision-making agility				0.946	0.94	0.74	0.474
	AGI5	0.85	0.728				
	AGI6	0.893	0.792				
	AGI7	0.911	0.766				
	AGI8	0.884	0.785				
	AGI9	0.866	0.762				
	AGI10	0.753	0.647				
Acting agility				0.943	0.97	0.9	0.474
	AGI11	0.799	0.679				
	AGI12	0.78	0.674				
	AGI13	0.833	0.689				
	AGI14	0.805	0.644				
	AGI15	0.836	0.686				
	AGI16	0.806	0.697				
	AGI17	0.779	0.662				
	AGI18	0.773	0.708				
	AGI19	0.753	0.691				
Sustainable competitive advantage				0.929			
Superior efficiency				0.903	0.91	0.72	0.482
	COM1	0.8	0.601				
	COM2	0.923	0.758				

(*continued*)

Table 1. (*continued*)

Variables	Items	Factor loading	Squared multiple (R)	A	CR	AVE	MSV > AVE
	COM3	0.88	0.705				
	COM4	0.758	0.629				
Superior quality				0.854	0.86	0.62	0.44
	COM5	0.837	0.612				
	COM6	0.836	0.6				
	COM7	0.877	0.641				
	COM8	0.653	0.692				
Superior innovation				0.847	0.86	0.61	0.506
	COM9	0.621	0.653				
	COM10	0.92	0.682				
	COM11	0.849	0.635				
	COM12	0.77	0.641				
Responding to customer needs				0.832	0.85	0.6	0.506
	COM13	0.647	0.772				
	COM14	0.819	0.681				
	COM15	0.896	0.698				
	COM16	0.842	0.602				

α,(> 0.7), CR (> 0.7), AVE (> 0.5), (MSV > AVE)

4.3 Structural Model Assessment

To test the direct and indirect effect's hypotheses, this study used structural equation modeling (SEM) technique. The obtained fit indices results were Chi-square (814.899, df = 451; P < .000); CFI = 0.947; TLI = 0.942; CMIN/DF = 1.807; GFI = 0.847; AGFI = 0.821; IFI = 0.948; NFI = .890; RMSEA = 0.055. These results were less than the acceptable criteria (Byrne, 2010; Hair et al., 2016). Hence, items 8, 9, and 13 were removed from the dependent variable (SCA) because their factor loadings were less than.60.

4.4 Hypotheses Testing

The relationship between the main variables was examined using the t-test values. In the second step, we assessed the direction of the relationship such as the positive or negative relationship between the independent variable and dependent variable. The nonparametric bootstrapping technique was applied with 2000 replications to test the structural model. Accordingly, the main hypothesis *H1*, strategic agility has a significant direct effect on sustainable competitive advantage (β = 0.79; p < 0.05; t-values = 7.023) in Yemeni large- and medium-sized manufacturing companies.

The coefficient of determination (Squared Multiple) represents the percentage of variance in which the dependent variable can be explained by the independent variable [40]. Strategic agility explains 62.7% of the variability of the sustainable competitive advantage. Thus, hypothesis H1 is supported and we can conclude that the independent variable (strategic agility) has a direct effect on the dependent variable (sustainable competitive advantage) as shown in Table 2.

Table 2. Direct effect of independent variable on dependent variable

H	Main Variables' Relationship	Path	S.E	C.R	P	Result
H1	Strategic Agility Sustainable Competitive Advantage	0.3	0.05	4.3	***	Supported

On the other hand, according to the SEM model in Fig. 1 which tests the direct sub hypotheses for the first main hypothesis, the obtained fit indices results were Chi-square result showed (933.513, df = 452; P < .000); CFI = 0.930; TLI = 0.954; CMIN/DF = 2.065; GFI = 0.874; AGFI = 0.804; IFI = 0.931; NFI = .0923; RMSEA = 0.063 which indicates that current results are acceptable [38, 41]. Thus, the results of goodness-of-fit indices indicated irrefutable evidence that the model exhibited an acceptable fit and a statistically significance fit.

Based on the results of the sub-hypotheses' *Ha* test, the direct effect of sensing agility on sustainable competitive advantage is positive and significant (β = 0.32; P-value < 0.05; t-values = 4.266) in Yemeni large- and medium-sized manufacturing companies. Furthermore, the result of this study demonstrates that decision-making agility has a significant direct effect on sustainable competitive advantage (β = 0.28; P-value < 0.05; t-values = 3.756) in Yemeni large- and medium-sized manufacturing companies. There is also a significant direct effect between acting Agility on sustainable competitive advantage (β = .51; p < 0.05; t-values = 6.183) in Yemeni large- and medium-sized manufacturing companies (Table 3).

The result of squared multiple correlations confirmed that the dimensions of strategic agility explained 43.8% of the variance of sustainable competitive advantage. The result is considered as a medium correlation score [40]. The model's fit indices results were Chi-square result showed (1021.656, df = 445; P < .000); CFI = 0.916; TLI = 0.907; CMIN/DF = 2.296; GFI = 0.814; AGFI = 0.780; IFI = 0.917; NFI = .0862; RMSEA = 0.069. These scores are considered to be acceptable. Thus, the results of goodness-of-fit indices indicated irrefutable evidence that the model exhibited an acceptable fit and a statistical significance fit to the data.

Consequently, the results of the sub-hypotheses Hb tests that belong to hypothesis 1 confirmed that sensing agility, decision-making agility, and acting Agility have a significant effect on superior efficiency. The model showed significant values of (β = .24; β = .15; β = .68) for the previous relationship respectively, with p-values of (*p*-value = .000; .020; .000) sequentially. Moreover, the *t*-value for those relationships were (3.434; 2.024; 6.749) respectively. Also, the results of the study showed that there were

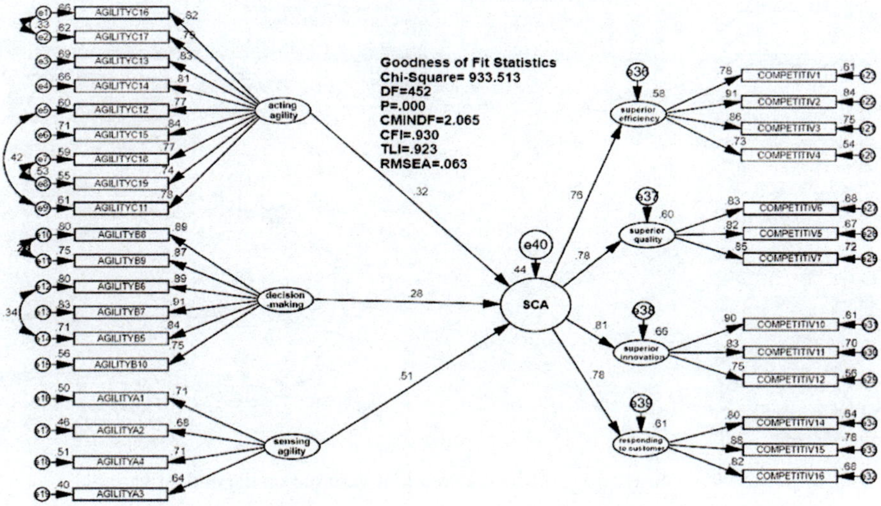

Fig. 1. Impact of dimensions of the independent variable on the dependent variable

Table 3. Testing of sub-hypotheses with the main hypothesis for direct effect

H*a*	IV Dimensions with DV	Path	S.E	C.R.	*P*	Result
Ha1	Sensing agility → SCA	0.3	0.05	4.3	***	Supported
Ha2	Decision-making agility → SCA	0.3	0.05	3.8	***	Supported
Ha3	Acting Agility → SCA	0.5	0.08	6.2	***	Supported

**SCA (Sustainable Competitive Advantage)*

significant direct effects of sensing agility, decision-making agility, and acting Agility on superior quality. These relationships have the values of (β = .20; β = .16; β = .73), *p*-values of (.004; .043; .000), and *t*-value (2.915; 2.062; 6.667) respectively. There are **also significant direct impacts of sensing agility, decision-making agility, and acting Agility on superior innovation. These relationships have coefficient's values of** (β = .25; β = .12; β = .74), *p*-values of (.000; .039; .000), and *t*-values of (3.695; 2.012; 7.209) respectively. In addition to that, the results of the study showed that there were **significant direct effects of sensing agility, decision-making agility, and acting Agility** on responding to customer needs. These relationships have the values of (β = .15; β = .30; β = .69), *p*-values of (.020; .047; .000), and *t*-value (2.318; 4.00; 6.557) respectively, in Yemeni large and medium manufacturing industries.

Figure 2, and Table 4 **summaries these results. The coefficient of determination shows** very good scores for these relationships (.581; .626; .593; .541) respectively.

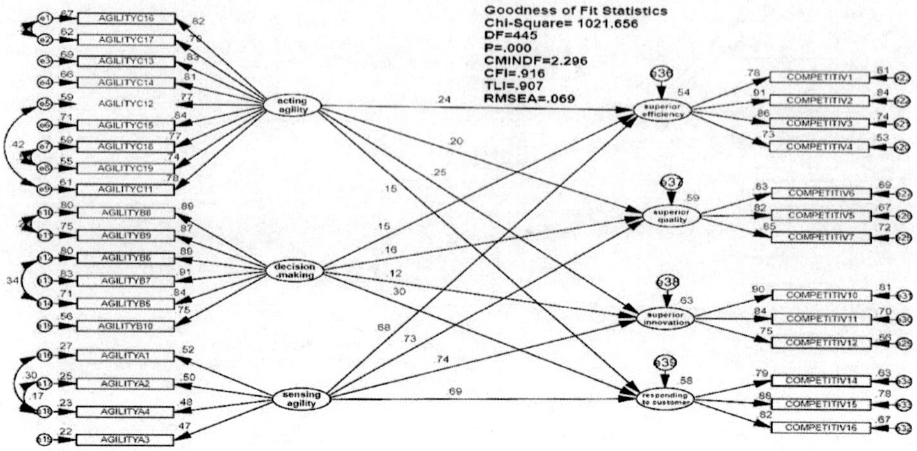

Fig. 2. Effect of dimensions of the independent variable on dependent variable.

Table 4. The direct effect of IV's sub-hypotheses on DV's sub-hypothesis

Hb	IV's Dimensions on DV's Dimensions	Path	S.E	C.R.	P	Result
Hb1	Sensing agility → Superior efficiency	0.24	0.05	3.434	0.001***	Supported
Hb2	Sensing agility → Superior quality	0.20	0.06	2.915	0.004**	Supported
Hb3	Sensing agility → Superior innovation	0.25	0.05	3.695	0.001***	Supported
Hb4	Sensing agility → Responding to customer needs	0.15	0.06	2.318	0.020*	Supported
Hb5	Decision-making agility → Superior efficiency	0.15	0.05	2.024	0.020*	Supported
Hb6	Decision-making agility → Superior quality	0.16	0.06	2.062	0.043*	Supported
Hb7	Decision-making agility → Superior innovation	0.12	0.04	2.012	0.039*	Supported
Hb8	Decision-making agility → Responding to customer needs	0.30	0.05	4.00	0.047*	Supported
Hb9	Acting Agility → Superior efficiency	0.68	0.13	6.749	0.001***	Supported
Hb10	Acting Agility → Superior quality	0.73	0.12	6.667	0.001***	Supported
Hb11	Acting Agility → Superior innovation	0.74	0.15	7.209	0.001***	Supported
Hb12	Acting Agility → Responding to customer needs	0.69	0.12	6.557	0.001***	Supported

*P ≤ 0.05 ** P ≤ 0.01 *** P ≤ 0.001

5 Conclusion

This study used the structural equations technique to test the relationship between strategic agility and its dimensions (sensing agility, decision-making agility, acting Agility) on sustainable competitive advantage and its dimensions (superior efficiency, superior quality, superior innovation and responding to customer needs) of Yemeni large and medium manufacturing industries. Sensing agility has a positive and significant impact on superior efficiency, superior quality, superior innovation and responding to customer needs. The findings are consistent with the findings of the previous studies [10]. Decision-making agility has a positive and significant impact on superior efficiency, superior quality, superior innovation and responding to customer needs [12]. Action agility has a positive and significant impact on superior efficiency, superior quality, superior innovation and responding to customer needs. Therefore, the findings are consistent with the findings of the previous studies that show the effective impact of strategic agility in accelerating responding to changes and developments [33, 34].

The findings of this study are unique. This study enriches literature by investigating the direct role of the sustainable competitive advantage in a very constrained context such as Yemen. Besides, the current study contributed to more in-depth understating of the interaction between the dimensions of strategic agility and the dimension of the sustainable competitive advantage in a more comprehensive model. This study created more in-depth analysis of the relationship between strategic agility and the creation of competitive advantage.

5.1 Theoretical and Practical Implications

Organizations managers can use the fining of this study to create competitive advantages over their competitors by concentrating on their efficiency, quality, and innovation. Managers can be ready with more than one model to create their competitive advantage and penetrate more lucrative markets. Moreover, managers could plan more sustainable strategies through strategic agility utilizations that would create more sustainable and unimputable products and services. Managers should not think about their strategies as a static endeavor, they should think of how to create a strategy and at the same time how to be agile in applying it.

Theoretically, this paper enhances the understanding of strategic agility and sustainable competitive advantage and uses them as multidimensional concepts. Scholars also now use the relationship of strategic agility and the sustainable competitive advantage under a dynamic organizational capability instead of static resources. The finding of the study has added a theoretical contribution to the previous body of knowledge such as the RBV theory by Barney [7] and the theory of dynamic capacity.

5.2 Limitations and Future Research

It is acknowledged that the limitations of the present study avail the opportunity for further research to direct the attention on variables that are not addressed in this article and to fill in the gap in the Yemeni context on the determinants of sustainable competitive advantage. This study covered only the large and medium-sized manufacturing companies in Yemen. The study concentrated on strategic agility in creating competitive advantages, yet it ignored many other important factors such as human resource management practices, performance management practices, employees' quality of life.

Future research should address the relationship between strategic agility and sustainable competitive advantage in different industries and different types of organizations such as small organizations. More future researches are needed to reveal the effect of many other factors that could affect the creation of sustainable competitive advantage.

Thus, the study on industrial companies (in general), and Yemeni industrial companies (in particular), should pay more attention to the concept of strategic agility. Moreover, to apply the two components in scientific manners to be able to achieve the sustainable competitive advantage to overcome the highly competitive situations resulted by the opening of markets and maintaining its position within the competitive environment, to improve its current situation and overcome the current and future problems resulting from rapid and complex changes in the surrounding environment.

References

1. Sheel, S., Sindhwani, B.K., Goel, S., Pathak, S.: Quality of work life, employee performance and career growth opportunities: a literature review. Int. J. Multi. Res. **2**, 291–300 (2012)
2. Alwehabie, A.M.: Criteria for measuring the efficiency and effectiveness of human resources management strategy and its relation to institutional performance at Al Rajhi Bank at Al Qassim in Saudi Arabia from the employees' point of view. IJBM **12**, 111 (2017)
3. Pfeffer, J.: Competitive advantage through people. Calif. Manage. Rev. **36**, 9–28 (1994)
4. Barahma, M.M., Ali, K.B., Nashief, M., Al-Awlaqi, M.A.: The impact of human resources management strategies on the strategic agility: a field study in large and medium yemeni manufacturing sector. COES&RJ-JSS **8**, 611–29 (2019)
5. Barahma, M., Ali, K.B., Nashief, M.: The use of measurement model of strategic agility among large and medium-size manufacturing sector in yemen. IJASOS- Int. E-J. Adv. Soc. Sci. **5**, 1036–1049 (2019)
6. AlQershi, N., Bin Abas, Z., Mokhtar, S.S.M.: The mediating effect of human capital on the relationship between strategic innovation and the performance of manufacturing smes in yemen. Organ. Market. Emerg. Econ. Vilniaus Universiteto Leidykla **10**, 57–77 (2019)
7. Barney, J.: Firm resources and sustained competitive advantage. J. Manag. **17**, 99–120 (1991)
8. Teece, D.J., Pisano, G., Shuen, A.: Dynamic capabilities and strategic management. Strateg. Manag. J. **18**, 509–533 (1997)
9. Tseng, Y.-H., Lin, C.-T.: Enhancing enterprise agility by deploying agile drivers, capabilities and providers. Inf. Sci. **181**, 3693–3708 (2011)
10. Chen, X.: Impact of Business Intelligence and IT Infrastructure Flexibility on Competitive Advantage: An Organizational Agility Perspective [Dissertation]. University of Nebraska-Lincoln, Lincoln (2012)
11. Almahamid, S., Awwad, A., McAdams, A.C.: Effects of organizational agility and knowledge sharing on competitive advantage: an empirical study in jordan. Int. J. Manag. **27**, 387–404 (2010)
12. Dabiri, S.A., Gholami, A.: Surveying the effect of strategic agility on gaining competitive advantage in Gachsaran oil and gas exploitation company. J. Renew. Nat. Resour. Bhutan. **3**, 196–207 (2015)

13. Al-Awlaqi, M.A., Aamer, A.M., Barahma, M.M., Battour, M.: The interaction between leadership styles and their followers' human capital: a correspondence analysis approach applied to micro-sized businesses. Journal of Management Development [Internet]. 2021 [cited 2021 Jan 14]; ahead-of-print (2021). https://doi.org/10.1108/JMD-05-2019-0172
14. Nejatian, M., Zarei, M.H., Rajabzadeh, A., Azar, A., Khadivar, A.: Paving the path toward strategic agility: a methodological perspective and an empirical investigation. J. Enterp. Inform. Manag. Emerald Publishing Limited **32**, 538–562 (2019)
15. Felipe, C.M., Roldán, J.L., Leal-Rodríguez, A.L.: An explanatory and predictive model for organizational agility. J. Bus. Res. **69**, 4624–4631 (2016)
16. Saha, N., Gregar, A., Sáha, P.: Organizational Agility and HRM Strategy: Do They Really Enhance Firms' Competitiveness? [Internet]. Rochester, NY: Social Science Research Network; 2017 Feb. Report No.: ID 3335720. https://papers.ssrn.com/abstract=3335720
17. Doz, Y., Kosonen, M.: The dynamics of strategic agility: nokia's rollercoaster experience. Calif. Manage. Rev. **50**, 95–118 (2008)
18. Sharifi, H., Zhang, Z.: A methodology for achieving agility in manufacturing organisations: an introduction. Int. J. Prod. Econ. **62**, 7–22 (1999)
19. Sambamurthy, V., Bharadwaj, A., Grover, V.: Shaping agility through digital options: reconceptualizing the role of information technology in contemporary firms. MIS Q. **27**, 237–263 (2003)
20. Jaworski, B.J., Kohli, A.K.: Market orientation: antecedents and consequences. J. Mark. **57**, 53–70 (1993)
21. Porter, M.E.: Competitive Advantage: Creating and Sustaining Superior Performance. Free Press, New York, NY (1985)
22. Beal, R.M.: Competitive Advantage: Sustainable or Temporary in Today's Dynamic Environment? Tallahassee. Florida A&M University, Florida (2001)
23. Dickson, P.R.: Toward a general theory of competitive rationality. J. Mark. **56**, 69–83 (1992)
24. Hoffman, N.P.: An examination of the "sustainable competitive advantage" concept: past, present, and future. Acad. Mark. Sci. Rev. **2000**(4), 1–16 (2000)
25. Coyne, K.P.: Sustainable competitive advantage—What it is, what it isn't. Bus. Horiz. **29**, 54–61 (1986)
26. Harrison, D.M.I. (ed.) Diagnosing Organizations: Methods, Models, and Processes. 3rd edn. SAGE Publications, Inc. Thousand Oaks, Calif (2004)
27. Day, G.S., Wensley, R.: Assessing advantage: a framework for diagnosing competitive superiority. J. Mark. **52**, 1–20 (1988)
28. Prahalad, C.K., Hamel, G.: The core competence of the corporation. In: Hahn, D., Taylor, B., (eds.) Strategische Unternehmungsplanung / Strategische Unternehmungsführung: Stand und Entwicklungstendenzen [Internet]. Heidelberg: Physica-Verlag HD, pp. 969–987 (1997). https://doi.org/10.1007/978-3-662-41482-8_46
29. Hunt, S.D., Morgan, R.M.: The comparative advantage theory of competition. J. Mark. **59**, 1–15 (1995)
30. Hill, C.W.L., Jones, G.R., Schilling, M.A.: Strategic Management: Theory: An Integrated Approach. Cengage Learning, Boston (2014)
31. Wójcik, P.: Exploring links between dynamic capabilities perspective and resource-based view: a literature overview. Int. J. Manag. Econ. Sci. **45**, 83–107 (2015)
32. Teece, D.J.: Explicating dynamic capabilities: the nature and microfoundations of (sustainable) enterprise performance. Strateg. Manag. J. **28**, 1319–1350 (2007)
33. Hemmati, M., Feiz, D., Jalilvand, M.R., Kholghi, I.: Development of fuzzy two-stage DEA model for competitive advantage based on RBV and strategic agility as a dynamic capability. J. Modell. Manag. **11**, 288–308 (2016)
34. Baškarada, S., Koronios, A.: The 5S organizational agility framework: a dynamic capabilities perspective. Int. J. Org. Anal. **26**, 331–342 (2018)

35. Al-Romeedy, B.S.: Strategic agility as a competitive advantage in airlines - case study: egypt air. J. Fac. Tourism Hotels-Univ. Sadat City **3**, 1–15 (2019)
36. Thompson, S.K.: Simple random sampling. In: Shewhart, W.A., Wilks, S.S. (eds.) Sampling [Internet]. Third Edition. Hoboken, NJ, USA: John Wiley & Sons, Inc.; 2012 [cited 2019 Sep 21], pp. 9–37 (2012). https://doi.org/10.1002/9781118162934.ch2
37. Sekaran, U., Bougie, R.: Research Methods for Business: A Skill-Building Approach, 7th edn. Wiley, Chichester, West Sussex (2016)
38. Hair, J.F., Black, W.C., Babin, B.J., Anderson, R.E.: Multivariate data analysis. Eighth edition. Cengage Learning EMEA, Andover, Hampshire (2019)
39. Kline, R.B.: Convergence of Structural Equation Modeling and Multilevel Modeling. The SAGE Handbook of Innovation in Social Research Methods [Internet]. 1 Oliver's Yard, 55 City Road, London EC1Y 1SP United Kingdom: SAGE Publications Ltd; 2011 [cited 2019 Jul 10], pp. 562–89 (2011). http://methods.sagepub.com/book/sage-hdbk-innovation-in-social-research-methods/n31.xml
40. Cohen, J.: Statistical Power Analysis for the Behavioral Sciences [Internet]. 2nd ed. Routledge; 2013 [cited 2019 Jul 26]. https://www.taylorfrancis.com/books/9780203771587
41. Byrne, B.M., van de Vijver, F.J.R.: Testing for measurement and structural equivalence in large-scale cross-cultural studies: addressing the issue of nonequivalence. Int. J. Test. **10**, 107–132 (2010)

The Role of Personality Traits on User' Continues Intention to Use Mobile Wallet

Suha Fouad Salem(✉), Khairiyah Salleh, and Nurulaini Shariffudin

Faculty of Business Management and Professional Studies, Management and Science University, Shah Alam, Malaysia
Suha_fksalem@msu.edu.my

Abstract. Mobile wallet has become an alternative method of making payments to purchase goods and services in the recent years. This study investigates the effects of antecedents of continues intention to use mobile wallet, by adapting the Personality Traits theory. The partial least squares structural equation modelling (PLS SEM) technique is used to evaluate the research model with a sample size of 290. Findings reveal that the three most personality traits that has strong connection to the attitude are agreeableness, extroversion, and openness to experience. Additionally, attitude has a positive impact on the continues intention to use mobile wallet in Malaysia. We hope that the results of this study can in several way benefit target individuals and stakeholders. It will also help managers and commerce industries in this field to learn more about consumer attitude toward mobile wallet adoption.

Keywords: Big five personality trait · Continues intention to use · Mobile wallet · Malaysia

1 Introduction

Mobility and new technologies nowadays shaping the life of customer and user doing business all around the world, especially on Malaysia [1]. The behavior of most consumers reflected directly by the emerging of this mobile technology. Finding the benefits and advantages of this new technology have become essential to most companies and business. Changes and adaptation of new ways of process from traditional transaction to newest payment technology have also become an important aspect among all companies and business entity. Thus, the necessity for mobile devices is increasing around the globe. The existences of mobile device already create multi dimension of value and opportunities on various aspects. Technology has dominated every aspect in human lives, and it is constantly changing the way we work, learn and shop. The newest trends in e-commerce sites has in-built with mobile wallet besides the online transfer, debit card or credit card as part of payment methods offered [2, 3]. Mobile wallet has given an alternative way to customer to shop, spend and collect points at the same times [4].

Online banking or mobile banking concept is the foundation of mobile wallet. It has become such a phenomenon in Malaysia, and there are various applications and systems

M. Al-Emran et al. (Eds.): ICETIS 2021, LNNS 299, pp. 419–428, 2022.
https://doi.org/10.1007/978-3-030-82616-1_36

available such as GrabPay, MPay, Boost and AirAsia's BigPay, Touch and Go and more. A mobile wallet will have user's information and details and will store money digitally through mobile devices. It acts as an extension system that allow user to do financial transaction by using their smartphone [5].

The emergence of mobile commerce started slowly with the increasing and adoption usage of mobile devices among consumer all around the world. Almost everyone in Malaysia use and own a smartphone. Though, younger generation has contributed to a higher percentage compared to older generation based on Statista Research Department, July 30, 2020. This number show, the necessity to have a smartphone in a modern life is due to work and lifestyle.

According to Mastercard Impact Study 2020, Malaysia was the lead in Southeast Asia in the usage of mobile or digital wallet. The result showed, COVID-19 is the main reason to the rapid implementation of e-commerce, online payment during online activities.

A number of researchers examined the Big Five Personality traits as predictor of technology adoption and internet functions [6–9]. However, there are only few studies that inspected and examined the effect of personality traits on user attitude towards mobile wallet. Therefore, the purpose of this paper is to investigate in depth the effect of personality traits on user attitude in regard to mobile wallet. Since mobile wallet is a new and emerging service in the market, this study could help to identify the issues and opportunities for this technology to be well adopt and continually used in Malaysia.

2 Theoretical Background

2.1 Personality Traits

One of the prominent theories for exploring and studying different personalities is the personality trait theory which is created by [10]. Personality has been defined as the combination of psychophysiological sequence of thinking, feelings, behaving and interacting which has an impact on human behavior [11]. These personality traits have different reaction when it comes to logical thinking and emotional behavior. According to [12], personality traits can be separated to five categories, which are known as the Big Five Model. These categories are: agreeableness, conscientiousness, extraversion, neuroticism as well as openness to experience (intellect), which have been used to successfully predict human behavior [13, 14].

Openness to Experience

People with this openness trait are more willing and open to accept new activities that can be risky; they are also attentive to artistic expressions, interested about others' ideas and adventurous. This is supported by [15] who stated that people with high level of openness expected to be more willing to try new ideas and tend to be more creative. This also shows the flexibility of thought and tolerance to accept new ideas, as this trait of openness is constantly related with engaging in learning experiences [9]. Thus, [9] believed that more openness to experience have strongest drive of online brand community engagement. Therefore, consistent with these findings, we hypothesize:

H1: Openness to experience is positively associated with user attitude toward mobile wallet.

Extraversion
People with extraversion trait are known to be cordial, affable, energetic, exuberant and positive [16]. Supported by [6], there are a variety of social activities that can be found on the internet, and since extraverts are more sociable, they are more likely to use social networking sites to follow, socialize and connect with each other than any other personality trait. This can also open the door for them to show their creative skills [17]. Consistent with this finding, we hypothesize:

H2: Extraversion is positively associated with user attitude toward mobile wallet.

Conscientiousness
Conscientiousness reflects people who have a task and goal directed behavior. They tend to be a hard-working, well-organized and motivated people [6] with more planning and self-discipline ability than others [6]. Thus, individual with conscientious personal traits will most likely have positive effect on perceived usefulness of social networking sites [8]. Consequently, we hypothesize:

H3: Conscientiousness is positively associated with user attitude toward mobile wallet.

Neuroticism
People who have neuroticism traits are more inclined to negative mood states, such as depression, anxiety, and insecurity [18]. They are frequently changing their opinions and have difficulty making quick decisions. Furthermore, neurotics are constantly changing emotions, where they tend to be nervous, disappointed and desperate, with a negative view toward issues. For that reason, they can be seen as cynical people [13]. Thereupon, individual with neuroticism personal traits will most likely have negatively influences continuance intention of social networking sites [8]. With those findings, we hypothesize:

H4: Neuroticism is negatively associated with user attitude toward mobile wallet.

Agreeableness
Agreeableness reflects people who are being cooperative, friendly, considerate, trusting, humane, and thoughtful [19]. These are usually straightforward and avoid exaggeration [20]. Furthermore, this type incline to be more adaptive and cooperating with others. These individuals are responsible, honest, friendly and compliant. Based on few previous studies on personality traits and intention to adopt a mobile application, it was found that agreeableness was positively related to the intention on the use of the application or software [21]. Contingent with those findings, we hypothesize:

H5: Agreeableness is positively associated with user attitude toward mobile wallet.

2.2 Attitude and Continuance Intention Towards Mobile Wallet

According to [22] attitude is a physiological state which capture favorable or unfavorable actions toward people, things or situation. Existing research and studies supports user attitude toward technology and intention to use in which are both positively associated [23]. Furthermore, [24] stated that significant impacts of perceived usefulness, satisfaction, and users' attitude effect continuance usage intention of mobile banking. It is supported by [25] that mobile banking continuance usage intention was positively influenced by perceived usefulness, satisfaction, users' attitude, self-efficacy and channel preferences. However, few studies have studied the effect of attitude on continuance use of mobile payments. Based on those findings, we hypothesize:

H6: Attitude towards mobile payment is positively associated with continuance usage (Fig. 1).

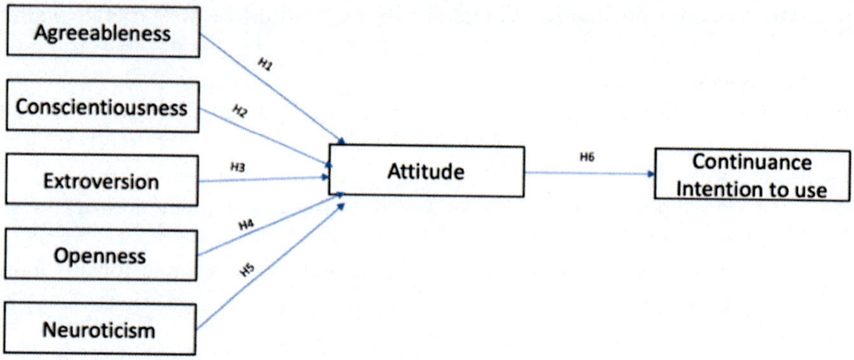

Fig. 1. Research framework

3 Methodology

This research followed a quantitative method to measure the hypothesized factors with an online self-administrative survey. The survey has three main sections: the first section comprises the introduction, instructions, and research confirmation using of data; the second section questioned the demographics (age, material status, occupation, and ethnicity) of the respondents, along with filter question to ensure that these respondents have owned a smart- phone and had at least a brief experience of using the m-wallet; and the third section presents the survey items that measure seven main variables defined as "Agreeableness" (measured with three items), "Conscientiousness" (measured with two items), "Extroversion" (measured with three items), "Openness" (measured with three items), "Openness" (measured with three items), "Attitude" (measured with Four items), and lastly, Continuance intention (measured with three items). The items used were formulated from previous studies to ensure the validity of the survey [26–29]. A five-point Likert scale was utilized to quantify independent variables (between 1 = strongly disagree and 5 = strongly agree) [30]. A pilot study was conducted to assess the validity of the survey. On the basis of feedback, minor modifications were done to improve the item comprehension. A total of 350 responded to the online questionnaire, with only 290 been filtered as usable responses were collected, representing 82.8% percent response rate. A convenient sampling was used to collect data during May to June 2020.

3.1 Structural Equation Modelling (SEM)

One of the SEM approaches is PLS which for many researchers have become the prime alternatives of CB-SEM and is deployed in fields such as business strategies, behavioral science and marketing [31]. Generally, existing relationships between variables of a research model are examined using PLS-SEM [32]. In addition to evaluating hypothesized relationships in a complex design [33]. By using Smart PLS application 3.2.0, the structural equation model is used to assess the measurement model and structural model.

4 Results

4.1 Measurement Model

Th measurement model was performed to assess the reliability of the study constructs. Values greater than 0.7 are considered acceptable [33]. The obtained results showed this criterion was met as the composite reliability ranged from 0.864 (Agr) to 0.932 (Con). The discriminant and convergent validities of the study constructs were also examined. Convergent validity was evaluated using the average variance extracted (AVE), in which values higher than 0.5 should be obtained [14]. The results of the measurements are shown in Table 1.

Table 1. Reliability and validity

Construct	Item	Loading	AVE	Composite Reliability
Agreeableness (Agr)	Agr 1	0.819	0.679	0.864
	Agr 2	0.859		
	Agr 3	0.793		
Conscientiousness (Con)	Con1	0.947	0.873	0.932
	Con2	0.922		
Extroversion (Ext)	EXT1	0.829	0.721	0.885
	EXT2	0.854		
	EXT3	0.864		
Openness (Ope)	Ope1	0.859	0.725	0.888
	Ope 2	0.847		
	Ope 3	0.847		
Neuroticism (Neu)	Neu1	0.873	0.755	0.903
	Neu2	0.875		
	Neu 3	0.858		
Attitude (Atu)	Atu1	0.809	0.707	0.903
	Atu2	0.864		
	Atu3	0.852		
	Atu4	0.828		
Continuance intention (CI)	CI1	0.904	0.814	0.929
	CI2	0.897		
	CI3	0.906		

Heterotrait-monotrait correlations (HTMT) ratio were also conducted to measure the discriminant validity [34]. Results showed that the values for all variables were below the critical value of 1.0, establishing discriminant validity (Table 2).

Table 2. Heterotrait–Monotrait Ratio (HTMT)

Construct	Agree-ableness	Attitude	Conscienti-ousness	Conscienti-ousness	Extra-version	Neuroti-cism;	Openness
Agreeableness							
Attitude	0.659						
Conscientiousness	0.681	0.5					
Continuance intention	0.679	0.764	0.479				
Extraversion	0.509	0.493	0.345	0.51			
Neuroticism	0.308	0.225	0.181	0.089	0.211		
Openness	0.691	0.6	0.708	0.575	0.443	0.131	

4.2 Structural Model

The structural model presented in Fig. 2 illustrates the value of total variance (R2) of the study model. The R2 value for Continuance intention is 0.459, indicating that all predictors explain approximately 45.9% of Continuance intention toward mobile wallet.

Hypotheses Testing: The bootstrapping with a resampling of 5000 procedure was performed to assess the significance of the hypothesized structural relationships. As shown in Table 3, the personality traits of agreeableness, extroversion and openness demonstrated a strong connection with Attitude toward mobile wallet with the value of T-statistics at 4.120, 4.157, 3.433 respectively. Therefore, the first, third and fourth hypotheses in this study was accepted (H1, H3, H4). However, the personality traits of Conscientiousness and Neuroticism did not show any association with attitude toward mobile wallet, hence rejecting the second and fifth hypothesis (H2, H5).

Finally, the sixth hypothesis (H6) can be accepted because there was a concrete relationship between attitude and continues intention toward mobile wallet, with the t-value above the threshold of 1.96 (t-value = 16.157, standard error = 0.041).

Table 3. Construct reliability and validity

Hypothesis	Path	Path coefficient (β)	SE	T-statistics	Decision
Hypothesis1	Agreeableness → Attitude	0.278	0.067	4.120	Supported
Hypothesis 2	Conscientiousness → Attitude	0.075	0.061	1.233	Not Supported

(continued)

Table 3. (*continued*)

Hypothesis	Path	Path coefficient (β)	SE	T-statistics	Decision
Hypothesis 3	Extroversion → Attitude	0.205	0.049	4.157	Supported
Hypothesis 4	Openness → Attitude	0.232	0.068	3.433	Supported
Hypothesis 5	Neuroticism → Attitude	0.049	0.045	1.091	Not Supported
Hypothesis 6	Attitude → Continuance intention	0.677	0.041	16.157	Supported

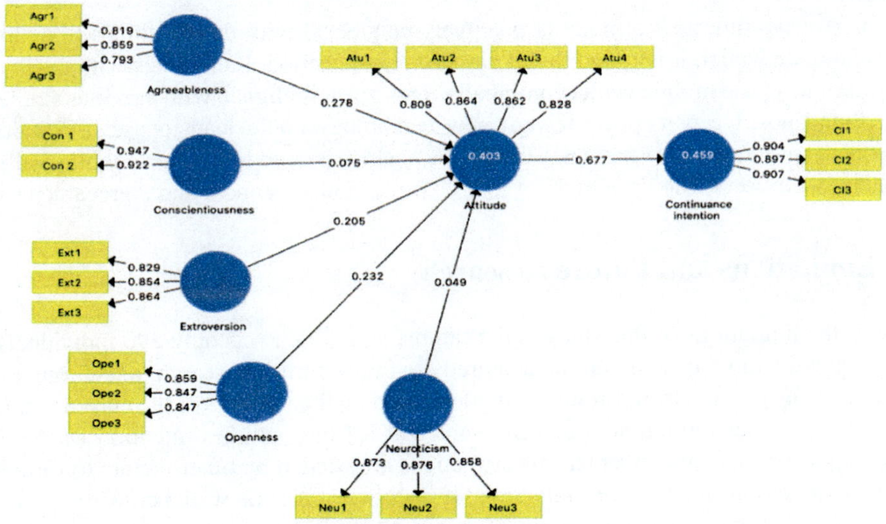

Fig. 2. PLS algorithm results

5 Discussion and Managerial Implications

This study mainly aims to identify the impact of personality traits on user attitude and determine how attitude could lead to continues intention toward mobile wallet. Among the five personality traits and their impact on user attitude (Agreeableness, Conscientiousness, Extroversion, Openness, and Neuroticism), Extroversion was found to have the highest impact on attitude. Extroverts are assumed to be sociable, outgoing, talkative [35] and generally have a higher personality to perceive and shape favourable perception toward mobile wallet. As a result, the continues intention to use mobile wallet by extroverts is high over time. However, service providers have to provide more features in their wallet applications to increase Extroverts' motivation and engagement toward the mobile wallet.

Moreover, results show that those who have strong traits of openness to experiences and agreeableness have strong relationships with Attitude. Individuals with a high level of agreeableness and healthy openness to experiences traits have a higher propensity to perceive new technology as innovative idea and useful [36] which in turn could lead to have positive attitude toward it. Service providers need to highlight the benefits and innovation of using mobile wallet to ensure that personality traits continue to have positive attitude toward it.

However, Conscientiousness and Neuroticism did not show any association with attitude toward continues intention to use mobile wallet. Individuals who are Conscientious, are logical intellectuals, with a high sense of ethical commitments and an inclination to have their own opinions and belief. This personality trait could tend to the action directly rather than controlled by the perception toward a mobile wallet. Similarly, Individuals who have high levels of neuroticism incline to experience anxiety, fear, mood swings and negative emotion more often than others; and these emotions lead to the actions directly.

In investigating how attitude is positively connected with the continues intention to use mobile wallet, it is found that it has strong connection. Individuals with positive attitude tend to use mobile wallet continually. This result is aligned with previous studies that found positive perception toward new technology could lead to use it. Service providers have to develop strategies that shape positive perception toward mobile wallet especially for users who have high level of Extroversion, Openness and Agreeableness.

6 Limitations and Future Research

One of the limitations of this study is the sample size which was only 290 individuals. Consequently, more data should be acquired for future further research; and a longitudinal investigational design should be implemented on the findings of the current study lengthily. Another limitation is that the values of R2 of continues intention is 45.9% which indicate that supplementary investigations are needed on other factors to further explain the reasons to continues intention towards use of mobile wallet construct.

This study has investigated the role of personality traits on users' attitude and continues intention toward mobile wallet. Future research should study the role of personality traits on users' actual use of mobile wallet.

References

1. Au, Y.A., Kauffman, R.J.: The economics of mobile payments: understanding stakeholder issues for an emerging financial technology application. Electron. Commer. Res. Appl. **7**(2), 141–164 (2008)
2. Al-Saedi, K., Al-Emran, M., Ramayah, T., Abusham, E.: Developing a general extended UTAUT model for M-payment adoption. Technol. Soc. **62**, 101293 (2020)
3. Singh, N., Sinha, N., Liébana-Cabanillas, F.J.: Determining factors in the adoption and recommendation of mobile wallet services in India: Analysis of the effect of innovativeness, stress to use and social influence. Int. J. Inf. Manage. **50**, 191–205 (2020)

4. Madan, K., Yadav, R.: Behavioural intention to adopt mobile wallet: a developing country perspective. J. Indian Bus. Res. **8**(3), 227–244 (2016). https://doi.org/10.1108/JIBR-10-2015-0112
5. Chawla, D., Joshi, H.: Consumer attitude and intention to adopt mobile wallet in India – an empirical study. Int. J. Bank Mark. **37**(7), 1590–1618 (2019). https://doi.org/10.1108/IJBM-09-2018-0256
6. Roos, J.M.: Five factor model of personality and the use of multiple internet functions. Open J. Soc. Sci. **05**(10), 109–116 (2017). https://doi.org/10.4236/jss.2017.510010
7. Xu, R., Frey, R.M., Vuckovac, D., Ilic, A.: Towards understanding the impact of personality traits on mobile app adoption – a scalable approach. In: ECIS 2015 Proceedings, pp. 1–12 (2015)
8. Mouakket, S.: The role of personality traits in motivating users' continuance intention towards Facebook: Gender differences. J. High Technol. Manag. Res. **29**(1), 124–140 (2018). https://doi.org/10.1016/j.hitech.2016.10.003
9. Yasin, M., Porcu, L., Abusharbeh, M.T., Liébana-Cabanillas, F.: The impact of customer personality and online brand community engagement on intention to forward company and users generated content: palestinian banking industry a case. Econ. Res. Istraz. **33**(1), 1985–2006 (2020). https://doi.org/10.1080/1331677X.2020.1752277
10. Allport, G.W.: Pattern and Growth in Personality. Holt, Rinehart & Winston, New York, NY (1961)
11. Maria Balmaceda, J., Schiaffino, S., Godoy, D.: How do personality traits affect communication among users in online social networks. Online Inf. Rev. **38**(1), 136–153 (2014)
12. Lin, L.Y.: The relationship of consumer personality trait, brand personality and brand loyalty: an empirical study of toys and video games buyers. J. Prod. Brand Manag. **19**(1), 4–17 (2010). https://doi.org/10.1108/10610421011018347
13. Liu, D., Campbell, W.K.: The big five personality traits, big two metatraits and social media: a meta-analysis. J. Res. Pers. **70**, 229–240 (2017). https://doi.org/10.1016/j.jrp.2017.08.004
14. Salem, S.F., Alanadoly, A.B.: Personality traits and social media as drivers of word-of-mouth towards sustainable fashion. J. Fash. Mark. Manag. (2020). https://doi.org/10.1108/JFMM-08-2019-0162
15. Watjatrakul, B.: Intention to adopt online learning : the effects of perceived value and moderating roles of personality traits. Int. J. Inf. Learn. Technol. **37**(1–2), 46–65 (2020). https://doi.org/10.1108/IJILT-03-2019-0040
16. De Raad, B., Peabody, D.: Cross-culturally recurrent personality factors: analyses of three factors. Eur. J. Personal. Publ. Eur. Assoc. Personal. Psychol. **19**(6), 451–474 (2005)
17. Kocak, E., Nasir, V.A., Turker, H.B.: What drives Instagram usage? user motives and personality traits. Online Inf. Rev. **44**(3), 625–643 (2020). https://doi.org/10.1108/OIR-08-2019-0260
18. Allen, M.S., Walter, E.E.: Personality and body image: a systematic review. Body Image **19**, 79–88 (2016)
19. Tang, J.H., Chen, M.C., Yang, C.Y., Chung, T.Y., Lee, Y.A.: Personality traits, interpersonal relationships, online social support, and Facebook addiction. Telemat. Inform. **33**(1), 102–108 (2016). https://doi.org/10.1016/j.tele.2015.06.003
20. Marques, B.P., Eng, D., Isep, I., Villate, J.E., Sim, E.: Applying the UTAUT model in Engineering Higher Education : Teacher ' s Technology Adoption
21. Shropshire, J., Warkentin, M., Sharma, S.: Personality, attitudes, and intentions: predicting initial adoption of information security behavior. Comput. Secur. **49**, 177–191 (2015)
22. Ajzen, I.: The theory of planned behaviour: reactions and reflections. Taylor & Francis, Milton Park (2011)

23. Rivera, M., Gregory, A., Cobos, L.: Mobile application for the timeshare industry: the influence of technology experience, usefulness, and attitude on behavioral intentions. J. Hosp. Tour. Technol. **6**(3), 242–257 (2015). https://doi.org/10.1108/JHTT-01-2015-0002
24. Lew, S., Tan, G.W.-H., Loh, X.-M., Hew, J.-J., Ooi, K.-B.: The disruptive mobile wallet in the hospitality industry: an extended mobile technology acceptance model. Technol. Soc. **63**, 101430 (2020)
25. Nikbin, D., Iranmanesh, M., Foroughi, B.: Personality traits, psychological well-being, Facebook addiction, health and performance: testing their relationships. Behav. Inf. Technol. **0**, 1–17 (2020). https://doi.org/10.1080/0144929X.2020.1722749
26. Hassim, N.: Glocalizing'the Hijab: a malaysian perspective. In: SHS Web of Conferences, vol. 33, p. 16 (2017)
27. Hui, S.L., Yee, R.W.Y.: Relationship among interpersonal relationship, customer satisfaction and brand loyalty in fashion retailing industry. Research Journal of Textile and Apparel (2015). https://doi.org/10.1108/RJTA-19-01-2015-B007
28. Sanchez Torres, J.A., Arroyo-Cañada, F.J.: Building brand loyalty in e-commerce of fashion lingerie. Journal of Fashion Marketing and Management: An International Journal (2017). https://doi.org/10.1108/JFMM-05-2016-0047
29. Sarkar, A., Sarkar, J.G.: Validating fashion brand centrality scale amongst young adults. Journal of Fashion Marketing and Management: An International Journal (2017). https://doi.org/10.1108/JFMM-05-2016-0045
30. Seo, S., Lang, C.: Psychological antecedents to customized apparel purchases. J. Fash. Mark. Manag. **23**(1), 66–81 (2019). https://doi.org/10.1108/JFMM-10-2018-0128
31. Henseler, J., Ringle, C.M., Sinkovics, R.: The use of partial least squares path modeling in international marketing. Advan Int. Mark. **20**(2009), 277–319 (2009). https://doi.org/10.1108/S1474-7979(2009)0000020014
32. Reinartz, W., Haenlein, M., Henseler, J.: An empirical comparison of the efficacy of covariance-based and variance-based SEM. Int. J. Res. Mark. **26**(4), 332–344 (2009)
33. Hair, J.F.J., Hult, G.T.M., Ringle, C., Sarstedt, M.: A Primer on Partial Least Squares Structural Equation Modeling (PLS-SEM), vol. 46, no. 1–2 (2014)
34. Henseler, J., Ringle, C.M., Sarstedt, M.: A new criterion for assessing discriminant validity in variance-based structural equation modeling. J. Acad. Mark. Sci. **43**(1), 115–135 (2014). https://doi.org/10.1007/s11747-014-0403-8
35. Cho, M., Auger, G.A.: Extrovert and engaged? Exploring the connection between personality and involvement of stakeholders and the perceived relationship investment of nonprofit organizations. Public Relat. Rev. **43**(4), 729–737 (2017)
36. Zhou, T., Lu, Y.: The effects of personality traits on user acceptance of mobile commerce. Intl. J. Hum.-Comput. Interact. **27**(6), 545–561 (2011)

Predicting the Adoption of Mobile Banking Practices Among Bangladeshi Millennial's

Swapnil Violet Dias[1], Abdullah Al Mamun[1](✉), Md Kausar Alam[2], and Noor Raihani Zainol[3]

[1] Faculty of Business and Management, UCSI University, 56000 Kuala Lumpur, Malaysia
abdullaham@ucsiuniversity.edu.my

[2] South Asia School of Business, University of South Asia, 1212 Dhaka, Bangladesh

[3] Faculty of Entrepreneurship and Business, Universiti Malaysia Kelantan, 16100 Kota Bharu, Malaysia

Abstract. This study explored the effect of performance expectancy (PE), effort expectancy (EE), social influence (SI), lifestyle compatibility (LC), and perceived trust (PT) on intention to adopt mobile banking (IMB), as well as the effect of facilitating conditions (FC) and intention on the adoption of mobile banking (AMB) among Bangladeshi millennials. Additionally, this study investigated the moderating effect of age, gender, and education level of Bangladeshi millennials under the premises of the unified theory of acceptance and use of technology (UTAUT) model. Upon adopting the cross-sectional research design, quantitative data were gathered from 312 Bangladeshi millennials via online survey. The study outcomes revealed that SI, LC, and PT displayed significantly positive effect on IMB, as well as for the effect of FC and intention on AMB. Additionally, IMB appeared to mediate the linkages of SI, LC, and PT with AMB. The findings revealed that gender moderated the relationship between SI and IMB. In promoting green financial services and fostering economic growth through mass AMB practices in Bangladesh, the government and its financial service providers should highlight on the benefits of using MB practices, as well as information security related details, to build trust among Bangladeshi millennials.

Keywords: Mobile banking · Intention and adoption · UTAUT · Millennials · Bangladesh

1 Introduction

The emerging technologies have adopted a range of financial inventions that make the financial system more extensive, rapid, and convenient. Mobile banking (MB) also refer as m-payment, m-transfer, and m-finance that denotes collectively applications that allow people to use their mobile or electronic devices to operate their bank accounts, connect their bank account with handsets, transfer money or funds or access credit and products [1]. MB is a key delivery platform for developing countries to boost economic growth and enable economic integration. In 2012, Mobile Financial Services was established in

M. Al-Emran et al. (Eds.): ICETIS 2021, LNNS 299, pp. 429–441, 2022.
https://doi.org/10.1007/978-3-030-82616-1_37

Bangladesh by the Dutch Bangla Bank. At present, 18 banks offer MB services through their traditional banking services to the people. Based on the Bangladesh Telecommunication Regulatory Commission (BTRC), internet users have escalated from 30.48 million to 90.05 million in 2013–2018 [2]. As the number of mobile phone users increases, interaction with their financial institutions has increased as well. The BTRC officials claimed that although the total number of mobile subscribers neared 16 cores, the MB segment is still in its infancy stage in Bangladesh. Hence, studies are called to probe into the behavioural intention towards mobile banking (IMB), particularly amidst Bangladeshi millennials. Millennials are the significant market segment that should be understood owing to its vast sector size and potential strong buying or spending power [3]. Due to the massive growing purchasing power, as well as usage of mobile devices and their services, millennial consumers have become an essential category in industries related to technology [4]. Millennials are described as technology-friendly, heavy mobile phone users, and are always attached to certain forms of machine [5].

Prior studies focused on the general consumers of banking services. As a result, only a handful of studies have assessed the adoption of mobile banking (AMB) practices amongst millennial clients. Past studies have revealed that altering technologies and users' preferences have a remarkable influence on the achievement of banking administrative and service distributive strategies [6]. Thus, it is important for bank retailers to comprehend the needs of millennial consumers if they want to appeal to the millennial customers to apply MB as they are different from the other generations in light of expectations and perceptions [7]. As the key players are the millennial generation, this study assessed their practices in adopting MB. As aforementioned above, there are numerous literatures related to the mobile banking practices, it is scarce in the Bangladeshi context. There were only few studies focus on the mobile banking among Bangladeshi millennials [8, 9]. This study therefore focused on examining the effect of PE, EE, SI, LC, and PT on IMB; as well as the effect of FC and IMB on the adoption of mobile banking (AMB) among Bangladeshi millennials.

2 Literature Review

2.1 Theoretical Foundation

Prior studies focused on MB application among consumers due to the vast opportunity in mobile commerce. Several researchers investigated the behavioural intention of adopting MB services by using Technology Acceptance Model (TAM) and Unified Theory of Acceptance and Use of Technology (UTAUT) [10–14]. The UTAUT looks into the intention of behavioural adoption and it is a general conceptual model to assess intention [12]. This model is mostly used to examine users' acceptance of information systems, as well as to understand the causes and likely problems faced by users. The UTAUT model composed of four variables; performance expectancy (PE), effort expectancy (EE), social influence (SI), and facilitating conditions (FC) that serve as the main components of behavioural intention and moderated by gender, age, experience, voluntariness of use [15]. As UTAUT is one of the most influential models that assess the aspects that affect the technology adoption decision of MB services, this study intended to examine the intention and adoption of MB under the premises of UTAUT.

2.2 Performance Expectancy (PE)

PE denotes the level of applying technology to provide services for customers in performing banking functions [16]. In online banking, PE emerged as the main factor for consumers to receive their services [17]. MB enables bank consumers to perform their financial transactions without any barrier using mobile devices. These probable benefits can affect users' behavioural IMB [14]. Brown et al. [18] found that PE positively affected IMB. Thus, the following is proposed:

Hypothesis (**H1**): *PE has a positive effect on IMB*

2.3 Effort Expectancy (EE)

Experts in the technology adoption model highlighted that EE denotes the individual competence in technology [15]. Easy and hassle-free technology can enhance MB implementation [19]. If users come to realize that MB is unchallenging, it will enhance their readiness to use MB proceedings. Most studies concluded that EE had a positive effect on behavioural IMB [20]. Thus, the following is hypothesized:

Hypothesis (**H2**): *EE has a positive effect on IMB*

2.4 Social Influence (SI)

SI denotes the level of one's understanding to implement another's behavior on a certain program [15]. Moral constraint is vital in shaping behavioural purpose in MB services [20]. Making decisions about using a new product or service, as well as recommendations from friends and family, are significant [21]. If users have no past experience of using specific technologies, the effect of the normative standard would be stronger [22]. In Bangladesh, when buying technologies related to financial or banking services, generation Y clients seemed to be the most dependent on the perceptions of family and peers. Concerning the above discussion, the following hypothesis is proposed:

Hypothesis (**H3**): *SI has a positive effect on IMB*

2.5 Lifestyle Compatibility (LC)

Compatibility is characterized by how much development is seen as being reliable with the existing esteems, past encounters, and the requirements of expected adopters A customer, upon seeing lifestyle incompatibility between the item and him/her, would not progress further in the adoption procedure [23]. Millennials seek products and labels that align their character, lifestyle, finances, and culture. They depend largely more on the content of social networks of their companions than on the range of the brands. An optimistic outlook on the innovation can be made through the interactive corroboration for appreciation and admiration from the peer group [13]. Based on the above depiction, the following is proposed:

Hypothesis (**H4**): *LC has a positive effect IMB*

2.6 Perceived Trust (PT)

Trust denotes possible customers' personal belief regarding the services offered by a service provider to fulfil their expectations. Trust is an important mechanism in any business transaction and MB as it enhances customers' creditability and minimizes uncertainty [24]. At times, customers face problems from the service providers due to longitudinal and temporal parting from their online service providers, as online services or transactions are not similar to direct transactions. This situation poses risk opportunity for the service providers as they do not have entire control of their actions due to privacy attacks and hackers. Based on the above depiction, the following hypothesis is proposed:

Hypothesis (**H5**): *PT has a positive effect on IMB*

2.7 Facilitating Conditions (FC)

According to Venkatesh et al. [15], FC serve as support in the application of such technology. The UTAUT model was applied to study the usage among young generation, wherein FC emerged as an essential factor [8]. Besides, clients from Jordanian banks for AMB showed that their behavioural intention was remarkably influenced by FC [25]. Hence, the following is hypothesized:

Hypothesis (**H6**): *FC has a positive effect on IMB*

2.8 Intention and Adoption of Mobile Banking (IMB)

Behavioural intention is the willingness of an individual to apply new technology. Prior investigations have reported behavioural intention as a vital component in assessing technology acceptance [15, 26]. Actual usage relies on the willingness of individuals to accept or reject the technology. Besides, intention was reported to exhibit a significantly positive relationship with actual usage [27]. As such, the hypothesis below is formulated:

Hypothesis (**H7**): *IMB practices has a positive effect on AMB*

2.9 Mediating Effect of Intention to Adopt Mobile Banking

A consumer's adoption intention is the most significant arena of investigation as it determines one's real usage behaviour of technology [27]. Behavioural intents incline to mediate the relationship between behavioural variables and the actual application of the behavior itself, thus intentions to estimate future actions properly. Past studies reported that usage of technology was mainly initiated by the behavioural intentions and assessments of an individual [16]. Lack of individual contact in MB is a significant aspect for users' faithfulness, mainly because absence of personal interaction can lead to higher opportunity for consumers to choose another bank [12]. With that, the following hypothesis is proposed:

Hypothesis (**H8**): *IMB mediates the relationships of PE, EE, SI, FC, and PT with AMB*

2.10 Moderating Effect of Age, Gender and Education

Age is concerned with the adoption and uses of new technologies, such as e-government and social media application. The study conducted by Alalwan et al. [25] provided the evidence that older users display high effort on intention compared to young users. Furthermore, study confirmed that age as a moderator of effort expectancy and intention to adopt mobile banking among Arab people [28]. Therefore, the following hypothesis formulated:

Hypothesis (**H9**): *Age moderates the effects of PE, EE, and SI on IMB*

Extensions of models influencing technology implementation have been performed by scholars to clarify the deployment of numerous procedures of technologies. Gender emerged as a significant moderator to address consumers' purchasing intention and shopping patterns via online business [29]. Males were more persuaded to use bank technology, internet banking, and MB, in comparison to females [20]. Female group exhibited stronger moderating effects of social rules, while the influence of enjoyment was stronger in the male group [30]. As such, gender can be posited as a moderator of mobile banking practices among Bangladesh millenials. Thus, the following is proposed:

Hypothesis (**H10**): *Gender moderates the effect of PE, EE, SI, and FC on IMB*

Educational background refers to the perceived education and the obtained degrees. Education is a vital component to determine applications related to information and communication technology. Studies have assessed education as a moderating factor in mobile government adaptation [31, 32]. Yousafzai and Yani-de-Soriano [33] noted that education moderated the association between perceived usefulness of internet banking and behavioural intention to use internet banking. Educational level also moderates the relationship between facilitating conditions and intention towards mobile banking such as technical and infrastructural supports to use a new system [15]. As such, the following is hypothesized:

Hypothesis (**H11**): *Education level moderates the effect of EE, SI, and FC on IMB*

3 Methodology

3.1 Sampling and Data Collection Method

This quantitative study adopted the cross-sectional research design to comprehend how behavioural factors predicted Bangladeshi millennials in adopting and using MB. The study questionnaire was designed based on previous researches pertaining to AMB. The partial least square structural equation modeling (PLS-SEM) was used to estimate complex cause-effect relationship models with latent variables. Since the study sample exceeded 100 respondents, the PLS SEM technique via Smart-PLS was suitable for this study to test the causal-effect relationships proposed in this study model.

3.2 Description of Respondent Profile

As this study assessed the AMB practices in Bangladesh by millennials, the population in this study refers to a group of people aged between 25 and 39 years. A convenience sample comprises of individuals who occur to be the most available to the researcher. Data were collected using an online survey conducted on the social media platform. Complete data were collected from a total of 312 respondents.

3.3 Instrument

PE, EE, SI, LC, PT, and FC were measured on a five-point Likert scale: 5 (strongly agree), and 1 (strongly disagree). Both IMB and AMB were measured on a seven-point Likert scale from 7 (strongly agree), and 1 (strongly disagree). Items that measured PE were adapted from Oliveira et al. [10], and AbuShanab and Pearson [11]. The five items of PE assessed productivity improvement, as well as time and cost saving in conducting banking transactions via MB. The level of EE was measured using five items adapted from Oliveira et al. [10], and AbuShanab and Pearson [11]. These items measured if MB is easy and hassle-free to understand and learn effectively. Items of SI assessed the effects of important others' opinions if they should adopt MB services. The items were adapted from Farah et al., [12] and AbuShanab and Pearson [11]. Items of LC, which evaluated experiences, preferences values, and lifestyle with AMB, were retrieved from Chawla and Joshi [13]. Scales of PT were adapted from Giovanis et al. [14]. The ensuing three-item scale that measured FC was adapted from Oliveira et al. [10] and Farah et al. [12]. The scale determined the degree to which one's MB usage was facilitated by his/her living and working environment [27].

4 Research Findings

The 312 Bangladeshi millennial respondents dwelled in Dhaka city. Among them, 33.3% were females and 66.7% were males. About 86.9% of the respondents were 25–29 years old (due to the focus on millennial group), while 7.4% and 5.8% were 30–34 and 35–39 years of age, respectively. Most of the respondents (37.0%) worked in various sectors, while 27.3% were businessmen, and 25.5% were students. In total, 75.6% of the respondents were still single, 23.1% were married, 1.0% had been divorcees, and 0.3% widow. As for the education background of respondents, 60.6% and 25.6% earned bachelor degree or equivalent and master's degree, respectively. Those with secondary school certificate and diploma/technical school certificate were represented by 9.9% and 2.9%, respectively; signifying a huge difference between undergraduate and postgraduate levels. A total of 40.4% of the respondents were full-time employees, 22.8% and 23.1% were self-employed and unemployed, respectively, and 12.8% were part-time employees. Most of them (46.5%) earned below BDT 20,000 a month, followed by 19.6% who earned BDT 20,001–30,000, 10.9% brought home BDT 30,001–40,000, 4.5% earned BDT 40,001–50,0001, 3.8% had an income of BDT 50,001–60,000, and 14.7% earning more than BDT 60,000. As expected, most of the respondents were urban dwellers (87.8%), while only 12.2% resided in rural areas.

4.1 Reliability and Validity

Cronbach's alpha presumes that all the used indicators are equally reliable [34]. As noted in Table 1, the PE, EE, SI, LC, PT, FC, IMB, and AMB scored above 0.7; indicating reliable constructs. As for the average variance extracted values for all the variables exceeded 0.50; indicating acceptable convergent validity [35].

Table 1. Reliability and validity

Code	Loading	Mean	SD	CA	DG *rho*	CR	AVE	VIF
PE – 1	0.853	4.356	0.83348	0.899	0.903	0.926	0.715	4.412
PE – 2	0.846							
PE – 3	0.872							
PE – 4	0.901							
PE – 5	0.748							
EE – 1	0.869	4.340	0.84117	0.925	0.927	0.943	0.769	3.583
EE – 2	0.865							
EE – 3	0.885							
EE – 4	0.881							
EE – 5	0.885							
SI – 1	0.894	4.125	0.91777	0.915	0.918	0.937	0.749	3.104
SI – 2	0.878							
SI – 3	0.904							
SI – 4	0.892							
SI – 5	0.749							
LC – 1	0.896	4.348	0.88118	0.875	0.877	0.923	0.800	3.187
LC – 2	0.893							
LC – 3	0.894							
PT – 1	0.895	4.141	0.92669	0.929	0.931	0.947	0.780	3.706
PT – 2	0.883							
PT – 3	0.934							
PT – 4	0.884							
PT – 5	0.816							
FC – 1	0.814	4.224	0.90385	0.807	0.809	0.887	0.723	2.232
FC – 2	0.903							
FC – 3	0.832							

(*continued*)

Table 1. *(continued)*

Code	Loading	Mean	SD	CA	DG *rho*	CR	AVE	VIF
IMB – 1	0.875	5.958	1.33047	0.957	0.958	0.966	0.824	2.268
IMB – 2	0.913							
IMB – 3	0.915							
IMB – 4	0.905							
IMB – 5	0.919							
IMB – 6	0.921							
AMB – 1	0.920	5.184	1.37045	0.877	0.938	0.916	0.669	–
AMB – 2	0.885							
AMB – 3	0.913							
AMB – 4	0.909							
AMB – 5	0.830							

Note: PE: Performance Expectancy, EE: Effort Expectancy, SI: Social Influence, LC: Lifestyle Compatibility; PT: Perceived Trust; FC: Facilitating Condition; IMB: Intention to Adopt Mobile Banking, AMB: Adoption of Mobile Banking, SD: Standard Deviation; CA: Cronbach's Alpha; DG rho - Dillon-Goldstein's rho; CR - Composite Reliability; AVE - Average Variance Extracted; VIF - Variance Inflation Factors
Source: Author's data analysis

4.2 Path Analysis

Path coefficients are assessed by path relationships in the structural model (Hair et al., 2013). Table 2 lists the path coefficients of SI, LC, and PT, which displayed positive and statistically significant (5% level of significance) effect on IMB. Next, the path coefficients of FC and intention with AMB appeared to be positive and statistically significant (5% level of significance) for Bangladeshi millennials. However, the effect of PE and EE on IMB was statistically insignificant (5% level of significance) among millennials in Bangladesh.

Table 2. Path coefficient

Hypo		Beta	CI - Min	CI - Max	t	p	r^2	f^2	Q^2
Factors effecting intention to adopt mobile banking									
H_1	PE → IMB	0.143	−0.006	0.275	1.634	0.051		0.022	
H_2	EE → IMB	−0.040	−0.151	0.084	0.544	0.293		0.002	
H_3	SI → IMB	0.285	0.169	0.380	4.514	0.000	0.786	0.122	0.635
H_4	LC → IMB	0.317	0.216	0.430	4.710	0.000		0.147	
H_5	PR → IMB	0.225	0.121	0.341	3.320	0.000		0.064	

(continued)

Table 2. (*continued*)

Hypo		Beta	CI - Min	CI - Max	t	p	r^2	f^2	Q^2
Factor effecting adoption of mobile banking									
H_6	FC → AMB	0.273	0.185	0.364	5.069	0.000		0.145	
H_7	IMB → AMB	0.656	0.569	0.734	12.818	0.000	0.771	0.826	0.510

Note: PE: Performance Expectancy EE: Effort Expectancy, SI: Social Influence, LC: Lifestyle Compatibility; PT: Perceived Trust; FC: Facilitating Condition; IMB: Intention to Adopt Mobile Banking, AMB: Adoption of Mobile Banking
Source: Author's data analysis

4.3 Moderating and Mediating Effect

The mediational effect presented in Table 3 reveals that IMB mediated the relationships between SI and AMB (β = 0.187, p = 0.000), between LC and AMB (β = 0.208, p = 0.000), as well as between perceived risk and AMB (β = 0.187, p = 0.000).

Table 3. Mediating and moderating effects

Associations	Beta	CI - Min	CI - Max	t	p	Decision
Mediating effect of intention to adopt mobile banking						
PE → IMB → AMB	0.094	−0.004	0.187	1.586	0.057	Reject
EE → IMB → AMB	−0.026	−0.102	0.054	0.543	0.294	Reject
SI → IMB → AMB	0.187	0.107	0.257	4.135	0.000	Accept
LC → IMB → AMB	0.208	0.137	0.282	4.616	0.000	Accept
PT → IMB → AMB	0.148	0.075	0.225	3.199	0.001	Accept
Moderating effect of age						
PE → IMB	−0.111	−0.231	0.020	1.411	0.079	Reject
EE → IMB	0.049	−0.075	0.139	0.729	0.233	Reject
SI → IMB	−0.018	−0.110	0.078	0.315	0.376	Reject
Moderating effect of gender						
PE → IMB	0.015	−0.087	0.094	0.259	0.398	Reject
EE → IMB	−0.092	−0.176	0.025	1.527	0.064	Reject
SI → IMB	0.140	0.019	0.223	2.211	0.014	Accept
FC → AMB	−0.002	−0.045	0.038	0.063	0.475	Reject

(*continued*)

Table 3. (*continued*)

Associations	Beta	CI - Min	CI - Max	t	p	Decision
Mediating effect of education						
EE → IMB	0.014	−0.021	0.069	0.505	0.307	Reject
SI → IMB	−0.016	−0.067	0.029	0.534	0.297	Reject
FC → AMB	−0.033	−0.090	0.014	1.050	0.147	Reject

Note: PE: Performance Expectancy, EE: Effort Expectancy, SI: Social Influence, LC: Lifestyle Compatibility; PT: Perceived Trust; FC: Facilitating Condition; IMB: Intention to Adopt Mobile Banking, AMB: Adoption of Mobile Banking
Source: Author's data analysis

The respondents behaved differently based on their age. It is critical to explore the effect of age on the model paths, as proposed in Hypothesis 9. However, no significant effect of age was found on the study model paths. This showed that the performance expectancy and effort expectancy of the respondents were not based on their age. They also behaved differently based on their gender. Thus, it is critical to assess the effect of gender on the model paths, as depicted in Hypothesis 10. The findings revealed that the effect of SI on IMB was moderated by gender. As aforementioned above, the gender significantly influenced the intention of the mobile banking among millennials via social rules, enjoyment, liveliness, and etc. Finally, the study results showed that the effect of respondents' education level on all the study paths appeared to be insignificant. The education level of sample had no essential role in the study paths.

5 Discussion

Overall, the results showed that most of the path coefficients in the proposed model were statistically significant, except for paths from PE to IMB and EE to IMB. In precise, the millennials' IMB was found to be significantly affected by SI, LC, and PR, while PE and EE displayed otherwise. Adoption of MB (AMB) was significantly affected by FC and IMB. Social influence (SI) exhibited a positive effect on IMB among Bangladeshi millennials. The results stipulated that SI had a significantly positive effect on behavioural intention, which is in line with prior studies. In this particular situation, it is prompted that professionals need to convince previous adopters of the framework to help in uplifting it to diverse users. Individuals may be influenced by positive word-of-mouth from their referent companions. As mobile users in this study were mostly between the ages of 25 and 32, they can be measured as a lead segment as they are easily influenced by mass media or peers to use MB. Bangladeshi banks should focus on the effective use of social media platforms, such as Facebook, Twitter, YouTube, and SMS services, through mobile phones to engage more users. The LC had a positive effect on IMB among Bangladeshi millennials; reflecting a significant part of exploration in innovation adoption. Millennials seek technology with automation and flexibility. According to Salesforce, more than a quarter of millennials were entirely dependent on MB application on their smart devices. Banks need to understand the LC of a user, such as how customers perform banking transactions, when they are required to do so, what assumptions they have from MB services, and in which way they use it; so that banks can offer easy guidance to provide proper services, thus resulting in effective and efficient AMB.

The outcomes exemplified that PT had a positive effect on IMB among Bangladeshi millennials, which agrees with prior studies. Banks with various operating systems, devices, and platforms to use should assure mobile security to fulfil all the necessary demands and obligations so that millennials can trust MB. Many MB service providers must give instructions to their consumers about conducting safe transactions through MB. The findings of this study showed that FC had a positive effect on AMB among Bangladeshi millennials. External factors, such as availability of resources regarding mobile or internet banking, can influence users, and thus the positive effect on the AMB. When MB is ready to be operated and the users have both adequate knowledge and appropriate mobile phone, FC can instruct, assist, and train to support the users. Finally, IMB practices displayed a positive effect on AMB among Bangladeshi millennials. The results suggest that IMB influenced the AMB of mobile payment or MB systems, which is in line with earlier studies. Users with expectation of adopting mobile payment are bound to utilize MB more frequently.

6 Conclusion

Bangladesh is an emerging potential market for MB and this study can deliver better understanding of user-centric intention that eventually affects AMB. For both academic and financial segments, this research can furnish both the perceptions from both the education and banking industries in devising viable marketing strategies and selling activities for MB services adoption. Bangladeshi millennials can be motivated to adopt MB if they succeed in marketing and promotional offerings. The MB services should be designed in a way that the user might like their ubiquity and personalization, useful content, and clear instruction, to name a few. Banks and policymakers have a central role to ascertain the security and trustworthiness of MB. To mitigate users' concerns regarding MB security, providers need to give proper payment documentation and briefly give instructions on the errors in the transaction process. Using motivational policies, a bank must provide reliable, secure, and trustworthy MB services to millennials. Mobile banking (MB) can help to adopt proper method of employing renewable energy, automated operating system, and other activities that lower carbon footprint from banking activities. Upon identifying the IMB of Bangladeshi millennials, this study offers several theoretical and practical implications. This study contributes to Bangladeshi millennials to increase their understanding on the usage and AMB technology. This study prescribes a viable way to millennial users in Bangladesh to accept MB services by investigating the significant aspects. Essentially, consumers' AMB motivates millennial clients to embrace this technology as their practices.

References

1. Donner, J., Tellez, C.A.: Mobile banking and economic development: linking adoption, impact, and use. Asian J. Commun. **18**(4), 318–332 (2008)

2. Kader, M.: The mobile and internet penetration growth continues, internet's deployment phase (2019). https://futurestartup.com/2019/01/31/mobile-and-internet-penetration-updates-internets-deployment-phase/. Accessed 25 July 2020
3. Wolburg, J.M., Pokrywczynski, J.: A psychographic analysis of generation Y college students. J. Advert. Res. **41**(5), 33–52 (2001)
4. Kim, J., Hahn, K.H.: Effects of personal traits on generation Y consumers' attitudes toward the use of mobile devices for communication and commerce. Hum. Technol. **8**(2), 133–156 (2012)
5. Leon, S.: Service mobile apps: a millennial generation perspective. Ind. Manag. Data Syst. **118**(9), 1837–1860 (2018)
6. Byers, R.E., Lederer, P.J.: Retail bank services strategy: a model of traditional, electronic, and mixed distribution choices. J. Manag. Inf. Syst. **18**(2), 133–156 (2001)
7. Tan, E., Lau, J.L.: Behavioural intention to adopt mobile banking among the millennial generation. Young Consum. **17**(1), 18–31 (2016)
8. Islam, M.M., Hossain, M.E.: Consumers' attitudes towards mobile banking in Bangladesh. In: The Second International Conference on E-Technologies and Business on the Web, Asia Pacific University of Technology and Innovation, Kuala Lumpur, Malaysia, March 18–20 (2014)
9. Parvin, A.: Mobile banking operation in Bangladesh: prediction of future (2015). https://www.icommercecentral.com/open-access/mobile-banking-operation-in-bangladesh-prediction-of-future-1-15.php?aid=37948. Accessed 25 July 2020
10. Oliveira, T., Faria, M., Thomas, M.A., Popovic, A.: Extending the understanding of mobile banking adoption: when UTAUT meets TTF and ITM. Int. J. Inf. Manage. **34**(5), 689–703 (2014)
11. Abushanab, E., Pearson, J.: Internet banking in Jordan. J. Syst. Inf. Technol. **9**(1), 78–97 (2007)
12. Farah, M.F., Hasni, M.J., Abbas, A.K.: Mobile-banking adoption: empirical evidence from the banking sector in Pakistan. Int. J. Bank Market. **36**(7), 1386–1413 (2018)
13. Chawla, D., Joshi, H.: Consumer attitude and intention to adopt mobile wallet in India – an empirical study. Int. J. Bank Market. **37**(7), 1590–1618 (2019)
14. Giovanis, A., Assimakopoulos, C., Sarmaniotis, C.: Adoption of mobile self-service retail banking technologies. Int. J. Retail Distrib. Manage. **47**(9), 894–914 (2019)
15. Venkatesh, V., Morris, M., Davis, G., Davis, F.: User acceptance of information technology: toward a unified view. MIS Q. **27**(3), 425–478 (2003)
16. Venkatesh, V., Thong, J.Y., Xu, X.: Consumer acceptance and use of information technology: extending the unified theory of acceptance and use of technology. MIS Q. **36**(1), 157–178 (2012)
17. Tarhini, A., Hone, K., Liu, X., Tarhini, T.: Examining the moderating effect of individual level cultural values on users' acceptance of E-learning in developing countries: a structural equation modeling of an extended technology acceptance model. Interact. Learn. Environ. **25**(3), 306–328 (2017)
18. Brown, I., Cajee, Z., Davies, D., Stroebel, S.: Cell phone banking: predictors of adoption in South Africa—an exploratory study. Int. J. Inf. Manage. **23**(5), 381–394 (2003)
19. Koenig-Lewis, N., Palmer, A., Moll, A.: Predicting young consumers' take up of mobile banking services. Int. J. Bank Market. **28**(5), 410–432 (2010)
20. Amin, H., Supinah, R., Aris, M.M., Baba, R.: Receptiveness of mobile banking by Malaysian local customers in Sabah: an empirical investigation. J. Internet Bank. Commer. **17**(1), 1–12 (2012)
21. Riquelme, H.E., Rios, R.E.: The moderating effect of gender in the adoption of mobile banking. Int. J. Bank Market. **28**(5), 328–341 (2010)

22. Hartwick, J., Barki, H.: Explaining the role of user participation in information system use. Manage. Sci. **40**(4), 440–465 (1994)
23. Saaksjarvi, M.: Consumer adoption of technological innovations. Eur. J. Innov. Manag. **6**(2), 90–100 (2003)
24. Slade, E., Williams, M., Dwivedi, Y., Piercy, N.: Exploring consumer adoption of proximity mobile payments. J. Strateg. Mark. **23**(3), 209–223 (2014)
25. Alalwan, A.A., Dwivedi, Y.K., Rana, N.P.: Factors influencing adoption of mobile banking by Jordanian bank customers: extending UTAUT2 with trust. Int. J. Inf. Manage. **37**(3), 99–110 (2017)
26. Teo, T., Milutinovic, V.: Modelling the intention to use technology for teaching mathematics among preservice teachers in Serbia. Australas. J. Educ. Technol. **31**(4), 363–380 (2015)
27. Yu, C.: Factors affecting individuals to adopt mobile banking: empirical evidence from the UTAUT model. J. Electron. Commer. Res. **13**(2), 104–121 (2012)
28. Arenas, G.J., Peral, B., Ramon, J.M.: Elderly and internet banking: an application of UTAUT2. J. Internet Bank. Comm. **20**(1), 1–23 (2015)
29. Zhang, X., Prybutok, V.R.: TAM: the moderating effect of gender on online shopping. J. Int. Inf. Manage. **12**(2), 99–118 (2003)
30. Hwang, Y.: The moderating effects of gender on e-commerce systems adoption factors: an empirical investigation. Comput. Hum. Behav. **26**(6), 1753–1760 (2010)
31. Mandari, H.E., Chong, Y.-L.: Gender and age differences in rural farmers' intention to use m-government services. Electron. Gov. Int. J. **14**(3), 217–239 (2018)
32. Zaidi, Z.H., Abro, A.A., Messo, M.S., Mahesar, G.A.: Demographic variables' effect on male and female awareness to use of e-government service in Pakistan. Women-Ann. Res. J. Gender Stud. **9**(9), 22–47 (2017)
33. Yousafzai, S., Yani-de-Soriano, M.: Understanding customer-specific factors underpinning internet banking adoption. Int. J. Bank Market. **30**(1), 60–81 (2012)
34. Hair, J., Sarstedt, M., Hopkins, L., Kuppelwieser, V.: Partial least squares structural equation modeling (PLS-SEM): an emerging tool in business research. Eur. Bus. Rev. **26**(2), 106–121 (2013)
35. Hair, J.F., Ringle, C.M., Sarstedt, M.: PLS-SEM: indeed a silver bullet. J. Market. Theor. Pract. **19**(2), 139–152 (2011)

Social Media Marketing Activities, Brand Equity and Loyalty Towards Fashion Accessories Among Chinese Youth

Liu Shuang, Mcxin Tee, and Abdullah Al Mamun(✉)

Faculty of Business and Management, UCSI University, 56000 Kuala Lumpur, Malaysia
abdullaham@ucsiuniversity.edu.my

Abstract. The research studied on effect of social media marketing activities (SMMA), i.e. customization, entertainment, interaction, trendiness, and word of mouth, on mediating effect of brand equity and brand loyalty towards fashion accessories among Chinesc youth. Cross-sectional research and quantitative approach are implemented. A total of 300 data were collected through online survey. Results proved that entertainment, trendiness, and word of mouth have significant positive impact on brand equity; and brand equity has significant positive impact on customer loyalty. Findings also revealed brand equity mediates association between entertainment, trendiness and word of mouth on customer loyalty. With the aim of increasing brand loyalty, Chinese fashion accessories industry should consider SMMA as important marketing strategy by focusing on the aspects of entertainment, trendiness and word of mouth.

Keywords: Social media marketing activities · E-commerce · Brand equity · Brand loyalty · Chinese youth

1 Introduction

Mobile devices become crucial part of current society for communicating, shopping and social networking purposes [1]. Popularity of social media is mirror of people's need for more convenient interaction. Social media marketing activities (SMMA) is commercial marketing process that attempt to generate positive influence on consumer purchasing behavior by using social media. SMMA is significant element on corporate branding. So, corporations have incorporated customer to social media for reaching excellent customer relationships [2]. Companies can convey brands concepts and values in attractive way, and offer customization in simple, inexpensive, and sustainable manner. Increasing demand for fashion industry has appeared in emerging market like China and there are more opportunities than ever to expand the business. Fashion industry includes different sectors, such as apparel, jewelry, accessories, footwear etc. Fashion industry can be categorized further based on market segmentation. Most fashion brands create social media official accounts on YouTube, Facebook, Instagram, or Weibo to post brands relevant contents due to possible advantages of creative two-way direct

M. Al-Emran et al. (Eds.): ICETIS 2021, LNNS 299, pp. 442–454, 2022.
https://doi.org/10.1007/978-3-030-82616-1_38

communication with customers [3]. Social media is appropriate online instrument for consumers' interaction and self-expressiveness in fashion industry [4]. Before making purchasing decision, customers will actively evaluate information through social media. Social media need to continuously update latest information due to possibility to enhance customer relationship. Hence, it is important to study on influences of SMMA in fashion industry.

Brand equity attracts lots of interest from marketing experts due to its strategic role in management decisions and competitive advantages [5]. Brand equity increase company value via increased popularity, brand loyalty and perceived quality. Besides, most fashion brands still adopt social media as a communication instrument rather than strategic means for marketing. They thoughtlessly prefer number of *likes* and comments received from followers, they are not aware of purpose of owning them. They spent a lot of money on traditional resources to increase brand values. In fact, no one is clearly aware of social media marketing's strategic role in taking brands to new level [6]. Brand loyalty can be key solution for customer relationship management. In return, brand loyalty generates profit, reputation, and market share to corporations. Hence, interaction between company and customers through social media can develop higher customer satisfaction and stronger brand loyalty. Meanwhile, companies need to enhance brand loyalty to develop competitive position in market. Brand loyalty is experiencing transformation because companies are losing control in handling brand recognition on traditional media. Though some companies have performed SMMA to earn market share, there is remaining uncertainty on how online marketing events on social media sites will directly affect customer loyalty. Objective of marketing events is to grow benefits of all stakeholders and achieving better customer relationship management [7]. However existing literature still lacks empirical proof on how SMMA influence brand equity and brand loyalty. In consideration of the growing use of social media for online marketing activities by fashion brands, it has developed to imperative to quantitatively investigate impact of social media. Although there are several sectors in fashion industry, this study will focus on fashion accessories sector as fashion accessories receive little attention from existing studies. Fashion accessories are decorative objects, such as sunglasses, handbags, watches, wallets, and jewelry that enable customers to supplement outfits in order to boost self-esteem, express self-image, and show social status. Accessories have longer life cycle compared to clothing. Furthermore, this research will study influence of SMMA on customer loyalty and brand equity in organizational environment. Findings can deliver recommendations to fashion accessories industry in building brand loyalty and brand equity through online marketing events on social media platform.

2 Literature Review

2.1 Theoretical Foundation

Mehrabian and Russell [8] conceptualised Stimuli-Organism-Response (S-O-R) model based on psychological research which is extension to Stimulus-Response (S-R) theory of behavioural psychology. Originally, S-O-R model concentrated only on emotional responses which stimuli mean factors that influence individual inner state. Bitner [9] integrated physiology and cognition within S-O-R model by expanding its application to explain behaviour of customers. Briefly, S-O-R model explains impact of environmental stimuli (S) on cognitive and emotional status of individual (O), such as consumers, and then causes certain consumers' reactions (R) in the form of behaviour [10]. It is commonly applied in retail and marketing context to uncover customers' behaviour [11] and relationship between emotions and impulsive purchasing behaviour [12]. This study proposes extended S-O-R model to predict Chinese youth's behaviour in fashion accessories industry. The proposed framework clearly connects the influences of SMMA as trigger on consumers' response, which is mediated by brand equity. Brand equity works as mediator to the extent that customers identify and remember a brand and how much added brand value that is generated by the marketing stimuli affects customers' loyalty [13]. Specifically, this paper utilizes SMMA as external stimuli (S), proposing a study on the impact of brand loyalty towards Chinese youth' (R) in fashion accessories industry through brand equity (O), targeting to enhance literature and deliver managerial insights on youth customers' behaviour.

2.2 Customers' Brand Loyalty

Brand loyalty refers to commitment of consumers on repeatedly buying products or services from same company, regardless of competitors' action or other conditions that make consumers have chance to change their purchasing option [14]. Drivers of brand loyalty on social media platform, for example customization, accessibility, connectivity, entertaining and interactivity, are different from traditional way to enhance customer loyalty [15]. High customer satisfaction on perceived value and positive emotional feeling are the two main factors that can increase loyalty of consumers. Customers' brand loyalty is important to maintain connectivity in long-term This study will explore customers' brand loyalty as a response to online marketing events on social media platform for fashion accessories industry.

2.3 Social Media Marketing Activities and Brand Equity

Entertainment and Brand Equity

Entertainment is intentional purpose of users to participate in online social media events. Entertainment is significant element that fosters users' response and continually stimulate users' positive affection on brand recognition of company [16]. Amusement, enjoyment, and relaxation are causes for social media users to search for entertainment by viewing company brand relevant contents [7]. Hence, corporations should provide more brand relevant entertaining contents on online platform to encourage more users to view, support and share contents which the actions are capable to transfer into competitive advantage for the company. Entertainment offered in online marketing events on social media platform can strengthen customers' brand equity. Ismail [17] has stated sharing of entertained contents through social media platform can be practicable method to attract attention of users to the brand and can lead to higher engagement and interaction, then eventually facilitate brand awareness in online social media community. Companies with luxury brands should assign higher amount of resources in entertaining consumers through social media platform to create positive customer-brand association [18]. Moreover, SMMA of luxury brands have highlighted that interactivity and entertainment can stimulate affection connection and contribute to co-creation of brand stories with consumers [12]. Hence, entertained marketing and fashion accessories brand relevant contents posted on social media platform can provide fun and exciting experience to online users that will stimulate them to share the content to others. This leads us to pose our first hypothesis:

Hypothesis (H1): *Entertainment has a significant positive impact on brand equity*

Customization and Brand Equity

Customization is tailoring of products or services to satisfy personal preferences and needs. Zhu and Chen [19] have categorized online content to two forms: 1. broadcasting which focuses on any users; 2. customized message, which focuses on targeted consumers. Customization is offered by understanding preferences of target customers. It is more cost-effective while is presented through social media compared to traditional media platform. Ding and Keh [20] have mentioned that customization is strategy to stimulate customer satisfaction and positive brand affection because it can generate exclusively marketing content to target consumers. Customization of SMMA is a tool for corporations to provide uniqueness of their brands to improve customer brands' preferences with the aim of boosting customer loyalty to the brands. Furthermore, consumers are more likely to choose brands that meet their own preference. Customization service is good strategy for companies to enhance customer loyalty and corporate reputation [16]. Besides, customization on e-service have direct and significant effect on brand equity [5]. Customization is believed to be as sign of respect from companies to customers with the aim of enhancing understanding on customer value. Hence, customization can affect brand equity positively in fashion accessories industry. We propose the second hypothesis:

Hypothesis (H2): *Customization has a significant positive impact on brand equity*

Trendiness and Brand Equity

Godey et al. [7] have defined trendiness as continuously presenting of state-of-the-art information about products, services or brands. Social media can deliver latest trends and popular topics. For today's world, social media become primary searching tool for consumers because they perceive information delivered by social media is trustworthy compared to corporations' traditional one-way controlled communication. Yadav and Rahman [2] have mentioned interactivity of social media to follow trendiness can strengthen relationship with customers and deliver positive influence on brand equity. Trendiness as one of SMMA features encourages consumers to monitor latest information and constantly pay attention to brand dynamics which the action can rise brand awareness. Trendiness is the most considered feature for consumers because it facilitates consumers' concentration and positive perception toward corporate brands [16]. Based on previous studies, trendiness is the highest significant measurement and it positively influences on corporate brand image and brand awareness. Moreover, trendiness feature can meet consumers' expectation on luxury fashion brands [21]. Since fashion brands dedicates to bring value to consumers, companies should allocate attention and resources to provide the latest trendiness of products and services by offering information on social media with the aim of positively enhancing brand image and brand awareness. We propose the third hypothesis:

Hypothesis (H3): *Trendiness has a significant positive impact on brand equity*

Interaction and Brand Equity

Interaction means boosting consumers to communicate and share contents. Due to social media provides platform to voice out and switch opinions, interaction on social media able to provide valuable insights and comments to corporations [22]. Kim and Ko [21] have mentioned fashion brands can attract consumers' brand affection to facilitate developing greater brand image and customer loyalty by communication through social media. Marketing events on social media should specifically deliver communication objectives and attach important information to stimulate active interaction with customers. Furthermore, interaction can build trust and remove uncertain reasons that might cause restraints for consumers to connect with companies [23]. SMMA can encourage active interaction which lead to enhancement of brand equity through providing brand recognition to consumers [18]. Godey et al. [7] have found out interaction between SMMA and Chinese consumers can deliver customer value and reinforce brand image for luxury fashion industry. Therefore, we propose our fourth hypothesis:

Hypothesis (H4): *Interaction has a significant positive impact on brand equity*

Word of Mouth and Brand Equity

Yadav and Rahman [2] have defined word-of-mouth (WOM) as communication among consumers. It is recognized as the most significant impact on consumers' affection, perception, and behaviour. Social media can be perfect instrument for electronic WOM because consumers create and share brands relevant information without difficulties or limitations [21]. WOM represents consumers' perspective towards brands. WOM is behavioural reaction of consumers to brands' marketing events [24]. This type of behavioural expression will influence on brand image. One of the best methods to improve consumers' perception and attitude towards brands is stimulating WOM through social media [25]. Information sharing of customers induces diffusing process of information which involves personal feeling and opinions will create impact on brand image. Use of social media is valuable marketing technique because it allows consumers to exchange perspectives as WOM in line with products, services and brands will boost better brand equity and relationship equity. Hence, we propose the fifth hypothesis:

Hypothesis (H5): *Word of mouth has a significant positive impact on brand equity*

2.4 Brand Equity and Customer Loyalty

Yadav and Rahman [2] have explained equity theory means perceived equity influences affective condition which will subsequently affect customer satisfaction and loyalty. If consumers perceive brand can meet their expectation, then they will hold positive brand image, high brand preference and high chance to purchase products or services from the company. Equity-based marketing technique is an imperative feature to increase customer loyalty. Higher brand equity lead to higher willingness of consumers to repurchase [26]. Positive customers' perception towards a brand can influence affective commitment positively. Brand equity can affect willingness of consumers to participate in brands' marketing activities, repurchase from the same brand, and recommend others about the brand. Brand equity is one of the significant factors to boost customer loyalty [27]. Hence, we propose following research hypothesis:

Hypothesis (H6): *Brand equity has a significant positive impact on brand loyalty*

2.5 Mediating Effect of Brand Equity

Online marketing events on social media reinforce relationship between brands and consumers, and among consumers. Reinforced relationship is dedicated enhancing customer loyalty [28]. Based on literature review, this research anticipates brand equity has mediating impact to SMMA on customer loyalty. Social media provide opportunities for brands to grow, and also improve brand awareness and brand image. While customers are highly aware of brand, stronger brand loyalty will exist [17]. Hence, this leads us to pose the eighth hypothesis:

Hypothesis (H7): *Brand equity mediates the relationship between entertainment, customization, trendiness, interaction and word of mouth on brand loyalty*

3 Research Methodology

3.1 Research Design

This research measured effect of SSMA on brand equity and brand loyalty towards fashion accessories among Chinese youth. All relationships hypothesized are presented in Fig. 1. This study adopted cross-sectional design and collected quantitative data through online survey. Then, data analysis was conducted by following all thresholds suggested by Hair et al. [29].

3.2 Population and Sample

Chinese youth in China who were from 18–35 years old and use social media in daily life would be targeted as sample population. Sample size was estimated by using G * Power. This study adopted convenience sampling method and collected data by sharing the link of online survey questionnaire through various social media platforms.

3.3 Survey Instrument

Self-completed questionnaire through online distribution was applied. Measurement items were adapted from previous literature (Table 1). Questionnaire design was 5-point Likert scale (from strongly disagree to strongly agree), close-ended questions.

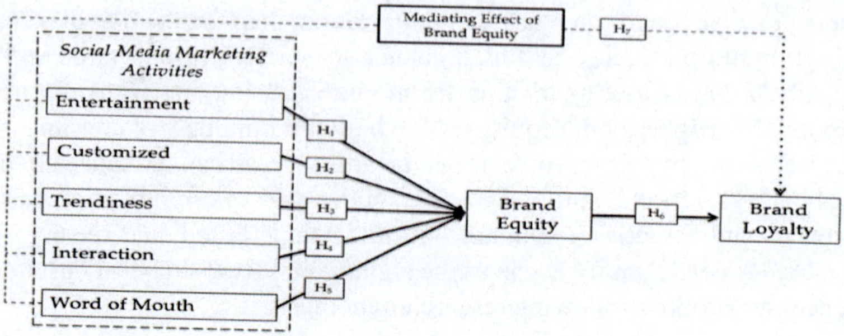

Fig. 1. Research framework

Table 1. Survey instrument

Variables	No. of Items	Sources
Entertainment (EN)	4	Kim and Ko [21]
Customization (CU)	4	Kim and Ko [21]; Srinivasan et al. [30]
Trendiness (TR)	4	Yadav and Rahman [2]; Kim and Ko [21]
Interaction (IN)	4	Zhang et al. [31]; Kim and Ko [21]
Word of Mouth (WOM)	3	Hwang et al. [32]; Kim and Ko [21]
Brand Equity (BE)	4	Kim and Hyun [33]
Customer Loyalty (CL)	5	Wang et al. [34]; Laroche et al. [28]

3.4 Common Method Bias

In social science, researchers adopted common method for measurement of variables which can result in common method variance (CMV). Harman's single-factor analysis was used. Recommended guiding principle was a single factor must be less than 50% variance. Result was 43.418% which indicated there is no single factor explained most of variance, means common method variance is not an issue for this research.

4 Data Analysis

4.1 Demographic Characteristics

Table 2 exhibits demographic profile of respondents, with a total of 300 respondents.

Table 2. Demographic characteristics

	N	%		N	%
Age Group			*Education*		
18–20 years	30	10.0	Secondary school	13	4.3
21–25 years	131	43.7	Diploma	72	24.0
26–30 years	91	30.3	Bachelor's degree	153	51.0
31–35 years	48	16.0	Master's degree	54	18.0
Total	300	100.0	Doctoral degree	8	2.7

4.2 Reliability and Validity

As shown in Table 3, Cronbach's alpha and composite reliability values for all constructs was more than 0.7. Hence, it indicates all constructs were reliable. AVE for all items were between 0.577 and 0.729, signifying good convergent validity. Fornell-Larcker criterion tested discriminant validity. Results in Table 4 display there is no issue with discriminant validity as all values were below 0.90.

Table 3. Reliability and validity

Variables	No. Items	Mean	SD	CA	DG *rho*	CR	AVE	VIF
EN	4	3.721	0.686	0.791	0.795	0.864	0.614	2.030
CU	4	3.768	0.651	0.755	0.756	0.845	0.577	2.362
TR	4	3.733	0.675	0.794	0.797	0.866	0.618	2.713
IN	4	3.742	0.660	0.784	0.790	0.860	0.606	2.773
WOM	4	3.763	0.730	0.809	0.810	0.887	0.724	2.298
BE	4	3.673	0.706	0.831	0.834	0.888	0.664	1.000
CL	5	3.749	0.646	0.818	0.822	0.873	0.579	-

Note: SD: Standard Deviation; CA: Cronbach's Alpha; DG *rho*: Dillon-Goldstein's *rho*; CR: Composite Reliability; AVE: Average Variance Extracted; VIF: Variance Inflation Factors

4.3 Path Analysis

Next step is to analyze relationship between variables. Path coefficient results are shown in Table 5, which suggested H1, H3, H5 and H6 are positive and statistically significant. However, results proved H2 and H4 are positive but statistically have insignificant impact on brand equity.

Effect size (f^2) can be found in Table 5. Entertainment, trendiness, customization and WOM have small effect on brand equity. Interaction has no effect on brand equity. Brand equity presents large effect on customer loyalty. Meanwhile, r^2 shows moderate

Table 4. Discriminant validity: Fornell-Larcker criterion

	EN	CU	TR	IN	WOM	BE	CL
EN	0.783						
CU	0.585	0.760					
TR	0.637	0.694	0.786				
IN	0.659	0.682	0.688	0.778			
WOM	0.559	0.625	0.680	0.686	0.851		
BE	0.567	0.568	0.650	0.581	0.655	0.815	
CL	0.637	0.645	0.722	0.703	0.693	0.703	0.761

Table 5. Path coefficients

Hypo		Beta	t	p	r^2	f^2	Q^2	Decision
H_1	EN → BE	0.162	2.472	0.007		0.028		Accept
H_2	CU → BE	0.075	0.968	0.167		0.005		Reject
H_3	TR → BE	0.257	3.461	0.000	0.529	0.052	0.341	Accept
H_4	IN → BE	0.021	0.321	0.374		0.000		Reject
H_5	WOM → BE	0.328	5.735	0.000		0.099		Accept
H_6	BE → CL	0.703	18.646	0.000	0.494	0.974	0.281	Accept

variance in dependent variable and is acceptable. The predictive relevance (Q^2) value is larger than zero which indicating that path model's accuracy is acceptable.

4.4 Mediation

Bootstrapping procedures were adopted to acquire estimation of indirect effects and using confidence intervals (CI) to identify significance. If estimated 95% confidence interval does not contain zero, it implies that indirect effect is statistically significant. Results in Table 6 confirmed that brand equity significantly mediates effect of entertainment, trendiness, and WOM on customer loyalty.

Table 6. Mediating effects

Associations	Beta	CI - Min	CI - Max	*t*	*p*	Decision
EN → BE → CL	0.114	0.044	0.189	2.545	0.006	Accept
CU → BE → CL	0.053	−0.041	0.145	0.970	0.166	Reject
TR → BE → CL	0.181	0.085	0.269	3.161	0.001	Accept
IN → BE → CL	0.014	−0.055	0.096	0.318	0.375	Reject
WOM → BE → CL	0.231	0.155	0.298	5.412	0.000	Accept

5 Discussion

This study empirically investigates effect of SMMA on brand equity and brand loyalty in fashion accessories industry. Findings suggested brand equity is positively and significantly affected by entertainment, trendiness, and WOM, which are consistent with results of existing literature. Entertainment can stimulate users' positive affection on brands. Hence, companies should provide more fashion accessories' entertaining contents on social media to create positive customer-brand association. Besides, customers positively perceive fashion accessories' information shared on social media is up-to-date and trustworthy. Therefore, trendiness can deliver positive influence on brand equity. Furthermore, WOM through social media can stimulate consumers' perception and attitude towards fashion accessories brands through process of exchanging opinions and feelings. Moreover, findings also implied brand equity significantly mediated relationships between entertainment, WOM, trendiness, and customer loyalty. In addition, brand equity is positively significant to build customer loyalty. The findings are accordant to results of previous literature. Brand equity can increase customers' loyalty via SMMAs' characteristics of entertainment, trendiness, and WOM. If consumers perceive fashion accessories brands can meet their expectation, then they will hold positive brand image, high brand preference and high loyalty to the company. However, customization is insignificant to brand equity. The reason may be customization on SMMA requires technological advancement. Web 3.0 technology is more suitable to offer customization on online platform [35]. Interaction are insignificant to brand equity too. There is one possible explanation. While increasing brand equity, interaction on social media should include emotional and cognitive dimensions, not only just focus on social brand engagement.

6 Conclusion

This research studied on empirical relationship between SMMAs' properties (entertainment, customization, trendiness, interaction, and WOM), brand equity and brand loyalty in the context of fashion accessories industry among Chinese youth in China. Findings support significant positive relationships between 1) entertainment, trendiness, WOM and brand equity; 2) brand equity and customer loyalty. Moreover, brand equity is significant mediator between entertainment, WOM, trendiness, and brand loyalty. The research contributes to present marketing literature in context of social media. Most of existing literature on SMMA didn't study on mediating effect of brand equity. This study proposed a model that identified mediating effect of brand equity in relationship between SMMA and customer loyalty. If consumers perceive fashion accessories brands can meet their expectation, then they will hold positive brand image, high brand preference and high loyalty to the company. Secondly, the study introduced S–O–R model to connect SMMA of fashion accessories industry to brand loyalty via brand equity which is seen barely in existing literature. It offers a powerful theoretical foundation. Findings of this study deliver important implications for marketers regarding SMMA. Marketers in fashion accessories industry should focus on significance of entertainment, trendiness and WOM to enhance brand equity and boost up brand royalty. Marketers can actively use social media to increase entertained brand relevant contents, grab the newest industry trends in time and design activities targeted on emphasizing on brand reputation. SMMA should be a tool to exert positive impact on brand equity so that can be beneficial to brand values. Fashion accessories industry should attract customers to use social media more frequently and initiatively via SMMA. This study assists marketers to identify SMMA as crucial influence on brand equity and customer loyalty. Consequently, to successfully enhance and maintain customer loyalty, SMMA as well as brand equity must be taken into consideration.

References

1. Tan, G.W.-H., Lee, V.H., Lin, B., Ooi, K.-B.: Mobile applications in tourism: the future of the tourism industry? Ind. Manag. Data Syst. **117**(3), 560–581 (2017)
2. Yadav, M., Rahman, Z.: Measuring consumer perception of social media marketing activities in e-commerce industry: scale development & validation. Telematics Inform. **34**(7), 1294–1307 (2017)
3. Ananda, A.S., Hernández-García, Á., Acquila-Natale, E., Lamberti, L.: What makes fashion consumers "click"? Generation of eWoM engagement in social media. Asia Pac. J. Mark. Logist. **31**(2), 398–418 (2019)
4. White, R.L.: 6 Reasons why fashion marketing rules social media (2016). http://trackmaven.com/blog/2016/02/6-reasons-why-fashion-marketing-rules-social-media/. Accessed 1 June 2020
5. Atashfaraz, M., Abadi, M.H.H.S.: Impact of E-Service innovation on brand equality and customer loyality in Samsung international corporation. Proc. Econ. Fin. **36**, 327–335 (2016)
6. Perera, G.R., Perera, I.: Influence of social media marketing on the brand image of organizations in the hospitality industry of Sri Lanka. Int. J. Asian Bus. Inf. Manage. **7**(1), 30–41 (2016)
7. Godey, B., et al.: Social media marketing efforts of luxury brands: influence on brand equity and consumer behavior. J. Bus. Res. **69**(12), 5833–5841 (2016)
8. Mehrabian, A., Russell, J.A.: An Approach to Environmental Psychology. MIT Press, Cambridge (1974)
9. Bitner, M.J.: Servicescapes: the impact of physical surroundings on customers and employees. J. Mark. **56**(2), 57–71 (1992)
10. Amarnath, D.D., Jaidev, U.P.: Toward an integrated model of consumer reactance: a literature analysis. Manage. Rev. Q. **71**(1), 41–90 (2020). https://doi.org/10.1007/s11301-020-00180-y
11. Zhu, B., Kowatthanakul, S., Satanasavapak, P.: Generation Y consumer online repurchase intention in Bangkok. Int. J. Retail Distrib. Manage. **48**(1), 53–69 (2020)

12. Liu, X., Shin, H., Burns, A.C.: Examining the impact of luxury brand's social media marketing on customer engagement: using big data analytics and natural language processing. J. Bus. Res. (2019). https://doi.org/10.1016/j.jbusres.2019.04.042
13. Wu, Y.-L., Li, E.Y.: Marketing mix, customer value, and customer loyalty in social commerce. Internet Res. **28**(1), 74–104 (2018)
14. Oliver, R.L.: Whence consumer loyalty? J. Mark. **63**(4), 33–44 (1999)
15. Ebrahim, R.S.: The role of trust in understanding the impact of social media marketing on brand equity and brand loyalty. J. Relat. Market. **19**(4), 287–308 (2019)
16. Seo, E.-J., Park, J.-W.: A study on the effects of social media marketing activities on brand equity and customer response in the airline industry. J. Air Transp. Manag. **66**, 36–41 (2018)
17. Ismail, A.R.: The influence of perceived social media marketing activities on brand loyalty. Asia Pac. J. Mark. Logist. **29**(1), 129–144 (2017)
18. Arrigo, E.: Social media marketing in luxury brands. Manage. Res. Rev. **41**(6), 657–679 (2018)
19. Zhu, Y.-Q., Chen, H.-G.: Social media and human need satisfaction: implications for social media marketing. Bus. Horiz. **58**(3), 335–345 (2015)
20. Ding, Y., Keh, H.T.: A re-examination of service standardization versus customization from the consumer's perspective. J. Serv. Mark. **30**(1), 16–28 (2016)
21. Kim, A.J., Ko, E.: Do social media marketing activities enhance customer equity? An empirical study of luxury fashion brand. J. Bus. Res. **65**(10), 1480–1486 (2012)
22. Muntinga, D.G., Moorman, M., Verlegh, P.W.J., Smit, E.G.: Who creates brand-related content, and why? The interplay of consumer characteristics and motivations. In: Rodgers, S., Thorson, E. (eds.) Digital Advertising - Theory and Research, pp. 259–284 (2017)
23. Khadim, R.A., Hanan, M., Arshad, A., Saleem, N., Khadim, N.: Revisiting antecedents of brand loyalty: impact of perceived social media communication with brand trust and brand equity as mediators. Acad. Strateg. Manage. J. **17**(1), 1–14 (2018)
24. Kunz, W., et al.: Customer engagement in a Big Data world. J. Serv. Mark. **31**(2), 161–171 (2017)
25. Leung, X.Y., Bai, B., Stahura, K.A.: The marketing effectiveness of social media in the hotel industry. J. Hosp. Tour. Res. **39**(2), 147–169 (2015)
26. Alavijeh, M.R.K., Esmaeili, A., Sepahvand, A., Davidaviciene, V.: The effect of customer equity drivers on word-of-mouth behavior with mediating role of customer loyalty and purchase intention. Inzinerine Ekonomika-Engineering Economics **29**(2), 236–246 (2018)
27. Liu, C.-R., Wu, T.-C., Yeh, P.-H., Chen, S.-P.: Equity-based customer loyalty mode for the upscale hotels—alternative models for leisure and business travels. Tour. Manage. Perspect. **16**, 139–147 (2015)
28. Laroche, M., Habibi, M.R., Richard, M.-O.: To be or not to be in social media: how brand loyalty is affected by social media? Int. J. Inf. Manage. **33**(1), 76–82 (2013)
29. Hair, J.F., Jr., Hult, G.T.M., Ringle, C., Sarstedt, M.: A Primer on Partial Least Squares Structural Equation Modeling (PLS-SEM). Sage Publications, Newbury Park (2016)
30. Srinivasan, S.S., Anderson, R., Ponnavolu, K.: Customer loyalty in e-commerce: an exploration of its antecedents and consequences. J. Retail. **78**(1), 41–50 (2002)
31. Zhang, K.Z.K., Benyoucef, M.: Consumer behavior in social commerce: a literature review. Decis. Support Syst. **86**, 95–108 (2016)
32. Hwang, I.J., Lee, B.G., Kim, K.Y.: Information asymmetry, social networking site word of mouth, and mobility effects on social commerce in Korea. Cyberpsychol. Behav. Soc. Netw. **17**(2), 117–124 (2014)
33. Kim, J.-H., Hyun, Y.J.: A model to investigate the influence of marketing-mix efforts and corporate image on brand equity in the IT software sector. Ind. Mark. Manage. **40**(3), 424–438 (2011)

34. Wang, T., Yeh, R.K.-J., Yen, D.C.: Influence of customer identification on online usage and purchasing behaviors in social commerce. Int. J. Hum. Comput. Interact. **31**(11), 805–814 (2015)
35. Idrees, S., Vignali, G., Gill, S.: Technological advancement in fashion online retailing: a comparative study of Pakistan and UK fashion e-commerce. Int. J. Econ. Manage. Eng. **14**(4), 318–333 (2020)

Adoption of Cashless Payment Practices Among Malaysian Millennials

Tan Zixin, Hao Siyu, Lai Wan Fei, and Abdullah Al Mamun(✉)

Faculty of Business and Management, UCSI University, 56000 Kuala Lumpur, Malaysia
abdullaham@ucsiuniversity.edu.my

Abstract. Cashless payment is a growing trend particularly during coronavirus pandemic but little attention from past studies paid to its adoption among Malaysian Millennials. This study examined the effect of Performance Expectancy (PE), Social Influence (SI), Effort Expectancy (EE), Facilitating Condition (FC) and Perceived Trust (TR) and the moderating effect of age, gender and education level on the intention and adoption of cashless payment among Malaysian Millennials. This study collected quantitative data from a total of 301 Millennials from Malaysia and analyzed by PLS-SEM method. The findings revealed that PE, EE, FC and TR have significant positive effect on intention to adopt cashless payment (ICP), and ICP has a positive and significant effect on adoption of cashless payment (ACP). To promote economic growth through the mass ACP practices among Malaysian, the government and financial service providers should therefore highlight the benefits of using cashless payment and inform security related details to build TR on cashless payment.

Keywords: Cashless payment · Intention and adoption · UTAUT · Millennials · Malaysia

1 Introduction

Cashless payment is the money transaction through internet [1]. Due to technological transformation nowadays, the trend of cashless payment such as E-wallet and digital money had been increased and widely used in daily activities and economic sectors. The intense competition of banking industry to provide more effective and high efficiency banking services in globalized business environment has led to the importance of implementation of digital banking service platform which this enabled worldwide business transactions to be completed in a least time. However, according to the statistical report published by Bank Negara Malaysia, the amount of cashless payment transaction was

M. Al-Emran et al. (Eds.): ICETIS 2021, LNNS 299, pp. 455–468, 2022.
https://doi.org/10.1007/978-3-030-82616-1_39

only RM203, 827.8 million, whereas the cash transaction was RM427, 481.7 million [2]. Less than 33% of total payment transaction was cashless payment [2]. Besides that, the adoption rate of cashless payment services by Malaysian consumers is still lower compared to other developing countries. Even though mobile transaction is considered a new model of payment and becomes a trend in Malaysia, the availability of cashless payment service is still limited to certain geographical areas like rural as compared to urban area. Safety and trust issues such as the leakage of personal information and scam with the use of cashless payment service has also become the spotlight in Malaysia. Therefore, further study should be carried out to increase the adoption of cashless payment.

Nowadays, as millennial or generation Y, aged between 1981 and 1996 that have grown in the era major technology disruptive and known as the "digital natives". They play a crucial role in economic development and more likely to adopt the service of cashless payment in daily money transaction. Therefore, in order to promote the ACP among Malaysian millennials that they are a crucial generation in the contribution to Malaysia economic growth nowadays, the purpose of this study will be focus on identify the determinants that will affect the Malaysian millennials intentions to adopt cashless payment. This study also served as a proposal for government and financial service providers such as banks and merchants to emphasize the benefits of cashless payment that enabled Malaysia to gain higher competitiveness in global business environment and also always in aware of the security related details to build and enhance the trust aspect in cashless payment service.

2 Literature Review

2.1 Theoretical Foundation

The Unified Theory of Acceptance and Use of Technology (UTAUT) Model underlie the importance of the perceive ease of use and perceived usefulness which will be a primary factor that will bring positive or negative impacts of the customer to accept the new technology in this era. UTAUT reported that there are four significant factors that affect the behavioral intentions such as the importance of performance expectancy (PE), social influence (SI), effort expectancy (EE) and facilitating condition (FC) which will influence the behavior of a user. In addition, UTAUT is a comprehensive model that can be applied to detect various technologies and has been proven to be an effective tool for predicting adoption behavior in various technology-based systems [3]. This study used UTAUT to study the impact of many factors such as PE, SI, EE and FCs on the behavioural intentions of Malaysian millennials, which will ultimately affect the behavior of actual use of cashless payment of Malaysian millennials.

2.2 Determinants of Intention to Adopt Cashless Payment (ICP)

Performance Expectancy (PE)
PE referred to the extent where a person realizes a system will benefit one in doing the daily activities [4]. This explained that with the improvement of technology, consumers will perceive cashless payment would provide greater benefits in term of convenience that can maximize their performance [5]. The research of Al-Qeisi and Hagazy [6] on consumers intentions towards usage of online banking has mentioned that PE has the most significant relationship with the usage intention, when consumers believed the payment method is more useful to them, there will be higher intention to adopt the cashless payment system. Therefore, based on the prior studies of several researchers that concluded PE was an important determinant on the adoption intention of cashless payment, this study proposed the following hypothesis:

Hypothesis (**H$_1$**): *PE has a positive effect on ICP among Malaysian Millennial*

Effort Expectancy (EE)
EE is referred to the required effort paid by a person to use a specific system [7]. EE is the main driver of adopting the online banking system which is believed by Jordanian customers that the system is ease to use with minimum effort to make the online money transactions [8]. However, the sample size will make a difference on the adoption intention particularly the variable of EE [9]. They explained that the degree of intention of online banking adoption would not be strong for Lebanese in different sample size of study even though they observed a significant result of intention due to EE. Therefore, this study proposed the following hypothesis:

Hypothesis (**H$_2$**): *EE has a positive effect on ICP among Malaysian Millennials*

Social Influence (SI)
SI is referred to one's adoption of a system is affected by another person [10]. This influence of adoption could come from a person's family, friends, spouse or communities. It is noteworthy that this influence of adoption would be getting less to a person who has more capability and knowledge on the adoption of a latest technology [7]. This assertion is further confirmed by Al-Qeisi and Hegazy [6] who found that SI is not a significant variable in the study of behavior intention to use online banking system in Jordan, Saudi Arabia and Egypt. Conversely, SI plays an important role in the adoption of a technology [3]. The awareness of using online banking system could be higher due to the positive influence by one's family and friends. Hence, this study proposed the following hypothesis:

Hypothesis (**H$_3$**): *SI has a positive effect on ICP among Malaysian Millennials*

Facilitating Condition (FC)
FC referred the adoption of a technology is supported by the technical infrastructure [7]. A good technical infrastructure will increase the intention of adoption of a new technology because it certainly assist and facilitate the users to understand and perform the tasks by the technology [9]. Therefore, the advanced technical infrastructure available in online banking system may encourage more Malaysian millennials to use the cashless payment system. Therefore, this study proposed the following hypothesis:

Hypothesis (H_4): *FC has a positive effect on ICP among Malaysian Millennials*

Perceived Trust (TR)
TR is explained as the level of confidence towards a system as expected [11]. In this context, it referred as the level of confidence of an individual to perform the cashless payment using an online system. TR is quite important for new users as they are more likely to use a new technology if it has less security threats [3]. Findings of Madan and Yadav [12] further confirmed TR has significant negative influence on the adoption of mobile wallet given that it fails to provide a safe security system on mobile transactions and further affecting the adoption intention of more users. As so, this study proposed the following hypothesis:

Hypothesis (H_5): *TR has a positive effect on ICP among Malaysian Millennials*

ICP and Adoption of Cashless Payment (ACP)
Adoption intention referred to people willingness to use a certain product or technology. Behavioural intentions have been constantly identified as the most influential determinants of personal behavior during technology acceptance. This is very important for investigating the intentions and behavior of individuals and reflecting their level of likelihood of using cashless payments. From prior research, it can be clearly proved that behavioural intention has a significant positive impact on adoption behavior [8]. EE can directly or indirectly promote behavioural intention to use technology by promoting the effects of perceived usefulness [8]. Therefore, this study proposed the following hypothesis:

Hypothesis (H_6): *ICP has a positive effect on ACP among Malaysian Millennials*

2.3 Mediating Effect of ICP

Past researchers have confirmed the consumers' intention of adoption of a technology is the key criteria leading the actual adoption of the technology [13–15]. The intention of adoption was demonstrated having the likelihood on the adoption of mobile banking services [16]. The consumers' intention was also found as the mediator on the relationship between behavioural variables and the actual behavior [17]. Similarly, the ICP may possibly mediate the relationships of the variables of EE, FC, PE, SI, TR on the ACP with the following explanations.

The study of Al-Qeisi and Hagazy [6] indicated that the higher ICP system was due to stronger believe on its benefits on online banking. The greatest feature of mobile banking service is the cashless payment system that brings the quick transaction and convenience to consumers. The belief of consumers on the benefits of a particular technology service or products does not necessary lead to the adoption directly without having intention of adoption in advance. Besides that, consumers normally possessed high intention for adoption due to high EE during early times of the adoption of a product or service. With the evolvement of artificial technology and machine learning, people will even pay less effort using cashless payment system.

Also, Gupta, Manrai and Goel [3] and Nysveen [18] have a same finding that SI will drive the intention of adoption prior to the adoption and acceptance of a technology, especially mobile banking applications. With more incentives given by Malaysian government in 2020 through cashless payment system under the economic stimulus, people are more likely to use cashless payment system to execute the cashless payment for shopping or even request the cashless payment from friends and family. As for the concern of the mediator of intention on the relationship between FC and ACP practice, Oliveira, Faria, Thomas and Popovic [19] provided their research model to support of the adoption of mobile banking was due to the behavioral intention. The adequate technical infrastructure would lead the adoption of mobile banking. Lastly, innovative cashless payment requires TR to increase the intention of adoption before the decision of adoption [20]. Therefore, higher security system is needed to reduce the possibility of less cybercrime in order to build up the TR on cashless payment. As a result, it provides the sense of security and confidence leading to the increase the intention of adoption. Therefore, this study proposed the following hypothesis:

Hypothesis (**H_7**): *ICP mediates the relationship between PE, EE, SI, FC and TR on ACP among Malaysian Millennials*

2.4 Moderating Effect of Age, Gender and Education

Age, Gender and Education are the three common moderating variables on behavioral intention investigated by a lot of past studies [7, 13, 21, 22]. According to UTAUT model by Venkatesh et al. [3], age and gender moderate the relationship between the behavioral variables and the intention of interest. Park, Amendah, Lee, and Hyun [23] also found the significant effect by education levels. However, past researchers also found the contradicted insignificant results of the moderating effects of age [19], gender [22] and education levels [24]. Therefore, this study proposed the following hypothesis:

Hypothesis (**H_8**): *Age, gender and education level moderates the effects of PE, EE, SI, FC and perceived TR on ICP among Malaysian Millennials*

3 Research Methodology

3.1 Research Design

Cross-sectional study is conducted for this study due to fast and easy to obtain large information from the respondents. Questionnaire was distributed through Google forms. Data are analyzed by SmartPLS software. Malaysian millennials which aged between

24 to 39 will be the targeted population to examine the adoption intention of cashless payment in this study. The sampling method adopted was convenience sampling as this is due to the time constraints for this study. As it is a large population of Malaysian millennials, therefore this study will used G-power version 3.1 to obtain the minimum sample size. Based on the calculation with level of confidence was set at 0.95 with an effect size (f^2) of 0.15 for 6 predictors, the minimum sample size of 74 was required. In order to achieve greater accuracy and avoid any problems that arise during data analyzation stage due to small sample size, 300 sample size was set as a target. The data from 294 Malaysian millennials was finally collected for analysis.

3.2 Survey Instrument

Questionnaire is used as the survey instrument in this study and questions were adapted from the earlier studies of researchers with a minor modification to match current research topic. A total of four questions adopted from Abrahão, Moriguchi and Andrade [4] and Dewi, Mohaidin and Murshid [10] to measure PE. A total of five questions adopted from Boonsiritomachai and Pitchayadejanant [25], Abrahão, Moriguchi and Andrade [4] and Dewi, Mohaidin and Murshid [10] to measure EE. A total of five questions adopted from Dewi, Mohaidin and Murshid [10] to measure SI. A total of five questions adopted from Sobti [26], and Boonsiritomachai and Pitchayadejanant [25], to measure FC. To measure TR, five questions adopted from Park et al. [23]. A total of six questions adopted from Sobti [26], Makanyeza [16], and Gupta, Manrai and Goel [3], to measure ICP. Finally, to measure ACP, this study adopted six questions from Makanyeza [16] and Al-Qeisi and Hegazy [6]. In this study, five-point Likert scale ranging from 1 (strongly disagree) to 5 (strongly agree) was used and to determine independent variables. A seven-point Likert scale ranging from 1(strongly disagree) to 7 (strongly agree) were used to determine ACP.

3.3 Common Method Bias

The Harman's single-factor test is normally adopted to examine any possible common method bias found in the dataset [27]. Only 45.5% of total variance was explained by single factor. Therefore, the dataset in this study has no significant common method bias.

4 Data Analysis

4.1 Demographic Characteristics

Table 1 depicted the data of demographics profile of respondents collected from 294 Malaysian millennials which included six questions, gender, age group average monthly income (RM), marital status, education and living areas.

In Table 2, most respondents (270 or 91.8%) were using cashless payments. Only 24 (8.2%) respondents had never use cashless payment. Among all the respondents, 129 respondents used online banking, while 30 respondents preferred mobile banking. The number of respondents using credit/debit cards were 39 whereas 71 respondents chosen mobile wallets or eWallet.

Table 1. Demographic characteristics

	N	%		N	%
Gender			*Marital Status*		
Male	83	28.2	Single	273	92.9
Female	211	71.8	Married	20	6.8
Total	294	100.0	Divorced	1	0.3
			Total	294	100.0
Age group			*Education*		
24–28 years	253	86.1	Secondary school certificate	20	6.8
29–33 years	26	8.8	Diploma certificate	28	9.5
34–39 years	15	5.1	Bachelor degree or equivalent	223	75.9
Total	294	100.0	Master's degree	21	7.1
			Doctoral degree	2	0.7
			Total	294	100.0
Average Monthly Income (RM)			*Living areas*		
Below 2500	158	53.7	Urban	267	90.8
2501 to 5000	107	36.4	Rural	27	9.2
5001 to 7500	19	6.5	Total	294	100.0
7501 to 10,000	5	1.7			
10,001 to 12500	1	0.3			
More than 12500	4	1.4			
Total	294	100.0			

4.2 Reliability and Validity

To ensure reliability of data, Cronbach's Alpha that is above 0.7 is acceptable and defined stronger consistency according to Cronbach's Alpha rule of thumb [28]. From Table 3, Cronbach's Alpha of all variables are above .7 which indicated all items are reliable. Another approach to measure internal consistency reliability is examine composite reliability, value that is above 0.7 is considered reliable. The composite reliability value in of all variables in Table 3 are greater than 0.8 which indicated that it is reliable. Furthermore, Dillon-Goldstein *rho* values of all variables in Table 3 are above 0.7 which this further confirmed the data collected is reliable.

Lastly, the average variance extracted (AVE) that is more than 0.5 is considered as achieved the convergent validity. Referring to Table 3, the AVE values for all variables has exceeded 0.5 and this indicated acceptable convergent validity. In this study, the variance inflation factors (VIF) are tested to identify multicollinearity issues. According to Diamantopoulos and Siguaw [29], variables that being tested that have VIF values

Table 2. Cashless payment

	N	%		N	%
Using cashless payment			*Types of cashless payment*		
No	24	8.2	None	24	8.2
Yes	270	91.8	Online banking	129	43.9
Total	294	100.0	Mobile banking	31	10.5
			Credit card/Debit card	39	13.3
Frequency of Using Cashless Payment Per Week			Mobile wallets/eWallet	71	24.1
			Total	294	100.0
Never	24	8.2			
1 to 5 times	183	62.2			
6 to 10 times	50	17.0			
11 to 15 times	8	2.7			
More than 15	29	9.9			
Total	294	100.0			

Table 3. Reliability and validity

Variables	No. Items	Mean	SD	CA	DG *rho*	CR	AVE	VIF
PE	4	4.304	0.601	0.816	0.827	0.878	0.643	2.600
EE	5	4.325	0.590	0.849	0.854	0.892	0.624	3.078
SI	5	3.844	0.769	0.853	0.860	0.896	0.636	1.721
FC	5	4.127	0.617	0.782	0.808	0.850	0.536	2.503
TR	5	3.710	0.889	0.922	0.947	0.943	0.769	1.693
ICP	6	5.611	1.073	0.953	0.953	0.962	0.810	1.000
ACP	6	5.495	1.119	0.940	0.945	0.952	0.769	–

Note: PE: Performance Expectancy; EE: Effort Expectancy; SE: Social Influence; FC: Facilitating Condition; TR: Perceived trust; ICP: Intention to Adopt Cashless Payment; ACP: Adoption of Cashless Payment; SD: Standard Deviation; CA: Cronbach's Alpha; DG *rho* - Dillon-Goldstein's *rho*; CR - Composite Reliability; AVE - Average Variance Extracted; VIF - Variance Inflation Factors

Source: Author's data analysis

lower than 3.3 indicated the multicollinearity issues is not serious. In Table 3, the multicollinearity issues are less significant as the VIF values of all variables are below 3.3.

4.3 Path Analysis

According to Table 4, the PE for the effect of ICP (H_1) was 0.164 with a p-value of 0.008. It indicated that PE had a positive effect on the ICP. The f^2 value of 0.027 indicated a small effect of PE on Malaysian millennials' ICP. Besides, the coefficient value for EE on the ICP (H_2) was 0.203 with a p-value of 0.003. In other words, EE had a positive effect on the ICP, with the f^2 value of 0.035. In contrast, the path coefficient value for effects of SI on ICP (H_3) was −0.001 with the p-value of 0.495. This implied there was no significant effect of SI on ICP, with the f^2 value of 0.000.

Table 4. Path coefficients

		Beta	t	p	r^2	f^2	Q^2	Decision
H_1	PE → ICP	0.164	2.439	0.008		0.027		Accept
H_2	EE → ICP	0.203	2.729	0.003		0.035		Accept
H_3	SI → ICP	−0.001	0.013	0.495	0.615	0.000	0.457	Reject
H_4	FC → ICP	0.215	3.256	0.001		0.048		Accept
H_5	TR → ICP	0.348	6.554	0.000		0.186		Accept
H_6	ICP → ACP	0.844	48.810	0.000	0.712	2.476	0.540	Accept

Note: PE: Performance Expectancy; EE: Effort Expectancy; SE: Social Influence; FC: Facilitating Condition; TR: Perceived trust; ICP: Intention to Adopt Cashless Payment; ACP: Adoption of Cashless Payment
Source: Author's data analysis

Furthermore, the path coefficient value for the effect of FC on the ICP is 0.215 with a p-value of 0.001 (H_4). This revealed a positive and statistically significant effect of FC on the ICP. Next, the path coefficient values recorded a positive ($\beta = 0.348$) and significant (p-value of $0.000 < 0.05$) effect of TR on the ICP (H_5), with the f^2 value of 0.186. It then indicated a moderately high effect of TR on the ICP. Finally, the path coefficient values for ICP shown a positive ($\beta = 0.844$) and significant (p-value of 0.000 < 0.05) effect on the ACP (H_6). A substantially significant effect of ICP towards ACP among Malaysian millennials was indicated.

The r^2 value that defined the degree of variance for the ICP is 0.615, indicating that 61.5% of the variation in the ICP can be explained by PE, EE, SI, FC and TR. Besides, the r^2 value for the ACP was 0.712, showing that 71.2% of the variation in the ACP can be explained by the ICP. Moreover, the Q^2 value of 0.457 indicated that PE, EE, SI, FC and TR have high predictive relevance for the ICP. Similarly, the Q^2 value of 0.540 indicated that ICP has high predictive relevance for the ACP.

4.4 Moderation and Mediation

As for the mediating effect, findings presented in Table 5 highlighted that TR mediates the relationship between PE, EE, FC and TR on ACP. The moderating effects of age, gender and education however are not statistically significant.

Table 5. Mediating and moderating effects

Associations	Beta	CI - Min	CI - Max	*t*	*p*	Decision
Mediating effect of intention to adopt cashless payment						
PE → ICP → ACP	0.138	0.052	0.241	2.431	0.008	Accept
EE → ICP → ACP	0.172	0.060	0.274	2.705	0.004	Accept
SI → ICP → ACP	−0.001	−0.067	0.071	0.013	0.495	Reject
FC → ICP → ACP	0.182	0.089	0.264	3.268	0.001	Accept
TR → ICP → ACP	0.294	0.214	0.361	6.518	0.000	Accept
Moderating effect of age						
PE → ICP	−0.081	−0.217	0.078	0.891	0.187	Reject
EE → ICP	−0.001	−0.185	0.129	0.014	0.495	Reject
SI → ICP	−0.033	−0.143	0.069	0.501	0.308	Reject
FC → ICP	0.042	−0.111	0.207	0.438	0.331	Reject
TR → ICP	−0.068	−0.169	0.001	1.316	0.094	Reject
Moderating effect of gender						
PE → ICP	0.043	−0.084	0.145	0.625	0.266	Reject
EE → ICP	0.060	−0.066	0.186	0.788	0.215	Reject
SI → ICP	−0.066	−0.169	0.021	1.125	0.131	Reject
FC → ICP	0.009	−0.084	0.137	0.122	0.451	Reject
TR → ICP	−0.015	−0.089	0.090	0.276	0.391	Reject
Mediating effect of education						
PE → ICP	−0.021	−0.184	0.140	0.226	0.411	Reject
EE → ICP	−0.059	−0.231	0.081	0.636	0.263	Reject
SI → ICP	0.060	−0.045	0.184	0.886	0.188	Reject
FC → ICP	0.049	−0.066	0.187	0.628	0.265	Reject
TR → ICP	0.020	−0.110	0.126	0.282	0.389	Reject

Note: PE: Performance expectancy, EE: Effort Expectancy, SI: Social Influence, FC: Facilitating Condition; TR: Perceived trust, ICP: Intention to Adopt Cashless Payment, ACP: Adoption of Cashless Payment
Source: Author's data analysis

5 Discussion

Rapid change in business environment and development of technology recently had substantially encouraged the use of cashless payment by consumers worldwide. Therefore, this finding will be described on the factors effecting ICP and actual ACP among Malaysian millennials with UTAUT model and extended TR construct based on data analysed. This study had completely supported the UTAUT theory as the results four predictors, PE, EE, SI and FC shown consistent positive effects [7] in the model. Based on the findings, both H_1: PE and H_2: EE was supported as it revealed that PE and EE had a significant positive effect on Malaysian millennials ICP. This result shown is consistent with the study of Chakraborty and Mitra [11] in India. Being said that, Malaysian millennials' strong ICP can be triggered when the performance of cashless payment service was useful for their day-to day activities and easier to perform, as the millennials are generation that highly prioritize freedom and efficiency, transactions system that provided clear and friendly interface can uphold their productivity in accomplished tasks related to financial transactions promptly will greatly motivate them to use it.

Subsequently, H_3: SI was not supported as it had the insignificant p-value showing its weak effect on the Malaysian millennials' intention of cashless payment adoption. This result was parallel with UTAUT model [7], where influence of society will not have a direct relationship on adoption intention of individual, in the context that individual had equipped sufficient knowledge and capability to use technology. To further discuss with UTAUT model, as millennials are the generation of tech-savvy and digital native that possessed high ability and knowledge to utilize the information technology devices, this implied that less guidance from family or friends will be needed when they have high ICP. H_4: The significant p-value from FC had revealed a positive effect on ICP among Malaysian millennials. In other words, FC towards Malaysian millennials ICP improved with infrastructure and facilities that supported greater convenience, efficient and effortless in transaction process. This finding was consistent with the research of Madan and Yadav [11] n India and Teo, Tan, Hew and Yew [30] in Malaysia. They supported that greater accessibility and high availability towards internet connection and cashless payment service providers can result in strong adoption intention as the transaction process can be completed more efficiently.

TR, the extended indicator in this finding was accepted and had a significant influence on the adoption intention of Malaysian millennials in cashless payment (H_5). To further justified, when Malaysian millennials have higher degree of TR towards the transparency of cashless payment where they possessed it will have less security and privacy threats, the more probable it will be triggered strong intention to adopt. This result was supported by the prior finding of Park et al. [23] that confirmed TR was a significant factor in determined consumer intention to use m-payment when they aware of the service provided are secure and trustworthy to complete transaction. Lastly, H_6: intention of adoption had disclosed a significant positive effect with the ACP among Malaysian millennials. This explained that when millennials were aware of the benefits of cashless payment in providing effective, efficient and reliable transaction process that will trigger their desire towards the use of cashless payment service, it will be more likely for millennials plan to use cashless payment as they believed it is worthwhile for them to use cashless payment in day-to-day activities and will directly be led to frequent actual adoption of them for the transaction process. Prior study of Al-Qeisi and Hegazy [6] also confirmed positive behavior perceived by consumers about internet banking will directly contribute to the internet banking adoption.

6 Implications

Findings of this study can contribute several significant theoretical and practical implications for researchers, Malaysia government and financial service providers. In term of theoretical implications, as limited prior study on millennials ICP using UTAUT model [7] was conducted by researchers in Malaysia, therefore this study can further support the applicability of UTAUT model in different context for future research purpose. In predicting Malaysian millennials ICP, the four variables, PE, EE, SI and FC adopted from UTAUT model, has shown consistent results with the UTAUT model recommended in which this had confirmed the precision of model. Moreover, UTAUT model was extended to incorporate TR as an indicator to examine the adoption intention of millennials in Malaysia with an attempt to have deeper understanding of millennials adoption behavior, that has revealed a significant positive effect in this context.

In term of practical implications, this finding provided valuable insight for Malaysia government and financial service providers to promote ACP of millennials together with an effort to foster economic growth and thus improve living standard of Malaysian. Government can encourage the ACP of millennials by enforced stringent policy that required cashless payment service providers abide and operates under the standards and regulations imposed in order to protect consumers that use the service. In addition, for cashless payment service provider to boost the level of confidence of millennials trustworthy towards using cashless payment, it is encouraged to issue their payment systems under Payments Network Malaysia (Paynet) that certified functional and security of cashless payment in Malaysia (Paynet, 2019). As so, the possibility of fraud done by cashless payment service provider can be mitigated due to the stringent law enforcement and credible certification services that secure millennial as a third party in transaction process. Moreover, in order to spur millennials ICP, retailers should offer multiple payment options such as online payment, digital wallet and credit or debit cards in transaction process as this could maximize the speed and convenience in conducting business activities which not only cap domestic transaction but worldwide that eventually promotes economic growth of Malaysia. Financial service provider and retailers can make an effort in improving or design the cashless payment system or application to present a user-friendly interface that required less knowledge and capability to function the cashless payment system. As most of the millennials respondents in this finding were lived in urban area, therefore, to continuously promote the cashless payment in rural area, improvement in infrastructure is an important way to do it, government should take action to improve the internet coverage or accessibility all around the nation as this able contribute to the development of country's economic where more Malaysian could have the opportunity to adopt cashless payment.

7 Conclusion

This study investigated the influencing factors of Malaysian millennials' ICP. UTAUT was employed and incorporated TR variable in finding. In accordance to the results analyzed, all variables studied demonstrated a significant positive effect on the Malaysian millennials ICP that will eventually led to actual usage except that SI was revealed to had weak effect on adoption intention among Malaysian millennials, this had fully supported the UTAUT model. As in Malaysia context, the main means to complete transaction is still the paper money, and adoption rate of cashless payment in Malaysia still considered low. The rapid transformation of technology and globalization in business markets had been moving towards millennials era, it will be important for Malaysian government and financial service providers to promote the benefits of cashless payment adoption to a greater extent as it can contribute to Malaysia economic growth. Therefore, this finding served as a tool for government and financial service providers to take initiative for them to implement strategic approaches that not only encouraged and build trust of millennial but all the Malaysian to realize the effectiveness and efficiency way in adopting cashless payment for day-to-day transactions.

References

1. Sfenrianto, J.: A model of factors influencing consumer's intention to use e-payment system in Indonesia. Procedia Comput. Sci. **59**, 214–220 (2015)
2. BNM – Bank Negara Malaysia (Online). https://www.bnm.gov.my/index.php?ch=34&pg=163&ac=2&bb=file. Accessed 20 June 2020
3. Gupta, K.P., Manrai, R., Goel, U.: Factors influencing adoption of payments banks by Indian customers: extending UTAUT with perceived credibility. J. Asia Bus. Stud. **13**(2), 173–195 (2019)
4. Abrahãoa, R.S., Moriguchi, S.N., Andrade, D.F.: Intention of adoption of mobile payment: an analysis in the light of the Unified Theory of Acceptance and Use of Technology (UTAUT). RAI Revista de Administraçãoe Inovacão **13**, 221–230 (2016)
5. Gholami, R., Ogun, A., Koh, E., Lim, J.: Factors affecting e-payment adoption in Nigeria. J. Electron. Commer. Organ. **8**(4), 51–67 (2010)
6. Al-Qeisi, K., Hegazy, A.: Consumer online behaviour: a perspective on internet banking usage in three non-western countries. Procedia Econ. Fin. **23**, 386–390 (2015)
7. Venkatesh, V., Morris, M.G., Davis, G.B., Davis, F.D.: User acceptance of information technology: toward a unified view. MIS Q. **27**(3), 425–478 (2003)
8. Alalwan, A.A., Dwivedi, Y.K., Rana, N.P., Algharabat, R.: Examining factors influencing Jordanian customers' intentions and adoption of internet banking: extending UTAUT2 with risk. J. Retail. Consum. Serv. **40**, 125–138 (2018)
9. Tarhini, A., El-Masri, M., Ali, M., Serrano, A.: Extending the UTAUT model to understand the customers' acceptance and use of internet banking in Lebanon. Inf. Technol. People **29**(4), 830–849 (2016)
10. Dewi, C.K., Mohaidin, Z., Murshid, M.A.: Determinants of online purchase intention: a PLS-SEM approach: evidence from Indonesia. J. Asia Bus. Stud. **14**(3), 283–306 (2019)
11. Chakraborty, S., Mitra, D.: A study on consumers' adoption intention for digital wallets in India. Int. J. Cust. Relat. **6**(1), 38–57 (2018)
12. Madan, K., Yadav, R.: Behavioural intention to adopt mobile wallet: a developing country perspective. J. Indian Bus. Res. **8**(3), 227–244 (2016)
13. Yu, C.S.: Factors affecting individuals to adopt mobile banking: empirical evidence from the UTAUT model. J. Electron. Commer. Res. **13**(2), 104–121 (2012)
14. Arahita, C.L., Hatammimi, J.: Factors affecting the intention to reuse mobile banking service. Int. J. Res. Bus. Soc. Sci. **4**(4), 15–22 (2015)
15. Venkatesh, V., Thong, J.Y., Xu, X.: Consumer acceptance and use of information technology: extending the unified theory of acceptance and use of technology. MIS Q. **36**(1), 157–178 (2012)
16. Makanyeza, C.: Determinants of consumers' intention to adopt mobile banking services in Zimbabwe. Int. J. Bank Market. **35**(6), 997–1017 (2017)

17. Baker, E.W., Al-Gahtani, S.S., Hubona, G.S.: The effects of gender and age on new technology implementation in a developing country: testing the theory of planned behavior. Inf. Technol. People **29**(4), 352–375 (2007)
18. Nysveen, H.: Intentions to use mobile services: antecedents and cross-service comparisons. J. Acad. Mark. Sci. **33**(3), 330–346 (2005)
19. Oliveira, T., Faria, M., Thomas, M.A., Popovič, A.: Extending the understanding of mobile banking adoption: when UTAUT meets TTF and ITM. Int. J. Inf. Manage. **34**(5), 689–703 (2014)
20. Gao, L., Waechter, K.A.: Examining the role of initial trust in user adoption of mobile payment services: an empirical investigation. Inf. Syst. Front. **19**(3), 525–548 (2015). https://doi.org/10.1007/s10796-015-9611-0
21. Wang, Y.S., Wu, M.C., Wang, H.Y.: Investigating the determinants and age and gender differences in the acceptance of mobile learning. Br. J. Edu. Technol. **40**, 92–118 (2009)
22. Koksal, M.H.: The intentions of Lebanese consumers to adopt mobile banking. Int. J. Bank Market. **34**(3), 327–346 (2016)
23. Park, J., Amendah, E., Lee, Y., Hyun, H.: M-payment service: interplay of perceived risk, benefit, and trust in service adoption. Hum. Factors Ergon. Manufact. Serv. Ind. **29**(1), 31–43 (2018)
24. Laforet, S., Li, X.: Consumers' attitudes towards online and mobile banking in China. Int. J. Bank Market. **23**(5), 362–380 (2005)
25. Boonsiritomachai, W., Pitchayadejanant, K.: Determinants affecting mobile banking adoption by generation Y based on the Unified Theory of Acceptance and Use of Technology Model modified by the Technology Acceptance Model concept. Kasetsart J. Soc. Sci. **40**(2), 349–358 (2017)
26. Sobti, N.: Impact of demonetization on diffusion of mobile payment service in India. J. Adv. Manage. Res. **16**(4), 472–497 (2019)
27. Podsakoff, P.M., MacKenzie, S.B., Lee, J.Y., Podsakoff, N.P.: Common method biases in behavioral research: a critical review of the literature and recommended remedies. J. Appl. Psychol. **88**, 879–903 (2003)
28. Pallant, J.: SPSS Survival Mannual. Mc Graw Hill, New York (2007)
29. Diamantopoulos, A., Siguaw, J.A.: Formative versus reflective indicators in organizational measure development: a comparison and empirical illustration. Br. J. Manage. **17**(4), 263–282 (2006)
30. Teo, A.C., Tan, G.W.-H., Ooi, K.B., Hew, T.-S., Yew, K.-T.: The effects of convenience and speed in m-payment. Ind. Manage. Data **115**(2), 311–331 (2015)

IoT Within the Saudi Healthcare Industry During Covid-19

Maram Alqahtani(✉)

Department of Management Information System, King Saud University, Riyadh, Saudi Arabia

Abstract. Internet of things (IoT) has improved quality of life and made life easier, more effective, and more productive. With the rapid spread of COVID-19, a need has emerged for technologies that enable countries and their healthcare sectors to confront the pandemic, reduce its spread and infection rate, and find a way to detect it as soon as possible. This research investigated the current status of adopting IOT technology within the Saudi healthcare sector during the COVID-19 pandemic. The research benefited from previous research that suggested solutions to leverage IOT technology to confront COVID-19. Structured interviews with experts from the field were conducted to collect the information for this research. The results indicated that all research participants are taking a clear initiative to adopt IOT technology and have taken some initial steps. It was observed that IOT technology was implemented in closed systems and not shared over the internet or outside a cloud. Privacy and security concerns were raised as major obstacles facing IOT adoption, in addition to strict standards imposed by the Ministry of Health (MOH). Moreover, the results show that COVID-19 can be seen as an enabler to promote IOT adoption. Moreover, participants highlighted many benefits of adopting IOT technology for example: improve advance patient care, provide real-time monitor and real-time data, provide proactive action, and homecare. Participants also emphasized the challenges of IOT adoption: poor infrastructure, shortage of staff and poor experience. Participants also highlighted IOT that awareness workshops can help in effective IOT adoption. This research should serve as a good starting point for further research on a greater scale.

Keywords: IOT adoption · Internet of Things · Saudi healthcare · COVID-19 · Coronavirus · Fight coronavirus · Saudi Arabia

1 Introduction

Information technology has been rapidly developing in recent times and has revolutionized the way we go about our lives. A major contributor to this revolution is the so-called Internet of Things (IoT). IoT has improved quality of life and made life easier, more effective, and more productive. IoT is defined as a network of connected objects or "things" that collect and send information from their software and sensors to central servers, all with a lower level of human involvement [1].

IoT is widely known for its impact on the personal level, as can be seen in the development of smart homes, smart lighting management systems, and smart parking,

M. Al-Emran et al. (Eds.): ICETIS 2021, LNNS 299, pp. 469–483, 2022.
https://doi.org/10.1007/978-3-030-82616-1_40

all of which can be easily controlled through remote devices. Fortunately, healthcare is also one sector that gets the most benefit out of IoT. The healthcare sector uses the IoT to monitor patient symptoms through sensors, to manage patient health, and to predict diseases. IoT can help with central patient care, which will lead to greater patient satisfaction. Moreover, IoT technology makes tracking staff, patients, and equipment much easier, faster, and more effective. Although the healthcare sector knows and trusts the IoT and its capabilities, there is a lot to be done before leveraging IOT to its full potential [2].

Currently, the coronavirus (COVID-19) pandemic is having a great impact on the world economy and on healthcare in particular [3]. Moreover, global public healthcare is also affected and there is a significant burden on their financial position and lives of patients and healthcare workers. With the rapid spread of COVID-19, a need has emerged for technologies that enable countries and their healthcare sectors to confront the pandemic, reduce its spread and infection rate, and find a way to detect it as soon as possible. Moreover, reaching patients is considered to be the second major issue after vaccine development; this is what the IoT can provide to the healthcare sector, in addition to other tools such as artificial intelligence, data analytics, and drones.

Smart wearable devices that use IoT technology can enable patients to track their own health and help healthcare providers deliver immediate care when needed [4]. These devices can also monitor and record symptoms and then analyze them to predict disease.

Many countries, such as China and Saudi Arabia, have implemented the IoT in their health sectors and leveraged its benefits to fight COVID-19. China has invested significantly in the IoT, especially in the healthcare sector. It uses IoT in conjunction with other technologies like robots, drones, and artificial intelligence (AI) to monitor patients wearing IoT-enabled devices to record any changes on the body. The Chinese health sector also use IoT technology to track infected people and make sure they are in quarantine. Saudi Arabia, with their promising Vision 2030, is also steadily moving forward to adopt new technologies in their healthcare sector. They know the power of the IoT and have started to adapt it into their health care.

This paper emphasizes the positive impact of the IoT in the healthcare sector and discusses the current situation in the healthcare sector of Saudi Arabia, especially during the coronavirus pandemic. Moreover, it can serve as a starting point to inform Saudi healthcare leader about the current status of IOT in healthcare and how IOT will help them to confront the current pandemic or any potential future crisis. Furthermore, it can serve as a foundation for future studies in addition to encourage and motivate healthcare leaders to fully adopt IOT technology in Saudi healthcare.

2 IoT Technology: Overview

Before discussing the research on the IoT, it would be helpful to discuss the architecture of IoT.

2.1 The Layers of IoT in Healthcare and Its Components

IoT technology is built on three layers. First, the physical or perception layer is where the devices and sensors are located. Second, the network layer is where the routers and gateways are used. Third, the application layer is where the cloud and servers are located. According to [1], the architecture of IoT in health care has six layers: perception, transport, processing, application, business, and network. The perception layer concerns objects and sensors that are available in the environment. It determines whether there is a smart object in the area and senses those objects in some "physical parameters" for information such as motion, temperature, and location. The transport layer transfers sensor data from the perception layer to the processing layer and reverse through RFID, Wi-Fi, and Bluetooth. The processing layer, or middleware layer, stores a large amount of information and analyses it. The application layer delivers the user an application-oriented service which IoT has different application on such as a smart home and smart health. The business layer manages all IoT systems, such as user privacy (Fig. 1).

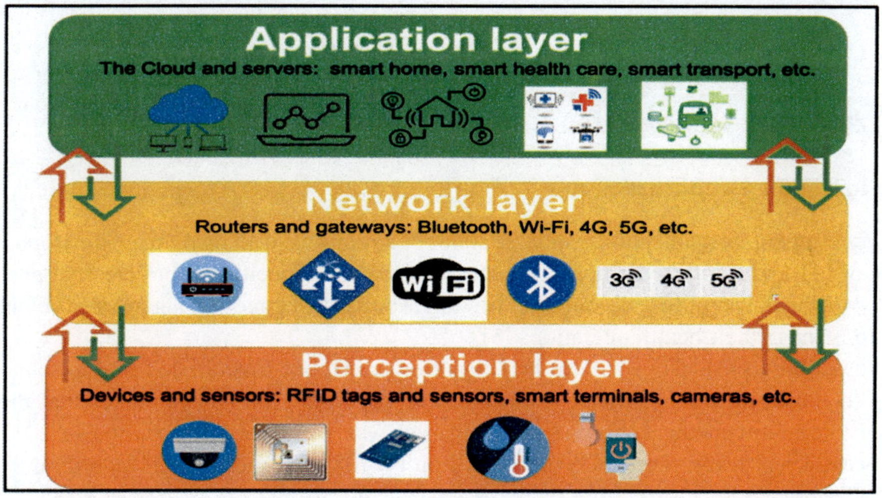

Fig. 1. IOT layers

[1] also explain the following five components of IoT in health care: 1) the sensor collects information; 2) the collected information goes through a smart network; 3) this collected information is saved in the cloud; 4) the information is investigated and analyzed using big data analysis to select the right choice; and 5) the clarified information is sent to a smart medical clinic, which takes appropriate actions toward diagnosis and treatment. This cycle then returns to step one to sustain the circularity of the system (Fig. 2).

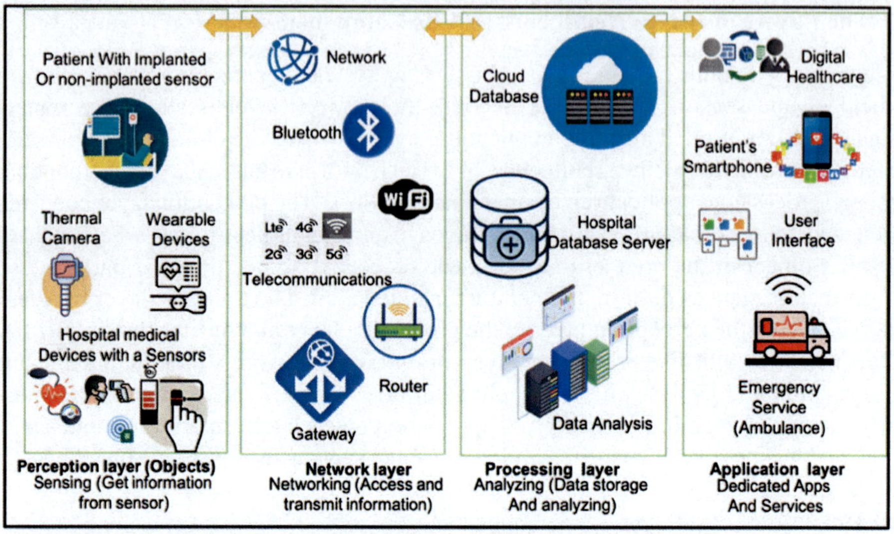

Fig. 2. IOT layers in healthcare

3 Literature Review

3.1 IoT in the Healthcare Sector

The governments, organizations, and initiatives of nations have all recognized the importance of IoT and its positive impact. The healthcare sector specifically has clearly benefited from the power of IoT, because IoT offers the best, most effective, and most efficient way to manage their health operations [5].

According to [3], about 76% of healthcare organizations trust the IoT and its capability to transform the healthcare industry. A report from Arab Health states that the global internet-of-medical-things (IoMT) market was worth US$41.2 billion in 2017 and is expected to increase to US$158.1 billion by 2022 [6] and to US$188 billion by 2024 [3].

IoT plays an important role in the healthcare sector by helping to manage its operations and prevent chronic disease. It helps monitor and track patients' health and sends alerts to healthcare provider that unusual changes happened by using radio-frequency identification (RFID) tags and Real-time locating system (RTLS) technology [2]. According to [7], older people (60 or older) are the patients that can benefit most from IoT. They benefit by using IoT-enabled devices like health trackers (e.g., a bracelet), glucometers, and skin sensors to monitor their blood pressure, blood sugar level, and heart rate. This can help save lives by recommending specific safety actions or by notifying healthcare providers and the patient's guardian of any unusual changes. In addition, [7] suggests using smart shirts that have numerous sensors that monitor the patient's vital signs and inform healthcare providers about the patient's condition in case of an emergency and even locate him or her via GPS. According to [1], IoT can be used in medical information management to detect and prevent falls, which are considered a serious problem for older people. Finally, [8] offers a fall-detecting system that is based on IoT and big data modelling, which uses machine-learning (ML) techniques based on a decision tree.

3.2 Impact of IoT on the Healthcare Sector During the COVID-19 Pandemic

From the studies discussed above, we can say that IoT can effectively help minimize the spread of the coronavirus. COVID-19 has changed the world and pushed countries to adopt new technology and search for "practical and cost-effective solutions" to manage the pandemic and reduce its negative impacts on the economy and on people's lives [9]. The healthcare sector has faced the following key critical issues: patient overload, safety of healthcare workers, and healthcare resource shortage. Moreover, [9] states that most of these problems occurred because of not effectively reaching the patient; however, IoT technology can improve patient tracking. IoT technology can ensure that all infected patients are quarantined and tracked using an internet-based network, and it can monitor and capture real-time data and information from infected patients. Moreover, [9] emphasizes that by successfully implementing IoT, there will be critical merits of IoT for the COVID-19 pandemic. It will reduce the "chance of mistakes, deliver superior treatment, lower expenses, effective control, and enhance patient diagnosis." This technology can be used to monitor not only infected patients but also older people and people with a high risk of developing severe complications from the disease [3].

Another suggested use for IoT technology in fighting COVID-19 is air cleaning systems. In [10], Terrence DeFranco said that one solution for managing and reducing COVID-19 infection is to use IoT technology to clean the air indoors and make sure the environment is free from viruses. Applying a program that manages indoor air quality will create a healthy environment and reduce infection. This can be done with an indoor virus detection sensor that controls contaminants and ensures the environment is getting enough ventilation [10].

In [11], the authors propose a home hospitalization system based on the IoT, cloud computing, and fog computing. This proposed system allows the patient to receive treatment and recover in their own home without going to the hospital and possibly getting infected with COVID-19. This home system will primarily benefit patients with chronic disease, especially older people. Moreover, this system will solve problems of resource shortages and overcrowding that hospitals have suffered during the pandemic. The system will monitor the patient's health and the room's environment during the home hospitalization period so healthcare providers can monitor patients remotely.

On the other hand, [12] suggests that implementing a centralized IoT system for diagnosing COVID-19 patients will be more effective if it is done by the government. The authors emphasize that integrating telemedicine benefits (e.g., accessing patient health records through software and with a communication channel between healthcare provider and patient) with their proposed IoT system will improve the system and make it suitable not just for the COVID-19 crisis, but for future situations.

According to [13], they proposed an IOT architecture with the use of smart sensors to measure and record body temperature that will identify infected individuals and maximize social distancing. Moreover, this IOT system is connected to a cloud computing and uses data analysis to make effective decisions based on real-time data.

Furthermore, [14] proposed a monitoring system that gathers real-time symptoms data from wearable sensors to identify potential coronavirus cases. This will be done via eight machine learning algorithms implemented within IoT infrastructure that tracks both potential and conformed cases in addition to the outcome of treatment of recovered patients. The goal of this system solution is to minimize mortality rates by early detection, tracking recovered cases and a better understanding of the nature of the disease. The result shows that five of these eight algorithms have received an accuracy of more than 90%.

Another system design proposed by [15] has the ability to detect and monitor coronavirus cases by IOT based smart glasses. The glasses will send a notification to the officer if system identify a suspected case among the masses of people based on the captured body temperature by thermal cameras. Moreover, the system can use face identification on suspected carriers of the virus who have high body temperature. The information of the visited location of the suspected cases via Google Location History (GLS) will be used to provide "reliable data on detection process". This system will help decrease screening time with minimum human interaction and as a result reduce the chance of the virus spreading.

During this pandemic, many countries have leveraged the use of the IoT in their healthcare sector. In China, Wuhan was the first city to identify a coronavirus outbreak. With its incredible Wuhan Wuchang Smart Field Hospital, China gives a great example on how to apply IoT technology in health care. This hospital uses several smart devices, such as bracelets, to monitor patients and record vital information like blood pressure, blood sugar level, blood oxygen saturation, and heart rate. If there is any sign of infection, the medical provider will be informed [16]. China also launched an app called the Close Contact Detector, which notifies people if they are near someone who is confirmed to have or is suspected of having the virus [9].

Hong Kong is using electronic tracker wristbands to track citizens and notify authorities when an individual, particularly one who recently arrived from an international destination, is not following compulsory home quarantine orders [17]. The passengers are given a bracelet to wear which has a unique QR code to track their movement. They then download an app called StayHomeSafe and scan the QR code, and once they reach their apartment, they should move around only to calibrate the device. The app uses geofencing, which is a location-based technology that uses RFID, GPS, Wi-Fi, or cellular data to create a virtual perimeter for an app. Russia, Poland, Singapore, and South Korea have all launched apps similar to the StayHomeSafe app used in Hong Kong [18].

With the Saudi Arabi promising vision 2030, there will be noticeable improvement in healthcare electronic services including IOT application. There are many Telehealth application and e-services in Saudi healthcare that helps in managing patient information, deliver advanced care remotely and effectively, and increase treatment outcomes. As [19] explain, Telemedicine (SEHA App) used to deliver care at a distance and enable patient to manage their healthcare journey. An appointment app called Mawid, which is a central appointment system that patients can use to manage their appointments. Moreover, recently, The Ministry of Health (MOH) lunch Tabaud App to help in fighting and contain COVID-19 pandemic. MOH comes with strategy to use those Mobile applications, in addition to Tawakklna and Tetamman, to contain and mitigating the risk of the disease [19].

4 Methodology

4.1 Methodological Approach and Data Collection

The goal of this research is to investigate and gain a better understanding of the current status of IOT adoption in Riyadh, Saudi Arabia, and if coronavirus pandemic affects its adoption. A qualitative research design was adopted. To gather the qualitative data, this research conducted structured interviews due to the limited time, the shortage of experts in the field, and the limited adoption of IOT technology in the Saudi healthcare sector. The interviews were conducted through phone, Microsoft Teams, or Zoom and lasted approximately 30 min each. It has eight open-ended questions and was answered by five participants. The participants were selected as they have the most expertise in the field and have knowledge of IOT adoption in Saudi Arabia's healthcare sector. The participants' roles and work locations are shown in Table 1.

Five out of the six participants that met the criteria were interviewed. The other participant refused to publish or use the information he provided.

Table 1. Participant role and working place

Participant	Role	Working place
Participant 1	Director of healthcare technology development	King Abdulaziz Medical City NGHA
Participant 2	Group Chief Information Technology Officer	Saudi German Hospital Group - BAB Medical Education Company Limited
Participant 3	Director of e-Services and e-Channels	Dr. Sulaiman AL Habib Medical Group
Participant 4	Integration Services Manager	King Faisal Specialist Hospital and Research Center
Participant 5	- Executive Vice President - Asst. Professor	- Taibah Valley company - Taibah University

4.2 Data Analysis

Thematic analysis was performed to analyze the qualitative data. During the interview, notes were taken, and the interviews were recorded in case of any missing notes that were not captured. To analyze the data, different steps were taken. First, the notes and data from the interview records were transcribed into a document. Second, the data were coded by highlighting similarities and differences between the data (text). Third, themes were generated by identifying patterns among created codes and then reviewing them. Finally, themes were defined and named for brevity and clarity. Moreover, the level of IOT adoption and the degree of obstacles hindering IOT adoption was measured on scale of one to five as shown in Table 2.

5 Result

After analyzing the data, eight themes were identified: (1) adoption of IOT technology in the Saudi healthcare sector, (2) obstacles to the application of IOT technology, (3) supported IOT technology devices, (4) effect of the COVID-19 pandemic on IOT application, (5) desired benefits of IOT adoption, (6) challenges faced when trying to adopt IOT technology, (7) raising leadership and end-user awareness of the power of IOT technology, and (8) criteria for measuring the success of adopting IOT technology.

1. Adoption of IOT technology in the Saudi healthcare sector

All participants agreed that Saudi Arabia's healthcare sector is moving forward to adopt IOT technology and has initiated steps in this action. Four of the participants discussed that they use IOT technology within closed systems and store data in local datacenters. They do not share information through the Internet and to an external cloud. Participant (P1) said, "No, we don't have full adoption of IOT technology. We don't have a device that collects data by sensors for example, and then feeding bigdata database or cloud over the Internet. However, we have solutions over a closed system (semi-IOT technology) that do some certain function, but not sharing the data to an open platform." On the other hand, P3 said that they are using IOT technology in their hospital, and they have started using smart wearable devices such as the Apple Watch, which is integrated with their applications to collect patients' vital signs and stores them directly within their electronic health records (EHR). He added that they have a new department that mainly focuses on emerging technology such as IOT, AI, and deep learning.

P4 highlighted that some critical departments such as cardiovascular diseases are required to collect some parameters when a patient is discharged from the hospital, so they give them a device to collect those vital signs. After that, there are two options: the device company either transmits the data to the hospital through a secure channel or stores the data within the device, and when the patient comes back to the hospital, they upload the patient's data to the hospital system. The adoption results are summarized in Table 2.

2. Obstacles to the application of IOT technology

As showing in Table 2, there are many obstacles that prevent Saudi healthcare from adopting IOT technology. Three participants agreed that privacy and security issues are the biggest obstacles they face. However, P2—who has experience in data security and privacy—argued that "any new technology comes out, people used to mention privacy and security as an issue." He added that "All technologies are supported with standards and controls that will make sure we use the technology correctly and protect our data in case if their integrity, availability, and confidentiality."

Table 2. Adoption results

Themes	Related Questions (Patterns)	Point-scale from (1 to 5) where: 1 Strongly disagree, 2 Disagree, 3 Nature, 4 Agree, and 5 Strongly agree				
		P1	P2	P3	P4	P5
1. Adoption of IOT technology in Saudi healthcare sector	Does Saudi healthcare sector has initiated steps in IOT technology?	5	5	5	4	3
	Have full IOT adaption	1	1	5	4	1
	Use IOT technology within closed systems (Semi-IOT technology)	5	5	1	5	3
2. Obstacles that prevented the application of IOT technology	Do you have any kind of obstacles that prevent you from adopting IOT technology?	5	2	1	5	3

Two participants (P1 and P2) stated that they are not sure about the accuracy and safety of the devices that collect data. However, P4 disagreed and said that the accuracy of the devices depends on the manufacturers, and nowadays these companies are more advanced and provide the market with excellent devices at different prices.

Four participants mentioned that there are no standards or regulations from the Ministry of Health (MOH) that guide the adoption of IOT technology in healthcare. Moreover, the MOH prevents sharing patient information over the Internet or storing it in the cloud. P3 mentioned that they don't have any obstacles.

3. Supported IOT technology devices

P1 referred to devices that patients can use to adjust the room temperature and lights and call the nurse. However, he emphasized that all those devices are linked to closed systems that perform certain functions. He also mentioned that on the hospital gates, they use thermal cameras that sense body temperature. In addition to a thermal camera, P2 added that they are using a smart analyzer in their laboratory. P3 said they used Apple Watches, Accu-Chek for testing blood glucose, smart refrigerators for medicines, and point of care (PoC) technology. Three participants mentioned that they are using bedside devices that are linked to their closed system. Moreover, P4 added that they are using Philips Patient Monitoring, Welch Allyn vital signs monitor, and MAQUET ventilators.

4. Effect of the COVID-19 pandemic on IOT application

P3 highlighted that COVID-19 did not affect their adoption of IOT technology. However, all other participants stressed that COVID-19 has had a positive impact on their hospitals' adoption of IOT technology, and it works as an enabler to promote IOT adoption. P1 said, "COVID-19 made everyone change their healthcare delivery. Hospitals want to minimize the exposure of their healthcare provider and patients to the virus, monitor infected patients inside and outside the hospital, and deliver the healthcare service to them. All of that can be done by applying IOT technology." P4 added that during the COVID-19 pandemic, there was a specific budget for handling the pandemic, and there was a high demand for devices that support IOT technology.

5. Desired benefits of IOT adoption

All participants agreed that IOT technology will improve patient care and increase patient engagement and satisfaction. Moreover, all of them discussed the benefit of IOT in terms of how it will help in home care especially for elder patients. They all pointed out that IOT technology will provide them with real-time monitoring and data which will help take proactive action before situations become critical. Three participants stressed that with real-time data, IOT technology will improve disease management.

Four participants believed that IOT technology will help reduce operating costs. P3 said, "Adopting IOT technology will reduce the usage of power like electricity. Moreover, with the real-time data, the health provider will deliver the service to the patient without unnecessary visits to the hospital which will save doctors time and hospital resources used in the reception."

Three participants agreed that IOT will improve data accuracy and reduce errors, mainly human errors. They also indicated that adopting IOT technology will provide them with massive amounts of data which will improve treatment outcomes and help in decision-making.

6. Challenges faced when trying to adopt IOT technology

Participants highlighted many challenges they faced when they trying to adopt IOT technology. Four participants stressed that patient privacy and security are the biggest challenges to IOT adoption. Moreover, they highlighted that that MOH standards and regulations prevent sharing patient information and saving it on the cloud, which will make full IOT adoption more difficult.

P3 noted that infrastructure is one of the big challenges faced by the Saudi healthcare sector. P1 and P2 agreed that many hospitals have poor healthcare infrastructures, but they did not face this problem in their hospitals. In addition, P1 mentioned that 5G technology can help solve many IOT challenges faced by the Saudi healthcare sector. P3 added that end users (doctors and nurses) are not technical oriented, and they need training and education programs to facilitate IOT application.

Three participants agreed that staff shortage and poor experience are two of the barriers they faced, as P4 said, "Lot of people knows the concept of the IOT and the broad idea behind it. However, less of them knows the details of IOT adoption and how much effort it will take to implement it effectively."

P4 expressed concerns about the availability and maintenance of the devices in the Saudi market. He said, "We don't have local manufacture in Saudi Arabia that provide the Saudi market with devices that support IOT technology. Which means that we have to buy spare parts from outside the country in case of devices' maintenance." He added that the companies raised their prices—especially after COVID-19—which may reduce the demand for such devices.

The main challenge P2 faced in their hospital is time. "We are moving forward in IOT application, but we need more time to fully implement it effectively in our hospital."

P5 highlighted that resistance to change—from both top management and end users—in medium to small size hospitals is one of the biggest challenges to Saudi healthcare's adoption of IOT technology.

7. Raising leadership and end-user awareness of the power of IOT technology

Four participants emphasized and highlighted the importance of education and awareness programs for end users and how they will help raise their awareness of the power of IOT technology and the added value it can potentially provide. They all agreed that they do not have problems with top management since they are the ones who guide and lead the hospital's digital transformation. As P4 said, "In our hospital, top management and healthcare provider are the ones who asked for more IOT solutions. They are aware of the added value that IOT can deliver and saw some of its potentiality. However, we have some concern about the patients."

All participants agreed and discussed how running demos help in raising leadership and end-user awareness of the power of IOT technology. As P2 said, "We can run a demo (proof of concept (POC)) help in raising leadership and end user awareness."

8. Criteria for measuring the success of adopting IOT technology

All participants agreed that decreased response time—which will improve customer service—will be one of the criteria for measuring the success of adopting IOT technology. They all discussed that optimizing and utilizing resources along with reducing operating costs will be considered as success factors. In light of the current COVID-19 pandemic, they all agreed that infection control and minimizing patients' hospital stay will be considered as a criterion for success.

Three participants mentioned patient satisfaction as another criterion for evaluating progress. The other two participants argued that having accurate results in real-time and the validity of devices are two important requirements. P4 and P5 added that increasing productivity and reducing errors can be considered as critical criteria.

6 Discussion

Studies have shown the importance of IOT technology in the healthcare sector and how its application will enhance the effectiveness of hospital performance, improve the patient's health and care, and increase their satisfaction. With the current situation, solutions to overcome the COVID-19 pandemic and minimize its spread are crucial. Results indicate that COVID-19 works as an enabler to promote IOT adoption in the Saudi healthcare sector.

Results show that IOT technology is still a nascent phenomenon especially in the Saudi healthcare sector, and its adoption is still at an early stage. Moreover, there are some emerging IOT applications, but they are done in closed systems (semi-IOT technology). Thermal cameras, wearable devices, and other devices that use sensors to gather data such as body temperature, blood pressure, heart rate, and other body parameters are used in Saudi hospitals but not shared over the Internet or stored in an external cloud. The current Saudi Ministry of Health (MOH) regulations prevent sharing patient information through the Internet or storing it in the cloud because of concerns about data privacy and security. Electronic health records (EHR) are still not stored in a central datacenter for all Saudi hospitals. This is one of obstacles to widespread IOT adoption by Saudi hospitals.

A central datacenter for storing all EHRs and sharing information among hospitals and over the Internet is important to adopt IOT successfully within the Saudi healthcare sector. In addition, [12] suggested that having a centralized IOT system built by the government and integrates telemedicine benefits will improve the system and make it suitable not just for the COVID-19 crisis but also for future situations.

The infrastructures in Saudi healthcare sector are undergoing a huge improvement, especially in big hospitals. On the other hand, medium to small size hospitals are still facing some infrastructure challenges in addition to resistance to change and a shortage of expert staff. However, these challenges can be overcome by launching demos (proof of concept (POC)) and conducting education and training programs.

Using 5G will solve many other challenges in IOT application and boost its adoption. 5G mobile networks can significantly provide a lot of capabilities compared with previous networks to effectively adopt IOT in healthcare. It can enhance the reliability of wireless connections, provide real-time data transmission, real-time monitoring, long-distance coverage, and strong security. 5G can reduce latency, which is the time required for data to be transmitted from point to point, to less than 5 ms, which will enhance the efficiency of the network and enable the availability of data in real-time. Moreover, it will provide a large bandwidth (more spectrum), which means faster throughput (faster transfer data) [20, 21].

7 Conclusion

This research investigated and obtained a better understanding of the current situation of IOT adoption in the Saudi Arabian healthcare sector in Riyadh and whether the COVID-19 pandemic affected its adoption. Moreover, this research was interested in finding the obstacles that prevent the Saudi healthcare sector from adopting IOT technology and the desired benefits from applying it. It also shed light on the usefulness of IOT in confronting COVID-19 and the proposed solutions to face it.

It can be concluded that IOT adoption within the Saudi Healthcare sector is still at infancy but there is clear initiative from Healthcare leaders to take advantage of the value it can provide. Therefore, this research should serve as a good starting point for further research on a greater scale. Because of limited time, the shortage of expert professionals in IOT adoption in Saudi healthcare, and access to these professionals, future research should cover a larger sample size of hospitals spreading across the country. Moreover, the Ministry of Health (MOH) should champion an IOT initiative and find solutions on how it can expedite IOT adoption across the Kingdom.

8 Managerial and Theoretical Implication

8.1 Managerial Implication

Research proved that IOT can help in managing healthcare operations and reduce chronic disease. It also shows that implementing IOT properly in healthcare will minimize the chance of errors, provide excellent treatment, minimize operation cost, and improve patients care and diagnosis. Moreover, many critical problems like patient overload, safety of healthcare workers, and healthcare resource shortage can be solved by adopting IOT in healthcare. For example, IOT can help healthcare provider remotely monitor and track patient's health and make an alert if unusual changes happened. It also can reduce operation cost by managing hospital utility cost to use it only when needed.

The results indicate that IOT within Saudi Healthcare still in infancy but there is a clear initiative from healthcare leaders to take advantage of the value IOT can provide. The results also confirmed that there are some emerging IOT applications however, they are done in closed systems. The Ministry of Health (MOH) prevent sharing information over the internet or store it in outside cloud for security and privacy concern. Therefore, the ministry of health can use international standard and regulation to make sure patient's information are protected. They also may use a framework that is developed by the Communication and information technology Commission (CITC) which is called Internet of Things (IoT) regulatory framework to ensure patient's privacy and security.

The results show that medium to small hospitals in Saudi Arabia didn't adopt IOT because of insufficient infrastructure and resistance to change. Therefore, lunching a demo (proof of concept (POC)) and conducting education and training programs may raise the awareness of the importance of IOT to the hospitals and the value it can add.

Regarding security issues, using 5G mobile network can solve many challenges in IOT application and boost its adoption. It can enhance the reliability of wireless connections, provide real-time data transmission, real-time monitoring, long-distance coverage, and strong security.

This paper should serve as a starting point for more future studies about IOT in Saudi healthcare and cover a larger sample size including medium and small hospitals spreading across the country.

8.2 Theoretical Implication

Based on the research [22], it can be said that covid-19 accelerate the adoption of digital technology in the healthcare sector. In contrast, this paper shows that COVID-19 can be seen as an enabler to promote IOT adoption in Saudi healthcare because IOT can help in effectively reaching the patient and deliver quality healthcare remotely without human interaction. Moreover, it is crucial to have a central datacenter for storing all EHRs and sharing information among hospitals and over the Internet to successfully adopt IOT in Saud healthcare. Research [12] suggested that having a centralized IOT system built by the government and integrates telemedicine benefits will improve the system and make it suitable not just for the COVID-19 crisis but also for future situations.

References

1. Mohammed, M.N., Desyansah, S.F., Al-Zubaidi, S., Yusuf, E.: An internet of things-based smart homes and healthcare monitoring and management system: review. In: Journal of Physics: Conference Series, vol. 1450 (2019)
2. Schweizer, E.: How IoT delivers better patient care and optimizes healthcare facility operation. In: Schneider Electric White Paper (2017)
3. Dahri, A.S., Massam, S,R, Thebo, L.A.: An overview of AI enabled M-IOT wearable technology and its effects on the conduct of medical professionals in public healthcare in Pakistan. In: 3C AQ4 Tecnología. Glosas de innovación aplicadas a la pyme, pp. 87–111, June 2020
4. AbdElnapi, N., Omran, N., Ali, A. Omara, F.: A survey of Internet of Things Technologies and Projects for Healthcare Services, pp. 48–55 (2018). https://doi.org/10.1109/ITCE.2018.8316599
5. Masmali, F., Miah, S.: Adoption of IoT based innovations for healthcare service delivery in Saudi Arabia (2019). https://doi.org/10.1109/CSDE48274.2019.9162390
6. Williams, R.: Arab Health, Dubai, rep. (2020)
7. Karahoca, A., Karahoca, D., Aksöz, M.: Examining intention to adopt to internet of things in healthcare technology products. Kybernetes **47**(4), 742–770 (2018). https://doi.org/10.1108/K-02-2017-0045
8. Yacchirema, D., de Puga, J.S., Palau, C., Esteve, M.: Fall detection system for elderly people using IoT and Big Data. Procedia Comput. Sci. **130**, 603–610 (2018)
9. Singh, R.P., Javaid, M., Haleem, A., Suman, R.: Internet of things (IoT) applications to fight against COVID-19. Diabetes Metab. Syndr. **14**(4), 521–524 (2020)
10. DeFranco, T., Roschelle, J.: Detecting/preventing infections, and moving instruction online. Commun. ACM **63**, 8–9 (2020)
11. Ben Hassen, H., Ayari, N., Hamdi, B.: A home hospitalization system based on the Internet of things, Fog computing and cloud computing. Inform. Med. Unlocked **20**, 100368 (2020)
12. Cacovean, D.Q., Ioana, I.U.Q., Nitulescu, G.U.O.: IoT system in diagnosis of Covid-19 patients. Informatica Economica **24**, 75–89 (2020)
13. Kumar, K., Kumar, N., Shah, R.: Role of IoT to avoid spreading of COVID-19. Int. J. Intell. Netw. **1**, 32–35 (2020)
14. Otooma, M., Otoumb, N., Alzubaidia, M.A., Etoomcde, Y., Banihanicf, R.: An IoT based framework for early identification and monitoring of covid 19 cases. Biomed. Sig. Process. Control **62**, 102149 (2020)
15. Abdulrazaq, M., et al.: 2019 novel coronavirus disease (Covid-19): detection and diagnosis system using IoT based smart glasses. Int. J. Adv. Sci. Technol. **29**, 954–960 (2020)
16. ETHealthWorld: As Covid-19 hounds the world, AI and IoT lead the charge in Healthcare 4.0. ETHealthworld.com. https://health.economictimes.indiatimes.com/news/health-it/as-covid-19-hounds-the-world-ai-and-iot-lead-the-charge-in-healthcare-4-0/77349407. Accessed 18 Oct 2020
17. Choudhary, M.: How IoT can help fight COVID-19 battle. Geospatial World. https://www.geospatialworld.net/blogs/how-iot-can-help-fight-covid-19-battle/. Accessed 18 Oct 2020

18. Mishra, S.: The growing role of IoT in COVID-19 response. IoT for All. https://www.iotforall.com/the-growing-role-of-iot-in-covid-19-response. Accessed 18 Oct 2020
19. Alghamdi, S.M., Alqahtani, J.S., Aldhahir, A.M.: Current status of telehealth in Saudi. J. Fam. Community Med. **27**(3), 208–211 (2020)
20. Varga, P., et al.: 5G support for industrial IoT applications, challenges, solutions, and research gaps. Sensors **20**(3), 828 (2020)
21. Bai, L., et al.: Chinese experts' consensus on the Internet of Things-aided diagnosis and treatment of coronavirus disease 2019. Clinical eHealth **3**, 7–15 (2020). https://doi.org/10.1016/j.ceh.2020.03.001
22. Clipper, B.: The influence of the COVID-19 pandemic on technology: adoption in health care. Nurse Lead. **18**(5), 500–503 (2020)

Check for updates

Determinants of Acceptance of m-payment in Morocco

Dounia Dahab and Nada Soudi(✉)

Institut Supérieur de Commerce et d'Administration des Entreprises (ISCAE), Casablanca, Morocco
{ddahab,nsoudi}@groupeiscae.ma

Abstract. As the first m-payment services are launched in Morocco, banks and payment institutions are wondering about the acceptance of these new payment solutions. This study focuses on the identification of the determinants of m-payment acceptance among Moroccan citizens.

The literature review highlighted several models with the main determinants in different countries. The study was developed in collaboration with a Moroccan bank, the data were collected through a questionnaire administered either face to face and on the Internet from February to May 2019 before the launching of m-payment's solutions. The field study were conducted in Morocco on 288 individuals who showed individual experience with bank services, Innovativeness, perceived risk, usefulness and ease of use are key determinants in the individual intention to use m-payment.

The Data were analyzed using SmartPLS. Experience was tested as antecedents of perceived usefulness, while innovativeness and perceived risk as antecedents of perceived ease of use. Thus, perceived usefulness was verified as a mediating variable of the relationship of experience with the intention to use m-payment. However, the role of ease of use has not been verified. Our research also focuses on the role of certain antecedents to beliefs, namely: experience with payment methods, innovativeness and perceived risk

Keywords: Mobile payment · Innovativeness · Perceived risk · TAM · PLS

1 Introduction

Cell phones first allowed the communication through voice, text, and then evolve into gaming, photos, videos, browse the web and geolocalisation. Today, in many countries, the mobile phone enables several transactions, from the simple payment of a transport ticket to other more complex forms. These transactions through the mobile phone are called Mobile Payment or m-payment. Recent research published by Allied Market Research (AMR[1]) predicts that the global m-payment market could reach $3.388 billion by 2022 with a global growth rate of 33.4% per year. Mobility, high penetration rate of smartphones, changes in consumption habits and the need for immediacy in transactions

[1] https://www.alliedmarketresearch.com/mobile-payments-market.

M. Al-Emran et al. (Eds.): ICETIS 2021, LNNS 299, pp. 484–497, 2022.
https://doi.org/10.1007/978-3-030-82616-1_41

are perceived as the main determining factors in the growth of this market. Today, mobile payment solutions are reaching people on all continents, however adoption rates vary from country to another. In developing countries specially in Africa, the Kenyan model represents a successful case study as 75% of payments are made through the mobile phone.[2] In these countries, the most popular models are the models offered by telecom operators as is the case in Kenya where the M-PESA system offered by Safaricom is very successful with more than 17 million users.

However, one of the essential brakes of the weak development of m-payment in Africa remains end user behavior, which is cited in a report on payments innovation as the first factor inhibiting the take-off of electronic money in Africa before capacity to pay or regulation.[3] However, the perceived ease of using mobile phones even for illiterate people could be a factor for success.

In Asia, the mobile wallet has become a disruptive innovation in several sectors, particularly in the Malaysian hotel industry. Indeed, a recent study has shown that the intention to adopt the mobile wallet is important (the borrowed research model, Extended TAM, was able to explain 61% of the variance in behavioral intention to adopt this service [1]). Thus, mobile payment has also had a significant impact on many sectors of the commercial sphere, particularly the retail sector in Malaysia. A study conducted on the intention to adopt m-payment, using the QR code technique, through the Technology Acceptance Model showed that the intention to adopt m-payment is influenced by two factors which are user optimism and the perceived usefulness of mobile payment [2]. Another study of the antecedents that affect users' behavioral intent to use mobile payment revealed, through an extended TAM, several determinants (utility, ease of use, perceived aesthetics, availability) important to the adoption of this technology [3] in Malaysia.

In Morocco, a banking law that entered into application in June 2016 offers a more flexible legal framework for the launch of m-payment offers. The first solutions of m-payment were then launched.

The high rate of use of mobile phones (mobile phone penetration rate of 130%) and the low rate of banking of Moroccans (estimated at 30%) encourage banks and payment institutions to launch m-payment solutions. They would then constitute a factor to reduce the circulation of cash. However, there is still some uncertainty about the acceptance of this service by the end user. The issue of accepting new payment methods has been the subject of numerous studies without reaching a consensus on the factors of acceptance. Regarding the specific case of mobile payment, most previous research has been conducted mainly on non-Arab samples and in developed countries [4]. Moreover, although these new payment methods have been available earlier, their usage rates remain low even in some pioneering countries such as Finland.

Thanks to open access in general and voluntary use by users, they do not always meet with the expected success and their levels of acceptance vary according to different countries [5, 6]. The study of user behavior is necessary to offer mobile payment solutions

[2] According to a KPMG report published in 2015 available at https://assets.kpmg.com/content/dam/kpmg/za/pdf/2016/09/Payment-Developments-in-Africa-2015.pdf.

[3] According to the Innovation Payments jury report available at http://www.paymentscardsandmobile.com/innovation-essential-currency-payments-africa/.

that meet their expectations [7, 8]. This research aims to better understand the behavior of the Moroccan user of m-payment. It is particularly interested in the role that variables such as experience, innovativeness, and perceived risk play in determining end user beliefs and intention to use this service. As the use of cash is a strong transaction mode in Morocco, m-payment would be a means of reducing the circulation of cash and more traceability of payments. In the following sections, will be studied respectively the theoretical and conceptual framework of the study, the context and methodology adopted and the main results of the research.

2 Theoretical and Conceptual Framework

First of all, m-payment should be distinguished from M banking: M banking refers to access via the Internet to banking services offered by a bank. The user can then carry out transactions [9] from his mobile or accede to other financial services (savings, credit), or even limit himself to the simple consultation of his accounts [5]. M-payment can be defined as any form of payment that can be made through a mobile terminal whether it involves a banking operator or not [10]. [11] define mobile payment as a transfer of funds in exchange for a good or service, where the cell phone is involved in both triggering and confirming the payment.

In this study we refer to m-payment as "the fact of using mobile phone to complete a transaction". This positioning joins many studies of the last decade on the acceptance of m-payment [12]. Several articles aimed to list and classify published m-payment studies [13, 14]. Others addressed the determinants of the adoption of this service. For example, [15] studied the influence of habits on Mobile Payment Acceptance. With the increase in the use of various mobile devices, mobile payments have become a crucial driver for commerce success. However, the percentage of consumers who use or continue using mobile payments in the US is low. This study adopts information technology (IT) ecosystem view and transfer of learning theory and explores the effects of five types of technology use habits on consumers' intention to continue using mobile payments. Results indicate that consumers' online shopping, mobile service use, and cell phone use habits have a positive relationship with their mobile payment use habit, positively affecting their intention to continue using mobile payments. [16] studied the disruptive technologies and abundance in the service sector. The main highlights are:

- Technology-specific characteristics influence mobile payment intention.
- Perceived personal innovativeness is related to behavioral intention.
- Intention to use mobile payment is explained by perceived social influence.
- Perceived risk reduces individuals' intention to use mobile payment.
- Perceived costs mitigate the positive impact of several characteristics.

A study [17] synthesized the main contextual variables impacting the adoption of mobile payments. These variables have been broken down into adoption factors and barriers in a developing country context. The former, who are the facilitators, have a positive influence on adoption or use, while the obstacles have a negative influence. [17] divided the contextual factors into two categories: Environmental factors (such

as nation-specific regulatory factors or region-specific tariff advantages) and cultural factors (the collective beliefs of members of a society or nation. The vast majority of research conducted have mobilized behaviorist theories. Nevertheless, the Technology Acceptance Model (TAM) is proving to be the most widely used and strongest model to explain m-payment acceptance [6]. **The TAM proved his efficiency over the past two decades, in the field of work on "acceptance", it was one of the most used** [5, 6], and above all, the most powerful in explaining the use of different ICTs [14]. Indeed, TAM has been used in many countries to explain different ways to use m-payment. [14] listed a large number of studies using behavioral theories to explain the adoption of Mobile banking and/or Mobile payment. Our summary (see Table 1) concerns the work that mobilized the TAM to explain the adoption of m-payment and validated it empirically. This work is presented below.

Table 1. Results of studies using TAM to explain the acceptance of m-payment (adapted from Slade et al. 2014)

Source	Country	Results
Hanudin (2008) [18]	Malaysia	Explain 63,8% of intention of adoption of m-payment
Chandra et al. (2010) [19]	Singapore	Explain 53% of intention of adoption of m-payment
Chen (2008) [20]	Unites States	Show only model suitability
Cheong et al. (2004) [21]	Korea	Explain 55,1% of intention of adoption of m-payment
Dastan and Gurler (20160 [22]	Turkey	Explain 62% of intention of adoption of m-payment
Goeke and Pousttchi (2010) [23]	Germany	Explain 75,7% of intention of adoption of m-payment
Kim et al. (2010) [24]	Korea	Explain 76% of intention of adoption of m-payment
Mallat et al. (2009) [25]	Finland	Explain 55% of intention of adoption of m-payment
Phonthanukitithaworn et al. (2015) [26]	Thailand	The model explains 63% of variance of the intention to adopt m-payment
Schierz et al. (2010) [8]	Germany	Explain 84% of intention of adoption of m-payment
Shin (2010) [27]	USA	Explains 72% of the variance in intention and 81% of use effective du m-payment

(*continued*)

Table 1. (*continued*)

Source	Country	Results
Liébana-Cabanillas et al. (2015) [28]	Spain	The model explains respectively 55.7% of the variance and 65.4% of the variance depending on the technology (NFC or SMS) used to make the m-payment
Pal et al. (2019) [17]	India	Correlations confirmed between environmental and cultural factors
Jia et al. (2020) [15]	China	Correlations confirmed between habits, mobile payment use habit, and intention to continue using
Schmidthuber et al. (2020) [16]	Europe	Highlighted determinants are: Technology-specific characteristics, personal innovativeness, intention to use mobile payment, perceived risk and costs
Lew et al. (2020) [1]	Malaysia	The extended model was able to explain 61% of the variance in behavioural intention to adopt mobile wallet
Lee et al. (2020) [3]	China	Perceived aesthetics, technology readiness, mobile usefulness, and mobile ease of use impact behavioral intention
Yan et al. (2020) [2]	Malaysia	Mobile usefulness and optimism have direct impacts on intention to adopt (Yan et al. 2020)

TAM explains user behavior based on two basic beliefs, perceived usefulness and ease of use of ICT [29]. These factors would play a key role in the acceptance by the end user of any ICT. Furthermore, TAM suggests, that perceived usefulness as well as perceived ease of use can be affected by many external variables.

2.1 The Determining Role of Usefulness and Perceived Ease of Use

Perceived usefulness is defined by [29] as "the degree of thinking that usage of a system improves the performance at work". Perceived usefulness can also be defined in the context of m-payment as the belief that m-payment would have added value for the individual [30]. Perceived ease of use refers to "the degree of a person believes that using a system does not require effort". This definition suggests that if m-payment is perceived to be easier to use, it will have a better chance of being accepted by users. Several research recognize the central role of perceived ease of use and perceived usefulness in the acceptance of m-payment, even if they do not always verify their positive influence on the adoption of the service. [20] in a study of 299 respondents demonstrates the direct link between UP and FUP and the intention to use m-payment as well as the subjective norm and user mobility. [24] conducted a study in schools, universities, businesses, research institutes and internet cafes among individuals who have used m-payment. They analyze 269 responses and verify the existence of a significant link between perceived ease and usefulness and adoption of m-payment. In their review of the literature on adoption of m-payment, [14] reveal that, respectively, perceived usefulness and perceived ease of use, lead the constructs that have shown positive relationships with the intention to adopt m-payment.

Few studies do not find a significant relationship between perceived ease of use, perceived usefulness, and intention to adopt m-payment. This is the case for the study conducted in Turkey by [22] which analyzed the responses of 225 individuals regarding their intention to adopt m-payment. [26] in their study of 256 respondents in Thailand were surprised about the results. They don't find a significant relationship between perceived ease of use, perceived usefulness and intention to use m-payment. The literature review leads us to formulate the following two hypotheses:

H1: Perceived usefulness positively influences the acceptance of m-payment

H2: Perceived ease of use positively influences the acceptance of m-payment

2.2 The Antecedents of Beliefs

As discussed above, perceived usefulness and perceived ease of use of an ICT can be influenced by many external variables. Some of the variables that have been found to be critical in the intention to use are individual experience, innovativeness, and perceived risk.

Individual Experience

According to the theory of [31], the habitual nature of a behavior influences an individual's response to a given situation. The author says that for many behaviors, habits are more important than intention. In ICT, habits are generally measured by user experience. [32] argues that user experience is a factor that would reduce the level of perceived risk of an electronic purchase, for example. The results regarding the influence of habits on online shopping intentions are not always obvious. In a comparison of two samples of students in France and in Hong Kong, [33] fail to demonstrate the link between user experience and purchase intention. In addition, users can generalize the experience related to a service provided to all others services offered. The link between experience and perceived usefulness of a technology has been demonstrated in numerous studies [34–36]. There may also be a contagion effect between technologies or services perceived as similar. [37] find that the user's experience and training have "no direct effect" on the intention to use of ICT but only "indirect effect" by acting on the user's beliefs. The results of this research also show that the user experience positively influences the perception of the usability of IT but does not influence the perception of its usefulness. In our study, we will focus on the influence of user experience on their intention to use m-payment. The assumptions are as follows:

H3a: Individual's experience with payment methods positively influences perceived usefulness of m-payment

H3b: Individual's experience with payment methods positively influences perceived ease of use of m-payment.

Innovativeness

Innovativeness can be defined as the speed degree of adoption of an innovation by an individual compared to other members of his social [38]. [37] validated the construct innovativeness in the particular context of ICT and named it (PIIT) 'personal innovativeness in the domain of IT'. Innovativeness is thus defined as "the willingness of an individual to try a new technology". It is a characteristic specific to a given individual like any other personality trait. [39] find that innovativeness has a significant impact on the intention to use location-based services (LBS). Innovativeness is a direct determinant of perceived usefulness and perceived ease of use. In the context of m-commerce and m-payment, the supposed effect of innovativeness on the intention to use m-payment has been demonstrated. [40] found that innovativeness was the most significant factor in the intention to use m-payment. We therefore propose that:

H4a: Innovativeness of positively influences perceived usefulness of m-payment

H4b: Innovativeness positively influences perceived ease of use of m-payment.

Perceived Risk

The perception of the risk associated with the use of technology is mentioned by many authors [8, 20, 41, 42] as an essential determinant of the acceptance of m-payment. However, the relationship between the perceived risk of the system and the intention to adopt m-payment has not been consistently demonstrated. Several authors have noted that due to the virtual aspect of mobile payment systems, risks and uncertainties are accentuated for the user who perceives a lack of control over the technology [43], which negatively impacts his confidence in this payment method. Users could also fear the intervention of hackers and the use of their personal payment data.

The higher the perceived risk associated with using m-payment, the less likely users will be to use m-payment. In a study on the adoption of m-payment [8, 20], demonstrate that perceived risk has a negative effect on the intention to adopt M-Payment. They suggest that users are less motivated to use m-payment when they consider the risk associated with their use is bigger than the usage of more common payment methods. Other authors also show that the perception of risk discourages users from adopting m-payment [44, 45]. In Moroccan context, users more accustomed to use cash payment might consider this new payment method to be more uncertain and therefore associate it with a higher level of risk. This leads us to formulate the following hypothesis:

H5a: the perceived risk of using m-payment negatively influences the perceived usefulness

H5b: the perceived risk of using m-payment negatively influences the perceived ease of use

2.3 M-Payment Acceptance

In TAM, technology acceptance is measured through intent to use. In the context of this study, the dependent variable is the intention to use mobile payment. Intention is usually measured by items such as: if I had the chance, I would use…., I think I would use, etc. The actual use of m-payment could not be measured since the service was not experienced yet.

3 Context and Methodology

3.1 Data Collection Context

In Morocco, the first m-payment solutions were launched by Telecom operators in 2010, Mobicash for Maroc Telecom and Medicash by Meditel (now Orange). Both of these products were unsuccessful. One of the reasons often given to explain this failure is the necessary route through banks to carry out the transactions. Indeed, in 2010, banks were the only institutions approved to make payments, which made the solution difficult to launch and decrease the interest of going through telecom (since in all cases the passage through the bank was mandatory). In other words, unbanked users ultimately had to open a bank account in order to access the m-payment service. This was a first break for the usage of the service. A new banking law (law 103–12 called "New banking law") promulgated in 2004 and entered into application in June 2016 offered a more flexible legal framework for telecom operators and Fintechs[4] for the launch of m-payment solutions. Since then, communication actions claimed the first m-payment solutions, then the contactless bank cards using NFC technology. Apart from the legal and regulatory framework, the high rate of mobile phones penetration and the very low rate of banking (not exceeding 40%) are two determining factors in the development of m-payment in Morocco. Although a majority of Moroccans do not have a bank account, practically all of them are equipped with mobile phones. Note that the penetration rate of mobile telephony has continued to increase in recent years to reach 130%[5] today. The rate of equipment in smartphones of the population exceeds 70%. M-payment would allow a large part of the Moroccan population to enter into the financial system. This study were developed in collaboration with a Moroccan bank which wanted to know the behavioral intentions of Moroccans regarding m-payment in order to launch its m-payment service. It followed a qualitative study through focus groups.

3.2 Data Collection

The data were collected through a questionnaire administered either face to face or on the Internet (Facebook page was shared with the respondents). Data collection spanned from February to May 2019 before m-payment's solutions in their current format were launched in Morocco. It mainly aimed to find out the intentions to use the service. Respondents were distributed throughout Morocco with a majority of respondents located in the region of Casablanca (68.1% of respondents) and the others distributed between Rabat, Fes and Marrakech. Incomplete questionnaires were eliminated. 288 valid questionnaires were thus analyzed.

The first part of the questionnaire was dedicated to general questions on the name of the bank, the telephone operator, the second part included questions relating to experience, perceived risk, innovativeness and the intention to use m-payment, then the

[4] Startups operating in Finance.

[5] According to National agency of reglementation of telecoms.

questions relating to the respondent's informations. The concepts were measured on the basis of scales adapted for most Anglo-Saxon research translated and adapted to the context of m-payment in Morocco. These are Likert scales ranging from 1 to 5 measuring the degree of agreement with respect to propositions except for the perceived risk scale which was a scale with semantic supports ranging from very low to very high. Their psychometric qualities were checked and found to be satisfactory. The items making up these scales are presented in Appendix 1. Among the 288 respondents, 145 are women (50.3%) and 143 are men (49.7%). 2.8% are under 18, 26.4% between 18 and 22, 20.5% between 23 and 29, 18.4% between 30 and 39, 13.5% between 40 and 49 and 17.4% were over 50 years old. Regarding the income of respondents, 18.1% have a monthly income below 2000 MAD, 16.7% between 2001 and 5000 MAD, 20.8% between 5001 and 10,000, 19.8% between 10,001 and 15,000 and 21.5% have an income of more than 15,000 MAD. The participants in the study are 32.8% of executives, 18.4% of employees, 27.5% of students, 3.1% are inactive and 1.9% of small business owners.

4 Results Analysis

4.1 Reliability, Validity and Adjustments

The partial least squares (PLS) method was adopted because it does not require a multinormal distribution of variables and is not sensitive to multi-collinearity issues [46]. Moreover, we preceded to a bootstrap of 500 iterations to ensure the stability of the model [47–49].

Table 2. Reliability and validity of measurements

Latent variable	Convergent validity (AVE[1])	Reliability (Rho of DG)
Perceived usefulness	0,577	0,822
Perceived ease of use	0,644	0,865
Expérience with payment methods	0,401	0,645
Innovativness	0,497	0,718
Perceived risk	0,354	0,600

First, we looked at the validation of measurement scales. As shown in Table 2, the reliability indicator is very satisfactory (Rho of DG greater than 0.6), while the convergent validity indicator is moderately satisfactory for certain variables for which the AVE is equal to or less than 0,4. The reliability and validity of the measure of intention to use m-payment could not be studied since it is a single item scale.

To check the discriminant validity, we checked that the HTMT (heterotrait-monotrait) criterion is much less than 0.85. The discriminant validity is well verified.

Secondly, we verified the quality of the overall model with the SRMR (Standardized Root Mean Square Residual) (Henseler and Sarstedt 2013). The indicator of 0.126 leads us to conclude that the quality of the proposed model is satisfactory. Thus, we can appreciate the structural model presented in Fig. 1, and look at the reading of the results. To accept or refute a hypothesis, we rely on the significant relationships provided by the SmartPLS software, that is, when reading the structural links (path coefficients) significant at 5%.

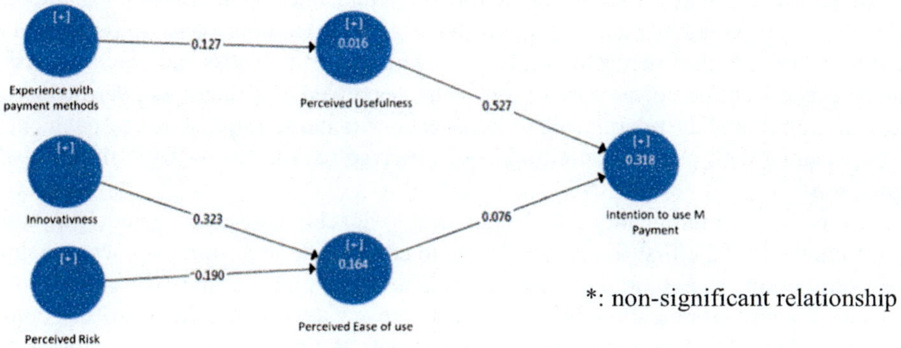

Fig. 1. PLS structural model

4.2 Hypothesis Tests

First, we tested at the role of usefulness and perceived ease of use in explaining the intention to use m-payment. This approach will allow to test hypotheses H1 and H2, based on the path coefficients provided by SmartPLS. The results shows that the perceived usefulness is indeed an antecedent of the intention to use the m-payment (path coefficient: +0.527). The H1 hypothesis is then confirmed and this result approves the studies by [14, 22, 24]. Tention to use m-payment. Hypothesis H2 is then rejected. This result was unpredictable. It could be explained by the fact that the intention to use the m-payment is a single-item variable. Second, we studied the link of each antecedent with usefulness and ease of use. Thus, the user's experience in using various payment methods influences their intention to use m-payment (path coefficient: +0.127). The H3a hypothesis is retained while the H3b hypothesis is rejected. As a reminder, it has been shown in numerous studies that there is a link between experience and perceived usefulness of technology [34–36]. Regarding the effects of innovativeness, our research confirms the existence of a significant association with ease of use (path coefficients of +0.323). Furthermore, there is no significant link with perceived usefulness. Hypothesis H4b is then confirmed and hypothesis H4a rejected. This result supports the research [51] who suggest the role of innovativeness in explaining beliefs in the ease of use of mobile wireless networks. Finally, the perceived risk negatively influences the perceived ease (path coefficient; −0.190). The higher the perceived risk associated with using m-payment, the less users perceive the ease of using this payment method. Furthermore, no significant relationship emerges with perceived usefulness. Hypothesis H5b is confirmed and hypothesis H5a rejected. Moreover, by studying the low contributions to R^2 (1.6% for perceived usefulness, 16.4% for ease of use), we can highlight the absence in the model studied, of other antecedents with significant explanatory power.

5 Conclusion

The results of the study have important implications. First, we focused on the antecedents of intending to use m-payment. Our research confirms the role of perceived usefulness in explaining the intention to use m-payment. However, the role of ease of use has not been verified. Our research also focuses on the role of certain antecedents to beliefs, namely: experience with payment methods, innovativeness and perceived risk. Experience is tested as a antecedent of perceived usefulness, while innovativeness and perceived risk as an antecedents of perceived ease of use. Thus, perceived usefulness was verified as a mediating variable of the relationship of experience with the intention to use m-payment:

Experience with payment methods → Perceived usefulness → Intention to use m-payment

As a managerial implication, the banks need to develop consumer experiences and to push clients for "the first time usage" even to collaborate with stores to have special offers, promotions and communication campaigns to enhance the trials.

As a theoretical implication, this study adapt the framework into the Moroccan context and test hypothesis according to multiple models. A first extension of this research will therefore consist in studying the impact of other individual variables (age, level of education, culture, resistance to innovation, satisfaction, enthusiasm, involvement, etc.), the impact of variables linked to the provider. Service (trust in the telephone operator, trust in the bank, diversity of services, quality of service, etc.), or variables linked to the technology itself (trust in technology, risk linked to data confidentiality, cost, etc.). Moderating variables, such as "Age" or "Level of education" on intentions to adopt m-payment could also be tested. Secondly, as the first m-payment services will be offered by payment institutions, it will be interesting to test a model that includes as a dependent variable not the user's intention to adopt m-payment but its current use (number of services paid using m-payment, variety of services, amounts spent, etc.). Another line of research will consist in carrying out a comparative study between a banked population and an unbanked population in Morocco as to its intention to use m-payment. It would indeed be interesting to verify whether the behavioral influence variables such as perceived usefulness, perceived ease of use, compatibility, experience with electronic banking, etc. impact the two populations of the same way. These results would be of great use for the marketing campaigns of future providers of this service in order to adapt their offers to both segments. In future research concerning m-payment, it would also be interesting to explore the impact of the type of technology offered by the service provider of m-payment (NFC, SMS or other) on user behavior in the continuity of the study conducted by [51].

Appendix 1. Items

PERCEIVED USEFULNESS (Adapted from Davis,1989)
It is necessary to use my mobile phone to make payments.
Using my cell phone to make payments would save time.
Using my cell phone to make payments would avoid using cash.
Using my cell phone to make payments would make many payments easier
Using my cell phone to make payments would help me a lot..
PERCEIVED EASE OF USE (Adapted from Davis,1989)
Learning to use my cell phone to make payments seems easy.
I find it easy to handle my cell phone for different tasks.
My interaction with my cell phone is clear and understandable.
I find it easy to get good at using my cell phone to make payments.
Overall, I find it easy to use my mobile phone to make payments
INNOVATIVNESS (Adapted from Mao & al.2005)
When I hear about a new technology, I usually find a way to try it.
In my entourage, I am generally one of the first people to try new technologies
In general, I hesitate before using a new technology
In general, I like to try new technologies
PERCEIVED RISK (Adapted from Phonthanukitithaworn, C. et al., 2015)
The level of confidentiality of information passing through the mobile phone is: □ Very low □ Rather low □ Medium □ Rather high □ Very high
The level of reliability (security) of the mobile telephone network for making payments is:
The probability of making mistakes while making a bill payment with my mobile phone is:
The probability that the phone battery is discharged or that the connection is interrupted when paying with my mobile phone is:
The risk associated with using the mobile phone to make payments is:
INTENTION (Adapted from Kim et al., 2010)
If I had the option, I would use my cell phone to make payments: □ Most definitely □ Probably □ Maybe yes / no □ Probably not □ Certainly not

References

1. Lew, S., Tan, G.W.H., Loh, X.M., Hew, J.J., Ooi, K.B.: The disruptive mobile wallet in the hospitality industry: an extended mobile technology acceptance model. Technol. Soc. **63**, 101430 (2020)
2. Yan, L.Y., Tan, G.W.H., Loh, X.M., Hew, J.J., Ooi, K.B.: QR code and mobile payment: the disruptive forces in retail. J. Retail. Consum. Serv. **58**, 102300 (2020)
3. Lee, V.H., Hew, J.J., Leong, L.Y., Tan, G.W.H., Ooi, K.B.: Wearable payment: a deep learning-based dual-stage SEM-ANN analysis. Expert Syst. Appl. **157**, 113477 (2020)
4. Ramadan, R., Aita, J.: A model of mobile payment usage among Arab consumers. Int. J. Bank Market. **36**(7), 1213–1234 (2018)
5. Chaix, L.: Le paiement mobile: perspectives économiques, modèles d'affaires et enjeux concurrentiels. Thèse de Doctorat en Sciences économiques, Université de Nice Sophia Antipolis (2013)
6. Shaikh, A.A., Karjaluoto, H.: Mobile banking adoption: a literature review. Telematics Inform. **32**(1), 129–142 (2014)
7. Mallat, N.: Exploring consumer adoption of mobile payments - a qualitative study. J. Strateg. Inf. Syst. **16**(4), 413–432 (2007)
8. Schierz, O., Schilke, O., Wirtz, B.: Understanding consumer acceptance of mobile payment services: an empirical analysis. Electron. Commer. Res. Appl. **9**(3), 209–216 (2010)

9. Varshney, G., Madan, P.: A study of functionality dilemma and barriers to optimal usage of m-commerce. Curie J. **3**(1), 60–73 (2010)
10. Kadhiwal, S., Zulfiquar, A.U.S.: Analysis of mobile payment security measures and different standards. Comput. Fraud Secur. **2007**(6), 12–16 (2007)
11. De Bel, J., Gaza, M.: Mobile payments 2012. In: Director (2011). http://www.mobiltarca.com/media/documents/mobey-forum-mobile-payments
12. Al-Saedi, K., Al-Emran, M., Ramayah, T., Abusham, E.: Developing a general extended UTAUT model for m-payment adoption. Technol. Soc. **62**, 101293 (2020)
13. Diniz, A., Albuquerque, J.P., Cernev, A.: Mobile money and payment: a literature review based on academic and practitioner-oriented publications. In: Proceedings of SIG Glob Dev Fourth Annual Workshop, Shanghai, China (2011)
14. Slade, E.L., Dwivedi, Y.K., Williams, M.D.: Devising a research model to examine adoption of mobile payments: an extension of UTAUT2. Market. Rev. **14**(3), 311–336 (2014)
15. Jia, L., Song, X., Hall, D.: Influence of habits on mobile payment acceptance. Inf. Syst. Front. (2020). https://doi.org/10.1007/s10796-020-10077-6
16. Schmidthuber, L., Maresch, D., Ginner, M.: Disruptive technologies and abundance in the service sector - toward a refined technology acceptance model. Technol. Forecast. Soc. Change **155**, 119328 (2020)
17. Pal, A., De', R., Herath, T., Rao, H.R.: A review of contextual factors affecting mobile payment adoption and use. J. Bank. Fin. Technol. **3**(1), 43–57 (2019). https://doi.org/10.1007/s42786-018-00005-3
18. Hanudin, A.: Factors affecting the intentions of customers in Malaysia to use mobile phone credit cards. Manage. Res. News **31**(7), 493–503 (2018)
19. Chandra, S., Srivastava, S., Theng, Y.-L.: Evaluating the role of trust in consumer adoption of m-payment systems: an empirical analysis (Article 29). Commun. Assoc. Inf. Syst. **27**(1), 561 (2010)
20. Chen, L.-D.: A model of consumer acceptance of mobile payment. Int. J. Mobile Commun. **6**(1), 32–52 (2008)
21. Cheong, J., Park, M., Hwang, J.: Mobile payment adoption in Korea: switching from credit card. In: 15th Europe Regional Conference of the International Telecommunication Society, Berlin (2004)
22. Daştan, İ., Gürler, C.: Factors affecting the adoption of mobile payment systems: an empirical analysis. Emerging Market J. **6**(1) (2016). https://doi.org/10.5195/emaj.2016.95, http://emaj.pitt.edu. ISSN 2158-8708
23. Goeke, L., Pousttchi, K.: A scenario-based analysis of mobile payment acceptance. In: Mobile Business and 2010 Ninth Global Mobility Roundtable (ICMB-GMR), International Conference, pp. 371–378 (2010)
24. Kim, C., Mirusmonov, M., Lee, I.: An empirical examination of factors influencing the intention to use mobile payment. Comput. Hum. Behav. **26**(3), 310–322 (2010)
25. Mallat, N., Rossi, M., Tuunainen, V., Oorni, A.: The impact of use context on mobile services acceptance: the case of mobile ticketing. Inf. Manage. **46**(3), 190–195 (2009). https://doi.org/10.1016/j.im.2008.11.008
26. Phonthanukitithaworn, C, Sellito, C, Fong, M.: User intentions to adopt Mobile Payment services: a study of early adopters in Thailand. J. Internet Bank. Commer. **20**(1) (2015)
27. Shin, D.-H.: Modelling the interaction of users and mobile payment system: conceptual framework. Int. J. Hum. Comput. Interact. **26**(10), 917–940 (2010). https://doi.org/10.1080/10447318.2010.502098
28. Liébana-Cabanillas, F., Ramos de Luna, I., Montoro-Ríos, F.: Intention to use new mobile payment systems: a comparative analysis of SMS and NFC payments. Econ. Res.-Ekonomska Istraživanja **30**(1), 892–910 (2017)

29. Davis, F.D.: Perceived usefulness, perceived ease of use, and user acceptance of information technology. MIS Q. **13**(3), 329–340 (1989)
30. Tzou, R.C., Lu, H.P.: Exploring the emotional, aesthetic, and ergonomic facets of innovative product on fashion technology acceptance model. Behav. Inf. Technol. **28**(4), 311–322 (2009)
31. Triandis, H.C.: Attitude and Attitude Change. Wiley, New York (1971)
32. Nantel, J., Tellier, C.: L'achat par services télématisés: Qui en sont les utilisateurs et pourquoi. Technologies de l'information et Société **4**(3), 333–350 (1992)
33. Limayem, M., Rowe, F.: Comparaison des facteurs influençant les intentions d'achat à partir du lneb à hong kong et en france: influence sociale, risques et aversion pour la perte de contact. Revue Française du Marketing **209**, 25 (2006)
34. Taylor, S., Todd, P.: Assessing IT usage: the role of prior experience. MIS Q. **19**(4), 561–570 (1995)
35. Karahanna, E., Straub, D.W.: The psychological origins of perceived usefulness and ease of use. Inf. Manage. **35**(4), 237–250 (1999)
36. Venkatesh, V.: Determinants of perceived ease of use: integrating control, intrinsic motivation, and emotion into the technology acceptance model. Inf. Syst. Res. **11**(4), 342–365 (2000)
37. Agarwal, R., Prasad, J.: Are individual differences germane to the acceptance of new information technologies? Decis. Sci. **30**(2), 361–392 (1999)
38. Rogers, E.M., Shoemaker, F.F.: Communication of Innovations. The Free Press, New York (1971)
39. Gupta, S., Xu, H., Zhang, X.: Balancing privacy concerns in the adoption of location-based services: an empirical analysis. Int. J. Electron. Bus. **9**(1/2), 118–137 (2011)
40. Zarmpou, T., Saprikis, V., Markos, A., Vlachopoulou, M.: Modeling users' acceptance of mobile services. Electron. Commer. Res. **12**(2), 225–248 (2012)
41. Bauer, H., Reichardt, T., Barnes, S., Neumann, M.: Driving consumer acceptance of mobile marketing: a theoretical framework and empirical study. J. Electron. Commer. Res. **6**(3), 181 (2005)
42. Laforet, S., Li, X.: Consumers' attitudes towards online and mobile banking in China. Int. J. bank Market. **23**(5), 362–380 (2005)
43. Yan, H., Yang, Z.: Examining mobile payment user adoption from the perspective of trust. Int. J. U-and EService Sci. Technol. **8**(1), 117–130 (2015)
44. Pham, T.T., Ho, J.C.: The effects of product-related, personal-related factors and attractiveness of alternatives on consumer adoption of NFC-based mobile payments. Technol. Soc. **43**, 159–172 (2015)
45. Yang, Q., Pang, C., Liu, L., Yen, D.C., MichaelTarn, J.: Exploring consumer perceived risk and trust for online payments: an empirical study in China's younger generation. Comput. Hum. Behav. **50**, 9–24 (2015)
46. Chin, W.W.: How to write up and report PLS analyses. In: Handbook of Partial Least Squares, Concepts, Methods and Applications, pp. 655–690. Springer, Heidelberg (2010). https://doi.org/10.1007/978-3-540-32827-8_29
47. Chin, W.W.: The partial least squares approach to structural equation modeling. In: Marcoulides, G.A. (ed.) Modern Business Research Methods, pp. 295–336. Lawrence Erlbaum Associates, Mahwah (1998)
48. Efron, B., Tibshirani, R.J.: An Introduction to the Bootstrap. Chapman and Hall, London (1993)
49. Tenenhaus, M., Vinzi, V.E., Chatelin, Y.-M., Lauro, C.: PLS path modeling. Comput. Stat. Data Anal. **48**, 159–205 (2005)
50. Henseler, J., Sarstedt, M.: Goodness-of-fit indices for partial least squares path modeling. Comput. Stat. **28**, 565–580 (2013)
51. JuneLu, J.E., Yaob, C.-S.: Personal innovativeness, social influences and adoption of wireless Internet services via mobile technology. J. Strateg. Inf. Syst. **14**(3), 245–268 (2005)

Predicting Intention to Use Mobile Banking Services among Yemeni Banks' Clients: Is Perceived Value Important?

Ahmed M. Mutahar[1](✉), Adnan Aldholay[1], Osama Isaac[2], Abdullah Nabeel Jalal[1], and Sharaf Alkibsi[3]

[1] Faculty of Business Management and Professional Studies, Management and Science University, Shah Alam, Malaysia
ahmed_mutahar@msu.edu.my

[2] Guangdong University of Finance, Guangzhou, Guangdong Province, China

[3] School of Advanced Studies, University of Phoenix, Tempe, AZ, USA

Abstract. While the advancements in technology have led the providers of financial services to seek new mediums to offer banking services, the perception of their clients regarding their usefulness varies, especially with respect to mobile banking. Despite various factors have been identified to gain a strong comprehension of the mobile banking services usage, limited focus has been given to the what contributes to the value of engage in mobile banking services. This study aims to identify that significant success elements that influence the perceived significance of mobile banking use in Yemen by employing revised TAM. The mediating influence of perceived value and the intent to use mobile banking services are also explored in this study. This model was evaluated through survey data gathered from 482 validated responses from persons who did not use mobile banking. SEM via PLS version 3.0 software was used to identify the significance of the relationships among the investigated parameters. The developed model in this study explained 58% of variation in mobile banking's perceived value and 51% of the intention to utilise it as indicated by the model fitness with the data. Results from the multivariate analysis revealed that awareness (AW), perceived usefulness (PU), and perceived ease of use (PEOU) are significant predictors of perceived value, whereas perceived risk (PR) was identified to minimise customers' perception of the value. Moreover, perceived value determined the correlations among these parameters of PU, PEOU, AW, and PR as well as the intention. The outcomes of this research give understanding of the mobile banking strategies.

Keywords: Technology acceptance model · Perceived value · Perceived risk · Mediation · Behavioural intention

1 Introduction

Technology today has altered all aspects of human life [1]. There has been a drastic change in the nature of selling and buying banking services and the use of wireless

M. Al-Emran et al. (Eds.): ICETIS 2021, LNNS 299, pp. 498–520, 2022.
https://doi.org/10.1007/978-3-030-82616-1_42

devices has turned into one of the banking domain's newest mediums of providing their services as well. This development is moving customers from conventional service encounters to a directly accessible technology-based service, a mechanism which is useful to both the clients and the bank, as it is more convenient and reduces the transaction cost [2, 3]. Mobile banking enables fast, smooth, anywhere and anytime banking. It is a groundbreaking and novel technology that enables the customers to efficiently and independently carry out transactions like fund transfer, balance enquiry, bill payment, etc. [4].

The current global population is estimated to be 7.10 billion, with a mobile phone subscription of 6.835 billion and mobile banking users 0.8 billion in 2014 (expected to reach 1.8 billion by 2019) [5]. Despite the global spread of this technology, in both the developing and developed countries, the increase in the mobile banking adoption rate is still insufficient, with great possibility for growth [6].

Recent years have witnessed a consistent development in Yemen's mobile technology industry as evidence by the improving frequency of mobile phone use which scaled to approximately 70% by 2014 [7], instigating a competitive rivalry in the market. In view of this development, the banks of Yemen have given special attention to launch mobile banking services, with 10 banks out of 18 now offering mobile banking, but this mobile banking evolution is yet to reach the height the mobile technology use has scaled. Although, the Yemeni government has been encouraging businesses and workers to adopt mobile banking and electronic money transfers, two technologies which are considered to hold the most promise so far [8], the intention of Yemeni bank clients to take them up is still low. Data offered by the two largest financial institutions in Yemen (IBY and CAC banks) advocate that merely 27% of the customers have participated in mobile banking up to 2014 (Quality Assurance reports on CAC bank, and IBY bank, 2014). Financial institutions in Yemen have started to articulate their concern on the minimal adoption rate of mobile banking among customers. In an unstructured survey conducted in 2014, the president of CAC bank, Ahmed Abdullah, mentioned that mobile banking is offered to customers of the banks in Yemen, though the mobile banking adoption is lower in Yemen than other Arab nations and the world at large.

Several models and hypotheses have been presented and developed with respect to IS in order to determine and justify the behaviour of the users with regards to technology, including Diffusion of Innovation Theory (DOI) [9]. However, Technology Acceptance Model (TAM) [10] is the most commonly acknowledge and employed model among IS researchers to investigate the adoption of a system and users' intention [11]. Thus, this research uses TAM as a supporting concept. Although numerous scholars such as, [4] performed studies in different countries and determined that there has not been any research on the factors that impact the adoption of mobile banking services among Yemeni banking customers [12]. Limited literature of this issue necessitates this study in order to identify the causes that contribute to the low adoption rate of mobile service in Yemen. Hence, it is crucial to explore the factors that influence customers' intention to use mobile banking services. The current literature has observed that a majority of research in the information technology industry has concentrated on innovation attributes and user characteristics instead of concentrating customer value–driven qualities and their criticality in the adoption of e-services [13]. However, the majority of these studies was conducted in the conventional context using value as the key element to predict customer purchase decision-making [14].

Some earlier studies which used TAM have reported diverse results with respect to the effect of PEOU (perceived ease of use) and PU (perceived usefulness). Others have examined the significance of the correlation between intention, PU and PEOU [15, 16], while some of them also reported a trivial correlation among these parameters and behavioural intention [17–19]. These inconsistent results may indicate that other factors can intervene in the relationship, depending on the context and the application under examination. Therefore, in order for convenience and efficiency parameters to positively affect intention of using mobile banking, they are first required to go through the mediating parameter that moderates the correlation between PEOU, PU and behavioural intention.

Those studies which applied TAM, ignore the role of any essential factor that may predict, moderate or mediate the relationship between variables with regard to adopting and accepting mobile banking services. Beyond this, there is the challenge to produce an environment that engages more clients in utilising mobile banking services, one that will consequently improve productivity, performance, return on investment, and income per capita. A vital factor in this study is the conceptualisation of 'perceived value', referred to as 'the overall assessment of the consumer regarding the utility of a service or product based on the perceptions of what is given and what is received' [20], while [14] inferred that the factors that influence value are also elements of value. Despite the suggestion of various factors as significant predictors of the mobile banking adoption, minimal attention has been given to fully explore what influence the perceived value of mobile banking has among the users, [21] stressed that future research must explore client value perceptions to examine the criticality of it in boosting client's intention to adopt mobile banking services. It recommends an extended TAM framework that uses the perceived value parameter as a mediator for each of the PU, PEOU, AW, SE, and PR with reference to the mobile banking services in Yemen. The results may potentially help in gaining significant insights for banks and strategists to respond to the challenges they encounter in the banking sector of Yemen.

2 Literature Review

2.1 Perceived Usefulness

[22] describe PU (perceived usefulness) as the subjective possibility that technology use will increase the performance of the individual. TAM postulated that PU is viewed as the magnitude of which a certain system would boost an individual's job performance [23]. Researchers have also claimed that PU plays a significant role with respect to technology [24]. Various technology acceptance and adoption applications have investigated the influence of PU on the intention to adopt and use technology. A research by [15] declared that PU plays a significant role with respect to technology. Internet adoption and mobile services for tourism were investigated by [25] in Korea and Germany in that order, and examined the correlation of PU and intention to utilise such technology. Other scholars have reported the insignificance of PU on other applications of technology. For example, in Yemen, [26] studied IT in public sector and noted that PU does not affect

the intention of usage. Also, a study in China and Malaysia was conducted by [27] on mobile commerce and their results agreed with the unimportance of the PU on the intention of using such technology. In the same way, in different cultural contexts, studies on the effect of 3G mobile services, E-learning, health services using smartphones, ICT adoption, GST application, E-invoice services, and internet shopping in Taiwan, Korea, Thailand, Ghana, Malaysia, and Jordan also supported a finding of the pettiness effect of PU on intention [17].

Also, there are several researchers who have explored mobile banking adoption and use. [28] have studied the factors affecting the intention to use mobile banking services in Malaysia, and revealed that PU especially influenced the intention to utilise mobile banking services. In Germany, [29] explored the same application and backed the Malaysian results. Scholars from Australia, China, the US, Indonesia, and Ghana [30] have also noted the significant influence exerted by PU on the intent to use mobile banking, remarking that as much as the clients perceive that the technology is useful, their intention to use mobile banking would be noteworthy.

H1: Perceived usefulness has a positive effect on perceived value.

2.2 Perceived Ease of Use

[22] defined PEOU (perceived ease of use) as "the extent of user's readiness to adopt the system where their effort is not required". PEOU in TAM signifies users' perception on the technology's ease of use. [31, 32] have coined the term 'effort expectancy', which is defined as 'the level of ease with respect to system use'. Empirical investigations on the PEOU concept and its uses have been carried out throughout the world. In Taiwan, a research by [33] stated the significant impact of PEOU on the intention of using E-tail, similar to a [34] research on internet banking services in Taiwan. [35] inspected mobile internet services in Latvia and also revealed that PEOU significantly influences the intention to use mobile internet.

Likewise, a research in Korea concerned with mobile payments also affirmed the importance of the influence of PEOU on the intention of customers to adopt and use a mobile service, and there are several other studies on various applications that emphasise the significance of PEOU. These include mobile financial services, Chinese eBay, internet banking, mobile payments, E-booking, online auction, E-book services, mobile learning, library application, higher education, mobile banking and social networking learning, studies conducted in Lebanon, Saudi Arabia, and Uganda by [36, 37], respectively. Moreover, mobile banking application usage has also been investigated in other contexts. [28] explored the influence of PEOU on intention of using mobile banking by Bank Islam customers in Malaysia, concluding that, PEOU is crucial in defining the intent of the users to utilise mobile banking services. A research in Korea conducted by [38] produced similar results as that of [39] research.

On the other hand, four studies carried out on various applications in diverse cultural contexts stated the triviality of PEOU with respect to the intention of the customers to adopt and use these services. In their study on mobile buying and selling in Malaysia, and Hong Kong [40, 41] observed that PEOU does not notably influence the intention of using mobile commerce in these diverse contexts. Moreover, [42] described the irrelevance of the PEOU concept on the intent of using mobile banking services in Malaysia.

Although a number of studies have indicated the criticality of PEOU, several studies have also stressed that PEOU is not significant in influencing users' intention to engage in of mobile banking services. Also, the literature has reported mixed results on the influence of PEOU in the mobile banking services setting. [43] in Tanzania, [44] in Iran, and [45] in Yemen observed that the concept of PEOU is significant for the same application. Meanwhile, there are certain research works in China, Taiwan, and Germany which have observed the insignificance of PEOU [46]. Therefore, the following hypotheses are postulated:

H2: Perceived ease of use has a positive effect on perceived value.
H6: Perceived ease of use has a positive effect on perceived usefulness.

2.3 Self-efficacy

Referring to the social cognition concept designed by [47], self-efficiency (SE) is the concept that an individual has the competence to complete a specific behaviour [48]. Research on SE has been conducted in many countries by examining different applications of technology. [49], in Taiwan, reported the significant SE influence on the intention of using the E-services, while [97] in Hong Kong, substantiated the SE importance on the intention of using mobile commerce. In Taiwan, [49] also identified the significant influence of SE on the intention to utilise mobile services. Moreover, Higher education context was studied in Serbia by [50], who also revealed the significance of SE in technology usage. In Yemen, [51] observed SE importance to explain intention of the users to learn through mobile technology.

The banking sector has also been engaged in the research related to E-banking services like internet banking as well as mobile banking. [52] examined the factors impacting the intention of using internet banking in Singapore and identified that SE significantly impact the intention of to use such services. In Yemen, [12] also observed similar findings for the same service but in another context. In Taiwan, mobile banking services were studied by [53], who were looking for an insight into the intention of using these services, and the outcomes of this research showed that SE was an important determinant of behavioural intention. [54], in Ghana, a study was conducted by to determine the factors to use mobile banking, and revealed that SE is a main factor for influencing the use of mobile banking services.

On the other hand, a research was carried out by [55] in Malaysia to find out the determinants of mobile entertainment use, and they discovered that SE was not too significant in affecting the intention of using mobile entertainment services in Malaysia. Therefore, the following hypotheses are postulated:

H3: Self-efficacy has a positive effect on perceived value.

2.4 Awareness

[56] have defined awareness of service as the degree bank customers know the existence of the service and its availability for their use. Similarly, [57] defined mobile banking

awareness as the awareness about the existence of a mobile banking practice and its benefits. The awareness of the customers regarding mobile banking can be determined by the amount of knowledge they possess about these services [28].

AW has been studied in several contexts including the adoption and usage of Internet banking services. [28, 42], in Malaysia, have examined the predictors that impact the mobile banking usage among Malaysian customers, and supported the significant influence of the awareness about these services among the customers on their intent of using mobile banking services. Correspondingly, in Taiwan, [58] revealed that the intent of using mobile banking is significantly affected by its awareness among the customers. [54] conducted a comparable research on the adoption and use of mobile banking services in Ghana and obtained similar outcomes.

Researchers in the field of internet banking have explored the awareness of internet banking services as a factor of intention of using these services. [59] substantiated the implication of awareness of such services in influencing the intent of using Internet banking services in Oman, while in another study that employed linear regression by [60] in Australia, established that awareness was an important predictor of internet banking usage. It agreed with Hamilton, However, [61] who inferred that awareness is a significant aspect for intention to use telephone banking services in the UK. Therefore, the following hypotheses are postulated:

H4: Awareness has a positive effect on perceived value.

2.5 Perceived Risk

Recent years have witnessed the modification in description of perceived risk (PR) which is attributed to the deviations in customer behaviour and the inclination to use online services. Formerly, PR was limited to item quality or fraud, but now it has become 'the potential for damage in seeking a desired result in using an e-service' [62]. A total of seven categories of risk were determine which are (1) the overall risk, (2) performance risk, (3) social risk, (4) privacy risk, (5) psychological risk, (6) time risk, and (7) financial risk. [62] mentioned that it was crucial to include PR in TAM since customers value and associate risk when evaluating services/products for adoption/use, and it may create distress and anxiety.

Numerous studies were conducted to inspect the influence of PR on the intention of adoption and technologies usage. In the United States of America, [62] examined online bill payment services and discovered the major negative impact of PR on its usage intention. [63] observed that PR is significant with respect to the intention of using mobile transactions in India, and a different Indian research reported a major negative impact on the intention of using online government services [64]. Similarly, studies carried out in Taiwan on various applications across several years by [65], on mobile e-invoice, commerce, and online auction, respectively, confirmed the significant impact of PR on the intention of using these applications.

Mobile banking adoption was examined in Malaysia by [42], but they revealed that perceived risk was not a main determinant of the intention of using mobile banking services. Therefore, it did not align with the findings of previous literature that established the implication of PR as a determine influencing the intent to use mobile banking. This inconsistency suggests that this concept needs further analysis; hence, it was included in the current research. Consequently, the following hypothesis is proposed:

H5: Perceived risk has a positive effect on perceived value.

2.6 Perceived Value

PV (perceived value) is described as 'the customer's overall judgement about the product/service being useful based on the opinion of what is given and what is received' [20]. [14] put forward that the perceived value can be 'a comparison among expenditure and utility and it can also be employed to derive the antecedents of perception of the value'. They deduced that the parameters that affect the value are also elements of value. The value perceived by the consumer is regarded as one of the most significant determinants in examining the intention for consumer adoption. In this regard, PR is a robust determinant in attracting new clients and holding on to current ones, and thus value perception should be a significant concept for e-services promoters seeking to create a competitive advantage. For achieving this, they require to reposition their services and products to meet predefined customer needs and thus increase the customer satisfaction [66].

Through a meticulous review of appropriate empirical studies related to PV, various technology applications of PV have been examined to assess the impact on intention to adopt and use in various cultural contexts. In Thailand, experts [67] studied E-booking services and observed that PV significantly affected the intention of using such services. A study conducted in Thailand by [68] linked to smartphone-based health services revealed that PV is a key factor that influences the intention to use such applications. In Taiwan, [69] examined the use of e-commerce as the centre of research, and discovered a significant correlation between PV and the intention of using mobile commerce services, while [70], in Singapore, investigated the construct with respect to mobile internet and observed that PV has a significant impact on the intention of using it. Other study on banking e-services has examined the determining capability of PV with respect to intention to use. [52] investigated internet banking adoption in Taiwan, and found out that PV significantly affected the intention of using internet banking. Correspondingly, [71] examined mobile banking usage in China and supported the results of previous literature which reported that PV is a main predictor in influencing the intention of using these services.

Nonetheless, in their study in Saudi Arabia regarding adoption and usage of mobile internet services by local populace, [72] discovered that PV is, in fact, an unimportant determinant of the intention of using mobile internet.

Thus, some studies claim that the influence customer PV has on the adoption intention of a service is mediated by customer satisfaction levels. Certain other studies have regarded PV as a cognitive notion with a direct effect on behavioural outcomes [73]. This characteristic has been subjected to empirical examination in many studies. For example, it is so in studies conducted by [74]. As far as the domain of mobile technology services is concerned, a limited PV conceptualisation shows it has a direct impact with respect to the intention to accept those services. Therefore, the following hypotheses are postulated:

H7: Perceived value has a positive effect on behavioural intention.
H8: Perceived value mediates the relationship between perceived usefulness, perceived ease of use, self-efficacy, awareness, perceived risk and behavioural intention.

2.7 Intention to Use Mobile Banking

BI implies readiness to present a certain behaviour [16, 75]. It can be also described as the intention of a customer to use an innovative technology [76]. That is to say, it is thc possibility of using a specific system. The intention to accept assesses the possibility of prospective bank customers to accept mobile banking one day [77].

In the previous research works, several scholars have revealed that individual behaviour can be described by investigating behavioural intention [10]. Intention here implies a purposeful and deliberate aim to adopt a new undertaking, and is therefore associated with the idea of using, when intentions directly affect the usage rate. Thus, usage or actual acceptance is intimately subject to the intention. Consumer intentions are also subject to their understanding of service or product attributes, their possible benefits and PR [78]. For the sake of this study, intention to accept will distinctively target the advancement of mobile banking technology. Intention to adopt and use various technological innovations have been well-researched in numerous settings the findings demonstrate that the intention for adoption is influenced by several parameters that are associated with consumer behaviour. In a proper review of past research, the researcher has encountered several experts who have examined intention as a dependent parameter in their research works. [67] both verify the importance of PV as an effective predictor of adoption intention besides usefulness and PEOU. Conversely, [59] has substituted awareness for PV which is also verified to exert a strong influence on intention of utilising online banking services in Oman.

3 Research Method

3.1 Research Model Development

The present study developed a model based on TAM [10] that simultaneously inspects factors that influence intention of using mobile banking services as dependent parameters related to mobile banking in Yemen. This study examined Perceived Ease of Use, Perceived Usefulness, Perceived Risk, and Self-Efficacy, and the mediating influence of PV to explain the intention of mobile banking usage. Correlations among the parameters and the complete model are shown in Fig. 1.

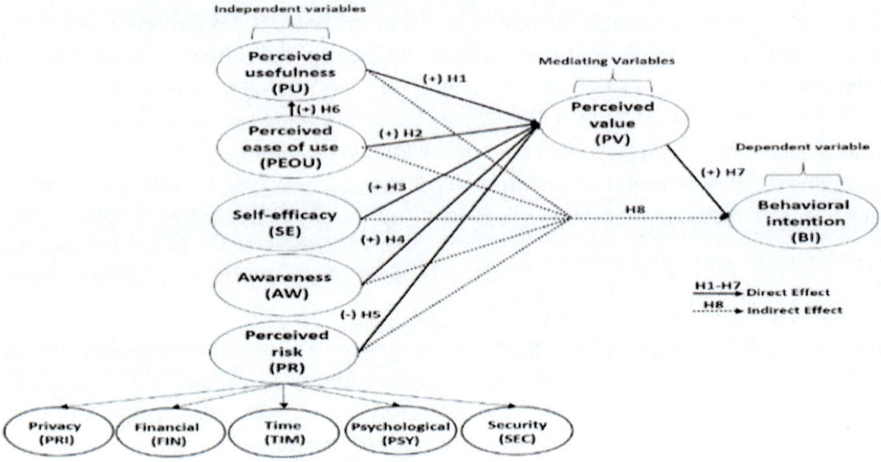

Fig. 1. Conceptual framework

3.2 Questionnaire Development of the Study and Data Collection

A survey questionnaire was used in this study to collect data designed specifically to gauge all the primary factors proposed in the study model. Survey questionnaire were administered to collect the necessary data for this study. The questionnaire contained two sections and encompassed close-ended statements that were translated to the Arabic Language to ensure a good understanding among the targeted Yemeni respondents. Strongly disagree to strongly agree are the two pillars for A 7-point Likert scale in the survey.

The respondents of this study were customers from a Yemeni bank that provides mobile banking services, but the customers opt to not use the offered services despite possessing a mobile phone. Snowball sampling was employed to recruit the respondents among the clienteles of Yemeni banks in Sana'a. This strategy is appropriate as the targeted population is difficult to be contacted [79]. This sampling method was used in this study because of the uncertainty regarding the identity of customers who do not use e-banking services, and mainly due to the difficulty in the region of Middle East of confronting people about a sensitive matter, such as their banking details [36]. this study successfully gathered a total of 428 complete and valid questionnaires. Analysis of Data and Outcomes PLS (Partial Least Squares) method was incorporated in this study to assess the developed research model with the use of Smart PLS software version 3.0 [80].

3.3 Descriptive Analysis of Demographic Profiles

Seven demographic categories of 482 respondents were analysed. The categories were marital status, gender, education, age, occupation, income, and banking experience. In terms gender, 71.0% of the respondents were male and 29.0% were female. 300 respondents were married, and 127 respondents were single. Concerning age, 13 were less than 20 years old, 449 were in the age group of 20–49 years and only 20 were

50 years old and above. With regards to banking experience, 20.3% had used banking services for a year, 57.0% had been bank clients for 2 to 7 years, and 22.6% had been users of banking services for more than 7 years.

3.4 Standard Deviation and the Mean

Respondents were asked to expressed their agreement with the statement according to a seven-point scale which ranged between 1 (i.e., strongly disagree) and 7 (i.e., strongly agree). PEOU documented the highest mean value of 4.945 out of 7, and a standard deviation of value 1.764, signifying that the respondents opine that mobile banking services are easy to use and flexible for simple interaction. PU (perceived usefulness) documented a mean value of 4.408 out of 7 and a standard deviation of value 1.778, signifying that the respondents view mobile banking would permit them to complete more financial transactions and would improve their experience of payments. PR (perceived risk) documented a mean value of 3.870 out of 7 and a standard deviation value of 1.618, implying that the respondents acknowledged there are existing risks when using mobile banking such as privacy, financial, time, security and psychological risks. PV (perceived value) mean value was 4.732 out of 7 and a standard deviation value of 1.728, suggesting that the candidates believe that mobile banking is useful and advantageous.

3.5 Measurement Model Assessment

Cronbach's alpha coefficients was employed to determine the reliability of each parameter in the measurement model. The results recommend that the Cronbach's alpha coefficients ranged between 0.854 and 0.966, exceeding the recommended level of 0.7 [81]. Moreover, in examining the reliability of the parameters, all the CR (composite reliability) values ranged from 0.912 to 0.9976, greater than 0.7 [82]), adequately suggesting that construct reliability was satisfied. Therefore, the survey achieved Cronbach's Alpha and CR for every construct which were recognised to be sufficiently error-free.

Factor loadings were used to inspect the reliability of the variables. High loadings on a parameter signify that the linked indicators are very common, and it was verified by the construct [83]. Factor loadings surpassing 0.50 are viewed as very important [21, 84, 85] the loadings for every items exceeded the recommended value of 0.5. In the model, therefore, the loading for each of the items have met all conditions, with the exception of FIN3, PSY3 and AW4 which were removed from the measurement due to their low loading value.

AVE (average variance extracted) was used to observe convergent validity, and all AVE values were higher than the advised 0.50 value [84, 86, 87] ranging between 0.776 and 0.932. The convergent validity of all constructs was effectively fulfilled and adequate convergent validity.

Table 1 Discriminant validity (referred to as the level to which items differentiate among parameters or assess distinct concepts) of measurement model was scrutinised by employing the Fornell-Larcker measurement. As displayed in Table 2, each AVE's square root, as shown by bold values diagonally, was higher than the corresponding relationships among the constructs (matching row and column values). This indicates that these constructs are significantly related with their specific indicators than other constructs in the model [88–90], therefore demonstrating respectable discriminant validity [83].

Table 1. Results of discriminant validity by Fornell-Larcker criterion

	Factors	1	2	3	4	5	6	7
		AW	BI	PEOU	PR	PU	PV	SE
1	AW	**0.920**						
2	BI	0.401	**0.966**					
3	PEOU	0.294	0.720	**0.899**				
4	PR	−0.351	−0.629	−0.486	**0.905**			
5	PU	0.410	0.837	0.682	−0.569	**0.949**		
6	PV	0.434	0.717	0.644	−0.573	0.693	**0.943**	
7	SE	0.282	0.718	0.561	−0.451	0.698	0.542	**0.912**

The Fornell-Larkcer measure has been subject to certain criticism. [91] mentioned that the measurement was unable to identify the non-existence of discriminant validity in universal research settings effectively, and have demonstrated an alternative which is (HTMT (heterotrait-monotrait) ratio of relationships based on multitrait-multimethod matrix). This research also assessed discriminant validity using HTMT as depicted in Table 2, and all values were shown to be lower than the suggested value of 0.90 [92], establishing discriminant validity.

Table 2. Results of discriminant validity by HTMT

	Factors	*1*	*2*	*3*	*4*	*5*	*6*	*7*
		AW	**BI**	**PEOU**	**PR**	**PU**	**PV**	**SE**
1	**AW**							
2	**BI**	0.415						
3	**PEOU**	0.304	0.773					
4	**PR**	0.363	0.651	0.517				
5	**PU**	0.424	0.869	0.731	0.590			
6	**PV**	0.451	0.753	0.700	0.601	0.728		
7	**SE**	0.297	0.770	0.622	0.482	0.750	0.589	

3.6 Structural Model Assessment

[83] suggested assessing a structural model by calculating beta (β), R^2 and related t-values using bootstrapping, with 5,000 resamples. Moreover, they mention that the

effect sizes (f^2) and predictive relevance (Q^2) should also be assessed. Nonetheless, [93] claim that while the presence of an effect is determined by p-value, it does not determine the effect size (Fig. 2).

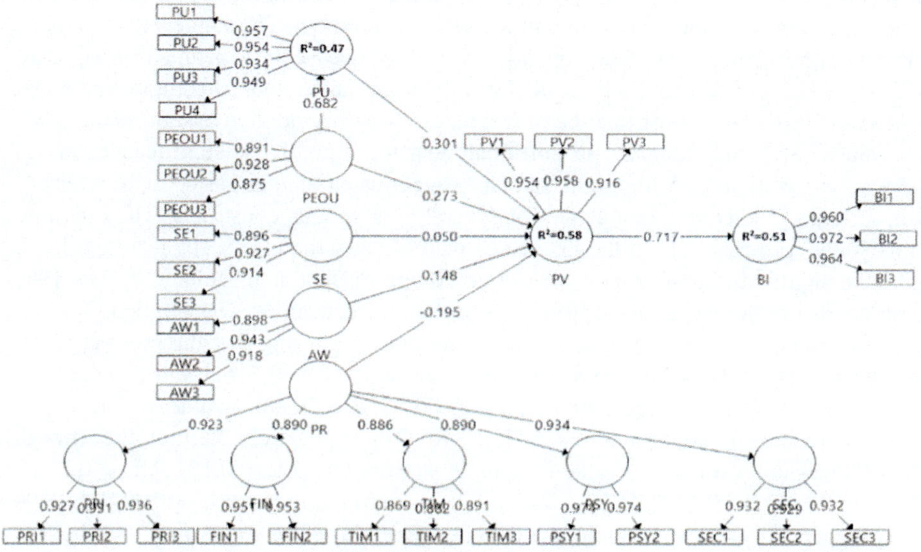

Fig. 2. PLS algorithm results

Hypotheses Tests

The structural model analysis as displayed in Table 3 provides results of the tests of the hypotheses. PU (perceived usefulness), AW (awareness) and PEOU (perceived ease of use) positively affect PV (perceived value), thus H1, H2 and H4 are accepted (β = 0.30, t-value = 5.224, $p < 0.001$), (β = 0.15, t-value =4.002, $p < 0.001$) and (β = 0.27, t-value = 5.794, $p < 0.05$), respectively. PR (perceived risk) negatively affects the PV, and hence H5 is also accepted (β = 0.20, t-value = 4.669, $p < 0.001$). PEOU also positively affects PU, and thus H6 is also accepted ($\beta = -0.68$, t-value = 22.747, $p < 0.001$). Similarly, H7 is accepted as PV significantly determines behavioural intention (β = 0.72, t-value = 26.561, $p < 0.001$). Nonetheless, H3 is not accepted. It is noted that the uniformed path coefficient (Beta) postulates the strength of the correlations between the independent and dependent variables, so the direct influence of PU on PV was stronger than other parameters.

PU, PEOU, AW and PR accounts for 58% of the variation in PV. PEOU accounts for 47% of the difference in the PU, while PV accounts for 51% of the variation in BI. The values of R^2 were larger than the 0.35 value as suggested by [94], signifying a substantial model, and R^2 is adequately high for the model to attain a satisfactory explanatory power. This research also analysed effect sizes (f^2). [83] mention that there must be an analysis of the variation in the R^2 values. [94] suggests an approach to assess the size of effect according to a scale (0.02, 0.15, and 0.35, which signify small, medium, and large effects in order). The result of f^2 as detailed in Table 3 is that there were 5 relationships that demonstrated small effect sizes and 1 relationship with a large effect size.

By employing the blindfolding method, this study evaluated the predictive significance of the recommended model. [83] suggested that this technique should only be used on the dependent variable which has a reflective measurement. The suggested model has predictive significance for several dependent variables in case the Q^2 value is larger than 0 [83, 95]. As depicted in Table 3, all Q^2 values are larger than 0 (ranging between 0.392 and 0.480). This infers that the model has satisfactory predictive significance. For v Q^2 values, [83] guide that as a proportional measure of predictive significance, values 0.02, 0.15, as well as 0.35 highlight that an exogenous parameter has a small, intermediate, or large predictive significance for several endogenous constructs. The outcome of this study confirms that all the Q^2 values were greater than 0.35, suggesting a large predictive significance. In order to test the mediation effect as postulated by H8, [96] recommends employing bootstrapping where the indirect impact was applied. Table 3 shows the outcomes of this bootstrapping technique, signifying that the indirect influence of H8.1, H8.2, H8.4 and H8.5 was significant with ($\beta = 0.216$, t-value = 4.857, $p < 0.001$), ($\beta = 0.343$, t-value = 9.429, $p < 0.001$), ($\beta = 0.106$, t-value = 4.427, $p < 0.001$) and ($\beta = -0.140$, t-value = 4.537, $p < 0.001$) respectively. Further, Preacher & Hayes (2008) suggest that the value of beta, 95% Boot CI: [LL = 0.131, UL = 0.301], [LL = 0.274, UL = 0.412], [LL = 0.055, UL = 0.161], and [LL = −0.199, UL = −0.077] respectively, does not include a 0 in between, therefore suggesting that there is mediation. Nonetheless, H8.3 is not accepted. Thus, this research can suppose that a mediation effect of the PV parameter is present with respect to the correlation between PU, PEOU, and AW and also between PR and BI.

Table 3. Structural path analysis result

Hypothesis	Relationship	Std Beta	t-value	Decision	R^2	f^2	Q^2	VIF
H1	PU → PV	0.301	5.224***	Supported	0.58	0.075	0.488	2.943
H2	PEOU → PV	0.273	5.794***	Supported		0.092		1.963
H3	SE → PV	0.050	1.146	N-supported		0.003		2.012
H4	AW → PV	0.148	4.002***	Supported		0.043		1.233
H5	PR → PV	−0.195	4.669***	Supported		0.059		1.566
H6	PEOU → PU	0.682	22.747***	Supported	0.47	0.868	0.392	1.000
H7	PV → BI	0.717	26.561***	Supported	0.51		0.452	1.000
H8.1	PU → PV → BI	0.216	4.857***	Supported				
H8.2	PEOU → PV → BI	0.343	9.429***	Supported				
H8.3	SE → PV → BI	0.036	1.096	N-supported				
H8.4	AW → PV → BI	0.106	4.427***	Supported				
H8.5	PR → PV → BI	−0.140	4.537***	Supported				

***p < .001; **p < .01; *p < .05

4 Discussion and Implications

4.1 Discussion

The main aim of this study is to examine the effect of PU, PEOU, PR, AW, and SE on the intention of using mobile banking through the mediation influence of PV. One of the special goals of this research is to investigate the parameters (PU, PEOU, PR, AW, and SE) that determine the PV (perceived value) of utilising mobile banking has been fulfilled by checking the hypotheses H1 to H6. PEOU, PU, and AW (perceived ease of use, perceived usefulness, and awareness) are observed to have noteworthy positive relationships with PV (perceived value). With regards to the β values, PU has the greatest influence on PV (β .301), while AW had the lowest beta value (.148); for PEOU, it is (β .273). Consequently, it is conceivable that as the perception of ease of use and usefulness rises among customers, and there is an intensification in the awareness concerning value and benefits of mobile banking usage, the perception of value will similarly rise. This discovery is consistent with the previous discoveries of [97], who realised that PU and PEOU of internet banking are associated positively with client value. Furthermore, this discovery is in line with another research by [70], in which PU and PEOU of the mobile internet are observed to have a positive correlation with PV, especially in the initial adoption phase.

Similar to past empirical studies regarding technology adoption, PU is the biggest concern of the possible users of mobile banking services when determining their value. A reasonable explanation for this inference is that banking clients feel assured with the ease of use and benefits of mobile banking services compared to different e-banking media and their needs are satisfied. Hence, the customers have developed a perspective that the service is advantageous which contributed to a larger PV of using these services.

Future clients are worried regarding the effort and time required to learn and use mobile banking services. If the mobile banking use is complex, has a sluggish response and is difficult to comprehend and learn, then it is not going to have a very good advantage. As a result, bank customers need to consider the technical knowledge needed in using those services when forming viewpoints regarding its value. The findings of this study refer to the technology acceptance model (TAM), which displayed that the ease-of-use parameter is more substantial in the early adoption phases when complications are issued by process concerns and need to address them.

When clients of financial institutions obtain aid and knowledge from banking employees concerning these services, proper method of utilisation of these services and their benefits became more apparent. This notion is supported by the results of this study. Additionally, the research outcomes confirm that the greater the awareness regarding values and benefits of mobile banking services (reliable and adequate information), the greater the perception about its value among the clients. Since this research target future customers, the non-existence of mobile banking data and mindfulness indicates that there are minimised benefits of these services.

H3 postulated self-efficacy positively influences perceived value and the findings of the study did not support this hypothesis. The reason for this might be that a potential user of a new technology, who has the ability and confidence that he/she got from prior experience and training to use similar technology, would know how to utilise such a technology. This would not influence the value perceived by clients and therefore the intention of using mobile banking. The higher the ability and confidence of clients in using a new technology will directly affect directly the behavioural intention without the intervention of other factors, such as perceived value. The reason could a result of an experience using other or similar technology.

Alternatively, the results obtained in this research show that PR (perceived risk) does have a noteworthy negative effect on PV (perceived value) with $\beta = -.195$. This is in line with the outcomes of [71] in which it is found that that there is a negative effect of PR on PV. That is to say, the greater the risk perception in the context of mobile banking, the lower the customers' value perception and this definitely affects the intention of using these services. That is why decision-makers as well as developers of the bank should keep in mind that they are required to make their services safer secure and reduce PR to increase the use of these services.

Moreover, Perceived value is found to mediate the relationship between perceived ease of use, perceived usefulness, awareness and perceived risk one side and the intention to use mobile banking services on the other, so supporting H.8.1, H8.2, H8.4 and H8.5. inconsistent with the finding of [70] study that revealed the mediating role of perceived value between the aforementioned antecedents and intention to use mobile internet service. This finding shed light on the significance of ensuring the easiness of use associated with the use of mobile banking services. Besides their direct effect, they have an indirect effect through perceived value. Further, H8.3 was not tested since it has failed to fulfill the three roles baron and Kenny mediation test also used by [98] and [99]. But self-efficacy has a significant positive direct effect on the intention to use mobile banking but no indirect effect. Mediation role of perceived value between the PU, PEOU, AW, PR and intention is found significant, which makes this result as an addition to the body of knowledge.

In a nutshell, the findings of the research are justified, and correspond to the past studies related to technology adoption and use, especially that PR, PEOU, PU, and AW are crucial for assessing PV.

4.2 Theoretical Implications and Implication for Practice

The fundamental concept in the current study is the TAM (technology acceptance model). The model demonstrates the function of PU, PEOU, AW, SE, and PR, in defining the PV and, more prominently, the intention to use mobile banking facilities.

The theoretical outcome of this study is that the examination of PV has contributed to the present literature in the BI (behavioural intention) field to accept mobile banking technology. The proposed theoretical model of this study has established additional correlations such as those between PEOU, PU, SE, and AW of a mobile banking platform and thus contributes to the initial models. Moreover, PV is a strong determinant of behavioural intention and a mediator between PR, PU, PEOU, AW, and BI. The result of this study also supported that the approximation offered by TAM received support as demonstrated by its ability to identify the relationship among the key parameters of the constructed model regarding E-services, i.e. mobile banking services.

The determination coefficient for the research displays that PEOU explained 47% of the variation in PU. PU, PEOU, AW, SE, and PR explained 58% of the variation in

PV to use mobile banking facilities, while PV explained 51% of the intention to utilise these services. In line with [94], the values of $\mathbf{R}^2$ for PV and BI in using mobile banking technology in the present research are viewed as essential.

This study has encouraged the PV parameter to be incorporated into an electronic setting, an unfamiliar field in the past. It has combined contemporary theories and frameworks and provide support for TAM. The psychological accounting theory and categorisation of effect model integration, as well as SE and PR in forecasting intention to use mobile banking facilities. The findings of this research should be added in the present mobile banking and PV literature which are not sufficiently addressed by previous studies, with a critical foundation of influence on pre-adoption behaviour by take no notice of the effect of PV and AW in IS.

Further, the conclusions and findings deduced from this study provide several possible implications for banking industry practitioners who are planning to jump into the market of mobile banking. First of all, mobile phones are technological devices and their adoption depends on a perception of usability and performance. The greater the perception that they make life easier, the better the adoption rate. Currently, the rate of adoption of mobile phone gadgets is very high and they contain many applications, including mobile banking capability. Present research shows that mobile phones are widely adopted [100], users have knowledge and experience in using them and many of their applications. This suggests that financial organisations (including banks) are unburdened of the trouble to expose customers to a mobile banking platform, and they need to make sure that the banking facilities, especially mobile banking services with mobile-based platforms for improved usefulness and ease of use, are easily understood and readily accessible. Technological features of applications like having an easy-to-use UI (user interface), easy Internet access, security issues, and so on, are crucial. This indicates that banks should endeavour to provide well-founded technological framework and service environments to the mobile telephony platform in order to expand and encourage the usage of their mobile banking facilities. Bank clients will expect that mobile banking services be more effective and freer of effort, and meeting this requirement will encourage them to accept and use such a service. Bank policy must then add usability of the technology within the banking context to enhance safety and security of customers banking information and reduce risk, make it easier for customers to transact online via a mobile banking platform and enhance transaction convenience through mobile banking as opposed to using conventional banking transactions channels.

According to the past research works, this research concludes that bank clients are more probable to use mobile banking facilities if these services are useful and easy to use [54, 101]. There is a robust correlation between a mobile banking platform and technological aspects with respect to acceptance and usage of the service offered. Financial institutions and mobile banking technologies developers must cooperate to exercise many initiative and resources into producing effective mobile banking platforms that are beneficial and convenient. Usefulness and ease of use will improve the likelihood of clients in adopting mobile banking services [36, 46]. These parameters are considered crucial in making the decisions such as whether to use these services, whether these services are easy to use, and whether they offer convenience, efficiency, and quality.

Perceived risk is assessed on the basis of mobile banking usage risks such as privacy concerns, financial issues, psychological effect, time, and security-related risks. The greater the perception of potential threats in any of the above-mentioned concerns, the lower its perceived value and thus the lower intention of using these services [101]. Clients are less likely to use a service that they do not trust to protect their privacy or money and interests, and they are less willing to use a service that will make take them out of their comfort zone and waste their time. High-grade safety characteristics must be incorporated to protect personal data such as personal identification or customer bank account numbers and other important details. The design and implementation of the framework of mobile banking facilities platforms must protect customer sensitive information. Guaranteeing a higher amount of usage of these services among bank customers depends on the capability of the technology to function with a degree of efficiency and trust similar to the other mediums of banking services.

The risk perceived by individuals towards the use of the mobile banking system is a significant roadblock towards technology adoption [13]. The market for mobile banking services possesses immense potential. Nevertheless, there is inadequate information about the services that lessens the adoption and behavioural intention. To boost the count of mobile banking consumers, useful steps like awareness programs highlighting the value and benefits need to be undertaken, while educating the masses and steps towards sensitisation may also be helpful. Further, since people are the initial target market, advertisement of mobile banking services and their value and benefits should be offered widely. When awareness is raised, bank clients would have a greater understanding of the mobile banking technology [54, 58].

Lastly, there are no parallels between the research conducted on mobile banking and internet-based banking. Marketing mobile banking services is an excellent approach to increase and improve the cognizance of the aspects of the banking sector, especially for individuals who have not adopted mobile banking services. Further, the current trends indicate that an increasing number of the younger generation are more comfortable with the latest innovations and technology. Consequently, the banks have an obligation to offer more innovative financial products in tandem with mobile banking. Furthermore, females observe mobile banking more positively than males. Therefore, females have a higher intention to utilise mobile banking. Therefore, it is recommended that financial institutions should concentrate on promoting mobile services to female customers to improve the adoption rate.

5 Limitations of the Research and Future Work Suggestions

The findings of this current study should be viewed in the light of certain limitations. There are three areas of limitation in this study. Given that there is no sampling frame present for this study, sample selection is made from the banks that currently provide a mobile banking platform in Yemen, and this aspect may affect the conclusions of the study. Generalising to a similar set of users than to all bank clients in Yemen, also consider the customers from other banks that currently do not provide mobile banking services. Moreover, the sample chosen in was from banks in the city of Yemen (Sana'a) while other provinces were not adequately represented. Another aspect causing hindrance is the cross-sectional research design, where all the included variables of the hypothesised model are assessed at one point in time. As per [102], such an approach does not provide for definite conclusions to be drawn, especially relating to the causality of the relationships between variables. Moreover, another limitation for the current study has been due to the inherent nature of the extended TAM concept. Though TAM has been established as a useful framework for the prediction of technology adoption, it is not inclusive enough to assess the entirety of factors that affect mobile banking even after the original TAM model has been augmented using newer constructs. An unidentified number of confounding variables may be present which could influence mobile banking adoption.

On the other hand, the current study improves the existing understanding on TAM (perceived ease of use and perceived usefulness), self-efficacy, perceived risk, awareness, intention to use mobile banking, and perceived value. Nevertheless, given that mobile banking in Yemen has transformed the way individuals handle banking transactions, additional research may explore additional areas from different points of view, for instance, market orientation, a more inclusive assessment of relationship marketing, and steps to better promote mobile banking as a concept.

Further research could also assess the potential of other variables as a moderator such as cultural values, trust, and demographic variables on the relationship between perceived value and the intention of mobile banking usage to improve the model's predictive power. It is recommended that additional research in this aspect may address this complexity.

Lastly, it is suggested that future researchers step further and integrate real behaviour (usage) with the proposed framework instead of solely focusing on behavioural intention. Though assessing actual behaviour is considered expensive from a time and monetary perspective and also logistically complicated [103], an understanding of actual behaviour shall increase the validity of the studies that are based on behavioural intention [104]. Although behavioural intention has been recognised as a sufficient determinant of actual behaviour in different behavioural-related fields, studies assessing the predictive power of intention on the actual behaviour is still minimal in the literature.

6 Conclusion

This study aims to determine the factors that impact the intention to utilise mobile banking services and test the mediating effect of perceived value for Yemini banks' customers. Despite the several limitations of the current study, there have been encouraging results since they provide information on new variables of intention or behaviour. An understanding of the value obtained by using mobile banking shall increase client intention to use mobile banking. Perceived risk (financial, time, privacy, security, and psychological) may be a significant roadblock towards the use of mobile banking because it may potentially adversely affect the perceived value and, therefore, the intention to use it. For the bank client, a mobile banking service should offer something new when compared to competing services in similar categories (e-banking services). This study reported that bank clients evaluate mobile banking services based on their perception of the ease of use, related perceived risk, and usefulness, and then elect whether or not to engage in those services.

The critical characteristics the influence perceived value are perceived ease of use, perceived value, awareness, and perceived usefulness. Perceived value is measured to influence the intention to use mobile banking services. Moreover, perceived ease of use is concluded as a significant factor of perceived usefulness. The present study recommends that it is vital to comprehend that perceived risk and TAM factors (perceived ease of use and perceived usefulness) have been regularly identified as important in impacting intention and perceived value for the mobile banking industry. Hence, there is a need for the practitioners to dedicate adequate resources towards these aspects to increase the probability of better returns on investment. Another interesting finding is that perceived value significantly mediates perceived risk, perceived ease of use, awareness, and perceived usefulness, and the intention in using mobile banking services. Finally, the general conclusion to be drawn from the evidence is that perceived value can affect client intention not only on fixed context but also in an e-services environment, especially when clients intend to adopt a new service because of its usefulness and ease of use.

References

1. Kumar, T.A., Ferdous, A.S.M., Nabeel, J.A.: Operational or strategic benefits: empirical investigation of internet adoption in supply chain management. Manag. Res. Rev. **40**(1), 28–52 (2017)
2. Wessels, L., Drennan, J.: An investigation of consumer acceptance of M-banking. Int. J. Bank Mark. **28**(7), 547–568 (2010)
3. Hammood, W.A., Abdullah, R., Hammood, O.A., Mohamad, S., Al-Sharafi, M.A., Hasan, A.M.: A review of user authentication model for online banking system based on mobile IMEI number. In: The 6th International Conference on Software Engineering & Computer Systems, p. 769 (2020)
4. Alalwan, A.A., Dwivedi, Y.K., Rana, N.P.: 7.pdf. Int. J. Inf. Manag. **37**, 99–110 (2017)
5. KPMG & UBS Evidence Lab: Global Mobile Banking Report, The Financial Brand (2015)
6. Hanafizadeh, P., Keating, B.W., Khedmatgozar, H.R.: A systematic review of Internet banking adoption. Telemat. Informatics **31**(3), 492–510 (2014)
7. World Development Indicators: Mobile Subscription Among Arab Countries, World Bank. The World Bank, May 2014
8. Al-Dailami, A.: Yemen Uses Mobile Banking & Electronic Money Transfers. Yemen (2013)
9. Rogers, E.M.: Diffusion of Innovations, 4th edn. The Free Press, New York, London (1995)
10. Davis, F.D.: Perceived usefulness, perceived ease of use, and user acceptence of inforamtion technology. MIS Q. **13**(3), 319–340 (1989)
11. Oliveira, T., Martins, M.F., De Lisboa, U.N.: Literature review of information technology adoption models at firm level. Electron. J. Inf. Syst. Eval. **14**(1), 110–121 (2011)
12. Zolait, A.H.S.: An examination of the factors influencing Yemeni Bank users' behavioural intention to use Internet banking services. J. Financ. Serv. Mark. **15**(1), 76–94 (2010)
13. Alsheikh, L., Bojei, J.: Customer's perceived value to use mobile banking services. In: International Conference on Management, Behavioral Sciences and Economics Issues (2012)
14. Gupta, S., Kim, H.-W.: Value-driven Internet shopping: the mental accounting theory perspective. Psychol. Mark. **27**(1), 13–35 (2010)
15. Anisur, M., Qi, X., Islam, T.: Banking access for the poor: adoption and strategies in rural areas of Bangladesh. J. Econ. Financ. Stud. **04**(03), 1–10 (2016)
16. Talib, Z.M., Mosbah, A., Al-Jubari, I.: Intention to pursue postgraduate studies in Malaysian universities. Int. J. Innov. Creat. Chang. **5**(2), 591–609 (2019)

17. Faqih, K.M.S.: An empirical analysis of factors predicting the behavioral intention to adopt Internet shopping technology among non-shoppers in a developing country context: does gender matter? J. Retail. Consum. Serv. **30**, 140–164 (2016)
18. Al-Jubari, I., Hassan, A., Liñán, F.: Entrepreneurial intention among University students in Malaysia: integrating self-determination theory and the theory of planned behavior. Int. Entrep. Manag. J. **15**(4), 1323–1342 (2018)
19. Al-Jubari, I.: College students' entrepreneurial intention: testing an integrated model of SDT and TPB. SAGE Open **9**(2), 1–15 (2019)
20. Zeithaml, V.A.: Consumer sperceptions of price, quality, and value-a means-end model and snthesis of evidence. J. Mark. **52**(1), 2–22 (1988)
21. Salem, S.F., Alanadoly, A.B.: Personality traits and social media as drivers of word-of-mouth towards sustainable fashion. J. Fash. Mark. Manag. (2020)
22. Davis, F.D., Bagozzi, R.P., Warshaw, P.R.: User acceptance of computer technology: a comparison of two theoretical models. Manag. Sci. **35**(8), 982–1003 (1989)
23. Al-Gahtani, S.S.: Testing for the applicability of the TAM model in the Arabic context: exploring an extended TAM with three. Inf. Resour. Manag. J. **21**(4), 177–204 (2008)
24. Ramayah, T.: Course website usage: does prior experience matter?, vol. 11, no. 2 (2006)
25. Krogstie, J.: The Adoption of Mobile Tourism Services: An Empirical Study, pp. 3–5 (2012)
26. Al-Haderi, S.M.S.: The effect of self-efficacy in the acceptance of information technology in the public sector. Int. J. Bus. Soc. Sci. **4**(9), 188–198 (2013)
27. Chong, A.Y.-L., Chan, F.T.S., Ooi, K.-B.: Predicting consumer decisions to adopt mobile commerce: cross country empirical examination between China and Malaysia. Decis. Support Syst. **53**(1), 34–43 (2012)
28. Amin, H., Rizal, M., Hamid, A., Lada, S., Anis, Z.: The adoption of mobile banking in Malaysia: the case of Bank Islam Malaysia Berhad (BIMB), vol. 9, no. 2, pp. 43–53 (2008)
29. Koenig-Lewis, N., Palmer, A., Moll, A.: Predicting young consumers' take up of mobile banking services. Int. J. Bank Mark. **28**(5), 410–432 (2010)
30. Purwanto, E., Deviny, J., Mutahar, A.M.: The mediating role of trust in the relationship between corporate image, security, word of mouth and loyalty in m-banking using among the millennial generation in Indonesia. Manag. Mark. Challenges Knowl. Soc. **15**(2), 255–274 (2020)
31. Venkatesh, V., Morris, M.G., Davis, G.B., Davis, F.D., Walton, S.M.: User acceptance of information technology: toward a unified view 1, vol. 27, no. 3, pp. 425–478 (2003)
32. Purwanto, E., Mutahar, A.M.: Examine the technology of acceptance model among mobile banking users in Indonesia. Technol. Reports Kansai Univ. **62**(7), 3969–3979 (2020)
33. Chiu, Y.-B., Lin, C.-P., Tang, L.-L.: Gender differs: assessing a model of online purchase intentions in e-tail service. Int. J. Serv. Ind. Manag. **16**(5), 416–435 (2005)
34. Wang, Y.-S., Wang, Y.-M., Lin, H.-H., Tang, T.-I.: Determinants of user acceptance of Internet banking: an empirical study. Int. J. Serv. Ind. Manag. **14**(5), 501–519 (2003)
35. Binde, J., Fuksa, M.: Mobile technologies and services development impact on mobile internet usage in Latvia. Procedia Comput. Sci. **26**, 41–50 (2013)
36. Koksal, M.H.: The intentions of Lebanese consumers to adopt mobile banking. Int. J. Bank Mark. **34**(3), 327–346 (2016)
37. Badwelan, A., Drew, S., Bahaddad, A.A.: Towards acceptance m-learning approach in higher education in Saudi Arabia. Int. J. Bus. Manag. **11**(8), 12–30 (2016)
38. Gu, J.-C., Lee, S.-C., Suh, Y.-H.: Determinants of behavioral intention to mobile banking. Expert Syst. Appl. **36**(9), 11605–11616 (2009)
39. Amin, H., Supinah, R., Aris, M.M., Baba, R.: Receptiveness of mobile banking by Malaysian local customers in Sabah: an empirical investigation. J. Internet Bank. Commerce **17**(1), 1–12 (2012)

40. Khalifa, M., Shen, K.N.: Explaining the adoption of transactional B2C mobile commerce. J. Enterp. Inf. Manag. **21**(2), 110–124 (2008)
41. Wu, J.-H., Wang, S.-C.: What drives mobile commerce? Inf. Manag. **42**(5), 719–729 (2005)
42. Norzaidi, M.D., et al.: Determining critical success factors of mobile banking adoption in Malaysia. Aust. J. Basic Appl. Sci. **5**(9), 252–265 (2011)
43. Mawona, A., Mpogole, H.: ICT and Financial Inclusion: Adoption of Mobile Phone Banking Among Small Business Owners in Iringa, Tanzania, pp. 1–9 (2013)
44. Hanafizadeh, P., Behboudi, M., Koshksaray, A.A., Tabar, M.J.S.: Mobile-banking adoption by Iranian bank clients. Telematics Informatics **31**(1), 62–78 (2014)
45. Mutahar, A.M., Daud, N.M., Ramayah, T., Putit, L., Isaac, O.: Examining the effect of subjective norms and compatibility as external variables on tam: mobile banking acceptance in Yemen. Sci. Int. **29**(4), 769–776 (2017)
46. Yen, H.R., Wang, W., Wei, C.P., Hsu, S.H.Y., Chiu, H.C.: Service innovation readiness: dimensions and performance outcome. Decis. Support Syst. **53**(4), 813–824 (2012)
47. Bandura, A.: Social cognitive theory, vol. 6, pp. 1–60 (1986)
48. Compeau, D.R., Higgins, C.A.: Application of social cognitive theory to training for computer skills. Inf. Syst. Res. **6**(2), 118–143 (1995)
49. Hsu, M.-H., Chiu, C.-M.: Internet self-efficacy and electronic service acceptance. Decis. Support Syst. **38**(3), 369–381 (2004)
50. Kocaleva, M.: Research on UTAUT application in higher education institution. In: International Conference on Information Technology and Development of Education, no. June (2014)
51. Alrajawy, I., Daud, N.M., Isaac, O., Mutahar, A.M.: Examine factors influencing the intention to use mobile learning in Yemen public universities. Asian J. Inf. Technol. **16**(2), 287–297 (2017)
52. Lee, M.-C.: Factors influencing the adoption of internet banking: an integration of TAM and TPB with perceived risk and perceived benefit. Electron. Commer. Res. Appl. **8**(3), 130–141 (2009)
53. Luarn, P., Lin, H.-H.: Toward an understanding of the behavioral intention to use mobile banking. Comput. Hum. Behav. **21**(6), 873–891 (2005)
54. Cudjoe, A.G., Anim, P.A., Nyanyofio, J.G.N.T.: Determinants of mobile banking adoption in the Ghanaian banking industry: a case of access bank Ghana limited. J. Comput. Commun. **3**, 1–19 (2015)
55. Leong, L.-Y., Ooi, K.-B., Chong, A.Y.-L., Lin, B.: Modeling the stimulators of the behavioral intention to use mobile entertainment: does gender really matter? Comput. Hum. Behav. **29**(5), 2109–2121 (2013)
56. Tibenderana, P., Ogao, P., Ikoja-Odongo, J., Wokadala, J.: Measuring levels of end-users' acceptance and use of hybrid library services Prisca Tibenderana Kampala International University, Uganda. Int. J. Educ. Dev. Inf. Commun. Technol. **6**(2) (2010)
57. Al-Somali, S.A., Gholami, R., Clegg, B.: An investigation into the acceptance of online banking in Saudi Arabia. Technovation **29**(2), 130–141 (2009)
58. Chen, C.: Perceived risk, usage frequency of mobile banking services. Manag. Serv. Qual. **23**(5), 410–436 (2013)
59. Riffai, M.M.M.A., Grant, K., Edgar, D.: Big TAM in Oman: exploring the promise of online banking, its adoption by customers and the challenges of banking in Oman. Int. J. Inf. Manag. **32**(3), 239–250 (2012)
60. Sathye, M.: Adoption of internet banking by Australian consumers: an empirical investigation. Int. J. Bank Mark. (1999)
61. Howcroft, B., Hamilton, R., Hewer, P.: Consumer attitude and the usage and adoption of home-based banking in the United Kingdom (2002)

62. Featherman, M.S., Pavlou, P.A.: Predicting e-services adoption: a perceived risk facets perspective. Int. J. Hum.-Comput. Stud. **59**(4), 451–474 (2003)
63. Thakur, R., Srivastava, M.: Customer usage intention of mobile commerce in India: an empirical study (2013)
64. Rana, N.P., Dwivedi, Y.K., Williams, M.D., Weerakkody, V.: Investigating success of an e-government initiative: validation of an integrated IS success model. Inf. Syst. Front. **17**(1), 127–142 (2014)
65. Chang, S., Lee, M.: The linkage between knowledge accumulation capability and organizational innovation. J. Knowl. Manag. **12**(1), 3–20 (2008)
66. Day, G., Fahey, L.: Valuing market strategies. J. Mark. **52**(3), 45 (1988)
67. Bhatiasevi, V., Yoopetch, C.: The determinants of intention to use electronic booking among young users in Thailand. J. Hosp. Tour. Manag. **23**, 1–11 (2015)
68. Boontarig, W., Chutimaskul, W., Chongsuphajaisiddhi, V., Papasratorn, B.: Factors influencing the Thai elderly intention to use smartphone for e-Health services. In: 2012 IEEE Symposium on Humanities, Science and Engineering Research, pp. 479–483, June 2012
69. Wang, Y.S.: Assessing e-commerce systems success: a respecification and validation of the DeLone and McLean model of IS success. Inf. Syst. J. **18**(5), 529–557 (2008)
70. Kim, H.-W., Chan, H.C., Gupta, S.: Value-based adoption of mobile internet: an empirical investigation. Decis. Support Syst. **43**(1), 111–126 (2007)
71. Xiong, S.: Adoption of mobile banking model based on perceived value and trust. In: 2013 6th International Conference on Information Management, Innovation Management and Industrial Engineering, pp. 632–635, November 2013
72. Alwahaishi, S., Snášel, V.: Consumers' acceptance and use of information and communications technology: a UTAUT and flow based theoretical model. J. Technol. Manag. Innov. **8**(2), 61–74 (2013)
73. Dodds, W.B., Monroe, K.B., Grewal, D.: Effects of price, brand, and store information on buyers' product evaluations. J. Mark. Res. **28**(3), 307–319 (1991)
74. Toften, K., Olsen, S.O.: The relationships among quality, cost, value, and use of export market information: an empirical study. J. Int. Mark. **12**(2), 104–131 (2004)
75. Ajzen, I.: Understanding Attitudes and Predicting Social Behaviour (1980)
76. Seymour, L., Makanya, W., Berrangé, S.: End-Users' Acceptance of Enterprise Resource Planning Systems: An Investigation of Antecedents (2007)
77. Lin, H.-F.: An empirical investigation of mobile banking adoption: the effect of innovation attributes and knowledge-based trust. Int. J. Inf. Manag. **31**(3), 252–260 (2011)
78. Upshaw, L.B.: Building Brand Identity: A Strategy for Success in a Hostile Marketplace. Wiley (1995)
79. Al-Qeisi, K.I.: Analyzing the Use of UTAUT Model in Explaining an Online Behaviour: Internet Banking Adoption A (2009)
80. Ringle, C.M., Wende, S., Becker, J.-M.: SmartPLS 3. Bonningstedt: SmartPLS (2015)
81. Kannana, V.R., Tan, K.C.: Just in time, total quality management, and supply chain management: understanding their linkages and impact on business performance. Omega Int. J. Manag. Sci. **33**(2), 153–162 (2005)
82. Kline, R.B.: Principles and Practice of Structural Equation Modeling, 3rd edn. The Guilford Press, New York (2010)
83. Hair, J.F., Hult, G.T.M., Ringle, C., Sarstedt, M.: A Primer on Partial Least Squares Structural Equation Modeling (PLS-SEM), 2nd edn. SAGE, London; Thousand Oaks (2017)
84. Hair, J.F., Black, W.C., Babin, B.J., Anderson, R.E.: Multivariate Data Analysis, 7th edn. Prentice Hall (2010)
85. Jalal, A.N., Bahari, M., Musa, M.W., Sherief, S.R.: Social media and traditional customer relationship management: an overview of theoretical challenges and future directions in healthcare context. Int. J. Adv. Trends Comput. Sci. Eng. **8**(1.6), 76–82 (2019)

86. Jalal, A.N., Bahari, M., Tarofder, A.K., Musa, W.M.N.M.W.: Factors influencing customer social relationship management implementation and its benefits in healthcare industry. Polish J. Manag. Stud. **19**(2), 196–205 (2019)
87. Salem, S.F., Salem, S.O.: Effects of social media marketing and selected marketing constructs on stages of brand loyalty. Glob. Bus. Rev. **1**(24) (2019)
88. Fornell, C., Larcker, D.F.: Evaluating structural equation models with unobservable variables and measurement error. J. Mark. Res. **18**(1), 39–50 (1981)
89. Aldholay, A., Abdullah, Z., Isaac, O., Mutahar, A.M.: compatibility perspective of Yemeni students on use of online learning extending the information systems. Inf. Technol. People **33**(1), 106–128 (2019)
90. Aldholay, A., Abdullah, Z., Isaac, O., Mutahar, A.M.: Perspective of Yemeni students on use of online learning: extending the information systems success model with transformational leadership and compatibility. Inf. Technol. People **33**(1) (2019)
91. Henseler, J., Ringle, C.M., Sarstedt, M.: A new criterion for assessing discriminant validity in variance-based structural equation modeling. J. Acad. Mark. Sci. **43**(1), 115–135 (2014)
92. Gold, A.H., Malhotra, A., Segars, A.H.: Knowledge management: an organizational capabilities perspective. J. Manag. Inf. Syst. **18**(1), 185–214 (2001)
93. Sullivan, G.M., Feinn, R.: Using effect size - or why the p value is not enough. J. Grad. Med. Educ. **4**(3), 279–282 (2012)
94. Cohen, J.: Statistical Power Analysis for the Behavioral Sciences, 2nd edn. Lawre Associatesnce Erlbaum (1988)
95. Fornell, C., Cha, J.: Partial least squares. In: Bagozzi, R.P. (ed.) Advanced Methods in Marketing Research, pp. 52–78. Blackwell, Cambridge (1994)
96. Preacher, K.J., Hayes, A.F.: SPSS and SAS procedures for estimating indirect effects in simple mediation models. Behav. Res. Methods Instruments Comput. **36**(4), 717–731 (2004)
97. Ho, S.-H., Ko, Y.-Y.: Effects of self-service technology on customer value and customer readiness the case of internet banking. Internet Res. **18**(4), 427–446 (2008)
98. Field, A.: Discovering Statistics Using SPSS, 4th edn (2013)
99. Hayes, A.: Introduction to Mediation, Moderation, & Conditional Process Analysis (2013)
100. World Development Indicators: Online users as Percent. Popul. Yemen vs. world Aver. (2014)
101. Gutierrez, D.M., Scavarda, L.F., Fiorencio, L., Martins, R.A.: Evolution of the performance measurement system in the logistics department of a broadcasting company: an action research. Int. J. Prod. Econ. **160**, 1–12 (2015)
102. De Wulf, K.: The Role of Seller in Enhancing Buyer-Seller Relationship. Unversity of Central Florida (1999)
103. Mitchell, R., Hall, C.M.: The post-visit consumer behaviour of New Zealand winery visitors. J. Wine Res. **15**(1), 39–49 (2004)
104. Cronin, J.J., Brady, M.K., Hult, G.T.M.: Assessing the effects of quality, value, and customer satisfaction on consumer behavioral intentions in service environments. J. Retail. **76**(2), 193–218 (2000)

The Predictors of Electric Vehicles Adoption: An Extended Theory of Planned Behavior

Ling Ming Wu(✉), Jason Wai Chow Lee, Yet Mee Lim, and Chuen Khee Pek

UCSI Graduate Business School, Faculty of Business and Management, UCSI University, Kuala Lumpur, Malaysia
{wclee,LimYM,pekck}@ucsiuniversity.edu.my

Abstract. As environmental problems become more serious, carbon dioxide emission from vehicles deserves attention from the global economies as it is the main cause of air pollution. Although the Chinese government formulated a series of policies for promoting electric vehicles, fuel vehicles still dominate the auto market. Based on the theory of planned behavior (TPB), this study aims to explore the factors that may influence electric vehicles purchase intention of people in China. Data were collected through the biggest online survey site named "Wen Juan Xing" in China. Four hundred and forty-eight participants responded the survey yielded a sample size of n = 448. PLS-SEM technique was used to analyze the data and to test the hypotheses. The findings in this study show that attitude towards buying electric vehicles, subjective norms, and environmental knowledge positively affect the electric vehicles purchase intention of people. Meanwhile, environmental concern mediated the relationship between environmental knowledge, subjective norms, collectivism, perceived behavioral control and electric vehicles purchase intention; while attitude towards green advertising moderated the relationship between environmental knowledge and electric vehicles purchase intention. In addition, female consumers were found to have higher electric vehicle purchase intention than male consumers. Implications of the research findings are discussed.

Keywords: Purchase intention · Electric vehicles · Green products · Collectivism · Environmental concern · Green advertisement · Environmental knowledge

1 Introduction

The protection of environment is a global concern and people all over the world are paying attention to green products. An international survey showed that more than 70% of Hong Kong consumers bought green products with labels and 80% of Taiwanese spent 10% more on green products; while 83% of Americans not only participated in green activities but also bought green products [7]. Nowadays, human health is affected as the deterioration of air quality. Past study indicated that the major contribution of air pollution comes from fuel vehicles [12]. The global vehicle sales reached to nearly 85 million units

M. Al-Emran et al. (Eds.): ICETIS 2021, LNNS 299, pp. 521–532, 2022.
https://doi.org/10.1007/978-3-030-82616-1_43

in 2019, of which China contributed more than 28% of the sales [11]. China's total vehicle inventory is expected to increase to more than 400 million in 2030 [16]. Thus, more serious environmental pollution will continue to rise if there are no alternative initiatives are taken. However, electric vehicles are introduced as an alternative to fulfill the demand for transport, which are less harmful to the environment [14]. Consumption of electric vehicles is a viable solution to global environmental pollution and should be promoted to replace some fuel vehicles. The purpose of this research is to understand the key factors that explain the electric vehicles purchase intention (EVPI) based on the TPB model. This research examines the influence of attitude towards green purchasing, subjective norms, perceived behavioral control, environmental knowledge and collectivism on EVPI. In addition, the moderating effect of attitude towards green advertising and the mediating effect of environmental concern were also be examined. This study would contribute to the body of knowledge from several aspects. Firstly, an extended TPB model was introduced with green media and cultural factors. Secondly, the research findings would help the government formulate policies to promote electric vehicles for environmental protection and the development of green gross domestic products. Finally, this study would provide useful information to help marketing managers to set strategy to attract consumers to purchase electric vehicles.

2 Literature Review

2.1 Attitude Towards Green Purchasing and Purchase Intention

Purchasing intention (PI) refers to the tendency of an individual to perform an action of buying an item in a certain situation, which indicates the preposition of this person to perform a specific behaviour [6]. PI is the core of the TPB model as it is used to predict a wide variety of purchase behaviors [10]. In the green market, green consumerism is closely related to green purchase intentions (GPI). GPI is defined as the motivation of people to choose green products. Past study indicated that EVPI mainly depends on the perception and attitude of people, not energy attributes [31]. Attitude (ATT) is considered a psychological path that determines the feelings and preferences of people about something—whether a person likes or dislikes a particular item or object. If a person has a positive attitude towards a specific behavior, the greater the tendency to perform that behaviour [27]. Several studies showed that a positive attitude towards the environment increases the likelihood of engaging environmental behaviors such as staying in green hotels during travel, eating green foods and recycling of waste [1]. In terms of green purchasing, studies such as Liu, Teng and Han (2020) indicated that a favourable attitude towards green is a strong predictor of GPI. Indeed, green attitude is found to be the strongest predictor of GPI of consumers among all of the TPB predictor constructs [29]. Hence, first hypothesis was formulated as follows:

H1: Attitude towards green purchasing has a positive impact on electric vehicles purchase intention.

2.2 Subjective Norm and Purchase Intention

Subjective norm (SN) is considered to be a social influence that may arouse the intention of individual to perform a certain behaviour [27]. Opinions and suggestions from friends, family and relatives may increase the intention to engage in something among the individuals. When an important person around an individual agrees or supports a certain behavior, the greater the probability that person will perform such behavior and vice versa [23]. For example, SN has been found to significantly influence BI of people to stay in green hotels [29]. The role of SN is more obvious in green buying behavior. According to Dong et al. (2020), SN can be a source of pressure that people feel from their surroundings when they make a decision of purchasing electric vehicles. Studies found that SN is a significant predictor of electric vehicles purchase intention include [10] and [31]. Consequently, the following hypothesis was developed:

H2: Subjective norm has a positive impact on electric vehicles purchase intention.

2.3 Perceived Behavioral Control and Purchase Intention

According to Sreen, Purbey and Sadarangani (2018), perceived behavioral control (PBC) is defined as the perceived ease or difficulty in conducting a behavior; i.e., people who thinks that he has the ability to afford the consequences. But the level of PBC will determine the probability of people to perform the behavior [31]. The main reasons for the low demand for green products are declining supply, perceived higher prices and insufficient understanding of these products. These are some factors negatively affect the level of PBC in relation to green buying. Nevertheless, studies showed that PBC is positively related to the PI of people to buy energy-saving appliances [23]. PBC has a significant effect on EVPI [28]. When people feel that it is easy to buy and use electric vehicles, they will have more motivation to purchase it [10]. Based on the above discussion, the third hypothesis was proposed:

H3: Perceived behavioral control has a positive impact on electric vehicles purchase intention.

2.4 Environmental Knowledge

Environmental knowledge (EK) refers to people's general knowledge of the environment and overall understanding of the environment issues [8]. It is a good predictor of green consumerism when consumers are aware of the causes and influence of environmental problems, possible solutions to environmental problems, and the relationship between people and the environment [24]. It plays an important role in studying consumers behaviors as it has a significant influence on their intention [17]. The more knowledgeable of consumers about the environment, the higher the tendency to buy green products [30]. Specifically, Tanwir and Hamzah (2020) found that EK has a positive effect on the intention of people to buy green car. Therefore, the fourth hypothesis was postulated as follows:

H4: Environmental knowledge has a positive impact on electric vehicles purchase intention.

2.5 Collectivism

According to Hofstede (1993), collectivism is one of the dimensions that can be used to describe the cultural differences between nations. Collectivism (COL) explains why people in a country are more willing to act as members of a group rather than individuals. In the context of this research, collectivism refers to the belief that group is necessary to protect the environment. People in the collectivist culture are more aware of environmental issues and they believe that cooperative spirits would solve environmental problems more solidly [3]. Several studies provided evidences that COL is positively correlated to green intention and behavior. Consumers in a collectivist society are willing to buy goods with higher price if they perceive that their acts are beneficial to the society [22]. Furthermore, COL was found to have a positive influence on PI towards green hotel selection [30]. This discussion forms the basis of the following hypothesis:

H5: Collectivism has a positive impact on electric vehicles purchase intention.

2.6 The Moderating Role of Attitude Towards Green Advertising

Green advertising (GA) refers to the publicity of green products [20]. Its main purpose is to use mass media message to educate the public with regards to the meaning and value of green products and to encourage them to take practical actions to reduce environmental damage. Alamsyah et al., (2020) indicated that people's attitude, perception and behavior towards the environment can be effectively affected by GA. This is probably because GA can enhance consumers' awareness and knowledge of green products. In addition, GA is found to have moderating effect on the relationship between ATT and GPI [20]. Alamsyah, Othman and Mohammed (2020) found that consumers' trust in organic products can be increased through GA, thereby improving their purchase intention of these products. The moderating role of green advertising leads to the following set of hypotheses:

H6a-H6e: Attitude towards green advertising moderates the relationships between attitude towards green purchasing, subjective norms, perceived behavioral control, environmental knowledge, collectivism and EVPI respectively.

2.7 The Mediating Role Environmental Concern

Environmental concern (EC) refers to the concern of an individual for the environment and environmental-related issues, i.e., how much people worry about the environment [8]. Due to people's attention to the environment has generally increased, experts in environment strongly recommend that more studies are needed to examine this important variable [25]. Recent studies found that EC plays a mediating role in various relationships. It was found to have a mediation effect on the relationships between COL, EK and PI [3]. EC also mediates the relationship between EK and ATT on environmental responsibility as well as the relationship between attitude on green products and PI [19]. Although EC is a significant mediator, virtually no studies have fully explored the mediating role of it in TPB, i.e., whether it plays a mediating role between SN, PBC and GPI of a specific product. Based on the above discussion, further testing on the mediating role of environmental concern is warranted. Hence, the following set of hypotheses were proposed:

H7a-H7e: Environmental concern mediates the relationship between attitude towards green purchasing, subjective norms, perceived behavioral control, environmental knowledge, collectivism and EVPI respectively (Fig. 1).

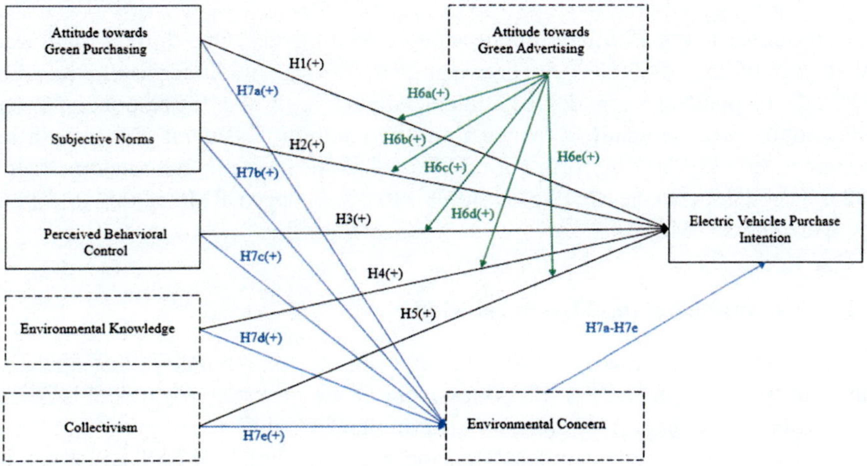

Fig. 1. Research model

3 Research Methodology

This study was conducted in Sichuan of China, which has a population of 83 million. The target population of this study consisted of those who were potential car buyers. Survey questionnaire was used to collect primary data from March to April of 2020 among the potential respondents via the biggest online survey site named "Wen Juan Xing" in China. Four hundred and forty-eight participants responded the survey yielded a sample size of n = 448. This sample size is considered large for the study of consumer behavior [21]. PLS-SEM technique was used to analyze the data and to test the hypotheses. The questionnaire consisted of two parts. The first part collected demographic data with respect to gender, age, occupation, education level, and monthly income. The second part included items measuring the variables of study. Attitude towards green purchasing was measured using five items adapted from [27]. Subjective norms were measured using four items adapted from [16]. Four items adapted from [26] and [29] were used to measure perceived behavioral control. Similarly, four items adapted from [18] were used to measure environmental knowledge. Collectivism was operationalized by six items adapted from [7]. Environmental concern was operationalized based on five items adapted from [6]. Attitude towards green advertising was measured using five items adapted from [5]. Lastly, purchase intention was measured using six items adapted from [23]. All variables were measured using a 5-point Likert scale, ranged from 1 (strongly disagree) to 5 (strongly agree). A pre-test was done to ensure that all questions were clear and easy to understand before the actual data collection process.

4 Results and Findings

4.1 Respondents Demographic Information

The sample consisted of 58.5% of females and 41.5% of males. One-third of them were below the age of 25 and 53.8% were between 25 to 39 years old. In terms of education, 52.5% of the respondents were degree holders. About two-third of the respondents were working adults with occupations ranged from drivers to professionals. The proportion of income level was fairly uniform with 23.9% of the respondents with income below RMB2,000; 4.8% between RMB2,000-4,999; 26.8% between RMB5,000-7,999; and 24.6% were above RMB8,000.

4.2 The Measurement and Structural Model

The measurement model was assessed to test the validity and reliability of all the latent variables in the research framework. Assessment of the measurement model is necessary to establish convergent validity and discriminant validity. Convergent validity is established when the factor loadings of all the indicator items for each construct are above 0.70 and the composite reliability (CR) are more than 0.70 while average variance extracted are more than 0.50. The convergent validity test showed that all the factor loadings are above the 0.70 threshold, the composite reliability (CR) of the latent variables are greater than 0.70, and the average variance extracted (AVE) are above the 0.50 benchmark. Since all of the factor loadings, CR and AVE have satisfied the conditions for convergent validity, the measurement model is deemed to have established the convergent validity. In establishing discriminant validity, the constructs must be unique and distinct from each other. To meet this requirement, the heterotrait-monotrait (HTMT) test was used. The cut off point for all HTMT values should be less than 0.85 to establish discriminant validity [13]. The results showed that all the HTMT values are less than 0.85, hence discriminant validity is established in the model.

The structural model was assessed using PLS-SEM technique which applies algorithm to determine the construct scores, path coefficients, and the R^2 values. To determine the t-values and to conduct hypotheses testing, bootstrapping was used. Results of R^2 of 0.78 for PI and 0.53 for EC indicated that all the independent variables could explain about 78% of the variance in purchase intention and 53% of variance in environmental concern respectively. The results showed that attitude towards green purchasing ($\beta = 0.468$, $p < 0.000$), subjective norms ($\beta = 0.256$, $p < 0.000$), and environmental knowledge ($\beta = 0.149$, $p < 0.002$) were positively related to purchase intention. Therefore, H1, H2, and H4 were supported, while H3 ($\beta = -0.034$, $p < 0.267$) and H5 ($\beta = 0.060$, $p < 0.189$) was not supported. The effect sizes as indicated by the f^2 values range from 0.020 to 0.275 suggesting that effect sizes range from small to medium following Cohen (1988) (Table 1).

Table 1. Hypotheses testing in relation to purchase intention

Hypothesis	Relationship	Std Beta	Std Error	t-values	p-values	BCI LL	BCI UL	f^2
H1	ATT -> PI	0.468	0.053	8.891	0.000	0.365	0.571	0.275
H2	SN -> PI	0.256	0.044	5.763	0.000	0.162	0.347	0.077
H3	PBC -> PI	−0.034	0.031	1.111	0.267	−0.095	0.025	0.004
H4	EK -> PI	0.149	0.049	3.041	0.002	0.052	0.242	0.039
H5	C -> PI	0.060	0.046	1.315	0.189	−0.036	0.146	0.006

4.3 Mediation and Moderation Analysis

The results of mediating effects of EC on EVPI showed that the effect of perceived behavioral control ($\beta = 0.020$, BCI LL = 0.005, BCI UL = 0.043), collectivism ($\beta = 0.044$, BCI LL = 0.009, BCI UL = 0.088), subjective norm ($\beta = -0.020$, BCI LL = −0.049, BCI UL = −0.003) and environmental knowledge ($\beta = 0.026$, BCI LL = 0.007, BCI UL = 0.054) on EVPI were mediated by EC. Therefore H7b, H7c, H7d and H7e are supported. However, no mediating effect of environmental concern was found on the relationship between attitude towards green purchasing and electric vehicles purchase intention. Hence, H7a ($\beta = 0.012$, BCI LL = −0.001, BCI UL = 0.042) was not supported. As for the moderation analysis, the findings indicate that there is a significant positive moderation effect on the strength of the relationship between environmental knowledge and purchase intention ($\beta = 0.096$, $p < 0.05$). Hence, H6d is supported. It means that when attitude towards green advertising is more positive, the strength of the relationship between environmental knowledge and purchase intention will be higher. H6a, H6b, H6c, and H6e were not supported with p values greater than 0.05 (Tables 2, 3 and 4).

Table 2. Results of the mediating effect of environmental control

Hypothesis	Beta	t values	p values	BCL LL	BCL UL	Decision
ATT -> EC -> PI	0.012	1.212	0.226	−0.001	0.042	No mediation
C -> EC -> PI	0.044	2.146	0.032	0.009	0.088	Mediation
EK -> EC -> PI	0.026	2.04	0.042	0.007	0.054	Mediation
PBC -> EC -> PI	0.020	2.111	0.035	0.005	0.043	Mediation
SN -> EC -> PI	−0.020	1.631	0.104	−0.049	−0.003	Mediation

Table 3. Moderation effects of attitude towards green advertising

Moderation	Std Beta	Std Error	t-values	p-values	BCI LL	BCI UL	Decision
AGA*ATT -> PI	0.001	0.052	0.017	0.987	−0.096	0.109	Not supported
AGA*SN -> PI	−0.006	0.054	0.107	0.915	−0.118	0.091	Not supported
AGA*PBC -> PI	−0.035	0.030	1.176	0.24	−0.085	0.026	Not supported
AGA*EK -> PI	0.096	0.046	2.075	0.038	0.0030	0.178	Supported
AGA*C -> PI	−0.003	0.043	0.066	0.948	−0.085	0.083	Not supported

Table 4. Measurement scale items

Items	Questions	Source
Attitude towards green purchasing	Purchasing electric vehicle is a good idea	Sreen et al. (2018)
	Purchasing electric vehicle is a wise idea	
	I believe that electric vehicles help to reduce pollution (water & air)	
	I think buying an electric vehicle is a good choice	
	Given a choice, I will prefer an electric vehicle over a fuel vehicle	
Subjective norms	People who are important to me think that I should buy electric vehicles	Huang and Ge (2019)
	My close friends think that I should purchase electric car rather than fuel car	
	People that I know or familiar would want me to buy electric vehicles	
	If people around me use electric vehicles, this will prompt me to buy	
Perceived behavioral control	Whether or not to buy electric vehicles is completely up to me	Ru et al. (2018); Verma and Chandra (2018)
	I believe I have the ability to purchase electric vehicle	
	I require a lot of time to search for electric vehicles	
	It is entirely my decision to buy electric vehicles	
Environmental knowledge	I know about environmental issues	Jaiswal and Kant (2018)

(*continued*)

Table 4. (*continued*)

Items	Questions	Source
	I know that electric vehicles are environmentally safe	
	Using electric vehicle is a primary way to reduce pollution	
	I know that traditional vehicles are damaging our environment	
Collectivism	The well-being of my group members (family, friends) is important to me	Chen et al. (2018)
	I respect the majority's wish	
	I maintain decisions made by my group	
	I will share with others the benefits of electric vehicles	
	I respect decisions made by my group	
	Whatever be the circumstances, I always support my group	
Environmental concern	I am very concerned about the environment	Chaudhary and Bisai (2018)
	I would be willing to reduce my consumption to help protect the environment	
	I am emotionally involved in environmentally protection issues in country	
	I believe Anti-pollution laws should be enforced more strongly	
	I often think about how the environmental quality of country can be improved	
Attitude towards green advertising	I think green advertising is good	Attia (2014)
	Overall, green advertising is important	
	Overall, green advertising is useful	
	Green advertising is valuable to society	
	Green advertising leads people to be more socially responsible	
Purchase intention	I will consider to buy electric vehicles	Li et al. (2019)
	I am very likely to buy electric vehicles in the future	
	I plan to buy electric vehicles	
	I will consider switching to electric vehicles for ecological reasons	
	I expect to purchase electric vehicles in the future because of its positive environmental contribution	
	I am willing to buy electric vehicles	

5 Implications of Research Findings

The findings of this study have some theoretical and practical implications. Firstly, this study has included these external factors–environmental knowledge, environmental concern, collectivism, and attitude towards green advertising—to gain a more in-depth understanding of the influencing factors in relation to green purchase intention. Then, the Chinese government may provide incentives for green vehicles production and creation of environmental knowledge among the car manufacturing companies. These car producers may educate the public to raise their environmental awareness through green advertising. Employees could also be trained to provide green services to the potential buyers. Thirdly, the marketing managers should invest more on advertising to raise consumers' awareness about electric vehicles. According to Wang et al. (2017), modern people who like to share information online have great possibilities to accept electric vehicles. Therefore, managers may formulate some preferential schemes to encourage old customers to deliver information of electric vehicles through word of mouth. Additionally, managers may set different strategies based on gender as females were found to have a higher electric vehicle purchase intention than male. For instance, in the early stage, marketers could place more ads on TV shows that are popular with women.

6 Study Limitations and Conclusion

This research is subject to a couple of limitations. Firstly, the sample used in this study was based on convenience sampling technique; hence, the results may not be generalized across the population. Secondly, this study examined EVPI without taken the influence of other factors into consideration such as vehicle ownership, the budget for automobile purchase, and whether the individual has already owned an electric vehicle. Therefore, future research may investigate whether the EVPI will also be affected by these personal factors. Thirdly, a qualitative study should also be conducted to explore the true feelings about purchasing green vehicles among the consumers as it able to discover much insights regarding the intention and the possibility of actual buying green vehicles. Finally, a longitudinal study is needed to examine the relationship between purchase intention and actual buying behaviour as PI can only be supposed to explain the variance of actual purchase behaviour [8]. In conclusion, this study provides a new foothold for predicting consumers' purchase intentions for electric vehicles. Nowadays, electric vehicles are increasingly becoming one of the choices that people buy cars. But its popularization still takes a long time to realize. To make the consumption of electric vehicles successful, both enterprises and the government need to do more for the development and supporting of electric vehicles.

References

1. Alamsyah, D.P., Othman, N.A., Bakri, M.H., Adjie, A.N., Salsabila, K., Syarifuddin, D.: Confirmatory factor analysis of green advertising and its impact on green awareness. Manag. Sci. Lett. **10**, 3899–3906 (2020)

2. Alamsyah, D.P., Othman, N.A., Mohammed, H.A.A.: The awareness of environmentally friendly products: the impact of green advertising and green brand image. Manag. Sci. Lett. **10**(9), 1961–1968 (2020)
3. Arisal, I., Atalar, T.: The exploring relationships between environmental concern, collectivism and ecological purchase intention. Procedia Soc. Behav. Sci. **235**(10), 514–521 (2016)
4. Arshad, H.M., Saleem, K., Shafi, S., Ahmad, T., Kanwal, S.: Environmental awareness, concern, attitude and behavior of university students: a comparison across academic disciplines. Pol. J. Environ. Stud. **30**(1), 1–10 (2020)
5. Attia, S.T.: The effect of green advertising as a moderator on green purchase attitude - green purchase intentions relationship: the case of young Egyptian consumers. J. IMS Group **11**(1), 1–15 (2014)
6. Chaudhary, R., Bisai, S.: Factors influencing green purchase behavior of millennials in India. Manag. Environ. Quality: Int. J. **29**(5), 798–812 (2018)
7. Chen, C.C., Chen, C.W., Tung, Y.C.: Exploring the consumer behavior of intention to purchase green products in Belt and Road countries: an empirical analysis. Sustainability **10**(3), 854–872 (2018)
8. Choi, D., Johnson, K.K.P.: Influences of environmental and hedonic motivations on intention to purchase green products: an extension of the theory of planned behavior. Sustain. Prod. Consumption **18**, 145–155 (2019)
9. Cohen, J.W.: Statistical Power Analysis for the Behavioural Sciences. Lawrence Erlbaum Associates, Hillsdale (1988)
10. Dong, X.Y., Zhang, B., Wang, B., Wang, Z.H.: Urban households' purchase intentions for pure electric vehicles under subsidy contexts in China: do cost factors matter? Transp. Res. Part A: Policy Pract. **135**, 183–197 (2020)
11. Focus2move. https://www.focus2move.com/global-car-industry-the-ranking-of-top-countries-in-2019/. Accessed 04 Jan 2020
12. Habich-Sobiegalla, S., Kostka, G., Anzinger, N.: Electric vehicle purchase intentions of Chinese, Russian and Brazilian citizens: an international comparative study. J. Clean. Prod. **205**, 188–200 (2018)
13. Hair, J.F., Hult, G.T.M., Ringle, C., Sarstedt, M.: A Primer on Partial Least Squares Structural Equation Modeling (PLS-SEM). SAGE Publications, Los Angeles (2017)
14. Hamilton, B., Terblanche-Smit, M.: Consumer behavioural intention to purchase green vehicles: a South African perspective. In: 16th International Association of Public and Non-Profit Marketing (IAPNM) Congress, pp. 1–7 (2017)
15. Hofstede, G.: Cultural constraints in management theories. Acad. Manag. Exec. **7**, 81–94 (1993)
16. Huang, X., Ge, J.: Electric vehicle development in Beijing: an analysis of consumer purchase intention. J. Clean. Prod. **216**, 361–372 (2019)
17. Indriani, I.A.D., Rahayu, M., Hadiwidjojo, D.: The influence of environmental knowledge on green purchase intention the role of attitude as mediating variable. Int. J. Multicultural Multireligious Understanding **6**(2), 627 (2019)
18. Jaiswal, D., Kant, R.: Green purchasing behaviour: a conceptual framework and empirical investigation of Indian consumers. J. Retail. Consum. Serv. **41**(11), 60–69 (2018)
19. Janmaimool, P., Chudech, S.: Effect of domestic and global environmental events on environmental concern and environmental responsibility among university students. Sustainability **12**(4), 1–20 (2020)
20. Kim, W.H., Malek, K., Roberts, K.R.: The effectiveness of green advertising in the convention industry: an application of a dual coding approach and the norm activation model. J. Hosp. Tour. Manag. **39**(4), 185–192 (2019)
21. Krejcie, R.V., Morgan, D.W.: Determining sample size for research activities. Educ. Psychol. Meas. **30**(3), 607–610 (1970)

22. Lee, Y.K.: A comparative study of green purchase intention between Korean and Chinese consumers: the moderating role of collectivism. Sustainability **9**(10), 1930–1946 (2017)
23. Li, G., Li, W., Jin, Z., Wang, Z.: Influence of environmental concern and knowledge on households' willingness to purchase energy-efficient appliances: a case study in Shanxi, China. Sustainability **11**(4), 1–18 (2019)
24. Liu, P., Teng, M., Han, C.: How does environmental knowledge translate into pro-environmental behaviors? The mediating role of environmental attitudes and behavioral intentions. Sci. Total Environ. 728–741 (2020)
25. Onurlubaş, E.: The mediating role of environmental attitude on the impact of environmental concern on green product purchasing intention. Emerg. Markets J. **8**(2), 5–18 (2019)
26. Ru, X., Wang, S., Chen, Q., Yan, S.: Exploring the interaction effects of norms and attitudes on green travel intention: an empirical study in eastern China. J. Cleaner Prod. **197**(96), 1317–1327 (2019)
27. Sreen, N., Purbey, S., Sadarangani, P.: Impact of culture, behavior and gender on green purchase intention. J. Retail. Consum. Serv. **41**(6), 177–189 (2018)
28. Tanwir, N.S., Hamzah, M.I.: Predicting purchase intention of hybrid electric vehicles: evidence from an emerging economy. World Electric Veh. J. **11**(2), 35–53 (2020)
29. Verma, V.K., Chandra, B.: An application of theory of planned behavior to predict young Indian consumers' green hotel visit intention. J. Clean. Prod. **172**, 1152–1162 (2018)
30. Wang, L., Wong, P.P.W., Alagas, N.E.: Antecedents of green purchase behaviour: an examination of altruism and environmental knowledge. Int. J. Cult. Tourism Hospitality Res **14**(1), 63–82 (2020)
31. Xu, Y., Zhang, W., Bao, H., Zhang, S., Xiang, Y.: A SEM-neural network approach to predict customers' intention to purchase battery electric vehicles in China's Zhejiang Province. Sustainability **11**(11), 3164–3182 (2019)

The Relationship Between Product Quality, Brand Image, Purchase Decision, and Repurchase Intention

Dam Tri Cuong(✉)

Industrial University of Ho Chi Minh City, Ho Chi Minh City, Vietnam
damtricuong@iuh.edu.vn

Abstract. Repurchase intention relates to the client's behavior to rebuying because of the pleasure with the prior buying experience. Little research has measured the connection between product quality, the brand image on purchase decision, and repurchasing intent. This research seeks to fill this space within the literature. The author suggests the study model from prior studies and then tested it using empirical data from 249 customers at fast-food restaurants in Ho Chi Minh City. We use the Partial Least Squares approach to test measuring and structural models. Partial least squares (PLS) path modeling is a variance-based SEM (structural equation modeling) method commonly implemented in marketing and social sciences. The findings indicate that product quality has positively linked the brand image, purchasing decision, and rebuying intent. The results display brand image has positively correlated to buying decisions and rebuying intentions. Moreover, the study's findings illustrate buying decisions has entirely connected to repurchase intention.

Keywords: Brand image · Product quality · Purchase decision · Repurchase intention · PLS-SEM

1 Introduction

Repurchase intention is a client's behavior to rebuying because of the pleasure with the prior buying experience [1]. Repurchase intention links to feelings and emotions. When clients are delighted and satisfied in purchasing goods/services, clients will reinforce repurchase intention. It can be an input for marketers to improve appropriate marketing strategies to enhance the offered products/services' repurchase intention to boost profitability for the business's sustainability [2]. Some researches focus on fundamental antecedent factors of rebuying intent [3, 4]. Some researchers disclosed elements (the quality of goods and brand image) were vital antecedent variables of repurchase intention [2, 5].

Besides, in today's dynamic business context, the business needs to understand individual buyers' buying decisions and how customers employ their existing resources such as time, money, and effort to get a product/service [5]. Prior studies proved the quality of goods and brand image were also crucial antecedent buying decisions [6].

M. Al-Emran et al. (Eds.): ICETIS 2021, LNNS 299, pp. 533–545, 2022.
https://doi.org/10.1007/978-3-030-82616-1_44

The fast-food industry in Vietnam is dominated by brands (KFC, Lotteria, Jollibee). Despite sales growth, the growth rate has slowed down compared to two years ago, partly because convenience stores have combined casual goods with fast food. Besides, the customer's mindset's change with priority for healthy products has also made the fast-food industry face greater growth challenges [9]. Therefore, to survive in a dynamic competitive environment, particularly in the fast-food industry in Vietnam, restaurants need to note consumer retention determinants (e.g., product quality and brand image, etc.) [7].

In the fast-food industry, previous studies declared that determinant impacting repurchase intention such as customer satisfaction, promotion, service quality, alternatives, brand, product quality, physical design, and price [3]; brand preference, and client satisfaction [8]; the relation between shopper satisfaction and re-buying intent [9]; effects of service quality and buyer satisfaction to the rebuying intention [10], etc. Nevertheless, little research has analyzed the relation between brand image, product quality, purchase decision, and rebuying intent. This research seeks to fill this space in the literature. Therefore, this research's principal purpose is to empirically research the correlation linking product quality, brand image, purchase decision, and rebuying intent at fast-food restaurants in Vietnam.

2 Literature Review

2.1 Consumer Behavior and Consumer Decision-Making Process

"Consumer behavior" is explained as the customer's purchasing, using, and disposing of products/services [11]. Other scholars argued that consumer behavior relates to clients' action to select products/services using their possible resources (e.g., money, effort, time, etc.) [12]. In this investigation, clients' purchasing decision is clients' actions in searching, giving, utilizing, judging, and using goods and services expected to meet their needs [12].

"Consumer decision-making process" has been researched by some researchers [11, 13]. Scholars indicate that customers will make five main steps of decision-making before buying goods/services. These are needed recognition, information search (externally and internally), evaluation of choices, purchase decision, and post-buying [11, 13, 14]. The buying decision is the significant determinant of repurchase intention [15]. In this investigation, repurchase intention relates to a client's client's behavior to rebuying because of the pleasure with the prior buying experience [1].

2.2 Product Quality

Product quality is regarded as the product's ability to deliver performance outcomes that meet or even surpass what consumers expect [16]. Product quality is one of the essential factors for consumers' intentions to buy and repurchase a product [17]. Product quality includes the features of a product/service that can perform the stated needs [18]. Product quality is affected through its visual appearances, such as sound colors, differential product shapes, logos, sizes, and detailed appearance. The visual appearance will affect the initial thought of purchasers about the quality product. Product superiority and good overall quality indicate the quality product [6, 19].

Besides, the quality of the goods will influence clients in the formation of the brand image. If the product quality is right in the clients' eyes, then a positive brand image is formed automatically in clients' eyes [1]. Furthermore, Aaker (1991) declared as perceived quality could explain the remarkable differentiation and become a particular brand in consumers' minds [20]. Prior studies recommended that commodity quality significantly influences brand image [1, 17].

Moreover, some authors proved that the more the goods quality, the more the purchasers' decision to buy [6, 21]. Previous researches said that commodity quality was an antecedent of purchasing decisions [6, 21, 22].

Meanwhile, Susskind and Chan (2000) pointed out that commodity quality is an essential determinant for visiting a restaurant from the client's perspective [23]. Other studies also revealed that commodity quality was a vital influence factor for repurchase goods/services indicates the quality level of that product's performance [17]. Likewise, prior studies revealed that the product's quality was a precursor of repurchase intention [1, 24].

Thus, we proposed the after hypotheses:

H1: Product quality relates positively to brand image.
H2: Product quality is positively associated with purchasing decisions.
H3: Product quality is positively linked to repurchase intention.

2.3 Brand Image

Brand image has been an essential notion in consumer behavior studies since the 1950s. Brand image is described as the rational or emotional perceptions buyers connect to specific brands [25]. Brand image was considered to collect associations related to a brand represented in consumers' minds [26]. Aaker (1991) evaluated the brand image as the collection of associations that brand managers want to build or preserve [20]. Brand image was also an essential intangible asset, which was difficult to imitate, leading to achieving sustainable financial higher performance [27]. Hanaysha (2014) stated that a positive brand image gave several advantages, such as higher profits and enhanced consumer loyalty [17]. The brand image's relative edge is represented as the entry barrier to any market, as consumers mainly prefer brands with a strong image during their buying decisions [17, 27]. Customers have employed the brand image to arrange and remember the brand's data on their minds and help purchase decisions [20, 28]. Earlier empirical examinations reported that brand image had a sign of buying decision and positively related to purchase decision [6, 29]. Moreover, other scholars demonstrated that brand image would become the essential selection of specific products in the rebuying intent and positively influenced repurchase intention [24, 30, 31]. Therefore, we propose the following hypotheses:

H4: Brand image associates positively with purchasing decisions.
H5: Brand image links positively to repurchase intention.

2.4 Linking between Purchasing Decision and Rebuying Intention

The purchasing decision is customers' actions in searching, giving, utilizing, judging, and using a product and service expected to meet clients' needs [12]. Purchase decisions include a chain of options made by purchasers before making a buying to meet their demand. The buyer should decide where to purchase, wanted brands, model, purchase quantity, time to purchase, a sum of money to be paid, and a payment method. Marketers could affect these decisions by giving information about their products/services that may inform the purchaser's evaluation process [5]. Purchasing decision was the actual buying process customers purchase a product [16].

Rebuying intent is regarded as the individual's judgment about rebuying the same firm, based on his prior experience [32]. In the similarly above point of view, another scholar declares that rebuying intention is the possibility of using the former company again in the future [33]. Other scholars disclose that rebuying intent is the consumer's inclination to buy goods of the same restaurant for a long time [34, 35]. The previous study illustrated that buying decisions were an essential antecedent of repurchase intention [15]. It also announced that purchase decisions had a link with repurchase intention [15].

Therefore, hypothesis 6 is recommended as follows:

H6: Purchase decision is positively linked to repurchase intent.

Figure 1 displays the recommended research model.

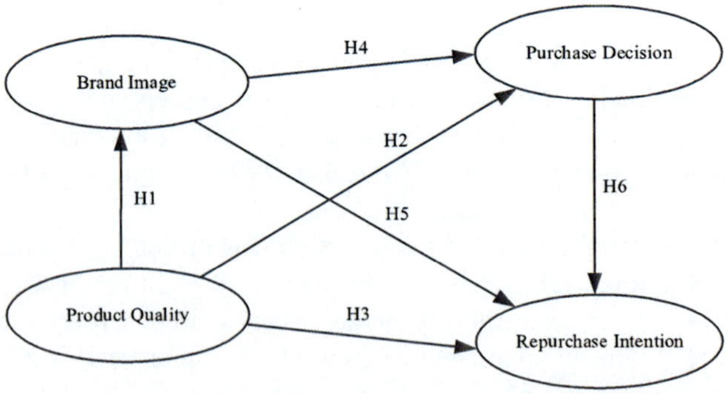

Fig. 1. The recommended research model

3 Methodology of Research

3.1 Measures

The constructs measure indicators drawn from prior studies; then, these measure indicators are slightly adjusted to accommodate the research context. We apply a five-point Likert scale to estimate the indicators. We adopt four indicators of product quality from [36], four items of the brand image from [37], five indicators of purchase decision from [5, 38], and four indicators of repurchase intention from [39, 40].

3.2 Sample and Data Collection

We gather the research data from customers at fast-food restaurants in Ho Chi Minh City, Vietnam. This research uses the convenience sampling technique. We give a sum of 300 questionnaires, and 51 responses were removed because there was not full information. We use 249 answers for this analysis. The sample included 94 male consumers (37.8%) and 155 female consumers (62.2%). Buyers were category ages below 18 made 13.3%; 63.5% were category ages 18 to 25; 14.9% in category ages 26 to 35; and 8.3% were older than 36.

3.3 Analytical Technique

We usage the PLS technique in this investigation as this method has the most meaningful reasons for using this approach was: (1) nonnormal data, (2) small sample quantities, and (3) formatively measured variables [41]. The PLS path modeling is a variance-based structural equation modeling (SEM) method that is widely performed in the business and social sciences. It is a new statistical tool of the research method [42]. The PLS technique is applied to examining the suggested research model and hypotheses.

4 Results and Discussion

4.1 Results

(1) Measurement Model

Table 1 shows the findings of concepts and sub-variables scale. We adopt Cronbach's alpha and composite reliability (CR) to estimate the study's reliability. Cronbach's alpha (α) criteria of the concepts should be over 0.70, and composite reliability (CR) values more numerous than 0.70, which signifies that there is satisfactory internal consistency among the notions [43]. Table 1 indicated Cronbach's alpha points of the constructs are over 0.70. So, these concepts get inside consistent reliability.

We implement the items' factor loading and the average variance extracted (AVE) to estimate convergent validity. The AVE and factor loading should be larger than 0.50 [43, 44]. In this research, the items' factor loading and the AVE points are over 0.50. Therefore, the constructs' convergent validity is suitable.

Table 1. The analysis results of constructs and sub-variables

Constructs and sub-variables	Factor loading	α	CR	AVE
Product quality		0.880	0.872	0.632
1. The goods display of the fast-food restaurant is visually charming	0.861			
2. The goods of the fast-food restaurant is delicious	0.902			
3. The goods of the fast-food restaurant is fresh	0.849			
4. The product of the fast-food restaurant provides healthy options	0.820			
Brand image		0.835	0.890	0.669
1. The fast-food restaurant brand is reliable	0.842			
2. The fast-food restaurant brand is attractive	0.793			
3. The fast-food restaurant brand is pleasing	0.836			
4. The fast-food restaurant brand recognizes a fine reputation	0.801			
Purchase decision		0.895	0.897	0.705
1. I feel right about my decision to purchase goods from the fast-food restaurant	0.857			
2. I often purchase from this fast-food restaurant	0.801			
3. I have a want to buy goods from this fast-food restaurant	0.824			
4. I would still buy from this fast-food restaurant, even if another fast-food restaurant has a promotion	0.824			
5. Overall, I am happy with buying goods at this fast-food restaurant	0.889			
Repurchase intention		0.804	0.872	0.632
1. I will purchase products from this fast-food restaurant next time	0.703			
2. If I need to buy the product from this fast-food restaurant next time, I will still choose this fast-food restaurant	0.826			
3. I will recommend this fast-food restaurant to someone who seeks my advice	0.835			
4. I will spread positive word-of-mouth about this fast-food restaurant	0.808			

Furthermore, we measure discriminant validity by the Fornell-Larcker standard [45]. It describes the square root of the AVE points with the latent variable. Exactly, the AVE's square root should be more potent than its highest correlation with any other concepts [43]. Table 2 declares that AVE's square root is reflective of concept product quality, brand image, purchase decision, and rebuying intention is more potent than the corresponding latent variables correlation. Consequently, the discriminant validity of these concepts is confirmed.

Table 2. The analyzing outcomes of discriminant validity

	Band image	Product quality	Purchase decision	Repurchase intention
Brand image	0.818			
Product quality	0.672	0.858		
Purchase decision	0.651	0.712	0.840	
Repurchase intention	0.712	0.716	0.789	0.795

(2) The Structural Model Evaluation and Hypotheses Testing

(2.1) The Model Fit Evaluation

Figure 2 and Table 3 illustrate the proposed model fit results. Table 3 shows that the Chi-square = 577.858 was meaningful at the 0.05 level (p = 0.00). SRMR (standardized root mean square residual) is a suitable measure standard of the recommended research model. A model is suitable by custom when SRMR is below 0.08 [46]. The summary outcomes in Table 3 reveal that this model has an SRMR value of 0.068 < 0.08. So, the suggested research model fits the research data. Besides, checking off a multicollinearity concern declares that all VIF points were under the cut-off of 5. Therefore, there is no multicollinearity matter in the structural model [43].

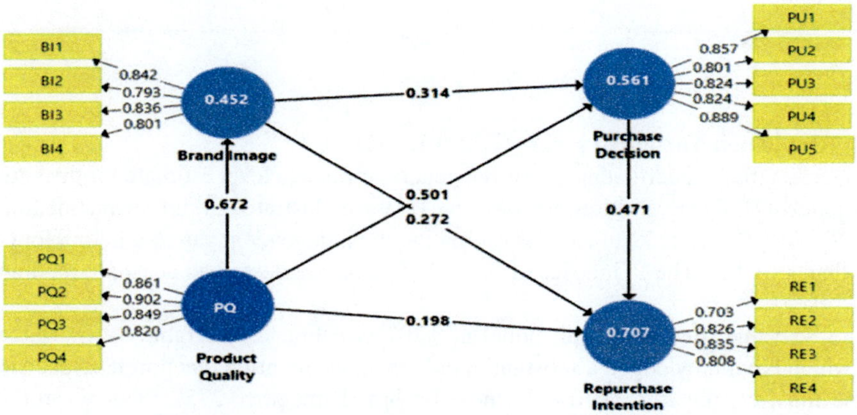

Fig. 2. The structural model' analyzing results (PLS-SEM)

Table 3. The summary of the model fit

	SRMR	d_ULS	d_G1	d_G2	Chi-square	NFI
Saturated model	0.068	0.710	0.441	0.389	577.858	0.806

(2.2) Hypotheses Testing

Table 4 displays the results of hypothesis testing. We apply 5000 resamplings bootstrapping results with a corresponding 95 percent confidence interval (CI) for the proposed model shows that all parameter estimates of the links are in the confidence interval. Besides, these confidence interval weights of these relations do not cover the zero value. In other words, all parameter estimates of these links in the recommended model are significant at the 5% level. Therefore, these hypotheses H1-H6 are approved.

Table 4. Hypothesis testing' results

Relationships	Hypotheses	Path coefficients	95% CI	Results
Product quality → Brand image	H1	0.672	[0.576–0.756]	Approved
Product quality → Purchase decision	H2	0.501	[0.390–0.612]	Approved
Product quality → Repurchase intention	H3	0.198	[0.071–0.315]	Approved
Brand image → Purchase decision	H4	0.314	[0.192–0.430]	Approved
Brand image → Repurchase intention	H5	0.272	[0.161–0.388]	Approved
Purchase decision → Repurchase intention	H6	0.471	[0.348–0.598]	Approved

(3) R^2 (Explained Variance) and f^2 (Effect Level)

The determination's coefficient (R^2) is the general influence level estimate for the structural model [47]. The R^2 indexes of 0.67, 0.33, 0.19 are illustrated as significant, medium, low [48]. The (f^2) effect level enables estimating the independent variable giving for the dependent variable. The f^2 indexes of 0.02, 0.15, 0.35 are described as small, medium, high [49].

The R^2 value for the overall model is 0.707 (see Fig. 2 and Table 5), more than 0.67, which is acknowledged a substantial influence; we mention that purchase decision has the most crucial impact (0.471), next by brand image (0.272), then by product quality (0.198). Besides, goods quality and brand image explain 56.1% of the variance on purchase decision; we also reveal that the product quality has a more significant effect

(0.501) than the brand image (0.314). Furthermore, the product quality interprets 45.2% of the variance on brand image, and it has a strong influence (0.672).

Table 5 describes the f^2 affecting sizes. The high f^2 effect levels happen for the linking of product quality → brand image (0.824), purchase decision → repurchase intention (0.333), and goods quality → purchase decision (0.313). The medium-weak f^2 effect sizes appear for the linking of brand image → purchase decision (0.123) and brand image → repurchase intention (0.124). The small f^2 effect level happens for the connection of product quality → repurchase intention (0.056).

Table 5. R^2 and f^2 results

Relationships	Path coefficients	f^2	Construct	R^2
Product quality → Brand image	0.672	0.824	Brand image	0.452
Product quality → Purchase decision	0.501	0.313		
Product quality → Repurchase intention	0.198	0.056	Purchase decision	0.561
Brand image → Purchase decision	0.314	0.123		
Brand image → Repurchase intention	0.272	0.124	Repurchase intention	0.707
Purchase decision → Repurchase intention	0.471	0.333		

4.2 Discussion

This research has explained the link between product quality, brand image on purchase decision, and rebuying intention in Vietnam's circumstances. The research results confirmed that the six hypotheses in the model were accepted.

The research finding demonstrated that product quality has significantly positively related to brand image. Product quality was an antecedent of the brand image. The f^2 effect extent of the linking between product quality and brand image was large (0.824). The prior empirical research confirmed this research [17]. The research results also explained that product quality correlated positively with purchasing decisions. Product quality was an antecedent of purchase decisions; the f^2 influence size of linking product quality and buying decisions was large (0.313). The previous authors proved this research [6, 21, 22]. The investigation findings also demonstrated that product quality positively influenced rebuying intention. Product quality was a predictor of repurchase intention, yet the f^2 impact size of the association between commodity quality and rebuying intent was small (0.056). The early empirical investigations verified this research [1, 24].

Besides, the research results also proved that brand image has positively connected to buying decisions. The brand image was a predecessor of the purchase decision, and the f^2 effect level of the linking of brand image and product quality was average-weak (0.123). The early empirical studies demonstrated this research result [6, 29]. Next, the research outcomes further explained that brand image has positively linked to repurchase intention. The brand image was a precursor of rebuying intent, and the f^2 impact level of the linking of brand image and rebuying intent was average-weak (0.124). The past investigations verified this research [24, 30, 31].

Moreover, the study's findings also reinforced that buying decisions have positively influenced repurchase intention. Buying decisions was a predictor of repurchase intent, and the effect size of the relationship of the purchase decision and repurchase intention was large (0.333). The f^2 previous empirical study supported this research [15].

5 Implications and Future Research

5.1 Theoretical Implications

This study has analyzed the link between product quality, the brand image on purchase decision, and rebuying intent. So, academics can consider this paper as the reference document when studying consumer behavior, especially concepts as product quality, band image, purchase decision, and rebuying intention. Besides, the scholars can use the model of this paper and applying it to another context.

5.2 Managerial Implications

In light of prior research, this study has revealed the linking between product quality, brand image on purchase decision, and rebuying intention at fast-food restaurants in Vietnam's circumstances. Therefore, managers should build plans that enhance the consumer's thought of product quality and brand image to improve buying decisions and rebuying intentions.

This study guides managers to see the critical of product quality on brand image, purchase decision, and repurchase intention. Thus, restaurant managers should create the product to match or even exceed what customers. Managers should note some of the product's points, such as the product display, visually charming, delicious, fresh, and healthy options. If consumers have positively known about the restaurant's product quality, they will tend to a favorable brand image and tend to buy decisions and rebuying intentions.

Likewise, managers should give marketing plans to develop the brand image, such as advertising through social media, promotions, events, etc., and other marketing activities. If clients positively link to brand image, clients will also positively tend to purchase decisions and rebuying intent.

Finally, this study will support practitioners to see the critical of buying decisions on rebuying intention. So, managers could give a restaurant atmosphere (inside and outside) to match consumers' needs and wants. Creating a positive restaurant atmosphere will help consumers feel good and happy about purchasing products. When consumers have the right purchasing choices, they will tend to be repurchase intention.

5.3 Future Research

The study could not represent all disciplines; therefore, future research could be continued research in various industries, such as online shopping, washing powder, etc. Second, this research only examines the link between brand image, product quality on purchase decision, and rebuying intention; thus, future investigations should focus on other concepts such as satisfaction, trust, etc.

References

1. Ketut, Y.I.: The role of brand image mediating the effect of product quality on repurchase intention. Russ. J. Agric. Socio-Econ. Sci. **83**, 172–180 (2018)
2. Paramananda, N., Sukaatmadja, I.P.G.: The impact of price perception and brand image on customer satisfaction and repurchase intention. Int. J. Econ. Commer. Manag. **12**, 59–73 (2018)
3. Fungai, M.: Factors influencing customer repurchase intention in the fast food industry. A case study of Innscor - Mutare, Zimbabwe. Bus. Soc. Sci. J. **2**, 113–133 (2017)
4. Shalehah, A., Trisno, I.L.O., Moslehpour, M., Lin Cor, P.K.: The effect of Korean beauty product characteristics on brand loyalty and customer repurchase intention in Indonesia. In: 2019 16th International Conference on Service Systems and Service Management, ICSSSM 2019, pp. 1–5 (2019)
5. Hanaysha, J.R.: An examination of the factors affecting consumer's purchase decision in the Malaysian retail market. PSU Res. Rev. **2**, 7–23 (2018)
6. Arjuna, H., Ilmi, S.: Effect of brand image, price and quality of the product on the smartphone purchase decision. EkBis: Jurnal Ekonomi dan Bisnis **3**, 294–305 (2020)
7. Chi, Q.: KFC, Lotteria, Jollibee collected trillions of VND from the Vietnamese market (Vietnamese). https://doanhnhan.vn/kfc-lotteria-jollibee-thu-ca-nghin-ty-dong-tu-thi-truong-viet-nam-33323.html
8. Awi, Y.L., Chaipoopirutana, S.: A study of factors affecting consumer's repurchase intention toward XYZ restaurant, Myanmar. In: International Conference on Trends in Economics, Humanities and Management (ICTEHM 2014), pp. 181–184 (2014). https://doi.org/10.15242/icehm.ed0814093
9. Wiranto, I.I.J., Husin, H.A.: Relationship between customer satisfaction and repurchase intention on Mcdonald's Kuala Lumpur. **6**, 70–79 (2016)
10. Mensah, I., Mensah, R.D.: Effects of service quality and customer satisfaction on repurchase intention in restaurants on University of Cape Coast Campus. J. Tourism Heritage Serv. Mark. **4**, 27–36 (2018)
11. Blackwell, R., Miniard, P., Engel, J.: Consumer Behavior. Thomson/South-Western, Mason (2006)
12. Schiffman, L., Kanuk, L.: Consumer Behavior. Prentice Hall, Upper Saddle River (2000)
13. Kotler, P., Armstrong, G.: Principles of Marketing. Pearson Education Limited, Harlow (2016)
14. Oke, A.O., Kamolshotiros, P., Popoola, O.Y., Ajagbe, M.A., Olujobi, O.J.: Consumer behavior towards decision making and loyalty to particular brands. Int. Rev. Manag. Mark. **6**, 43–52 (2016)
15. Ali, H.: Purchase decision and repurchase models: product quality and process analysis (case study of house ownership credit financing in Permata Sharia Bank Jakarta). Scholars Bull. **05**, 526–535 (2019)
16. Kotler, P., Keller, K.L.: Marketing Management. Pearson Education Inc., Upper Saddle River (2016)

17. Hanaysha, J., Hilman, H., Abdul-Ghani, N.H.: Direct and indirect effects of product innovation and product quality on brand image: empirical evidence from automotive industry. Int. J. Sci. Res. Publ. **4**, 2250–3153 (2014)
18. Yee, W., Sidek, Y.: Influence of brand loyalty on consumer satisfaction towards sportswear. Int. J. Econ. Manag. **2**, 221–236 (2008)
19. Wang, E.S.T.: The influence of visual packaging design on perceived food product quality, value, and brand preference. Int. J. Retail Distrib. Manag. **41**, 805–816 (2013)
20. Aaker, D.A.: Managing Brand Equity. The Free Press, New York (1991)
21. Rohman, E.S.I.D.: The model of product quality, promotion, price, and purchase decisions. Jurnal Ekonomi **23**, 260–271 (2018)
22. Brata, B.H., Husani, S., Ali, H.: The influence of quality products, price, promotion, and location to product purchase decision on nitchi at PT. Jaya Swarasa Agung in Central Jakarta. Saudi J. Bus. Manag. Stud. **2**, 433–445 (2017)
23. Susskind, A.M., Chan, E.K.: How restaurant features affect check averages. Cornell Hotel Restaurant Adm. Q. **41**(56–63), 8 (2000)
24. Girsang, N.M., Rini, E.S., Gultom, P.: The effect of brand image and product quality on re-purchase intention with customer satisfaction as intervening variables in consumers of skincare oriflame users – a study on students of north sumatra university, faculty of economics and business. Eur. J. Manag. Mark. Stud. **5**, 40–57 (2020)
25. Dobni, D., Zinkhan, G.M.: In search of brand image: a foundation analysis. In: Goldberg, M.E., Gorn, G., Pollay, R.W. (eds.) Advances in Consumer Research, Provo, UT, pp. 110–119 (1990)
26. Keller, K.L.: Conceptualizing, measuring, and managing customer-based brand equity. J. Mark. **57**, 1–22 (1993)
27. Roberts, P.W., Dowling, G.R.: Corporate reputation and sustained superior financial performance. Strateg. Manag. J. **23**, 1077–1093 (2002)
28. Low, G., Charles, W., Lamb, J.: The measurement and dimensionality of brand associations the measurement and dimensionality of brand associations. J. Product Brand Manag. **9**, 350–370 (2010)
29. Suyanti, N., Khuzaini, K.: The role of brand image in the relationship of product quality and price perceived on purchasing decisions for oppo smartphone consumers. In: Proceeding Book 7th Asian Academic Society International Conference, pp. 140–144 (2019)
30. Ranjbarian, B., Sanayei, A., Kaboli, M.R., Hadadian, A.: An analysis of brand image, perceived quality, customer satisfaction and re-purchase intention in iranian department stores. Int. J. Bus. Manag. **7**, 40–48 (2012)
31. Shah, S.H.H., Lei, S., Hussain, S.T., Mariam, S.: How consumer perceived ethicality influence repurchase intentions and word-of-mouth? A mediated moderation model. Asian J. Bus. Ethics **9**(1), 1–21 (2019). https://doi.org/10.1007/s13520-019-00096-1
32. Hellier, P.K., Geursen, G.M., Carr, R.A., Rickard, J.A.: Customer repurchase intention: a general structural equation model. Eur. J. Mark. **37**, 1762–1800 (2003)
33. Upamannyu, N.K., Gulati, C., Chack, A., Kaur, G.: The effect of customer trust on customer loyalty and repurchase intention: the moderating influence of perceived CSR. Int. J. Res. IT Manag. Eng. **5**, 1–31 (2015)
34. Chinomona, R., Maziriri, E.T.: The influence of brand awareness, brand association and product quality on brand loyalty and repurchase intention: a case of male consumers for cosmetic brands in South Africa. J. Bus. Retail Manag. Res. **12**, 143–154 (2017)
35. Gounaris, S., Dimitriadis, S., Stathakopoulos, V.: An examination of the effects of service quality and satisfaction on customers' behavioral intentions in e-shopping. J. Serv. Mark. **24**, 142–156 (2010)
36. Namkung, Y., Jang, S.C.: Does food quality really matter in restaurants? Its impact on customer satisfaction and behavioral intentions. J. Hosp. Tourism Res. **31**, 387–409 (2007)

37. Lien, C.H., Wen, M.J., Huang, L.C., Wu, K.L.: Online hotel booking: the effects of brand image, price, trust and value on purchase intentions. Asia Pac. Manag. Rev. **20**, 210–218 (2015)
38. ALHuwaishel, N.S., AL-Meshal, S.A.: The impact of perceived value, quality, and loyalty on purchase decision in the accessories department: study on saudi females. Br. J. Mark. Stud. **6**, 21–31 (2018)
39. Aslam, W., Ham, M., Farhat, K.: Influencing factors of brand perception on consumers' repurchase intention: an examination of online apparel shopping. Manag. (Croatia) **23**, 87–101 (2018)
40. Goh, S.K., Jiang, N., Hak, M.F.A., Tee, P.L.: Determinants of smartphone repeat purchase intention among Malaysians: a moderation role of social influence and a mediating effect of consumer satisfaction. Int. Rev. Manag. Mark. **6**, 993–1004 (2016)
41. Hair, J.F., Sarstedt, M., Hopkins, L., Kuppelwieser, V.G.: Partial least squares structural equation modeling (PLS-SEM): an emerging tool in business research. Eur. Bus. Rev. **26**, 106–121 (2014)
42. Henseler, J., Hubona, G., Ray, P.A.: Using PLS path modeling in new technology research: updated guidelines. Ind. Manag. Data Syst. **116**, 2–20 (2016)
43. Hair, J.F., Hult, G.T.M., Ringle, C.M., Sarstedt, M.: A Primer on Partial Least Squares Structural Equation Modeling (PLS-SEM). SAGE Publications Inc, Los Angeles (2017)
44. Bagozzi, R.P., Yi, Y.: On the evaluation of structural equation models. Acad. Mark. Sci. **16**, 74–94 (1988)
45. Fornell, C., Larcker, D.F.: Evaluating structural equation models with unobservable variables and measurement error. J. Mark. Res. **18**, 39–50 (1981)
46. Hu, L., Bentler, P.: Fit indices in covariance structure modeling: sensitivity to underparameterized model misspecification. Psychol. Methods **3**, 424–453 (1998)
47. Garson, G.D.: Partial Least Squares: Regression & Structural Equation Models. Statistical Associates Publishers, Asheboro (2016)
48. Chin, W.W.: The partial least squares approach to structural equation modeling. In: Macoulides, G.A. (ed.) Modern Methods for Business Research, pp. 295–336. Lawrence Erlbaum Associates, Mahwah (1998)
49. Henseler, J., Ringle, C.M., Sinkovics, R.R.: The use of partial least squares path modeling in international marketing. Adv. Int. Mark. **20**, 277–319 (2009)

Millennials' Intention to Use Digital Real Estate Platform During the COVID-19 Pandemic: The Stimulus-Organism-Response Approach

Chang Mui Ling Dyana[1](✉), Rizal Hamid[2], Nelson Lajuni[3], and Norazah Mohd Suki[4]

[1] Centre for the Promotion of Knowledge and Language Learning, Universiti Malaysia Sabah, Jalan UMS, 88400 Kota Kinabalu, Sabah, Malaysia
dyana@ums.edu.my

[2] Labuan Faculty of International Finance, Universiti Malaysia Sabah, Labuan International Campus, 87000 Labuan, Malaysia

[3] Faculty of Business, Economics and Accountancy, Universiti Malaysia Sabah, Jalan UMS, 88400 Kota Kinabalu, Sabah, Malaysia

[4] Othman Yeop Abdullah Graduate School of Business, Universiti Utara Malaysia, 50300 Kuala Lumpur, Malaysia

Abstract. Malaysia's real estate industry is undergoing its digital transformation as COVID-19 takes effect. Both forms of physical property transactions have been digitalised, which prompted many opportunities for the sector. In response, the real estate industry acts as an enterprise that provides more streamlined offerings to buyers than ever before. Much of the digital technology has offered new opportunities to realty business to target specific market segments such as millennials who have a unique characteristic compared to other generations. However, few studies have investigated this area concerning the millennials. Thus, this study examines the intention to use the digital real estate platform among millennials in Malaysia by utilising the Stimulus-Organism-Response (S-O-R) model as the guiding principle. Partial Least Squares-Structural Equation Modelling (PLS-SEM) approach was employed to test the hypotheses. The effectiveness factor strongly influenced millennials' intention to use the digital real estate platform because they believed that by using the technology, the real estate platform would be free of effort and enabled them to save time when making important decisions whether to purchase or rent any of the real estate product or services. Digital technology also allows millennials to search at their pace and convenience, learning a great deal about the property and related issues before they begin any actual negotiations. The findings highlight several factors that substantially affect millennial consumers' intention to use the real estate platform. Suggestions for further study are also discussed.

Keywords: Millennial intention · Informativeness · Effectiveness · Entertainment · Stimulus-Organism Response (SOR) model

M. Al-Emran et al. (Eds.): ICETIS 2021, LNNS 299, pp. 546–557, 2022.
https://doi.org/10.1007/978-3-030-82616-1_45

1 Introduction

Malaysia's real estate industry is undergoing its digital transformation in response to the COVID-19 pandemic. Both forms of physical property transactions have been digitalised, which has created endless new possibilities for the sector, therefore, necessitated the need for best practises. Therefore, the real estate industry plays a crucial role as an enterprise that provides more streamlined offerings to buyers than ever before [19, 20, 25, 34, 43]. The recent COVID-19 pandemic has steered the real estate market industry to change their norm of conducting business by inculcating businesses through online platforms by creating a marketplace of sorts for buyers and sellers and utilising the digital real estate platforms in finding opportunities for technology to establish itself in the real estate market [3, 16]. The digital platform is growing important, but the online platforms have yet been optimised [36] by the real estate sector. Much of the digital technology on offer seems geared towards the millennial generation who has a unique characteristic compared to other generations. Despite this, the millennials using the real state platform has depreciated by one percent, from 35% last year to 34% this year. This percentage drop happened even though there are nearly four billion dynamic internet users, and almost three billion web-based clients in the year 2017 [20]. This marks a growth of an encouraging seven percent for social media websites in 2016 [32].

In the field of marketing, management, and tourism, social media advertising via ads-platform has received sufficient attention from researchers [47]. On the contrary, little investigation has been accomplished on the application of real estate platforms for the property industry, especially that of developing countries, concerning the millennials [6]. For that reason, this research aims to investigate factors that affect millennial's intention to use the digital real estate platform by utilising the Stimulus-Organism-Response (S-O-R) model as the guiding principle. The findings from this study can contribute to foster understanding of the behavioural study of the millennial's intention to use the online property website, particularly customer intention behaviour, stimulus criteria, interjecting factors, and behavioural responses.

2 Literature Review

2.1 Stimulus-Organism-Response

The Stimulus-Organism-Response (hereinafter referred to as S-O-R) model which was developed by Mehrabian and Russell (1974) "incorporates three aspects: the atmosphere or stimulus (S) that initiates customer reactions, the organism (O) that responds, and the bona fide reaction (R)". The model aims to align customers' reactions to make clear individual perceptions and emotions to the outer stimuli, and the good or poor practices which are formed in the process. Using the S-O-R model to investigate customers' actions can help in defining the ecological enhancement and buyers' outward and inward responses behaviours while using real estate online platform.

Following some prior studies, the S-O-R model was employed in this study to reveal that digital technology's outline affects the consumer's emotions and initiates acts that influence the number of items purchased and the quantity spent at the store [46, 55]. Moreover, several studies have also employed this model to investigate several variables such as online booking, auction, purchase intention, impulse buying behaviour, flow experience, and mobile purchasing [11, 12, 22]. S-O-R is used in this current study to investigate one's intention in using the real estate platform for advertisement or to find information about the real estate in terms of informativeness, effectiveness, and entertainment.

2.2 Informativeness

Informativeness is explained by [5] as "ease and dismantles the limitations on mobility, but also improves user understanding and skill by instant appliance services in mobile manufacturing" [51]. Informativeness affects consumers' cognitive satisfaction during their time on social media. [13, 14] asserted that the informativeness of e-commerce affects customers who are engaged in sending, getting information, and conducting online links on shopping platforms whenever, and wherever instantly. Besides, past studies noted that informativeness has a strong impact on students' personal lives concerning the students' acceptance of ICT technology in an educational setting [18, 21, 45, 52–54]. Also, such studies have demonstrated that female students are more strongly affected than their male counterparts in Chinese business schools. They reported that using elaborate business application software does not require effort and that the technology could improve their educational output. Therefore, it is hypothesised that:

H1: Informativeness has a positive effect on millennials' intention to use the digital real estate platform.

2.3 Effectiveness

Effectiveness refers to the reflection of the degree to which information is accurate, up-to-date, available, wide-ranging, and relevant to Internet users [4, 40]. Moreover, [31] pointed out that the "perception of site content can be measured by how informative it is if it provides detailed and specific information on products or other relevant topics". However, [17] argued that "information content of Web pages, per se, does not appear to attract visitors". Consequently, consumers are prone to recognize and reminiscence websites that satisfy their desires and provide them with the effective information needed [1, 28, 44]. Therefore, the second hypothesis was derived:

H2: Effectiveness of use has a good influence on millennials' intent to use the digital real estate platform.

2.4 Entertainment

The construct of entertainment has been extensively used in advertising settings. [41] pointed out that entertainment is entirely associated with (a) site attitudes and (b) site involvement. [48] found that entertainment can be measured as a low task-relevant atmospheric cue and significantly impact site attitudes in the S-O-R framework. Meanwhile, [26] pointed out that entertainment in social media marketing can influence the millennials' online purchasing behaviour of low involvement products. As discussed earlier, entertainment is one of the new constructs integrated into the S-O-R framework in this study. This construct was adopted from a literature review by [10]. The study shows that entertainment has a positive influence on the millennials' intention to use digital real estate platforms in Malaysia. The entertainment variable has a positive influence on intention to use the digital real estate platforms due to millennials unique characteristics that can be defined as trendy and technologically savvy [26]. Hence, the subsequent hypothesis is put forth:

H3: Entertainment exchange positively influences the millennial's intention to use the digital real estate platform.

2.5 Flow Experience

Flow experience needs a measured quality in providing situations (e.g., service counter) to execute certain behaviours (e.g., digital technology exploration through an online platform) [56]. The previous scholar emphasises digital technology such as digital real estate platforms that the flow experience can trigger behavioural consequences [30, 39]. Thus, the intention to use the real estate platform is affected by purchasing behaviour [27, 58]. Moreover, the flow experience affects behavioural intention, such as creating a higher tendency to purchase from the digital real estate platform [15]. An Internet user who lacks experience has a strong inclination to perform an unintended online transaction and has extended online usage [35]. Likewise, [29] indicated that flow experience positively influences one's intention to play online computer games. Therefore, it is postulated that:

H4: Flow experience has a positive influence on millennial intention to use online property websites (Fig. 1).

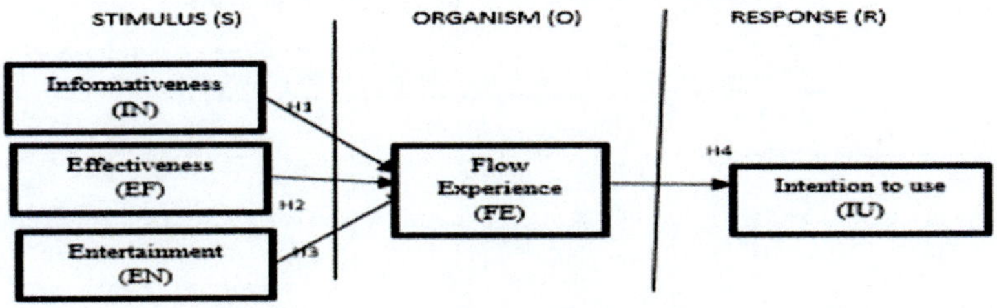

Fig. 1. Proposed conceptual framework

3 Methodology

The samples of this study constitute millennials in Malaysia in the age bracket of 18 to 38 years old which were approached using non-probability purposive sampling. The sample size was estimated using G*power 3.0 software, with the effect size of the f square 0.15, α error pro-0.05, and power Gf 0.95 with three tested predictors. Consequently, the minimum sampling necessary for this study was 119 respondents. Of the

400 online surveys emailed to the respondents, 354 surveys were sent to the respondents considered appropriate for data analysis as the balance 46 were incomplete responses. Data collection using an online survey was carried out from 28th July 2020–30th August 2020.

The capability of online consumers to respond fast and minimise costs has encouraged researchers to use online surveys to conduct research topics of interest [57]. Using a five-point Likert scale from 1 = disagree to 5 = agree, the respondents scored the items in the 3-section online questionnaire. The measurement of items for informativeness and entertainment was adapted from [10], while the item's effectiveness was based on studies from [4].

The collected data was firstly analysed using Statistical Package for Social Sciences (SPSS) Version 23. The data were tested with frequency and descriptive statistics to understand the respondents' demographic information and the descriptive data of each item used in the questionnaire. Subsequently, reliability analysis and confirmatory factor analysis tests were conducted to test each variable's reliability and validity. Partial Least Squares-Structural Equation Modelling (PLS-SEM) approach was employed to analyse the data which was used to test the hypotheses [8, 9, 49].

4 Data Analysis

Table 1 displays the socio-demographic profile of the respondents. The results show that, more than half of the respondents are female (58.8%) while the rest being male (41.2%) with age between 18 and 30 years old. Additionally, in terms of races, 53.7% are Chinese followed by Bumiputra (31.4%), Malay (8.5%), and others (5.1%), respectively. In terms of respondents' occupation most of them are students (25.4%), professional workers

Table 1. Socio-demographic profile of respondents

Variable		Frequency	Percentage
Gender	Male	146	41.2
	Female	208	58.8
Races	Chinese	190	53.7
	Bumiputra	111	31.4
	Malay	30	8.5
	Others	18	5.1
Occupation	Student	90	25.4
	Professional	86	24.3
	Housewives	48	13.6
	Clerks	44	12.4
	Manager	44	12.4
	Retirees	23	6.5
	Front liners	19	5.4
Age (year old)	18–30	354	100

(24.3%), housewives (13.6%), managers (12.4%), clerks (12.4%), retirees (6.5%) and front liners (5.4%).

4.1 Measurement Model Assessment

In the measurement model assessment, the convergent and discriminatory validity of 25 indicators is analysed. Both the convergent and discriminatory validity are validated and satisfactory as each factor loading is beyond 0.50 and none of the factor items share significant degrees of residual variance with other constructs. Table 2 reveals the results of reliability and validity testing. The findings confirmed that the variables have high internal consistency and adequate average variance extracted (AVE) to confirm the convergent validity [24]. The results were summarised in Table 2.

Table 2. Measurement model assessment

Construct	Item	Loadings	CA	CR	AVE	Convergent validity (AVE > 0.5)
Informativeness (IN)	EC1a	0.779	0.903	0.926	0.675	Yes
	EC1b	0.783				
	EC2	0.819				
	EC3	0.836				
	EC4	0.831				
	EC5	0.877				
Flow Experience (FE)	FE1	0.676	0.847	0.886	0.566	Yes
	FE2	0.723				
	FE3	0.722				
	FE4	0.774				
	FE5	0.774				
	FE6	0.834				
Intention to Use (IU)	IU1	0.912	0.941	0.958	0.850	Yes
	IU2	0.929				
	IU3	0.941				
	IU4	0.907				
Effectiveness (EF)	WE1	0.877	0.924	0.944	0.770	Yes
	WE2	0.914				
	WE3	0.898				
	WE4	0.913				
	WE5	0.780				
Entertainment (EN)	WI1	0.763	0.919	0.943	0.805	Yes
	WI2	0.776				
	WI3	0.765				
	WI4	0.743				

No item was deleted as loading Composite Reliability > .708 (Hair et al., 2017)

Table 3 displays the HTMT criterion to evaluate discriminant validity [49]. The result specifies that discriminant validity is strongly established at HTMT 0.85 [24]. The findings indicate that it is appropriate to proceed for structural model assessment to assess the research hypotheses as no concerns of multi-collinearity emerged between the items loaded on separate constructs in the outer model.

Table 3. HTMT criterion

	IN	FE	IU	EF	EN
IN	0.821				
FE	0.437	0.752			
IU	0.574	0.580	0.822		
EF	0.634	0.577	0.554	0.878	
EN	0.746	0.340	0.583	0.552	0.897

Criteria: Discriminant validity is established at HTMT0.90

To evaluate the formative measurement models, the formative construct must highly correlate with a reflective measure of the same construct. This type of analysis is known as redundancy analysis [9]. Specifically, [23] mentioned that redundancy analysis can be achieved by using formative construct as an exogenous latent variable predicting the same construct operationalized by reflective indicators or a global single item, which summaries the essence of the construct that the formative indicators are intended to measure. Table 4 details the formative construct of path coefficients of 0. 728, which surpassed a cut-off value of 0.70, signifying convergent validity is satisfactory [24]. Furthermore, the VIF values are below the threshold value of 5.0. Therefore, the measured constructs have sufficient degrees of convergent validity [33].

Table 4. Formative measurement model assessment

Construct	Items	Convergent validity	Weight	VIF	t-value weights	Sig	
Informativeness (IN)	IN1	0.728	0.111	3.351	15.417	0.000	**
	IN2		0.099	3.104	13.578	0.000	**
	IN3		0.096	2.567	16.516	0.000	**
	IN4		0.096	2.879	15.351	0.000	**
	IN5		0.091	3.016	18.449	0.000	**
	IN6		0.100	3.058	17.635	0.000	**
Effectiveness (EF)	EF1		0.072	3.424	12.890	0.000	**

(*continued*)

Table 4. (*continued*)

Construct	Items	Convergent validity	Weight	VIF	t-value weights	Sig	
	EF2		0.072	4.219	13.250	0.000	**
	EF3		0.079	4.148	14.259	0.000	**
	EF4		0.073	4.550	12.654	0.000	**
	EF5		0.077	2.072	14.494	0.000	**
Entertainment (EN)	EN1		0.113	4.333	15.698	0.000	**
	EN2		0.119	4.241	16.932	0.000	**
	EN3		0.124	3.603	17.315	0.000	**

4.2 Structural Model Assessment

Table 5 presents the assessment of the path coefficient by using 5000 bootstrap re-sampling. The results reveal that the informativeness variable has a positive effect on millennials' intention to use the digital real estate platform ($\beta = 0.115, p < 0.01$). Thus, H1 is supported. Likewise, the effectiveness variable reveals a strong positive significant impact on the millennials' intention to use the digital real estate platform. Hence, retaining H2 as expected. Subsequently, the informativeness variable has a positive effect on millennials' intention to use the digital real estate platform ($\beta = 0.163$, $p < 0.01$). In addition, the results of H3 indicated that millennial's intention to use the digital real estate platform was significantly influenced by entertainment as standardized beta coefficient is 0.163 at $p < 0.01$. Therefore, maintaining the postulated H3. The ensuing hypothesis, H4 put forward that flow experience has a positive influence on millennial intention to use online property websites. This relationship produced standardized beta coefficient of 0.163 and $p < 0.01$. Consequently, H4 is also strengthened. The predictive relevance values for the dependent variable are greater than 0, showing that the independent variables can predict the millennial's intention to use the property websites as anticipated by Q^2 using the blindfolding procedure [24].

Table 5. Path Coefficients and Model Quality Assessment

Paths	Beta	S.E	t-value	p-value	LLCI	ULCI	Decision
H1 Informativeness ---> Intention to Use	0.115	0.014	8.014	0.000	0.142	0.084	Supported
H2 Effectiveness ---> Intention to Use	0.163	0.021	7.710	0.000	0.201	0.120	Supported
H3 Entertainment ---> Intention to Use	0.156	0.022	7.220	0.000	0.198	0.116	Supported
H4 Flow ---> Intention to Use	0.095	0.041	2.319	0.021	0.014	0.175	Supported

Notes: R2 = 0.337; f2 = 0.507; Q2 = 0.265; VIF = 2.447

5 Discussion

This study has examined the factors that affect millennial's intention to use the digital real estate platform by utilising the Stimulus-Organism-Response (S-O-R) model as the guiding principle. The PLS-SEM has revealed that millennials' intention to use digital real estate platforms is influenced by informativeness, effectiveness, and entertainment variables as predicted by the S-O-R model. Specifically, informativeness has a significant positive relationship on the millennials' intention to use digital real estate platforms. The links are predominantly explained by its ability to address the problem of connectivity between potential trading partners and the consumers. This proves that digital technology can attract the attention of the global population and has impacted the way business is conducted today, which is gradually moving from conventional methods of advertising (e.g., sale boards, property bulletins, magazines) to the use of the digital real estate's platforms such as Property Guru, Durian Property, and Property Circles.

The Effectiveness factor strongly influenced millennials' intention to use the digital real estate platform because they are attracted by the simplicity of the digital technology to purchase online [38, 50]. Moreover, millennials believe that by using the technology, assessing the real estate platforms is effortless and it enables them to save time when making important decisions whether to purchase or rent any of the real estate product or services. Digital technology also allows millennials to search at their pace and convenience, learning a great deal about the property and related issues before they begin any actual negotiations. Moreover, the entertainment factor affects the millennial's intention to use the real estate platforms in retrieving related information via the Internet [2]. Furthermore, it allows millennials a pleasant online experience in attaining information about the targeted property and its surrounding neighbourhood [19].

6 Conclusion

The findings of this study have shed light on the important implications of informativeness, effectiveness, entertainment, and flow experience on millennials intention to use the digital real estate platforms. This study has several implications for the management of online property and online marketing in general. The primary managerial implication is that the web site atmospheric cues can be used purposely to encourage the flow experience. Therefore, it may enhance millennials intention to use the websites and improve their satisfaction with the digital real estate platforms. In addition, this study dealt with flow experience in the context of online searching for real estate information. Thus, this study generalizes the flow theory in the real estate context. Furthermore, the findings suggested that millennials intention behaviour are influenced by informativeness, effectiveness, and entertainment as predicted by the S-O-R model. This study indicates that flow as an optimal experience has significant effects on millennials intention to use digital real estate platforms.

Future studies are recommended to perform a multigroup analysis to examine the differences among groups of different generations. Moreover, to understand the future of property markets, future studies can consider examining the effect of other demographic factors such as gender, races, and education levels. Also, it is recommended that future studies to investigate the green buying behaviour in the property industry, facilities perspectives such as internet coverage real estate platform [37], and the use of social networking services with up-to-date interactivity features. This study has several limitations. For instance, the data collection procedure employed in the study was online survey in Malaysia. Thus, the findings may different if the S-O-R model was retested in different cultural environment or in a different industrial context. Thus, it is encouraged that the property investors to consider this limitation when interpreting the findings of this current study and validate the findings based on different cultural and industrial contexts.

References

1. Ahuja, M., Gupta, B., Raman, P.: An empirical investigation of online consumer purchasing behavior. Commun. ACM **46**(12), 145–151 (2003)
2. Baen, J., Guttery, R.: The coming downsizing of real estate: Implications of technology. J. Real Estate Portfolio Manag. **3**(1), 1–18 (1997)
3. Bardhan, A.D., Jaffee, D., Kroll, C.: The internet, e-commerce, and the real estate industry (2000)
4. Bell, H., Tang, N.K.: The effectiveness of commercial Internet Web sites: a user's perspective. Internet Research (1998)
5. Boonsiritomachai, W., Sud-On, P.: Increasing purchase intention and word-of-mouth through hotel brand awareness. Tourism Hosp. Manag. **26**(2), 265–289 (2020)
6. Chang, M.L.D., Suki, N.M.: Understanding consumers intention to use online property websites: a conceptual model. Labuan Bull. Int. Bus. Finan. (LBIBF) **17**, 139–146 (2019)
7. Chin, W.W.: Overview of the PLS Method (2018). http://plsgraph.com/
8. Chin, W.W., Marcolin, B.L., Newsted, P.R.: A partial least squares latent variable modelling approach for measuring interaction effects: results from a Monte Carlo simulation study and electronic-mail emotion/adoption study. Inf. Syst. Res. **14**(2), 189–217 (2003)
9. Chin, W.W.: The partial least squares approach to structural equation modelling. Mod. Methods Bus. Res. **295**(2), 295–336 (1998)
10. Chen, Q., Clifford, S.J., Wells, W.D.: Attitude toward the site: new information. J. Advert. Res. **39**(5), 28–38 (1999)
11. Chen, C.C., Yao, J.Y.: What drives impulse buying behaviours in a mobile auction? The perspective of the stimulus-organism-response model. Telemat. Inform. **35**(5), 12491262 (2018)
12. Choi, Y.: Technology acceptance model and stimulus-organism response for the use intention of consumers in social commerce. Int. J. E-Bus. Res. (IJEBR) **15**(2), 93–101 (2019)
13. Clarke III, I.: Emerging value propositions for m-commerce. MIS Q. J. Bus. Strat. **25**(2), 319–340 (2008)
14. Davis, F.D.: Perceived usefulness, perceived ease of use, and user acceptance of information technology (1989)
15. Dailey, L.: Navigational web atmospherics: explaining the influence of restrictive navigation cues. J. Bus. Res. **57**(7), 795–803 (2004)
16. Del Giudice, V., De Paola, P., Del Giudice, F.P.: COVID-19 infects real estate markets: short and mid-run effects on housing prices in Campania region (Italy). Soc. Sci. **9**(7), 114 (2020)
17. Dholakia, U.M., Rego, L.L.: What makes commercial Web pages popular? An empirical investigation of Web page effectiveness. Eur. J. Mark. (1998)

18. Durkaya, B.: Examining the helpfulness of online customer reviews based on review related factors: the moderating effect of product type (Doctoral dissertation, Institute of Science and Technology) (2020)
19. Dyana, M.L., Hamid, R., Lajuni, N., Suki, N.M.: What drives consumers intention to use online property websites: a conceptual model. J. Soc. Sci. Res. 630–633 (2018)
20. Dyana, C.M.L., Suki, N.M., Lajuni, N., Hamid, R.: Towards industry revolution 4.0 practice: millennial's intention to use online property websites by applying the Stimulus-Organism-Response (SOR) model. Int. J. Supply Chain Manag. **8**(4), 1032–1038 (2019)
21. Edmunds, R., Thorpe, M., Conole, G.: Student attitudes towards and use of ICT in course study, work, and social activity: a technology acceptance model approach. Br. J. Edu. Technol. **43**(1), 71–84 (2012)
22. Ettis, S.A.: Examining the relationships between online store atmospheric colour, flow experience and consumer behaviour. J. Retail. Consum. Serv. **37**(2017), 43–55 (2017)
23. Hair, J.F., Hult, G.T.M., Ringle, C.M., Sarstedt, M., Thiele, K.O.: Mirror, mirror on the wall: a comparative evaluation of composite-based structural equation modelling methods. J. Acad. Mark. Sci. **45**, 616–632 (2017)
24. Hair, J.F., Sarstedt, M., Ringle, C.M., Mena, J.A.: An assessment of the use of partial least squares structural equation modelling in marketing research. J. Acad. Mark. Sci. **40**(3), 414–433 (2012)
25. Hamzah, A., Yazid, M.F., Shamsudin, M.F.: Post Covid-19: what next for real estate industrial sector in Malaysia? J. Postgraduate Curr. Bus. Res. **1**(1) (2020)
26. Harun, A., Husin, W.H.R.: Is the purchasing behavior of suburban millennials affected by social media marketing? Empirical Evidence from Malaysia (2019)
27. Hoffman, D.L., Novak, T.P.: Flow online: lessons learned and future prospects. J. Interact. Mark. **23**(1), 23–34 (2009)
28. Hoffman, D.L., Novak, T.P., Peralta, M.: Building consumer trust online. Commun. ACM **42**(4), 80–85 (1999)
29. Hsu, C.L., Lu, H.P.: Why do people play online games? An extended TAM with social influences and flow experience. Inf. Manag. **41**(7), 853–868 (2004)
30. Hsu, C.L., Chang, K.C., Chen, M.C.: Flow experience and internet shopping behavior: investigating the moderating effect of consumer characteristics. Syst. Res. Behav. Sci. **29**(3), 317–332 (2012)
31. Huizingh, E.K.: The content and design of web sites: an empirical study. Inf. Manag. **37**(3), 123–134 (2000)
32. Joia, L.A., Gutman, L.F.D., Moreno, V., Jr.: The intention of use of home broker systems from the stock market investors' perspective. J. High Technol. Managem. Res. **27**(2), 184–195 (2016)
33. Klassen, R.D., Whybark, D.C.: Environmental management in operations: the selection of environmental technologies. Decis. Sci. **30**(3), 601–631 (1999)
34. Kim, Y.: Cultural difference in motivations for using social networking site, a comparative study of American and Korean college students. Comput. Hum. Behav. **27**(1), 365–372 (2011)
35. Koufaris, M.: Applying the technology acceptance model and flow theory to online consumer behaviour. Inf. Syst. Res. **13**(2), 205–223 (2002)
36. Kurniawati, N.: Creating low-cost animation video using online platform: a learning media user review. Jurnal Pendidikan Kedokteran Indonesia: Indonesian J. Med. Educ. **9**(1), 26–31 (2020)
37. Lim, I.: Reality for Malaysia's university students: online learning challenges, stress, workload; possible solutions for fully digital future until Dec (2020). www.malaymail.com
38. Lin, J.-J., Chung, X.-J., Yang, C.-Y., Lau, H.-L.: A meta-analysis of trials using the intention to treat principle for glutamine supplementation in critically ill patients with burn. Burns **39**(4), 565–570 (2013)

39. Lee, S.M., Chen, L.: The impact of flow on online consumer behaviour. J. Comput. Inf. Syst. **50**(4), 1–10 (2010)
40. Marjerison, R.K., Hu, Y.A.: Exploring the impact of peer influence on online shopping: the case of Chinese Millennials. In: Quality Management for Competitive Advantage in Global Markets, pp. 196–210. IGI Global (2021)
41. Mazaheri, E., Richard, M.O., Laroche, M.: The role of emotions in online consumer behavior: a comparison of search, experience, and credence services. J. Serv. Mark. **26**(7), 535–550 (2012)
42. Mehrabian, A., Russell, J.A.: An Approach to Environmental Psychology. The MIT Press (1974)
43. Michaelidou, N.: Usage, barriers and measurement of social media marketing, an exploratory investigation of small and medium B2B brands. Ind. Mark. Manag. **40**(7), 1153–1159 (2011)
44. Ni, A.Y., Ho, A.T.K.: Challenges in e-government development: lessons from two information kiosk projects. Gov. Inf. Q. **22**(1), 58–74 (2005)
45. Park, S.Y., Nam, N.W., Cha, S.B.: University students' behavioural intention to use mobile learning: evaluating the technology acceptance model. Br. J. Edu. Technol. **43**(4), 592–605 (2012)
46. Peterson, S., Bredow, T.S.: Middle Range Theories: Application to Nursing Research and Practice. Lippincott Williams & Wilkins (2019)
47. Rangaswamy, A., Moch, N., Felten, C., van Bruggen, G., Wieringa, J.E., Wirtz, J.: The role of marketing in digital business platforms. J. Interact. Mark. **51**, 72–90 (2020)
48. Richard, M.O.: Modelling the impact of internet atmospherics on surfer behavior. J. Bus. Res. **58**(12), 1632–1642 (2005)
49. Ringle, C.M., Wende, S., Becker, J.M.: SmartPLS 3.2.7 (2015). http://www.smartpls.com. Accessed 4 Feb 2017
50. Shin, J.I., Chung, K.H., Oh, J.S. Lee, C.W.: The effect of site quality on repurchase intention in Internet shopping through mediating variables: the case of university students in South Korea. Int. J. Inf. Manag. **33**(3), 453463 (2013)
51. Suki, N.M., Lajuni, N., Hamid, R.: Towards industry revolution 4.0 practice: Millennial's intention to use online property websites by applying the Stimulus-Organism-Response (SOR) model. Int. J. Sup. Chain. Mgt. **8**(4), 1032 (2019)
52. Teo, T., Fan, X., Du, J.: Technology acceptance among preservice teachers: does gender matter? Australas. J. Educ. Technol. **31**(3), 235–251 (2015)
53. To, W.M., Lai, L.S., Leung, V.W.: Technology acceptance model for the intention to use advanced business application software among Chinese business school students. Australas. J. Educ. Technol. **34**(4), 160–173 (2019)
54. To, W.M., Tang, M.N.F.: Computer-based course evaluation: an extended technology acceptance model. Educ. Stud. **45**(2), 131–144 (2019)
55. Verhagen, T., Van, D.W.: The influence of online store beliefs on consumer online impulse buying a model and empirical application. Inf. Manag. 48(8), 320327 (2011)
56. Venkatesh, V.: Determinants of perceived ease of use: integrating control, intrinsic motivation, and emotion into the technology acceptance model. Inf. Syst. Res. **11**(4), 342–365 (2000)
57. Wright, K.B.: Researching internet-based populations: advantages and disadvantages of online survey research, online questionnaire authoring software packages, and web survey services. J. Comput.-Mediated Commun. **10**(3) (2005)
58. Zhao, L., Lu, Y., Wang, B., Huang, W.: What makes them happy and curious online? An empirical study on high school students' Internet use from a self-determination theory perspective. Comput. Educ. **56**(2), 346–356 (2011)

Transformational Leadership Using IT to Increase the Success Rate of SMEs in Malaysia

Sujendren Dewarajan(✉) and Dania Sammani

International University of Malaya Wales, Kuala Lumpur, Malaysia
daniasammani@iumw.edu.my

Abstract. There is constantly a need to improve the success rate of SMEs in Malaysia. In this time of Covid 19, the whole world is working remotely. It's inevitable of using IT to support, lead, and coach the team from remote locations. This study uses Literature survey to understand the existing transformational leadership (TL) behaviors that could be used by SMEs with the help of IT tools. In Malaysia even there is a huge gap between the SMEs and the success rate, its stated that we have an estimate of 60% failure of SMEs. With this Covid19 situation the SMEs are struggling to survive the lockdowns and managing projects remotely. There are proof that almost any SMEs in Malaysia are not able to withstand the first 5 years of start-up. Therefore, there is a voracious need to improve and Increase the success rate of a Small Medium Enterprise (SME). There is a need to Increase the success rate of a Small Medium Entrepreneur (SME) by integrating Leadership and IT tools as a support system. In this study, the link between the transformational leadership attributes and the IT tools that can be used to evaluate the leadership behaviours are mapped out. By using this IT tools, the leader would be able to enhance their personal attributes by reaching out to the wider team.

Keywords: Transformational leadership · ICT · IT Tools · Leadership behaviours · SME · SME success

1 Introduction

Over the years there are oceans of studies on leadership theories that has been developed throughout the years [1]. One of the most famous leadership styles is Transformational Leadership (TL). The concept or the idea of Transformational leadership was introduced by [2] as a process or direction in which "both leaders and followers elevate each other to greater levels of ethics, morality and motivation". A crucial element of transformation is the ability to cultivate the needs of the follower in a follower-centred (person-centred) organization or group of people. This study will not be focusing on the definition or what is leadership, instead focusing on how it has been evolving and what are the types of transformational leadership behaviors.

Transformational leaders are proactive. Rather than working within a particular organizational culture, they work to create an adaptive organizational culture by inspiring and

M. Al-Emran et al. (Eds.): ICETIS 2021, LNNS 299, pp. 558–567, 2022.
https://doi.org/10.1007/978-3-030-82616-1_46

implementing new ideas [3]. Researcher has further explained that "Transformational leadership is about bringing change, creating innovation, and establishing entrepreneurship". Transformational leadership can be divided into two major components of personal traits and behaviors, the first is the macro level Transformational Leadership and secondly is the micro level transformational leadership [2].

Focusing on the followers needs makes leaders accountable to the follower in general [4]. Conflict can be used to initiate a possible change for good using IT tools it supports in segregation and ensure clear decision and discussion is captured. The leap to the next level conflict and the process of is by having empathy, understanding, insights and humanly consideration, not any type of manipulation, power wielding or coercion. Burn's work on Transformational leadership lays the foundation transformational movement which we won't be neglecting in this study [4].

On the other hand, there is supporting study that proves leader can manage to encourage and stimulate employees whom have the "don't care" and "its ok" behavior on managing risk [2]. This can be done not only physically but through the support of IT tools, via conferencing and video camera to have a personal touch [5]. With Transformational Leadership, leaders can psychologically empower followers to do things in a better way and can develop trust in employees resulting in high commitment; highly committed employees with direct or indirect contacts [5].

While most leaders spend their time on planning and analyzing each step of the task or strategy. Some might take a more intuitive approach; there are a lot of tools to support the planning and sharing the documents and knowledge remotely, there are shared drive where all of the team members could share their input and update while the rest of the team are able to view the changes live [6].

At the moment, Malaysia SME is having the worst time of their life. In a research it shows that 25% of the SME failed due to the fact that they couldn't with stand the current situation [7]. The survival of the Small Medium enterprise resilience skills and the usage of the IT tools to support the business is insufficient to overcome the challenging times. This resulted with very poor performance to meet the customers' needs and end up closing the business [8].

1.1 Research Problem

This issue of SME failure has been brought up in various research and it's in the SME news. During the Launch of the entrepreneurship challenge, AmBank's Managing Director said that almost 80% of SMEs do not survive the first three years [2]. There is another research that shows similar issue identified also in Australia. In Australia, the SME failure rate is reported to be as high as 23% [8]. This has a clear conclusion that lack of skills and techniques are the main reason SMEs are not able to be resilient on Survival [9]. Two main factor of SME failures are business cycles are not managed well and firms faced internal problems due to poor leadership and mismanagement. [9] has indicated that the SMEs are not able to recover from a crisis due to the adoptability and the vision and leadership of the organization is not well defined and supported. Base on the evidences and proof collected above it is significant to understand the low success rate of SMEs in Malaysia. SMEs in Malaysia are not able to sustain the first 5 years of start-up.

Therefore, there is a need to Increase the success rate of a Small Medium Entrepreneur (SME) by integrating Leadership and IT tools as a support system.

2 Literature Review

2.1 Transformational Leadership Behaviors and IT

Based on the literature study by scholars, [10] it can be recapitulated into six behavior groups. It is crucial to understand this group of behaviors. Firstly, it's on identifying and articulating a vision. Followed by providing an appropriate model. The third behavior is fostering the acceptance of group goals. Fourth behavior would-be high-performance expectations. The fifth behavior is around providing individualized support to staff and finally the last behavior style is intellectual stimulation [11].

It can be concluded that there is no single definition on leadership, it can simply be defined in 1 sentence or a single meaning [12]. After 40 years of researchers trying to unravel the meaning of leadership, had list down more than 3,500 definitions and concluded, There are too many definitions of leadership as those who have attempted to define the concept of leadership [13, 14]. Hence this study will not aim to create or identify a new definition of leadership nor drill deeper on the definition of leadership. It will focus on TL using IT tools to bring the best out of a SME in Malaysia.

Leadership is how you bring that idea to the successful implementation and ensuring it reaches the right target group using the IT tools and ensuring the team is fully aligned. It is innovation and bringing team together differentiate a leader and entrepreneur [15]. Making the team accountable by assigning the task, and ensuring they updating it remotely gives the accountability to each and every team members [15].

Based on abundant studies on the theory of leadership, it can be grouped and summarized into main five broad theories. Among the five theories, the first is about Trait behaviors followed by contingency or situational approach theory. Third theory is contemporary integrative approach. Fourth is power and finally the fifth is about influence approach [3, 14].

The trait behaviors approach is an inherent attribute and personal characteristics of the leaders. As a human being we have our own personal trades, this is influence by few factors such as the leaders background, upbringing, the environment and their guru or the footstep of the person they follow. This behavior is built into their personality and very difficult to change. The behavioral approach is defined as behaviors of the leaders.

Behavioral approach can be broken down into, contingency or situational approach. It is a behavior that comes with a combination of situation and the leader's personal traits. The leaders will be managing using their behavior and they are more adoptable and flexible based on the situation [3]. The situation will either be giving them an opportunity to take control or to be controlled and manage the uncontrolled situations. The influencing capabilities and skills will be changing based on the situation. Hence it a trait that gives the strength of individual behaviors with adaptations capabilities [14].

Contemporary integrative approach is an approach where the leaders will be an innovator of their own behaviors and situations. This type of leadership will always be thinking about different ways or approach to deal with a situation or team member. The leader is designed to constantly venture and develop a more interactive way of having a discussion, for an example even now with Teams and google meet, there are a lot of exciting features. All that can only be done in a face to face scenario can be done remotely.

Power and Influence approach are two similar traits that the first is about gathering the power over the decision and the organization using multiple ways, most importantly the trust factor. As they have the trust in the organization the team will subconsciously obey and give the power to leader as a decision maker and the followers will obey. It is similar case for the influencing where influencing is an approach where the leaders will be able to win a debate and somehow make the other parties think and believe that the certain decision will be a win-win situation. This allow the leader to influence and stir the decision into his court.

Over time, there can be group of four central components of IT and Leadership. These components include idealized knowledge, motivation online, intellectual stimulation, and individualized discussion. The first two components can be clustered together as personality or charisma of the leader [14]. Each individual leader represents them self as a proud transformational leader displays these components in different levels. It is based on their interpersonal strength, desired outcome of the organization and also adopting to the outcomes of the followers. So there is no certain standard or equal traits of displaying these for components [13].

Therefore, it is important for everyone to appreciate and understand that idealizing the influence integrates two totally different characteristics of the follower and leader's relationship. First, the followers will admire certain characteristic towards the transformational leader, and they will try to imitate them, this happens in their subconscious mind. On the second point the leaders can impress and making the followers owe on every movement or gestures of the leaders [14].

The most important factor to enable the supporters/followers to reach their full potential is on the individualized consideration involves transformational leadership to coach and mentor them. This needs to be done in a structured and progressive way using the above behaviors [16].

As much as the team mates believe and trust on the leader they will support and dedicate themselves to ensure the success of the project [17]. Leadership items from measures produced by [18]. Those leadership attributes were selected on the simple basis of how well they evaluate the theoretical construct under any type of study. The leadership attributes are: Articulating a vision, Intellectual stimulation, Inspirational communication, Supportive leadership, Personal recognition [19].

2.2 Transformational Leadership Behavioral Trends in IT

"Times are changing" is a very common term and being used vastly for all topics, it is time to acknowledge the fact that times are evolving rapidly as the day goes by more than anybody ever imagined. Leaders must accommodate to these accelerated requirements and demands else we will be in the Jurassic world [20]. With the old approaches they are not able to keep abreast with the demand and the needs of current world. Using the IT tools to keep track on the actions, target, and ensuring the team is aligned is very crucial [20].

In this era we can't simply rely on the classical leadership style, the world needs a CEO is capable of meeting every person in the organization globally and later improved by [14] and [21]. Supports these elements of caring, visionary, role modelling and team building leadership through the transformational and transactional leadership model.

Transformational leaders raise the consciousness of followers by appealing to those trades which attract followers more and increases trust [20]. The four types of transformational leadership used by the executives in this study in order of most often to least often used are Individualized consideration, Inspirational motivation, Intellectual stimulation and Idealized influence [20].

People need feelings & ideas, and teams need leadership [20]. It has been a trend to have multiple leadership styles and names for it, despite all that it is important for them to keep in par in the changing world and be part of the change. IT would be there to support and elevate their capability. They encapsulate this by describing leadership as a combination of individual traits, abilities, skills, and behaviors demonstrated by the leaders within the professional relationships to achieve common goals.

Following the most current methods and ways of working and having a Transformational leadership behavior could potentially enhance the performance of subordinates, this is done by combining their motive with the right IT tools.

2.3 Effects of Transformational Leadership in Businesses With and Without IT Tools

Generally, everyone would agree to the point, that leadership is one of the most important criteria in any type of business especially in the SME and we need IT to support the business. Everything and anything about the performance are vital to be captured and tracked a business perspective especially when it's a start up business.

All strong leaders will consider their employees as their biggest and most important assets. In an organisation high turnover rate is considered to affect the performance and the quality of the product from the organisation. Equalling to that is the organisation IT system is efficient and capable to supporting the staff to deliver, the team will have the satisfaction of delivery [22].

One of the critical success factors of projects in a business is to use the leadership style of a project manager remotely. it is important for the project managers to use the project management tool virtually and assign the task to the members and ensuring each member understand and be a part of the project [23–25].

A very positive association between a manager's transformational leadership and an organisation's business success [17]. Base on their research, their finding shows that the manager's leadership style plays an important role in business success, this is not limited to type of projects or business. Essentially, a transformational leader motivates and inspires team members towards a direction to ensure business, with few important traits such as improving efficiency, producing high quality products and providing a reliable service to the customers. Leaders will make the team understand and strive to these days to day operations which defines the success of the organisation. This finding answers the call by [17], who underlined that the project management literature failed to give sufficient attention to the role of business managers' leadership styles. It is also clearly stated that team building is capable to increase the bond and the trust among the team members which will positively impact the organisation success. This finding confirms the meta-analysis by [26] and later on [17] supported this study.

The study by Jandhagi [27]that from the four most famous Transformational leadership style inspirational motivation criteria has the highest, followed by idealized influence, intellectual stimulation and individualized consideration [17]. It is indeed very interesting finding because most research conducted in Western countries by Toshi [28] and Waldaman's study [29] has proven that having a strong influence and followers to idealize the goals are the key element of transformational leadership in any business in western countries [14, 28]. However, in the context of SMEs in Malaysia, especially focusing on the IT sector, inspirational motivation is perceived to be the key attribute of TL that can reap the fruit of success. A similar finding was reported in another non-Western context study in Iran by aghi, GH, Matin, H and Farjami [14] and that study was strongly supported by Aga [27].

3 Methodology

This study will be focusing on Transformational leadership and their critical behaviours. It is followed by Malaysia SME failure and success Literature review. This study focuses on critical analysis and discussion to identify the most suitable transformational leadership behaviours that is can be supported by IT tools to support the SME business. The overall research is to critically review literature survey to understand the behaviours of transformational leadership that has been evolving over time. This study will be able to support SMEs and future studies on leadership improvement in utilising IT tools to lead an organisation. Hence this study will be focusing on the behaviours of transformational leadership using IT that could influence the success of a SME in Malaysia. It was based on constructivism and an interpretive theoretical perspective, beginning without hypothesis. This study will stress on the data collected from the literature review from various scholars. A critical review of literature was undertaken to answer the research question.

4 Result and Discussion

As specified in the beginning of this journey, the last 4 decades the topic transformational leadership has been one of the favourite studies and there are a lot more to understand and research on this topic. The vast benefit of Leadership and brining it out remotely is crucial at this point of time. Using IT tools for success is one of the most common topic and also important topic for the current fragile and ever evolving economic situation [12, 16, 30].

This implies that leaders believed in the concept of having the leaders as the main power source and formalizing the power of a leader and their practise of showing and using the power for influencing and getting things done. There is another study by [31] is equally important, which shows a very significant difference between the authentic transformational leadership and IT Project management leadership.

Transformational Leadership has proven to have a significantly stronger relationship with the organizational IT support tools. They feel that they must ensure that the organisation's values and quality deliverables are their responsibility. This is consistent with a study by [2] and also by [2] who proof that transformational leaders have capabilities to achieve and have at most higher level of commitments in achieving the organisation goals and be part of the organisation success if they have the right tools [2].

Based on a development experience inventory, for a leadership development experience exercise, it is scientifically proven that the transformational leadership was the most important trait and the leadership behaviours are crucial and important for an organisation. Hence this can be used in the SME in Malaysia as well. The scholar tested and proved that potentially the transformational leadership behaviours are most liked by the followers. They have also proved that there are linkage and relationship between the leadership and IT tools [32].

Having a latest way or in other word cutting age innovative tools and strong transformational leadership can be directly linked to inspire and stimulate the followed to deliver results beyond expectations while grooming and building their own leadership traits and their capabilities in leading [12, 33].

5 Conclusion

Base on the literature survey above, this study could conclude that the Leadership and its behavioral attributes will boost the capabilities and drive the employees of an organization towards a uniformed direction to achieve a common goal. In an IT Project environment, it is crucial to constantly evolve and be on the toes for changes and innovation. The day to day employees will get bored and tired on day to day routine, as a transformational leader it's their role to constantly motivate and empower every individual in supply chain to deliver extra ordinary results and keeping them connected through the IT tools that's is available. It is crucial to set up a home office and ensure the team is connected and feel that their part of the organizations. Therefor it's important for a mature business or start-up to cultivate the transformational leadership behaviors and provide with the efficient IT tools to run the show.

Therefor these findings have important theoretical implications; these behaviors will benefit and ensure both leaders and followers elevate each other to greater levels of ethics, morality, and motivation to achieve a common goal as a team. Based on the discussion above the 4 important behaviors are listed down in the table below (Table 1).

Table 1. Shows how each attribute of TL will impact the organization and its employees to deliver a common goal.

TL key attributes/behaviours	IT tools that can be used
Inspirational motivation criteria	Video conferencing, E-cards,
Idealised influence	Online whiteboard, idea generation tools,
Intellectual stimulation	Virtual breakout rooms, E-learning, online training
Individualised consideration	Action tracker, scheduling PM tools

With the above, this study can conclude that the behaviors of leadership combined with the right IT tool not in any ratio or order will definitely deliver a better working experience and more productivity in an individual or a team or in an organization. This will directly impact the success of a business in long term. Managerial implication from the study shows that the right leader needs to have the right tool to ensure the message and guidance is transfer accurately. They should not only rely on the Old ways of working but ensure the followers have the right tools to deliver their goal.

Therefore, the Leadership behaviors and trades should be considered by the SME's in Malaysia. It is crucial to have a leader who can demonstrate the above behaviors directly with his personal trait to bring the organization to the right direction and manage changes accordingly. Additional budget and plan on the IT tools will ensure that the entire team is fully equipped for success.

Suggestion for future research is on task-oriented strategies, change management by leadership for industries in Malaysia. A comprehensive study would enable greater understanding of the leadership and risk-taking processes as it could measure leadership effects on risk and issue management and the growth of small medium entrepreneur by tackling the risk and issues as a holistic.

References

1. Lo, M., Ramayah, T., Min, H.W.: Leadership styles and organizational commitment: a test on Malaysia manufacturing industry. Leadersh. Styles Organ. Commit. Test Malaysia Manuf. Ind. **1**(6), 133–139 (2009)
2. Hayward, T.R.Q., Goss, M. "Hayward, Q., Goss, M., Tolmay, R. (2004) The relationship between transformational and transactional leadership and employee commitment. Grahamstown: Rhodes University, Business Report". Rhodes Univ. Bus. J. **118**(1), 112–122 (2004)
3. Muenjohn, N., McMurray, A.: The Palgrave Handbook of Leadership in Transforming Asia. In: The Palgrave Handbook of Leadership in Transforming Asia (2017). https://doi.org/10.1057/978-1-137-57940-9
4. Tyssen, A.K., Wald, A., Spieth, P.: The challenge of transactional and transformational leadership in projects. Int. J. Proj. Manag. **32**(3), 365–375 (2013). https://doi.org/10.1016/j.ijproman.2013.05.010
5. Mittal, S.: Effects of transformational leadership on turnover intentions in IT SMEs. Int. J. Manpow. **37**(8), 1322–1346 (2016). https://doi.org/10.1108/IJM-10-2014-0202
6. **Badewi, A.: The impact of project management (PM) and benefits management (BM) practices on project success: towards developing a project benefits governance framework. Int. J. Proj.** Manag. **34**(4), 761–778 (2016). https://doi.org/10.1016/j.ijproman.2015.05.005
7. Lim, J.: 25 % of SMEs face closure risk amid new wave of Covid-19. Edge (October 2020). https://www.theedgemarkets.com/article/25-smes-face-closure-risk-amid-new-wave-covid19
8. **Osarumwense, O.R.: Influence of resilience on survival of small and medium enterprises. Int.** Bus. Educ. J. **13**(1), 127–132 (2020)
9. Çera, G., Belas, J., Zapletalikova, E.: Explaining business failure through determinist and voluntarist perspectives. Serbian J. Manag. **14**(2), 257–275 (2019). https://doi.org/10.5937/sjm14-23348

10. Podsakoff, P.M., MacKenzie, S.B., Moorman, R.H., Fetter, R.: Transformational leader behaviors and their effects on followers' trust in leader, satisfaction, and organizational citizenship behaviors. Leadersh. Q. **1**(2), 107–142 (1990)
11. Mudin, D.K.D., How, S.E., Rahman, M.M., Ibrahim, P., Jopony, M.: Industrial revolution 4.0: Universiti Malaysia Sabah perspective. In: E3S Web Conference, vol. 48, pp. 3–6 (2018). https://doi.org/10.1051/e3sconf/20184803005
12. Bass, B.M., Riggio, R.E.: Transformational Leadership, 2nd edn. Taylor & Francis, New Jersey (2012)
13. Bass, B.M., Avolio, B.J., Jung, D.I., Berson, Y.: Predicting unit performance by assessing transformational and transactional leadership. J. Appl. Psychol. **88**(18), 207–218 (2003)
14. Arham, A.F.: The relationship between leadership behaviour, entrepreneurial orientation and organisational performance in Malaysian small and medium enterprises, p. 431 (2014)
15. Sengupta, S., Sahay, A.: Social entrepreneurship research in Asia-Pacific: perspectives and opportunities. Soc. Enterp. J. **11**(1) (2017). https://doi.org/10.1108/sej-02-2015-0007
16. Avolio, B.J., Bass, B.M., Jung, D.I.: Re-examining the components of transformational and transactional leadership using the multifactor leadership questionnaire. J. Occup. Organ. Psychol. **72**, 441 (1999)
17. Turner, J.R., Muller, R.: The project manager's leadership style as a success factor on projects. Proj. Manag. J. **2**(2), 29–50 (2005)
18. House, R.J.: Leadersh. Mult. Lev. approaches Contemp. Altern. London JAI Press, vol. 24, Part B, pp. 23–30 (1998)
19. Rafferty, A.E., Griffin, M.A.: Dimensions of transformational leadership: conceptual and empirical extensions. Leadersh. Q. **15**(3), 329–354 (2004). https://doi.org/10.1016/j.leaqua.2004.02.009
20. Sarros, J.C., Santora, J.C.: The transformational-transactional leadership model in practice. Leadersh. Organ. Dev. J. **22**(8), 383–394 (2001). https://doi.org/10.1108/01437730110410107
21. Arham, A.F., Fadlisham, H., Hasan, A., Ridzuan, A.R., Sulaiman, N., Studies, M.: The effect of leadership behaviours on entrepreneurial orientation: an online survey on Malaysian SMEs. Int. J. Latest Res. Sci. Technol. **4**(5), 47–50 (2015)
22. Kontoghiorghes, K., Frangou, C.: The association between talent retention, antecedent factors, and consequent organisational performance. SAM Adv. Manag. J. **74**(1), 29–58 (2009)
23. Anantatmula, V.S.: Project manager leadership role in improving project performance. EMJ - Eng. Manag. J. **22**(1), 13–22 (2010). https://doi.org/10.1080/10429247.2010.11431849
24. Lindgren, J., Packendorff, M.: Project leadership revisited: towards distributed leadership perspectives in project research. Int. J. Proj. Organ. Manag. **1**(1), 285–308 (2009)
25. Hayat, N., Riaz, M.T.: The influence of the SMES top-level managers' leadership styles and their entrepreneurial orientation on the business performance. SSRN Electron. J. (2012). https://doi.org/10.2139/ssrn.1884069
26. Klein, L., Biesenthal, C., Dehlin, E.: Improvisation in project management: a praxeology. Int. J. Proj. Manag. **33**(2), 267–277 (2015)
27. Jandaghi, A., Matin, G.H., Farjami, H.: Comparing transformational leadership in successful and unsuccessful companies. Int. J. Soc. Sci. **4**(3), 211–216 (2009)
28. Tosi, H.L., Misangyi, V.F., Fanelli, A., Waldman, D.A., Yammarino, F.J.: CEO charisma, compensation, and firm performance. Leadersh. Q. **15**(1), 405–425 (2004)
29. Waldman, D.A., Ramirez, G.G., House, R.J., Puranam, P.: Does leadership matter? CEO leadership attributes and profitability under conditions of perceived environmental uncertainty. Acad. Manag. J. **44**, 134–143 (2001)
30. Lowe, K.B., Kroeck, K.G., Sivasubramaniam, N.: Effective correlates of transformational and transactional leadership: a meta-analytic review. Leadersh. Sage **10**, 385–425 (2006)

31. Sobh, R., Perry, C.: Research design and data analysis in realism research. Eur. J. Mark. **40**(11–12), 1194–1209 (2006). https://doi.org/10.1108/03090560610702777
32. Law, M.J.: The study of transformational leadership and leader development of public health directors in the Midwest. J. Appl. Christ. Leadersh. **7**, 127 (2013)
33. Kissi, J., Dainty, A., Tuuli, M.: Examining the role of transformational leadership of portfolio managers in project performance. Int. J. Proj. Manag. **31**(4), 485–497 (2013). https://doi.org/10.1016/j.ijproman.2012.09.004

Mobile Payment Adoption: Barriers for Baby Boomers in Malaysia

U Hao Lau[1], Lai Ee Lee[1], Di Kang Lew[1], Lian Sien Loo[1], Sze Xiong Ooi[1], Voon Hsien Lee[1], Xiu Ming Loh[1](✉), and Garry Han Wei Tan[2]

[1] Universiti Tunku Abdul Rahman, 31900 Kampar, Malaysia
{lauuhao1kp1,dikangboy123,sien096,jasonooi1996}@1utar.my, {leevh,lohxm}@utar.edu.my

[2] UCSI University, 56000 Cheras, Malaysia

Abstract. Mobile payment has been defined by many as the employment of wireless and other such technologies to pay for goods, services, and bills through mobile devices. While this method of payment has increased in popularity in Malaysia, its adoption rate among Malaysians categorized as baby boomers has not been as promising even though this group of people has high levels of spending. Unfortunately, the adoption of such payment methods has been the prime focus of many recent studies in this area, and this has consequently caused many scholars to overlook resistance. Hence, this study addresses the gap by utilizing an extended Innovation Resistance Theory. With this theory, variables such as usage, value, risk, tradition, image, as well as information barriers are used to comprehensively examine the behavioral intention of baby boomers to adopt mobile payment. A survey was conducted and 310 usable responses were collected from baby boomers across Malaysia. These participants had very limited to no experience with mobile payment. The data was then analyzed and it was found that barriers included in this study have significant relationships with the behavioral intention of baby boomers in Malaysia to adopt mobile payment. The present study provides contributions to many stakeholders to encourage Malaysian baby boomers to use mobile payment services.

Keywords: Mobile payment · Mobile commerce · Innovation Resistance Theory · Baby boomers

1 Introduction

Evolving from traditional payment methods that utilized credit cards, the development and growth of technology have enabled customers to make payments using mobile devices. In particular, mobile payment is defined as the utilization of mobile devices to pay for tangible products, services, and bills [1]. Mobile payment is classified into two types which are proximity and remote. More specifically, proximity mobile payment refers to the close interaction between the consumers' mobile device with the merchant's point of sale terminal. On the other hand, remote mobile payment refers to the non-direct

M. Al-Emran et al. (Eds.): ICETIS 2021, LNNS 299, pp. 568–579, 2022.
https://doi.org/10.1007/978-3-030-82616-1_47

interaction between the consumers' mobile device with the merchant's point of sale terminal [2].

In Malaysia, there are numerous mobile payment systems that consumers can choose from. For example, Boost, Grabpay, Touch n' Go eWallet, and many others [3]. Even with these various systems in place, it was found that only around 10% of payments made in Malaysia were through mobile means; indicating that this method of payment is still sparingly used [4]. Moreover, the percentage of baby boomers are still lacking behind when it comes to using mobile-related technology and service [5]. As such, cash is the most preferred payment method for this group of people despite the Malaysian government's efforts to transform the country into a cashless one [6]. Thus, a gap is present in terms of understanding the barriers that affect the behavioral intention of baby boomers to adopt mobile payment.

In view of the above-mentioned gap, this study utilizes the Innovation Resistance Theory as the foundation of the research model. This is because the focus is to look into the barriers rather than the motivators that affect the behavioral intention of baby boomers to adopt mobile payment. Therefore, other frameworks such as the Mobile Technology Acceptance Model [7], is not suitable for this study. Furthermore, this study extends the theory to include "information barrier" given its importance with the context of this study. Overall, the research objectives are to determine (1) the barriers that determine the behavioral intention of baby boomers to adopt mobile payment and (2) the robustness of incorporating "information barrier" into the Innovation Resistance Theory in relation to mobile payment resistance.

In general, this study is posited to contribute to a number of novel findings and insights. Practically, various stakeholders, such as government officials and business operators, can collectively gain significant value through this study. Theoretically, the present research extends the existing knowledge pool of mobile payment resistance from the perspective of a developing nation.

2 Literature Review

2.1 Innovation Resistance Theory

The Innovation Resistance Theory examines consumers' negative response towards a certain innovation; caused either by the potential changes it brings to the existing status quo that evoke functional barriers, or its mismatch with the consumer's belief structure which results in psychological barriers [8]. These negative responses are categorized as active resistance because they are the attitudinal outcomes developed by consumers as a result of negative product evaluation. Functional barriers are represented by usage barrier, value barrier, and risk barrier whereas psychological barriers include tradition barrier and image barrier. The application of this theory is in several research areas in the mobile context such as mobile payment [9] and mobile website purchasing [10].

It has been posited that there are other barriers present beyond those contained in the Innovation Resistance Theory. In particular, innovation adoption can be hindered by limited information as these innovations require new users to invest substantially in the form of learning effort [11]. As such, information barrier is integrated into this study's research model. Past studies have also shown that information barrier has a debilitating effect on the intention to adopt mobile services; further justifying this integration [12]. Moreover, service providers' failure to deliver comprehensive guidance will cause adults to become disoriented and lack knowledge [13]. Consequently, they will be unwilling to adopt such technologies. Therefore, this research extends the Innovation Resistance Theory with information barrier to examine the behavioral intention of baby boomers to adopt mobile payment.

3 Hypotheses Development

3.1 Usage Barrier

In terms of usage barrier, it arises when a certain innovation is incongruous with a consumer's preset lifestyle, habits, or practices [8]. This form of barrier has been shown by several past studies to have negative influences on the adoption of mobile services. In particular, in the process of adopting mobile financial services, users tend to face complications in the form of a small screen [14]. Usage barriers relating to mobile services also happen due to visual difficulty and the lack of keying in sensitive information [13]. Thus, the following hypothesis was developed:

H1: Usage barrier has a significantly negative relationship with the behavioral intention of baby boomers to adopt mobile payment.

3.2 Value Barrier

When a consumer perceives that the cost of learning is outweighed by the performance-to-price value, a certain resistance, known as value barrier, is formed towards the usage of the product or services [15]. A number of past studies have contributed empirical evidence that shows the significant effects of value barrier on the adoption of mobile services. More precisely, the existence of a significantly negative relationship between value barrier and innovation adoption were found [12]. This situation is also similar when it comes to mobile services [16]. Thus, the following hypothesis was developed:

H2: Value barrier has a significantly negative relationship with the behavioral intention of baby boomers to adopt mobile payment.

3.3 Risk Barrier

The level of uncertainty and unpredictable side effects that an innovation may bring due to its built-in characteristics is commonly known as risk barrier [8]. These risks can come in terms of economic, functional, physical, and social. Several past studies have proven that risk barrier is a significantly negative determinant of behavioral intention. For example, when adopting mobile technology, people become hesitant when they realize that there are unpredictability and unfavorable outcomes resulting from its adoption [17]. Particularly in mobile services, these risks can come in terms of privacy and security of a consumer's personal information [18]. Thus, the following hypothesis was developed:

H3: Risk barrier has a significantly negative relationship with the behavioral intention of baby boomers to adopt mobile payment.

3.4 Tradition Barrier

When an innovation contradicts a user's existing values, norms, and past experiences, tradition barrier is induced [8]. Its significance as a barrier in mobile services adoption has been empirically proven by past studies. These past studies include those that were conducted on mobile banking in Finland [19] and mobile shopping in India [20]. It is further postulated that this situation is such because consumers prefer sticking with the status quo when performing financial transactions [9]. Hence, the following hypothesis was developed:

H4: Tradition barrier has a significantly negative relationship with the behavioral intention of baby boomers to adopt mobile payment.

3.5 Image Barrier

As the term suggests, image barrier is formed when a consumer perceives a certain product or service, its image, the company stature, and the country of origin negatively [21]. Many studies have empirically proven that image barrier has significant negative influences on behavioral intention. For example, older individuals were found to be not as prone to adopt mobile banking as their younger counterparts [19]. This is because they perceive that the application provided are not user friendly. This situation is aligned with the findings of [22]. Therefore, the following hypothesis was developed:

H5: Image barrier has a significantly negative relationship with the behavioral intention of baby boomers to adopt mobile payment.

3.6 Information Barrier

The presence of information barrier is due to the limited information about a new product or service [23]. The significantly negative effect of information barrier on behavioral intention has been shown by numerous past studies. More specifically, service providers who fail to deliver comprehensive guidance will cause older adults to become disoriented due to the lack of knowledge required [13]. In general, consumers reported that they felt discouraged to adopt mobile banking services on the grounds of insufficient information and unfamiliarity with the use of these services [24]. Thus, the following hypothesis was developed (Fig. 1):

H6: Information barrier has a significantly negative relationship with the behavioral intention of baby boomers to adopt mobile payment.

Based on the hypotheses developed, the proposed conceptual model of this study is shown in Fig. 1

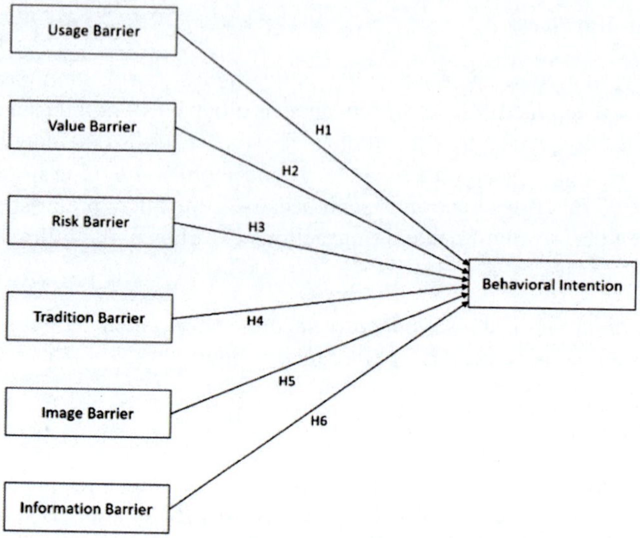

Fig. 1. Proposed conceptual model.

4 Methodology

Given the focus of this study, baby boomers in Malaysia who are new to mobile payment and have little to no experience with it were the target respondents for this study. Non-probability sampling was employed because there is no sampling frame of baby boomers with little to no experience with using mobile payment in Malaysia available. Therefore, this study employed purposive sampling as the respondents would have to possess certain attributes [25]. More precisely, they have to be within the age range of a baby boomer (born between 1946 and 1960) and have little to no experience with using mobile payment.

The responses were collected via a survey [26–28]. The survey used in this study has three sections which are the (1) cover page, (2) demographic, and (3) measurement items. The demographic section covered the age, gender, and other personal characteristics of the respondents. The measurement items section was for the factors of usage barrier (four items), value barrier (four items), risk barrier (five items), tradition barrier (five items), image barrier (four items), information barrier (four items), and behavioral intention (five items). Overall, the 31 questionnaire items were derived from past studies [7, 11] as shown in Appendix A. All measurement items were gauged through a seven-point Likert scale that ranges from (1) strongly disagree to (7) strongly agree. Next, the minimum sample size was calculated via the ratio of 1 item to 4 responses [29]. Therefore, the minimum sample size was established at 124 responses. Overall, 288 responses were obtained which were subsequently analyzed.

5 Analysis

5.1 Demographic Profile

Based on Table 1, a majority of the participants never used mobile payment services before, are female, between the ages of 57–61 years old, and possess a primary/secondary school level of education.

Table 1. Descriptive analysis.

Characteristics	Description	Count	Percentage
Experience with mobile payment	Never used it before	253	87.85
	Used it once or twice	35	12.15
Gender	Male	107	37.15
	Female	181	62.85
Age	57–61 years old	169	58.68
	62–66 years old	87	30.21
	67–71 years old	31	11.11
Highest level of education	Primary/Secondary School	189	65.62
	Diploma/Advanced Diploma	47	16.32
	Bachelor/Professional Qualification	41	14.24
	Master/Ph.D.	11	3.82

5.2 Reliability, Multicollinearity, and Normality

According to Table 2, all constructs are reliable as the Cronbach's alpha values are all above the threshold of 0.7 [30–33]. In addition, given that all the values for variance inflation factor are less than 5, this indicates that multicollinearity is not an issue [34–37]. Moreover, the data is found to be normally distributed as all the values for skewness and kurtosis are within ±3 and ±10 respectively [38].

5.3 Multiple Linear Regression

According to Table 3, usage barrier ($p = <0.001$), value barrier ($p = <0.001$), risk barrier ($p = 0.0140$), tradition barrier ($p = 0.001$), image barrier ($p = 0.026$), and information barrier ($p = <0.001$) all have significant relationships with behavioral intention as the p-values are all below 0.05. Moreover, all antecedents have a negative influence on behavioral intention whereby usage barrier ($\beta = -0.355$) has the greatest effect, followed by value barrier ($\beta = -0.221$), information barrier ($\beta = -0.171$), tradition barrier ($\beta = -0.169$), image barrier ($\beta = -0.106$), and risk barrier ($\beta = -0.094$). Overall, all six hypotheses are supported with a resultant R^2 of 0.5766 on behavioral intention. In

Table 2. Reliability, multicollinearity, and normality.

Construct	Cronbach's alpha	Variance inflation factor	Measurement item	Skewness	Kurtosis
Usage barrier	0.816	1.381	UB1	−0.002	−0.767
			UB2	0.097	−0.672
			UB3	0.024	−0.867
			UB4	−0.220	−0.165
Value barrier	0.791	1.275	VB1	0.203	−0.278
			VB2	0.275	0.399
			VB3	0.369	−0.305
			VB4	0.804	1.668
			VB5	0.024	−0.330
Risk barrier	0.864	1.230	RB1	−0.694	−0.021
			RB2	−0.303	−0.206
			RB3	−0.260	0.278
			RB4	−0.761	0.625
			RB5	−0.439	0.719
Tradition barrier	0.713	1.256	TB1	0.504	1.200
			TB2	0.080	−0.348
			TB3	0.366	0.043
			TB4	0.103	−0.338
Image barrier	0.766	1.266	IMB1	0.489	1.567
			IMB2	0.258	−0.177
			IMB3	0.018	−0.646
			IMB4	0.318	0.944
Information barrier	0.818	1.289	INB1	0.235	0.689
			INB2	−0.298	1.338
			INB3	−0.071	1.250
			INB4	−0.486	1.026
Behavioral intention	0.890		BI1	−0.315	−0.561
			BI2	−0.187	−0.666
			BI3	−0.127	−0.443
			BI4	−0.176	−0.443

(*continued*)

Table 2. (*continued*)

Construct	Cronbach's alpha	Variance inflation factor	Measurement item	Skewness	Kurtosis
			BI5	−0.126	−0.286

Table 3. Parameter estimates.

Hypothesis	Construct	Parameter estimates	Standardized estimates	T-value	p-value
	(Constant)	8.663		31.12	<0.001
H1	Usage Barrier → Behavioral Intention	−0.355	−0.408	−8.94	<0.001
H2	Value Barrier → Behavioral Intention	−0.221	−0.201	−4.58	<0.001
H3	Risk Barrier → Behavioral Intention	−0.094	−0.106	−2.47	0.014
H4	Tradition Barrier → Behavioral Intention	−0.169	−0.141	3.23	0.001
H5	Image Barrier → Behavioral Intention	−0.106	−0.098	−2.24	0.026
H6	Information Barrier → Behavioral Intention	−0.171	−0.162	−3.68	<0.001

other words, 57.66% of the dependent variable's variation can be accounted for by all six independent variables.

Overall, the equation of multiple linear regression is as follows.

$$BI = 8.663 - 0.355(UB) - 0.221(VB) - 0.094(RB) - 0.169(TB) - 0.106(IMB) - 0.171(INB)$$

6 Discussion

The results indicate that usage barrier is the most significant inhibitor on the behavioral intention to utilize mobile payment. This could be attributed to the complexity of using mobile payment as there are several technical jargons involved, including purchase authorization code and card code verification. Moreover, other significant barriers in this area are value barrier, risk barrier, and tradition barrier. These results show that people would still prefer to use traditional methods of payment as they have reservations regarding the risks and value that mobile payment will bring to their lives. In addition to that, behavioral intention to adopt mobile payment is also significantly influenced by image barrier. This is attributed to the perception that baby boomers who use mobile payment would be seen as different and stand out among others. Furthermore, information barrier has a significantly negative association with the behavioral intention to adopt mobile payment. This might be due to the insufficient or unclear information provided by the mobile payment service providers to baby boomers.

In line with the aforementioned analysis, several implications can be drawn by mobile payment service providers to look into overcoming the barriers affecting baby boomers to adopt mobile payment. Firstly, mobile payment service providers should look into designing their system in such a way that it is easy for baby boomers to understand and use. This can be in terms of navigation and readability. Additionally, in their marketing campaign, mobile payment service should inform the public by highlighting the key benefits and safeguards implemented to eliminate any risks of using mobile payment. This can be in the form of educational videos such as step-by-step tutorials. Moving on to the theoretical implications, this research has proven that the addition of "information barrier" into the Innovation Resistance Theory when studying mobile payment resistance is robust. Not only was "information barrier" found to be a significant inhibiting factor in the area of this study but it also better explains mobile payment adoption.

This study does not discount its limitations. Firstly, as this study was only carried out in Malaysia, the findings may not correctly indicate the state of mobile payment resistance in other countries. With this in mind, future researchers should look into conducting cross-country studies that will help to overcome this limitation. In addition, this study looked into mobile payment from a general perspective. As such, future studies can focus on a specific type of mobile payment or its applications in a particular sector [39].

Acknowledgement. This paper is part of a final year project that has been submitted to Universiti Tunku Abdul Rahman.

Appendix A (Questionnaire Items)

Construct	Measurement item
Usage barrier	UB1: I find that mobile payment is difficult to use
	UB2: I find that mobile payment is inconvenient to use
	UB3: I find that mobile payment is inefficient compare to paying in cash
	UB4: I find that instruction provided on the mobile payment platform is unclear
Value barrier	VB1: For me, using mobile payment is uneconomical
	VB2: For me, using mobile payment does not help to increase the ability to control my own financial matters
	VB3: For me, using mobile payment services does not offer any extra benefits when compared to cash payment
	VB4: For me, using mobile payment does not eliminate the constraint of time when conducting the transactions

(continued)

(*continued*)

Construct	Measurement item
	VB5: For me, mobile payment is not a good substitute for traditional cash payment
Risk barrier	RB1: I am afraid of making any mistakes in the process of using mobile payment
	RB2: I am afraid of entering wrong information during the payment process
	RB3: I am afraid of the exposure of private information by using the mobile payment platform
	RB4: I am afraid of any unreasonable or fraudulent charges if using the mobile payment services
	RB5: I am afraid of faultiness in the functions of mobile payment
Tradition barrier	TB1: I feel impatient with the mobile payment applications
	TB2: I prefer to face-to-face communication with the seller to purchase goods and services that I want
	TB3: I prefer to use physical forms of payment for my transactions
	TB4: I prefer to make payment through a computer rather than using mobile phones or tablets
Image barrier	IMB1: Mobile payment projected a very negative image
	IMB2: Mobile payments are perceived to be difficult to use
	IMB3: New technologies are always too complicated to use
	IMB4: The reputation of mobile payment service providers is not so good
Information barrier	INB1: I think it is difficult to get enough information about mobile payment services
	INB2: I think the information available on mobile payment services is unclear and unhelpful
	INB3: I think there is not enough guidance from service providers in relation to mobile payment service
	INB4: The information available concerning mobile payment services is not overwhelming
Behavioral intention	BI1: I will use mobile payment services in the near future
	BI2: I will use mobile payment services if the opportunity arises
	BI3: It is likely that I will use mobile payment services in the future
	BI4: I am planning to use mobile payment services
	BI5: I intend to learn how to use mobile payment services to carry out my transactions

References

1. Loh, X.M., Lee, V.H., Tan, G.W.H., Ooi, K.B., Dwivedi, Y.K.: Switching from cash to mobile payment: what's the hold-up? Internet Research (2020)
2. Slade, E.L., Williams, M.D., Dwivedi, Y.K.: Mobile payment adoption: classification and review of the extant literature. Mark. Rev. **13**(2), 167–190 (2013)
3. Lew, S., Tan, G.W.H., Loh, X.M., Hew, J.J., Ooi, K.B.: The disruptive mobile wallet in the hospitality industry: an extended mobile technology acceptance model. Technol. Soc. **63**, 101430 (2020)
4. Golingai, P.: Going cashless at a snail's pace (2019). https://www.thestar.com.my/opinion/columnists/one-mans-meat/2019/01/26/going-cashless-at-a-snails-pace-malaysians-through-force-of-habit-are-still-very-reliant-on-cash-whi
5. MCMC: Hand phone user survey (2018). https://www.mcmc.gov.my/skmmgovmy/media/General/pdf/HPUS2018.pdf
6. Nielsen: Cash or cashless? Malaysia's shifting payment landscape (2019). https://www.nielsen.com/my/en/insights/article/2019/cash-or-cashless-malaysias-shifting-payment-landscape/
7. Ooi, K.B., Tan, G.W.H.: Mobile technology acceptance model: an investigation using mobile users to explore smartphone credit card. Expert Syst. Appl. **59**, 33–46 (2016)
8. Ram, S., Sheth, J.N.: Consumer resistance to innovations: the marketing problem and its solutions. J. Consum. Mark. **6**(2), 5–14 (1989)
9. Leong, L.Y., Hew, T.S., Ooi, K.B., Wei, J.: Predicting mobile wallet resistance: a two-staged structural equation modeling-artificial neural network approach. Int. J. Inf. Manage. **51**, 102047 (2020)
10. Nel, J., Boshoff, C.: Status quo bias and shoppers' mobile website purchasing resistance. Eur. J. Mark. **54**(6), 1433–1466 (2020)
11. Laukkanen, T., Sinkkonen, S., Kivijärvi, M., Laukkanen, P.: Innovation resistance among mature consumers. J. Consum. Mark. **24**(7), 419–427 (2007)
12. Joachim, V., Spieth, P., Heidenreich, S.: Active innovation resistance: an empirical study on functional and psychological barriers to innovation adoption in different contexts. Ind. Mark. Manage. **71**, 95–107 (2018)
13. Kuerbis, A., Mulliken, A., Muench, F., Moore, A.A., Gardner, D.: Older adults and mobile technology: factors that enhance and inhibit utilization in the context of behavioral health. Mental Health Addict. Res. **2**(2), 1–11 (2017)
14. Sun, B., Sun, C., Liu, C., Gui, C.: Research on initial trust model of mobile banking users. J. Risk Anal. Crisis Response **7**(1), 13–20 (2017)
15. Rammile, N., Nel, J.: Understanding resistance to cell phone banking adoption through the application of the technology acceptance model (TAM). Afr. J. Bus. Manage. **6**(1), 86–97 (2012)
16. Awasthi, P., Sangle, P.S.: The importance of value and context for mobile CRM services in banking. Bus. Process Manage. J. **19**(6), 864–891 (2013)
17. Makanyeza, C.: Determinants of consumers' intention to adopt mobile banking services in Zimbabwe. Int. J. Bank Market. **35**(6), 997–1017 (2017)
18. Yu, C.S., Chantatub, W.: Consumers' resistance to using mobile banking: evidence from Thailand and Taiwan. Int. J. Electron. Commer. Stud. **7**(1), 21–38 (2016)
19. Laukkanen, T.: Consumer adoption versus rejection decisions in seemingly similar service innovations: the case of the Internet and mobile banking. J. Bus. Res. **69**(7), 2432–2439 (2016)
20. Gupta, A., Arora, N.: Understanding determinants and barriers of mobile shopping adoption using behavioral reasoning theory. J. Retail. Consum. Serv. **36**, 1–7 (2017)

21. Cruz, P., Neto, L.B.F., Munoz-Gallego, P., Laukkanen, T.: Mobile banking rollout in emerging markets: evidence from Brazil. Int. J. Bank Market. **28**(5), 342–371 (2010)
22. Priya, R., Gandhi, A., Shaikh, A.: Mobile banking adoption in an emerging economy. Benchmark. Int. J. **25**(2), 743–762 (2018)
23. Laukkanen, T., Kiviniemi, V.: The role of information in mobile banking resistance. Int. J. Bank Market. **28**(5), 372–388 (2010)
24. Yang, Y., Liu, Y., Li, H., Yu, B.: Understanding perceived risks in mobile payment acceptance. Ind. Manage. Data Syst. **115**(2), 253–269 (2015)
25. Loh, X.M., Lee, V.H., Tan, G.W.H., Hew, J.J., Ooi, K.B.: Towards a cashless society: the imminent role of wearable technology. J. Comput. Inf. Syst. (2019)
26. Foo, P.Y., Lee, V.H., Tan, G.W.H., Ooi, K.B.: A gateway to realising sustainability performance via green supply chain management practices: a PLS–ANN approach. Expert Syst. Appl. **107**, 1–14 (2018)
27. Ooi, K.B., Lee, V.H., Tan, G.W.H., Hew, T.S., Hew, J.J.: Cloud computing in manufacturing: the next industrial revolution in Malaysia? Expert Syst. Appl. **93**, 376–394 (2018)
28. Leong, L.Y., Hew, T.S., Ooi, K.B., Tan, G.W.H.: Predicting actual spending in online group buying–an artificial neural network approach. Electron. Commer. Res. Appl. **38**, 100898 (2019)
29. Hinkin, T.R.: A brief tutorial on the development of measures for use in survey questionnaires. Organ. Res. Methods **1**(1), 104–121 (1998)
30. Tew, H.T., Tan, G.W.H., Loh, X.M., Lee, V.H., Lim, W.L., Ooi, K.B.: Tapping the next purchase: embracing the wave of mobile payment. J. Comput. Inf. Syst., 1–9 (2021)
31. Ooi, K.B., Foo, F.E., Tan, G.W.H., Hew, J.J., Leong, L.Y.: Taxi within a grab? A gender-invariant model of mobile taxi adoption. Ind. Manage. Data Syst. **121**(2), 312–332 (2020)
32. Hew, J.J., Wong, L.W., Tan, G.W.H., Ooi, K.B., Lin, B.: The blockchain-based Halal traceability systems: a hype or reality? Supply Chain Manage. Int. J. **25**(6), 863–879 (2020)
33. Wong, L.W., Tan, G.W.H., Hew, J.J., Ooi, K.B., Leong, L.Y.: Mobile social media marketing: a new marketing channel among digital natives in higher education? J. Market. High. Educ., 1–25 (2020)
34. Tan, G.W.H., Lee, V.H., Hew, J.J., Ooi, K.B., Wong, L.W.: The interactive mobile social media advertising: an imminent approach to advertise tourism products and services? Telematics Inform. **35**(8), 2270–2288 (2018)
35. Hew, J.J., Leong, L.Y., Tan, G.W.H., Ooi, K.B., Lee, V.H.: The age of mobile social commerce: an artificial neural network analysis on its resistances. Technol. Forecast. Soc. Chang. **144**, 311–324 (2019)
36. Lee, V.H., Hew, J.J., Leong, L.Y., Tan, G.W.H., Ooi, K.B.: Wearable payment: a deep learning-based dual-stage SEM-ANN analysis. Expert Syst. Appl. **157**, 113477 (2020)
37. Wong, L.W., Tan, G.W.H., Lee, V.H., Ooi, K.B., Sohal, A.: Unearthing the determinants of Blockchain adoption in supply chain management. Int. J. Prod. Res. **58**(7), 2100–2123 (2020)
38. Saunders, M.N.K., Lewis, P., Thornhill, A.: Research Methods for Business Students, 8th edn. Pearson, London (2019)
39. Yan, L.Y., Tan, G.W.H., Loh, X.M., Hew, J.J., Ooi, K.B.: QR code and mobile payment: the disruptive forces in retail. J. Retail. Consum. Serv. **58**, 102300 (2020)

Driving Factors Towards Live-Stream Shopping in Malaysia

Li Qing Chan[1], Yi Ming Kong[1], Zhan Ye Ong[1], Jia Xing Toh[1], Yang Hui Von[1], Voon Hsien Lee[1], Xiu Ming Loh[1](✉), and Garry Wei Han Tan[2]

[1] Universiti Tunku Abdul Rahman, 31900 Kampar, Malaysia
{yiming0316,ongzhan.ye,tohjiaxing,zifeii}@1utar.my, {leevh, lohxm}@utar.edu.my
[2] UCSI University, 56000 Cheras, Malaysia

Abstract. Live-stream shopping is growing exponentially in Malaysia but limited research has been conducted to investigate the driving factors that lead consumers to shop via live-stream. Thus, this research is aimed to establish the factors that affect the purchase intention of Malaysian online shoppers on live-stream shopping to fill this research gap. An integrated research model of Uses and Gratification Theory and Source Credibility Theory was used to investigate the purchase intention in a live-stream context. The integration of these three theories would provide a more comprehensive result as compared to merely adopt only one theory. A self-administered questionnaire was distributed in several states in Malaysia which yielded 252 sets of usable responses. The data collected was analyzed to provide empirical support for the hypotheses developed. This research then provides vast practical and theoretical contributions from the results obtained. In particular, this study provides better comprehension to retailers in understanding the motivating factors that lead consumers to complete the checkout process and thus assist in designing their business strategy accordingly.

Keywords: Live-stream shopping · Online commerce · Uses and gratifications theory · Source credibility theory

1 Introduction

A notable occurrence in the realm of e-commerce is the integration of live-stream and online shopping [1]. Live-stream shopping is a new process of shopping through mixed media which has social commerce attributes and unique media attributes where real-time interaction is offered [2]. Individuals who sell their products or services in real-time videos are known as live-stream sellers [3], which includes social media influencers who are paid to promote merchants' products in live-stream.

In terms of research, live-stream shopping is still a relatively understudied area despite having been present for some time. This is especially true in the context of Malaysia. More specifically, many e-commerce platforms in the country such as Lazada and Shopee utilize this feature extensively [4]. Hence, this study aims to fill the practical

M. Al-Emran et al. (Eds.): ICETIS 2021, LNNS 299, pp. 580–591, 2022.
https://doi.org/10.1007/978-3-030-82616-1_48

and research gap by investigating the variables that will trigger consumers' purchase intention in the context of live stream shopping. In particular, this study integrates the Uses and Gratifications Theory with the Social Credibility Theory to develop the research framework. These two theories were initially developed for the context of media which are therefore suitable for this study given that live-stream is a form of media [5, 6].

As such, the research objectives are to determine (1) the antecedents of purchase intention in the context of live stream shopping as well as (2) the robustness of incorporating the Uses and Gratifications Theory with the Social Credibility Theory. Overall, this study is posited to contribute to numerous novel findings and insights. Practically, this study provides significant value to a number of stakeholders, especially e-commerce retailers. Theoretically, this study extends the literature of purchase intention in the context of live stream shopping from the perspective of a developing nation.

2 Literature Review

2.1 Uses and Gratifications Theory

The Uses and Gratification Theory is an audience-centered approach to understanding the reasons people actively seek out certain media to obtain gratification. It is positivistic in its approach and holds heuristic value [5]. Earlier on, this theory was used to look into the individual motivations behind the utilization of different traditional media such as newspapers, radio, and so on. However, in recent times, it has been widely used to explore new media and communication technologies such as social networking sites [7]. This theory suggests several categorical reasons individuals use technology such as for entertainment and to access information [8]. The Uses and Gratifications Theory was integrated into the research model given its suitability with the subject matter of this study. More specifically, it is a well-used and tested theoretical framework for understanding the reasons individuals engage with a certain media or platform. Moreover, it has been successfully utilized in the contexts of social media and other electronic retail segments [9] which are categorically similar to live-stream shopping.

2.2 Source Credibility Theory

The Source Credibility Theory [6] postulated that the source of a communication is the most significant variable for consumers to determine the effectiveness and reliability of an information. There are certain characteristics of a particular source that affect the persuasiveness and impact of information. In particular, they suggested that the credibility of information depends on the source's attractiveness, expertise, and trustworthiness. Given its suitability with the subject matter of this research, the Source Credibility Theory was incorporated into the research framework. This is because a number of past studies have found that this theory is significant in contexts that are categorically similar to live-stream shopping such as social media [10] and purchase intention [11].

3 Hypotheses Development

3.1 Entertainment

Entertainment is defined as the degree to which something is perceived as fun, enjoyable, and pleasurable [12]. Rationally, consumers feel satisfied when they feel pleasure and increase their willingness to acquire products or services via an online platform [13]. Several empirical studies have found a significantly positive relationship between entertainment and purchase intention [14, 15]. When it comes to live-stream shopping, entertainment is related to the fun and pleasurable experience which could be obtained via the interaction with the live-stream seller. Thus, the following hypothesis was developed:

H1: Entertainment has a significantly positive relationship with purchase intention.

3.2 Informativeness

Informativeness refers to the consumers' desire to obtain useful, timely, and accurate information from a particular source [14]. The significance of informativeness as an antecedent of purchase intention has been empirically proven by a number of past studies [14, 16]. In the context of this study, the source is referring to the live stream seller. If the live-stream seller is able to provide information regarding its products or trend that are accurate and complete in a timely manner, consumers are likely to make a purchase decision. Thus, the following hypothesis was developed:

H2: Informativeness has a significantly positive relationship with purchase intention.

3.3 Attractiveness

Attractiveness is the perception towards the source which includes physical appearance, personality, and similarity [17]. Consumers' acceptance of the information will increase if the information is deliberated by an attractive source [18]. In the context of live-stream shopping, if the information or the product is conveyed by attractive live-stream sellers, consumers will have higher acceptance towards the information [3]. Therefore, the following hypothesis was developed:

H3: Attractiveness has a significantly positive relationship with purchase intention.

3.4 Expertise

Expertise is the perceived extent of the understanding, skills, and knowledge of the source [18]. The information conveyed by the live-stream seller who is an expert tends to have more influence over the consumers' decision [3]. The ability to persuade and convince will subsequently prompt consumers' purchase intention [17]. This is because consumers would place more trust in the live-stream seller who is deemed to be an expert [19]. Hence, the following hypothesis was developed:

H4: Expertise has a significantly positive relationship with purchase intention.

3.5 Trustworthiness

Trustworthiness refers to the degree of consumers' confidence placed in the information delivered by a particular source [20]. In the case of this study, it relates to the live-stream seller and the level of acceptance of the consumers towards the information. Information conveyed by a trusted seller while promoting products via live-stream will be increasingly believed and accepted by consumers [3]. Moreover, consumers will tend to have higher self-assurance towards the products due to a higher level of trustworthiness which will then increase the purchase intention. Thus, the following hypothesis was developed:

H5: Trustworthiness has a significantly positive relationship with purchase intention.

Based on the hypotheses developed, the proposed conceptual model of this study is shown in Fig. 1.

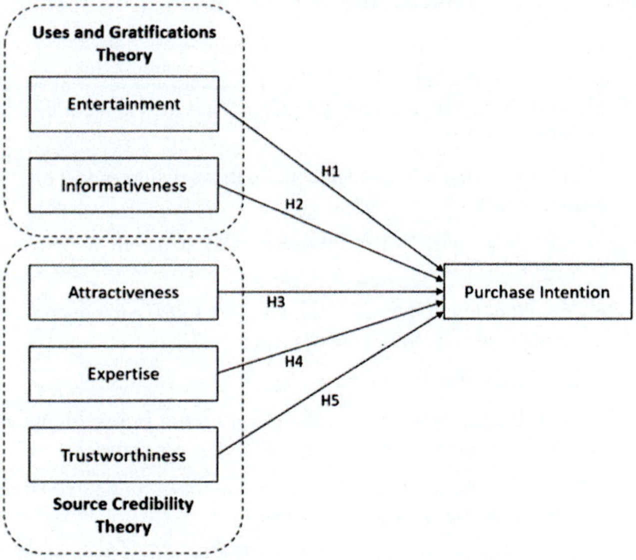

Fig. 1. Proposed conceptual model.

4 Methodology

The target respondents of this study are Malaysian internet users who have watched a promotional live-stream of a product or service. Non-probability sampling was selected as there is no sampling frame available for this group of people. Thus, this study utilized judgmental sampling as the respondents would have to fit certain requirements [21]. More specifically, a qualifier question ("Have you ever watched a live-stream video content about a product or service?") was included in the questionnaire to filter the non-target respondents out.

A survey was conducted in order to collect the data. This survey has three sections which are (1) cover page, (2) demographic, and (3) measurement items [22–24]. The demographic section included the age, gender, occupation, and other personal characteristics of the respondents. The measurement items section was for the factors of entertainment (five items), informativeness (five items), attractiveness (four items), expertise (five items), trustworthiness (five items), and purchase intention (four items). Overall, as shown in Table 1, there were a total of 28 questionnaire items which were adapted from past studies [25–27]. All measurement items were gauged using a seven-point Likert scale that ranges from (1) strongly disagree to (7) strongly agree. After that, the minimum sample size was calculated using the ratio of 1 item to 4 responses [28]. Hence, the minimum sample size was calculated to be 112 responses. Overall, 252 responses were collected and then analyzed.

Table 1. Measurement items.

Construct	Measurement item
Entertainment	EN1: I find it entertaining to watch a live video content that sells/promotes product/service
	EN2: I find that a live video content that sells/promotes product/service is fun to watch
	EN3: I feel excited when I watch a live video content that sells/promotes product/service
	EN4: Watching a live video content that sells/promotes product/service provides me with lots of enjoyment
	EN5: I have fun when interacting with live-stream seller
Informativeness	IN1: It is important that the live-stream seller is able to give me product/service information that is of interest to me
	IN2: Accurate product/service information from the live-stream seller improves my shopping effectiveness
	IN3: Timely information of product/service from the live-stream seller improves my shopping performance
	IN4: Product/service information from the live-stream seller that is useful in aiding my shopping decisions is appreciated
	IN5: Product/service information from the live-stream seller that makes it easier for me to compare product/service choices when shopping online is desirable
Attractiveness	AT1: The live-stream seller is good looking (pretty/handsome)
	AT2: The live-stream seller has a stylish image
	AT3: The live-stream seller is very attractive
	AT4: Others will want to look like the live-stream seller

(continued)

Table 1. (*continued*)

Construct	Measurement item
Expertise	EX1: The live-stream seller knows about the product/service very well
	EX2: It is an undeniable fact that the live-stream seller is an expert on the product/service
	EX3: The live-stream seller has a lot of experience with the product/service
	EX4: The live-stream seller is likely to see/read a lot of reference sources/materials related to the product/service
	EX5: The live-stream seller gives viewers information about the product/service
Trustworthiness	TR1: The live-stream seller will be sincere every time he/she promotes/sells the product/service
	TR2: The live-stream seller will not either exaggerate or lie about the product/service
	TR3: The live-stream seller will not pretend to know about what he/she does not know well about the product/service
	TR4: The live-stream seller will talk validly and reasonably about the product/service
	TR5: The live-stream seller will frankly present his/her position, thoughts and opinions about the product/service
Purchase intention	PI1: The probability that I would consider buying the product/service from live-stream seller is high
	PI2: If I were to buy product/service, I would consider buying it from the live-stream seller
	PI3: The likelihood of my purchasing the product/service from the live-stream seller is high
	PI4: My willingness to buy the product/service from the live-stream seller is high

5 Analysis

5.1 Demographic Profile

Based on Table 2, all of the respondents have previous experience with live-stream shopping given as it is a filter question. Moreover, a majority of the respondents are female, between the ages of 20–25 years old, and possess a bachelor degree/professional qualification.

Table 2. Descriptive analysis.

Characteristics	Description	Count	Percentage
Experience with Live-Stream	Yes	252	100.0
Shopping	No	0	0.0
Gender	Male	116	46.0
	Female	136	54.0
Age	20–25 years old	166	65.9
	26–30 years old	39	15.5
	31–35 years old	23	9.1
	Above 36 years old	24	9.5
Highest Level of Education	Primary/Secondary school	54	21.4
	Diploma/Advanced diploma	51	20.2
	Bachelor degree/Professional qualification	140	55.6
	Master/PhD degree	7	2.8
Occupation	Self-employed	44	17.5
	Employed	98	38.9
	Unemployed	16	6.3
	Student	93	36.9
	Retired	1	0.4
Monthly Income	Dependent (zero income)	97	38.5
	RM1,000 and below	21	8.3
	RM1,001–RM3,000	52	20.6
	RM3,001–RM5,000	49	19.4
	Above RM5,000	33	13.1

5.2 Reliability, Multicollinearity, and Normality

Based on Table 3, reliability has been confirmed as all values of Cronbach's alpha for each construct are above the threshold of 0.7 [29–33]. Moreover, the issue of multicollinearity is absent as all of the values for variance inflation factor are less than 5 [34–38] In addition, the data is considered to be normal given that all values for skewness and kurtosis are within ± 3 and ± 10 respectively [39].

Table 3. Reliability, multicollinearity, and normality.

Construct	Cronbach's alpha	Variance inflation factor	Measurement item	Skewness	Kurtosis
Entertainment	0.915	1.413	EN1	−0.380	−0.403
			EN2	−0.333	−0.369
			EN3	−0.047	−0.603
			EN4	−0.010	−0.592
			EN5	−0.163	−0.674
Informativeness	0.922	1.925	IN1	−0.771	0.569
			IN2	−0.617	0.069
			IN3	−0.399	0.096
			IN4	−0.604	0.264
			IN5	−0.559	0.235
Attractiveness	0.853	1.409	AT1	−0.349	−0.252
			AT2	−0.395	−0.302
			AT3	−0.397	−0.103
			AT4	−0.149	−0.500
Expertise	0.884	2.515	EX1	−0.736	0.090
			EX2	−0.220	−0.393
			EX3	−0.386	−0.104
			EX4	−0.266	−0.20
			EX5	−0.609	0.292
Trustworthiness	0.907	2.291	TR1	−0.609	0.090
			TR2	−0.005	−0.290
			TR3	0.114	−0.397
			TR4	−0.341	−0.040
			TR5	−0.261	−0.239
Purchase intention	0.949		PI1	−0.205	−0.358
			PI2	−0.248	−0.795
			PI3	−0.034	−0.400
			PI4	−0.007	−0.420

5.3 Multiple Linear Regression

With reference to Table 4, entertainment ($p = < 0.001$), informativeness ($p = 0.041$), attractiveness ($p = 0.022$), and trustworthiness ($p = 0.006$) have significant effects on purchase intention as the p-values are all below 0.05. In addition, these four constructs have a positive relationship with purchase intention whereby entertainment ($\beta = 0.445$)

has the greatest effect, followed by trustworthiness (β = 0.232), informativeness (β = 0.163), as well as attractiveness (β = 0.146). However, expertise ($p > 0.05$) was found to be insignificant in determining purchase intention. Therefore, all hypotheses are supported except H4. Furthermore, the model has an R^2 of 0.454 which indicates that 45.4% of the changes in the dependent variable can be accounted for by all the variables.

Table 4. Multiple linear regression.

Hypothesis	Construct	Parameter estimates	Standardized estimates	T-value	p-value
	(Constant)	−0.491		−1.407	<0.001
H1	Entertainment → Purchase Intention	0.445	0.415	7.406	<0.001
H2	Informativeness → Purchase Intention	0.163	0.134	2.049	0.041
H3	Attractiveness → Purchase Intention	0.146	0.129	2.304	0.022
H4	Expertise → Purchase Intention	−0.001	0.000	−0.006	0.995
H5	Trustworthiness → Purchase Intention	0.232	0.198	2.776	0.006

Overall, the multiple linear regression equation is as follows.

$$PI = -0.491 + 0.445\,(EN) + 0.163\,(IN) + 0.146\,(AT) - 0.001\,(EX) + 0.232\,(TR)$$

6 Discussion

The results show that entertainment is the most significant antecedent of purchase intention in the context of live-stream shopping. This could be attributed to the purpose of live-stream which is also as a form of entertainment via engagement [40]. Furthermore, purchase intention was found to be affected by attractiveness. This is because Asian consumers are more swayed if the live-stream personality is famous or good-looking. In addition, informativeness and trustworthiness were also found to be significant constructs in this context. These results show that the real-time features of live-stream enable sellers to provide information more accurately and timely. More precisely, sellers are able to perform a live demonstration and explanation while at the same time allow viewers to ask questions and get an immediate response. Conversely, expertise was discovered to be not significant in this context. This could be due to companies getting celebrities as ambassadors or spokespersons for the live-stream. Hence, viewers may perceive that celebrities may not be experts with the product but are still engaged given their popularity and attractiveness.

Based on the findings, there are several implications that live-stream sellers can look into to better develop their live-stream strategy to better sway the viewers' purchase intention. Firstly, live-stream sellers should take advantage of the features of live-stream which allows instant replies, comments, and reaction buttons to make the live-stream more entertaining to viewers by engaging them. Additionally, live-stream sellers should engage more with their audience by giving timely and accurate information. This can be in the form of honest demonstrations and promotions by describing the limitations of the product to increase the trustworthiness of the information. Finally, they should engage celebrities and social media influencers to feature in the live-stream to attract viewers.

In terms of theoretical implications, this study has found that the integration of Uses and Gratifications Theory and Source Credibility Theory is robust in the live-stream shopping context. This study's research model focused on the factors that are relevant to the features of live-stream shopping. More specifically, Uses and Gratifications Theory addressed the real-time and entertainment aspects of live-stream while Source Credibility Theory looked into the relationship of the live-stream sellers' attributes with the viewers' purchase intention.

Similar to other studies, this one also has its own limitations. Firstly, as this study was conducted purely in Malaysia, the findings may not accurately reflect the situation in other countries given the diverse differences [41]. Hence, future studies should carry out a comparative study among multiple countries in order to overcome this limitation. Moreover, the cross-sectional nature of this study does not allow for the analysis of differences over a period of time [42]. Therefore, future studies should carry out a longitudinal study given that live-stream shopping is still a novel phenomenon.

References

1. Chen, Z., Cenfetelli, R., Benbasat, I.: Grassroots internet celebrity live streaming activating it-mediated lifestyle marketing services at e-commerce websites. In: 38th International Conference on Information Systems, pp. 4954–4965. Association for Information Systems, Seoul (2017)
2. Cai, J., Wohn, D. Y., Mittal, A., Sureshbabu, D.: Utilitarian and hedonic motivations for live streaming shopping. In: 2018 ACM International Conference on Interactive Experiences for TV and Online Video, pp. 81–88. Association for Computing Machinery, Seoul (2018)
3. Todd, P.R., Melancon, J.: Gender and live-streaming: source credibility and motivation. J. Res. Interact. Mark. **12**(1), 79–93 (2018)
4. The Star. Steering Malaysia towards a new normal (2020). https://www.thestar.com.my/news/nation/2020/04/23/steering-malaysia-towards-a-new-normal
5. Katz, E., Blumler, J.G., Gurevitch, M.: Uses and gratifications research. Pub. Opin. Q. **37**(4), 509–523 (1973)
6. Hovland, C.I., Weiss, W.: The influence of source credibility on communication effectiveness. Pub. Opin. Q. **15**(4), 635–650 (1951)
7. Hossain, M.A.: Effects of uses and gratifications on social media use. PSU Res. Rev. **3**(1), 16–28 (2019)
8. Leong, L.Y., Hew, T.S., Ooi, K.B., Tan, G.W.H.: Predicting actual spending in online group buying–an artificial neural network approach. Electron. Commer. Res. Appl. **38** (2019)
9. Ray, A., Dhir, A., Bala, P.K., Kaur, P.: Why do people use food delivery apps (FDA)? a uses and gratification theory perspective. J. Retail. Consum. Serv. **51**, 221–230 (2019)

10. Lou, C., Yuan, S.: Influencer marketing: How message value and credibility affect consumer trust of branded content on social media. J. Interact. Advert. **19**(1), 58–73 (2019)
11. Chakraborty, U.: The impact of source credible online reviews on purchase intention. J. Res. Interact. Mark. **13**(2), 142–161 (2019)
12. Huang, T., Bao, Z., Li, Y.: Why do players purchase in mobile social network games? an examination of customer engagement and of uses and gratifications theory. Program **51**(3), 259–277 (2017)
13. Ramayah, T., Ignatius, J.: Impact of perceived usefulness, perceived ease of use and perceived enjoyment on intention to shop online. ICFAI J. Syst. Manag. **3**(3), 36–51 (2005)
14. Zamzuri, N.H., Kassim, E.S., Shahrom, M., Humaidi, N., Zakaria, N.: Entertainment gratification, informative gratification, web irritation and self-efficacy as motivational factors to online shopping intention. Manag. Account. Rev. **17**(3), 95–108 (2018)
15. Tan, P.K., Goh, H.B., Stany, W.L.F., Yeow, J.A.: Factors that influence the consumer purchase intention in social media websites. Int. J. Supply Chain Manag. **6**(4), 208–214 (2017)
16. Chaturvedi, S., Gupta, S., Hada, D.S.: Perceived risk, trust and information seeking behavior as antecedents of online apparel buying behavior in India: an exploratory study in context of Rajasthan. Int. Rev. Manag. Mark. **6**(4), 935–943 (2016)
17. Chekima, F.Z., Wafa, S.A.W.S.K., Sulong, R.S.: The impact of celebrity credibility on purchase intention of cosmetic products: the moderating role of ethnocentrism. Asian J. Econ. Bus. Account. **7**(1), 1–10 (2018)
18. Wang, S.W., Scheinbaum, A.C.: Enhancing brand credibility via celebrity endorsement. J. Adv. Res. **58**(1), 16–31 (2017)
19. Khan, S.K., Rukhsar, A., Shoaib, M.: Influence of celebrity endorsement on consumer purchase intention. J. Bus. Manag. **18**(1), 6–9 (2016)
20. Malik, H.M., Qureshi, M.M.: The impact of celebrity endorsement on consumer buying behavior. Adv. Soc. Sci. Res. J. **4**(3), 149–170 (2017)
21. Loh, X.M., Lee, V.H., Tan, G.W.H., Hew, J.J., Ooi, K.B.: Towards a cashless society: the imminent role of wearable technology. J. Comput. Inf. Syst. (2019)
22. Foo, P.Y., Lee, V.H., Tan, G.W.H., Ooi, K.B.: A gateway to realising sustainability performance via green supply chain management practices: A PLS–ANN approach. Expert Syst. Appl. **107**, 1–14 (2018)
23. Ooi, K.B., Lee, V.H., Tan, G.W.H., Hew, T.S., Hew, J.J.: Cloud computing in manufacturing: the next industrial revolution in Malaysia? Expert Syst. Appl. **93**, 376–394 (2018)
24. Leong, L.Y., Hew, T.S., Ooi, K.B., Wei, J.: Predicting mobile wallet resistance: A two-staged structural equation modeling-artificial neural network approach. Int. J. Inf. Manag. **51**, 102047 (2020)
25. Lim, W.M., Ting, D.H.: E-shopping: an analysis of the uses and gratifications theory. Modern Appl. Sci. **6**(5), 48 (2012)
26. Choi, W., Lee, Y.: Effects of fashion vlogger attributes on product attitude and content sharing. Fash. Text. **6**(1), 1–18 (2019). https://doi.org/10.1186/s40691-018-0161-1
27. Peng, L., Zhang, W., Wang, X., Liang, S.: Moderating effects of time pressure on the relationship between perceived value and purchase intention in social E-commerce sales promotion: considering the impact of product involvement. Inf. Manag. **56**(2), 317–328 (2019)
28. Hinkin, T.R.: A brief tutorial on the development of measures for use in survey questionnaires. Organ. Res. Methods **1**(1), 104–121 (1998)
29. Hew, J.J., Leong, L.Y., Tan, G.W.H., Ooi, K.B., Lee, V.H.: The age of mobile social commerce: an artificial neural network analysis on its resistances. Technol. Forecast. Soc. Chang. **144**, 311–324 (2019)
30. Loh, X.M., Lee, V.H., Tan, G.W.H., Ooi, K.B., Dwivedi, Y.K.: Switching from cash to mobile payment: What's the hold-up? Internet Res. (2020)

31. Ooi, K.B., Foo, F.E., Tan, G.W.H., Hew, J.J., Leong, L.Y.: Taxi within a grab? A gender-invariant model of mobile taxi adoption. Ind. Manag. Data Syst. (2020)
32. Wong, L.W., Tan, G.W.H., Lee, V.H., Ooi, K.B., Sohal, A.: Unearthing the determinants of Blockchain adoption in supply chain management. Int. J. Prod. Res. **58**(7), 2100–2123 (2020)
33. Tew, H.T., Tan, G.W.H., Loh, X.M., Lee, V.H., Lim, W.L., Ooi, K.B.: Tapping the next purchase: Embracing the wave of mobile payment. J. Comput. Inf. Syst., 1–9 (2021)
34. Hew, J.J., Leong, L.Y., Tan, G.W.H., Lee, V.H., Ooi, K.B.: Mobile social tourism shopping: a dual-stage analysis of a multi-mediation model. Tour. Manag. **66**, 121–139 (2018)
35. Tan, G.W.H., Lee, V.H., Hew, J.J., Ooi, K.B., Wong, L.W.: The interactive mobile social media advertising: an imminent approach to advertise tourism products and services? Telematics Inform. **35**(8), 2270–2288 (2018)
36. Hew, J.J., Wong, L.W., Tan, G.W.H., Ooi, K.B., Lin, B.: The blockchain-based Halal traceability systems: a hype or reality? Supply Chain Manag. Int. J. (2020)
37. Lee, V.H., Hew, J.J., Leong, L.Y., Tan, G.W.H., Ooi, K.B.: Wearable payment: a deep learning-based dual-stage SEM-ANN analysis. Expert Syst. Appl. (2020)
38. Wong, L.W., Tan, G.W.H., Hew, J.J., Ooi, K.B., Leong, L.Y.: Mobile social media marketing: a new marketing channel among digital natives in higher education? J. Mark. High. Educ. (2020)
39. Saunders, M.N.K., Lewis, P., Thornhill, A.: Research Methods for Business Students, 8th edn. Pearson, London (2019)
40. Yokoi, T.: Could livestream shopping provide next-generation entertainment for those stuck at home? Forbes (2020). https://www.forbes.com/sites/tomokoyokoi/2020/04/23/could-livestream-shopping-provide-next-generation-entertainment-for-those-stuck-at-home/?sh=3c9b6e7f5229
41. Yan, L.Y., Tan, G.W.H., Loh, X.M., Hew, J.J., Ooi, K.B.: QR code and mobile payment: the disruptive forces in retail. J. Retail. Consum. Serv. **58** (2020)
42. Lew, S., Tan, G.W.H., Loh, X.M., Hew, J.J., Ooi, K.B.: The disruptive mobile wallet in the hospitality industry: an extended mobile technology acceptance model. Technol. Soc. **63**, 101430 (2020)

Models and Frameworks for IS Outsourcing Structure and Dimensions
A Holistic Study

Eman Elnakeep(✉), Iman M. A. Helal, and Sherif A. Mazen

Faculty of Computers and Artificial Intelligence, Cairo University, Giza, Egypt
eman.elnakeep@pg.cu.edu.eg, {i.helal,s.mazen}@fci-cu.edu.eg

Abstract. Organizations can outsource information systems (IS) at various scales, from a little percentage of services to the whole range of services that they want in which an organization decides to separate the IS-related tasks and outsource them to another organization to produce, based on a defined contract. Different factors affect the organization's decision to outsource. After the decision to outsource is taken, an organization must decide on the type of outsourcing model to follow. There are various outsourcing models and frameworks and each of them focus on specific dimensions like strategic, capability, resources, financial, relationship, organizational structure, institutional arrangements, knowledge, management, security, computability, function characteristics and performance.

Also, there are some challenges/limitations in the existing outsourcing models and frameworks like a restriction to specific outsourcing environments, determining performance outcomes, risks and organization issues, knowledge integration and lack of factors that needed to be added in each outsourcing model to improve the performance of the model. There is no outsourcing model/framework that decides which of these dimensions should be included.

The aim of this paper is to present a comparative study of the existing outsourcing models and frameworks. The results of this study lead to insights into their capabilities, dimensions, factors, advantages, and limitations for building an enhanced and integrated outsourcing model.

Keywords: Outsourcing · Models · Frameworks

1 Introduction

Information systems (IS) have evolved over the years and become an important component of any organization. Thus, the organization must decide among few options to get new software. The first option is to develop in-house software which incurs a lot of money and effort. The second option is to buy ready-made software to solve a specific problem. Despite having many advantages, it still has a major setback as the organization will always be not aware of the inner components of the software that they are dealing with. Finally, the third option is to outsource the software development to an external party.

M. Al-Emran et al. (Eds.): ICETIS 2021, LNNS 299, pp. 592–607, 2022.
https://doi.org/10.1007/978-3-030-82616-1_49

Outsourcing is one of the possible solutions in which an organization decides to separate IS-related tasks and provide them to a different organization to supply. Outsourcing can play an important role in improving the performance and quality of IS development. It has several key advantages, such as cost reduction, usage of skilled personnel, focus on the core business area, management overhead of employing and maintaining an IS task force, keeping up-to-date technology [1]. There are several areas that utilize from outsourcing IS-related tasks. For example: application development and maintenance, managing network, testing, integration, and data entry [2].

After the decision to outsource is taken, an organization must decide on the suitable type of outsourcing model to follow. The previous studies show the importance of using outsourcing in achieving information system tasks. Also, it shows the dimensions and their related factors that affect outsourcing models and frameworks that contribute to achieving successful outsourcing.

In this paper, we aim to determine the main dimensions and their related factors affecting the outsourcing models and frameworks. Also, we aim to present a comparative study of the existing models and frameworks with their advantages and limitations.

The rest of this paper is organized as follows: an overview of IS outsourcing with its basic concepts and characteristics in Sect. 2. Section 3 discusses related work of outsourcing models and frameworks. Section 4 discusses the results that we derive. Finally, we conclude the paper in Sect. 5 with an overlook of the future work.

2 Background Study

In this section, we will cover basic concepts related to information systems outsourcing in Sect. 2.1. In Sect. 2.2 will present the most common benefits and drawbacks of IS outsourcing in the literature.

2.1 Definitions, Types and Levels of IS Outsourcing

Outsourcing is a common practice in many large organizations around the world. This practice can lead to the improvement of the organization's efficiency [1]. There are various outsourcing **definitions**. It can be defined as a decision-making process, where the management of the organization decides whether they would keep a specific activity in-house or buy it from an external subcontractor [3]. In [4], outsourcing is a production of services externally which are consumed by any enterprise. And in this paper, we define *outsourcing* as a strategy in which an organization contracts other organization to provide services externally, which may involve transferring employees and assets from one organization to another.

Outsourcing can also be categorized into three different **types** which are *total, selective,* and *transitional* outsourcing. The organization uses total outsourcing when moving more than 80% of the organization's business operating budget to an external service provider. When an organization outsources a preferred function to single or multiple vendors while managing between 20 to 80% of the business in-house, this is called selective outsourcing while transitional outsourcing occurs during a major transition for an organization to the new platform or system, such as bringing in new technology [5].

According to [5], there are three different **levels** of outsourcing available to firms which are *tactical*, *strategical* and *transformational*. An organization should choose tactical outsourcing when they are faced with a specific problem and want to quickly solve it. Strategic outsourcing is a more advanced approach taken to maintain an organization's primary value and its core competencies, whereas transformational outsourcing is the most developed form of cooperation with the service provider in outsourcing.

2.2 Advantages and Drawbacks of Outsourcing

There are several benefits and drawbacks for applying outsourcing in an organization. Each organization selects the suitable type and level of outsourcing according to them.

Benefits of Outsourcing
The most important benefits of IS outsourcing [6, 7] are stated as follows:

1. *Expertise and fast delivery*: It ensures better performance and quality with highly skilled domain experts, suitable equipment and technologies.
2. *Focus on core business*: It helps organizations to utilize their time on their core activities and gain a competitive advantage over other organizations.
3. Reduced costs: It reduces the need to hire individuals in-house. Thus, it minimizes set up, operational, and recruitment costs, as well as better, utilizes time.
4. *Risk-sharing*: It involves components that help the organization to transfer certain functions to the outsourced provider, who plans a risk-mitigating factor better.
5. *Improvements in service and quality level*: The outsourcing vendor can provide more proficient services via suitable technical skills and make quicker deliveries to customers that accommodate rapid changes in technology.
6. *Flexibility and control*: It aids in adapting to a rapidly changing market, supporting organizations with better flexibility and control to the changing environment.

Drawbacks of Outsourcing
The main risks of IS outsourcing [6, 8] are stated as follows:

1. *Danger of exposing sensitive data and technology*: An organization involves a risk of exposing confidential organization information in addition to technology to a third-party during the outsourcing process.
2. *Many Hidden costs*: Outsourcing includes hidden costs involved in switching vendors, signing a contract while signing a contract across international boundaries may pose a serious threat.
3. *Possibility of Weak Management*: It can be more difficult for organizations to deal with a new type of management, i.e., outsourcing management.
4. *Lack of customer focus*: An outsourced vendor may need to serve many organizations at a time. Therefore, they will lack an entire specialize in the client organization's tasks.

5. *Inexperienced Staff*: Clients are usually supported by their own staff transferred to the seller during the outsourcing process. This transfer can lose the experienced IT staff of the organization to the vendor organization.
6. *Business Uncertainty*: Organization's needs can change after the outsourcing decision. So strategic intent must be restated according to this change.
7. *Outdated Technology and Skills*: If the vendor cannot keep up with the changes in the market and specific sectors, the organization may have outdated technology at the end of the contract term.
8. *Loss of Innovative Capacity*: The innovation capacity of the organization may be impaired after outsourcing due to its dependency on the vendor(s) in outsourced operations/services.

3 Assessment of Outsourcing Models and Frameworks

There are several dimensions (Sect. 3.1) affecting outsourcing models and frameworks (Sect. 3.2), where each model and framework focus on specific concerns/dimensions.

3.1 Dimensions in Outsourcing Models and Frameworks

There are several dimensions that affect outsourcing models, frameworks, and system architectures, these dimensions are as follows:

1. **Strategic** dimension includes mergers and acquisitions, and business transitions and reduction in project delivery lead time [9].
2. **Financial** dimension includes cost reduction through reducing total assets and the number of employees involves in the project. This can also improve the financial performance index and cash flow [9].
3. **Capability** dimension includes an assembly of skills, techniques, technologies which enable an organization to acquire, deploy, and leverage investments in pursuit of business strategies and increase productivity [10].
4. **Resources** dimension includes access to external expertise and technical skills needed. This will help the organizations to achieve competitive advantages after delegating outsourced services to a vendor organization [9].
5. **Relationship** dimension is the main reason for a successful partnership between organizations because the interaction between the parties is based on human resources, good communication and compatibility to achieve the quality of the partnership between two parties [1].
6. **Knowledge** dimension includes transferring and utilizing knowledge which helps organizations to create advantages for competing in the market and improve efficiency and performance of processes [11].
7. **Organizational structure** dimension describes how certain activities are coordinated to realize the objectives of an organization. It includes roles, responsibilities, rules and layers in the hierarchy [10].
8. **Institutional arrangements** dimension involves policies, systems, and processes that organizations use to organize and manage their activities efficiently and to effectively collaborate with others [12].

9. **Function characteristics** include a business function that is delivered to business customers by business units. The function has some characteristics that affected it that should be considered to ensure the successful delivery of business services [13].
10. **Management** dimension includes legal contracts that can achieve project goals and create positive competition, coordination to ensure that interdependence of tasks and responsibilities is well managed between two parties [9].
11. **Security** dimension includes information security that is a set of practices, protocols, strategies, policies, procedures, standards, techniques and tools that protecting information by mitigating information risks, Information security's primary objective is the balanced protection of confidentiality, integrity and availability of data [14].
12. **Performance** dimension includes methodologies, processes and metrics to manage and optimize business performance and overcome potential problems and risks. It provides key performance indicators (KPI) that help organizations monitor the efficiency of projects against operational targets [10].

Figure 1 shows the relationship between dimensions that existed in the outsourcing structure. The outsourcing process should start with a strategic dimension. Strategic evaluation reflects the extent to which the outsourcing team carried out a comprehensive assessment of the strategic implications of outsourcing the business activity. The degree to which firms effectively perform a strategic evaluation is concurrently reflected by their evaluation from a capability and financial perspective [15]. So, the capability and financial dimensions are considered as output for the strategic evaluation process.

Capability is often combined with required resources, persons, organization's services, knowledge and technologies. So, we need to identify resources that will help to perform the business activity. So, the capability dimension is considered as a prerequisite for the resources dimension. There is a relationship between capability and organizational structure. It is used to describe the structure within an organization. It identifies each job, its function and where it reports to within the organization. A provider's organizational structure can be considered as an intermediary of the relationship between the client's actions and the provider's outsourcing capabilities that were designed to support the client [10].

So, organizational structure is used to assign roles to activities that are set by capability, hence we see capability dimension as a prerequisite for organizational structure dimension. There is a relationship between institutional arrangements and organizational structures as institutional arrangements refer to formal governmental organizational structures as well as informal standards in place in a country for arranging and undertaking its own policy actions. These arrangements are essential, as they provide the government, at all levels, with the framework for formulating and executing policies. So, the output of determining/making organizational structure process can be input for the institutional arrangements dimension as part of its process.

The relationship is the main reason for successful cooperation/partnership between organizations and IT providers. Clients should have a close relationship with the provider to achieve effective IT outsourcing because IT outsourcing relies on human resources [1]. Hence, we consider resources as prerequisites for the relationship dimension because building and sustaining a successful relationship depends on resources. Knowledge can be shared effectively through efficient use and utilization of capability and relationship dimensions [11].

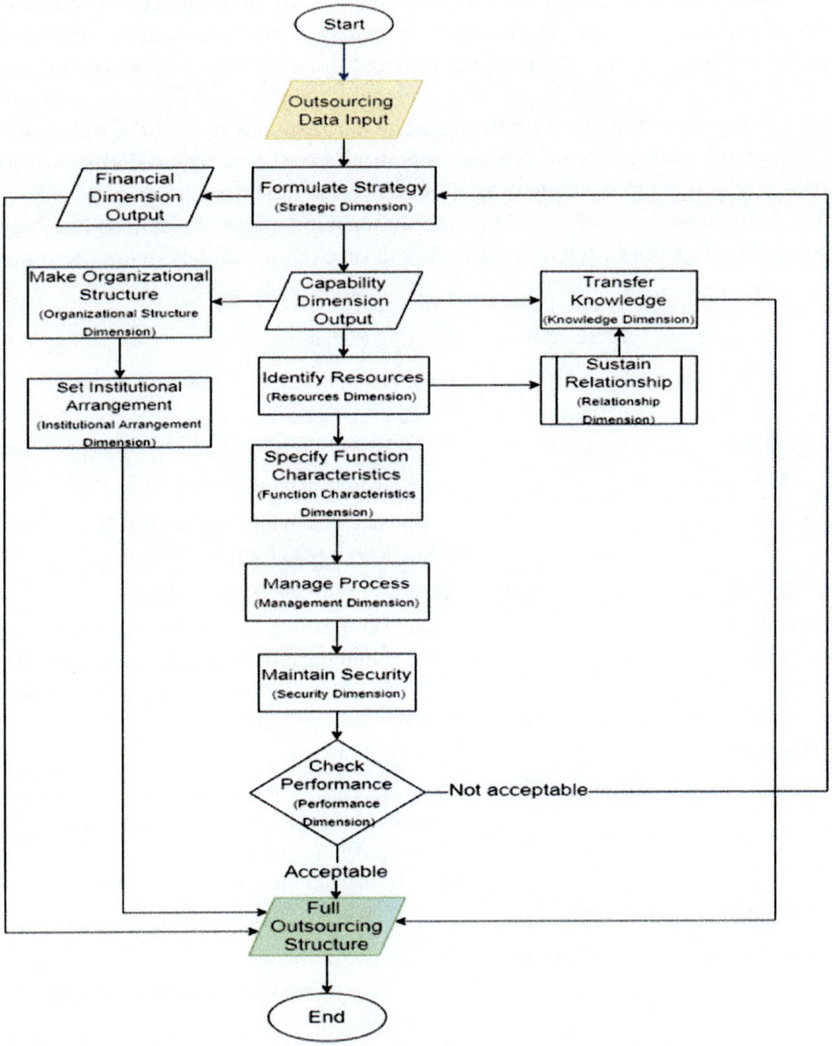

Fig. 1. Dimensions required to reach satisfying/acceptable outsourcing structure

Therefore, these two dimensions can be used as input for achieving successful knowledge transfer and utilization process. A function is a service that is delivered to business customers by business units. Functions have some characteristics that affected them as the complexity of function, function integration and structure, function control and equipment/tool availability [13]. So, we need first to allocate resources responsible for function implementation before set function characteristics. So, we see resources are considered as input for the set function characteristics process.

The management dimension is considered a wide umbrella that contains a lot of processes, activities and practices like resource allocation, set function characteristics, coordination and administration of tasks to achieve goals [13]. Hence, the set function characteristics process is the responsibility of management. The security dimension is an important dimension to protect information and mitigate risks. So, we need to ensure security to prevent any unexpected behavior that can arise and keep the outsourcing structure more secure. Finally, we can check the performance if it is acceptable, we can get an acceptable and satisfying outsourcing structure. If not, we should return to the first process which is the strategic evaluation to revise and check it again.

Table 1 illustrates a brief sample of the dimensions and their factors. It shows the most common factors per each dimension based on existing studies in the literature.

Table 1. Factors affecting each dimension

#	Dimension	Factor(s)
1	Strategic	Time, strategic partnership, organization strategy
2	Financial	Cost, savings, lending, investments
3	Capability	Technical, technology, skills, experience, user training and consulting, know-how
4	Resources	Technical staff, end-users, culture
5	Relationship	Culture compatibility, temporal, standard, loyalty, trust, commitment, interaction, quality, interface characteristics, information sharing, strategic compatibility, flexibility
6	Knowledge	Quality content, accuracy
7	Organizational structure	Competences, responsibilities, roles, degree of formalization, specialization, standardization, layers in the hierarchy, horizontal integration, and professionalism, business environment
8	Institutional arrangements	Policies, systems, processes, laws, regulations
9	Function/service characteristics	Complexity of function, function integration and structure, function control and equipment/tools availability
10	Management	Contract, leadership, coordination, organizational factors (shared values, norms, beliefs, development process), risk management
11	Security	Confidentiality, integrity, availability, access control, firewall, authenticity
12	Performance	Balanced scorecard, key performance indicators (KPIs), reliability, quality performance, supporting infrastructure, performance contract

3.2 Outsourcing Models, Frameworks and System Architectures

In this subsection, we will present a detailed description of each outsourcing model, framework and system architecture showing dimensions that affect them and gaps in each one.

Outsourcing Models

Table 2 illustrates the existing outsourcing models and their mapping with respect to one or more dimensions. In [1], the proposed model determined the factors that positively impact the continuity of their relationship. However, this study has missed other aspects that can influence partnership quality as risk, capability, and resources. In [2], the proposed model is generic and can be extended to different IT outsourcing arrangements. However, they missed other aspects while describing the alternatives as used resources, cost, and risk management. The model presented in [3], focused on outsourcing in SMEs instead of large enterprises. However, it did not consider financial and strategic dimensions and what factors are needed to determine the capability dimension.

If the proposed model in [4] and software tool are being followed, risks can be mitigated, and the success of the projects is assured. However, they missed how to develop and sustain a long-term relationship between the service provider and client. While in [10], Outsourcing capabilities and organizational structure dimensions are crucial factors in the achievement of quality performance, and that a fit between them is paramount, but missed the risks and cost aspects. Moreover, the model presented in [11] can aid practitioners to determine how to use outsourcing to improve knowledge management, but it depends on limited sample size.

In [12], the authors confirmed that effectively designing work practices and contracts achieve better public-private goals. However, they need to develop more cases in different countries to improve the generalizability of the findings and explore more factors. In [13], the model could be used for a specific maintenance activity in any generic location and specialty. However, they depend on the user to give the relative weight of the factors, which need more experienced users in the outsourcing model context. Meanwhile, in [15], the proposed model was empirically presented and has validated the effects of the three key practices on outsourcing performance. However, other elements that impact performance as advanced technology, financial and knowledge transfer aspects were skipped.

In [16], the model identified the factors that affect the knowledge transfer process in IT outsourcing, but missed other factors, e.g., knowledge integration mechanisms. The presented model in [17], is an effective and efficient decision-making tool for the selection of the best service provider for outsourcing and can be extended to other service sectors. However, it did not consider the reputation of the service provider, experience, resources, and capability dimensions in selecting an appropriate service provider. While in [18], the presented model ensures the efficiency of the IT infrastructure, achieves a high degree of innovativeness of the business and makes use of the resource advantages of service providers.

Table 2. Mapping existing models w.r.t dimensions

No	Models	Dimensions											
		Strategic	Financial	Capability	Resources	Relationship	Knowledge	Organizational structure	Institutional arrangements	Function characteristics	Management	Security	Performance
1	[1]					✓							
2	[2]			✓	✓								✓
3	[3]			✓		✓					✓		
4	[4]	✓	✓	✓	✓						✓		
5	[10]			✓				✓					✓
6	[11]			✓			✓						✓
7	[12]			✓		✓	✓		✓				
8	[13]	✓	✓	✓						✓	✓		
9	[15]	✓		✓		✓					✓		✓
10	[16]		✓	✓	✓		✓						
11	[17]		✓			✓	✓				✓		
12	[18]		✓								✓		
13	[19]	✓	✓										
14	[20]			✓				✓					
15	[21]			✓									
16	[22]		✓	✓	✓	✓							
17	[23]		✓		✓		✓				✓		
18	[24]										✓		
19	[25]	✓		✓	✓						✓		
20	[26]		✓	✓	✓	✓					✓		✓
21	[27]				✓							✓	

Meanwhile, in [19], the government not only saves costs but can also keep ahead in the development of IT, while the vendor can acquire returns from the construction and maintenance of the e-government systems. However, the proposed model did not consider risk management. The model in [20], investigated the factors affecting the outsourcing decisions of e-banking services, but the findings cannot be generalized to other industries. While in [21], the authors studied the moderating role of learning capability on the relationship between integration capability and performance in IT outsourcing while prior related studies neglecting this hypothesis. However, it will be more valuable if they can combine the data from both the vendor and their clients.

Meanwhile, in [22], many subcontracting SMEs can internationalize their business if they leverage their resources, capabilities, and customer networks proficiently. In [23], the proposed model is the first to consider subcontracting forms and how they have led to the sustainability of SMEs activities. Meanwhile, in [24], the authors proposed a generic approach, but it is a manual checklist-based method and can become time-consuming. They depend on documents for the assessment process, so problems can arise if documents are not updated regularly. In [25], the authors addressed the issues concerning outsourcing and reversibility as one unified process while others considered them as opposite phenomena. The key factor for reversibility should be considered and handled properly. Otherwise, reversibility will be difficult to implement.

Meanwhile, in [26], managers used this model to prepare an appropriate interaction design with the representatives to achieve successful outsourcing. While in [27], the proposed OPE model can be implemented by any programming language, and users can define their split methods and encrypt function. It achieved high security. Both [26, 28] did not consider cost management.

The models were validated in different approaches: either by a questionnaire [1, 3, 4, 10, 13, 15–17, 23, 26], using a case study [2, 4, 12, 13, 17, 24, 27], a survey [11, 20–23], or an experiment [27]. All the models presented in [16, 18, 19, 26] are dedicated to a specific sector, they are not considered generic models.

Outsourcing Frameworks

Table 3 illustrates the existing outsourcing frameworks and their mapping with respect to one or more dimensions. In [28], the authors confirmed that strong partnership and goal compatibility can be the foundation of a long-term sustainable relationship. The presented framework in [29] identified the critical success factors for IT outsourcing implementation. Meanwhile, in [30], the authors provide practitioners with pointers on the way to maximize benefits and minimize the obstacles caused by dependency. However, client dependence can affect other aspects as vendor performance.

In [31], the authors introduced the catalyst role of partnership dynamic capability in transforming IT investments into successful partnership value. However, they did not consider how the interfirm power structure impacts the effects of IT investments. Meanwhile, in [32], the proposed framework reduced time, cost and improve the security of the auditing process. However, they did not consider the confidentiality of the data. The framework in [33], covered different business and IT facets of the enterprise which empowers the efficiency and achieve the business/IT alignment. It should be applied to other cloud models to find other dimensions and factors to be assured that it is not restricted to one cloud service.

Table 3. Mapping existing frameworks w.r.t dimensions

No	Frameworks	Dimensions											
		Strategic	Financial	Capability	Resources	Relationship	Knowledge	Organizational structure	Institutional arrangements	Function characteristics	Management	Security	Performance
1	[28]					✓							
2	[29]	✓	✓	✓	✓						✓		
3	[30]				✓	✓							
4	[31]			✓	✓								✓
5	[32]	✓	✓									✓	
6	[33]	✓	✓	✓	✓								✓

There are several validation techniques to existing frameworks, either by questionnaire [14, 30, 31], prototype [32], or an experiment [32]. Moreover, the proposed frameworks in [28, 29] missed risk management.

Outsourcing System Architecture

Table 4 illustrates the existing outsourcing systems architectures and their mapping with respect to one or more dimensions. In [14], the authors proposed a design of a new architecture to achieve the confidentiality and integrity of query results on outsourced data. But they did not state how to optimally deploy these services over the cloud to reduce cost. While in [34] the authors provided optimized business process outsourcing through the CaaS model and enhanced business process deployment. However, they did not consider the integrity of outsourced stored data and did not show how to search and retrieve stored data on the cloud. They validated the proposed system architecture through a case study.

4 Results and Discussions

Figure 2 shows the total ratio and the number of each dimension that existed in the models, frameworks, and systems architectures. Through investigating the models and frameworks that existed in the literature review, we found 12 dimensions affecting outsourcing success which are strategic, financial, capability, resources, relationship, knowledge, organizational structure, institutional arrangements, function characteristics, management, security, performance. From Fig. 2, we notice that capability, resources, financial and management dimensions have the highest percentages over the other dimensions. Thus, we can consider them the four key players in most models and frameworks.

Other dimensions have lower percentages due to the context or applicability of outsourcing models and frameworks. For example, the institutional arrangements dimension considers determinants of collaborative interfaces in public-private IT outsourcing relationships [12]. Also, the function/service characteristics dimension is available in a multi-criteria decision-making model and applied as an approach for maintenance managers to consider before deciding on outsourcing [13]. We conclude that there are some dimensions with the same objective that we can merge them into one dimension like organizational structure, function characteristics and institutional arrangements dimensions. They have the same objectives with the management dimension so we can merge them into one dimension as *management*.

Table 4. Mapping existing system architectures w.r.t dimensions

No	System architectures	Dimensions											
		Strategic	Financial	Capability	Resources	Relationship	Knowledge	Organizational structure	Institutional arrangements	Function characteristics	Management	Security	Performance
1	[14]			✓	✓							✓	
2	[34]		✓	✓	✓						✓		✓

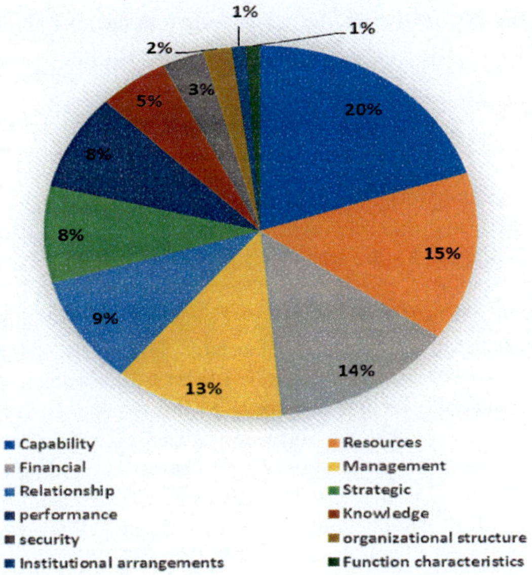

Fig. 2. Percentage of each dimension in models, frameworks, and system architectures

5 Conclusion and Future Work

Outsourcing is a common practice in many large organizations around the world. This practice transfers some functions of an organization to external parties for the improvement of the organization's efficiency. Among the most outsourced functions are information system (IS) and information technology (IT) functions. Several factors affect the organization's decision to outsource. After the decision to outsource is taken, an organization must decide on the suitable type of outsourcing model to follow. To achieve this target, we investigate several models, frameworks, system architectures and approaches that help in achieving successful outsourcing.

This research has important theoretical and managerial implications for several research directions. From the theoretical perspective, this research contributes to the outsourcing literature by focusing on the main dimensions and their related factors affecting the success of outsourcing models and frameworks. Also, it contributes to the literature by addressing the advantages and limitations of each of them. On the other hand, from the managerial perspective, this research supports managers in different organizations to decide on the suitable type of outsourcing model to follow after the decision to outsource is taken. It also helps end-users in different organizations to select an outsourcing third-party based on their system requirements.

As a result of this research, we can conclude that no outsourcing model or framework considers all the dimensions. Also, there is no consensus regarding the models and dimensions determining outsourcing success. Some dimensions/factors can be added or removed according to the context or environment that we design the model or framework for it. This can be very challenging to achieve successful IS outsourcing with good service quality, minimum cost, time, risks, and effort through the application of a comprehensive model or framework. For future work, we aim to create an enhanced and integrated outsourcing model to consider the available models and frameworks.

References

1. Wibisono, Y.Y., Govindaraju, R., Irianto, D., Sudirman, I.: Managing differences, interaction, and partnership quality in global inter-firm relationships: an empirical analysis on offshore IT outsourcing. Int. J. Manag. Proj. Bus. **12**(3), 730–754 (2019). https://doi.org/10.1108/IJMPB-04-2018-0074
2. Morais, D.C., Costa, A.P.C.S., de Almeida, A.T.: Group decision model for outsourcing IT services. Procedia Technol. **16**, 562–568 (2014). https://doi.org/10.1016/j.protcy.2014.10.004
3. Hamzah, A.K.: A review of IT outsourcing relevant and success determinants and a proposed model. J. Adv. Comput. Sci. Technol. Res. **3**, 18 (2013)
4. Kazmi, S.H.M., Hafeez, Y., Ali, S.: Software outsourcing model for risk mitigation. In: 2018 International Conference on Computing, Mathematics and Engineering Technologies – iCoMET 2018, p. 11 (2018). https://doi.org/10.1109/ICOMET.2018.8346317
5. Akbari, M., Hopkins, J.L.: The changing business landscape in Iran: establishing outsourcing best practices. Oper. Supply Chain Manag. Int. J. **9**, 184–197 (2016). https://doi.org/10.31387/oscm0250172
6. Somjai, S.: Advantages and disadvantages of outsourcing. Bus. Manag. Rev. **9**(1), 4 (2017)
7. Shumanskaya, M.: Improvement of outsourcing information technology services from other countries to Finland, p. 49 (2011)
8. Vural, L.: Success factors in public information systems outsourcing: a case study. JITCA **19**(1), 107 (2004)
9. Sabri, S.M., Sulaiman, R., Ahmad, A., Tang, A.Y.: A comparative study on IT outsourcing models for Malaysian SMEs e-business transformation. ARPN J. Eng. Appl. Sci. **10**(23), 8 (2015)
10. Plugge, A., Bouwman, H., Molina-Castillo, F.-J.: Outsourcing capabilities, organizational structure and performance quality monitoring: toward a fit model. Inf. Manag. **50**(6), 275–284 (2013). https://doi.org/10.1016/j.im.2013.04.006
11. Teo, T.S.H., Bhattacherjee, A.: Knowledge transfer and utilization in IT outsourcing partnerships: a preliminary model of antecedents and outcomes. Inf. Manag. **51**(2), 177–186 (2014). https://doi.org/10.1016/j.im.2013.12.001
12. Duhamel, F.B., Gutiérrez-Martínez, I., Picazo-Vela, S., Luna-Reyes, L.: Determinants of collaborative interfaces in public-private IT outsourcing relationships. Transform. Gov. People Process Policy **12**(1), 61–83 (2018). https://doi.org/10.1108/TG-07-2017-0042
13. Hassanain, M.A., Assaf, S., Al-Hammad, A.-M., Al-Nehmi, A.: A multi-criteria decision making model for outsourcing maintenance services. Facilities **33**(3/4), 229–244 (2015). https://doi.org/10.1108/F-01-2013-0003
14. Rady, M., Abdelkader, T., Ismail, R.: Integrity and confidentiality in cloud outsourced data. Ain Shams Eng. J. **10**(2), 275–285 (2019). https://doi.org/10.1016/j.asej.2019.03.002
15. Handley, S.M., Benton, W.C.: Unlocking the business outsourcing process model. J. Oper. Manag. **27**(5), 344–361 (2009). https://doi.org/10.1016/j.jom.2008.11.002
16. Arshad, N.H., Hanapi, R., Buniyamin, N.: IT outsourcing and knowledge transfer in Malaysia. In: 2010 2nd International Congress on Engineering Education, Kuala Lumpur, Malaysia, pp. 16–21 (December 2010). https://doi.org/10.1109/ICEED.2010.5940756
17. Hsu, C.-C., Liou, J.J.H.: An outsourcing provider decision model for the airline industry. J. Air Transp. Manag. **28**, 40–46 (2013). https://doi.org/10.1016/j.jairtraman.2012.12.009

18. Sosunova, L.A., Yakhneeva, I.V., Agafonova, A.N.: Outsourcing model evaluation for business process management in digital economy. Mediterr. J. Soc. Sci. (2015). https://doi.org/10.5901/mjss.2015.v6n6s3p328
19. Lu, J., Zhong, W., Mei, S.: BOO model for outsourcing e-government services. Syst. Eng. - Theory Pract. **29**(4), 75–80 (2009). https://doi.org/10.1016/S1874-8651(10)60016-6
20. Hanafizadeh, P., Zare Ravasan, A.: An empirical analysis on outsourcing decision: the case of e-banking services. J. Enterp. Inf. Manag. **31**(1), 146–172 (2018). https://doi.org/10.1108/JEIM-11-2016-0182
21. Wang, M.-M., Wang, J.-J.: How vendor capabilities impact IT outsourcing performance: an investigation of moderated mediation model. J. Enterp. Inf. Manag. **32**(2), 325–344 (2019). https://doi.org/10.1108/JEIM-07-2018-0167
22. Kim, J., Hemmert, M.: What drives the export performance of small and medium-sized subcontracting firms? A study of Korean manufacturers. Int. Bus. Rev. **25**(2), 511–521 (2016). https://doi.org/10.1016/j.ibusrev.2015.09.002
23. Nwankwo, C.: The effects of subcontracting forms on the sustenance of SMEs. World J. Entrep. Manag. Sustain. Dev. **15**(4), 15 (2019). https://doi.org/10.1108/WJEMSD-01-2019-0006
24. Kaur, K., Sharma, R.: Critical: threat model for an outsourcing business. In: 2017 8th International Conference on Computing, Communication and Networking Technologies (ICCCNT), Delhi, pp. 1–5 (July 2017). https://doi.org/10.1109/ICCCNT.2017.8204093
25. Al-Ahmad, W., Al-Oqaili, A.: Towards a unified model for successful implementation of outsourcing and reversibility of information systems. J. King Saud Univ. - Comput. Inf. Sci. **25**(2), 229–240 (2013). https://doi.org/10.1016/j.jksuci.2013.03.003
26. Sanchis-Pedregosa, C., Machuca, J.A.D., González-Zamora, M.-M.: Determinants of success in transport services outsourcing: empirical study in Europe. Int. J. Logist. Manag. **29**(1), 33 (2018). https://doi.org/10.1108/IJLM-09-2016-0207
27. Liu, Z., Chen, X., Yang, J., Jia, C., You, I.: New order preserving encryption model for outsourced databases in cloud environments. J. Netw. Comput. Appl. **59**, 198–207 (2016). https://doi.org/10.1016/j.jnca.2014.07.001
28. Chaurasia, S.S.: I can't but we can!: impact of goal compatibility on value co-creation in retailer–manufacturer outsourcing relationship. J. Glob. Oper. Strateg. Sourc. **11**(1), 123–144 (2018). https://doi.org/10.1108/JGOSS-04-2017-0010
29. Zhang, P., Zeng, Z., Huang, C.: A framework for IT outsourcing decision process. In: 2008 International Conference on Service Systems and Service Management, Melbourne, Australia, pp. 1–4 (June 2008). https://doi.org/10.1109/ICSSSM.2008.4598453
30. Gopalakrishnan, S., Zhang, H.: Client dependence: a boon or bane for vendor innovation? A competitive mediation framework in IT outsourcing. J. Bus. Res. **103**, 407–416 (2019). https://doi.org/10.1016/j.jbusres.2017.12.023
31. Chang, K.-H., Chen, Y., Huang, H.-F.: Information technology and partnership dynamic capabilities in international subcontracting relationships. Int. Bus. Rev. **24**(2), 276–286 (2015). https://doi.org/10.1016/j.ibusrev.2014.08.003
32. Guo, W., et al.: Outsourced dynamic provable data possession with batch update for secure cloud storage. Future Gener. Comput. Syst. **95**, 309–322 (2019). https://doi.org/10.1016/j.future.2019.01.009
33. Rekik, M., Boukadi, K., Ben-Abdallah, H.: A comprehensive framework for business process outsourcing to the cloud. In: 2016 IEEE International Conference on Services Computing (SCC), San Francisco, CA, pp. 179–186 (June 2016). https://doi.org/10.1109/SCC.2016.30
34. Boukadi, K., Grati, R., Rekik, M., Ben-Abdallah, H.: Business process outsourcing to cloud containers: how to find the optimal deployment? Future Gener. Comput. Syst. **97**, 12 (2019). https://doi.org/10.1016/j.future.2019.02.069

Impact of Total Quality Management on SMEs Sustainable Performance in the Context of Industry 4.0

Kashif Ali(✉) and Satirenjit Kaur Johl

University Teknologi PETRONAS, 32610 Seri Iskandar, Perak, Malaysia
kashif_18001526@utp.edu.my

Abstract. This empirical study analyzes the relationship of digital Total Quality Management (TQM) or Quality 4.0 practices on the sustainable performance in Malaysian small & medium (SMEs) manufacturing industries. Based on the literature review, a research model is constructed. The study hypotheses are formulated and analyzed by multiple regressions technique. The data was collected from 73 respondents at managerial levels from ISO 9000 certified small & medium manufacturing industries in Malaysia through a stratified sampling technique. The study results indicate positive and significant impact of soft TQM practices (top management commitment, customer focus and employee training) on financial, social, and environmental performance with R^2 values of 0.839, 0.739 and 0.656, respectively. This study also indicates that hard TQM practices (process management, quality information analysis and advance manufacturing technology) significantly enhances financial ($R^2 = 0.775$), social ($R^2 = 0.797$) and environmental ($R^2 = 0.712$) sustainable performance. In order to implement Quality 4.0 properly in SMEs manufacturing industries, industry personnel need to recognize the different role of hard and soft TQM. It is important that SMEs allocate resources to establish both type of TQM practices to achieve the effectiveness of the whole quality system.

Keywords: Industry 4.0 · TQM practices · Quality 4.0 · Sustainable performance

1 Introduction

In the past few decades, technological advancement and development transform human life in almost every area. Internet of Things (IoT), Artificial Intelligence (AI), big data and social media all represent modern technologies [1]. These new technologies transform the manufacturing models have that broadly reflect recent developments in industry as a response to the innovations in related technologies. These technologies can help to enhance product and service quality, and organizational efficiency [2]. Industry 4.0 (I4.0) is a modern new industrial model that is rapidly gaining importance and extensively debated research area among researchers and practitioners [3].

According to Lu [4], "Industry 4.0 is an automated, adaptive, optimized, service oriented and interoperable manufacturing process which is correlates with the use of big

M. Al-Emran et al. (Eds.): ICETIS 2021, LNNS 299, pp. 608–620, 2022.
https://doi.org/10.1007/978-3-030-82616-1_50

data, algorithms, and advance technologies". Industry 4.0 is "the integration of complex physical machinery and devices with networked sensors and software, used to predict, control and plan for better business and societal outcomes" [5]. Industry 4.0 was appeared firstly in 2011 in Germany as the fourth industrial revolution [6] but still in exploratory phase [7].

Quality management has developed since it was first introduced by quality experts in industrial applications [1]. The scope has expanded to a more comprehensive approach by applying TQM practices. Industry 4.0 has changed the way TQM practices are implemented [5]. Quality techniques are enhanced by the utilization of advanced sensors and instant analytical techniques. Traditional quality management practices shall be upgraded to meet such advancement in industry 4.0. Moreover, industry 4.0 significantly impacted on socio-economic and industrial issues [8]. There is need to analyze that how industry 4.0 transform the TQM practices and the improvement incurred on sustainable performance [8]. A literature review is conducted to identify the interface where TQM and industry 4.0 are interacting.

In Industry 4.0, the manufacturing sector gains most attention including sustainability and cyber-physical system (CPS) architecture because it affects process/system and human related factors [7]. Industry 4.0 affects different approaches of quality management like Lean Six Sigma lean manufacturing and increasingly total quality management (TQM) practises [6]. Since Industry 4.0 incorporates human and technological systems, therefore, on the basis of socio-technical system TQM becomes an integral part in industry 4.0 [9]. Furthermore, Sader, Husti and Daroczi [5] stated that successful implementation of TQM practices is supported by Industry 4.0. The study of Industry 4.0 is relatively new; therefore, many study areas are still unexplored particularly quality management and TQM in general. Some researchers have demonstrated conceptual grounding between Industry 4.0 and TQM [1, 6, 7]; however, the prior literature depicted that very few papers have been empirically analyzed Industry 4.0 and its relationship with quality management and sustainable performance. Therefore, this research paper empirically analyzes the association between TQM backed by Industry 4.0 and sustainable performance.

The objective of this research study is to empirically examine the effect of TQM practices on sustainable performance in the context of Industry 4.0. Firstly, a literature review of TQM in the context of Industry 4.0 and sustainable performance are discussed. Secondly, conceptual model and hypotheses development are discussed. Thirdly the research methodology, data analysis and research instruments are discussed. Finally, the paper will conclude the outcomes of this research work.

2 Literature Review and Hypothesis Formulation

2.1 TQM in the Context of Industry 4.0

Digitalization of businesses provides unique opportunities for product and service quality delivered by the organization. TQM is an evolving framework practises, tools and training to manage companies in an ever-changing environment to ensure customer loyalty and satisfaction [10]. Digitalization transforms the conventional TQM into Quality 4.0. According to Sisodia and Forero [11] Quality 4.0 refers to the "digitalization of TQM and its impact on quality management, process and people". According to American Society for Quality (2020) quality 4.0 means the "future of quality and organizational excellence within the context of industry 4.0".

Based on the prior literature and the identification of TQM in the context of Industry 4.0, the following analysis established the connection between TQM and Industry 4.0; a summary of this discussion is illustrated in Fig. 1.

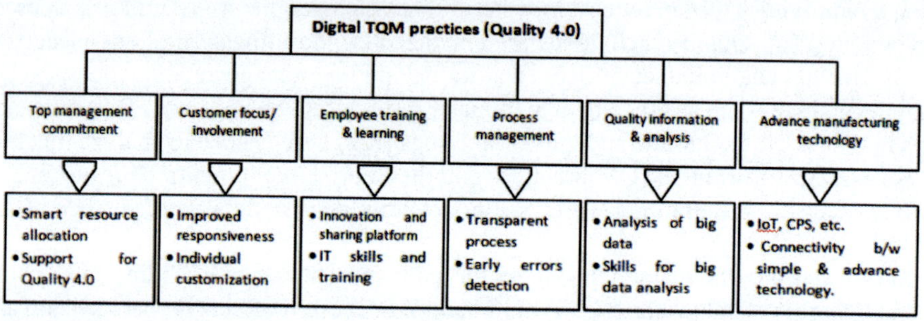

Fig. 1. Total quality management in the context of industry 4.0 (Source: self-developed according to literature)

Top Management Commitment/Leadership (TMC): Top management commitment means the top management support and involvement for quality planning and execution [12]. In industry 4.0, TMC refers that top management understand the importance of quality 4.0 and provide necessary support for implementation [3]. The success of quality 4.0 based on the acceptance of new technology. A fully committed TMC helps organization in the new technology adoption. Industry 4.0 provides a obvious production process, thus supporting the leadership capabilities in aligning and optimizing resources such as labor and machines to fulfill demand efficiently and effectively [1].

Customer Focus/Involvement (CF): Customer focus means how efficiently a business reacts to current and emerging customer needs and requirements to sustain effective customer relationship and determine customer loyalty [12]. Industry 4.0 will enable businesses to increase their customer loyalty and satisfaction through their goods and services, to produce and innovate new products that the requirements and expectations of consumers [1]. Additionally, Industry 4.0 will enable businesses to provide 'individual customization' of product and services at a regular time [5].

Employee Education and Training (EET): Employee training and learning is a vital activity in TQM; in fact three of Deming's out of 14-points directly relates with employee training [13] Industry 4.0 uses a range of technologies to improve the employee expertise and competency. These technologies are smart factory, drones, autonomous vehicles, 3D printing and other things. Furthermore, social media networks and tools can be used to exchange ideas and experience outside the organizations and within the organizations as well [3]. The technical skills required in Industry 4.0 skills include the installation and operation of IT equipment's, bar codes and big data analysis. In Industry 4.0, transformational skills such as critical thinking, adaptability, innovation and social skills will also be needed [14]. Hence, employee training and learning would play an important role in the successful adoption and implementation of digital TQM (Quality 4.0).

Process Management (PM): Process management means efficiently and effectively maintaining production process without breakdown and ensures that it run smoothly [15]. Industry 4.0 will support the transparency in production process; it will help to improve efficiency, resource allocation and process optimization [1]. Furthermore, Industry 4.0 will facilitate production defects, bottleneck and minimize production cost [1]. Moreover, Industry 4.0 will provide accurate and reliable processes information (risk, time, resources & critical constraints) and therefore help to maintain consistency and productivity at the planning level of key-processes [5].

Quality Information and Analysis (QIA): It refers to the availability of performance related quality data to identify quality problems quickly, which will help managers to make effective decisions. In Industry 4.0, big data is possible because technological development like CPS, sensing and data acquisition systems. It will help to gather information from different sources so that information will be processed in integrated databases and easily accessible to managers for decision making at any time [16].

Advance Manufacturing Technology (AMT): AMT refers as a group of integrated software and hardware-based technologies that can contribute to improving productivity and effectiveness of the organization if properly applied, controlled and evaluated [17]. In Industry 4.0, AMT integrate IoT with computer networks and big data to intelligently examine and exchange output information across all manufacturing facilities [18]. Furthermore, AMT will provide connectivity between simple and advance manufacturing equipment to leverage the correct information for decision making [19].

2.2 Soft and Hard TQM Practices

According to Socio-technical Systems (STSs) theory, there are reliable descriptions of soft and hard TQM practices exists [20]. The soft TQM practices are theoretical aspects relating to human resources and management principles and concepts. On the other hand, hard TQM practices referring to management practices, tools and techniques [21]. The STSs theory argues that there should be congruence between social (soft) and technical (hard) systems of process and adopted by many researchers in TQM studies [22, 23]. Based on the prior literature that TQM practices sub-divided into hard and soft as shown in Table 1.

2.3 TQM and Sustainable Performance

Sustainability is a very important criterion in modern day business in order to increase financial and social capital while reducing environmental pollution [27]. Sustainable performance is conceptualized as an outcome of sustainability management and defined as the combination of economic, social and environmental performance [28]. Nguyen, Phan and Matsui [27] empirically analyzed the association between TQM practices and

Table 1. Soft and hard TQM practices

Study	Soft TQM practices	Hard TQM practices
Samson and Terziovski [24]	Leadership; customer focus; human resources management	Information & analysis; process analysis; strategic planning
Rahman and Bullock [25]	Workforce commitment; personnel training; shared vision; cooperative relations; customer focus; use teams	Just-in-time principles; technology utilization; continuous improvement; computer based technologies
Khan and Naeem [26]	Customer focus; workforce commitment; shared vision; cooperative supplier relations; personnel training	Quality policy & target objectives, information & analysis, quality system processes, continuous improvement
Zeng, Zhang, Matsui and Zhao [23]	Employee suggestions; small group problem solving; employees' task-related training	Quality information, process management
Ershadi, Najafi and Soleimani [22]	Leadership; HR management; focus on customer	Process management; strategic management; quality management tools

sustainable performance. The results indicate that TQM practices have significant relationship with economic and social performance but insignificant association with environmental performance. Tasleem, Khan and Nisar [28] studied the relationship between TQM and sustainable performance. The outcomes depicted that TQM has positive and significant association with financial, social and environmental performance.

2.4 Hypotheses Development

In the prior literature, the association between soft TQM and performance was positive and significant. For instance, Abdallah [29] found that soft TQM improve firm performance. Furthermore, Sciarelli, Gheith and Tani [30] found that soft TQM directly and indirectly (innovation) improve the organization performance. Moreover, Khan and Naeem [26] and Rahman and Bullock [25] suggested that soft TQM has positive impact on performance. Hence, the following hypotheses are proposed.

H_1= Soft TQM has positive association with sustainable performance.

H_{1a} = Soft TQM has positive association with financial performance.

H_{1b} = Soft TQM has positive association with social performance.

H_{1c} = Soft TQM has positive association with environmental performance.

The prior literature suggests that hard TQM practices have relationship with performance. Sciarelli, Gheith and Tani [30] found the association between hard TQM practices and performance indicators. Likewise, Khan and Naeem [26] found that hard TQM practices established a positive and significant relationship with different constructs of organization performance. Therefore, this suggests the following hypotheses.

H_2= Hard TQM has positive association with sustainable performance.

H_{2a} = Hard TQM has positive association with financial performance.

H_{2b} = Hard TQM has positive association with social performance.

H_{2c} = Hard TQM has positive association with environmental performance.

All the hypothesized relationships are pictorial depicted in Fig. 2.

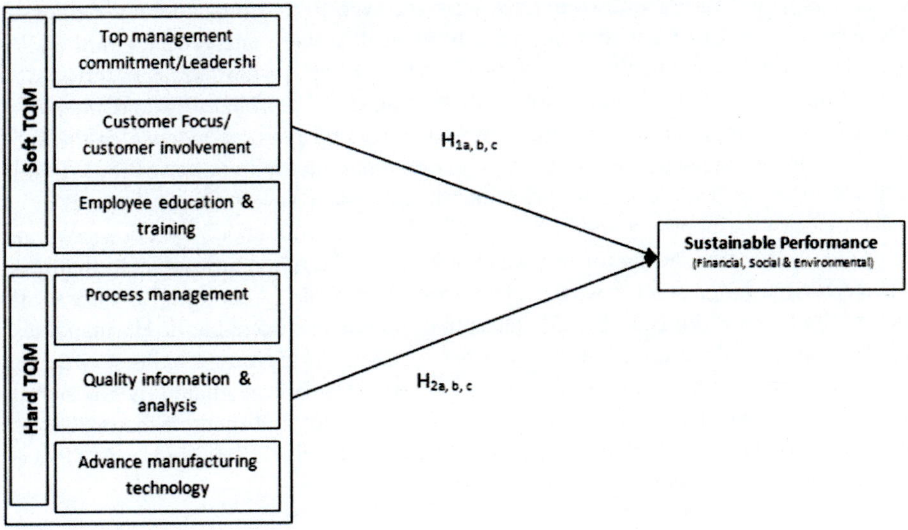

Fig. 2. Conceptual model (Source: self-developed according to literature)

3 Research Methodology

The section discusses the sample size, data collection methods and research instruments used in this study.

3.1 Sample Frame and Data Collection Method

The purpose of this study is to examine the association between TQM and sustainable performance in the context of Industry 4.0 from the view point of Malaysian SMEs manufacturing. This study employed survey research in Malaysia from August to September 2020. According to SMEcorp Malaysia [31] small manufacturing businesses have employees from 5 to not more than 75 and medium manufacturing businesses have full-time employees from 75 to not more than 200. Therefore, the study population is all the SME manufacturing businesses operating in Malaysia. The unit of analysis is at the organization levels and the respondents hold managerial positions in organizations under the umbrella of "Federation of Malaysian Manufacturers" (FMM) (FMM 51st Directory, 2019). The FMM directory is a well-known and prominent directory of active manufacturing companies in Malaysian and actively adopted in the prior literature [32]. There are 1480 SMEs manufacturing companies in Malaysia. The stratified sampling technique is applied to exclusively for the collection of research data only from SMEs ISO 9001 certified manufacturing companies which are listed in the 2019 FMM directory. The proposed model has six predictors; therefore, based on G-power software, the sample size is 98 with power at 0.80 and effect size at 0.15 (medium) and α at 0.05 [33]. The adapted questionnaire were distributed and collected for two months time long. In total 85 completed survey questionnaires were received from online sources. Out of 85 responses, 17 (15.7%) were missing data. Further, data were analyzed for outliers by SPSS. The final data for further analysis is 73 with 74.50% response rate.

The data were analyzed through SPSS version 23. The reliability and inter correlations were computed to determine the variability and interdependence of the subvariables. Multiple regressions were employed to examine the hypotheses [32]. In this study, all items were adapted from prior literature and on a seven-point Likert scale.

To measure the common method bias (CMB), as recommended by Podsakoff and Organ [34] Harman's one-factor test was performed. The outcomes depicted that total variance for the single factor was 33.98% (less than 50%). According to Podsakoff, MacKenzie, Lee and Podsakoff [35] stated that the power of one-factor Harman's test may be undermined by an increased number of items. Therefore, another CMB was adopted to examine the problem [36]. In this method, CMB is evaluated by comparing the variances of substantive factor loading ($R_1{}^2 = 0.790$) and method factor loading ($R_2{}^2 = 0.006$). The ratio between these two values is nearly 132:1. Hence, CMB was not a serious issue in this research study.

4 Results of the Survey

4.1 Constructs Reliability and Correlation

To measure the reliability of data for TQM practices and sustainable performance, scale reliability analysis were used using Cronbach's alpha. Table 2 depicted the results of reliability analysis. According to Taber [37] stated that Cronbach's offered to measure that different research items would give the same outcomes. The minimum acceptable value is .70. Table 2 all the values of Cronbach's alpha are above 0.70 [37], which indicated that the study constructs are reliable.

The correlation analysis in Table 2 depicted that relationship between research variables. It indicates the strength of relationship and significant if p-value is less than 0.01. As shown in Table 2, there is a significant correlation exist among all study variables. According to Hair et al. [38] stated that the correlation value above 0.90 indicate the multi-collinearity. In Table 2 all values are less than 0.90, hence, multi-collinearity was not found in the study data.

Table 2. Constructs reliability and correlation

Constructs	Reliability		Correlation								
	Items	Conbrach α	TMC	CF	EET	PM	QIA	AMT	FP	SP	EP
TMC	4	.894	1								
CF	6	.895	.820**	1							
EET	3	.728	.886**	.814**	1						
PM	5	.811	.677**	.718**	.658**	1					
QIA	8	.930	.853**	.836**	.845**	.679**	1				
AMT	5	.746	.771**	.700**	.758**	.621**	.710**	1			
FP	3	.844	.891**	.820**	.885**	.684**	.845**	.776**	1		
SP	4	.952	.848**	.751**	.830**	.652**	.794**	.856**	.847**	1	
EP	4	.813	.765**	.696**	.812**	.604**	.712**	.832**	.798**	.852**	1

**. Correlation is significant at the 0.01 level (2-tailed).

b. Listwise N = 73.

4.2 Hypotheses Testing

To analyze the relationship between TQM practices (soft & hard) and sustainable performance, the multiple regression analysis was chosen. Hair et al. [38] stated that it is a statistical tool that analyzes the association between set of independent constructs with one dependent variable. The outcome of multiple regressions is shown in Table 3 and 4. To prevent the existence of multi-collinearity, the tolerance values must be greater than 0.1 and VIF (variation inflation factors) no more than 10 (Hair et al. 1998). The Table 3 and 4 shows the tolerance and VIF values and depicted that there is no issues of multicollinearity.

From Table 3, it can be observed that soft TQM practices have positive and significant association with financial social and environmental performance. The value of R^2 is 0.846 representing that 84.6% of FP can be explained by soft TQM practices ($F = 125.96$, $p < 0.05$). The value of R^2 is 0.749 representing that 74.9% of SP can be explained by soft TQM practices ($F = 68.804$, $p < 0.05$). The value of R^2 is 0.670 representing that 67% of EP can be explained by soft TQM practices ($F = 46.677$, $p < 0.05$). Therefore, the hypotheses H1a, H1b and H1c are supported.

From Table 4, it can be observed that hard TQM practices have positive and significant association with financial social and environmental performance. The value of R^2 is 0.784 representing that 78.4% of FP can be explained by soft TQM practices ($F = 83.591$, $p < 0.05$). The value of R^2 is 0.805 representing that 80.5% of SP can be explained by soft TQM practices ($F = 94.977$, $p < 0.05$). The value of R^2 is 0.724 representing that 72.4% of EP can be explained by soft TQM practices ($F = 60.359$, $p < 0.05$). Therefore, the hypotheses H2a, H2b and H2c are supported.

Table 3. Regression analysis of soft TQM practices on sustainable performance

Constructs	Financial performance		Social performance		Environmental performance		Collinearity statistics	
	Std. Beta	Sig.	Std. Beta	Sig.	Std. Beta	Sig.	Tolerance	VIF
TMC	0.42	0.000	0.483	0.001	0.186	0.250	0.186	5.378
CF	0.171	0.055	0.082	0.463	0.049	0.700	0.292	3.421
EET	0.374	0.001	0.335	0.018	0.607	0.000	0.191	5.24
R	0.920		0.866		0.818			
R^2	0.846		0.749		0.670			
Adj. R^2	0.839		0.739		0.656			
F	125.96		68.804		46.677			
Sig.	0.000		0.000		0.000			

Table 4. Regression analysis of hard TQM practices on sustainable performance

Constructs	Financial performance		Social performance		Environmental performance		Collinearity statistics	
	Std. Beta	Sig.	Std. Beta	Sig.	Std. Beta	Sig.	Tolerance	VIF
PM	0.121	0.130	0.061	0.418	0.059	0.513	0.500	1.988
QIA	0.535	0.000	0.347	0.000	0.215	0.034	0.404	2.474
AMT	0.321	0.000	0.572	0.000	0.643	0.000	0.460	2.172
R	0.886		0.897		0.851			
R^2	0.784		0.805		0.724			
Adj. R^2	0.775		0.797		0.712			
F	83.591		94.977		60.359			
Sig.	0.000		0.000		0.000			

5 Discussion and Implications

The prior literature supports the outcome of the study. The results of H_1 depicted that soft TQM practices have positive and significant impact on sustainable performance; this finding is consistent with the findings of [25, 28, 30] and [39]. It is also interesting that soft TQM is slightly more impacted than hard TQM practices on sustainable performance. This result can be in line with the findings of [26] that unlike hard TQM practices which can be easily deployed in the organisation, soft TQM practices like intricacy and tastiness which are not easy to imitate and help organisations to achieve improved performance. The results of the study also indicate that hard TQM practices have significant and positive impact on sustainable performance (H_2); this outcome is in line with the studies conducted by [26, 30] and [27]. In a nutshell, the results provide support that soft and hard TQM practices or Quality 4.0 have significant impact on sustainable performance (financial, environmental & social).

This study contributes to the debate in the literature regarding quality management and sustainable performance relationship in industry 4.0 by providing information about the different impact of soft and hard TQM (Quality 4.0) on sustainable performance, which allows for more generalizability of findings in the SMEs manufacturing sectors. The multidimensional view of quality management is proven to be useful and important as there are two paths going through either soft or hard practices respectively leading to different influence on sustainable performance. The recent literature on Quality 4.0 has more concentrated on conceptualization, causing a limited understanding to the contribution of TQM to sustainable performance in industry 4.0. By applying quantitative technique and breaking down of TQM into soft and hard, that provide more detailed approach for the SMEs which could help them to efficiently allocated their resources according to a particular sustainable performance.

The empirical findings indicate that soft and hard TQM practices have a positive and significant impact on sustainable performance. This suggest that SMEs should give equal importance to soft (TMC, CF & EET) as well as hard (PM, QIA, & AMT) in order to have an effective quality management implementation and improved sustainable performance. It is worth noted that the effect of soft TQM is stronger than hard TQM practices on sustainable performance. SMEs should focus more on soft TQM practices over hard TQM practices; however, SMEs should not leave hard TQM practices unaddressed because both factors are stronger predicator of sustainable performance.

6 Conclusion

This empirical research study highlighted the key soft and hard TQM practices from an Industry 4.0 perspective and their impact on sustainable performance. The outcomes affirmed that soft and hard digital TQM or Quality 4.0 have positive and significant affect on financial, social and environmental performance. Furthermore, the results validate the significance of the social-technical system theory to Industry 4.0 and the effective implementations of soft and hard TQM practices in Industry 4.0 context. Industry 4.0 provides a strong support for the successful implementation of soft and hard TQM practices. This empirical study tries to fill a gap in adapting existing TQM practices and empirically analyze it in the context of Industry 4.0. The implications of this research will be beneficial to society, HR managers and TQM researchers and practitioners to their professions practices and training The theoretical contribution of this study involves, in particular, the expansion of the importance of STS theory from macro TQM to micro soft and hard TQM in order to close the theoretical and practical difference.

References

1. Daróczi, M., Husti, I., Sader, S.: Industry 4.0 as a key enabler toward successful implementation of total quality management practices. Periodica Polytechnica Soc. Manag. Sci. **27**(2), 131–140 (2019)
2. Radziwill, N.: Quality 4.0: let's get digital - the many ways the fourth industrial revolution is reshaping the way we think about quality, pp. 1–10 (2018)
3. Sony, M., Antony, J., Douglas, J.A.: Essential ingredients for the implementation of Quality 4.0. TQM J. **32**(4), 779–793 (2020)
4. Lu, Y.: Industry 4.0: a survey on technologies, applications and open research issues. J. Ind. Inf. Integr. **6**(1), 1–10 (2017)
5. Sader, S.S.A., Husti, I., Daroczi, M.: Suggested indicators to measure the impact of Industry 4.0 on total quality management. Int. Sci. J. **6**, 298–301 (2017)
6. Chiarini, A.: Industry 4.0, quality management and TQM world. A systematic literature review and a proposed agenda for further research. TQM J. **32**(4), 603–616 (2020)
7. Babatunde, O.K.: Mapping the implications and competencies for industry 4.0 to hard and soft total quality management. TQM J. **33**(4), 896–914 (2020)
8. Sader, S.: An experimental approach to total quality management in the context of Industry 4.0. Szent István University (2020)
9. Gunasekaran, A., Subramanian, N., Ngai, W.T.E.: Quality management in the 21st century enterprises: research pathway towards Industry 4.0. Int. J. Prod. Econ. **207**, 125–129 (2019)
10. Kiran, D.R.: Total Quality Management, pp. 1–14 (2017)
11. Sisodia, R., Forero, D.V.: Quality 4.0 - How to Handle Quality in the Industry 4.0 Revolution. Chalmers University of Technology (2020)

12. Hietschold, N., Reinhardt, R., Gurtner, S.: Measuring critical success factors of TQM implementation successfully – a systematic literature review. Int. J. Prod. Res. **52**(21), 6254–6272 (2014)
13. Jacob, D.: Quality 4.0 Impact and Strategy Handbook, pp. 1–28 (2017)
14. Schallock, B., Rybski, C., Jochem, R., Kohl, H.: Learning factory for Industry 4.0 to provide future skills beyond technical training. Procedia Manuf. **23**, 27–32 (2018)
15. Flynn, B.B., Schroeder, R.G., Sakakibara, S.: A framework for quality management research and an associated measurement instrument. J. Oper. Manag. **11**, 339–366 (1994)
16. Zonnenshain, A., Kenett, R.S.: Quality 4.0—the challenging future of quality engineering. Qual. Eng. **32**(4), 614–626 (2020)
17. Baldwin, J.R., Diverty, B.: Advanced technology use in Canadian manufacturing establishments. Statistics Canada Working Paper No. 85 (1995)
18. Moon, S.H.: Industry 4.0 for advanced manufacturing and its implementation. Eurasian J. Anal. Chem. **13**(6), 491–497 (2018)
19. Lee, J., Bagheri, B., Kao, H.-A.: Recent advances and trends of cyber-physical systems and big data analytics in industrial informatics (2014)
20. Appelbaum, S.H.: Socio-technical systems theory: an intervention strategy for organizational development. Manag. Decis. **35**(6), 452–463 (1997)
21. Saleh, R.A., Sweis, R.J., Mahmoud Saleh, F.I.: Investigating the impact of hard total quality management practices on operational performance in manufacturing organizations. Benchmarking: Int. J. **25**(7), 2040–2064 (2018)
22. Ershadi, M.J., Najafi, N., Soleimani, P.: Measuring the impact of soft and hard total quality management factors on customer behavior based on the role of innovation and continuous improvement. TQM J. **31**(6), 1093–1115 (2019)
23. Zeng, J., Zhang, W., Matsui, Y., Zhao, X.: The impact of organizational context on hard and soft quality management and innovation performance. Int. J. Prod. Econ. **185**, 240–251 (2017)
24. Samson, D., Terziovski, M.: The relationship between total quality management practices and operational performance. J. Oper. Manag. **17**, 393–409 (1999)
25. Rahman, S.-U., Bullock, P.: Soft TQM, hard TQM, and organisational performance relationships: an empirical investigation. Omega **33**(1), 73–83 (2005)
26. Khan, B.A., Naeem, H.: Measuring the impact of soft and hard quality practices on service innovation and organisational performance. Total Qual. Manag. Bus. Excell. **29**(11–12), 1402–1426 (2016)
27. Nguyen, M., Phan, A., Matsui, Y.: Contribution of quality management practices to sustainability performance of Vietnamese firms. Sustainability **10**(2), 375 (2018)
28. Tasleem, M., Khan, N., Nisar, A.: Impact of technology management on corporate sustainability performance: the mediating role of TQM. Int. J. Qual. Reliab. Manag. **36**(9), 1574–1599 (2019)
29. Abdallah, A.B.: The influence of "soft" and "hard" total quality management (TQM) practices on total productive maintenance (TPM) in Jordanian manufacturing companies. Int. J. Bus. Manag. **8**(21), 1 (2013)
30. Sciarelli, M., Gheith, M.H., Tani, M.: The relationship between soft and hard quality management practices, innovation and organizational performance in higher education. TQM J. **32**(6), 1349–1372 (2020)
31. https://smecorp.gov.my/index.php/en/policies/2020-02-11-08-01-24/sme-definition
32. Ooi, K.-B., Lin, B., Teh, P.-L., Chong, A.Y.-L.: Does TQM support innovation performance in Malaysia's manufacturing industry? J. Bus. Econ. Manag. **13**(2), 366–393 (2012)
33. Cohen, J.: Statistical Power Analysis for the Behavioral Sciences, 2nd edn. Lawrence Erlbaum Associates, Publishers (1988)
34. Podsakoff, P.M., Organ, D.W.: Self-reports in organizational research: problems and prospects. J. Manag. **12**, 69–82 (1986)

35. Podsakoff, P.M., MacKenzie, S.B., Lee, J.Y., Podsakoff, N.P.: Common method biases in behavioral research: a critical review of the literature and recommended remedies. J. Appl. Psychol. **88**(5), 879–903 (2003)
36. Liang, H., Saraf, N., Hu, Q., Xue, Y.: Assimilation of enterprise systems: the effect of institutional pressures and the mediating role of top management. MIS Q. **31**(1), 59–87 (2007)
37. Taber, K.S.: The use of Cronbach's alpha when developing and reporting research instruments in science education. Res. Sci. Educ. **48**, 1273–1296 (2018)
38. Hair, J.F.A., Anderson, R.E., Tatham, R.L., Blank, W.C.: Multivariate Data Analysis. Prentice Hall, Englewood Cliffs (1998)
39. Fotopoulos, C.B., Psomas, E.L.: The impact of "soft" and "hard" TQM elements on quality management results. Int. J. Qual. Reliab. Manag. **26**(2), 150–163 (2009)

Factors Affecting Customers' Intention to Use Mobile Payment Services in the Retailing Industry

Jin Yang[1], Garry Wei Han Tan[1], Keng Boon Ooi[1], Voon Hsien Lee[2], and Xiu Ming Loh[2](✉)

[1] UCSI University, 56000 Cheras, Malaysia
1001954816@ucsiuniversity.edu.my
[2] Universiti Tunku Abdul Rahman, 31900 Kampar, Malaysia
{leevh,lohxm}@utar.edu.my

Abstract. Recently, mobile payment services are becoming more and more popular. However, compared to other parts of the world, Malaysia's mobile payment services are still in the infancy stages. Hence, this study looks to study the variables that impact Malaysian customers' intention to adopt mobile payment services in the retailing industry. In achieving this objective, a conceptual model was developed via an integration of the Information System Success Model and Trust Transfer Theory. An online survey was conducted that yielded 359 usable responses for this study which was subsequently analyzed using PLS-SEM. The results show that customer satisfaction and trust in mobile payment have a significant relationship with the intention to adopt mobile payment services in the retailing industry. From the results of this study, there were several insights proposed to retailers, mobile payment technology companies, and mobile payment application developers. In addition, this study fills the theoretical gap of comprehending the main antecedents influencing customers' intention to adopt mobile payment services in the retailing industry via an integrated model.

Keywords: Mobile payment · Mobile commerce · Information system success model · Trust transfer theory

1 Introduction

The emergence of mobile devices has brought about significant changes to the way people carry out their daily activities. Mobile devices, especially smartphones, have now become a necessity in the lives of many people [1]. Given the development surrounding the use of smartphones, this has resulted in a new way of making payments known as mobile payment. More specifically, mobile payment refers to a service that enables users to initiate, authorize, and complete financial transfers using a mobile device through mobile networks or wireless communication technologies [2]. In Malaysia, there are a number of mobile payment systems that consumers can choose from. Among the more popular ones include Boost, Grabpay, and Touch n' Go eWallet [3].

M. Al-Emran et al. (Eds.): ICETIS 2021, LNNS 299, pp. 621–637, 2022.
https://doi.org/10.1007/978-3-030-82616-1_51

Mobile payment has the potential to be a competitive advantage for companies in the retailing industry. This is because mobile payment can lead to numerous benefits for both customers and merchants. On the customers side, using mobile payment in retail stores simplifies the payment process, speeds up services, and reduces the need to carry cash around [4]. For merchants, mobile payment helps in terms of automation as it can easily collect information and generate reports accordingly such as daily incomings and outgoings, average customer spending per transaction and so on in an instant. In view of the above-mentioned benefits, businesses are more inclined to enhance their business by providing mobile payment services at their stores.

Even as more and more merchants are integrating mobile payment into their business operations, it was found that only around 10% of payments made in Malaysia were through mobile means. This is a clear indication that this mobile payment is still sparingly used by Malaysians [5]. This is because cash still plays the dominant role as the most preferred payment method despite the Malaysian government's actions to transform the country into a cashless one [6]. However, in more recent times, as the COVID-19 virus can be transferred via surfaces and objects such as cash and cards, there is now an added health incentive for the use of mobile payment in the retailing industry. Hence, there is the presence of a gap in terms of comprehending the variables that affect the intention to adopt mobile payment in the retailing industry. In an attempt to fill the gap, this study utilizes a research framework that incorporates the Information System Success Model with the Trust Transfer Theory.

Thus, this study aims to determine (1) the constructs that influence the intention to adopt mobile payment in the retailing industry as well as (2) the robustness of incorporating the Information System Success Model with the Trust Transfer Theory. Overall, this study is postulated to contribute to numerous novel findings and insights. Practically, this study provides significant value to numerous stakeholders, especially mobile payment service providers. Theoretically, this study extends the knowledge on adoption intention for mobile payment in the retailing industry the perspective of a developing nation.

2 Literature Review

2.1 Information System Success Model

The Information System Success Model was developed by [7]. The model included six factors that are system quality, information quality, user satisfaction, use, individual impact, and organizational impact. However, due to the continuous progress of technology and society, the initial model gradually showed many deficiencies. Thus, an update for the model was proposed [8]. Among the major changes in the updated model include the addition of service quality as a variable and integration of individual and organizational impact into net benefit. These changes brought about improvements to the model by making it more relevant when it comes to understanding information system success in the 21st century.

The Information System Success Model is one of the most extensive models when it comes to evaluating information system success [9]. Given the comprehensiveness of the model, it has attracted the attention of many researchers in different areas of information system. These areas include mobile application [10], business intelligence system [11], online learning [12], and many others. As mobile payment is a form of information system, this study utilizes the Information System Success Model to look into the intention to adopt.

2.2 Trust Transfer Theory

Trust is an important predictor of an individual's use intention and behavior. Under the Trust Transfer Theory, it is posited that when two targets are connected, the trust in the source-target can be transferred to another related target [13]. When the trusted target and the unknown novel target possess a high degree of similarity, it can prompt users to transfer the trust to the unknown new target. With that said, trust transfer is a cognitive process that can promote the establishment of new trust [14]. People's attitudes toward familiar things will change or increase their trust towards unknown targets. In particular, this is because of the perceived connection between the familiar and new targets [15].

The presence of trust transfer has been empirically supported by past studies in different areas. One of these studies looked into the trust transfer from automated teller machines and online banking to mobile banking [16]. Another study investigated the transference of trust from online to mobile shopping [17]. Moreover, given that online payment precedes mobile payment, this study applies the Trust Transfer Theory to look into the dynamics between the different payment platforms.

3 Hypotheses Development

3.1 Information Quality

Information quality refers to the ability of mobile payment to provide fast and accurate payment information which makes it convenient for customers to use in retail stores [11]. **The significance of information quality as an antecedent of satisfaction has been** empirically proven by a number of past studies. These include those in the areas of mobile banking [18] and mobile application [10]. Hence, the hypothesis below was developed:

H1: Information quality has a significantly positive relationship with customer satisfaction.

3.2 System Quality

System quality refers to the correlation, ease of use, response speed, and visual appeal of the mobile payment service system [10]. System quality has been evaluated in terms of loading speed, response time, and ease of operation [12]. **Specifically for mobile** payment, system quality in the form of usability and responsiveness were found to **influence satisfaction** [19]. Thus, the hypothesis below was developed:

H2: System quality has a significantly positive relationship with customer satisfaction.

3.3 Service Quality

Service quality indicates that the system provides reliable and guaranteed services to users [20]. More precisely, mobile payment providers are to provide precise services, assured services, and tangible services provided to users [12]. A number of past studies have found service quality to be a significant determinant of satisfaction in the contexts of mobile shopping applications and mobile government services [21, 22]. Based on the above-mentioned, the hypothesis below was developed:

H3: Service quality has a significantly positive relationship with customer satisfaction.

3.4 Customer Satisfaction

Customer satisfaction refers to the customers' overall evaluation of the use of mobile payment services in retail stores [23]. In particular, it indicates whether a particular service meets the expectations and needs of customers. Customer satisfaction or otherwise is based on customers' sentiment which is either positive or negative respectively [24]. In the setting of this study, customer satisfaction was found to significantly encourage the intention to adopt mobile payment [19, 25]. Therefore, the following hypothesis was developed:

H4: Customer satisfaction has a significantly positive relationship with the intention to adopt mobile payment.

3.5 Trust in Online Payment

The trust transfer theory posits that an individual's trust is transferable from a known target to an unknown one [13]. The applicability of this theory has been shown by past studies to be valid from an offline and online to mobile context. More specifically, it was found that users' trust in mobile payment are significantly determined by their trust in both automated teller machines and online banking [16]. In this study's context, online payment is the source while mobile payment is the unknown target. This is because online payment precedes mobile payment [24]. Hence, the following hypothesis was developed:

H5: Trust in online payment has a significantly positive relationship with trust in mobile payment.

3.6 Functional Consistency

Functional consistency refers to the degree in which the source and target is similar in terms of their functions and operations [13]. The similarities are posited to facilitate the trust transfer between the known and unknown target. For instance, it was shown that the similarities between online and mobile shopping services have a significant influence on users' trust in mobile shopping services [17]. Particularly in this study, it refers to the similarities in the functions of online and mobile payment. This is because when comparing between both platforms, there are several similar functions that can be observed such as making payments, transferring funds, and checking balances. Thus, the hypothesis below was developed:

H6: Functional consistency has a significantly positive relationship with trust in mobile payment.

3.7 Perceived Entitativity

Perceived entitativity refers to the degree whereby a source and target object are deemed to be grouped in the same category [26]. Despite the difference in context from online to mobile, as long as the service is still the same, perceived entitativity is found to be present. Subsequently, this will increase users' trust in the mobile service [27]. In this study, it denotes the degree in which users believe that online and mobile payment belong to the same category. Particularly, the trust transfer of users from online to mobile payment is facilitated in the event that users perceive a high entitativity between both platforms [24]. Hence, the following hypothesis was developed:

H7: Perceived entitativity has a significantly positive relationship with trust in mobile payment.

3.8 Trust in Mobile Payment

Trust has been theorized as the belief that the other party will conduct according to the proper behavior [28]. Particularly, trust in the context of mobile payment refers to the willingness of users to make payment over a mobile device and network while expecting the payment platform to perform its duties [24]. Trust is essential in the setting of mobile payment as it heavily involves money. Past studies have indicated that trust is important in determining the intention to use mobile payment [1, 29]. Therefore, the hypothesis below was developed:

H8: Trust in mobile payment has a significantly positive relationship with the intention to adopt mobile payment.

Based on the hypotheses developed, the proposed conceptual model of this study is shown in Fig. 1.

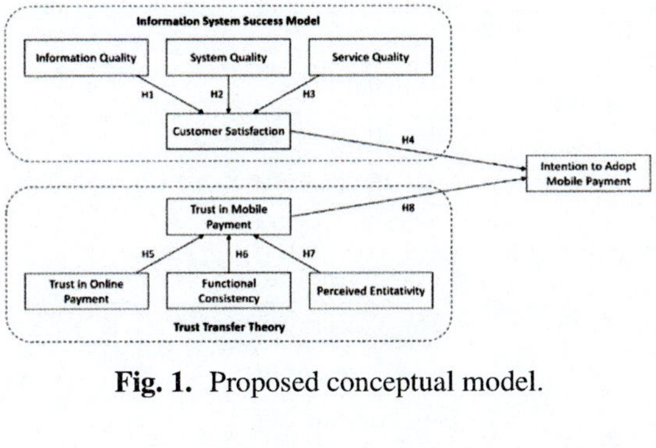

Fig. 1. Proposed conceptual model.

4 Methodology

Given the context of this study, the target respondents are Malaysian mobile device users who have experience in using online payment services. Besides that, non-probability sampling was used as there is no sampling frame available for this group of people. Thus, this study utilized purposes sampling as the respondents would have to fit certain requirements [30]. In particular, two qualifier questions ("I have experience utilizing online payment services." and "I have a mobile device.") were included to filter out non-target respondents. For the minimum sample size, G*Power was utilized with a setting of 8 predictors, 15% effect size, 5% alpha level, and 80% power. The software calculated that 109 responses were sufficient.

The data collection of this study was conducted via an online survey. The survey has three sections which are (1) cover page, (2) demographic, and (3) measurement items [31–33]. The demographic section covered the age, gender and other personal characteristics of the respondents. The subsequent section measured the variables of information quality (four items), system quality (four items), service quality (four items), customer satisfaction (four items), trust in online payment (four items), functional consistency (four items), perceived entitativity (four items), trust in mobile payment (three items) and intention to adopt mobile payment (four items). Overall, as shown in Appendix A, there were a total of 34 questionnaire items which were adapted from past studies [19, 24, 34]. All questionnaire items were measured with a 7-point Likert scale. Overall, this study collected 359 responses which were subsequently analyzed with PLS-SEM.

5 Analysis

5.1 Demographic Profile

Based on Table 1, the majority of the respondents are male (64.9%), possess a bachelor degree/professional qualification (56.8%), and uses mobile payment every day (54.9%).

Table 1. Descriptive analysis.

Characteristics	Description	Count	Percentage
Gender	Male	233	64.9
	Female	126	35.1
Age	16–20 years old	28	7.8
	21–25 years old	63	17.6
	26–30 years old	59	16.4
	31–35 years old	49	13.7
	36–40 years old	52	14.5
	41–45 years old	54	15.0
	46–50 years old	39	10.8
	51 years old and above	15	4.2

(continued)

Table 1. (*continued*)

Characteristics	Description	Count	Percentage
Highest level of education	Primary/Secondary school	29	8.1
	Diploma/Advanced diploma	80	22.3
	Bachelor degree/Professional **qualification**	204	56.8
	Master/PhD degree	46	12.8
Personal Income (per month)	Less than RM2,000	89	24.8
	RM2,001–RM5,000	156	43.4
	RM5,001–RM8,000	74	20.6
	RM8,001–RM10,000	30	8.4
	RM10,001 and above	10	2.8
Number of mobile devices owned	1–2	175	48.8
	3–5	151	42.0
	More than 5	33	9.2
Frequency of using mobile payment	Every day	197	54.9
	Every week	62	17.3
	Every month	41	11.4
	Every 3 months	38	10.6
	Every 6 months	13	3.6
	Every year	8	2.2

5.2 Measurement Model Assessment

Based on Table 2, **reliability has been confirmed as all values of Cronbach's alpha** for each construct are above the threshold of 0.7 [35, 36]. Furthermore, convergent validity was also established as all the values for average variance extracted and factor loading are greater than 0.5 and 0.7 respectively [37–39].

Table 2. Factor loading, reliability, and average variance extracted

Construct	Measurement item	Factor loading	Cronbach's Alpha	**Variance inflation** factor
Information quality	IQ1	0.847	0.891	0.754
	IQ2	0.852		
	IQ3	0.897		
	IQ4	0.877		

(*continued*)

Table 2. (*continued*)

Construct	Measurement item	Factor loading	Cronbach's Alpha	Variance inflation factor
System quality	SYQ1	0.894	0.896	0.764
	SYQ2	0.891		
	SYQ3	0.811		
	SYQ4	0.897		
Service quality	SEQ1	0.889	0.907	0.781
	SEQ2	0.889		
	SEQ3	0.873		
	SEQ4	0.884		
Customer satisfaction	CS1	0.901	0.911	0.789
	CS2	0.892		
	CS3	0.895		
	CS4	0.865		
Trust in online payment	TOP1	0.920	0.906	0.781
	TOP2	0.860		
	TOP3	0.899		
Functional consistency	FC1	0.894	0.912	0.790
	FC2	0.898		
	FC3	0.880		
	FC4	0.882		
Perceived entitativity	PE1	0.883	0.901	0.772
	PE2	0.865		
	PE3	0.859		
	PE4	0.907		
Trust in mobile payment	TMP1	0.894	0.836	0.751
	TMP2	0.847		
	TMP3	0.859		
Intention to adopt	IA1	0.905	0.908	0.784
	IA2	0.886		
	IA3	0.893		
	IA4	0.858		

In addition, discriminant validity was also found to be present as all values for the 2.5% and 97.5% confidence intervals are less than 1 as shown in Table 3 [40].

Table 3. Hetero-Trait-Mono-Trait Inference

Path	Original sample	Mean sample	Bias	Confidence interval	
				2.5%	97.5%
IQ → CS	0.881	0.881	−0.001	0.812	0.937
IQ → FC	0.799	0.799	0.000	0.697	0.876
IQ → IA	0.925	0.926	0.000	0.872	0.969
SYQ → CS	0.949	0.949	0.000	0.903	0.984
SYQ → FC	0.816	0.817	0.000	0.731	0.882
SYQ → IA	0.946	0.946	0.001	0.898	0.983
SYQ → IQ	0.958	0.958	0.000	0.923	0.988
SYQ → PE	0.871	0.872	0.000	0.794	0.928
SEQ → CS	0.715	0.715	0.000	0.624	0.794
SEQ → FC	0.797	0.798	0.000	0.697	0.879
SEQ → IA	0.796	0.795	0.000	0.703	0.870
SEQ → IQ	0.872	0.873	0.001	0.784	0.940
SEQ → PE	0.782	0.783	0.000	0.676	0.868
SEQ → SYQ	0.830	0.83	0.000	0.740	0.902
TOP → CS	0.733	0.733	0.000	0.622	0.822
TOP → FC	0.858	0.858	0.000	0.787	0.914
TOP → IA	0.816	0.817	0.000	0.712	0.897
TOP → IQ	0.891	0.892	0.001	0.787	0.961
TOP → PE	0.817	0.817	0.001	0.706	0.902
TOP → SYQ	0.834	0.834	0.000	0.733	0.910
TOP → SEQ	0.904	0.904	0.001	0.785	0.984
TOP → TMP	0.922	0.923	0.001	0.811	0.993
FC → CS	0.725	0.725	0.000	0.626	0.806
PE → CS	0.792	0.792	0.000	0.701	0.859
PE → FC	0.917	0.917	0.000	0.850	0.964
PE → IA	0.934	0.935	0.000	0.885	0.976
PE → IQ	0.868	0.868	0.000	0.775	0.936
TMP → CS	0.769	0.768	0.000	0.677	0.849
TMP → FC	0.716	0.716	0.000	0.606	0.809
TMP → IA	0.796	0.796	0.000	0.705	0.866
TMP → IQ	0.920	0.921	0.001	0.855	0.974

(*continued*)

Table 3. (*continued*)

Path	Original sample	Mean sample	Bias	Confidence interval	
				2.5%	97.5%
TMP → PE	0.794	0.795	0.001	0.692	0.878
TMP → SYQ	0.813	0.814	0.000	0.727	0.889
TMP → SEQ	0.863	0.864	0.001	0.774	0.934
IA → CS	0.881	0.881	0.001	0.806	0.934
IA → FC	0.835	0.835	0.000	0.752	0.897

5.3 Structural Model Assessment

Model fit is achieved as SRMR values for the saturated model (0.045) and estimated model (0.069) are less than 0.080. With regards to Table 4, support was provided for both H1 and H2 but not for H3. This is as information quality ($\beta = 0.26$, $p = 0.001$) and system quality ($\beta = 0.683$, $p < 0.001$) have significantly positive relationships with customer satisfaction whereas service quality ($\beta = -0.062$, $p = 0.195$) was discovered to be insignificant. In addition, H5 and H7 were supported while H6 was not. This is as trust in online payment ($\beta = 0.72$, $p < 0.001$) and PE ($\beta = 0.333$, $p = 0.005$) were significantly positive variables of trust in mobile payment while functional consistency ($\beta = -0.209$, $p = 0.013$) was not. Additionally, H4 and H8 were supported as customer satisfaction ($\beta = 0.61$, $p < 0.001$) and trust in mobile payment ($\beta = 0.284$, $p < 0.001$) were revealed to have significantly positive relationships with the intention to adopt mobile payment. Overall, except H3 and H6, all remaining hypotheses were supported. Also, the model can account for 68.6% of the changes in the intention to adopt mobile payment.

Table 4. Hypotheses testing

Hypothesis	Relationship	Path coefficient	Confidence interval		t-value	p-value	Remark
			2.5%	97.5%			
H1	IQ → CS	0.260	0.101	0.406	3.297	0.001	Supported
H2	SYQ → CS	0.683	0.543	0.818	9.826	<0.001	Supported
H3	SEQ → CS	−0.062	−0.164	0.024	1.295	0.195	Not supported
H4	CS → IA	0.610	0.479	0.720	9.961	<0.001	Supported
H5	TOP → TMP	0.720	0.532	0.857	8.739	<0.001	Supported
H6	FC → TMP	−0.209	−0.388	−0.059	2.492	0.013	Not supported
H7	PE → TMP	0.333	0.122	0.575	2.784	0.005	Supported
H8	TMP → IA	0.284	0.173	0.414	4.567	<0.001	Supported

F-square is used to assess the intensity of relationships in which values of 0.02 to 0.15 indicates a small effect, 0.15 to 0.35 specifies a medium effect, above 0.35 specifies a large effect, and less than 0.02 indicates no effect [41–43]. Therefore, as shown in Table 5, information quality and system quality have a small and large effect respectively whereas service quality has no effect on customer satisfaction. Furthermore, trust in online payment has a large impact on trust in mobile payment while functional consistency and perceived entitativity has small effects on trust in mobile payment. Lastly, customer satisfaction and trust in mobile payment has a large and small effect on intention to adopt mobile payment respectively.

Table 5. Effect size (f^2)

Construct	IQ	SYQ	SEQ	CS	TOP	FC	PE	TMP	IA
Information quality				0.060					
System quality				0.483					
Service quality				0.006					
Customer satisfaction									0.652
Trust in online payment								0.584	
Functional consistency								0.033	
Perceived entitativity								0.097	
Trust in mobile payment									0.141

Based on Table 6, the Q^2 values for customer satisfaction, trust in mobile payment and intention to adopt exceed 0 which indicates that the structural model has predictive relevance.

Table 6. Predictive relevance (Q^2)

Construct	SSO	SSE	Q^2 (= 1 − SSE/SSO)
Information quality	1436	1436	
System quality	1436	1436	
Service quality	1436	1436	
Customer satisfaction	1436	592.829	0.587
Trust in online payment	1436	1436	

(continued)

Table 6. (*continued*)

Construct	SSO	SSE	Q^2 (= 1 − SSE/SSO)
Functional consistency	1436	1436	
Perceived entitativity	1436	1436	
Trust in mobile payment	1077	544.719	0.494
Intention to adopt	1436	678.616	0.527

6 Discussion

For the antecedents of customer satisfaction, the results were mixed. More specifically, information quality and system quality were found to be significant facilitators of customer satisfaction. These findings show that users value mobile payment systems that brings about convenience and ease to them when making payments as well as provide additional information such as automated payment record. However, service quality was revealed to have an insignificant relationship with customer service. This could be attributed to the expectations that mobile payment should be of high quality which serves as a pre-requisite for consumers to use the service. Furthermore, there were varying results for the factors related to trust in mobile payment. Particularly, trust in online payment and perceived entitativity have positive effects whereas functional consistency has a negative impact on trust in mobile payment respectively. These results show that even though users believe that online and mobile payment are related, they also believe that they are functionally different. This is the case as mobile payment does not only have many similar functions with online payment, but mobile payment also has several additional functions such as the scanning of QR codes and others. Finally, customer satisfaction and trust in mobile payment are key determinants for the intention to adopt mobile payment in the retailing industry. In view that mobile payment is a financial service, users place emphasis on trust and satisfaction when deciding on its use.

Based on the findings, there are several implications that mobile payment service providers can look into to better encourage the public to adopt mobile payment. Firstly, mobile payment service providers should always ensure that the information provided on their services are accurate and timely. Moreover, mobile service system developers need to make sure that the mobile payment system is reliable, easy to use, and responsive. Besides, mobile payment service providers should also monitor the feedback given by users and take prompt actions to address any grievances that were brought up. Additionally, financial service providers that have both online and mobile versions of their services should provide equal emphasis on the development and user experience on both platforms as trust transference is present. Also, mobile payment service providers need to build trust with their users. This can be carried out by strengthening the reputation and security of the mobile payment services.

For theoretical implications, this study has successfully extended the literature of mobile payment in several ways. Firstly, this study emphasizes on the application of mobile payment in the retailing industry which has been relatively understudied. This is because past studies tend to focus on the general application of mobile payment [44, 45]. Furthermore, this study provides further empirical evidences for trust transference between similar services. In addition, this study reveals that the integration of Information System Success Model and Trust Transfer Theory is robust when it comes to the intention to adopt mobile payment.

There are several limitations in this study which were identified. Firstly, this study is not able to analyze the changes between different periods of time due to the cross-sectional nature. Thus, future studies should conduct a longitudinal study as consumer behavior on mobile payment is constantly changing. Furthermore, the findings of this study may not accurately reflect the situation in other countries. This is because this study was conducted purely in Malaysia and there are diverse differences present in other countries. Therefore, future studies should obtain data from multiple countries to overcome this limitation.

Appendix A (Questionnaire Items)

Construct	Measurement Item
Information quality	IQ1: The information provided by the mobile payment program is timely
	IQ2: The information provided by the mobile payment program is accurate
	IQ3: The information provided by the mobile payment program is relevant
	IQ4: The information provided by the mobile payment program is useful
System quality	SYQ1: It's easy and convenient to make payments using mobile payment applications
	SYQ2: The mobile payment application responds quickly
	SYQ3: The visual design of the mobile payment application looks great
	SYQ4: The mobile payment application can help me complete payment quickly
Service quality	SEQ1: Services provided by mobile payment is reliable
	SEQ2: Services provided by mobile payment is guaranteed
	SEQ3: The customer service center can effectively solve my problem
	SEQ4: When I encounter difficulties, the customer service center can provide a quick response

(continued)

(*continued*)

Construct	Measurement Item
Customer satisfaction	CS1: Mobile payment services meet my needs
	CS2: Mobile payment services meet my expectations
	CS3: Mobile payment services are useful during payment
	CS4: Mobile payment services are good for me
Trust in online payment	TOP1: I think it is safe to use online payment services
	TOP2: The use of online payment services is within my ability/control
	TOP3: The use of online payment services is guaranteed
Functional consistency	FC1: I think the functions of online payment services and mobile payment services are the same
	FC2: I think online payment services and mobile payment services have similar operating procedures
	FC3: I use online payment services and mobile payment services for the same purpose
	FC4: I think mobile payment applications have all the functions of online payment applications
Perceived entitativity	PE1: I think mobile payment services and online payment services are in the same category
	PE2: I think there is not much difference between mobile payment services and online payment services
	PE3: I think mobile payment services and online payment services are related
	PE4: I think mobile payment services and online payment services are very similar
Trust in mobile payment	TMP1: Using mobile payment is safe for me
	TMP2: Using mobile payment will not cause financial loss or personal information leakage
	TMP3: Mobile payment service providers will strengthen security measures
Intention to adopt	IA1: I plan to use mobile payment services in retail stores
	IA2: If I have the opportunity, I will use mobile payment services to make payment in retail stores
	IA3: I am willing to use mobile payment services in retail stores
	IA4: I will not refuse to use mobile payment services in retail stores

References

1. Loh, X.M., Lee, V.H., Tan, G.W.H., Ooi, K.B., Dwivedi, Y.K.: Switching from cash to mobile payment: what's the hold-up? Internet Res. (2020)
2. Slade, E.L., Williams, M.D., Dwivedi, Y.K.: Mobile payment adoption: Classification and review of the extant literature. Mark. Rev. **13**(2), 167–190 (2013)
3. Lew, S., Tan, G.W.H., Loh, X.M., Hew, J.J., Ooi, K.B.: The disruptive mobile wallet in the hospitality industry: an extended mobile technology acceptance model. Technol. Soc. **63** 101430 (2020)
4. Dinh, V.S., Nguyen, H.V., Nguyen, T.N.: Cash or cashless? Promoting consumers' adoption of mobile payments in an emerging economy. Strat. Direct. **34**(1), 1–4 (2018)
5. Golingai, P.: Going cashless at a snail's pace (2019). https://www.thestar.com.my/opinion/columnists/one-mans-meat/2019/01/26/going-cashless-at-a-snails-pace-malaysians-through-force-of-habit-are-still-very-reliant-on-cash-whi
6. Nielsen.: Cash or cashless? Malaysia's shifting payment landscape (2019). https://www.nielsen.com/my/en/insights/article/2019/cash-or-cashless-malaysias-shifting-payment-landscape/
7. DeLone, W.H., McLean, E.R.: Information systems success: the quest for the dependent variable. Inf. Syst. Res. **3**(1), 60–95 (1992)
8. Delone, W.H., McLean, E.R.: The DeLone and McLean model of information systems success: a ten-year update. J. Manag. Inf. Syst. **19**(4), 9–30 (2003)
9. Tam, C., Loureiro, A., Oliveira, T.: The individual performance outcome behind e-commerce. Internet Res. **30**(2), 439–462 (2019)
10. Hsiao, K.L., Lin, K.Y., Wang, Y.T., Lee, C.H., Zhang, Z.M.: Continued use intention of lifestyle mobile applications: the Starbucks app in Taiwan. Electron. Libr. **37**(5), 893–913 (2019)
11. Gonzales, R., Wareham, J.: Analysing the impact of a business intelligence system and new conceptualizations of system use. J. Econ. Finance Admin. Sci. **24**(48), 345–368 (2019)
12. Aldholay, A., Isaac, O., Abdullah, Z., Abdulsalam, R., Al-Shibami, A.H.: An extension of Delone and McLean IS success model with self-efficacy. Int. J. Inf. Learn. Technol. **35**(4), 285–304 (2018)
13. Stewart, K.J.: Trust transfer on the world wide web. Organ. Sci. **14**(1), 5–17 (2003)
14. Shi, S., Chow, W.S.: Trust development and transfer in social commerce: prior experience as moderator. Ind. Manag. Data Syst. **115**(7), 1182–1203 (2015)
15. Al-Htibat, A., Garanti, Z.: Impact of interactive eReferral on tourists behavioral intentions. Mark. Intell. Plan. **37**(5), 527–541 (2019)
16. Chaouali, W., El Hedhli, K.: Toward a contagion-based model of mobile banking adoption. Int. J. Bank Mark. **37**(1), 69–96 (2019)
17. Yang, S., Chen, Y., Wei, J.: Understanding consumers' web-mobile shopping extension behavior: a trust transfer perspective. J. Comput. Inf. Syst. **55**(2), 78–87 (2015)
18. Sharma, S.K., Sharma, M.: Examining the role of trust and quality dimensions in the actual usage of mobile banking services: an empirical investigation. Int. J. Inf. Manage. **44**, 65–75 (2019)
19. Ramadan, R., Aita, J.: A model of mobile payment usage among Arab consumers. Int. J. Bank Mark. **36**(7), 1213–1234 (2018)
20. Martins, J., et al.: Assessing the success behind the use of education management information systems in higher education. Telematics Inform. **38**, 182–193 (2019)
21. Kim, Y., Wang, Q., Roh, T.: Do information and service quality affect perceived privacy protection, satisfaction, and loyalty? Evidence from a Chinese O2O-based mobile shopping application. Telemat. Inform. **56** (2020)

22. Wang, C., Teo, T.S.: Online service quality and perceived value in mobile government success: an empirical study of mobile police in China. Int. J. Inf. Manag. **52** (2020)
23. Hossain, M.A., Hossain, M.S., Jahan, N.: Predicting continuance usage intention of mobile payment: an experimental study of Bangladeshi customers. Asian Econ. Financ. Rev. **8**(4), 487–498 (2018)
24. Cao, X., Yu, L., Liu, Z., Gong, M., Adeel, L.: Understanding mobile payment users' continuance intention: a trust transfer perspective. Internet Res. **28**(2), 456–476 (2018)
25. Humbani, M., Wiese, M.: An integrated framework for the adoption and continuance intention to use mobile payment apps. Int. J. Bank Mark. **37**(2), 646–664 (2019)
26. Gong, X., Zhang, K.Z., Chen, C., Cheung, C.M., Lee, M.K.: What drives trust transfer from web to mobile payment services? The dual effects of perceived entitativity. Inf. Manag. **57**(7) 103250 (2020)
27. Wang, N., Shen, X.L., Sun, Y.: Transition of electronic word-of-mouth services from web to mobile context: a trust transfer perspective. Decis. Support Syst. **54**(3), 1394–1403 (2013)
28. Shao, Z., Zhang, L., Li, X., Guo, Y.: Antecedents of trust and continuance intention in mobile payment platforms: the moderating effect of gender. Electron. Commer. Res. Appl. **33** 100823 (2019)
29. Zhou, T.: Understanding the determinants of mobile payment continuance usage. Ind. Manag. Data Syst. **114**(6), 936–948 (2014)
30. Leong, L.Y., Hew, T.S., Ooi, K.B., Wei, J.: Predicting mobile wallet resistance: a two-staged structural equation modeling-artificial neural network approach. Int. J. Inf. Manag. **51** (2020)
31. Foo, P.Y., Lee, V.H., Tan, G.W.H., Ooi, K.B.: A gateway to realising sustainability performance via green supply chain management practices: a PLS–ANN approach. Expert Syst. Appl. **107**, 1–14 (2018)
32. Ooi, K.B., Lee, V.H., Tan, G.W.H., Hew, T.S., Hew, J.J.: Cloud computing in manufacturing: the next industrial revolution in Malaysia? Expert Syst. Appl. **93**, 376–394 (2018)
33. Leong, L.Y., Hew, T.S., Ooi, K.B., Tan, G.W.H.: Predicting actual spending in online group buying–An artificial neural network approach. Electron. Commer. Res. Appl. **38** 100898 (2019)
34. Loh, X.M., Lee, V.H., Tan, G.W.H., Hew, J.J., Ooi, K.B.: Towards a cashless society: the imminent role of wearable technology. J. Comput. Inf. Syst., 1–11 (2019)
35. Hew, J.J., Leong, L.Y., Tan, G.W.H., Ooi, K.B., Lee, V.H.: The age of mobile social commerce: an Artificial Neural Network analysis on its resistances. Technol. Forecast. Soc. Chang. **144**, 311–324 (2019)
36. Ooi, K.B., Foo, F.E., Tan, G.W.H., Hew, J.J., Leong, L.Y.: Taxi within a grab? A gender-invariant model of mobile taxi adoption. Ind. Manag. Data Syst. **121**(2), 312–332 (2020)
37. Hew, J.J., Leong, L.Y., Tan, G.W.H., Lee, V.H., Ooi, K.B.: Mobile social tourism shopping: a dual-stage analysis of a multi-mediation model. Tour. Manag. **66**, 121–139 (2018)
38. Wong, L.W., Tan, G.W.H., Hew, J.J., Ooi, K.B., Leong, L.Y.: Mobile social media marketing: a new marketing channel among digital natives in higher education? J. Mark. High. Educ., 1–25 (2020)
39. Hew, J.J., Wong, L.W., Tan, G.W.H., Ooi, K.B., Lin, B.: The blockchain-based Halal traceability systems: a hype or reality? Supply Chain Manag. Int. J. **25**(6), 863–879 (2020)
40. Tan, G.W.H., Ooi, K.B.: Gender and age: do they really moderate mobile tourism shopping behavior? Telematics Inform. **35**(6), 1617–1642 (2018)
41. Tan, G.W.H., Lee, V.H., Hew, J.J., Ooi, K.B., Wong, L.W.: The interactive mobile social media advertising: an imminent approach to advertise tourism products and services? Telematics Inform. **35**(8), 2270–2288 (2018)
42. Lee, V.H., Hew, J.J., Leong, L.Y., Tan, G.W.H., Ooi, K.B.: Wearable payment: a deep learning-based dual-stage SEM-ANN analysis. Expert Syst. Appl. (2020)

43. Wong, L.W., Tan, G.W.H., Lee, V.H., Ooi, K.B., Sohal, A.: Unearthing the determinants of Blockchain adoption in supply chain management. Int. J. Prod. Res. **58**(7), 2100–2123 (2020)
44. Yan, L.Y., Tan, G.W.H., Loh, X.M., Hew, J.J., Ooi, K.B.: QR code and mobile payment: the disruptive forces in retail. J. Retail. Consum. Serv. **58** (2020)
45. Tew, H.T., Tan, G.W.H., Loh, X.M., Lee, V.H., Lim, W.L., Ooi, K.B.: Tapping the next purchase: Embracing the wave of mobile payment. J. Comput. Inf. Syst., 1–9 (2021)

The Impact of Sales Promotion on Consumer of GSM in Yemen: MTN-Yemen

Awsan R. Rassam[1], Kamal Tammim[1], Hussein Mohammed Esmail Abualrejal[2](✉), Fathey Mohammed[3], Nabil Hasan Al-kumaim[4], and Yousef Fazea[3]

[1] Faculty of Business and Finance, International University of Technology Twintech, Sana'a, Yemen

[2] Knowledge Science Research Lab, School of Technology Management and Logistics, College of Business, Universiti Utara Malaysia, 06010 Sintok, Kedah, Malaysia

abualrejal@uum.edu.my

[3] School of Computing, Universiti Utara Malaysia, 06010 Sintok, Kedah, Malaysia

[4] Faculty of Technology Management and Technopreneurship, Universiti Teknikal Malaysia Melaka (UTeM) Melaka, Hang Tuah Jaya, 76100 Malacca, Malaysia

Abstract. Consumers are observed as modest drivers in today's business and commercial world. There are many adoptions to be made among alternative services, and they demonstrate an extraordinary level of inspiration in the marketplace in terms of service quality and price. Thus, GSM providers must struggle to meet the requirements to stay a competitive supplier and achieve a higher level of business sustainability. Sales promotion is one of the best marketing strategies for persuading consumers to purchase and strain a precise service. It has been determined to be an effective weapon in GSM networks' influences to make a supplementary sale. Sales promotions continue to be exceedingly effective in introducing customers to services for the first time. In the early stages of new services performance, they can serve as vital promotional constituents. This paper aims to conclude the consequence of continuous sales promotion of GSM network consumers in Yemen. In this study survey, the technique remained used in collecting evidence commencing the respondents. The sample size was 100 participants selected using the simple random sampling technique and 95 used to do analysis. However, statistical analyses remained conducted using descriptive and inferential methods with the assistance of SPSS software. This paper originates that there is a positive relationship between sales promotion and consumer on GSM networks.

Keywords: GSM companies · Sales Promotion · Yemen

1 Introduction

Cell phones are the most innovative and appealing devices to consumers [1]. Consumers are more likely to buy cell phones with advanced features and a perceptive presence in the marketplace [2]. Many GSM manufacturers around the world are implementing various sales promotion schemes to meet the increasing demand for cell phones among consumers [3]. Various network service providers provide their customers with appealing

M. Al-Emran et al. (Eds.): ICETIS 2021, LNNS 299, pp. 638–645, 2022.
https://doi.org/10.1007/978-3-030-82616-1_52

compacts, cost-effective deals, and desirable devices [4]. Other providers may offer dual SIM features and complimentary headsets [5].

Furthermore, many GSM networks provide their customers with unique benefits, including free use of their most modern brands, provided that the client must comply with all requirements [6]. To get a deal, consumers have to sign different types of contracts with varying periods. The customers can choose the best plan for themselves. Consequently, customers can have an unlimited call and text access [7]. Furthermore, customers will only be billed every month for their call and text consumption [8]. The purpose of these attractive plans is to provide customers with various mobile models in different variations. Conversely, the plans are cheaper for the customers and have several incentives, such as text message applications, free talk time, and gift hampers for using mobiles. For making an eccentric return, the network service provider focuses on the best plans for customers [9]. In general, telcommunication technology is vital in information sharing adopted by many industries such as education [10, 11], manufacturing [12, 13], telehealth readness [14], E-Commerce readness [15], and achieving sustainable competitive advantages [16]. However, many companies still not adopt Green technology that preserve environment in there promoting efforts [17].

In Yemen, there are two GSM operators: MTN Yemen and Sabafon, which share the same market; in 2007, Y mobile company entered the market but was unsuccessful in gaining power. However, Yemen Telecom, which is owned by the 55% obligatory national PTC and operates a CDMA network, is a significant performer in the market, with a comparable marketplace segment to MTN and Sabafon [18]. Delta Partners, a consultancy firm, has investigated the different results by Dubai–based research. four mobile operators' market shares increased to 37% for MTN, 31% for Yemen Mobile, and 29% for Sabafon, while Y-Tel trailed on 5%. Delta Partners predicts that by 2014, these four operators will have market shares of 38%, 30%, 24%, and 8%, respectively [19]. In comparison to Etisalat, the UAE has achieved a market share of 67% due to increased sales and promotion of its services and offers and the consequences of decentralized government [20].The feedback from consumers will decide which of the sales promotion strategies work better and by what percentages [21]. This paper aims to investigates the impacts of sales promotion tactics on GSM network users and the relationship between sales promotion on GSM networks' consumers.

2 Literature Review

2.1 GSM Network

Initially, GSM (Global System for Mobile Communications) was designed to empower mobile peripatetics between participating countries [22]. GSM digital wireless service station stayed offered in approximate form in over 193 states as of March 2003. Almost 69% of entirely digital mobile subscriptions in June 2002 used GSM phones on GSM networks [23]. The persistence of GSM is to digitize and compress voice data before sending it down through the channel using two additional streams of user data, each in its time slot [24]. In the era of this 21st Century, the Global System for Mobile communications (GSM) employed digital organisms reasonably rather than the outdated analogue structures and used ordinal data [25]. GSM is a cellular network that describes the mobile devices bond in its direction by examining for partitions cells in the instant area.

2.2 Sales Promotion

Sales promotion is the collecting procedures used by salespeople to stimulate instant procurement if consistent products due to actual flexible demand [26]. In the most extreme cases, sales promotion is used in grouping through additional promotion procedures, specifically with advertising and direct sales [27]. In sales promotion standards, advertising is the leading competitive factor, but regarding demand, there are some limitations on behalf of product included for comparative advertising procedures [28]. Sales promotion can be defined as small-term inducements to encourage the purchase of a product or service [29]. Even the integrated supply chain can enhance sales [30].

3 Research Methodology

This study follows the quantitative approach, using a survey for data collection from three Yemeni GSM companies. The sample size consists of 100 participants who are working in GSM networks in Sana'a. A simple random sampling technique was employed. The participants hold carrying positions, such as first-line managers and middle-level managers. The first part of the distributed survey contains demographic information, and the second part contains (9) questions to measure sales promotion [31]. The third section also included (9) questions to assess consumer GSM networks [32]. In this study, the Five Likert Scale and SPSS were used.

H1: There is a significant relationship between sales promotion and consumers of GSM networks.

4 Research Finding

SPSS software was used to analyze the collected data. Various statistical analyses, such as descriptive analysis, frequency statistical analysis, correlation analysis, and regression analysis, were carried out to explain the relationship and inspiration of the self-determining variable (sales promotions) on the dependent variable (GSM networks).

4.1 Demographic Variables

The questionnaire distributed to 100 respondents; 95 of the responses were usable for analysis. The results Fig. 1 show the respondents' demographic variables, such as gender, age, education, job level and experience.

In the results of the gender variable, male respondents account for (79) of the total respondents (83.2%), while female respondents accounted for (16) with a percentage of (16.8%). The significance of age (30–40) years user's variables of (58) respondents represent their results (61.1%), and consumers with age 40–50 years represent the results (12. 6%). The percentage of customers below the age of 30 years is also shown in

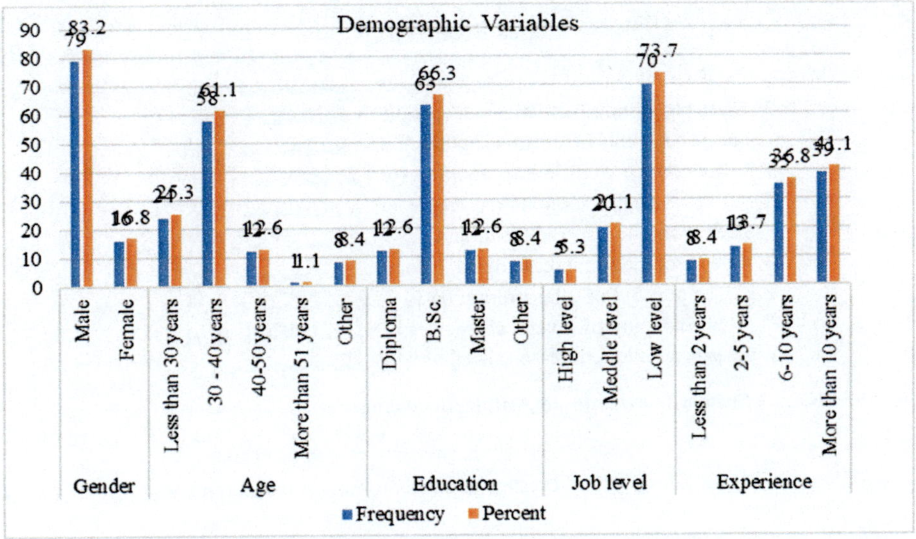

Fig. 1. Description of frequency analysis results of respondent demographic variables

the results (25.3%). Thus, the last result shows the group's percentage with more than 51 years old, representing (1.1%). According to the respondents' education, 66.3% had a bachelor's degree, and 12.6% had a master's degree. Furthermore, 73.7% of respondents were executives. According to the findings, 41.1% of respondents had more than ten years of experience, while 36.8% had 6–10 years of work experience.

4.2 Sales Promotion

The descriptive analysis includes the means and standard deviations expected to remain dependent and independent, as defined in the preceding sections. Figure 2 explains the verbal evaluation method for the descriptive statistics output from SPSS.

The primary rank is represented in Fig. 2 by a mean of 4.25 and a standard deviation of 0.922. The majority of defendants (85.05%) agree that sales promotion increases sales, which increases profit. The variable's overall average is 3.92, and the standard deviation is 0.59.

4.3 GSM Networks

The primary rank is represented in Fig. 3 by a mean of 4.46 and a standard deviation of 0.908. The majority of respondents strongly agree that wave coverage of your networks affects directly your sales activities.

Also, few of respondents agree that, not only the services and its quality what makes you able to increase your market share.ranked last with a mean of 2.96 and standard deviation of 1.22. The variable's overall average is 3.86, and the standard deviation is 0.872.

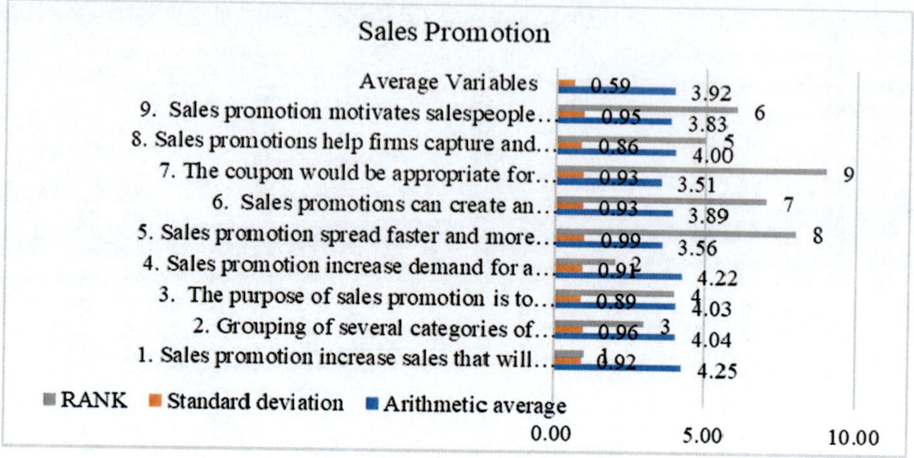

Fig. 2. Descriptive statistics of sales promotion measurement

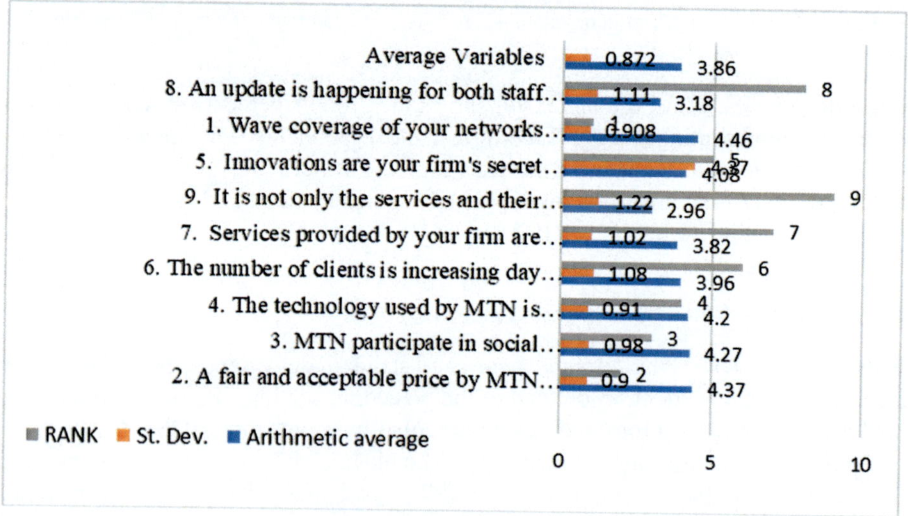

Fig. 3. Measurement of descriptive statistics of GSM networks

4.4 The Relationship Between Sales Promotion and GSM Networks

Table 1 show that correlation between Sales Promotion and GSM Networks is significant at 0.01 level with correlation 0.323 that represent Moderate relationship.

Table 1. Correlation analysis

		Sales Promotion Measurement	GSM Networks Measurement
Sales Promotion Measurement	Pearson Correlation	1	.323(**)
	Sig. (2-tailed)		.001
	N	95	95
GSM Networks Measurement	Pearson Correlation	.323(**)	1
	Sig. (2-tailed)	.001	
	N	95	95

** Correlation is significant at the 0.01 level (2-tailed).

5 Discussion

According to the study's findings, sales promotion repetition is a very projecting feature in GSM networks, and consumers are very aware of the various sales advertising approaches accomplished with the assistance of GSM suppliers. This finding supported by [32]. The numerous strategies recognized now the GSM networks by customers' provisions the most commonly used sales promotions in the global domain sphere. The GSM industries always stand in one advertising system or the other demanding to induce customers to support their services rather than their competitors. Today, a system comprises processers, big servers, supercomputers, network devices, peripherals, or additional devices associated with sharing the data to offer customers a better offers than competitors [33].

6 Conclusion

In conclusion, this paper has determined that GSM network suppliers have several facilities and products available to their customers through their innumerable collaboration and communication tools. Sales promotion is only a powerful tool used in marketing and advertising for companies services and products. In the long run, this may avoid the buyer beginning profitable concluded all the resolution platforms because of capability and available information to them. This has correspondingly perceived that the buyer would regularly ruminate which GSM operator is proposing the adequate money off and this compassionate service that fulfils their necessity. Consequently, sales promotion is an unavoidable promotional tool aimed at telecom companies to maintain or upturn their market share. Sales promotion should be a prominent feature in GSM networks providers' budget, year in, year out.

The analysis and results demonstrated a relationship between sales promotions and GSM networks, which significantly impacted Yemen's GSM networks. Sales promotions were found to be correlated with GSM networks ($r = .323$, $p = .001$, 2-tailed), and regression analysis revealed that sales promotions have a significant effect on GSM networks in Yemen ($\text{Beta} = 0.480$, $p = .001 > 0.05$), supporting the two hypotheses developed in this study. The researcher collected the data via a questionnaire that asked about each variable and the questionnaires..

References

1. Zachariah, T., Dutta, P.: Browsing the web of things in mobile augmented reality, pp. 129–134 (2019). https://doi.org/10.1145/3301293.3302359
2. Monahan, T.: Built to lie: investigating technologies of deception, surveillance, and control. Inf. Soc. **32**(4), 229–240 (2016). https://doi.org/10.1080/01972243.2016.1177765
3. Oladimeji, T.T., Akinwole, O.O.: Fifteen years of GSM services : technological issues. Adv. Soc.-Econ. Impact Niger. Econ. **6**(1), 14–18 (2018)
4. Zhang, X., et al.: Incentives for mobile crowd sensing: a survey. IEEE Commun. Surv. Tutor. **18**(1), 54–67 (2016). https://doi.org/10.1109/COMST.2015.2415528.Sample, http://www.springer.com/lncs. Accessed 21 Nov 2016
5. Shaaban, K., Gaweesh, S., Ahmed, M.M.: Characteristics and mitigation strategies for cell phone use while driving among young drivers in Qatar. J. Transp. Health **8**(October 2017), 6–14 (2018). https://doi.org/10.1016/j.jth.2018.02.001
6. Grewal, D., Bart, Y., Spann, M., Zubcsek, P.P.: Mobile advertising: a framework and research agenda. J. Interact. Mark. **34**(April), 3–14 (2016). https://doi.org/10.1016/j.intmar.2016.03.003
7. Hodge, S.K., Honeycutt, E.D., Shipley, D.: Toward an improved understanding of online customer service delivery to millennials. Serv. Mark. Q., 1–15 (2019). https://doi.org/10.1080/15332969.2019.1587865
8. Óskarsdóttir, M., Bravo, C., Sarraute, C., Vanthienen, J., Baesens, B.: The value of big data for credit scoring: enhancing financial inclusion using mobile phone data and social network analytics. Appl. Soft Comput. J. **74**, 26–39 (2019). https://doi.org/10.1016/j.asoc.2018.10.004
9. Gonzales, A.L., Ems, L., Suri, V.R.: Cell phone disconnection disrupts access to healthcare and health resources: a technology maintenance perspective. New Media Soc. **18**(8), 1422–1438 (2016). https://doi.org/10.1177/1461444814558670
10. Al-Kumaim, N.H., Hassan, S.H., Shabbir, M.S., Almazroi, A.A., Abu Al-Rejal, H.M.: Exploring the inescapable suffering among postgraduate researchers: information overload perceptions and implications for future research. Int. J. Inf. Commun. Technol. Educ. (IJICTE) **17**(1), 19–41 (2021)
11. Alamri, M.M., et al.: Towards adaptive e-learning among university students: by applying technology acceptance model (TAM). Int. J. Eng. Adv. Technol. **8**(6 Special Issue 3), 270–276 (2019)
12. Abualrejal, H.M.A., Udin, Z.M., Yusoff, R.Z.B.: The role of supply chain management toward performance of manufacturing companies in Yemen. Int. J. Supply Chain Manag. **7**(6), 92–99 (2018)
13. Abualrejal, H.M., Abu Doleh, J.D., Salhieh, L.M., Udin, Z.M., Mohtar, S.: Barriers of supply chain management practices in manufacturing companies in republic of Yemen: pre-war perspective. Int. J. Supply Chain Manag. **6**(3), 246–251 (2017)
14. Al-Fadhli, A.A., Othman, M., Al-Rejal, H.M.A., Al-Jamrh, B.A.: Developing a telehealth readiness assessment framework and tool for least developed countries: a case of yemen. Indian J. Public Health Res. Dev. **9**(11), 1489–1498 (2018)
15. Alqershi, N., Ismail, A.I., Abualrejal, H., Salahudin, S.N.: Competitive advantage achievement through customer relationship management dimensions. J. Distrib. Sci. **18**(11), 61–67 (2020). https://doi.org/10.15722/jds.18.11.202011.61

16. Makhloufi, L., Al-Rejal, H.M.E.A., Mohtar, S.: An investigation of the moderating effect of IT personnel capability on the relationship between intangible IT resources and IT infrastructure flexibility on the sustainable competitive advantages. J. Asian Sci. Res. **8**(9), 277 (2018)
17. Abu Al-Rejal, H.M.E., Udin, Z.M., Hassan, M.G., Sharif, K.I.M., Al-Rahmi, W.M., Al-kumaim, N.H.: Green information technology adoption antecedence: a conceptual framework. In: Saeed, F., Mohammed, F., Gazem, N. (eds.) IRICT 2019. AISC, vol. 1073, pp. 1098–1108. Springer, Cham (2020). https://doi.org/10.1007/978-3-030-33582-3_103
18. Ali, M.N., Alareqi, M.A., Zahary, A.T.: Opportunities and obstacles to deploy internet of things services by telecom operators in developing countries (case study telecom operators in Yemen). Asian J. Res. Comput. Sci. **3**(1), 1–14 (2019). https://doi.org/10.9734/ajrcos/2019/v3i130085
19. Tarapiah, S., Aziz, K., Atalla, S.: Third generation (3G) mobile network planning process and methodology - case study. Int. J. Comput. Appl. **143**(12), 1–6 (2016). https://doi.org/10.5120/ijca2016909881
20. Gielens, K., Steenkamp, J.B.E.M.: Branding in the era of digital (dis)intermediation. Int. J. Res. Mark., 1–18 (2019). https://doi.org/10.1016/j.ijresmar.2019.01.005
21. Felix, R., Rauschnabel, P.A., Hinsch, C.: Elements of strategic social media marketing: a holistic framework. J. Bus. Res. 9 (2016). https://doi.org/10.1016/j.jbusres.2016.05.001
22. Ali-Vehmas, T., Casey, T.R.: Examining possible value system transitions: the case of smart mobility services. In: Proceedings of the Annual Hawaii International Conference on System Sciences, 2015–March (August 2014), pp. 4366–4375 (2015). https://doi.org/10.1109/HICSS.2015.523
23. Simon, J.P.: How Europe missed the mobile wave. Info **18**(4), 12–32 (2016). https://doi.org/10.1108/info-02-2016-0006
24. Goud, D.S., Kumar, V.N.: Advanced handheld electronic banking system using Raspberry Pi. Adv. Res. J. Sci. Technol. 276–279 (2017)
25. Kadiri, K.O., Lawal, S.O., Adejuwon, A.A.: Analysis of congestion of mobile network in Offa. J. Sci. Res. Rep. **22**(6), 1–8 (2019). https://doi.org/10.9734/jsrr/2019/v22i630108
26. Fam, K.-S., Brito, P.Q., Gadekar, M., Richard, J.E., Jargal, U., Liu, W.: Consumer attitude towards sales promotion techniques: a multi-country study. Asia Pac. J. Mark. Logist. **31**(2), 437–463 (2019)
27. Vasile, T.C., Popescu, L., Dăniasă, C.I., Iancu, A., Popescu, V.: Business marketing, sales and promotion - field receptive to expert systems technology. Adv. Eng. Forum **27**, 280–289 (2018). https://doi.org/10.4028/www.scientific.net/aef.27.280
28. Kim, C., Takashima, K., Newell, S.: Asia Pac. J. Mark. Logist. **30**(3), 571–586 (2017). https://doi.org/10.1108/MRR-09-2015-0216
29. Familmaleki, M., Aghighi, A., Hamidi, K.: Analyzing the influence of sales promotion on customer purchasing behaviour. Int. J. Econ. Manag. Sci. **4**(4) (2015). https://doi.org/10.4172/2162-6359.1000243
30. Abualrejal, H.M., Abu Doleh, J.D., Mohtar, S.: Supply-chain management capabilities practices in the industrial organization in republic of Yemen. Int. J. Supply Chain Manag. **6**(4), 98–109 (2017)
31. Tan, H., Akram, U., Sui, Y.: An investigation of the promotion effects of uncertain level discount: evidence from China. Asia Pac. J. Mark. Logist. (March), APJML-05-2018-0191 (2019). https://doi.org/10.1108/APJML-05-2018-0191.
32. Opele, A.M., Afolabi, O.J., Onifade, T.: A Consumers' preference and satisfaction of GSM service providers among students of tertiary institutions in Lagos State, Nigeria. Niger. J. Technol. **37**(4), 1099 (2018). https://doi.org/10.4314/njt.v37i4.33
33. Qolomany, B., et al.: Leveraging machine learning and big data for smart buildings: a comprehensive survey. IEEE Access **7**, 90316–90356 (2019)

The Role of Supply Chain Integration on Project Management Success in Jordanian Engineering Companies

Ahmed Ali Atieh Ali(✉), Hussein Mohammed Esmail Abualrejal(✉), Zulkifli B. Mohamed Udin, Hassan Olayan Shtawi, and Amer Zaid Alqudah

Technology and Supply Chain Excellence Institute (TESCE), School of Technology Management and Logistics, University Utara Malaysia, Sintok, Kedah, Malaysia
abualrejal@uum.edu.my

Abstract. This study aims to identify the role of supply chain integration on project management success in Jordanian engineering companies. The study employed a quantitative approach. A questionnaire was developed to collect the data. The sample of the study consisted of 118 project managers in these companies. The researcher used the statistical package for the social sciences (SPSS) program to analyze the data. Based on the results, managers recommended that companies and their suppliers continue their partnership, coordination, and cooperation in decision-making, such as product design decisions, decisions to develop materials specifications, and decisions to improve production methods.

Keywords: Supply chain · Integration · Project success · Engineering companies · Jordan

1 Introduction

Delivering project output on time and at a low cost was the main focus of project managers from 1960 to the 1980s [1]. Although other factors impact a project's success (e.g. quality of project operations), most project managers have focused more on the iron triangle (i.e. cost, time, and scope) of project management. The importance of the time factor and its significant impact on a project's success has been highlighted by some authors, such as [2, 3]. In addition, [3] pointed that a project's success is judged by stakeholders, especially the primary sponsor, instead of the tangibles performance, which represents the conventional measurement of a project's success. Hence, there are many determinants to the project's success. This study focuses on the whole construction project's success. According to [4] the standard evaluation system for construction project success includes time, cost, quality, health and safety, environmental performance, participants' satisfaction, user satisfaction, and commercial value.

Recently, supply chains (S.C.s) and their impact on companies' performance have gained significant attention from both practitioners and researchers. The management

M. Al-Emran et al. (Eds.): ICETIS 2021, LNNS 299, pp. 646–657, 2022.
https://doi.org/10.1007/978-3-030-82616-1_53

of S.C.s refers to the systemic and strategic coordination of conventional market functions. The management of S.C.s involves the movement of resources, goods, knowledge, and money from a single business and through suppliers to customers, enhancing its performance [5]. Implementing construction supply chain integration is necessary for organizations to improve the construction business's response time. In this regard, organizations must acknowledge the importance of supply chain management toward business performance [6, 7]. Companies must increase the efficiency of storage management and operation management to improve their profitability [8]. Increasing the efficiency of S.C.s for companies is important to increase their profitability. Companies must deal with practical and sustainable operations in construction supply chain management [9]. For any project success, organizations have to adopt proper technology [10] for sustainable competitive advantages [11]. Therefore, this research focuses on the sustainability of construction material, supply chain integration and project success in Jordan.

2 Problem Statement

Project success has ceased to be just a matter of the triple constraints of budget, time and quality. It has become a detailed evaluation of stakeholders' benefits. The concept of project success has been revisited several times and still being revisited to make it fit with organizations' practical needs. This is due to the growing importance of the project's success in organizations and the increasing complexity of projects [12]. The change in market dynamics, market development, internationalization, and increasing competitiveness has led to increasing the importance of project success and increasing the importance of the notion of building and construction and its domain of studies [13]. According to [14] there is an imperative need to conduct more research on the project's success. In addition, there is a need to shed more light on the concept of S.C.s in general and on the concept of supply chain integration in specific. More studies to understand S.C.s practices' barriers are needed [15]. There is also a need to study the relationship between suppliers and customers and operations' efficiency [16].

The supply chain costs depend on the period of the implemented project [17]. Thus, the supply chain's integration is of great importance in developing the ideal approach to selecting suppliers in engineering projects, procurement and construction. Weak supply chain integration leads to project failure. The findings of [18] indicates that cost-based supplier selection leads to the prevalence of arm's length-type relationships. A possible explanation for these results is that projects supply chains are often temporary; hence, fostering strong partnerships and achieving full supplier integration might not always be feasible. Eventually, these practices may turn the supply market into an oligopolistic one with just a few large multinational suppliers; this could have dire consequences, such as harming pure competition and national economies. This research aims to describe the impact of supply chain integration on project management success.

3 Literature Review

3.1 Project Management Success

Just in Time. The introduction of the Just-in-Time (JIT) concept can be advantageous in managing the logistics of building materials and the workforce at construction sites. The construction industry is mainly project-based, renowned for being inefficient with low productivity, and the industry often encounters difficulties in keeping time schedules due to delays during the construction period. Concerning construction management, there is a significant difference between how JIT is implemented in the construction industry and the manufacturing industry. This is because construction is a different product that can resemble product development more closely than the manufacturing process [19].

Quality of Project Operations represents all the works and processes related to project completion. Therefore, companies have to keep continuous communication with the project's beneficiaries to ensure that their needs are well defined accurately [20].

Project Cost is a set of necessary processes for planning and estimating the costs allocated to the project's implementation, budgeting, and financing. The most significant challenges facing the success of projects and the integration of the supply chain are cost management [17].

3.2 Supply Chain Integration and Projects Success

An engineering companies follow the supply chain in their projects, they contribute to reorganizing their project management processes to improve and organize supply tasks and activities and improve their performance [21]. To creating customer satisfaction, the relationship between the use of integration in the supply chain and end-users may be positive, indicating one of the indicators for implementing the supply chain in engineering projects [22].

Suppliers Relationship. It refers to the degree of coordination between suppliers and manufacturers in inventory management decisions, collaborative planning, and sourcing. Sharing information is vital for any success [23]. The integration with the supplier involves a set of practices represented in partnership with the supplier, the participation in many activities such as design, reducing the number of supply sources, a commitment of resources, and sharing of information [24].

Storage. The storage process consumes a considerable amount of resources, time, and money; these processes must be carefully planned and managed by quality standards in the management sector to maintain the integrity of the inventory and the quality of the product. In addition, in the right place, the right condition at the right time, in a way that is cheap and in compliance with customer requirements [25].

Supply Integration. The supply integration process consists of sub-processes related to shipping, transporting, receiving, and testing goods to ensure that they comply with requirements in terms of forms, products, quantities, and specifications. Also, carry out the required research to evaluate the desired characteristics and requirements before carrying out any supplied process and those related to the selection of suitable sources of supply, how to negotiate with them, shipping terms, transport, insurance of products, inspection, and inspection procedures necessary for receiving [26].

The Structure of the Supply Network. Supply Chain Network Design is the discipline used to determine the facilities' optimal location and size through the facilities [27]. It is the most critical one is to locat the facilities in different tiers of the supply chain [28].

4 Methodology of Research

The study used a quantitative approach. The study population consisted of all 211 consulting engineering firms specialized in project management (Jordanian Engineers Syndicate, 2018). The selected sample consisted of 140 consulting engineering firms specialized in project management, depending on the completed size of the projects. The analysis unit consisted of project managers as respondents in each of the selected sampled companies, consisting of 140 project managers. Out of 124 questionnaires distributed to participants, six questionnaires were excluded because they were incompletely filled. Therefore, 118 questionnaires were valid and ready for conducting statistical analysis. The construct items adopted from previous studies were supply integration from [29], the structure of the supply network and the relationship with suppliers from [30], Just in Time from [31], and quality of project operations and cost of the project from [32].

5 Finding

The results describing the demographic appears in Fig. 1, which shows that the male project managers were 88.1%, while the female managers represented 11.9% of the study sample. The result shows that 61.9% of the respondents' ages fell in the range 41–50 years, followed by 22.9% of the respondents aged 51 years or more and representing 83.9% of the study sample. Regarding the educational level, 62.7% of the managers held a bachelor's degree, 26.3% holding a master degree, and 7.6% holding a PhD degree.

Moreover, the result shows that managers with experience ranging from 5 to 10 years were the most frequent, representing 37.3% of the sample, followed by those with experience less than five years (39 managers with 33%), and then those with experience of more than 15 years (19 managers with 16.1%). However, managers with 11–15 years of experience were the least frequent as 16 managers (13.6%).

Moreover, 31 managers with 11–15 years of experience came second with a percentage of 26.3, followed by 22 managers with experience of 16 years and over with a percentage of 18.6. Those with experience (less than five years) were less frequent. They were 17 with a percentage of 14.4 of the study sample.

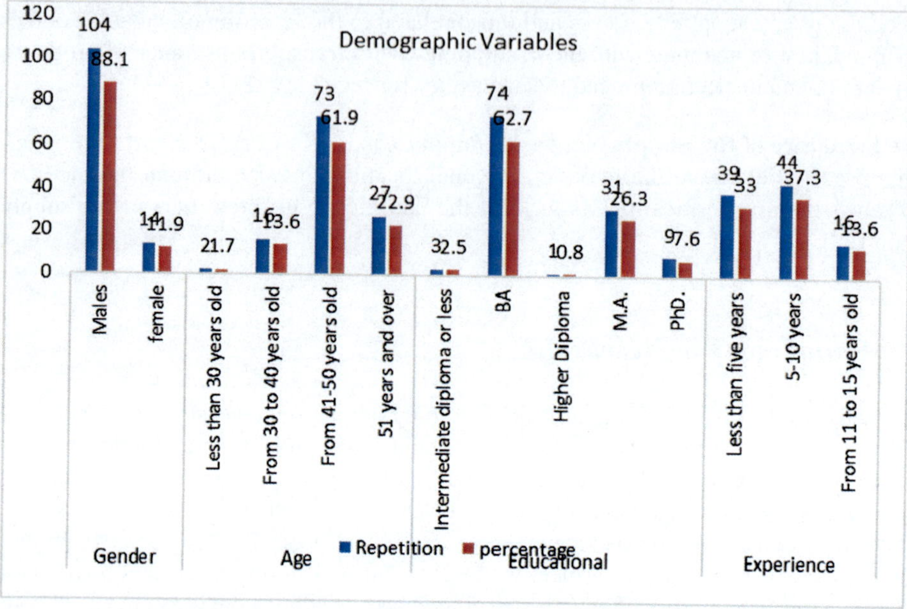

Fig. 1. A frequency and percent of respondents according to demographic and employment variables

5.1 Supply Chain Integration

Figure 2 indicates that the arithmetic means for the dimensions of the variable "supply chain integration" ranged between (4.09–4.32) with a high degree of appreciation of the minimum degree (1) and a maximum degree of integration of the supply chain (5). The dimension of supply integration with the highest arithmetic average (4.32), followed by a "storage" with an arithmetic average (4.24). The "relationship with suppliers" was the arithmetic average (4.22), and the dimension was the "supply network structure" with the lowest arithmetic average (4.09). On the other hand, the arithmetic mean for the variable as a whole was (4.22) with a high degree of appreciation.

The Relationship with Suppliers. Figure 3: The arithmetic means and the standard deviations of the dimensions of the relationship with the supplier and the dimension as a whole. Figure 3 shows a high level of estimation of the dimension of the relationship with the supplier where the arithmetic means of the dimension was used (4.22). The figure shows that the arithmetic mean of the study sample individuals' opinions on the axis of the study ranged from (4.65–3.78) to a high degree of appreciation for all paragraphs.

The fig. indicates that Paragraph (4) is the most appreciated, which stipulates that "suppliers are obligated to have flexible production and supply operations with the company", with an average of (4.65) and a high degree of appreciation, followed by paragraph (2) which states that "Post-supply services are compatible with the requirements of the company And its needs are " with an arithmetic average (4.53) and with a high degree of appreciation, and paragraph (5) which states" The company and suppliers share the

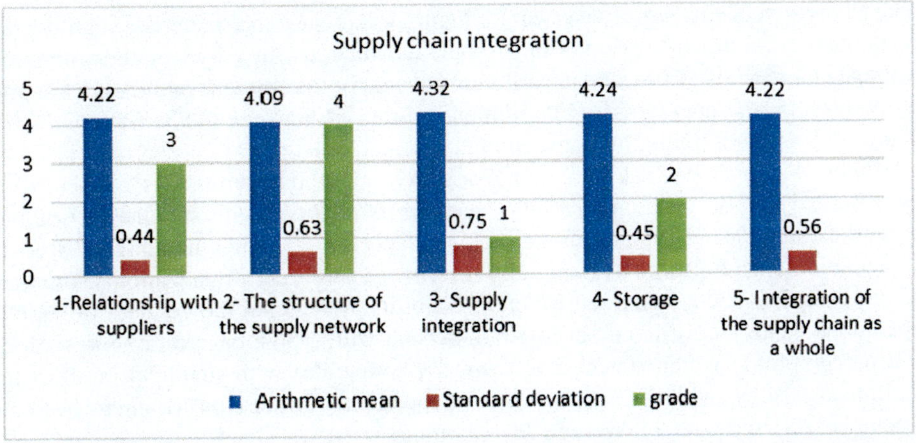

Fig. 2. A mean and standard deviation of the supply chain integration and its items

future vision of the workflow in all operations" with the lowest estimate with an average of (3.78) and with a high degree of appreciation.

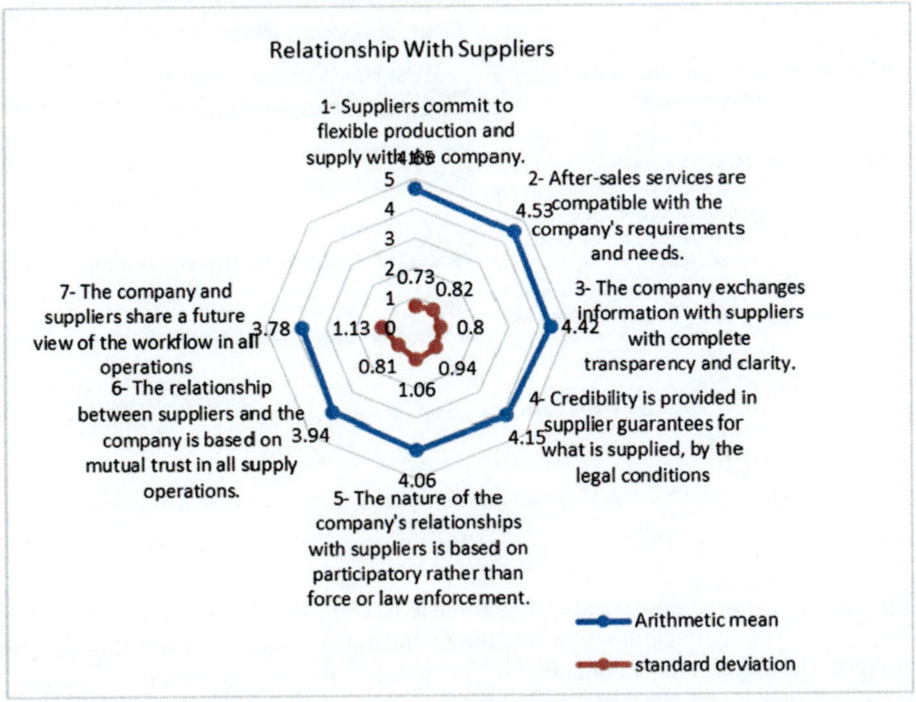

Fig. 3. Mean and standard deviations of the relationship with suppliers and their items

The Structure of the Supply Network. Figure 4 indicates that there is a high degree of estimation for the dimension of the "supply network structure", where the arithmetic average of it reached (4.06), and the arithmetic mean for the opinions of the study sample individuals on the paragraphs of the dimension ranged between (4.73–3.58) and with degrees of estimation ranging between high and medium.

Figure 4 also shows that Paragraph (5), which states "the company is characterized by a limited number of administrative levels in its relations with suppliers", was the highest estimate, where its arithmetic average reached (4.73) and with a high degree of appreciation, then followed by paragraph (2) which states " The organizational structure of the company is characterized by high flexibility to regulate the conduct of supply operations "with an arithmetic average (4.44) and with a high degree of appreciation, and paragraph (1) which states" The company cooperates with suppliers in all of its multiple activities "with the lowest degree of appreciation, where its arithmetic average reached (3.58) and with an Average degree rating.

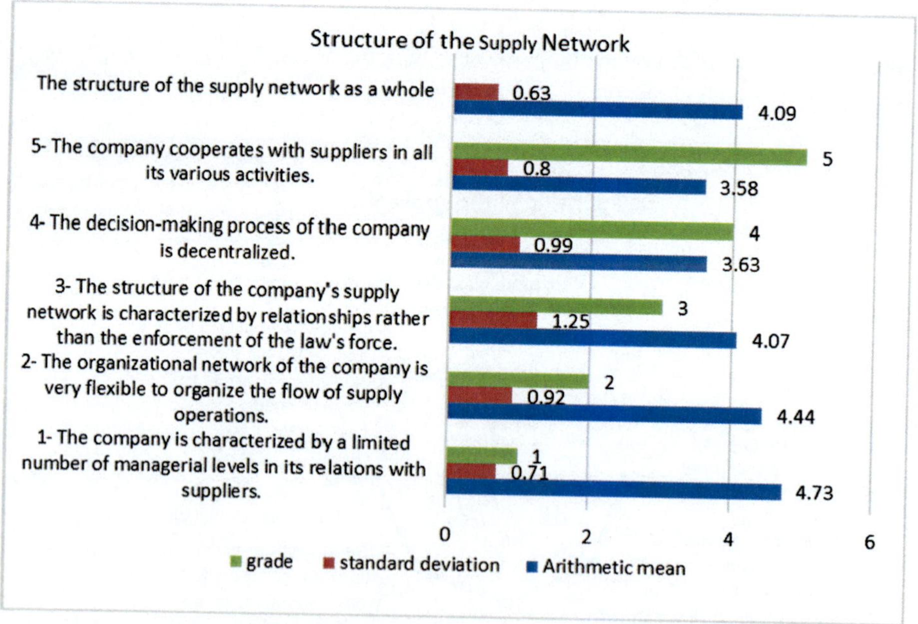

Fig. 4. A mean and standard deviations of the supply network structure its items

Supply Integration. Figure 5 indicates that there is a high degree of estimation for the dimension of "supply integration" where the arithmetic average of it reached (4.32). It also appears from Fig. 5 that Paragraph (4), which states "there is a smooth flow between the company and the suppliers in the flow of information and materials", was the highest estimate, where the arithmetic average of it reached (4.36) and with a high degree of appreciation, followed by Paragraph (1), which states " The company is in continuous coordination with suppliers continuously" with an arithmetic average (4.34) and with a high degree of appreciation, and Paragraph (2), which states" The company's logistics activities are integrated with the suppliers' logistics activities" with the lowest degree of appreciation, with an arithmetic average of (4.26) and with a high degree of appreciation.

Storage. Figure 6 indicates that there is a high degree of appreciation for the "storage" dimension where the arithmetic average of it reached (4.24). The arithmetic means for the study sample individuals' opinions on the dimension's paragraphs ranged between (4.43–4.14) and with high degrees of appreciation for all paragraphs. It also appears from Fig. 6. that Paragraph (1), which states "The company uses the most recent systems in warehouse management to secure suppliers' requests", was the highest estimate, as its arithmetic average reached (4.43) with a high degree of appreciation, then followed by paragraph (2), which states "The company seeks to allocate facilities and transportation related to storing materials between it and suppliers" with an average score of 4.23 and with a high degree of appreciation, and Paragraph (4), which states "The company seeks to adopt national and international standards related to the stored materials" with the lowest degree of appreciation as the average Arithmetic for her (4.14) and with a high degree of appreciation.

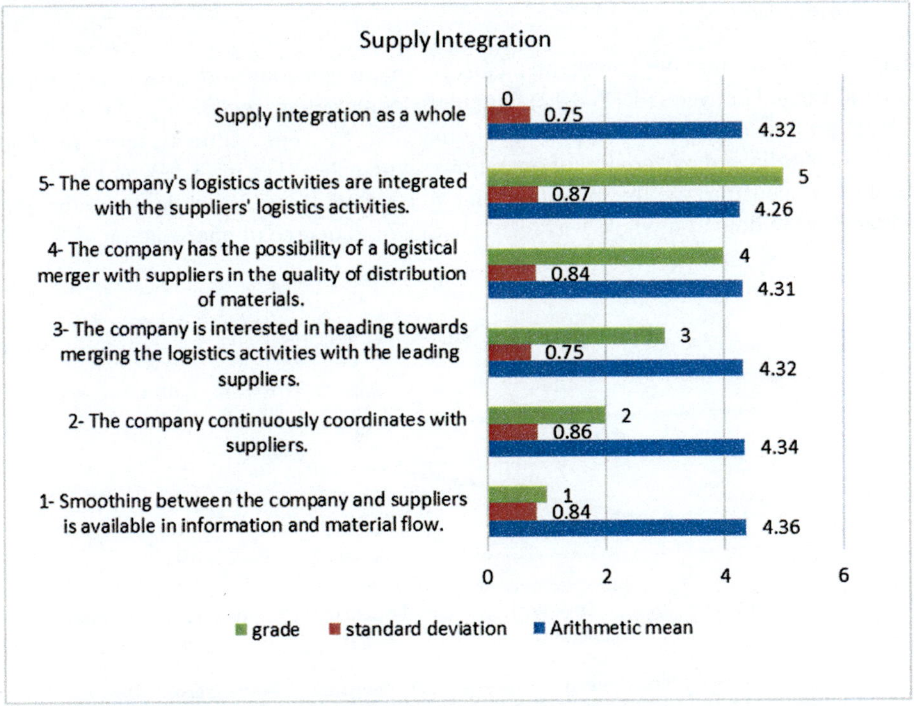

Fig. 5. The means and standard deviations of the supply integration and its items

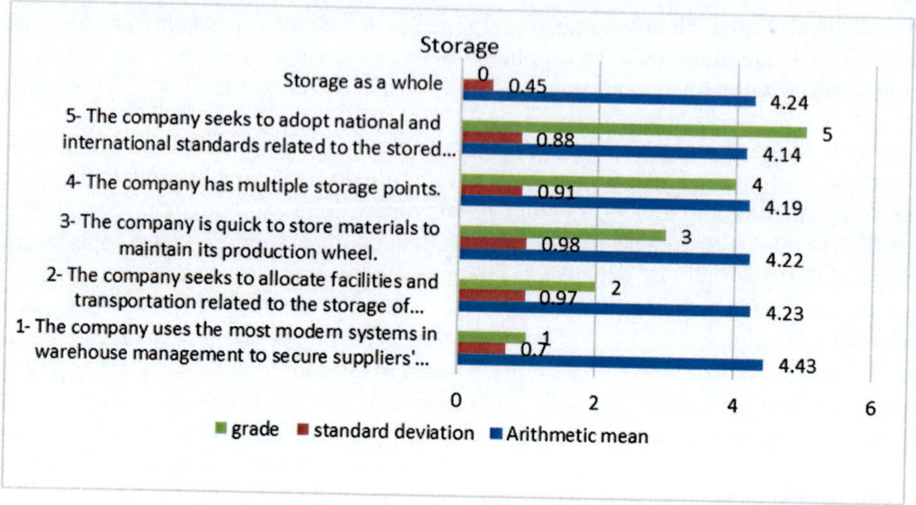

Fig. 6. The means and standard deviations of the storage and its items

5.2 Project Success

Figure 7 indicates that the arithmetic means for the dimensions of the variable "project success" ranged between (4.17–4.12) with high degrees of appreciation, and the dimension of "production on time" was with the highest average. The arithmetic terms reached (4.17), then followed by "quality of operations" with an average of (4.13), and the "project cost" dimension was the lowest arithmetic average of (4.12), while the arithmetic average for the field as a whole was (4.14) with a high degree of appreciation.

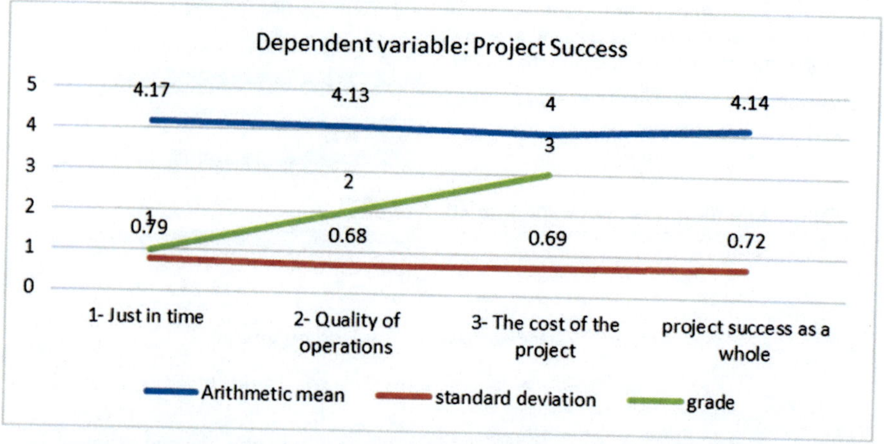

Fig. 7. The means and standard deviations of the project success dimensions

6 Conclusion

It is clear from these results that the integration of the supply chain has a significant contribution to the success of the projects, as it was found that the integration of the supply chain was the primary key to the long-term success of companies by maintaining the integration of supply chain practices based on establishing long-term relationships with suppliers, effective communication, and partnership with suppliers, and maintaining relationships with consumers by opening the way to customers to determine the nature of the services or products. Most of them showed a high level of relationship between the integration of the supply chain dimensions and the project's success dimensions. This result is in agreement with the study by [33], which showed that the success of the projects depends to a large extent on the quality of the project operations represented in strategic support, operational support, project budget achievement, quality of the project and project schedule. Besides, these days organizations must adopt green technology to achieve Green supply chain integration with their suppliers to protect the environment [34] and consider the green building index [35].

References

1. Ika, L.A.: Project success as a topic in project management journals. Proj. Manag. J. **40**(4), 6–19 (2009)
2. Shenhar, A.J., Levy, O., Dvir, D.: Mapping the dimensions of project success. Proj. Manag. J. **28**(2), 5–9 (1997)
3. Turner, R., Zolin, R.: Forecasting success on large projects: developing reliable scales to predict multiple perspectives by multiple stakeholders over multiple time frames. Proj. Manag. J. **43**(5), 87–99 (2012)
4. Chan, A.P., Scott, D., Chan, A.P.: Factors affecting the success of a construction project. J. Constr. Eng. Manag. **130**(1), 153–155 (2004)
5. Mentzer, J.T., et al.: G Defining supply chain management. J. Bus. Logist. **22**(2), 1–25 (2001)
6. Abu Al-Rejal, H.M., Udin, Z.M., Yusoff, R.Z.B.: The role of supply chain management toward performance of manufacturing companies in Yemen. Int. J. Supply Chain Manag. **7**(6), 92–99 (2018)
7. Abu Al-Rejal, H., Abu Doleh, J.D., Mohtar, S.: Supply-chain management capabilities practices in industrial organization in Republic of Yemen. Int. J. Supply Chain Manag. **6**(4), 98–109 (2017)
8. Pakurár, M., Haddad, H., Popp, J., Khan, T., Oláh, J.: Supply chain integration, organizational performance and balanced scorecard: an empirical study of the banking sector in Jordan. J. Int. Stud. **12**(2), 129–146 (2019)
9. Shaikh, F.A., Shahbaz, M.S., Din, S.U., Odhano, N.: The role of collaboration and integration in the supply chain of construction industry. Civ. Eng. J. **6**(7), 1300–1313 (2020)
10. Alamri, M.M., et al.: Towards adaptive e-learning among university students: by applying technology acceptance model (TAM). Int. J. Eng. Adv. Technol. **8**(6 Special Issue 3), 270–276 (2019)
11. Makhloufi, L., Al-Rejal, H.M.E.A., Mohtar, S.: An investigation of the moderating effect of IT personnel capability on the relationship between intangible IT resources and IT infrastructure flexibility on the sustainable competitive advantages. J. Asian Sci. Res. **8**(9), 277 (2018)
12. Shenhar, A.J., Dvir, D., Levy, O., Maltz, A.C.: Project success: a multidimensional strategic concept. Long Range Plan. **34**(6), 699–725 (2002)

13. Piccarozzi, M., Aquilani, B., Gatti, C.: Industry 4.0 in management studies: a systematic literature review. Sustainability **10**(10), 3821 (2018)
14. Luo, L., He, Q., Xie, J., Yang, D., Wu, G.: Investigating the relationship between project complexity and success in complex construction projects. J. Manag. Eng. **33**(2), 04016036 (2017)
15. Abualrejal, H.M., Abu Doleh, J.D., Salhieh, L.M., Udin, Z.M., Mohtar, S.: Barriers of supply chain management practices in manufacturing companies in republic of Yemen: pre-war perspective. Int. J. Supply Chain Manag. **6**(3), 246–251 (2017)
16. Han, Z., Huo, B.: The impact of green supply chain integration on sustainable performance. Ind. Manag. Data Syst. **120**(4), 657–674 (2020)
17. Panfilova, E., Dzenzeliuk, N., Domnina, O., Morgunova, N., Zatsarinnaya, E.: The impact of cost allocation on key decisions of supply chain participants. Int. J. Supply Chain Manag. **9**(1), 552–558 (2020)
18. Koufteros, X., Vickery, S.K., Dröge, C.: The effects of strategic supplier selection on buyer competitive performance in matched domains: does supplier integration mediate the relationships? J. Supply Chain Manag. **48**(2), 93–115 (2012)
19. Hybel Lasse 19 November 2018, at 14:35. The Just in Time concept in construction. http://apppm.man.dtu.dk/index.php/The_Just_In_Time_concept_in_construction. Accessed 3 Jan 2020
20. Wysocki, R.K.: Effective Project Management: Traditional, Agile Extreme. . Wiley, Hoboken (2011)
21. Briscoe, G., Dainty, A.: Construction supply chain integration: an elusive goal? Supply Chain Manag. **10**(4), 319–326 (2005)
22. Khalfan, M.M.A., McDermott, P.: Innovating for supply chain integration within construction. Constr. Innov. **6**(3), 143–157 (2006)
23. Al-Kumaim, N.H., Hassan, S.H., Shabbir, M.S., Almazroi, A.A., Abu Al-Rejal, H.M.: Exploring the inescapable suffering among postgraduate researchers: information overload perceptions and implications for future research. Int. J. Inf. Commun. Technol. Educ. (IJICTE) **17**(1), 19–41 (2021)
24. Han, J., Lu, H., Trienekens, J.H., Omta, S.W.F.: The impact of supply chain integration on firm performance in the pork processing industry in China. Chin. Manag. Stud. **7**(2), 230–252 (2013)
25. Singh, R., Sandhu, H.S., Metri, B.A., Kaur, R.: Relating organized retail supply chain management practices, competitive advantage and organizational performance. Vis. J. Bus. Perspect. **14**(3), 173–190 (2010)
26. Abu Al-Rijal, H.: The impact of supply chain management on institutional performance in Yemeni industrial companies. Unpublished Master thesis, Yarmouk University, Irbid, Jordan (2007)
27. Autry, T.G., et al.: The Definitive Guide to Modern Supply Chain Management (Collection). Pearson Education (2013)
28. Farahani, Z., Rezapour, S., Drezner, T., Fallah, S.: Competitive supply chain network design: an overview of classifications, models, solution techniques and applications. Omega **45**, 92–118 (2014)
29. McGinnis, M.A., Vallopra, R.M.: Purchasing and supplier involvement in process improvement: a source of competitive advantage. J. Supply Chain Manag. **35**(4), 42–52 (1999)
30. Monczka, R., Trent, R., Handfield, R.: Purchasing and Supply Chain Management. South-western College Publishing, Cincinnati (2002)
31. Basnet, C., Corner, J., Wisner, J., Tan, K.-C.: A Survey of Supply Chain Management Practice in New Zealand, January 2014 (2000)
32. Schroeder, D.M.: A dynamic perspective on the impact of process innovation upon competitive strategies. Strateg. Manag. J. **11**(1), 25–41 (1989)

33. Luo, L., He, Q., Xie, J., Yang, D., Wu, G.: Investigating the relationship between project complexity and success in complex construction projects. J. Manag. Eng. **33**(2), 04016036 (2016)
34. Abu Al-Rejal, H.M.E., Udin, Z.M., Hassan, M.G., Sharif, K.I.M., Al-Rahmi, W.M., Al-kumaim, N.H.: Green information technology adoption antecedence: a conceptual framework. In: Saeed, F., Mohammed, F., Gazem, N. (eds.) IRICT 2019. AISC, vol. 1073, pp. 1098–1108. Springer, Cham (2020). https://doi.org/10.1007/978-3-030-33582-3_103
35. Abualreja, H.M., Udin, Z.M.: Green building toward construction sustainability: energy efficiency with material and design aspects. J. Technol. Oper. Manag. (Special Issue), 100–109 (2017)

Intention to Adopt Robo-Advisors Among Malaysian Retail Investors: Using an Extended Version of TAM Model

Kai Wei Zheng(✉), Jin Hao Cheong, and Mohsen Jafarian

Faculty of Business and Management, UCSI University, 56000 Kuala Lumpur, Malaysia
1001645932@ucsiuniversity.edu.my

Abstract. The emergence of digital technology has accelerated the growth of the financial sector in Malaysia. Artificial Intelligence (AI) technology has been integrated into investing activities to create automated investment management services known as robo-advisors. Despite the fact that numerous efforts have been attempted to publicize robo-advisors, its adoption among Malaysian retail investors remains relatively low. Hence, the aim of this paper is to extend the existing Technology Acceptance Model (TAM) to examine the main drivers of intention to adopt robo-advisors that explain the robo-advisors adoption process among Malaysian retail investors. In this study, 300 valid primary responses from Malaysian retail investors aged 18 to 77 who have used robo-advisors were collected through online questionnaire and analyzed using Partial Least Square Structural Equation Modelling (PLS-SEM) method. All the hypotheses in this study were supported. In particular, attitude was found to be the strongest predictor of behavioral intention to adopt robo-advisors followed by perceived usefulness, subjective norms, perceived behavioral control, and lastly perceived risk. As such, robo-advisory vendors and the relevant regulatory authorities should focus on these key drivers to boost robo-advisors adoption in Malaysia.

Keywords: Robo-advisor · Artificial Intelligence · FinTech · TAM model

1 Introduction

The financial sector in Malaysia is experiencing drastic technological advancement in recent years that forms the FinTech industry. In this industry, robo-advisor has been introduced, where Day et al. [13] described it as a digital platform that offers investors financial service in terms of robotic investment management through the use of algorithms with inclusion of Artificial Intelligence (AI) in some robo-advisory companies. Robo-advisors outperform traditional investment vehicles by integrating technological advancements in the financial sector which significantly reduce the costs of portfolio management and increase the capability to make important investment decisions in the absence of emotional biases [28].

Robo-advisory companies have begun to emerge in the Malaysian market following the inauguration of Digital Investment Management framework by Securities Commission Malaysia (SC) in May 2017 [27]. According to the Electronic Licensing Application

M. Al-Emran et al. (Eds.): ICETIS 2021, LNNS 299, pp. 658–672, 2022.
https://doi.org/10.1007/978-3-030-82616-1_54

(ELA) by Securities Commission [14], there are five licensed robo-advisory platforms available in Malaysia, where the first to obtain the license was StashAway in 2018. In 2020, Akru Now was introduced as the first robo-advisory platform developed in Malaysia [31]. Despite the increasing attention given to such disruptive innovation, the adoption of robo-advisors appears to be sluggish. Nonetheless, the slow adoption of robo-advisors is apparently a worldwide issue [7, 10, 20]. According to Laukkanen and Pasanen [24], it is common for disruptive innovations to have small-scale willing early adopters of modern-day systems to substitute conventional practices in the financial services sector.

Although several studies and conceptual frameworks have been established in developed countries to examine the key determinants that affect robo-advisors adoption among consumers [7, 20], this research discipline is severely scarce in Malaysia. Given the fact that Malaysia is a developing country, the determinants of Malaysian consumers' intention to adopt robo-advisors may be different from the findings of researches based in developed countries such as United States and United Kingdom. Belanche et al. [7] suggested future studies in this area to incorporate other cultures such as Asian, whereas Cheng et al. [10] and Hohenberger et al. [20] acknowledged the need for more studies that explore different segments of population with different cultural backgrounds to be conducted to obtain a more thorough understanding of the robo-advisory services. The future development and long-term survival of robo-advisors are directly determined by consumers' perceptions that shaped their adoption intention, which ultimately leads to actual adoption; thus there is a necessity of conducting this study to close the research gap in the Malaysian robo-advisory industry which conforms to the local cultures. An in-depth understanding of whether and how the factors influence decision-making of retail investors to use robo-advisors can offer a clear direction for the relevant parties involved, specifically the robo-advisory vendors and regulatory authorities, to strive for a greater adoption rate of robo-advisors in Malaysia. Henceforth, the purpose of this study is to develop and design a comprehensive model that determines the key drivers of intention to adopt robo-advisors in order to describe the overall adoption process of robo-advisors from the Malaysian retail investors' perspective. Identification of the key drivers for the intention to adopt robo-advisors among the retail investors that prompt their current usage behaviour is essential in determining the future intention to adopt robo-advisors by potential customers as it described the exhaustive adoption process. To achieve this, the authors utilize a further extended version of Technology Acceptance Model (TAM) for this study as an extension to Belanche et al.'s [7] research.

The remainder of this paper constitutes four main sections including literature review where the conceptual framework and research hypotheses are proposed in this section. The subsequent section presents the methodological approach of this study followed by presentation of the research findings. Finally, the paper is concluded with discussion on major findings, implications, limitations and lines for future research.

2 Literature Review

2.1 Theoretical Foundation

Technology Acceptance Model (TAM). TAM is an information technology framework primarily designed by Davis et al. [12] to predict and understand users' adoption, acceptance and application of technology-enabled inventions. This theoretical framework has been commonly employed by researchers including those in the FinTech field. The model postulates that people's actual usage behavior of a technology is positively predicated by their behavioral intention to use; these are jointly determined by the positive effects of a person's salient belief about the technology's usefulness (perceived usefulness) and the person's attitude toward the use of a particular technology [11]. Attitude and perceived usefulness are in turn determined by the positive effect of perceived ease of use. As robo-advisor is a significant form of technology innovation, this study uses TAM as the fundamental framework.

Extended Technology Acceptance Model (TAM). There are, however, critiques of TAM on a number of grounds including the limited constructs of TAM that results in the failure to generate more meaningful information about the users' perceptions or opinions on a particular technology innovation. Henceforth, previous literatures have extended the original TAM model to include other variables [7, 32]. Similarly, this study incorporates additional variables including those related to individual perception (perceived risk, perceived trust, and perceived behavioral control), determinants of subjective norms (interpersonal influence and external influence), and subjective norms which could be vital for examining the intention and actual adoption of robo-advisors by retail investors in Malaysia.

2.2 Hypotheses Development and Research Model

Attitude Toward Robo-advisors. Attitude is an individual's favorable or unfavorable perception toward performing the behavior [3]. Based on the justifications in the theoretical frameworks of TAM, attitude is an essential antecedent of behavioral intention, which in turn predicts the usage behavior or adoption of a given object [11]. In this research, attitude toward behavior refers to the tendency of retail investors in Malaysia to react in a favorable or unfavorable manner toward the robo-advisors adoption. In the study by Belanche et al. [7], it is evidenced that attitude has positive effect on the behavioral intention to adopt robo-advisors. Thus, it is hypothesized that retail investors who have favorable attitudes toward adopting robo-advisors are more likely to develop intention to adopt robo-advisors.

Factors Affecting Attitude Toward Robo-advisors

Perceived Ease of Use. According to TAM developed by Davis [11], perceived ease of use refers to the extent by which the users of technology believes the particular system could be easily used in order to prevent any problems that they may face in the future; it is a key factor positively influencing perceived usefulness and attitude.

In this research, perceived ease of use is the opinion of retail investors that they could employ robo-advisor in their investment activities without taking much physical and mental efforts. The findings by Kulkarni and Patil [23] have highlighted the positive influence of perceived ease of use on both the perceived usefulness and attitude toward FinTech services. This is further supported by Belanche et al.'s [7] study on robo-advisors. Henceforth, this study proposes that robo-advisors which are viewed as simple to use and user friendly could increase their perceived usefulness and develop positive attitudes toward robo-advisors by retail investors.

Perceived Usefulness. Davis [11] stated that perceived usefulness is the extent to which users of technology expect the particular system to improve their performance. In this study context, perceived usefulness refers to the perception of retail investors that robo-advisors would be useful in improving their portfolio performance and effectiveness. In the robo-advisory sector, there are still insufficient evidence on the impact of perceived usefulness on both the user's attitude and behavioral intention. The positive effect of perceived usefulness on attitude has been evidenced by Belanche et al.'s [7] study. In the same study, however, the researchers found the effect of perceived usefulness on behavioral intention to be not significant. This study aims to re-examine the relationships and hypothesizes that retail investors who believe using robo-advisors will impact them positively tend to either indirectly develop intention to adopt robo-advisors through favorable attitude, or directly develop intention to adopt robo-advisors.

Perceived Risk. Tan et al. [30] defined perceived risk as the anticipation of losses that hinders consumers from acquiring a product or service. Kulkarni and Patil [23] further explained that perceived risk is the extent to which the users believe that the adoption of new technology innovation could involve various types of risks to them due to unknown situations. In this study, given that robo-advisor has limited popularity in Malaysia, retail investors may perceive that the adoption of robo-advisor will expose them to certain risks such as financial risk, identity theft, and risk of fraud. Previous literatures in other context have proven that perceived risk negatively influences the attitude and behavioral intention of users such as the study by Jantarakolica and Jantarakolica [21] which concluded that investors who are concerned about the risk and security of online trading systems tend to have negative attitude toward online trading, thus do not intend on adopting it. Hence, it is posited that perceived risk has a negative effect on both the attitude toward robo-advisors and the behavioral intention to adopt it.

Perceived Trust. Chandio et al. [9] described perceived trust as a "confident belief on the service provider". Perceived trust is typically based on one's previous experience and will affect his or her future expectations or willingness to engage in the activities. In this study, perceived trust is a vital element in determining the adoption of robo-advisors through attitude since the degree of uncertainty is high given that the monetary transactions are handled in a virtual context where users do not usually have direct control over their activities. As previous literatures [9, 21] have proven perceived trust to be a significant determinant in studies associated with technology acceptance, this study proposes that the higher level of trust perceived by retail investors with regards to the robo-advisors predispose them to developing positive attitude toward such innovation.

Perceived Behavioral Control. Perceived behavioral control is defined as the level of control or ability perceived by an individual to successfully execute a particular behavior [1]. In this research, it refers to the absence or existence of required skills and resources of the retail investors to employ robo-advisor in their investing activities. Ajzen [2] suggests in his theoretical framework of Theory of Planned Behavior (TPB) that there is an indirect effect of perceived behavioral control on the usage behavior through behavioral intention. Sun, Law, and Schuckert [29] demonstrate that perceived behavioral control positively impacts the repurchase intention of consumers using electronic payment. Moreover, Jantarakolica and Jantarakolica [21] has incorporated perceived behavioral control in their extended TAM and concluded that it is a significant determinant of the intention to use algorithm trading. Thus, perceived behavioral control is expected to develop the intention to adopt robo-advisors.

Factor Affecting Perceived Behavioral Control. Perceived trust is also a common antecedent of perceived behavioral control as examined by previous literatures. Mutual trust between the users and robo-advisory firms would increase the users' self-efficacies, which in turn increases their perceived behavioral control [35]. Since robo-advisors offer investment management service, the retail investors' trust on such innovation can have direct effect on their perception of whether they could efficiently control or use the robo-advisors [25]. Hence, perceived trust on robo-advisors is expected to increase perceived behavioral control of Malaysian retail investors.

Subjective Norms. Ajzen [1] explained in his TPB model that subjective norms signify a person's perception about others' preferences or opinions on him or her, such as whether or not others would approve his or her execution of a particular behavior, thereby posing some form of social pressure on the person as to whether the given behavior should be performed or not. In this research context, subjective norms describe how the expectations derived from the social environment surrounding the retail investors' impact their behavior [1]. If a person that is viewed as an exemplar by the investors think that they should use robo-advisors, they may be positively inclined to try out robo-advisors to comply with the expectation of others. Bhattacherjee [8] in his study pertaining to e-commerce adoption has pointed out that the importance of subjective norms must not be underrated, while existing literatures in the related field have proven that subjective norms positively affect behavioral intention [7, 21, 22]. Henceforth, it is expected that subjective norms will positively contribute to the behavioral intention to adopt robo-advisors.

Factors Affecting Subjective Norms. Both interpersonal and external influences are important for determining subjective norms. Bhattacherjee [8] has included interpersonal influence (e.g. word-of-mouth influences by family members, peers or friends, colleagues, or superiors) in his earlier studies, and later included external influence into the model (e.g. opinions of experts and reports of mass media). Belanche et al. [7] posits that interpersonal and external influences jointly determined the significance of subjective norms as a key determinant for robo-advisors adoption, since consumers generally require validation and social enticements when it comes to adopting robo-advisors. Thus, it is hypothesized that interpersonal and external influences are vital in explaining the subjective norms of robo-advisors adoption among Malaysian retail investors.

Behavioral Intention to Adopt Robo-advisors. Behavioral intention is defined as the extent by which an individual has planned to conduct or not to conduct some particular behavior in the future [33]. According to Ajzen [1], behavioral intention is the effort that individuals have prepared for their decision on the particular behavior. The higher the level of behavioral intention, the higher effort they prepared on their decision, and hence the higher probability that they will perform that behavior. Hence, it is proposed that the willingness of retail investors to adopt robo-advisors will positively influence usage behavior as supported by past studies; Weedige et al. [34] have justified that the tendency to adopt mobile banking is a significant factor of a person's actual adoption of mobile banking, whereas findings by Jantarakolica and Jantarakolica [21] show that the usage behavior of online and algorithm trading is positively predicated by the investors' behavioral intention.

Besides that, Ajzen [1] revealed in his TPB model about the importance of behavioral intention as a mediating variable that efficiently explains the effects of its antecedents on its consequent variable. As supported by Al Mamun et al. [4], behavioral intention is prompted by various perceptive determinants prior to the actual usage behavior. In this context, the behavioral intention to adopt robo-advisors mediates the effects of perceived usefulness, attitude toward robo-advisors, perceived risk, perceived behavioral control, and subjective norms on the usage behavior. This manifests the central role of behavioral intention in determining the adoption of robo-advisors in Malaysia.

Based on the premises of theory and extensive review of literatures, this study proposes hypotheses as follows:

H1a. Perceived ease of use has a positive effect on the perceived usefulness.
H1b. Perceived ease of use has a positive effect on the attitude toward robo-advisors among Malaysian retail investors.
H2a. Perceived usefulness has a positive effect on the attitude toward robo-advisors among Malaysian retail investors.
H2b. Perceived usefulness has a positive effect on the behavioral intention to adopt robo-advisors among Malaysian retail investors.
H3. Attitude toward robo-advisors has a positive effect on the behavioral intention to adopt robo-advisors among Malaysian retail investors.
H4a. Perceived risk has a negative effect on the attitude toward robo-advisors.
H4b. Perceived risk has a negative effect on the behavioral intention to adopt robo-advisors among Malaysian retail investors.
H5a. Perceived trust has a positive effect on the attitude toward robo advisors.
H5b. Perceived trust has a positive effect on the perceived behavioral control.
H6. Perceived behavioral control has a positive effect on the behavioral intention to adopt robo-advisors among Malaysian retail investors.
H7. Interpersonal influence has a positive effect on the subjective norms.
H8. External influence has a positive effect on the subjective norms.
H9. Subjective norms has a positive effect on the behavioral intention to adopt robo-advisors among Malaysian retail investors.
H10. Behavioral intention to adopt robo-advisors has a positive effect on the robo-advisors usage behavior of Malaysian retail investors.

The proposed conceptual framework is summarized in Fig. 1.

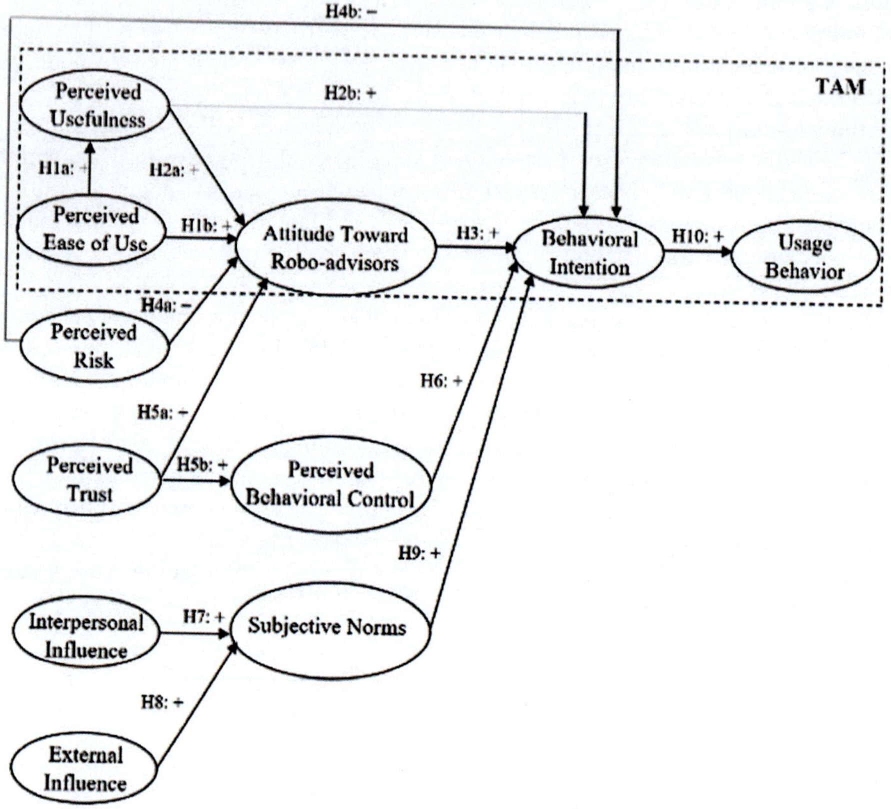

Fig. 1. Research model.

3 Methodology

The sample of this study comprised Malaysia's retail investors who were Internet users and have used robo-advisors, aged 18 to 77 or born between the years of 1943 to 2002, and who voluntarily invested or had the experience in trading financial instruments including, but not limited to stocks, bonds, exchange-traded funds (ETFs), and mutual funds. According to the PLS-SEM power tables adopted from Hair et al. [18], the recommended sample size for this study would be 205 in order to achieve a commonly used statistical power of 80% at a 1% level of significance. However, a total of 300 survey responses were collected to facilitate a robust PLS-SEM analysis. This study employed purposive sampling technique since the required data was collected deliberately from retail investors who have used robo-advisors so as to achieve the research objective [15].

In order to collect data, a structured online questionnaire had been developed based on extensive review of relevant research studies that employed TAM or extended TAM models. Online questionnaire was used as it is best suited for studies with large sample size, wide geographic coverage, and when interactions between interviewers and respondents are non-essential. In addition, the online survey is in tandem with the core of this study, robo-advisors, which is generally associated with technology adoption.

The questionnaire was segregated into two sections: Section A gathered data on demographics of respondents (age, gender, investment experience, employment status and income level) while Section B collected data on the 12 key constructs in the proposed model from the respondents' viewpoint. The 12 main constructs were measured through a total of 41 questionnaire items using a seven-point Likert scale ranging from 1 (strongly disagree) to 7 (strongly agree) so that all latent variables in the proposed model could be measured with optimized reliability in the study while reflecting the true judgement of respondents. The 41 questionnaire items were prepared based on an extensive review of relevant research studies that employs TAM or extended TAM models which revolve around technology adoption including those about robo-advisors, e-commerce services, algorithmic trading, as well as other FinTech innovations [7, 8, 10, 17, 21]. These questionnaire items adapted from previous literatures are modified to accommodate the context of this study which is robo-advisory services. After the collection of data, Smart-PLS 3.0 was used for the PLS-SEM analysis which includes the assessment of measurement model and the assessment of structural model. Instead of employing the CB-SEM method, PLS-SEM befitted the essence of this study as it will yield a higher statical power which is useful for exploring less developed or still developing theory [19].

4 Results

4.1 Assessment of Measurement Model

The assessment of measurement model involves convergent validity and discriminant validity. To assess convergent validity, item loadings, Cronbach's Alpha (α), Composite Reliability (CR), and Average Variance Extracted (AVE) were calculated using Smart-PLS 3.0. As suggested by Hair et al. [19], item loadings of 0.7 and above are recommended with the purpose of validating that the latent variables explain more than 50 per cent of the items' variances. On the basis of this recommendation, two items, PR1 and PR2, which had item loadings of 0.401 and 0.559 respectively, were removed from this study so that all the item loadings were higher than 0.7. To assess the construct reliability of the model, Cronbach's Alpha (α) and Composite Reliability (CR) were also measured. Table 1 illustrates that the α and CR values were all above the minimum value of 0.7. This indicated the high reliability of the measures as high levels of internal consistency were shown by all the constructs. To further ensure convergent validity, all the AVE values of the constructs were verified to be greater than 0.5, indicating that convergent validity was achieved by the model.

To assess discriminant validity, Fornell–Larcker criterion was used, where the comparison between the square root of Average Variance Extracted (AVE) and all inter-construct correlations was conducted. As stated by Fornell and Larcker [16], the square root of AVE has to be higher than all the inter-construct correlations in order to ensure that the respective item does not measure unrelated constructs. Table 2 shows that the discriminant validity was established since the square root of AVE for each construct (bolded) appeared to be higher than the correlations of any other latent variables. Hence, with these few assessments, it was concluded that the model had sufficient convergent and discriminant validity.

4.2 Assessment of Structural Model

Before assessing the structural model, assessment of collinearity among the constructs was conducted. Ringle, Wende, and Becker [26] suggested that the level of acceptable VIF should not be more than 5. As exhibited in Table 3, the overall inner VIF values were acceptable except for the influence of attitude toward robo-advisors on the intention to adopt robo-advisors, which had a slightly higher VIF of 6.099. To further guarantee that these variables were measuring different concepts and could be employed in the study, the discriminant validity assessment depicted in Table 2 could be used as a reference. Therefore, it was concluded that collinearity was not a concern in this study and the model could be considered free of common method bias.

Table 1. Convergent validity: Construct reliability and validity.

Construct	Item	Loading	Cronbach's Alpha (α)	Composite Reliability (CR)	AVE
Perceived Ease of Use (PEU)	PEU1	0.895	0.934	0.953	0.836
	PEU2	0.930			
	PEU3	0.951			
	PEU4	0.878			
Perceived Usefulness (PU)	PU1	0.943	0.963	0.973	0.899
	PU2	0.952			
	PU3	0.951			
	PU4	0.948			
Perceived Risk (PR)	PR3	0.955	0.887	0.946	0.898
	PR4	0.940			
Perceived Trust (PT)	PT1	0.953	0.957	0.972	0.921
	PT2	0.963			
	PT3	0.963			
Attitude Toward Robo-Advisors (ATR)	ATR1	0.958	0.965	0.974	0.905
	ATR2	0.966			
	ATR3	0.949			
	ATR4	0.932			

(continued)

Table 1. (*continued*)

Construct	Item	Loading	Cronbach's Alpha (α)	Composite Reliability (CR)	AVE
Perceived Behavioral Control (PBC)	PBC1	0.889	0.899	0.93	0.771
	PBC2	0.932			
	PBC3	0.752			
	PBC4	0.927			
Interpersonal Influence (INTP)	INTP1	0.918	0.936	0.959	0.887
	INTP2	0.949			
	INTP3	0.958			
External Influence (EXT)	EXT1	0.917	0.907	0.942	0.843
	EXT2	0.903			
	EXT3	0.934			
Subjective Norms (SN)	SN1	0.891	0.919	0.949	0.861
	SN2	0.954			
	SN3	0.938			
Intention to Adopt Robo-advisor (INT)	INT1	0.960	0.949	0.967	0.908
	INT2	0.940			
	INT3	0.958			
Usage Behavior (UB)	UB1	0.950	0.940	0.961	0.893
	UB2	0.961			
	UB3	0.923			

Note: Items with indicator loadings below 0.7 are removed: PR1, PR2.

Table 2. Discriminant validity: Fornell–Larcker criterion.

Construct	ATR	EXT	INT	INTP	PBC	PEU	PR	PT	PU	SN	UB
ATR	**0.951**										
EXT	0.754	**0.918**									
INT	0.894	0.753	**0.953**								
INTP	0.870	0.769	0.818	**0.942**							
PBC	0.897	0.715	0.821	0.786	**0.878**						
PEU	0.833	0.648	0.797	0.780	0.820	**0.914**					
PR	-0.232	-0.04	-0.222	-0.119	-0.223	-0.133	**0.947**				
PT	0.856	0.638	0.813	0.758	0.793	0.739	-0.182	**0.960**			
PU	0.867	0.684	0.849	0.819	0.807	0.849	-0.213	0.812	**0.948**		
SN	0.742	0.752	0.771	0.757	0.662	0.624	-0.050	0.742	0.719	**0.928**	
UB	0.855	0.752	0.879	0.785	0.788	0.702	-0.130	0.760	0.803	0.759	**0.945**

To evaluate the in-sample fit of the model, R^2 as well as adjusted R^2 values of the endogenous latent variables in the path model were considered as shown in Table 4.

As the adjusted R^2 values for attitude toward robo-advisors and robo-advisors adoption intention were greater than 0.75, the values were deemed substantial. The remaining endogenous constructs were considered as moderate given that they had adjusted R^2 values in between 0.5 and 0.75.

In order to test the research hypotheses, the researchers performed the standard procedure of bootstrapping of 5,000 resamples and examined the significance of the path coefficients. Indicators including standard beta, standard error, t-values and p-values had been computed to assess the significance of hypothesized relationships between the constructs. Table 5 exhibits the output of the hypothesis testing utilizing the PLS-SEM algorithm with two-tail test. The results revealed that all of the structural model relationships were significant. It was found that attitude toward robo-advisors had the most powerful impact on behavioral intention to adopt robo-advisors (t-value = 5.408, $p < 0.01$), vindicating that intention to adopt robo-advisors was heavily affected by attitude. The result also showed that perceived risk had a significant negative effect on attitude toward behavior with t-value of 2.579 ($p < 0.01$) and direct negative influence on behavioral intention to adopt robo-advisors (t-value = 2.045, $p < 0.1$). Lastly, the most significant predictor, behavioral intention to adopt robo-advisors construct was examined. It was found that it had a significant strong positive effect on usage behavior of robo-advisors (t-value = 63.150, $p < 0.01$).

Table 3. Collinearity: Variance Inflation Factor (VIF).

Construct	ATR	EXT	INT	INTP	PBC	PEU	PR	PT	PU	SN
Perceived Ease of Use (PEU)	3.698								1	
Perceived Usefulness (PU)	5.006		4.300							
Perceived Risk (PR)	1.057		1.101							
Perceived Trust (PT)	3.022				1					
Attitude Towards Robo-Advisors (ATR)			6.099							
Perceived Behavioral Control (PBC)			4.518							
Interpersonal Influence (INTP)										2.450
External Influence (EXT)										2.450
Subjective Norm (SN)			2.320							
Intention to Adopt Robo-advisor (INT)										
Usage Behaviour (UB)										

Table 4. Results of predictive accuracy (R^2 and Adjusted R^2).

Endogenous Latent Variables	R2	Adjusted R2
Attitude Toward Robo-advisors	0.822	0.820
Perceived Usefulness	0.720	0.719
Perceived Behavioural Control	0.629	0.628
Subjective Norm	0.644	0.642
Usage Behaviour	0.772	0.771
Intention to Adopt Robo-advisors	0.832	0.829

Note: Decision criteria to assess predictive accuracy (R2) according to Hair et al. [19]; R2 values: 0.25 (weak), 0.50 (moderate), 0.75 (substantial)

Table 5. Results of hypothesis testing.

Hypotheses	Casual Path (CP)	Std. Beta	Std. Error	*t*-Values	*p*-Values	Decision
H1a	PEU → PU	0.848	0.023	36.439	0.000*	Supported
H1b	PEU → ATR	0.235	0.054	4.402	0.000*	Supported
H2a	PU → ATR	0.290	0.074	3.901	0.000*	Supported
H2b	PU → INT	0.258	0.069	3.683	0.000*	Supported
H3	ATR → INT	0.337	0.063	5.408	0.000*	Supported
H4a	PR → ATR	-0.063	0.025	2.579	0.009*	Supported
H4b	PR → INT	-0.043	0.021	2.045	0.041**	Supported
H5a	PT → ATR	0.434	0.073	5.918	0.000*	Supported
H5b	PT → PBC	0.793	0.032	24.549	0.000*	Supported
H6	PBC → INT	0.140	0.051	2.771	0.006*	Supported
H7	INTP → SN	0.434	0.061	7.154	0.000*	Supported
H8	EXT → SN	0.420	0.062	6.719	0.000*	Supported
H9	SN → INT	0.252	0.045	5.526	0.000*	Supported
H10	INT → UB	0.878	0.014	63.150	0.000*	Supported

Note: $*p < 0.01$; $**p < 0.1$

5 Conclusions, Implications and Limitations

This research is an initial attempt to explore retail investors' intention to adopt robo-advisors and actual usage behavior of robo-advisors in Malaysia using an extended version of TAM. It was found that the influence of perceived ease of use on perceived usefulness and attitude toward robo-advisors are significantly positive, whereas the influence of perceived usefulness on attitude toward robo-advisors is also significant. An effective robo-advisory platform that features user-friendly interface and clear instructions is crucial to the development of favorable attitude toward such innovation by retail investors. Besides, as expected, perceived risk has a negative influence on the users' attitudes [21]. It is, however, imperative to note that perceived trust has the strongest effect on attitude toward robo-advisors. This could be due to the fact that investors may feel insecure if they make monetary transactions virtually using robo-advisors, especially if the amount is significant, thereby causing perceived trust to be highly regarded by investors before growing a favorable attitude toward robo-advisors. Perceived trust is also a strong positive antecedent of perceived behavioral control, as supported by existing researches such as Pavlou [25]. When users trust that robo-advisors will perform based on their expectations and meet their objectives, they will perceive that they can use and gain control of the robo-advisors efficiently. Moreover, subjective norm is positively influenced by both interpersonal and external influences [7, 8]. Nevertheless, the effect of interpersonal influence is slightly stronger than that of external influence. This implies that retail investors place more credence on opinions of family, peers, and superiors than on mass media coverage. Both Alfred [5] and Bhattacherjee [8] supported that word-of-mouth communication has stronger convincing capability than print media communication. Thus, robo-advisory vendors should consider tapping into favorable experiences of their existing clients in order to entice new customers instead of relying solely on mass media.

Subsequently, behavioral intention to adopt robo-advisors is influenced by five constructs in this study with attitude toward robo-advisors being the strongest predictor followed by perceived usefulness, subjective norms, perceived behavioral control, and lastly perceived risk. Undeniably, a positive attitude toward robo-advisors is a prerequisite of adoption intention especially when robo-advisors are still at the rudimentary phase of dissemination, which is in the case of Malaysia. Majority of the literatures attest attitude as an essential factor that can induce intention to adopt technological innovation [6]. This could be due to the essence of attitude which encompasses cognition, affect and behavior that ultimately leads to the positive reactions in retail investors' intention to adopt robo-advisors. As a strategy to boost the number of robo-advisors' adopters, the vendors and SC have to assuage the doubts of the consumers to better enhance their attitudes toward robo-advisors [7]. In addition, it is surprising to find that the impact of perceived usefulness slightly overshadows subjective norms given that the importance of subjective norms as the second most impactful antecedent of behavioral intention has been widely acknowledged [1, 10]. On top of that, Belanche et al. [7] found the direct effect of perceived usefulness on behavioral intention to be insignificant, while subjective norm was the second strongest predictor after attitude. This could be explained by the fact that the study by Belanche et al. [7] comprised participants that come from various backgrounds who may not have investment experience, whereas the sample in this study is delimited to solely retail investors. The traits of retail investors include logical, rational, and analytical thinking compared to the general public with no investment experience. Thus, the findings of this study indicated that Malaysian retail investors put more effort in rationally assessing the benefits offered by robo-advisors instead of soliciting views and advices from people around them or resorting to mass media information. Hence, vendors can either get feedback from or conduct surveys on existing clients to find out how the platforms can be enhanced or what new features clients would find beneficial. In turn, the minor influences of perceived behavioral control and perceived risk on behavioral intention suggest that they are inputs to, but not key drivers to robo-advisors adoption.

This study is not without limitations. As the sample of this study is merely limited to Malaysian retail investors, the findings may not readily apply to different countries [7]. Thus, future researches could cover different countries or cultures to comprehend the adoption intention of robo-advisors with a global perspective. Additionally, this study does not examine the effects of moderating variables. Individual characteristics such as familiarity with robots and self-enhancement motives were proven to be relevant moderators that alter the strengths of relationships among the variables [7, 20]. Hence, future researchers can incorporate certain individual characteristics as moderating variables in their respective studies to obtain more meaningful information.

References

1. Ajzen, I.: The theory of planned behavior. Organ. Behav. Hum. Decis. Process. **50**(2), 179–211 (1991)
2. Ajzen, I.: Perceived behavioral control, self-efficacy, locus of control, and the theory of planned behavior. J. Appl. Soc. Psychol. **32**(4), 665–683 (2002)
3. Ajzen, I., Fishbein, M.: Understanding Attitudes and Predicting Social Behavior. Prentice-Hall, Englewood Cliffs (1980)
4. Al Mamun, A., Mohamad, M., Yaacob, M., Mohiuddin, M.: Intention and behavior towards green consumption among low-income households. J. Environ. Manage. **227**, 73–86 (2018)
5. Alfred, O.: The significant effect of word of mouth (WOM) on the attitude of potential consumers. Int. J. Econ. Behav. Organ. **4**(6), 59–63 (2016)
6. Altawallbeh, M., Soon, F., Thiam, W., Alshourah, S.: Mediating role of attitude, subjective norm and perceived behavioural control in the relationships between their respective salient beliefs and behavioural intention to adopt e-learning among instructors in Jordanian universities. J. Educ. Pract. **6**, 152–159 (2015)
7. Belanche, D., Casaló, L., Flavián, C.: Artificial intelligence in FinTech: understanding robo-advisors adoption among customers. Ind. Manag. Data Syst. **119**(7), 1411–1430 (2019)
8. Bhattacherjee, A.: Acceptance of e-commerce services: the case of electronic brokerages. IEEE Trans. Syst. Man Cybern. Part A: Syst. Hum. **30**(4), 411–420 (2000)
9. Chandio, F., Burfat, F., Abro, A., Naqvi, H.: Citizens' acceptance and usage of Electronic-Government services: a conceptual model of trust and technological factors. Sindh Univ. Res. J.-SURJ (Sci. Ser.) **49**(3), 665–668 (2017)
10. Cheng, X., Guo, F., Chen, J., Li, K., Zhang, Y., Gao, P.: Exploring the trust influencing mechanism of robo-advisor service: a mixed method approach. Sustainability **11**(18), 4917 (2019)
11. Davis, F.: Perceived usefulness, perceived ease of use, and user acceptance of information technology. MIS Q. **13**(3), 319–340 (1989)
12. Davis, F., Bagozzi, R., Warshaw, P.: User acceptance of computer technology: a comparison of two theoretical models. Manage. Sci. **35**, 982–1002 (1989)

13. Day, M., Cheng, T., Li, J.: AI robo-advisor with big data analytics for financial services. In: IEEE/ACM International Conference on Advances in Social Networks Analysis and Mining (ASONAM), pp. 1027–1031. Institute of Electrical and Electronics Engineers (IEEE), Barcelona (2018)
14. Electronic Licensing Application. http://ers.seccom.com.my/public/Default.aspx?menu=2. Accessed 7 Mar 2021
15. Etikan, I.: Comparison of convenience sampling and purposive sampling. Am. J. Theor. Appl. Stat. **5**(1), 1 (2016)
16. Fornell, C., Larcker, D.: Evaluating structural equation models with unobservable variables and measurement error. J. Mark. Res. **18**(1), 39–50 (1981)
17. Gerlach, J.M., Lutz, J.K.: Evidence on usage behavior and future adoption intention of fintechs and digital finance solutions. Int. J. Bus. Finance Res. **13**(2), 83–105 (2019)
18. Hair, J., Hult, G., Ringle, C., Sarstedt, M.: A Primer on Partial Least Squares Structural Equation Modeling. Sage, Thousand Oaks (2014)
19. Hair, J., Risher, J., Sarstedt, M., Ringle, C.: When to use and how to report the results of PLS-SEM. Eur. Bus. Rev. **31**(1), 2–24 (2019)
20. Hohenberger, C., Lee, C., Coughlin, J.: Acceptance of robo-advisors: Effects of financial experience, affective reactions, and self-enhancement motives. Financial Plan. Rev. **2**(2), e1047 (2019)
21. Jantarakolica, K., Jantarakolica, T.: Acceptance of financial technology in Thailand: case study of algorithm trading. In: Banking and Finance Issues in Emerging Markets (International Symposia in Economic Theory and Econometrics), vol. 25, pp. 255–277 (2018)
22. Jerene, W., Sharma, D.: The adoption of financial technology in Ethiopia: a study of bank customers perspective. J. Bank. Financ. Technol. **4**, 53–63 (2020)
23. Kulkarni, M., Patil, K.: Artificial intelligence in financial services: customer chatbot advisor adoption. Int. J. Innov. Technol. Explor. Eng. **9**(1), 4296–4303 (2019)
24. Laukkanen, T., Pasanen, M.: Mobile banking innovators and early adopters: how they differ from other online users? J. Financ. Serv. Mark. **13**, 86–94 (2008)
25. Pavlou, P.: What drives electronic commerce? A theory of planned behavior perspective. In: Academy of Management Annual Meeting Proceedings, pp. 9–14 (2002)
26. Ringle, C.M., Wende, S., Becker, J.M.: SmartPLS 3. SmartPLS, Bönningstedt (2015)
27. Securities Commission Malaysia: SC Annual Report 2017. Securities Commission Malaysia, Bukit Kiara (2017)
28. Sironi, P.: Fintech Innovation: From Robo-Advisors to Goal Based Investing and Gamification. Wiley, West Sussex (2016)
29. Sun, S., Law, R., Schuckert, M.: Mediating effects of attitude, subjective norms and perceived behavioural control for mobile payment-based hotel reservations. Int. J. Hosp. Manag. **84**, 102331 (2020)
30. Tan, G.W.H., Ooi, K.B., Chong, S.C., Hew, T.S.: NFC mobile credit card: the next frontier of mobile payment? Telematics Inform. **31**(2), 292–307 (2013)
31. The Edge Malaysia. https://www.theedgemarkets.com/article/solutions-malaysias-first-homegrown-roboadvisory-platform. Accessed 7 Mar 2021
32. Venkatesh, V., Davis, F.: A theoretical extension of the technology acceptance model: four longitudinal field studies. Manage. Sci. **46**(2), 186–204 (2000)
33. Warshaw, P., Davis, F.: Disentangling behavioral intention and behavioral expectation. J. Exp. Soc. Psychol. **21**(3), 213–228 (1985)
34. Weedige, S.S., Ouyang, H., Gao, Y., Liu, Y.: Decision making in personal insurance: impact of insurance literacy. Sustainability **11**(23), 6795 (2019)
35. Wu, I., Chen, J.: An extension of trust and TAM model with TPB in the initial adoption of on-line tax: an empirical study. Int. J. Hum Comput Stud. **62**, 784–808 (2005)

JomFacial Recognition Attendance Systems

F. K. Tee[1], G. X. Loh[1], K. Sumaiya[1], Z. Y. Cha[1], P. S. JosephNg[1(✉)], K. Y. Phan[2], and J. T. Lim[2]

[1] Institution of Computer Science and Digital Innovation, UCSI University, UCSI Heights, 56000 Cheras, Kuala Lumpur, Malaysia
josephng@ucsiuniversity.edu.my

[2] Faculty of Information and Communication Technology, Universiti Tunku Abdul Rahman, Kampar, Perak, Malaysia
{kyphan,jtlim}@utar.edu.my

Abstract. Nowadays, most universities as well as offices need to keep the attendance of their students and employees. As technology is growing rapidly, machines are designed to reduce the daily workload of mankind. An automatic facial recognition system can make it easier for institutes to keep track of attendance. Introducing a facial recognition attendance system to a university will promote its branding, make them more efficient, and cost-effective. A mixed-method of data collection is used in this research. In this paper, we will be proposing a facial recognition attendance system to replace the traditional manual attendance system. Theoretical and management implication are discussed.

Keywords: Face recognition · Automatic attendance system · OpenCV · Authentication · PCA · Biometric · Web-Based · Machine learning · Real-Time monitoring

1 Introduction

As technology is growing fast day by day, machines had been built to replace the manual workload of mankind [1]. Under repetition and precise situations, such as manual attendance taking system, machines are always doing better than mankind [2, 3]. This is because machines would not be getting unmotivated or tired as compared to humans. Facial recognition technology is based on modern advances technology in Artificial Intelligent (AI) and neural network development [4, 5]. It uses humans' biometrics to outline the unique facial features, then verifies the identity through the key features of the face [6, 7]. Hence, the use of facial recognition technology for an automated attendance system is a very smart approach. Nowadays, most universities need an attendance system to keep a record of the attendance rate of the students. Some universities do it manually and some of them use attendance systems [8–11]. For example, private universities like University UTAR, Taylor's University, and University Technology Petronas are using the facial recognition system in Malaysia. Attendance systems can include RFID, iris scanning, GPS tracking, fingerprint scanning, face recognition [12–14]. Few universities still have obstacles to implementing the facial recognition attendance system, therefore

M. Al-Emran et al. (Eds.): ICETIS 2021, LNNS 299, pp. 673–681, 2022.
https://doi.org/10.1007/978-3-030-82616-1_55

they are still practicing the manual attendance system by using the attendance sheet given by the lecturer and pass it to the students to sign their signature on it [15]. Lecturers are required to print out the name list and bring the attendance sheet to the classroom. Lecturers have to manually key in the attendance record into the database management system after every class [16].

In this paper, we have proposed the automatic facial recognition attendance system to replace the traditional manual attendance system [18–21]. The traditional manual attendance system consumes a lot of time because the attendance sheet needs to be passed around and around until every student has signed [7]. Not just that, the traditional manual attendance system is easily causing chaos in the classroom, and the student's learning process might be interrupted [2]. Also, the traditional manual attendance system is a waste of resources for printing materials which is not environmentally friendly [22]. In the facial recognition attendance system, the flow process starts by capturing portrait from the input devices such as cameras which are placed in the classroom [15]. The system will start to verify and compare the captured portrait with the registered image in the database to determine the authentication of the person [23]. Therefore, we have proposed an enhancement on the system which builds a more perfect existing facial recognition attendance system.

2 Problem Statement and Hypothesis

The goal of this study is to develop a facial recognition attendance system that could reduce cost and eliminate fake attendance without attention needed from the students and lecturer. The hypothesis for every criterion extracted from the objectives is summarized in Fig. 1 below.

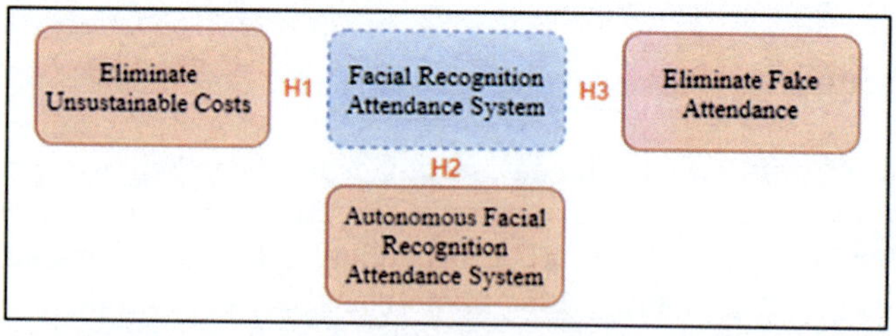

Fig. 1. Research model

H1: Eliminate unsustainable costs will bring more profit.
The cost of paper, printing, and pen is not only required frequently. As the capacity of student increase, these costs will also increase in proportion. By using the Facial Recognition Attendance System, these costs can be eliminated and capable of future student capacity growth.

H2: Autonomous Facial Recognition Attendance System will reduce the burden of lecturer and student.

Autonomous Facial Recognition Attendance System will record student attendance in front of the class which brings minimal to no distraction without any effort required from the lecturer. This system will reduce the burden of the lecturer to key in the attendance of the system. Students can pay full concentration in class without the concern of not signing the attendance sheet. This will significantly increase the lecturer efficiency on academic tasks and the efficiency of students to learn in class.

H3: Eliminate fake attendance will improve the branding of the university.
Fake attendance occurs frequently when using paper-based manual attendance and this may spread to the public which shows that the university didn't have a complete attendance taking system. With 0% of fake attendance, branding name will arise and this will give parents of students more confidence to send the children to the university. By using the Facial Recognition Attendance System, the student is impossible to create fake attendance as every attendance is done digitally. By creating a facial recognition system with all the features added, fake attendance can be avoided. This system will allow both lecturers and students to save time and money. So, the time can be used to teach and learn a lecture properly. As a result, Higher attendance rates and more efficiency can be achieved.

3 Methodology

Based on the research questions discussed earlier, we have created a few questions to be asked in a survey and interview. The survey method will consist of questions using a Likert scale. The interview questionnaire will go more on the depth of the client's point of view. Our target group is the education community, such as students, lecturers, and staff. Therefore, the answers can be more accurate and more effective. All surveys and interviews are going to be performed during the data collection process. The data will be collected using the following methodology as outlined in Table 1.

Table 1. Research methodology [24–36]

Research dimension	Explanatory sequential design
Research methodology	Quantitative & Qualitative perceptive
Research method	Survey & Interview

Based on Table 1, the research dimension will be illustrated in the sequential design which will clarify every step of the data collection. Qualitative reasoning is used to generalize the views of many respondents from the University. Using Survey as the tool allowed the research team to collect more data. And using interview, allowed the research team to go in-depth with their opinion and reasoning. The survey was completed by seventy-four respondents. The interview consisted of thirty-three respondents. All the data gathered from the interview was used to verify the reliability of the answers in the survey.

4 Results and Findings

According to Fig. 2, The percentage of the respondents who think the facial recognition system's cost is sustainable in the long run is 70% (31.1% + 39.2% = 70.3%). Few lecture classes have more than 250 students and few lecturers have more than one class. The lecturers have to print out more than one attendance sheet for every class they have and at least three attendance sheets for the classes with a vast number of students. They have to reprint the attendance sheet again after the add and drop period. If by any chance the lecturer forgets to bring the attendance paper, students have to find a piece of empty paper to write their names and id on and pass to the whole class. Hence, a lot of papers and ink is consumed. As a result, facial recognition is considered sustainable as there will be less to no wastage of paper for attendance and the information will be available 24/7. This system will save costs in the long run by not printing out materials for attendance. As the system will have certain costs related to maintenance and database storage. Whereas many of them believe that the existing university fees should be enough to implement a facial recognition attendance system.

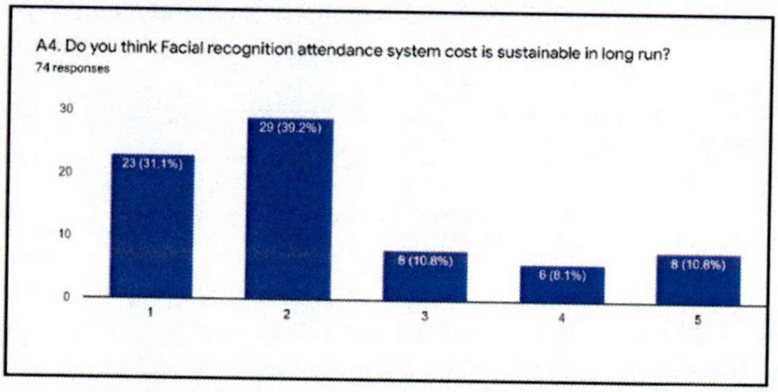

Fig. 2. Survey results related to cost

A higher percentage of the respondents (59.5% + 17.6% = 77.1%) agree or strongly agree that the traditional paper-based attendance system wastes a lot of time for the lecturer and the students. the traditional attendance system takes a lot of time for the lecturers as they have to print out the papers, at times pause their lectures so that the students can sign and once students have signed the attendance, the lecturers also have to update the student's attendance. The lecturers also have to reprint the attendance list once "add and drop" ends to get the updated list. For the students, they have to wait for the attendance list to come to them for them to sign. If the class has more than one hundred students, half of the class most probably have to stay after the class to sign for the attendance. The student also has to look for their name throughout the attendance list to sign. Overall a lot of time is consumed by both lecturers and students.

The higher percentage of respondents (52.7% + 17.6% = 70.3%) agree and strongly agree that students will be able to concentrate more in class rather than getting distracted by taking attendance. The facial recognition system will allow students to save time as the students don't have to concentrate on looking for the attendance paper to sign, look for their name on the list and they don't have to sign any paper or write down their name while the lecturer is teaching. Introducing a facial recognition system will force the students to come to the classes to keep up their attendance rate higher. So, they can't skip classes and will be forced to attend all the classes. The more the student attends the classes, the more possibility there is to get better results according to research [6].

As per the results of data collection shown in Fig. 3, a total of 71.6% of individuals who conducted the survey agree facial recognition attendance system will have better accuracy with less effort from the lecturer and students. The facial recognition attendance system can detect faces and compare the face with the database to approve the presence of the student in the class. The facial recognition system will work automatically and update the database. The students just have to show up for class and concentrate. Moreover, the efficiency of the lecturer will also increase due to a lesser workload for attendance matters. With the facial recognition attendance system, which can work autonomously, the lecturer will no longer need to bother with the attendance sheet and marking attendance on the lecturer portal and focus on academic tasks.

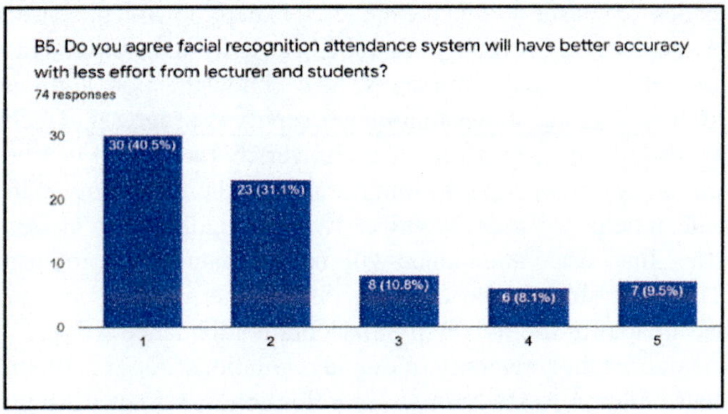

Fig. 3. Survey results related to the efficiency.

As can be seen in Fig. 4, Almost 68% (47.3% + 20.6% = 67.6%) of the respondents agree or strongly agrees that it's not nice to fallback as a university while there are primary schools out there with the feature of facial recognition attendance system. The respondents have agreed with that might be due to a few reasons. Firstly, the university students' education fee is much more expensive than the primary school, but the university's facilities are not good enough and worse than the primary school, therefore it would become an unbalance for them. Secondly, from primary school to university regardless of education, the knowledge provided, or facilities environment should be gradually enriched or diversify. However, the primary school had practiced this kind of technology advanced system, but not for the university, so, naturally, students feel bad for it. Thirdly, the impression for university students will be affected due to the facilities and efficiency compared to the rest of the university and even primary school which are applying the facial recognition attendance system. As a university who do not practice the facial recognition system will depress its student.

Due to the implementation of a facial recognition system which improves the branding for universities as one of the leading or trending universities. Facial recognition is a

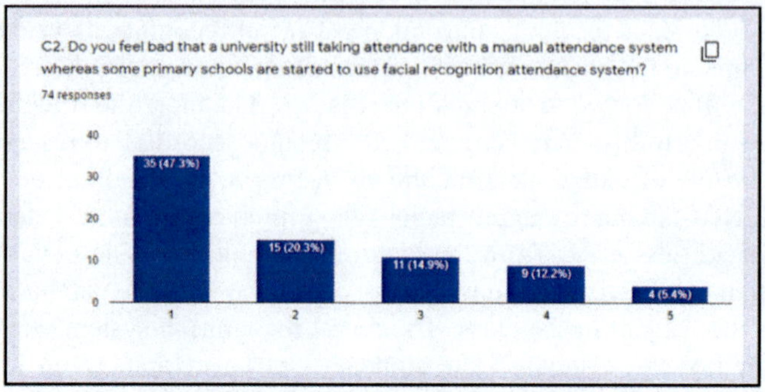

Fig. 4. Survey results related to branding

popular computer technology research field. If the universities introduce the system and use it to replace the traditional manual attendance system, will make the university more professional. Besides that, it shows that the university keeps up with the latest technology which boosts the image of the university in public in terms of the adoption of technology. Hence give a better reputation to the university. As we know the more automated systems are introduced to the university shows that the university is progressing technologically and heading towards industry 4.0. Moreover, a university that is making the digitalized attendance system will attract people to join because of the technology part of the university. Meanwhile, it helps to make the university be recognized as a modern university with great facilities. Inaccurate attendance will make the student more unsatisfied with the university. The more students become unsatisfied, the more the students will vent out creating a bad image for the university. Inaccurate attendance will create an image for the parents to not trust the university rules and regulations. As a result, they wouldn't want to allow their children to study in an unprofessional and not trustable university [37, 38]. Thus, practicing the facial recognition attendance system does provide the best environment for students and lecturers. The good image of the university will also allow graduate students to get work in popular companies. The introduction of a facial recognition attendance system will also show investors, students, and parents, that the university is taking a step further by digitalizing the university. The attendance page in Fig. 5, shows all the active class information such as class code, class title, student capacity, lecturer, and venue. Once the user clicks on the class title, it will redirect to the Take Attendance Page to perform take attendance. To view class attendance users just need to click the eye icon on the action column and it will redirect to View Class Enrolment Attendance.

Users can start taking attendance with the capture face button as shown in Fig. 6. The user able to set intervals which is the time between each image is taken. The number of taken images will increase every time a new image is taken. The number of taken images will affect the concentration rate of the student. For instance, the number of taken images is 5 but the student only appears on the image 3 times, so the concentration rate is 60%. Once the user decides to submit the attendance, the user just needs to click the submit button to finish the attendance taken procedure. Once clicked on the submit button, it will send an attendance email to the attended student and an absent email to the absent student and the total class conducted will increase.

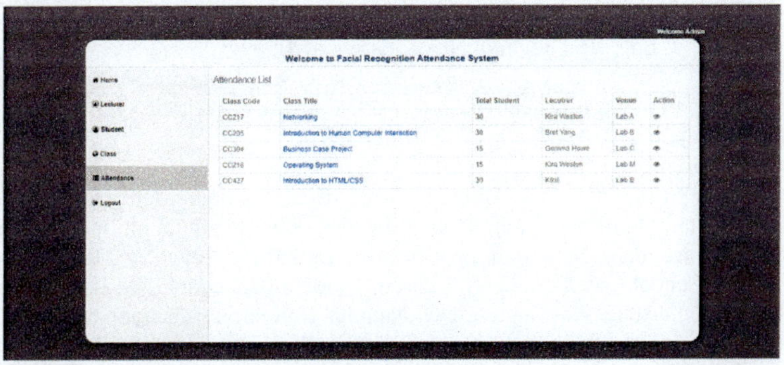

Fig. 5. Attendance page

Fig. 6. Take attendance page

5 Conclusion and Limitation

Facial recognition technology is considered mature technology as many devices are implementing facial recognition for identity ID purposes such as smartphone, laptop, immigration kiosk and other thus using it to identity purpose is completely reasonable and more accurate. In conclusion, the face recognition attendance system will make taking attendance more efficient, cost-effective, and boost the university image by taking attendance, verifying the face with the database, updating the system with the attendance, sending a notification email, and calculating the attendance percentage rate. More Universities can implement it to make everyone's life easier.

Acknowledgement. This work is a part of a project submitted to UCSI University.

References

1. Arsenovic, M., Sladojevic, S., Stefanovic, D.: FaceTime - deep learning-based face recognition attendance system. Int. Symp. Intell. Syst. Inform. **2017**, 53–58 (2017)

2. Bussa, S., Mani, A., Kaushik, S.: Smart attendance system using OPENCV based on facial recognition. Int. J. Eng. Res. Technol. **9**(3), 54–59 (2020)
3. Chakraborty, P., Muzammel, C.S., Khatun, M., Islam, S.F., Rahman, S.: Automatic student attendance system using face recognition. Int. J. Eng. Adv. Technol. (IJEAT) **9**(3), 93–99 (2020). https://doi.org/10.35940/ijeat.B4207.029320
4. Vica, V., et al.: JomImage SnapFudo: control your food in a snap. In: International Conference on Engineering Technologies and Applied Science, Kuala Lumpur, Malaysia, pp. 1–5 (2019)
5. Vivilyana, V., et al.: JomImage: weight control with mobile snapfudo. In: Intelligent Systems Conference, Amsterdam, Netherland, pp. 168–180 (2020)
6. NCES: Every School Day Counts: The Forum Guide to Collecting and Using Attendance Data (NFES 2009–804). Washington, DC: National Forum on Education Statistics (2020). https://nces.ed.gov/pubs2009/attendancedata/chapter1a.asp
7. Hammim, R.: SK Taman Perling 1 uses facial recognition scanners to mark pupils' attendance. (New Straits Times) (2020). https://www.thestar.com.my/news/nation/2020/01/02/jb-school-using-facial-scanner-for-attendance
8. Liow, K.M., et al.: JomDesignLab: bringing artwork design nearer. In: IEEE International Conference on Control Systems, Computing & Engineering, Penang, Malaysia, pp. 1–6 (2020)
9. Kang, C.M., et al.: A study on integrating penetration testing into the information security framework for Malaysian higher education institutions. In: International Symposium on Mathematical Science & Computing Research, Malaysia, pp. 156–161 (2015)
10. AdrianChin, Y.K., et al.: JomDataMining: learning behaviour affecting their academic performance, really? In: IEEE 6th International Conference on Engineering Technologies and Applied Science, Malaysia (2020)
11. AdrianChin, Y.K., et al.: JomDataMining: academic performance and learning behavior dubious relationship. In: International Journal of Business Information Systems, Forthcoming article (2020)
12. Hava, V., Kale, S., Bairagi, A., Prasad, C., Chatterjee, S., Varghese, A.: Free & generic facial attendance system using android. Int. Res. J. Eng. Technol. **06**(09), 1846–1850 (2019)
13. Helmi, R.A., Yusuf, S.S., Jamal, A., Abdullah, M.I.: Face recognition automatic class attendance system. In: IEEE International Conference on Automatic Control and Intelligent Systems, pp. 50–55 (2019). https://doi.org/10.1109/I2CACIS.2019.8825049
14. Vanitha, J.J.A., Adithya, S., Vaidehi, A.V.: Deep learning based automated attendance. In: International Conference on Recent Trends in Advanced Computing, pp. 307–313 (2019)
15. Mishra, A., Kumar, A., Pandey, A.: Student attendance system by face detection. Int. Res. J. Eng. Technol. (IRJET) **6**(4), 2206–2209 (2019)
16. Owandkar, M., Kolte, A., Peshave, D., Jadhav, M.S.: Attendance monitoring system using face recognition. Int. Res. J. Eng. Technol. **04**(05), 1163–1168 (2017)
17. Patel, S., Kumar, P., Garg, S., Kumar, R.: Face recognition based smart attendance system using IoT. Int. J. Comput. Sci. Eng. **6**(5), 871–877 (2018)
18. Pei, Z., Xu, H., Zhang, Y., Guo, M., Yang, Y.-H.: Face recognition via deep learning using data augmentation based on orthogonal experiments. Electronics **8**(10), 1088 (2019)
19. Nandhini, R., Duraimurugan, N., Chokkalingam, S.P.: Face recognition based attendance system. Int. J. Eng. Adv. Technol. **08**(03), 574–577 (2019)
20. Waweru, B.W., et al.: Gamesy: how videogames serve as a better replacement for school. In: Student Conference on Research and Development, Kuala Lumpur, Malaysia, pp. 10–15 (2020)
21. Yeo, J.L., et al.: Time scheduling and finance management: university student survival kit. In: Student Conference on Research and Development, Kuala Lumpur, Malaysia, pp. 1–6 (2020)
22. Salim, O.A., Olanrewaju, R.F., Balogun, W.A.: Class attendance management system using face recognition. In: 7th International Conference on Computer and Communication Engineering, pp. 93–98 (2018). https://doi.org/10.1109/ICCCE.2018.8539274

23. Tantak, A., Sudrik, A., Kale, A., Mehetre, R., Pophale, P.: Face recognition for E-Attendance for student and staff. IOSR J. Comput. Eng. **19**(02), 89–94 (2017)
24. JosephNg, P.S., Loh, Y.F., Eaw, H.C.: Grid computing for MSE during volatile economy. In: International Conference on Control, Automation and Systems, IEEE Explore, Busan, Korea, pp. 709–714 (2020)
25. JosephNg, P.S., et al.: EaaS: available yet hidden infrastructure inside MSE. In: 5th International Conference on Network, Communication, and Computing, ACM International Conference Proceeding Series, Kyoto, Japan, pp. 17–20 (2016)
26. JosephNg, P.S., Kang, C.M., Choo, P.Y., Wong, S.W., Phan, K.Y., Lim, E.H.: Beyond cloud infrastructure services in medium-size manufacturing. In: International Symposium on Mathematical Sciences Computing Research, IEEE Explore, Ipoh, Malaysia, pp. 150–155 (2015)
27. Joseph, N.P.S., Choo, P.Y., Wong, S.W., Phan, K.Y., Lim, E.H.: Malaysia SME ICT during economic turbulence. In: International Conference on Information & Computer Network, Singapore, pp. 67–71 (2012)
28. Ng Poh Soon, J., Yin, C.P., Wan, W.S., Nazmudeen, M.S.H.: Energizing ICT infrastructure for Malaysia SME during economic turbulence. In: Student Conference on Research and Development, IEEE Explore, Cyberjaya, Malaysia, pp. 328–322 (2011)
29. JosephNg, P.S., Eaw, H.C.: Making financial sense from EaaS for MSE during economic uncertainty. In: Advances in Intelligent Systems and Computing, Forthcoming (2021)
30. JosephNg, P.S., Eaw, H.C.: Still technology acceptance model? Reborn: exostructure as a service model. In: International Journal of Business Information Systems, Forthcoming (2021)
31. JosephNg, P.S.: EaaS infrastructure disruptor for MSE. Int. J. Bus. Inform. Syst. **30**(3), 373–385 (2019)
32. JosephNg, P.S.: EaaS optimization: available yet hidden information technology infrastructure inside medium-size enterprises. J. Technol. Forecast. Soc. Change **132**(July), 165–173 (2018)
33. JN, P.S., et al.: Exostructure services for infrastructure resources optimization. J. Telecommun. Electron. Comput. Eng. **8**(4), 65–69 (2016)
34. JosephNg, P.S., Kang, C.M.: Beyond barebone cloud infrastructure services: stumbling competitiveness during economic turbulence. J. Sci. Technol. **24**(1), 101–121 (2016)
35. Joseph, N.P.S., Mahmood, A.K., Choo, P.Y., Wong, S.W., Phan, K.Y., Lim, E.H.: Barebone cloud IaaS: revitalization disruptive technology. Int. J. Bus. Inform. Syst. **18**(1), 107–126 (2015)
36. Joseph, N.P.S., Mahmood, A.K., Choo, P.Y., Wong, S.W., Phan, K.Y., Lim, E.H.: IaaS cloud optimization during economic turbulence for malaysia small and medium enterprise. Int. J. Bus. Inform. Syst. **16**(2), 196–208 (2014)
37. Mainkar, V., Pednekar, P.A., Shinde, H.V.: Attendance monitoring system using image processing. Int. J. Innov. Res. Exp. **05**(02), 22–26 (2018)
38. Zion, G.D., Karthik, K., Reddy, D.M.: IoT based facial recognition biometric attendance. Int. Res. J. Eng. Technol. **06**(03), 5191–5198 (2019)

Crafting an Online Shopping Repurchase Decision Making Process: Does the Perceived Competitive Price Matters?

Linda Seduram(✉), Phillis Wong Mun Lum, and Lim Xim Yi

Faculty of Business and Management, UCSI University, Kuala Lumpur, Malaysia
linda@ucsiuniversity.edu.my

Abstract. Online shopping creates opportunities for businesses to reach consumers globally and directly. However, retaining consumers has been challenging for many online vendors as the e-environment is highly competitive. Thus, this study investigated the continuance intention and its determinants using a combination of the expectation-confirmation theory (ECT) with perceived competitive price, to explain online shopping continuance intention. Based on a sample of 370 students collected from the UCSI University Kuala Lumpur, this study found that the perceived competitive price influences expected service quality and the expectation-confirmation subsequently determines repurchase intention. This study demonstrates that investigating the perceived competitive price as key driver of online shopper's expectation amplifies the understanding of future researchers, considering that ECT's validity has been strongly established in online shopping. The results demonstrate the need for online vendors need to prioritize competitive pricing to stimulate repurchase intention of customers.

Keywords: Online shopping · Repurchase intention · Expectation-Confirmation Theory (ECT) · Perceived Competitive Price

1 Introduction

Over the last few years, e-commerce has become an indispensable part of the business. Brick-and-mortar stores are reducing as more than 2.14 billion people are expected to purchase goods and services online globally [1]. Malaysia is not an exception when it comes to online shopping. The survey conducted by Statista [2] on e-commerce shows that the number of online shoppers is estimated to reach 14.4 million in 2021. In addition, the revenue of the e-commerce market is projected to reach US$5,540 million in 2021. Although the increase in online shoppers presents excellent opportunities, many online vendors are facing difficulties in retaining their customers [3]. Trivedi and Yadav [4] pointed out that once a customer visits a particular e-marketplace, the challenge for the e-retailer lies in the strategy to persuade that customer to repurchase from the same platform. Furthermore, to ensure continuous profitability, online vendors need to attract new customers and convert them into repeat customers [5]. Thus, generating repeat

M. Al-Emran et al. (Eds.): ICETIS 2021, LNNS 299, pp. 682–690, 2022.
https://doi.org/10.1007/978-3-030-82616-1_56

customers in e-commerce markets is a critical challenge and considered highly important as compared to offline markets.

The ECT has been largely applied in the study of repurchase intention in e-commerce and information systems [6, 7]. However, Pee, Jiang, and Klein [8] pointed out that prior research have mostly investigated the consequences of expectation rather than its creation. Investigating the drivers of online shoppers' expectations can be beneficial for future research. On the other hand, given the nature of online shops, customers cannot evaluate and physically observe the products [7]. Thus, online shoppers will look for extrinsic cues that involves less sacrifice and more benefits to form an initial expectation about the service quality. Perceived competitive price can be a vital extrinsic cue that online shoppers can rely on as they can compare prices of the products via shopping websites without any sacrifice. Therefore, this study incorporated perceived competitive price as the predictor of expectation service quality into ECT.

2 Literature Review

Bhattacherjee [9] proposed the expectation confirmation theory to explain the formation of a customer's repurchase intention. According to past literature, this theory has been extensively used in service marketing, e-commerce, and the IT domain [10]. The theory proposes that customers will first form their own expectation of the goods or services they want to purchase. Customers will then use the specific goods or services and evaluate the actual performance of goods or services based on initial expectation and determine whether the expectation has been met. However, customer's expectation will be difficult to confirm if the expectation is too high. Subsequently, customers will be led to feeling satisfied or dissatisfied based on the level of confirmation and expectation. Eventually, satisfied customers will carry out repeat purchases while dissatisfied customers will cease purchase.

As explained in the introduction, the perceived competitive price is incorporated as the predictor of expectation service quality in this review section.

2.1 Perceived Competitive Price and Expected Service Quality

Kim, Xu, and Gupta [11] defined the perceived competitive price as an individual's subjective perception that the price offered by a particular online retailer is lower than the prices offered by others. Customers perceive the competitive price as cost-efficient. Previous research found that a product is assumed to be of high-quality if the price is high compared to products with a lower price. It was suggested that the perceived competitive price should be a negative indicator of perceived quality in online shopping setting [11, 12]. However, Sullivan and Kim [7] claimed that purchasing goods from an online shop with a highly competitive price encourages customers to pay less and therefore, decrease costs of product acquisition and its use. By subjectively adjusting the price point, consumers are likely to modify their expectations towards a service offered by a particular online shop. Therefore, it is possible that the adaptation level of the perceived competitive price influences customers' expectation toward a particular service. It is expected that customers would purchase a product within an acceptable level of quality weighted based on the lowest price point [7, 13]. In this regard, customers' expectations of product quality tend to be high with the lowest price they are willing to pay for the product. Thus, we hypothesize:

H1. The perceived competitive price positively associated with expected service quality.

Service Quality Expectation, Service Quality Confirmation and Repurchase Intention

Based on the discussion about ECT in Sect. 2.1, we hypothesized that customers' expectation about a service quality before purchase negatively influences service quality confirmation as high expectation is often difficult to meet. On the other hand, there is positive relationship between the perceived service quality and service quality confirmation. Expectation of a service quality is positively associated with satisfaction since a higher expectation sets a higher reference level to determine attainable satisfaction [14, 15]. Thus, we hypothesize:

H2. Service quality expectation is negatively associated with service quality confirmation.

H3. Service quality expectation is positively associated with satisfaction.

H4. The perceived service quality is positively associated with service quality confirmation.

Based on ECT, an online service quality that exceeds customer's expectation (service quality expectation is confirmed) will highly likely encourage customer's satisfaction. On the other hand, achieving customer's satisfaction will enhance customer's repurchase intention [16]. Previous studies have found that the service quality confirmation is positively associated with satisfaction, whereby satisfaction positively influences continuance intention e.g. [9, 15, 17–20]. Therefore, we posit the following hypothesis:

H5. Service quality confirmation is positively associated with satisfaction.

H6. Satisfaction is positively associated with repurchase intention.

3 Research Methodology

This study employed hypothesis testing as a method to investigate the nature of relationship between the variables. Investigation carried out in this study was of causal research as the focus was to delineate the causal sequence of the variables in forming online shopper repurchase intention, with minimum interference by researcher in a non-contrived setting (work proceeds as normal). The target population for this study was the students of UCSI University in Kuala Lumpur. The sample was collected from the students in classes and the library. A total of 370 completed responses were received and subjected to further data analysis.

The structured questionnaire was designed to collect data from the respondents. The perceived competitive price consists of two items which were adopted from Sullivana and Kim [7]. The items for expected service quality (5 items), perceived service quality (4 items), service quality confirmation (4 items), and customer repurchase intention (3 items) were adopted from Pee et al. [8]; the items for satisfaction (5 items) were adopted from Wu [21]. Two academic panels have reviewed the entire questionnaire to ensure the clarity and accurate representation of the items. The reliability test was conducted to determine the items' relative consistency and acceptability for the six variables investigated in this study.

This study employed PLS path modelling as suggested by Henseler, Ringle, and Sinkovics [22] to explain the variance (prediction of constructs). This study tested the structural model using PLS Smart 3.0 software.

4 Data Analysis

The values of Cronbach's alpha and composite reliability (CR) calculated for all constructs exceed the threshold values, demonstrating a strong internal consistency: CR > .60 [23] and Cronbach's alpha > .70. Table 1 shows the reliability of the study.

Table 1. Measurement model results.

Construct	Mean	SD	Alpha (α)	CR	AVE	VIF
Perceived Competitive Price (PCP)			0.721	0.877	0.781	1.000
Expected Service Quality (ESQ)			0.789	0.863	0.613	1.292
Perceived Service Quality (PSQ)			0.741	0.838	0.565	1.292
Service Quality Confirmation (SQC)			0.763	0.849	0.585	1.138
Satisfaction (S)			0.898	0.924	0.710	1.000
Repurchase Intention (RI)			0.886	0.929	0.825	

Both convergent and discriminant validity were established to determine the validity of the study. The average variance extracted (AVE) for all constructs are higher than the threshold value of 0.5, with the minimum value recorded at 0.613 (see Table 1). Therefore, the convergent validity is proven. On the other hand, the discriminant validity was examined using the heterotrait–monotrait (HTMT) ratio. Table 2 reveals the fulfilment of discriminant validity.

Table 2. Heterotrait-Monotrait Ratio (HTMT)

Construct	ESQ	PCP	PSQ	RI	S	SQC
ESQ						
PCP	0.549					
PSQ	0.619	0.625				
RI	0.295	0.524	0.672			
S	0.338	0.597	0.784	0.856		
SQC	0.444	0.661	0.979	0.700	0.882	

The HTMT ratio of correlations in Table 2 shows that all values, except for S on RI and SQC on PSQ, are below 0.85. The value has been identified as the most conservative threshold for establishing discriminant validity [24]. Further, Henseler et al. [24] stated that the HTMT inference criterion could be used for conceptually similar constructs, in which the HTMT ratio of correlation should be lower than 1. Thus, the HTMT ratio for S and SQC is acceptable. The results confirm the discriminant validity of the constructs.

In addition, this study analysed the possible multicollinearity among the constructs. The calculated variance inflation factor (VIF) is below the standard cut-off threshold of 5 (refer to Table 1). This study also checked the potential method bias. According to Harman's single factor test, a common method bias is identified when the first component of the data set variance is more than 50%. The variance of the first component in this study is 41.28%, which value demonstrates that there is no common method bias among the collected data.

4.1 Path Analysis

Based on the discussion, the measurement model has satisfied all criteria set. Thus, this study proceeded with developing the structural model. Bootstrapping of 5,000 resampled datasets were employed to determine the statistical significance of path coefficients. Table 3 shows the inspection results of inner structural model, and Fig. 1 shows the detailed structure of the structural model.

The structural equation modelling reveals that the perceived competitive price influences expected service quality ($\beta = 0.419$, $t = 7.264$). On the other hand, the expected service quality negatively related to service quality confirmation ($\beta = -0.002$, $t = 0.045$) while positively related to satisfaction ($\beta = 0.041$, $t = 1.029$). In addition, the perceived service quality influences service quality confirmation ($\beta = 0.738$, $t = 18.054$), and the service quality confirmation influences satisfaction ($\beta = 0.718$, $t = 21.462$). Eventually, satisfaction influences repurchase intention ($\beta = 0.763$, $t = 25.106$). Respectively, the findings confirm the hypotheses H1, H2, H4 and H5 and H6, except for H3.

The strength of a relationship is determined by evaluating the effect sizes $(f)^2$. A threshold value of 0.02, 0.15, and 0.35 demonstrates that an exogenous variable has a small, medium, and large effect size, respectively, on an endogenous variable [23]. Based on Table 3, expected service quality does not affect service quality confirmation and satisfaction. On the other hand, the perceived competitive price has a medium effect on the expected service quality, while the rest of the relationships demonstrate a large effect size.

In terms of coefficient determination (R2), repurchase intention has the most remarkable value (47.00%), followed by satisfaction (37.60%), service quality confirmation (30.90%), and expected service quality (10.4%). These values represent the model's adequate predictive capabilities based on the criteria published by Henseler et al. [22], which necessitate the values to be greater than zero. Thus, the proposed structural model is capable to offer an in-depth explanation about repurchase intention of online shoppers.

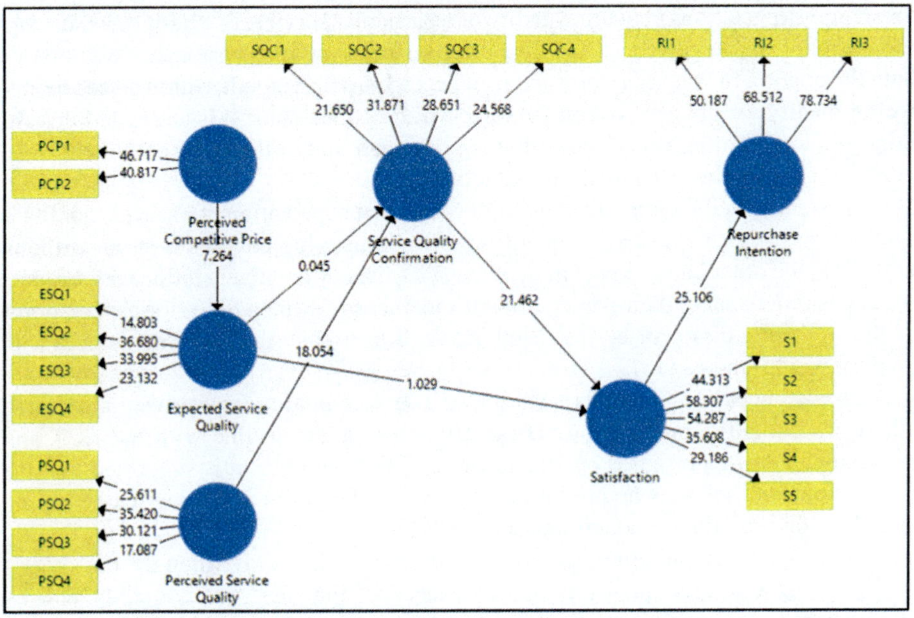

Fig. 1. Structural model

Table 3. Result of structural model inspection

Relationship	Standardized coefficient	CI (Min)	CI (Max)	t-value	*p*-value	f^2
H1: PCP → ESQ	0.419	0.305	0.529	7.264	0.000	0.213
H2: ESQ → SQC	−0.002	−0.101	0.106	0.045	0.964	0.000
H3: ESQ → S	0.041	−0.029	0.124	1.029	0.304	0.003
H4: PSQ → SQC	0.738	0.655	0.819	18.054	0.000	0.922
H5: SQC → S	0.718	0.645	0.775	21.462	0.000	0.978
H6: S → RI	0.763	0.699	0.813	25.106	0.000	1.396

5 Discussion of Results and Implications

The proposed model hypothesized that the perceived competitive price positively influences expected service quality. As expected, the empirical evidence from this study shows that there is significant positive relationship between the perceived competitive price and expected service quality. However, the findings contradict the prior studies, which have reported that the perceived competitive price negatively influences service quality e.g. [12, 25]. One of the possible explanations is consumers can compare prices of the products through shopping websites, leading customers to form a perception of competitive price. By subjectively adjusting the price point, customers are inclined to adjust their expectations about the quality of a particular service. Furthermore, the adaptation level of the perceived competitive price positively affects customers' expectation about the quality. In that sense, customers expect to purchase goods within an acceptable level of quality weighted based on the lowest price point. Thus, it is suggested that the online vendors to offer a service at a lower price than that of their competitors in order to encourage customer's repurchase intention.

Consistent with ECT, this study has also found that expectation of a service quality is negatively associated with service quality confirmation as high expectations are difficult to fulfil. On the other hand, expectation of a service quality positively influences satisfaction as customers adapt their perceptions of satisfaction in line with prior expectations, and thus reducing dissonance. The result obtained from this study is in agreement with that of previous studies [15].

It has also been found that expectation of a service quality is positively associated with service quality confirmation. These relationships are in line with the ECT and the previous finding presented by Bhattacherjee [9]. Therefore, it is suggested that the organizations and sellers who are interested to enter the online shopping market to put more effort on providing products and services of good quality; this includes the quality of the websites where the offerings are being displayed. It is only when the customer's expectation of a service quality is met or exceeded, that the service quality will be confirmed.

Based on ECT, individuals are inclined to be satisfied if the expectation of a service quality is confirmed, which will influence repurchase intention. The results obtained from this study are in agreement with the ECT. The results are also supported by previous studies, which have also found that the service quality confirmation influences satisfaction [8, 16]; in turn, satisfaction influences repurchase intention [8, 18, 20] Thus, to enhance repurchase intention, online vendors need to ensure an after-sale service quality that matches online shopper's expectation (i.e., the expected service quality is confirmed) and perception of satisfaction. By doing so, online vendors can increase the chance of turning new customers into returning customers.

As far as theoretical study is concerned, this study adds new knowledge in the field by integrating the perceived competitive price into the ECT. This study will help future researchers to better understand the ECT, and how does a service quality, price, and satisfaction affect repurchase intention of online shoppers. In terms of practicality, these findings offer direction for practitioners in online shopping industry to pursue in order to encourage continuance intention of online shoppers.

6 Limitation and Suggestions for Future Research

Thera are few limitations encountered in this study. Firstly, the sample is only composed of students at UCSI University, which may not be adequate to represent the whole population of online shoppers in Malaysia. Online shopping customers who come from different regions may have different views, attitudes or behaviour that determine their inclination toward repeat purchase, as compared to the respondents from UCSI University. Therefore, sample collection from wider population should be conducted to examine the potential different points of view.

On the other hand, independent variables investigated in this study are consists of only the perceived competitive price, expected service quality, service quality confirmation and satisfaction. However, the R-Square value obtained from the analysis shows that there are some other external factors that may also affect customer's repurchase intention. Thus, future research should integrate the possible independent variables into the model.

References

1. Statista: Number of digital buyers worldwide from 2014 to 2021 (In Billions) (2020)
2. Statista: e-commerce (2020)
3. Rita, P., Oliveira, T., Farisa, A.: The impact of e-service quality and customer satisfaction on customer behavior in online shopping. Heliyon **5**(10), 1–14 (2019)
4. Trivedi, S.K., Yadav, M.: Predicting online repurchase intentions with e-Satisfaction as mediator: a study on Gen Y. VINE J. Inf. Knowl. Manage. Syst. **48**(3), 427–447 (2018)
5. Shin, J.I., Chung, K.H., Oh, J.S., Lee, C.W.: The effect of site quality on repurchase intention in Internet shopping through mediating variables: the case of university students in South Korea. Int. J. Inf. Manage. **33**(3), 453–463 (2013)
6. Qazi, A., Tamjidyamcholo, A., Raj, R.G., Hardaker, G., Standing, C.: Assessing consumers' satisfaction and expectations through online opinions: expectation and disconfirmation approach. Comput. Hum. Behav. **75**, 450–460 (2017)
7. Sullivan, Y.W., Kim, D.J.: Assessing the effects of consumers' product evaluations and trust on repurchase intention in e-commerce environments. Int. J. Inf. Manage. **39**, 199–219 (2018)
8. Pee, L.G., Jiang, J., Klein, G.: Signaling effect of website usability on repurchase intention. Int. J. Inf. Manage. **39**, 228–241 (2018)
9. Bhattacherjee, A.: Understanding information systems continuance: an expectation-confirmation model. MIS Q. **25**(3), 351–370 (2001)
10. Park, E.: User acceptance of smart wearable devices: an expectation-confirmation model approach. Telematics Inf. **47**, 101318 (2020)
11. Kim, H.W., Xu, Y., Gupta, S.: Which is more important in Internet shopping, perceived price, or trust? Electron. Commer. Res. Appl. **11**(3), 241–252 (2012)
12. Chen, Z., Dubinsky, A.J.: A conceptual model of perceived consumer value in ecommerce: a preliminary investigation. Psychol. Mark. **20**(4), 323–347 (2003)
13. Teas, R.K., Agarwal, S.: The effects of extrinsic product cues on consumers' perceptions of quality, sacrifice, and value. J. Acad. Mark. Sci. **28**(2), 278–290 (2000)
14. Brown, S.A., Venkatesh, V., Goyal, S.: Expectation confirmation in technology use. Inf. Syst. Res. **23**(2), 474–487 (2012)
15. Kim, D.J., Ferrin, D.L., Rao, H.R.: Trust and satisfaction, two steppingstone for successful e-commerce relationships: a longitudinal exploration. Inf. Syst. Res. **20**(2), 237–257 (2009)
16. Parasuraman, A., Zeithaml, V.A., Berry, L.L.: A conceptual model of service quality and its implications for future research. J. Mark. **49**(4), 41–50 (1985)
17. Bhattacherjee, A., Premkumar, G.: Understanding changes in belief and attitude toward information technology usage: a theoretical model and longitudinal test. MIS Q. **28**(2), 229–254 (2004)
18. Gupta, A., Dhiman, N., Yousaf, A., Arora, N.: Social comparison and continuance intention of smart fitness wearables: an extended expectation confirmation theory perspective. Behav. Inf. Technol. 1–14 (2020)
19. Fu, X.M., Zhang, J.H., Chan, F.T.: Determinants of loyalty to public transit: a model integrating satisfaction-loyalty theory and expectation-confirmation theory. Transp. Res. Part A Policy Pract. **113**, 476–490 (2018)

20. Tam, C., Santos, D., Oliveira, T.: Exploring the influential factors of continuance intention to use mobile apps: extending the expectation confirmation model. Inf. Syst. Front. **22**(1), 243–257 (2018). https://doi.org/10.1007/s10796-018-9864-5
21. Wu, I.-L.: The antecedents of customer satisfaction and its link to complaint intentions in online shopping: an integration of justice, technology, and trust. Int. J. Inf. Manage. **33**(1), 166–176 (2012)
22. Henseler, J., Ringle, C., Sinkovics, R.: The use of partial least squares path modelling in international marketing. Adv. Int. Mark. **20**(1), 277–319 (2009)
23. Hair, J.F., Hult, G.T.M., Ringle, C., Sarstedt, M.: A Primer on Partial Least Squares Structural Equation Modeling, 2nd edn. Sage, Thousand Oaks (2014)
24. Henseler, J., Ringle, C.M., Sarstedt, M.: A new criterion for assessing discriminant validity in variance-based structural equation modelling. J. Acad. Mark. Sci. **43**(1), 115–135 (2015)
25. Dodds, W.B., Monroe, K.B., Grewal, D.: Effects of price, brand, and store information on buyers' product evaluations. J. Mark. Res. **28**(3), 307–319 (1991)

Critical Success Factors for ERP Implementation: A Study on Mobile Telecommunication Companies in Egypt

Ahmed Noureldeen[1], Mohamed Salaheldeen[2](✉), and Mohamed Battour[3]

[1] Faculty of Commerce, Menoufia University, Shibin Al Kawm, Egypt
[2] Faculty of Economics and Muamalat, Universiti Sains Islam Malaysia (USIM), Nilai, Malaysia
[3] College of Business Administration, University of Sharjah, Sharjah, UAE
mbattour@sharjah.ac.ae

Abstract. The success of the ERP system implementation has a great importance in Egyptian companies. This research seeks to know the relationship between the Critical Success Factors of the ERP system (strategic, tactical, and technological) and support decisions in mobile telecommunication companies in Egypt (Vodafone, Orange, Etisalat, WE). Data were collected through a survey list distributed to 443 respondents from the top and middle management mobile telecommunication companies in Egypt. The results indicate that there is a significant impact of the Critical Success Factors of the ERP system on supporting the decisions of top and middle management in mobile telecommunication companies. Also, differences were found in the level of perception of top management and middle management of critical success factors according to their different demographic characteristics (Experience Years, Education Level).

Keywords: ERP success · Support decisions · Mobile telecommunication companies

1 Introduction

Information plays a critical role in companies' success for both product and service industries [1]. Companies sought to adopt an information system based on modern technology called the Enterprise Resource Planning System (ERP), which works on the full flow of data and information between all departments in the company through subsystem that have databases that aggregate in one system that has standardized databases for tightening organizational control [2, 3]. Hence, the scientific development in information technology must be exploited to promote companies and support their decision [4, 5], especially when it possesses a system such as the ERP system [6]. This research tried to explain critical success factors for ERP implementation (strategic, tactical, and technology) to support the decisions in the mobile telecommunication companies in Egypt.

M. Al-Emran et al. (Eds.): ICETIS 2021, LNNS 299, pp. 691–701, 2022.
https://doi.org/10.1007/978-3-030-82616-1_57

2 Literature Review

Information plays an important role in industrial or service companies [7]. Companies sought to adopt an information system based on modern technology called the Enterprise Resource Planning System (ERP) [8], which works on the full flow of data and information between all departments in the company through subsystem that have databases that aggregate in one system that has standardized databases for tightening organizational control [9]. The scientific development in information technology must be exploited to promote companies and support their Decision, especially when it possesses a system such as the ERP system [10].

ERP systems generally comprise different software modules, which allow organizations to automate and integrate the majority of business functions by accessing, and sharing common information, data, and practices across the enterprise in real-time [11]. ERP system has the ability to help organizations to handle internal resources (human resources, financial resources, material resources, information resources, etc.) more easily and efficiently [12]. Enterprise resource planning (ERP) has become an effective IT tool and a driver of performance enhancement in organizations all over the world as a result of these different attributes [13]. The implementation of enterprise resource planning (ERP) is loaded with difficulties. A greater understanding of these problems and how to resolve them during the implementation process [14].

Shatat [15] found that the main factors for the success of implementing the ERP system are: support for top management, user participation, clear goals, strategic information technology planning, user training, and education, supplier support, teamwork, project leader, performance monitoring, and evaluation, Education on new business processes by application to Omani organizations. While Arvidsson and Kojic [16] found four main factors: training and education, change management, appropriate supplier selection, and risk management. And in the study of Moh'd Anwer [17] found that technical challenges, technological readiness, company size, and top management support have a major impact on the approval of the companies under study to implement the project resource planning system.

ERP systems have increased in value as a means of exploiting transaction data for decision support and analytics applications. There is a relationship between the success of the ERP system and decision support applications [18]. Guo, Wang [19] aimed to apply the intelligent decision support system based on data mining technology to enterprises to establish an internet of things-based intelligent DSS for the manufacturing industry, thereby supporting the decision-makers. There is interest from many researchers and studies regarding identifying the success factors for implementing an ERP system. However, there is not any research link these success factors to support decisions in mobile companies in Egypt. Consequently, this research mainly seeks to identify the success factors for implementing the ERP system, represented by strategic factors, tactical factors, and technological factors to support decisions in mobile companies operating in Egypt, according to the different demographic characteristics of the investigated, represented in experience years and academic qualification (Fig. 1).

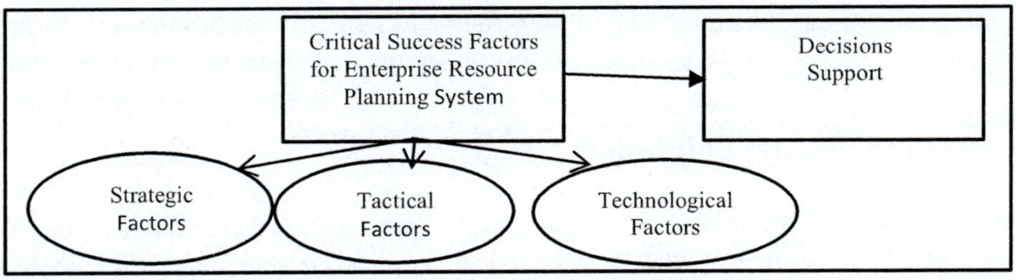

Fig. 1. The relationship between the Critical Success Factors of ERP and Decision Support

3 Research Methodology

The research sample is represented by members of the top management and middle management of mobile telecommunication companies in Egypt, on the basis that this category is responsible for ERP implementation, and they also have decision-making in those companies according to their demographic characteristics Experience Years (less than 5, from 5 years to less than 10 years, from 10 years to less than 15 years, and 15 years and more), and Education Level (Intermediate qualification, higher qualification, and MA/Ph.D.). The Instrumentation of collecting data for this study was questionnaires that were distributed to 443 individuals four mobile telecommunication companies.

The questionnaire is divided into the following sections: Section one includes a set of questions to identify the critical success factors for ERP implementation. a five-point scale is used ranging from 1 = strongly disagree to 5 = strongly agree. The second section: It includes a set of questions directed to identify the dimension of decision support. The answer could be given by using a five-point scale ranging from 1 = strongly disagree to 5 = strongly agree. The third section covers the demographic characteristics for top and middle management, which are divided into experience years and education level.

The data were subjected to statistical analysis and to test the validity of the hypotheses using the computer through some methods provided by the statistical program for social sciences (SPSS) as reliability analysis (alpha correlation coefficient) is used to make sure about the degree of reliability in multi-scales items employed in this study [20]. It focuses on the degree of internal consistency among items that formed the scale. Factor analysis is used to reduce the number of variables in this study, isolate variables that do not affect the results of the study, and identifying the key dimensions that express the actual characteristics that are measured [21]. The study was depended on one of the dispersion measures which is the standard deviation. Standard deviation is a measure of how precise the mean of a population or sample [22]. It is used to indicate the trends in the objects in a given data set concerning the mean, i.e., the spread of these objects from the mean [23].

Multiple regression analysis is to investigate the existence of the relationship between the independent variable and the dependent one. So it is used to identify the relationship between Critical Success Factors for ERP system implementation as the independent variable, and decision support as dependent variables. One-way analysis of variance (one - way ANOVA) is used to assure the relationship between a dependent variable and an independent variable, and to identify the degree of similarity or difference between two groups of persons or things based on several factors as independent variables [24].

4 Research Model Validation

In this part, we will briefly talk about how valid questionnaires were obtained and the percentage of the responses that the researchers got. The researcher distributed 443 questionnaires to top management and middle management in four Mobile Companies (Vodafone – Orange – Etisalat – WE) in Egypt. The valid questionnaires were 420 which were completed in all questions. On the other hand, the invalid questionnaires were 23 ones because it was uncompleted and misleading in answers, The response rate was 95%.

The reliability assessment includes the measures of the critical success factors for ERP implementation and the decision support. Reliability analysis was conducted to investigate the internal consistency of the questionnaire. The internal consistency reliability means the extent to which there is cohesiveness among the questionnaire's items that have been used in the field study. the acceptable values of Cronbach Alpha (α) coefficients range from 0.60 to 0.95.

The results of reliability analysis revealed that most of the item-total correlations were above 0.30 for all items in the questionnaire. On the other hand, alpha coefficients were more than .60 for all scales used in the questionnaire which is acceptable and assures the internal consistency between items in the questionnaire (Table 1).

Table 1. Reliability analysis of the Scales used in Study.

Variables	Items number	Alpha coefficient
Critical success factors for ERP implementation	37	.851
Decision Support	12	.778

After evaluating the reliability of scales and related measures used in the study, the validity of these scales was also assessed to assure that the items included in the questionnaire are valid to measure what is supposed to measure. Two methods were used in validity assessment: Face validity, 13 professors of business have made their comments about the questionnaire and the researcher responded to their light modifications. And the content validity, according to this method, the different items for each scale have been carefully identified and defined, and previous studies conducted in the enterprise resource planning system have been reviewed to help develop the various measures. Also, an exploratory study was conducted on some mobile companies in Egypt, in a way that helps to accurately identify the study variables. The researcher was also keen on having open questions about the critical success factors for implementing the system. The survey list was field-tested by presenting it to (21) company managers, and they were informed that this procedure aims to test the list so that they do not hesitate to mention any notes or write any comments that are useful in the final design of the list.

Accordingly, some amendments were made to the list, some of them were linguistic and others were interpretations.

5 Results

Research Hypothesis 1
There is a difference between top management and middle management perceptions regarding the critical success factors for ERP implementation based on demographic characteristics (Experience years – Education Level) in the mobile companies.

The measure of critical success factors for implementing the ERP system contains twenty-seven variables that measure the difference in the perception of top and middle management of these factors. It was found from the results of the descriptive analysis that the technological factors occupy the first place with an arithmetic mean (4.60) and a standard deviation (.72) then the tactical factors come in second place with arithmetic mean (3.50) a standard deviation (.40) and finally the strategic factors. It comes in third place with an arithmetic mean (3.20) and a standard deviation (.31). It is clear to the researcher that the level of awareness of top management and middle management is relatively high of technological success factors, including the company must have the technology infrastructure necessary to implement the system, The ability to collect and analyze data to access information and make decisions, The IT team can solve technical problems related to the system, The IT team has the advantage of its ability to reactivate the system at a high speed in the event of a technical failure, capable of appropriate software coding for the ERP system.

The level of awareness is relatively average of tactical success factors, including the availability of a suitable training program for both technical staff and end-users, whether internal or external training, Consultants should share their expertise with relevant internal staff to assist in the implementation process, analyzing and describing the current system applied in the company in terms of its processes, outputs, problems and technology used in it. Monitor and evaluate performance in managing the ERP system. As well as their level of awareness is relatively average of strategic success factors, including the availability of sustainable administrative support, whether from the upper or middle levels, change management ensures complete harmony between the employees, the new technology, and the processes. Business processes must be re-engineered to align business processes with the ERP system business model, developing a plan and a clear vision for the application that specifies the expected benefits of the system and the challenges of its implementation. For more clarification, see the following table. The researcher attributes the difference in the perception of both the top and middle management towards the success factors of implementing the ERP system due to several reasons, including the low level of academic qualification of managers, their lack of experience in administrative work, as well as the high cost of implementing the system, where the cost of the application reaches more than 2 million dollars. In addition to the weak ability to overcome the challenges of implementing this system, the most important of which is resistance to change.

The one-way method of analysis of variance was applied, to determine the extent of the difference in the perceptions of upper and middle management in the companies under study towards the critical success factors of implementing the ERP system according to the different demographic characteristics of the investigated from them (Experience Years and Education Level).The results showed that there are statistically significant differences between the level of perception of the Top and Middle management towards the critical success factors of implementing the ERP system according to their demographic characteristics (Experience Years and Education Level), as the value of (q) is significant at the level of statistical significance (0,01, 0,05).

Experience Years*
The results of the analysis of variance show that there are differences in the perceptions of top and middle management towards the critical success factors of implementing an ERP system with different years of experience (less than 5 years, from 5 years to less than 10 years, from 10 years to less than 15 years, 15 years and more) Whereas the value of (q) is significant at the level of statistical significance (0,01, 0,05) and then the null hypothesis must be rejected and the alternative hypothesis should be accepted which states that there are statistically significant differences between the level of awareness of top and middle management towards the success factors of implementing the ERP system with different years of experience, as shown in the following table (Table 2).

Table 2. ANOVA of critical success factors for ERP (Experience years)

Items		Sum of squares	df	Mean square	F	Sig
Strategic factors	Between groups	14.157	1	14.157	78.65	.000
	Within groups	75.442	419	.180		
	Total	89.599	420			
Tactical factors	Between groups	18.781	1	18.781	79.245	.000
	Within groups	99.343	419	.237		
	Total	118.124	420			
Technology factors	Between groups	20.898	1	20.898	102.441	.000
	Within groups	85.361	419	.204		
	Total	102.259	420			

Education Level*
The results of the analysis of variance show that there are differences in the perceptions of top and middle management towards the critical success factors of implementing an ERP system with different education levels (Intermediate qualification, higher qualification "BA", MA/Ph.D.) Whereas the value of (q) is significant at the level of statistical significance (0,01, 0,05), and then the null hypothesis must be rejected and the alternative hypothesis should be accepted which states that there are statistically significant differences between the level of awareness of top and middle management towards the critical success factors of implementing the ERP system with different education level, as shown in the following table (Table 3).

Table 3. ANOVA of critical success factors for ERP (Education Level)

Items		Sum of squares	df	Mean square	F	Sig.
Strategic factors	Between groups	37.041	2	18.520	97.989	.000
	Within groups	79.347	418	.189		
	Total	116.388	420			
Tactical factors	Between groups	38.915	2	19.458	92.218	.028
	Within groups	88.199	418	.211		
	Total	127.114	420			
Technology factors	Between groups	39.949	2	19.974	107.968	.000
	Within groups	77.347	418	.185		
	Total	117.296	420			

Research Hypothesis 2

There is a difference between top management and middle management perceptions regarding the decision support based on demographic characteristics (Experience years – Education Level) in the companies.

The decision support scale contains ten variables that measure the extent of the difference in the perception of the top and middle management of them. It was found from the results of the descriptive analysis that the quality of the information ranks first with arithmetic mean (4.40) and a standard deviation (.39), and the quality of the content ranked second with arithmetic mean (3.99) and a standard deviation (1.08). It is clear to the researchers that there is a difference in the level of awareness of both the top and middle management of the variables of decision support.

The one-way method of analysis of variance was applied, to determine the extent of the difference in the perceptions of top and middle management in the companies under study towards the decision support according to the different demographic characteristics of the investigated from them (Education Level, years of experience). The results showed that there are statistically significant differences between the level of perception of the top and middle management towards the decision support according to their demographic characteristics (Experience years – Education Level), as the value of (q) is significant at the level of statistical significance (0,01, 0,05).

Experience Years*

The results of the analysis of variance show that there are differences in the perceptions of top and middle management towards the decision support with different experience years (less than 5 years, from 5 years to less than 10 years, from 10 years to less than 15 years, 15 years and more) Whereas the value of (q) is significant at the level of statistical significance (0,01, 0,05), and then the null hypothesis must be rejected and the alternative hypothesis should be accepted which states that there are statistically significant differences between the level of awareness of top and middle management towards the decision support with different years of experience, as shown in the following table (Table 4).

Table 4. ANOVA of decision support (experience years)

Items		Sum of squares	df	Mean Square	F	Sig.
Information quality	Between groups	23.186	1	23.186	97.831	.000
	Within groups	99.432	419	.237		
	Total	122.618	420			
Content quality	Between groups	25.781	1	25.781	106.975	.000
	Within groups	101.323	419	.241		
	Total	123.094	420			

Education Level*

The results of the analysis of variance show that there are differences in the perceptions of top and middle management towards the decision support with different Education Level (Intermediate qualification, higher qualification "BA", MA/Ph.D.) Whereas the value of (q) is significant at the level of statistical significance (0,01, 0,05), and then the null hypothesis must be rejected and the alternative hypothesis should be accepted which states that there are statistically significant differences between the level of awareness of top and middle management towards the decision support with different education level, as shown in the following table (Table 5).

Table 5. ANOVA of decision support (education level).

Items		Sum of squares	df	Mean Square	F	Sig.
Information quality	Between groups	47.043	2	23.522	109.916	.000
	Within groups	89.357	218	.214		
	Total	136.4	220			
Content quality	Between groups	55.395	2	27.698	51.579	.028
	Within groups	117.165	218	.537		
	Total	172.56	220			

Research Hypothesis 3

There is a relationship between the critical success factors for ERP implementation and the decision support in the companies.

There is a positive linear relationship with statistical significance between the critical success factors of implementing the ERP system and decision support in the mobile companies under study (taken in total), and that this relationship represents 85% (according to the multiple correlation coefficient in model R) and this relationship is positive, as the more the factors of successful implementation such as strategic factors, tactical factors, and technological factors are taken into account, the more that supports decisions at the level of the company as a whole. The critical success factors of implementing the ERP system in the companies under study can explain about 68% (according to the coefficient of determination in the R2 model) of the total variance in the level of decision support of the companies under study.

It is found that the strategic, tactical, and technological factors have a positive linear relationship with statistical significance between them and support decisions in the mobile companies under study. This means that mobile companies in Egypt can support decisions by taking into consideration the aforementioned success factors. as shown in the following table (Table 6).

Table 6. The type and degree of relationship between critical success factors for implementing the ERP system and decision support. (Multiple regression analysis output).

Critical success factors for ERP implementation	Regression coefficient Beta	Correlation coefficient R	Determination coefficient R^2
Strategic factors	.28*	.32	.10
Tactical factors	.31**	.36	.13
Technology factors	.45**	.50	.25
Multiple correlation coefficient of the R model	.85		
Determination coefficient of the model R2	.68		
Computed f value	36.91		
Level of statistical significance	0.000		

* Indicates T-Test significance at .05 level.
**Indicates T-Test significance at .01 level.

6 Results and Discussion

The results of the current study revealed several key findings. The strategic, tactical, and technological factors have a positive linear relationship with statistical significance between them and support decisions in the mobile companies under study. This means that mobile companies in Egypt can support decisions by taking into consideration the aforementioned critical success factors. There is a difference between top management and middle management perceptions regarding the critical success factors for ERP implementation based on demographic characteristics (Experience Years – Education Level) in the mobile companies in Egypt. The level of awareness of top management and middle management is relatively high of technological success factors, including The company must have the technology infrastructure necessary to implement the system, The ability to collect and analyze data to access information and make decisions, The IT team can solve technical problems related to the system, The IT team has the advantage of its ability to reactivate the system at a high speed in the event of a technical failure, capable of appropriate software coding for the ERP system.

The level of awareness of top management and middle management is relatively average of tactical success factors, including Availability of a suitable training program for both technical staff and end-users, whether internal or external training, Consultants should share their expertise with relevant internal staff to assist in the implementation process, Analyzing and describing the current system applied in the company in terms of its processes, outputs, problems, and technology used in it. Monitor and evaluate performance in managing the ERP system. As well as their level of awareness is relatively average of strategic success factors, including the availability of sustainable administrative support, whether from the upper or middle levels, change management ensures complete harmony between the employees, the new technology, and the processes. Business processes must be re-engineered to align business processes with the ERP system business model, developing a plan and a clear vision for the application that specifies the expected benefits of the system and the challenges of its implementation. For more clarification, see the following table.

The researcher attributes the difference in the perception of both the top and middle management towards the critical success factors of implementing the ERP system due to several reasons, including: the low level of Education Level of some managers, their lack of experience in administrative work, as well as the high cost of implementing the system, where the cost of the application reaches more than 2 million dollars. In addition to the weak ability to overcome the challenges of implementing this system, the most important of which is resistance to change.

References:

1. Jones, P., et al.: The sustainable development goals and information and communication technologies. Indonesian J. Sustain. Account. Manag. **1**(1), 1–15 (2017)
2. Zarei, B., Naeli, M.: Critical success factors in enterprise resource planning implementation a case-study approach. Int. J. Enterpr. Inf. Syst. (IJEIS) **6**(3), 48–58 (2010)
3. Ahmad Suhaimi, N.S., Nawawi, A., Puteh Salin, A.S.A.: Impact of enterprise resource planning on management control system and accountants' role. Int. J. Econ. Manag. **10**(1) (2016)
4. Kovalenko, A.S., et al.: Concept of information technology of monitoring and decision-making support. In: Photonics Applications in Astronomy, Communications, Industry, and High Energy Physics Experiments 2017. International Society for Optics and Photonics (2017)
5. Housel, T.J., et al.: A comparative analysis of advanced methodologies to improve the acquisition of information technology in the Department of Defense for optimal risk mitigation and decision support systems to avoid cost and schedule overruns. In: Acquisition Research Symposium, Monterey, CA (2019)
6. Zhezhnych, P., Tarasov, D.: Methods of data processing restriction in ERP systems. In: 2018 IEEE 13th International Scientific and Technical Conference on Computer Sciences and Information Technologies (CSIT). IEEE (2018)

7. Battor, M., Battor, M.: The impact of customer relationship management capability on innovation and performance advantages: testing a mediated model. J. Mark. Manag. **26**(9–10), 842–857 (2010)
8. Seethamraju, R.: Adoption of software as a service (SaaS) enterprise resource planning (ERP) systems in small and medium sized enterprises (SMEs). Inf. Syst. Front. **17**(3), 475–492 (2015)
9. Faccia, A., et al.: Accounting information systems and ERP in the UAE: an assessment of the current and future challenges to handle big data. In: Proceedings of the 2019 3rd International Conference on Cloud and Big Data Computing (2019)
10. Li, H., et al.: Organizational mindfulness towards digital transformation as a prerequisite of information processing capability to achieve market agility. J. Bus. Res. **122**, 700–712 (2021)
11. Davenport, T.H.: Putting the enterprise into the enterprise system. Harvard Bus. Rev. **76**(4), 121–131 (1998)
12. Nah, F.F.H., Lau, J.L.S., Kuang, J.: Critical factors for successful implementation of enterprise systems. Bus. Process Manag. J. (2001)
13. Alam, M.S., Uddin, M.A.: Adoption and implementation of enterprise resource planning (ERP): an empirical study. J. Manag. Res. **6**(1), 1–33 (2019)
14. Ali, M., Miller, L.: ERP system implementation in large enterprises–a systematic literature review. J. Enterpr. Inf. Manag. (2017)
15. Shatat, A.S.: Critical success factors in enterprise resource planning (ERP) system implementation: an exploratory study in Oman. Electron. J. Inf. Syst. Eval. **18**(1), 36 (2015)
16. Arvidsson, J., Kojic, D.: Critical Success Factors in ERP Implementation: The Perspective of the Procurement System User (2017)
17. Moh'd Anwer, A.-S.: Towards better understanding of determinants logistical factors in SMEs for cloud ERP adoption in developing economies. Bus. Process Manag. J. (2019)
18. Holsapple, C., Sena, M., Wagner, W.: The perceived success of ERP systems for decision support. Inf. Technol. Manage. **20**(1), 1–7 (2017). https://doi.org/10.1007/s10799-017-0285-9
19. Guo, Y., et al.: The internet of things-based decision support system for information processing in intelligent manufacturing using data mining technology. Mech. Syst. Signal Process. **142**, p. 106630 (2020)
20. Green, S.B., Yang, Y.: Evaluation of dimensionality in the assessment of internal consistency reliability: Coefficient alpha and omega coefficients. Educ. Meas. Issues Pract. **34**(4), 14–20 (2015)
21. Brown, T.A.: Confirmatory factor analysis for applied research. Guilford publications (2015)
22. Boos, D.D., Brownie, C.: Comparing variances and other measures of dispersion. Stat. Sci. **19**(4), 571–578 (2004)
23. Walther, B.A., Moore, J.L.: The concepts of bias, precision and accuracy, and their use in testing the performance of species richness estimators, with a literature review of estimator performance. Ecography **28**(6), 815–829 (2005)
24. Moder, K.: Alternatives to F-test in one way ANOVA in case of heterogeneity of variances (a simulation study). Psychol. Test Assess. Model. **52**(4), 343–353 (2010)

Correction to: Blockchain Adoption Intention in Higher Education: Role of Trust, Perceived Security and Privacy in Technology Adoption Model

Nishant Kumar, Mitushi Singh, Kamal Upreti, and Divya Mohan

Correction to:
Chapter "Blockchain Adoption Intention in Higher Education: Role of Trust, Perceived Security and Privacy in Technology Adoption Model" in: M. Al-Emran et al. (Eds.): ***Proceedings of International Conference on Emerging Technologies and Intelligent Systems*, LNNS 299,**
https://doi.org/10.1007/978-3-030-82616-1_27

The affiliation "Inderprastha Engineering College, Ghaziabad, India" of the author "Kamal Upreti" in the original version of the book has been removed from the chapter "Blockchain Adoption Intention in Higher Education: Role of Trust, Perceived Security and Privacy in Technology Adoption Model". The correction book has been updated with the change.

The updated version of this chapter can be found at
https://doi.org/10.1007/978-3-030-82616-1_27

M. Al-Emran et al. (Eds.): ICETIS 2021, LNNS 299, p. C1, 2022.
https://doi.org/10.1007/978-3-030-82616-1_58

Author Index

M. Al-Emran et al. (Eds.): ICETIS 2021, LNNS 299, pp. 703–705, 2022.
https://doi.org/10.1007/978-3-030-82616-1